Withdrawn From
University of Florida Libraries

AF709432

QELS '95

Summaries of papers presented at the

Quantum Electronics and Laser Science Conference

May 22–26, 1995
Baltimore Convention Center
Baltimore, Maryland

Volume 16
1995 Technical Digest Series
Conference Edition

Sponsored by
Optical Society of America
IEEE/Lasers and Electro-Optics Society
Laser Science Topical Group
of the American Physical Society

In cooperation with
Quantum Electronics Division
of the European Physical Society
Japanese Quantum Electronics Joint Group

Optical Society of America
2010 Massachusetts Avenue NW
Washington DC 20036-1023

UNIVERSITY OF FLORIDA LIBRARIES

Articles in this publication may be cited in other publications. To facilitate access to the original publication source, the following form for the citation is suggested:
Name of Author(s), "Title of Paper," in *Quantum Electronics Conference,* Vol. 16, 1995 OSA Technical Digest Series(Optical Society of America, Washington, D.C., 1995), pp. xx-xx.

Optical Society of America

ISBN

Conference Edition	1-55752-401-7
Postconference Edition	1-55752-402-5
(Note: Postconference Edition includes postdeadline papers.)	
1993 Technical Digest Series	1-55752-368-1

Library of Congress Catalog Card Number

Conference Edition	95-68678
Postconference Edition	95-68679

Institute of Electrical and Electronics Engineers

ISBN

Casebound	0-7803-2661-X
Microfiche	0-7803-2662-8
Catalog Number	95CH35799

Copyright © 1995, Optical Society of America

Individual readers of this digest and libraries acting for them are permitted to make fair use of the material in it, such as to copy an article for use in teaching or research, without payment of fee, provided that such copies are not sold. Copying for sale is subject to payment of copying fees. The code 1-55752-368-1/95/$6.00 gives the per-article copying fee for each copy of the article made beyond the free copying permitted under Sections 107 and 108 of the U.S. Copyright Law. The fee should be paid through the Copyright Clearance Center, Inc., 21 Congress Street, Salem, MA 01970.

Permission is granted to quote excerpts from articles in this digest in scientific works with the customary acknowledgment of the source, including the author's name and the name of the digest, page, year, and name of the Society. Reproduction of figures and tables is likewise permitted in other articles and books provided that the same information is printed with them and notification is given to the Optical Society of America. In addition, the Optical Society may require that permission also be obtained from one of the authors. Address inquiries and notices to Director of Publications, Optical Society of America, 2010 Massachusetts Avenue, NW, Washington, DC 20036-1023. In the case of articles whose authors are employees of the United States Government or its contractors or grantees, the Optical Society of America recognizes the right of the United States Government to retain a nonexclusive, royalty free license to use the author's copyrighted article for United States Government purposes.

Printed in the U.S.A.

FOREWORD

Welcome to the 1995 Quantum Electronics and Laser Science Conference. This fifth QELS conference will be held May 21–26 in Baltimore, Maryland. QELS'95 is organized under the aegis of the Joint Council for Quantum Electronics, which is sponsored by the American Physical Society (Laser Science Topical Group), Institute of Electrical and Electronics Engineers (Laser and Electro-Optics Society, and the Optical Society of America.

As in previous years, QELS'95 will be held concurrently with the Conference on Lasers and Electro-Optics (CLEO'95). The complementary nature of the meeting is highlighted with a joint plenary session held on Monday afternoon featuring talks by three internationally recognized researchers on topics of importance to both basic and applied science. The topics chosen this year—near-field optical microscopy, optoelectronics with conducting polymers, and applications of laser spectroscopy in atmospheric and environmental problems—epitomize the scope and vitality of the field covered by the conference. The common interest of the participants and contributors is further seen in the joint symposia arranged between QELS and CLEO to highlight areas of strong mutual interest. In addition, both meetings schedule a series of tutorials designed to enable people with an interest in an emerging field to obtain a basic introduction to the concepts. This year 9 tutorials have been arranged.

The excitement and activity of the field are reflected in the eight parallel sessions for oral presentations taking place from Monday through Friday and the three separate sessions for posters, allowing for a total of more than 1000 carefully selected papers to be presented. Another unique feature of the meeting is the comprehensive exhibit and demonstration of equipment organized by CLEO.

In the 1995 call for papers, QELS identified five primary subject categories for the meeting: laser spectroscopy, nonlinear optical phenomena, quantum optics, optical interactions with condensed matter and ultrafast phenomena, and the physics of coherent light sources. Papers in each category have been reviewed and selected by a subcommittee composed of scientists active in the area of research. Through this process, a total of 206 papers for oral presentation and 103 papers for poster presentation were selected to make up the contributed portion of the meeting. In additional 46 scientists were invited to present papers to provide in-depth discussion of significant new developments.

A successful meeting requires both outstanding submissions and a capable and diligent program committee. QELS'95 was fortunate to have both. We would particularly like acknowledge the key contributions to the meeting of the subcommittee chairs: Wendell T. Hill, Robert W. Boyd, Prem Kumar, Arto V. Nurmikko, and Kerry J. Vahala. It is a special pleasure to acknowledge the leadership and hard work of the QELS Program Chairs, Jeffrey Bokor and Richard E. Slusher. We also wish to acknowledge on behalf of the entire QELS technical committee, the exceptional support provided by the staff of the OSA. In particular, it was a pleasure to work with CLEO General Chairs Thomas F. Deutsch and John E. M. Goldsmith and CLEO Program Chairs Dennis K. Kilinger and George C. Valley. We also are indebted to the 1995 CLEO/QELS liaison, John E. Bowers, for his highly effective contribution to coordination of the two meetings.

This year's QELS meeting promises to provide a forum for exciting new results over the gamut of laser science and quantum electronics. We hope that the meeting will prove to be enjoyable and stimulating for all of the participants.

QELS'95

General Chairs

Tony F. Heinz
IBM T.J. Watson Research Center

George I. Stegeman
University of Central Florida/CREOL

Technical Program Chairs

Jeffrey Bokor
University of California, Berkeley

Richard E. Slusher
AT&T Bell Laboratories

QELS '95 Committees

Program Committee

General Chairs

Tony F. Heinz, *IBM T.J. Watson Research Center*
George I. Stegeman, *University of Central Florida/ CREOL*

Program Chairs

Jeffrey Bokor, *University of California, Berkeley*
Richart E. Slusher, *AT&T Bell Laboratories*

CLEO/QELS Technical Program Liaison

John E. Bowers, *University of California, Santa Barbara*

1. **Laser Spectroscopy**
   Wendell T. Hill, *University of Maryland, Chair*
   Charles W. Clark, *National Institute of Standards and Technology*
   Eric Cornell, *Joint Institute for Lab Astrophysics*
   Phillip L. Gould, *University of Connecticut*
   Kenneth C. Kulander, *Lawrence Livermore National Laboratory*
   Daniel J. Larson, *University of Virginia*
   Sune R. Svanberg, *Lund Institute of Technology, Sweden*
   Hiroshi Takuma, *University of Electro-Communication, Japan*

2. **Nonlinear Optical Phenomena**
   Robert W. Boyd, *University of Rochester, Chair*
   Alexander L. Gaeta, *Cornell University*
   Peter Gunter, *ETH Honggeberg Institut fur Quantumelectronik, Switzerland*
   Jean-Pierre Huignard, *Thompson CSF, France*
   Paul L. Kelley, *Tufts University*
   Barry Luther-Davies, *Australian National University, Australia*
   John F. Reintjes, Jr., *U. S. Naval Research Laboratory*
   John E. Sipe, *University of Toronto, Canada*

3. **Quantum Optics**
   Prem Kumar, *Northwestern University, Chair*
   Izo I. Abram, *Centre Natl. d'Etudes (CNET), France*
   Girish S. Agarwal, *University of Hyderabad, India*
   Roy S. Bondurant, *Massachusetts Institute of Technology*
   Luigi A. Lugiato, *Universita degli Studi di Milano, Italy*
   Gerard J. Milburn, *University of Queensland, Australia*
   Michael G. Raymer, *University of Oregon*
   Malvin C. Teich, *Columbia University*
   Bernard Yurke, *AT&T Bell Laboratories*

4. **Optical Interactions with Condensed Matter and Ultrafast Phenomena**
   Arto V. Nurmikko, *Brown University, Chair*
   Arthur H. Guenther, *Sandia National Laboratories*
   Erich P. Ippen, *Massachusetts Institute of Technology*
   Karl Leo, *Technical University of Dresden, Germany*
   Jason Stark, *AT&T Bell Laboratories*
   Duncan Steel, *University of Michigan*
   Henry van Driel, *University of Toronto, Canada*
   Masamichi Yamanishi, *Hiroshima University, Japan*
   Jeff F. Young, *University of British Columbia, Canada*

5. **Physics of Coherent Light Sources**
   Kerry J. Vahala, *California Institute of Technology, Chair*
   Luiz da Silva, *Lawrence Livermore National Laboratories*
   George Harvey, *AT&T Bell Laboratories*
   Thomas J. Kane, *Lightwave Electronics*
   Eli Kapon, *Swiss Federal Institute of Technology*
   Robert J. Lang, *Spectra Diode Laboratories*
   Howard M. Milchberg, *University of Maryland*
   Paul Morton, *AT&T Bell Laboratories*
   Margaret M. Murnane, *Washington State University*
   Klaus Streubel, *Royal Institute of Technology, Sweden*

Joint Council on Quantum Electronics

APS/Laser Science Topical Group

Thomas J. McIlrath, *University of Maryland, Chair*

Roger Falcone, *University of California, Berkeley*

William C. Stwalley, *University of Connecticut*

IEEE/Lasers &Electro-Optics Society

Arto V. Nurmikko, *Brown University*

John F. Reintjes, Jr., *U.S. Naval Research Laboratory*

William Silfvast, *University of Central Florida/CREOL*

OSA

Tony F. Heinz, *IBM T. J. Watson Research Center*

Jagdeep Shah, *AT&T Bell Laboratories*

Duncan G. Steel, *University of Michigan*

JCQE Publicity Chair

Robert W. Boyd, *University of Rochester*

JCQE Exhibits Chair

James J. Snyder, *Blue Sky Research*

JCQE Treasurer

Norman A. Kurnit, *Los Alamos National Laboratory*

CLEO Liaison to JCQE

Jagdeep Shah, *AT&TBell Laboratories*

Ex-Officio Members

Jeffrey Bokor, *University of California, Berkeley*
Richard R. Freeman, *AT&T Bell Laboratories*
James G. Fujimoto, *Massachusetts Institute of Technology*
Richart E. Slusher, *AT&T Bell Laboratories*
George I. Stegeman, *University of Central Florida/ CREOL*

Contents

Thursday, May 25

Friday, May 26

QMA 8:00 am
Hyatt A

Organic Nonlinear Optical Materials

Anthony J. Campillo, *U.S. Naval Research Laboratory, Presider*

QMA1 (Invited) 8:00 am

All-optical poling of polymers

F. Charra, C. Fiorini, J.-M. Nunzi, F. Kajzar, LETI, *Centre d'Etudes de Saclay, F-91191 Gif-sur-Yvette Cedex, France*

The orientation, under the influence of an electric field, of the dipoles of chromophores embedded in polymer matrices is now a widely used poling technique for the preparation of noncentrosymmetric polymer materials. However, in many cases, the use of light to achieve such vectorial orderings would be advantageous from both basic and practical points of view.

Now, whereas the electric-field poling permits a vectorial ordering of the chromophores, a linearly-polarized monochromatic light can only induce axial ordering, as exemplified by the dichroism induced by light in azo-dye doped polymers via a *trans* to *cis* isomerization (Ref. 1). Similarly, an optical poling can be expected to result from a photoselection of azo chromophores, based on their vectorial (and not only axial) orientation. The lowest-order polar optical effects are those resulting from a nonzero time average of the third power of the field, $\langle E^3\rangle_t$. The combination of a light field and its second harmonic possesses such a nonzero $\langle E^3\rangle_t$. Such a field superposition can break the intrinsic centrosymmetry of a material, as evidenced by the observation of light-induced second-harmonic-generation (SHG) in silica optical fibers.[2] It can also yield polar photoselective excitations, as we have demonstrated in a dye solution.[3] In azo-dye-doped polymers this interaction yields the expected vectorial ordering (poling) of the nonlinear optical chromophores.[4] Although two monochromatic optical fields of different frequencies do not interfere directly, the simultaneous 1-photon excitation of a quantum system by a field $E_{2\omega}$ at frequency 2ω and 2-photon excitation by a field E_ω at frequency ω yield quantum-interference effects, which results in a polar photoselection.[4]

We have used second-harmonic generation (SHG), which is forbidden in centrosymmetric media, as a probe for monitoring this light-induced noncentrosymmetry. Optical poling is achieved by simultaneous excitation of the sample by both fundamental (1064 nm) and second-harmonic (532 nm) light from a mode-locked *Q*-switched Nd^{3+}:YAG laser. A SHG susceptibility ($\chi^{(2)}(2\omega;\omega,\omega)$) of 90 pm/V has been reached, which is of the same order of magnitude as that obtained by electric-field (Corona) poling. Contrary to what has been observed in silica,[5] the spatial variations of SHG efficiency around the prepared spot reveal a single poled domain. This confirms that the optical-poling of this material results from a local effect. By controlling the phase between fundamental and second-harmonic preparation beams we have shown that the poling efficiency depends periodically on this phase, with a period of 180° as expected from a $E^*_{2\omega}E^2_\omega$ dependence of the optical poling. This dependence has important consequences as concerns nonlinear optical applications of optical poling. As a matter of fact, on longer propagation lengths, $\chi^{(2)}$ oscillate in space so as to fulfill the quasi-phase-matching conditions for efficient SHG. This property also permits the recording of holographic $\chi^{(2)}$ gratings.[6]

The possibility of inducing optically a vectorial ordering of organic molecules offers a simple way to realize new kinds of organized structures. For example, it permits the patterning of poled microstructures with controlled 3D arrangements. It can be applied to electrically conducting materials, *e.g.* for the polar ordering of ions.[7] In addition, the rich tensor properties of the $E^*_{2\omega}E^2_\omega$-optical preparation permits the preparation of organized structures with various kinds of noncentrosymmetric geometries, such as octupolar symmetries.[7]

1. C. Jones and S. Day, Nature **351**, 15 (1991).
2. R. H. Stolen and H. W. K. Tom, Opt. Lett. **12**, 585 (1987).
3. F. Charra, F. Devaux, J. M. Nunzi, and P. Raimond, Phys. Rev. Lett. **68**, 2440 (1992).
4. F. Charra, F. Kajzar, J. M. Nunzi, P. Raimond, and E. Idiart, Opt. Lett. **18**, 941 (1993).
5. T. J. Driscoll and N. M. Lawandy, J. Opt. Soc. Am. **B 11**, 355 (1994).
6. N. B. Baranova and B. Ya. Del'dovitch, JETP Lett. **45**, 716 (1987).
7. J. M. Nunzi, F. Charra, C. Fiorini, and J. Zyss, Chem. Phys. Lett. **219**, 349 (1993).

QMA2 8:30 am

Second-harmonic generation in new polymeric Langmuir–Blodgett films

Marguerite-Marie Denariez-Roberge, Michel St-Hilaire, Tarik Rahem,* Anna M. Ritcey,* *Center for Optics, Photonics, and Lasers, Department of Physics, Faculty of Science and Engineering, University of Laval, Québec, Québec G1K 7P4, Canada*

Media with strong second-order susceptibilities have always been sought for wave-mixing and phase-dependent amplification applications. Their interest is still increasing with new developments in $\chi^{(3)}$ phase nonlinearities obtained by cascading $\chi^{(2)}$ (Ref. 1). The Langmuir–Blodgett (LB) deposition technique is very attractive because it gives us the opportunity to build new media with specific nonlinear segments adequately oriented.

Recently, a new cellulose derivative, 6'-(*N*-methyl-*N*-*p*-nitrophenylamino)hexyl cellulose, was prepared from *N*-methyl-*p*-nitroaniline and cellulose acetate.[2] Stable monolayers of this polymer can be spread at the air–water interface and can be successfully transferred to hydrophillic glass substrates by vertical dipping at 20°C and a constant surface pressure of 2.5 mN/m. Transfer occurs only during substrate withdrawal with a ratio close to unity. No transfer is observed in the downstroke. Such behavior suggests that this derivative forms Z-type multilayers, with a head-to-tail organization. The LB films exhibit an absorption band with a maximum at 402 nm. The absorbance at this wavelength is a linear function of the number of transferred layers. Second-harmonic generation of a dye laser operating near 0.7 μm has been obtained in these samples. Its quadratic variation with the number of layers confirms a Z-type deposition. The variation of signal with the fundamental beam polarization allows us to estimate a mean orientation of 20° with respect to the sample normal for the nitroaniline group. The magnitude of the $\chi^{(2)}$ has been obtained relative to a crystalline quartz plate. This technique seems very promising: the density of chromophores along the polymer backbone can be increased, new noncentrosymmetric groups with stronger nonlinear response can eventually be introduced, and the number of layers can be increased.

**Department of Chemistry, Faculty of Science and Engineering, University of Laval, Québec, Québec G1K 7P4, Canada*

1. G. Assanto, D. Neher, G. I. Stegeman, W. E. Torruellas, M. B. Marques, W. H. G. Horthuis, and G. R. Mohlmann, Mol. Cryst. Liq. Cryst. **222**, 33 (1992).
2. P. Basque and A. Ritcey, submitted to Polymer Prepr.

QMA3 8:45 am

Origin of the two-dimensional optical nonlinearity of in-plane poled polymer films

P. M. Lundquist, S. Yitzchaik, T. J. Marks, G. K. Wong, R. Cohen,* G. Berkovic,* S. DiBella,** *Materials Research Center, Northwestern University, Evanston, Illinois 60208-3113*

Organic materials are widely studied for photonic applications because of their large nonlinear-optical (NLO) coefficients and their ease of fabrication.[1] Poled polymer films are particularly attractive candidates for photonic applications, such as optical switches and modulators. NLO effects arise from hyperpolarizable constituent chromophores that are macroscopically aligned by an applied electric field.[2] Corona poling and contact poling in the standard sandwich geometry result in optically uniaxial films; however, contact poling in nonuniform electric fields has been shown to result in surprisingly large NLO coefficients for polarizations normal to the direction of the applied electric field.[3]

The samples studied in the present work are polymethylmethacrylate thin films containing dispersed DANS dye molecules. Electrodes 300-Å thick were deposited 2 mm apart on a glass substrate that was then coated with the polymer film. In this geometry, the electric field inside the polymer

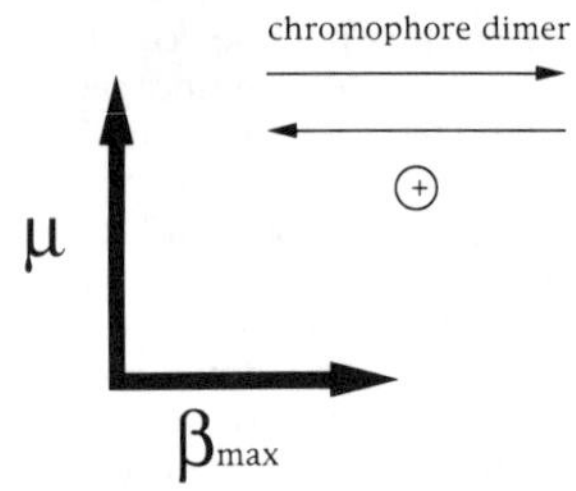

QMA3 Fig. 1. Schematic diagram of a two-chromophore aggregate (or dimer). The arrows denote the chromophore long axis.

has been shown to be principally in the plane of the film, pointing from one electrode to the other (defined as the x direction), with a small gradient normal to the film plane (defined as the z direction).[4] The field gradient in the z direction has been postulated to arise from charge injection and diffusion in the polymer film.[5] However, second-harmonic-generation studies reveal that $\chi^{(2)}_{zzz}$ is ~2 orders of magnitude larger than $\chi^{(2)}_{xxx}$. This type of response cannot be explained by the simple thermodynamic alignment of isolated chromophore molecules in the field. The frequency dependence of these NLO coefficients was examined by using an optical parametric amplifier[6] to determine whether the $\chi^{(2)}_{xxx}$ and $\chi^{(2)}_{zzz}$ responses originate from similar mechanisms. We find that the large magnitude of $\chi^{(2)}_{zzz}$, the coefficient perpendicular to the poling field, extends to longer wavelengths than does the parallel coefficient.

The formation of aggregates of chromophore molecules that trap injected charge has been proposed as a mechanism for the unusually large perpendicular response in such materials.[7] Figure 1 displays a schematic model of dimers that have trapped an injected hole. Recent calculations on these aggregates based on the CFG *ab initio* formalism indicate that the dimer electric-dipole moment is directed normal to the chromophore long axes but that the principle direction of β, the aggregate hyperpolarizability, is perpendicular to the dipole moment. This result could explain the large perpendicular $\chi^{(2)}_{zzz}$ that is observed. The predictions of this model are discussed in terms of the frequency dependence of the parallel and perpendicular NLO response.

*Department of Materials and Interfaces, Weizmann Institute of Science, Rehovot, Israel 76100
**Department of Chemistry, University of Catania, Via le A. Doria, I-95125 Catania, Italy

1. P. N. Prasad, *Contemporary Nonlinear Optics* (Academic, San Diego, Calif., 1992), p. 265.
2. K. D. Singer, J. E. Sohn, and S. L. Lalama, Appl. Phys. Lett. **49**, 248 (1986).
3. S. Yitzchaik, V. Krongauz, and G. Berkovic, Nonlinear Opt. **4**, 265 (1993).
4. S. Yitzchaik, G. Berkovic, and V. Krongauz, Chem. Mat. **2**, 162 (1990).
5. G. Berkovic, R. Cohen, and A. Donval, Proc. Soc. Photo-Opt. Instrum. Eng. **2285**, 76 (1994).
6. P. Lundquist, S. Yitzchaik, T. Zhang, D. R. Kanis, M. A. Ratner, T. J. Marks, and G. K. Wong, Appl. Phys. Lett. **64**, 2194 (1994).
7. R. Cohen, G. Berkovic, S. Yitzchaik, and V. Krongauz, Mol. Cryst. Liq. Cryst. **240**, 169 (1993).

QMA4 9:00 am

Novel nonlinear phenomena originating from nonlocality in optical response in ultrathin film

Hajime Ishihara, *Semiconductor Research Laboratory, Advanced Device Technology Department, PRESTO, Research Development Corporation of Japan, Mitsubishi Electric Corp., 1-1, Tsukaguchi Honmachi 8-Chome, Amagasaki, Hyogo 661, Japan*

A study of nonlocality in the nonlinear optical response shows that control of the intensity and the spatial distribution of the excitonically resonant internal field drives various novel nonlinear phenomena.

Recent studies have revealed the essential importance of nonlocality in the nonlinear optical response of confined electronic systems.[1–3] In nonlocal systems the motion of the internal field is determined self-consistently with that of the induced dipole, and this causes the resonant size enhancement of the internal field with a particular spatial pattern of mesoscopic scale. Because the strong size, shape, and internal-structure dependence of the nonlinear response appears through this effect, it is expected that the control of the internal field by the design of the sample structure will provide the remarkable nonlinear effects peculiar to mesoscopic systems. By nonlocal theory[1] based on the model of one-dimensional Frenkel excitons, two examples of new nonlinear responses in ultrathin films are presented.

As the first example, Fig. 1 shows transmitted intensity as a function of incident intensity calculated for a 226.8-Å-thick CuCl thin film, where the beam energy is 3.20322 eV, which is in the vicinity of the second one-exciton level. The notable points are the decrease of the transmitted intensity with an increase of input intensity and the bistable characteristic. In the case of strong nonlocality, the spatial distribution of the

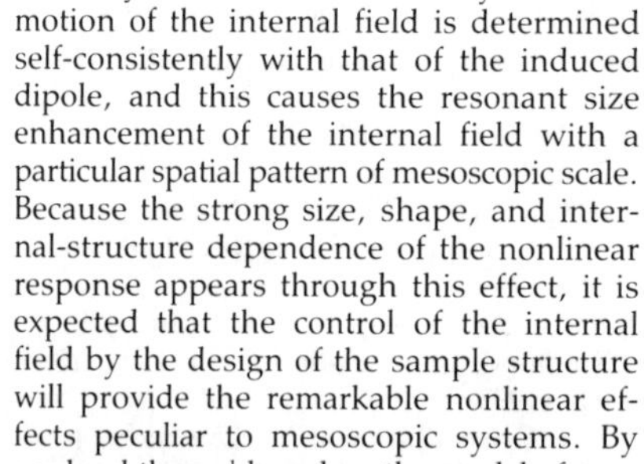

QMA4 Fig. 1. Transmitted intensity versus incident intensity for an ultrathin film with a 226.8-Å thickness. Normal incidence of the beam with 3.20322-eV energy is assumed. The material parameters of the Z_3 exciton of CuCl are chosen. The phase and population decay constants are 0.06 and 0.02 meV, respectively.

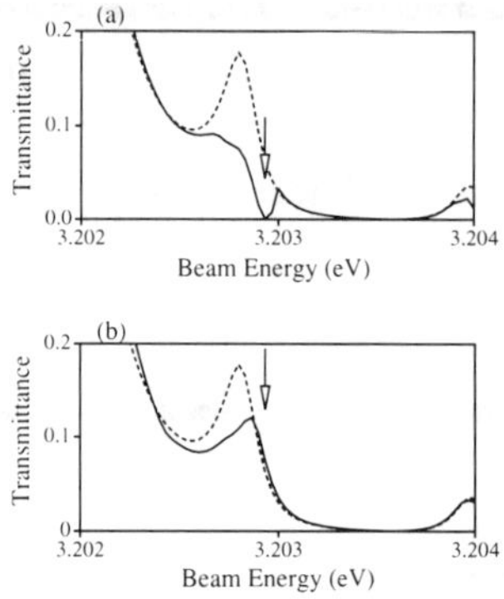

QMA4 Fig. 2. Transmittance spectra for an ultrathin film of CuCl with a 378.0-Å thickness containing the planar defect with infinite potential at 270.0 Å from a front surface. The beam intensity is 430 kW/cm², and the material parameters are the same as those in Fig. 1. Parts (a) and (b) are for normal incidence from the front surface and back surface, respectively. In each part the dashed curve denotes the linear response and the solid curve denotes the nonlinear response.

internal field varies strongly as the input intensity increases in the nonlinear regime. For the thickness and the beam energy used here, the intensity of the internal field near the back (front) surface of the film decreases (increases) with an increase of the input intensity. This causes the inverted input–output characteristic. On the other hand, the strong spatial dependence of the internal field indicates the existence of a feedback effect owing to multiple reflections of the resonant internal field with a very short wavelength at the surface of the sample. This brings about bistability, even in such a thin film.

The second example is a nonlinear response in a composite thin film that contains a planar defect, an artificially inserted layer or a delta-doped layer that acts as a potential wall for the translational motion of free excitons. The potential wall also causes multiple interference of the internal field through the coherent scattering of excitons. If this wall is not located at the center of the film, the intensities of the internal field for the incidences from the front surface (case F) and from the back surface (case B) are generally different. Making use of this fact, we can realize a condition for which the strength of the nonlinearity for case F is very different from that for case B. Figure 2 shows the nonlinear transmitted spectra for a 378.0-Å-thick CuCl thin film containing the planar defect with infinite potential at 270.0 Å from the front surface. For the beam energy indicated by arrows, the transmittance for case F is very different from that for case B, in which the thin film acts as an optical isolator.

1. H. Ishihara and K. Cho, Phys. Rev. B **48**, 7960 (1993).
2. K. Cho, H. Ishihara, and Y. Ohfuti, Solid State Commun. **87**, 1105 (1993).
3. H. Ishihara and K. Cho, Solid State Commun. **89**, 837 (1994).
4. H. Ishihara and K. Cho, J. Phys. Soc. Jpn. **60**, 4431 (1991).

QMA5 **9:15 am**

Two-photon states of two phthalocyanines

Maria A. Diaz-Garcia,
Fernando Agullo-Lopez,
William E. Torruellas,*
George I. Stegeman,* *Department of Materials C-IV, Autonomous University of Madrid, Cantoblanco, Madrid 28049, Spain*

Phthalocyanines have received a large amount of attention because of their potential for nonlinear optical applications.[1–4] However, little is known about the origin of their nonlinear optical response. Because they belong to a family of two-dimensional centrosymmetric organic molecules with extensively delocalized conjugated π-electrons, one can predict the existence and the crucial role of two-photon states in their nonlinear optical response.[5] Symmetry selection rules make these two-photon states invisible to single-photon transitions. We have therefore investigated spectroscopically the existence of such two-photon states with third-harmonic generation (THG) in metal-free phthalocyanine (PcH_2) and copper-substituted phthalocyanine (PcCu) thin films.

THG probes the ultrafast third-order response of the two phthalocyanine molecular systems and does not suffer from the contributions of excited populations, known to play a dominant role in optical limiters.[4] We show in Fig. 1 the spectrum of the magnitude of the third-order susceptibility $\chi^{(3)}(3\omega)$ for the metal-free molecule. When one compares this spectrum with the linear absorption spectrum, also shown in Fig. 1, one could think that only states accessible through one-photon transitions are significant in the third-order nonlinear optical response. Our analysis shows otherwise.

Indeed, when a model that includes only the ground state and the well-known visible bands (*Q* band at 670 nm and *B* band at 340 nm) is compared with our experimental results, a poor fit is obtained, particularly for the phase of $\chi^{(3)}(3\omega)$; see Fig. 2. One should notice that such a model does not have any adjustable parameters;

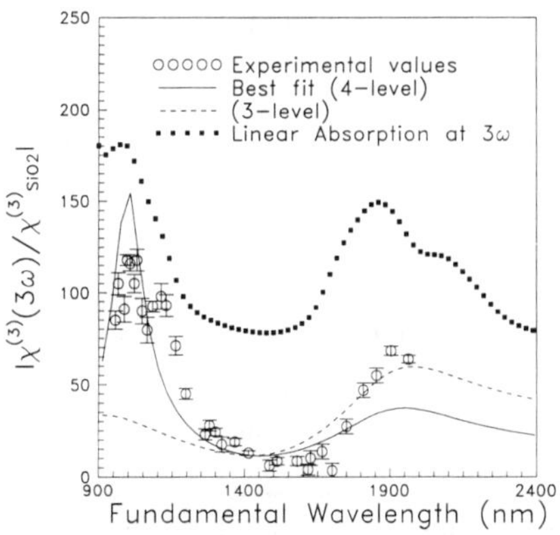

QMA5 Fig. 1. Relative magnitude (with respect to fused silica) of $\chi^{(3)}(3\omega)$ for sublimated PcH_2 thin films: open circles, THG experimental results; solid curve, best fit to the four-level model; dashed curve, best fit to a model in which only the *Q* and *B* bands are included; dotted curve, linear absorption (shown for comparison).

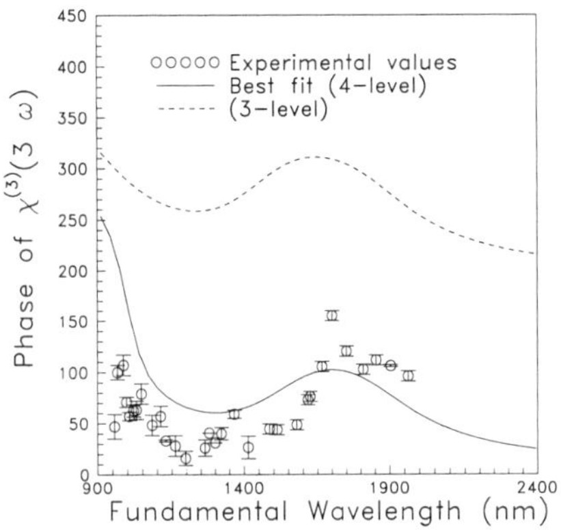

QMA5 Fig. 2. Relative phase of $\chi^{(3)}(3\omega)$ for sublimated PcH_2 thin films. Dashed curve clearly indicates the inadequacy of a model in which two photon states are omitted.

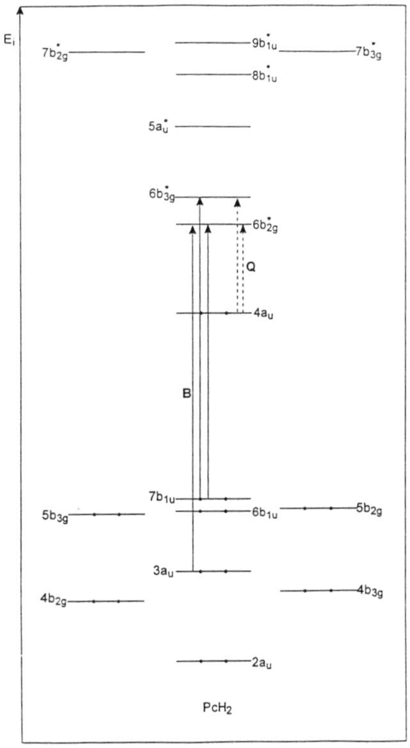

QMA5 Fig. 3. Valence effective Hamiltonian one-electron energy levels calculated for PcH_2.[6]

they are all obtained from the linear absorption spectrum. The poor fit in the region of the *B* band is due to quantum interferences of different pathways present in the nonlinear susceptibility. Because such a model includes only one- and three-photon resonances, a phase of approximately 180° is expected for $\chi^{(3)}(3\omega)$ in the spectral region under investigation. On the other hand, when a two-photon band is empirically added to the previous model, a much better agreement can be obtained, as shown in Figs. 1 and 2. Such a model is in good agreement with effective Hamiltonian calculations on PcH_2.[6] The inclusion of an additional two-photon state at 500 nm between the *B* and *Q* bands corresponds to transitions from the $4a_u$ to any of the $5a_u$*, $8b_{1u}$*, and $9b_{1u}$* states predicted by the valence-electron Hamiltonian calculation (Fig. 3) in the 400–600-nm spectral range. Similar results on PcCu and a comparison with the previously mentioned molecular calculations are presented.

**Center for Research and Education in Optics and Lasers, University of Central Florida, Orlando, Florida 32826*

1. Z. Z. Ho, C. Y. Ju, and W. M. Hetherington, J. Appl. Phys. **62**, 716 (1987).
2. R. A. Norwood, J. R. Sounik, D. Holcomb, J. Popolo, D. Swanson, R. Spitzer, and G. Hansen, Opt. Lett. **17**, 577 (1992).
3. H. S. Nalwa, Adv. Mater. **5**, 341 (1993).
4. J. Perry, K. Mansour, S. Marder, K. J. Perry, D. Alvarez, and I. Choong, Opt. Lett. **19**, 625 (1994).
5. M. A. Diaz-Garcia, I. Ledoux, J. A. Duro, T. Torres, F. Agullo-Lopez, and J. Zyss, J. Chem. Phys. **98**, 8761 (1994); M. A. Diaz-Garcia, F. Agullo-Lopez, M. G. Hutchings, P. F. Gordon, and F. Kajzar, Proc. Soc. Photo-Opt. Instrum. Eng. **2285**, 227 (1994).
6. E. Orti, J. L. Bredas, and C. Clarisse, J. Chem. Phys. **92**, 1228 (1990).

QMA6 **9:30 am**

Linear and nonlinear optical properties of phthalocyanine glasses

Steven R. Flom, James S. Shirk,
Richard G. S. Pong, F. J. Bartoli,
Robert D. George,* Arthur W. Snow,*
Optical Sciences Division, U.S. Naval Research Laboratory, Washington, D.C. 20375

Peripherally substituted phthalocyanines are of considerable interest for nonlinear-optical[1] applications. In particular, these materials have proven to be useful in optical-limiting applications[2] because the excited-state absorption coefficient exceeds that of the ground state over a large portion of the visible region. Soluble phthalocyanines have been synthesized and used for nonlinear-optical studies in solution. These materials tend to stack cofacially in organized structures that influence the photodynamics and the nonlinear-optical properties. In addition, pure thin films of these materials show optical losses due to the scattering and birefringence associated with the tendency to form liquid-crystal-like stacks.

Recently, a class of phthalocyanine glasses has been synthesized.[3] In these materials, bulky substituents with irregular symmetry and the capability to form intermolecular hydrogen bonds are introduced onto the periphery of the phthalocyanine. These groups inhibit the formation of cofacially stacked structures. In addition, the glass transition temperature T_g is depressed to the point that the materials can be processed into isotropic thin films. These thin films have high concentrations of the phthalocyanine moiety, low light scattering, and no birefringence. The phthalocyanines can also be incorporated into polymeric epoxy hosts, with little apparent phase separation.

Here we report the nonlinear-optical

properties of metal-free and Pb-substituted tetrakis (3-(*N*,*N*-bis(3-phenoxy-2-hydroxypropylether)aminophenoxy)phthalocyanine, which are examples of such phthalocyanine glasses. Degenerate four-wave-mixing (DFWM) nonlinear-transmission (NLT) experiments and picosecond transient-absorption experiments were performed on each of the films.

The transient-absorption spectra of both films after optical pumping at 590 nm have broad regions of induced absorption in the visible region. The excited-state spectrum of the lead material was broader and stronger than that of the metal-free material. Transient absorption demonstrates that $PbPc(PGE)_4$ has a lifetime substantially longer than 1 ns, whereas the metal-free films decay within tens of picoseconds.

Time-resolved DFWM signals at 590.5 nm for the films at various fluences are shown in Figs. 1 and 2. A fluence-dependent exciton–exciton annihilation is observed in the metal-free film, as shown in Fig. 1. The observed lifetimes of the metal-free material are significantly longer than those of $H_2Pc(CP)_4$ (Ref. 4) and other substituted-phthalocyanine thin films (Ref. 5). In contrast, the Pb-substituted material showed no fluence dependence at similar fluences over the time scale studied.

We conclude that these phthalocyanine glasses have excited states with absorption cross sections similar to those of other soluble phthalocyanines. The molecular structure inhibits cofacial stacking, which, in turn, inhibits the exciton–exciton annihilation rate in the excited state. The thin films also show improved optical qualities.

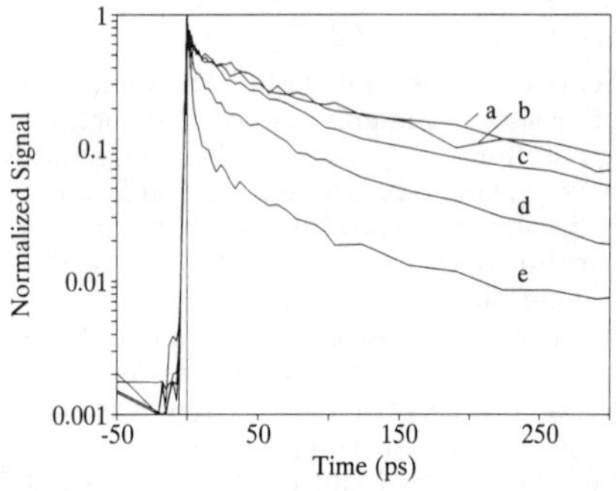

QMA6 Fig. 1. Normalized DFWM signal observed from a glassy $H_2Pc(PGE)_4$ film at fluences of (a) 0.16, (b) 0.45, (c) 1.4, (d) 2.9, and (e) 6.8 mJ/cm².

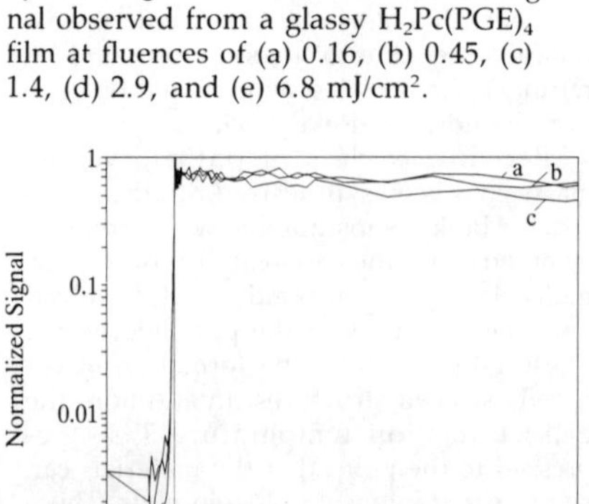

QMA6 Fig. 2. Normalized DFWM signal observed from a glassy $PbPc(PGE)_4$ film at fluences of (a) 0.45, (b) 0.91, and (c) 4.5 mJ/cm².

**Chemistry Division, U.S. Naval Research Laboratory, Washington, D.C. 20375*

1. J. S. Shirk, J. R. Lindle, F. J. Bartoli, Z. H. Kafafi, A. W. Snow, and M. E. Boyle, Intl. J. Nonlin. Opt. Phys. **1**, 699 (1992).
2. J. S. Shirk, R. G. S. Pong, and F. J. Bartoli, Appl. Phys. Lett. **63**, 1880 (1993).
3. R. D. George and A. W. Snow, submitted to Chem. Mater.
4. S. R. Flom, J. S. Shirk, J. R. Lindle, F. J. Bartoli, Z. H. Kafafi, R. G. S. Pong, and A. W. Snow, Mater. Res. Soc. Symp. Proc. **247**, 271 (1992).
5. B. I. Greene and R. R. Millard, Phys. Rev. Lett. **55**, 1331 (1985); M. K. Casstevens, M. Samoc, J. Pfleger, and P. N. Prasad, J. Chem. Phys. **92**, 2019 (1990); Z. Z. Ho and N. Peyghambarian, Chem. Phys. Lett. **148**, 107 (1988).

QMA7 **9:45 am**

Dispersion of third-order optical susceptibility of epitaxial phthalocyanine films prepared by organic MBE

J. P. Jiang, H. Tajalli,* J. Murray, S. Mazumdar,** A. Schmidt,† N. R. Armstrong,† N. Peyghambarian, *Optical Sciences Center, University of Arizona, Tucson, Arizona 85721*

Phthalocyanine molecules have been the subject of vigorous studies because of their two-dimensional delocalized electronic structures and the possibility of forming epitaxial films by molecular-beam-epitaxy (MBE) techniques.[1–3] Although the third-order nonlinear optical properties of phthalocyanines have been of interest for awhile, existing experimental data are for specific wavelengths only. One of the more definitive means for probing the underlying mechanism of nonlinearity is the acquisition of spectra of the third-order nonlinear susceptibility over a wide spectral range. We have obtained what we believe to be the first wavelength-dependent third-harmonic-generation (THG) spectrum of an epitaxially prepared metallophthalocyanine film over a wide wavelength range (900–1800 nm).

Thin films of Cl-In-Pc containing different epitaxial phases of the molecular arrangement were used in our experiment. The linear absorption spectra of an epitaxial phase and a polycrystalline phase are shown in Fig. 1. The choice of different epitaxial forms of phthalocyanine resulted from our desire to see the effect of intermolecular interaction in metal-substituted phthalocyanines on their nonlinear optical properties. The two distinct bands of Soret (or *B*) and *Q* bands are clearly observable. Wavelength-tunable THG was used to measure the third-order optical nonlinear coefficient of Cl-In-Pc. The tunable laser source was an optical parametric generator–amplifier (OPG-OPA) using two BBO crystals, which was pumped by the frequency-tripled (355-nm) output from a mode-locked Nd:YAG laser with a 35-ps pulse width (Continuum model PY61). Maker fringes were measured and were used

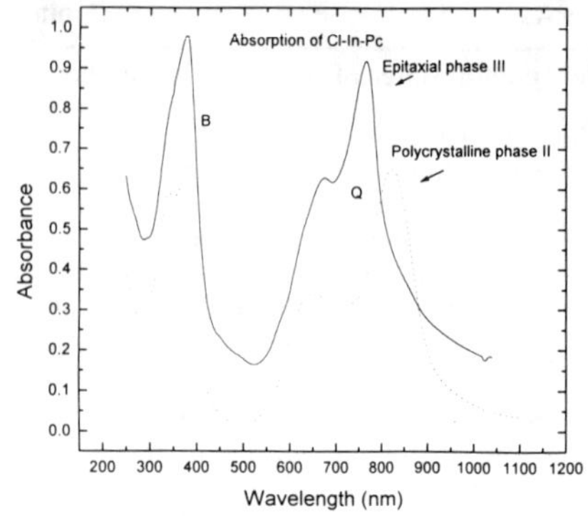

QMA7 Fig. 1. Linear absorption spectra of epitaxial phase III and polycrystalline phase II of Cl-In-Pc.

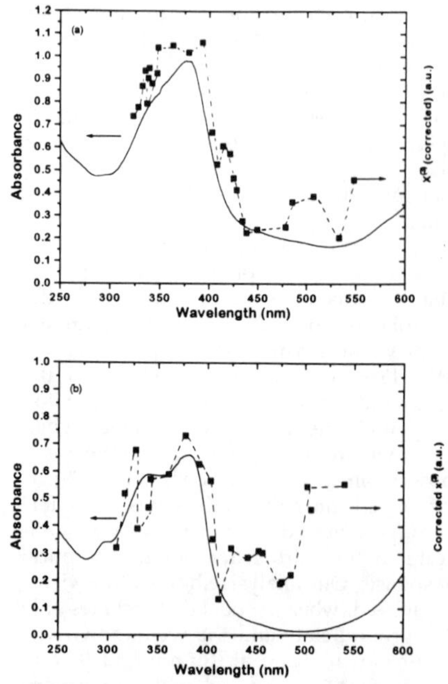

QMA7 Fig. 2. (a) Spectrum of third-order nonlinear susceptibility $\chi^{(3)}$ of epitaxial phase III of Cl-In-Pc (points); solid curve is a portion of its linear absorption curve. (b) Same as in (a) except polycrystalline phase II of Cl-In-Pc (points).

to extract the sample contribution to the THG from the convolution with the substrate.

Figures 2(a) and 2(b) show the measured $\chi^{(3)}$ spectra of two Cl-In-Pc films superimposed on their respective linear absorption curves. There are two three-photon-resonance peaks that occur in the 370-nm region. Because the wavelength plotted on the *x* axis is the THG signal measured at one third of the fundamental wavelength, these two peaks correspond to three-photon resonance to the Soret band. More significantly, in each case two other resonances between 400 and 600 nm are seen below the intense three-photon resonance. These resonances have no counterparts in the linear absorption. These could be (a) two-photon resonances due to even-parity two-photon accessible states or (b) three-photon resonances due to an odd-parity state that is weakly dipole coupled to the ground state but that is significant in nonlinear absorption. We have carried out Hartree–Fock and single-CI calculations that indicate that the above resonances are not three-photon resonances; hence they are both two-photon resonances. However, although the calculations show two two-photon states between the Q and Soret bands, the higher-energy two-photon state is too close to the states giving rise to the Soret band to appear below the three-photon resonance in THG. This indicates that, as with other π-conjugated systems, many-electron interactions lower the energies of the two-photon states relative to those of the one-photon states. Interestingly, the two-photon resonances shift according to the epitaxial phase of the phthalocyanine molecule; as the Q band is red shifted in epitaxial phase III, so are the two-photon absorption levels. These features provide an interesting opportunity to study the effect of intermolecular interaction in molecules with a highly delocalized two-dimensional electron system.

Work is in progress to understand the role of Coulomb interaction and the detailed mechanism of optical nonlinearity in metallophthalocyanines.

The authors would like to acknowledge support from the U.S. Air Force Office of Scientific Research and the National Science Foundation.

*On leave from Department of Physics, University of Tabriz, Tabriz, Iran
**Also with Department of Physics, University of Arizona, Tucson, Arizona 85721
†Also with Department of Chemistry, University of Arizona, Tucson, Arizona 85721

1. A. M. Schaffer, M. Gouterman, and E. R. Davidson, Theor. Chim. Acta **30**, 9 (1973).
2. Z. Z. Ho, C. Y. Yu, and W. M. Hetherington, J. Appl. Phys. **62**, 716 (1987).
3. L. K. Chau, C. D. England, S. Chen, and N. R. Armstrong, J. Phys. Chem. **97**, 2699 (1993).

QMB 8:00 am
Hyatt B

Entangled States and Bell's Inequalities

Roy S. Bondurant, *MIT, Presider*

QMB1 (Invited) 8:00 am

Spatiotemporal coherence properties of parametric downconverted light

Bahaa E. A. Saleh, Adel Joobeur,*
Department of Electrical, Computer and Systems Engineering, Boston University, Boston, Massachusetts 02215

The coherence properties of spontaneous parametric downconverted light are governed by the energy and phase-matching conditions. Ideally, this implies a strict one-to-one correspondence between each direction (signal) and a matching direction (idler). In the type-I configuration the light takes the form of concentric rings and is monochromatic everywhere, but the frequency varies with the polar angle. It is spatially incoherent, but the coherence time is infinite. The fourth-order coherence function (rate of photon coincidence) at pairs of directions vanishes, except for directions satisfying perfect matching. The entanglement time is infinite.

We have examined these properties under practical conditions for which the pump has a finite spectral width and the interaction region is of limited extent, so that there is some flexibility in satisfying the energy and phase-matching conditions. The coherence area is then finite, but the Van Cittert-Zernike theorem is not applicable. The coherence time is also finite. Photons in one direction are entangled with photons in directions within a cone surrounding the perfect matching direction, whose angles are the entanglement angles. The entanglement time is finite.

We have developed an approximate multi-dimensional Gaussian expression for the wave function describing the downconverted light for finite pump spectral width and beam waist, and finite crystal width. The magnitude of the second-order coherence function is a product of Gaussian functions of the azimuthal angles, the polar angles, and the time delay. Its phase is a linear function of time delay, representing a spectral shift that varies with the polar angles. The coherence angles generally decrease as the pump beam width or the crystal width increases. The coherence time increases as the length of the crystal or the beam width increases, and also as the pump spectral width is reduced. Although the pump beam possesses complete spatial coherence, the downconverted beam has an elliptically shaped coherence area. Even if the pump is perfectly monochromatic, the downconverted light is not.

The photon coincidence rate is also separable as a product of Gaussian functions of the azimuthal angles, the polar angles, and the time delay, with widths representing the entanglement angles and time. The azimuthal entanglement angle increases with decrease of the pump beam width, but is independent of the crystal width and the pump spectral width. The polar entanglement angle is independent of the pump beam width, but increases with reduction of the crystal width or increase of the pump spectral width. The entanglement area is also elliptical, but it does not match the coherence area, so that the Siegert relation is not satisfied. The entanglement time increases with increase of the pump beam width or with decrease of the pump spectral width. When the signal and idler photons are collected by apertures of finite areas, the photon coincidence rates drop sharply if the aperture areas mismatch the entanglement areas in size or location. These effects have been demonstrated experimentally.

*Department of Electrical and Computer Engineering, University of Wisconsin, Madison, Wisconsin 53706

QMB2 8:30 am

Delayed-choice quantum eraser: theory and experiment

Thomas Herzog, Paul Kwiat,
Harald Weinfurter, Anton Zeilinger, *Institute for Experimental Physics, University of Innsbruck, 6020 Innsbuck, Austria*

The complementary nature of wavelike and particlelike behavior is commonly interpreted as follows: Because of the uncertainty principle, any attempt to measure the position (particle aspect) of a quantum will lead to an uncontrollable, irreversible disturbance in its momentum, thereby washing out any interference pattern (wave aspect).[1] However, in some circumstances no state reduction is necessary for destroying wavelike behavior, and measurements can be reversible in a certain sense. To fully understand this phenomenon, one must view the loss of coherence as arising from an entanglement of the system wave function with that of the measuring apparatus (MA). Through this entanglement, previously interfering paths can become distinguishable such that no interference is observed. The mere possibility of obtaining welcher-Weg ("which-path") information is sufficient to destroy interference. It may be regained, however, if one somehow manages to erase the distinguishing information through a suitable measurement on the MA and correlates these results with measurements on the original quantum. This is the physical content of quantum erasure.[2] There are four basic measurements possible on the MA—two of which yield which-path information, one of which recovers the initial interference fringes, and one which yields interference antifringes. The decision to measure wavelike or particlelike behavior may be delayed until after *detection* of the quantum, an irreversible process. In all cases one must correlate the results of measurements on the MA with this detection.

In the previous demonstration of the quantum-erasure phenomenon,[3] the which-path information was carried by the same particles that interfered and a delayed-choice implementation was difficult. Here we report new results obtained by using a system that permits such an implementation. The signal and idler modes associated with a right-going UV pump photon are superposed with the signal and idler modes associated with a left-going pump photon (Fig. 1). In the absence of the wave plate in the idler arm (lower right), the singles rate at D_S displays fringes as the mirror M_S is translated.[4] These fringes are possible because one cannot tell, even in principle, whether the signal-and-idler pair was created from a left-going or a right-going pump photon. If a wave plate is used to rotate the polarization of the idler photon by 90°, then these two possibilities are distinguishable and interference at D_S vanishes. However, by using a polarizer at 45° before the idler detector, one can erase the distinguishability and recover interference fringes, upon correlating detections at D_S and D_I (Fig. 2). Note that with a fast polarization switch in front of D_I, the decision to measure which-path information or to recover

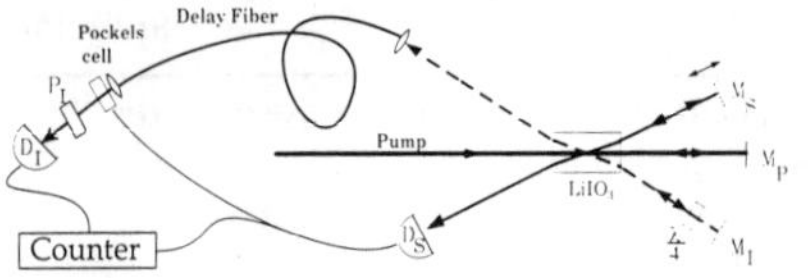

QMB2 Fig. 1. Schematic diagram of experiment to demonstrate a delayed-choice quantum eraser. In the absence of the wave plate in the idler arm (lower right), the singles rate at D_S displays fringes as the mirror M_S is translated. Interference vanishes if a wave plate is used to rotate the polarization of the idler photons by 90°. By using a polarizer P_I at 45° before the idler detector, one can erase the distinguishability and can recover interference, upon correlating detections at D_S and D_I. With a delay and a fast switch (Pockels cell) in front of D_I, the decision to measure which-path information or to recover interference may be made *after* the signal photon is detected.

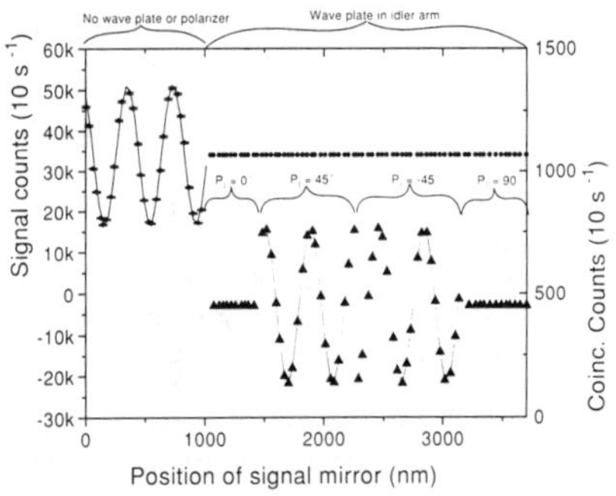

QMB2 Fig. 2. In the absence of any polarizers or wave plates, interference is observed in the singles rate D_S (circles, left axis) as a function of mirror M_S's position. With a polarization rotator (wave plate) in the idler arm, the interference should vanish. If one correlates the measurements of the signal photons with the measurement of the polarization of the idler photons, which-path information or interference fringes can be recovered (triangles, right axis).

interference may be made *after* the signal photon is detected.

As discussed in Ref. 5, another method of distinguishing the interfering paths is simply to increase the path length of the idler photon by moving mirror M_I a distance greater than the coherence length of the downconverted photons (about 30 μm). Again, interference fringes at D_S vanish, but may be recovered by coincidence counting with a narrow-bandwidth filter in front of D_I.

1. N. Bohr, in *Quantum Theory and Measurement*, J. A. Wheeler and W. H. Zurek, eds. (Princeton University Press, Princeton, N.J., 1983), p. 32.
2. M. O. Scully, B.-G. Englert, and H. Walther, Nature **351**, 111 (1991).
3. P. G. Kwiat, A. M. Steinberg, and R. Y. Chiao, Phys. Rev. A **45**, 7729 (1992).
4. T. J. Herzog, J. G. Rarity, H. Weinfurter, and A. Zeilinger, Phys. Rev. Lett. **72**, 629 (1994).
5. P. G. Kwiat, A. M. Steinberg, and R. Y. Chiao, Phys. Rev. A **49**, 61 (1994).

QMB3 8:45 am

Polarization-entangled states: the next generation

Paul Kwiat, Harald Weinfurter, Anton Zeilinger, A. V. Sergienko,* Y. H. Shih,* *Institute for Experimental Physics, University of Innsbruck, 6020 Innsbruck, Austria*

We have discovered theoretically and confirmed experimentally that a vast improvement over the traditional (type-I) downconversion sources (in which both correlated photons have the same polarization) is possible by using type-II downconversion (in which the correlated photons are orthogonally polarized).[1] In previous systems it was necessary to use a number of very carefully aligned extra optical elements, beam splitters, mirrors, etc., to obtain and measure the quantum-mechanical entangled states required for many fundamental experiments.[2,3] Even then, the actual state produced was usually a product state, and only by selecting out half of the correlated photons could an entangled subensemble be realized.[4] Whole classes of experiments (e.g., loophole-free tests of Bell's inequalities) are then impossible, even in principle.[5] With our new development, there are, in fact, special directions out of the crystal for which the photons are automatically in the desired quantum state, with no need for extra elements or for discarding any counts.

Figure 1 shows a schematic diagram of type-II downconversion (e.g., in BBO); only the degenerate case is shown, and so the frequencies of the two photons are identical. The correlated photon pairs are spontaneously emitted in cones, one extraordinary (*e*) polarized and the other ordinary (*o*) polarized; correlated photons lie dramatically opposed, on opposite sides of the pump. When the angle between the pump and the optic axis is 48.9°, the two cones are exactly tangent along the pump direction, allowing collinear phase matching. However, for slightly greater angles, there are two directions outside the crystal for which the *o*-cones and the *e*-cones intersect. Along these directions, the light appears unpolarized. Nevertheless, there are strong correlations in the polarizations of photons in the two directions; the state of the light

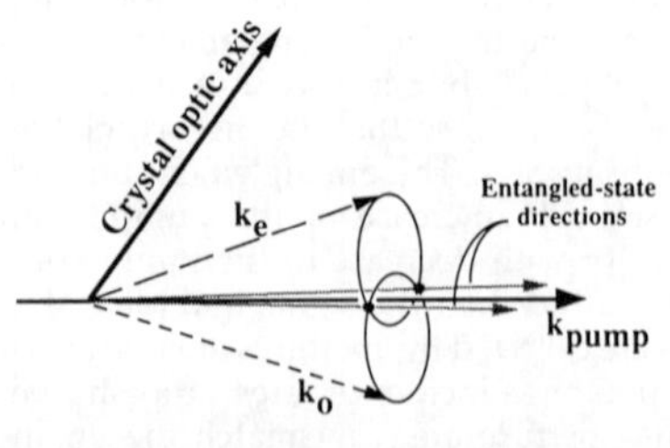

QMB3 Fig. 1. Degenerate downconversion process in type-II phase matching (in BBO). The *e*-polarized and the *o*-polarized photons of any correlated pair lie on opposite sides of the pump. Outside the crystal there are two directions for which the two cones intersect. The photons in these directions are polarization entangled. (Cone angles are exaggerated for clarity.)

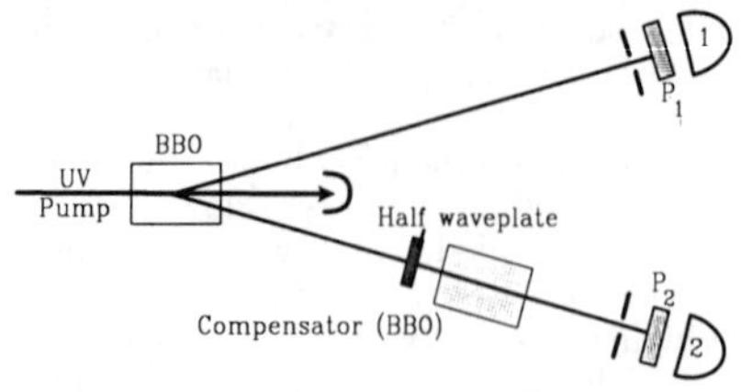

QMB3 Fig. 2. Simplified schematic diagram of experiment. Orientations of the polarizers P1 and P2 can be varied. The phase factor φ was adjusted by slight tilts of the compensating BBO crystal.

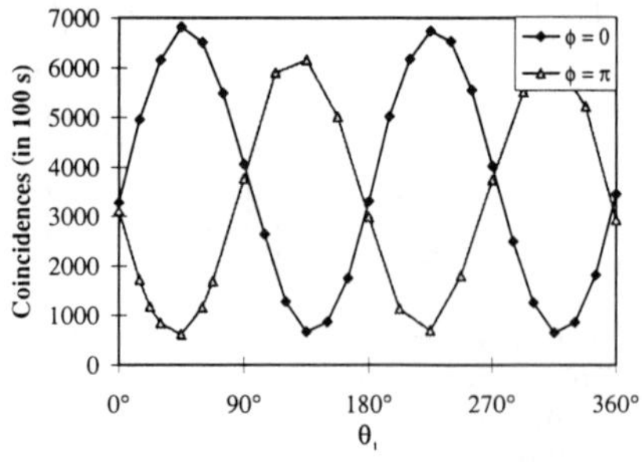

QMB3 Fig. 3. Coincidence fringes for the $\varphi = 0$ and $\varphi = \pi$ states when P2 is fixed at 45°. Visibilities in excess of 88% were observed.

in these directions (called 1 and 2) is

$$|\Psi\rangle \propto \frac{1}{\sqrt{2}}\left(|e_1, o_2\rangle + e^{i\varphi}|o_1, e_2\rangle\right),$$

where the phase φ is adjustable by small tilts of the downconversion crystal. One can thus readily switch between singletlike ($\varphi = 0$) and tripletlike ($\varphi = \pi$) states.

The birefringence of the 3-mm BBO crystal leads to a rather large longitudinal walk-off. We used a second 3-mm crystal in one of the arms to compensate this, after a half-wave plate to exchange the *o*-polarization and the *e*-polarization (see Fig. 2). The nonlocal polarization correlations were measured by observing the coincidence rate as a function of the orientation angles of the polarizers in front of the detectors. For the $\varphi = 0$ state, the rate displayed sinusoidal fringes depending only on the difference of the polarizer angles, with fringe visibilities in excess of 88%. For the $\varphi = \pi$ state, the fringes depended instead on the sum of the polarizer angles (see Fig. 3). Appropriate Bell's inequalities were violated by over 38 standard deviations for both states, modulo the usual supplementary assumptions.

The new method has another tremendous advantage: Whereas previous experiments required weeks or even months to properly align, the new system took just two days and is remarkably stable! We believe that it will be adopted as the standard for producing polarization-entangled states.

**Department of Physics, University of Maryland, Baltimore, Maryland 21228*

1. P. G. Kwiat, H. Weinfurter, A. Zeilinger, A. V. Sergienko, and Y. H. Shih, submitted to Phys. Rev. Lett.
2. Z. Y. Ou and L. Mandel, Phys. Rev. Lett. **61**, 50 (1988).

3. P. G. Kwiat, A. M. Steinberg, and R. Y. Chiao, Phys. Rev. A **47**, R2472 (1993).
4. T. E. Kiess, Y. H. Shih, A. V. Sergienko, and C. O. Alley, Phys. Rev. Lett. **71**, 3893 (1993).
5. P. G. Kwiat, P. H. Eberhard, A. M. Steinberg, and R. Y. Chiao, Phys. Rev. A **49**, 3209 (1993).

QMB4 (Invited) **9:00 am**

Implications of the recent experimental realization of the Einstein-Podolsky-Rosen paradox

M. D. Reid, *Physics Department, University of Waikato, Hamilton, New Zealand*

In this presentation I discuss the implications of the recent experiment by Ou *et al.*[1] This experiment is the first realization of the gedanken experiment considered by Einstein, Podolsky, and Rosen[2] (EPR) in 1935 as an argument for the incompleteness of quantum mechanics. I conclude that the Ou *et al.* realization of the EPR experiment suggests that there exists an even stronger paradox than that of EPR. Either: 1. We reject macroscopic local realism, where macroscopic local realism will be defined in this paper; or; 2. Accepting macroscopic local realism, we must accept as a consequence that quantum mechanics is incomplete.

The strength of this paradox is in the new feature that macroscopic local realism must now be rejected, if quantum mechanics is to be regarded as complete.

In the Ou *et al.* experiment one generates spatially separated pairs of fields $\hat{a}_1$, $\hat{a}_2$ and $\hat{b}_1$, $\hat{b}_2$ at A and B respectively. The observables measured are the Schwinger operators $S_x^A = \hat{a}_1\hat{a}_2^\dagger + \hat{a}_1^\dagger\hat{a}_2$ and $S_y^A = (\hat{a}_1\hat{a}_2^\dagger - \hat{a}_1^\dagger\hat{a}_2)/i$ at A, and $S_x^B = \hat{b}_1\hat{b}_2^\dagger + \hat{b}_1^\dagger\hat{b}_2$ and $S_y^B = (\hat{b}_1\hat{b}_2^\dagger - \hat{b}_1^\dagger\hat{b}_2)/i$, at B. Here $\hat{a}_i$, $\hat{a}_i^\dagger$, $\hat{b}_i$, $\hat{b}_i^\dagger$, are the usual boson creation and destruction operators. The fields $\hat{a}_1$ and $\hat{b}_1$ are generated via nondegenerate parametric oscillation, while $\hat{a}_2$ and $\hat{b}_2$ are intense coherent ("local oscillator") fields. Commutation relations imply the Heisenberg uncertainty relation $\Delta S_x^A \Delta S_y^A \geq |\langle S_z^A\rangle|$, where $S_z^A = \hat{a}_2^\dagger\hat{a}_2 - \hat{a}_1^\dagger\hat{a}_1$. Since $\langle\hat{a}_1^\dagger\hat{a}_1\rangle$ is negligible compared to $\langle\hat{a}_2^\dagger\hat{a}_2\rangle$, we may re-express the Heisenberg relation as:

$$\Delta S_x^A \Delta S_y^A \geq E^2 \quad (1)$$

where $\hat{a}_2$ is written as a classical amplitude E, assumed real, and such that $E^2 = \langle\hat{a}_2^\dagger\hat{a}_2\rangle$, the photon number of the local oscillator field. The nondegenerate parametric oscillator produces correlated fields, so that we may estimate S_x^A as $g_1 s_x^B$ where s_x^B is the result of the measurement of S_x^B. Similarly we may estimate S_y^A as $g_2 s_y^B$ where s_y^B is the result of the measurement of S_y^B. The variance of the error in our estimates is quantum mechanically given by $\Delta_1^2 = \Delta^2\hat{\delta}_x$ and $\Delta_2^2 = \Delta^2\hat{\delta}_y$ where $\hat{\delta}_x = S_x^A - g_1 S_x^B$ and $\hat{\delta}_y = S_y^A - g_2 S_y^B$. The experiment of Ou *et al.* achieved: $\Delta_1\Delta_2 = (0.70 \pm 0.01)E^2$. Since $\Delta_1\Delta_2 < E^2$, this is a realization of the EPR paradox.

In the original argument given by EPR, the fluctuations corresponding to the relevant quantum limit, that of position and momentum, were microscopic. Local realism is then used to establish the existence of "elements of reality" defined to a microscopic level. In contrast, the minimum quantum noise level $\Delta^2 S_x^A = \Delta^2 S_y^A = E^2$ corresponding to the equality (1) is macroscopic. We note that the quantities measured in the Ou *et al.* experiment are macroscopic. We note also that one can make the necessarily precise S_x^A measurements without ever having to know precisely the number of photons detected. Experimentally one needs never to resolve the number of photons detected to a microscopic number of photons. The "elements of reality" associated with the S_x^A and S_y^A are intrinsically macroscopic.

The essence of the EPR argument is that, provided the EPR correlations can be experimentally realized, local realism (as defined by EPR) implies that quantum mechanics is incomplete. Here I show that in the Ou *et al.* situation even "macroscopic local realism" is sufficient to imply that quantum mechanics is incomplete.

1. Z. Y. Ou, S. F. Pereira, H. J. Kimble, and K. C. Peng, Phys. Rev. Lett. **68**, 3663 (1992).
2. A. Einstein, B. Podolsky, and N. Rosen, Phys. Rev. **47**, 777 (1935).

QMB5 **9:30 am**

Experimental tests of Bell's inequalities based on space–time and spin variables

T. B. Pittman, Y. H. Shih, A. V. Sergienko, M. H. Rubin, *Department of Physics, University of Maryland, Baltimore, Maryland 21228*

Although the earliest tests of Bell's inequalities examined the spin correlations of two spatially separated particles, recent significant advances have led to the use of continuous variables, such as position and momentum or energy and time.[1] In all of these pioneering situations, the experiments relied on the production of two-particle states that were entangled with respect to only one type of observable. It is, however, possible to have two-particle states that are simultaneously entangled in more than one variable. In this paper we report an experiment that realizes a two-photon state that is entangled in both polarization and space–time variables. This double entanglement allows us to perform two different types of Bell's-inequality violations in a single experimental setup.

The experiment, which is shown schematically in Fig. 1, is based on the pairs of orthogonally polarized signal and idler photons produced in type-II spontaneous parametric downconversion. By using a series of quartz plates and Pockels cells, the initial two-photon state is projected into a polarization-entangled state. Furthermore, the birefringence of these optical elements essentially establishes a long and a short path leading to each detector, in analogy to the experiments proposed by Franson.[1] Coincidence measurements therefore realize the state

$$|\Psi\rangle \simeq \cos\theta_1\cos\theta_2|S\rangle_1|S\rangle_2 + \sin\theta_1\sin\theta_2 e^{-i\phi}|L\rangle_1|L\rangle_2 \quad (1)$$

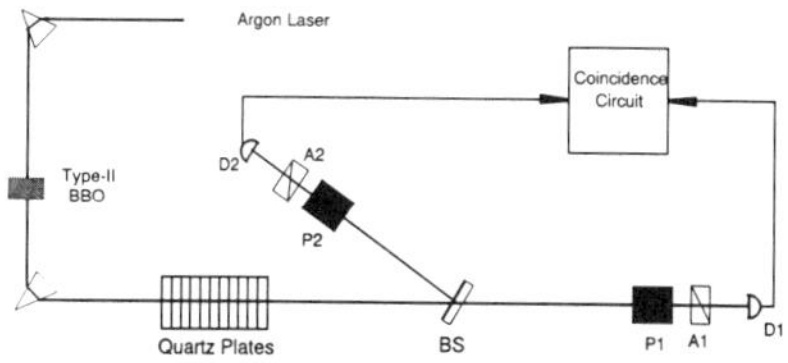

QMB5 Fig. 1. Experimental setup. The photons constituting a downconverted pair emerge collinearly from the BBO nonlinear crystal and are directed to a beam splitter BS through a series of quartz plates. Coincidence detections are measured between detectors D1 and D2, in front of which are analyzers A1 and A2 and Pockels cells P1 and P2.

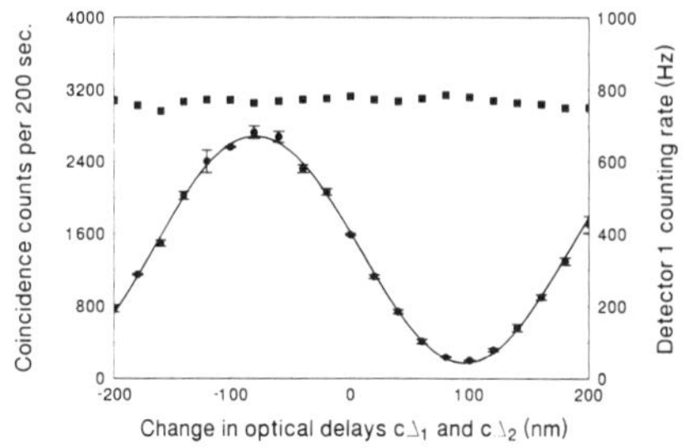

QMB5 Fig. 2. Coincidence counting rate as a function of the optical delay $\phi = \omega_1\Delta_1 + \omega_2\Delta_2$.

where θ_i ($i = 1, 2$) is the angle of the analyzer in front of detector i with respect to the fast axes of the quartz, and $|S\rangle_i$ and $|L\rangle_i$ denote the short and long space–time paths, with phase difference ϕ, respectively.

The remarkable features of the double entanglement are clearly manifest in state (1). For example, when both analyzers are set to 45° we can observe a space–time two-photon interference pattern in the coincidence counting rate, such as that shown in Fig. 2, by varying ϕ. From the experimentally measured visibility of (88.2 ± 1.2)%, we may infer a violation of Bell's inequality by more than 14 standard deviations. Likewise, we can go to a space–time minimum or maximum (i.e., $\phi = 0$ or π) and observe two-photon polarization interference by rotating the analyzers. As an example, we accumulated data at specific analyzer angles and found a violation of the Freedman–Clauser inequality[2] by more than 6 standard deviations.

Finally, it should be noted that this type of experiment is not just a simple new twist, or modification, of previous experiments. The curious nature of the double entanglement has useful applications in demonstrations of, for example, Hardy's nonmaximal entanglement[3] and the Greenberger–Horne–Zeilinger multiparticle interference.[4]

1. J. D. Franson, Phys. Rev. Lett. **62**, 2205 (1989).
2. S. J. Freedman and J. F. Clauser, Phys. Rev. Lett. **28**, 938 (1972).
3. L. Hardy, Phys. Rev. Lett. **71**, 1665 (1993).
4. D. M. Greenberger, M. A. Horne, A. Shimony, and A. Zeilinger, Am J. Phys. **58**, 1131 (1190).

QMB6 **9:45 am**

Tunable Bell-inequality violations in parametric downconversion

T. E. Kiess, Y. H. Shih,* A. V. Sergienko,* M. H. Rubin,* C. O. Alley, *Department of Physics, University of Maryland, College Park, Maryland 20742*

We prepare two-photon quantum states by using type-II phase matching in parametric downconversion to exhibit an interference term of variable strength. This interference, set by the selection of bandwidth of interference filters used in front of two detectors, is of quantum origin. We document this by exhibiting Bell-inequality violations.

In previous work (atomic-cascade,[1–3] type-I,[4–5] and type-II[6] experiments) with optical photon violations of Bell-type inequalities in polarization variables, the quantum state of interest has been a photon polarization analog of the spin-1/2 singlet state. Such analogs provide the maximum theoretical quantum violation for two particles of a Bell-type inequality. The experimental tests[1–6] have confirmed this, up to experimental nonidealities and detection-efficiency losses.

In this work we generate states that are no longer singlet-state analogs. They lack the rotational symmetry of singlet-state analogs and exhibit weaker quantum interference. We show with data that we can tune both the interference and the degree to which two different Bell-type inequalities in polarization variables are violated.

Specifically, we show Bell-inequality violations of greater than three standard deviations for quantum states characterized by values of $\rho = 0.84$, 1.53, and 1.73. Here ρ is a dimensionless parameter (equal to 2 for a singlet-state analog) useful[7] in quantifying the magnitude of the tunable interference.

* *Department of Physics, University of Maryland, Baltimore, Maryland 21228*

1. W. Perrie, A. J. Duncan, H. J. Beyer, and H. Kleinpoppen, Phys. Rev. Lett. **54**, 1790 (1985).
2. A. Aspect, J. Dalibard, and G. Roger, Phys. Rev. Lett. **49**, 1804 (1982) and references therein.
3. For a review of earlier work, see J. F. Clauser and A. Shimony, Rep. Prog. Phys. **41**, 1881 (1978).
4. C. O. Alley and Y. H. Shih, in *Proc. Second International Symposium on Foundations of Quantum Mechanics in the Light of New Technology*, Tokyo, M. Namiki et al., eds. (Physical Society of Japan, Tokyo, 1987), p. 47; Y. H. Shih and C. O. Alley, Phys. Rev. Lett. **61**, 2921 (1988).
5. Z. Y. Ou and L. Mandel, Phys. Rev. Lett. **61**, 50 (1988).
6. T. E. Kiess, Y. H. Shih, A. V. Sergienko, M. H. Rubin, and C. O. Alley, Phys. Rev. Lett. **71**, 3893 (1993).
7. Y. H. Shih, A. V. Sergienko, M. H. Rubin, T. E. Kiess, and C. O. Alley, Phys. Rev. A **50**, 23 (1994).

QMC 8:00 am
Hyatt C/D

Short Pulse Laser Physics and Applications

Paul A. Morton, *AT&T Bell Laboratories, Presider*

QMC1 (Invited) **8:00 am**

Optical sources for soliton transmission experiments

George T. Harvey, *AT&T Bell Laboratories, P.O. Box 900, Princeton, New Jersey 08542*

Soliton transmission systems have used a variety of pulsed optical sources. This talk will address differences between different types of sources and compare their performance in a series of recent experiments.

QMC2 **8:30 am**

Stabilization of a mode-locked ring laser

Briggs Atherton, Scott Diddams, Jean-Claude Diels, *Department of Physics and Astronomy, University of New Mexico, Albuquerque, New Mexico 87131*

In a mode-locked ring laser the counterpropagating pulses are used to measure nonreciprocal phase differences of 10^{-6}. When the pulses are combined outside the cavity on a detector, a beat note that represents the difference in the longitudinal mode spacing (or phase difference) for the two counterpropagating fields is observed. This difference in longitudinal mode spacing can be realized by having nonreciprocal effects, such as rotation,[1] or by placing matter in the cavity.

Because of mechanical vibrations and thermal gradients, the longitudinal modes of the cavity move about, which is not desirable in high-resolution spectroscopy. The Pound–Drever–Hall[2,3] stabilization scheme is implemented to stabilize the longitudinal modes to a reference cavity with a 3-MHz bandwidth (Fig. 1). The reference cavity is an ultralow-expansion block of quartz with mirrors optically contacted on the ends to produce a confocal cavity. This stabilization technique is now being applied to a mode-locked ring cavity in which many modes are present. The cavity lengths must be integer multiples of one another for sufficient energy to build up in the reference cavity for stabilization to occur.

If the mode-locked ring laser is not stabilized, the beat note described above has a bandwidth of 100 Hz. This bandwidth corresponds to the rotation rate of the earth. With the longitudinal modes stabilized to the reference cavity, the uncertainty of the bandwidth[4] can be as small as 0.02 Hz. This reduction by 4 orders of magnitude means that rotation rates as small as 6×10^{-7} deg/s can be obtained.

The laser is also used for high-resolution intracavity spectroscopy on an atomic beam of samarium. The need arises in this work to move one set of longitudinal modes with respect to the other. The shifting of the longitudinal modes is accomplished by taking a pulse from an output coupler and using it to modulate the Pockels cell at the cavity round-trip time (Fig. 1). When the delay is adjusted correctly, one of the interactive pulses passes through the Pockels cell at the peak of the sinusoidal wave while the other is at the trough. This produces a difference in the index of refraction that is thereby used to control the difference of longitudinal-mode spacing. The longitudinal modes of one of the pulses are fixed on the transition from $m_j = +1$ to $m_j = 0$ at 571 nm (σ^- transition); see Fig. 2. The longitudinal modes of the other pulse are

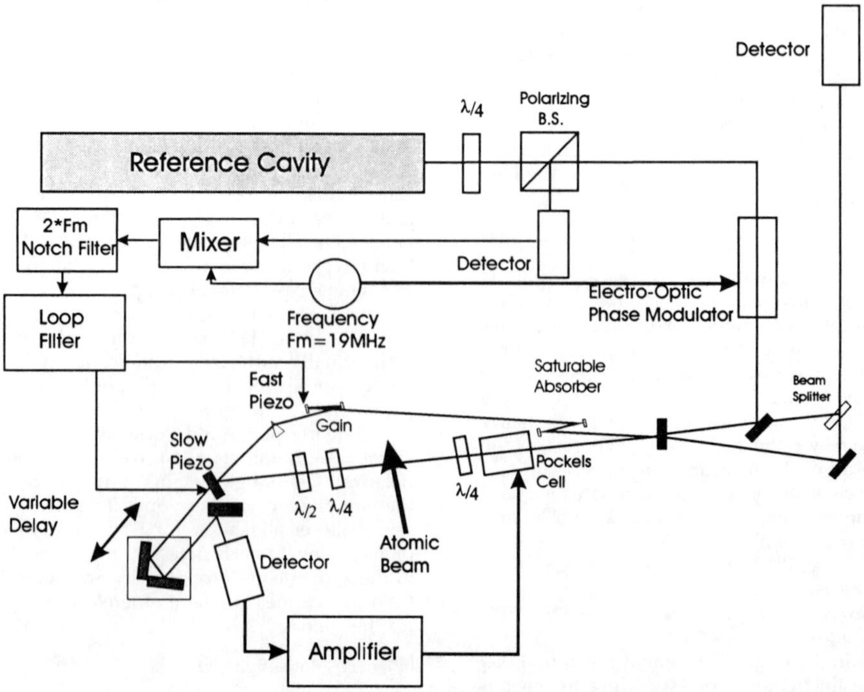

QMC2 **Fig. 1.** Schematic diagram of the cavity, stabilization scheme, and beat note.

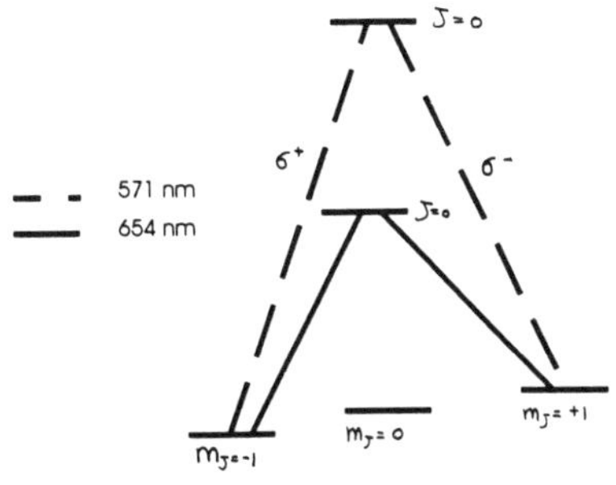

QMC2 **Fig. 2.** Schematic diagram of the relevant energy levels of samarium.

shifted by use of the Pockels cell described above to scan across the $m_j = -1$ to $m_j = 0$ (σ^+) transition at 571 nm to map out the index of refraction. This technique can sense index-of-refraction changes as small as 10^{-9} over a 1-cm path length.

1. M. Lai, J.-C. Diels, and M. L. Dennis, Opt. Lett. **17**(21) (1992).
2. R. W. P. Drever, J. L. Hall, F. V. Kowalski, J. Hough, G. M. Ford, A. J. Munley, and H. W. Ward, Appl. Phys. B **31**, 97 (1983).
3. I. Steiner, V. Enders, G. Elsner, W. Neujauser, P. E. Toschek, R. Blatt, and J. Helmche, Appl. Phys. B **49**, 251 (1989).
4. T. A. Dorschner, H. A. Haus, H. Holtz, I. W. Smith, and H. Statz, IEEE J. Quantum Electron. **QE-16**, 1376 (1980).

QMC3 **8:45 am**

Čerenkov emission from self-guided femtosecond pulses in air

E. T. J. Nibbering, P. F. Curley, B. S. Prade, G. Grillon, M. A. Franco, A. Mysyrowicz, *Laboratory of Applied Optics, National Center for Scientific Research, URA 1406, ENSTA, École Polytechnique, F-91120 Palaiseau, France*

Recently, it has been reported that intense femtosecond pulses can self-channel through air over distances of tens of meters.[1] High-intensity ultrashort pulses induce the nonlinear Kerr effect in air, leading to self-focusing and thereby to an increase of the pulse intensity in a confined region. Eventually, the peak intensity exceeds the ionization threshold of air, and the generation of free electrons then produces a negative contribution to the refractive index. Long-range channeling (or filamentation) is a consequence of the balancing of two effects: self-focusing due to the Kerr effect and defocusing due to the presence of the free electrons. The propagation process of the filaments can thus be regarded as self-guiding. This can be contrasted with the situation in which a laser beam can undergo channeling in a preformed plasma.[2]

Using pulses of a 150-fs duration at a wavelength of 800 nm, energies ranging from 20 to 40 mJ, and a beam diameter of 15 mm FWHM,[3] we observed that the onset of the filamentation occurs after a propagation distance of 20 m in air and continues up to 50 m, with a pulse energy of 0.7–1 mJ confined within a region of 100 μm. By

QMC3 **Fig. 1.** Photograph of the weakly diverging conical emission recorded at a distance of 30 m after the onset of filamentation. The diameter of the outer ring is about 10 cm.

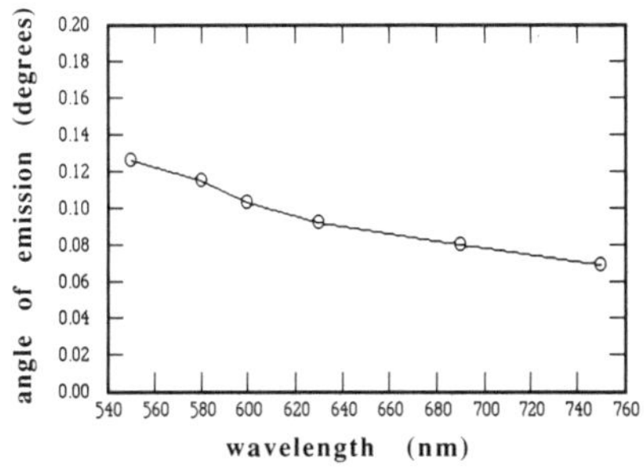

QMC3 **Fig. 2.** Angle Θ of the conical emission as a function of wavelength λ.

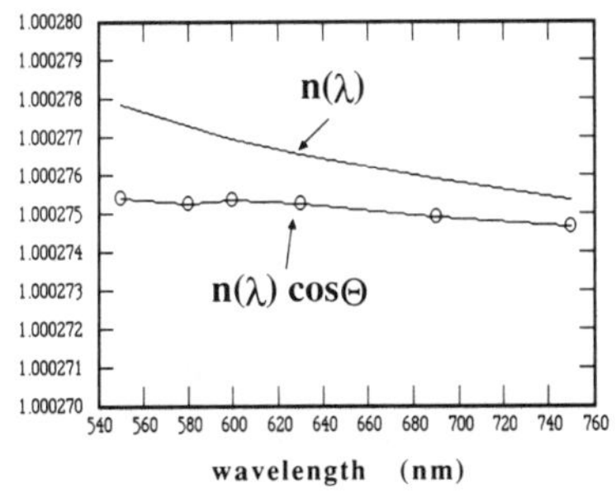

QMC3 **Fig. 3.** Wavelength dependence of the normal refractive index of air $n(\lambda)$ and of the product $n(\lambda)\cos\Theta$.

assuming the self-channeled pulse duration remains in the region of 100 fs, an intensity ranging from 10^{13} to 10^{14} W/cm^2 can be deduced, which is above the ionization threshold of air.

The spectrum of the filament exhibits characteristics of self-phase modulation and continuum generation. A striking feature accompanying the channeling is the conical emission of rings with different colors in the visible part of the spectrum originating from the filament (see Fig. 1). Note that the outer rings correspond to shorter wavelengths. This ordering is the opposite to that observed in normal diffraction. The weakly diverging conical emission occurs in the forward direction under a small wavelength-dependent angle (see Fig. 2). In Fig. 3 we plot the product of refractive index of air $n(\lambda)$ and the cosine of the angle of emission Θ as a function of the wavelength λ. Since this product is a constant, we interpret the conical emission as a signature of a leakly quasi-stable mode. This mode propagates in an antiguiding structure consisting of a core with a lower refractive index (due to the free electrons) and a surrounding cladding with a higher refractive index (due to the normal index of air and the Kerr effect). In such an antiguiding structure total internal reflection cannot occur, and, consequently, losses occur for all wavelengths. Because the phase velocity of the pulse in this leaky quasi-mode is larger than the speed of light in the cladding, its propagation is accompanied by Čerenkov-like radiation. From Fig. 3 we deduce a flat dispersion curve for the quasi mode by applying the Snell–Descartes law for refraction at the core–cladding interface. As a consequence, no modification of the temporal shape of a self-guided pulse is expected.

1. G. Korn, A. Braun, X. Liu, D. Du, J. Squier, and G. Mourou, in *High Field Interactions and Short Wavelength Generation*, 1994 Technical Digest Series, Vol. 16 (Optical Society of America, Washington, D.C., 1994), p. 140.
2. C. G. Durfee and H. M. Milchberg, Phys. Rev. Lett. **71**, 2409 (1993).
3. C. Le Blanc, G. Grillon, J. P. Chambaret, A. Migus, and A. Antonetti, Opt. Lett. **18**, 140 (1993).

QMC4 (Invited) **9:00 am**

Prospects for the miniaturization of femtosecond solid-state lasers

W. Sibbett, *I. E. Allen Physics Research Laboratories, University of St. Andrews, St. Andrews, Fife, KY16 9SS, Scotland*

There has been extensive research and development of femtosecond laser systems in which the self-mode-locking technique, first reported at CLEO '90,[1,2] is utilized. As a consequence, femtosecond lasers are now available with unprecedented tunabilities and peak-pulse powers. When such oscillator outputs are combined with regenerative and chirped-pulse amplifier arrangements, multi-teraWatt optical powers can be generated with excellent reliability and reproducibility. Access to extended tuning ranges can be readily and efficiently obtained by using such femtosecond lasers either directly or with regenerative amplification as pump sources for optical parametric oscillators and generators (amplifiers) respectively. Moreover, the exploitation of further intracavity or extracavity nonlinear frequency-mixing techniques, opens up additional ultraviolet/visible and mid-infrared spectral regions while x-ray generation in laser-produced plasmas provides ultrashot pulses in the high photon-energy regime.

The versatility and practicality of modern femtosecond lasers are such that an increasingly wide range of applications in photophysics, photochemistry, photobiology, and photomedicine is being identified and pursued. Although in some instances, the most appropriate laser (-amplifer) arrangements are elaborate and quite extensive, there are other situations when compact and more efficient, femtosecond-pulse lasers are required for integration into system configurations. For this reason there is a growing demand for all-solid-state versions of tunable vibronic lasers in particular.

In this presentation I review the recent advances that have been made in the diode-based laser pumping schemes and in the optimization of resonator designs that permit low self-model-locking thresholds in solid-state systems. With a view to miniaturization, emphasis will be placed upon direct laser-diode pumping considerations for Cr:LiSAF, Cr:YAG, etc. and on cavities that permit higher pulse repetition rates. Where possible, comparisons will be made with the minilaser-pumped alternatives within the context of simplified, self-starting, self-mode-locking arrangements. The suitability of such all-solid-state, femtosecond lasers and master-oscillator drivers for higher-power amplifier chains or as synchronous pumps for low-threshold optical parametric oscillators will also be discussed.

1. D. E. Spence, P. N. Kean, and W. Sibbett, Technical Digest CLEO '90, paper CPDP10 (May 21–25, 1990, Anaheim, Cal.), p. 619.
2. D. E. Spence, P. N. Kean, and W. Sibbett, Opt. Lett. **16**, 42 (1991).

QMC5 **9:30 am**

Soliton mode locking with slow saturable absorbers for ultrashort-pulse generation

F. X. Kärtner, I. D. Jung, U. Keller, *Institute of Quantum Electronics, Swiss Federal Institute of Technology, ETH Hönggerberg–HPT, CH-8093 Zürich, Switzerland*

Recently, we have shown that in the presence of solitonlike pulse shaping, a slow saturable absorber alone with a recovery time much longer than the soliton width can stabilize the pulse.[1] This is in contrast to the traditional concepts of ultrashort-pulse generation, which rely either on a fast saturable absorber, as is the case for APM or KLM systems, or on the interplay between a slow saturable absorber and gain saturation, as is the case for dye lasers. This is possible in the soliton regime because for the soliton the nonlinear effects due to SPM and the linear effects due to the negative group-velocity dispersion are in balance. In contrast, the noise or instabilities that would like to grow are not intense enough to experience the nonlinearity and are therefore spread in time. However, when they are spread in time, they experience higher absorption due to the slowly recovering absorber after passage of the solitonlike pulse. Thus the instability modes see less gain per round trip than does the soliton, and they will decay with time. Our experiments supporting this theory include Nd:glass lasers with sub-100-fs pulses[2] and a diode-pumped Cr:LiSAF laser producing 40–50-fs pulses.[3] The pulses are always transform limited because of their solitonlike nature.

We present a theoretical justification of this simple picture based on soliton perturbation theory, and we investigate further the limits on ultrashort-pulse generation by pure soliton mode locking with a slow absorber. We simulate a Ti:sapphire laser by the split-step FFT in the framework of the master equation of mode locking.[4] We assume cavity losses of 5% per round trip in the cavity, a strong saturable absorber that has a slow component with a relaxation time of 10 ps and 10% peak absorption, and a fast component with a relaxation time of 100 fs with 30% peak absorption. This absorber models the typical response of our semiconductor absorbers; however, we simulate large amounts of absorption to push for the shortest pulses. We also assume a constant intracavity group-velocity dispersion of $-7\ fs^2$, a self-phase-modulation coefficient of $0.057\ MW^{-1}$, and an intracavity pulse energy of 100 nJ. The saturation energy of the absorber is assumed to be 10 nJ, so that the pulse can saturate the absorber completely. Because we use large amounts of saturable absorption, the tendency of the laser to Q switch is rather strong. However, because the pulse energy saturates the absorber completely, the laser is stable against Q switching. Figure 1 shows the results of the numerical simulation in the time domain when the laser is in steady state. The pulse width is 8.25 fs, and the pulse has a slight chirp on it. Figure 2 shows the perfectly sech-shaped spectrum of the solitonlike pulse. However, because of the response of the absorber, the center frequency of the pulse is not exactly at the maximum of the gain. Figure 3 shows the intensity of the pulse over the full simulated window in time of 1-ps length. We clearly see a very good

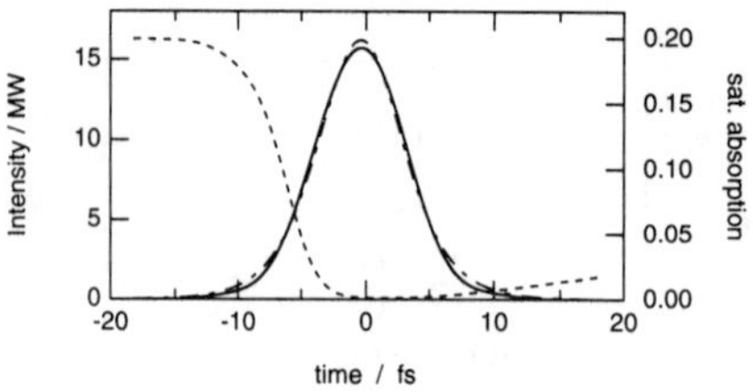

QMC5 Fig. 1. Steady-state intensity profile by numerical solution of the master equation for a Ti:sapphire laser: solid curve, pulse shape; dashed curve, loss modulation of the absorber; dashed-dotted curve, fit to a sech^2. The deviation from the soliton shape is due to a small chirp that arises from discrete action of the self-phase modulation when the pulse passes the laser crystal and then exits the cavity.

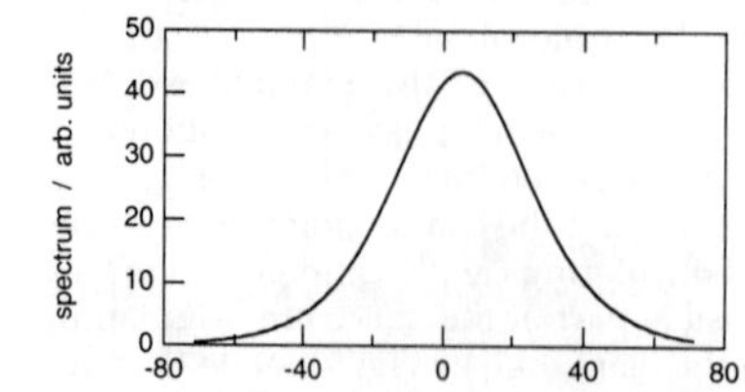

QMC5 Fig. 2. Steady-state intensity spectrum (solid curve) that is perfectly sech^2 shaped (dashed curve) but not located at the peak of the gain center because of the nonsymmetric response of the absorber.

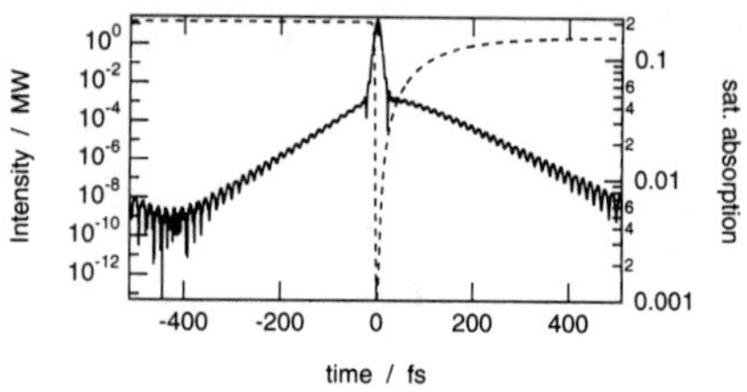

QMC5 Fig. 3. Intensity profile over the full simulated window in time (solid curve) and response of the saturable absorber (dashed curve). The simulation shows that the slow absorber and soliton-like pulse shaping suppress the continuum well by 40 dB.

suppression of the continuum of about 40 dB below the peak, despite the slow recovery of the absorber.

In conclusion, we have demonstrated that semiconductor absorbers with a fast recovery time due to intraband relaxation processes of 100 fs or even below are fast enough to allow for generation of sub-10-fs pulses without the need of Kerr-lens mode locking in which strong self-focusing of the laser beam is used to produce an equivalent fast saturable-absorber action. The strong self-focusing imposes strong limitations on the cavity design and leads to an undesired strong space–time coupling of the pulses in the laser crystal.[5] The soliton mode locking demonstrated here allows for decoupling of the nonlinearity responsible for self-phase modulation and therefore soliton formation from the saturable-absorber nonlinearity that is necessary for stable pulsed operation. This approach allows the design of a laser that reliably self-starts, without overdriving the nonlinearities when the laser reaches sub-10-fs pulses.

1. F. X. Kärtner and U. Keller, Opt. Lett., in press.
2. F. X. Kärtner, D. Kopf, and U. Keller, in *Ultrafast Phenomena*, 1994 Technical Digest Series, Vol. 7 (Optical Society of America, Washington, D.C., 1994), paper MA5.
3. D. Kopf, K. J. Weingarten, L. Brovelli, M. Kamp, and U. Keller, submitted to 1995 Conference on Lasers and Electro-Optics (Optical Society of America, Washington, D.C., 1995).
4. H. A. Haus, J. Appl. Phys. **46**, 3049 (1975).
5. I. P. Christov, H. C. Kapteyn, M. M. Murnane, C. P. Huang, and J. Zhou, Opt. Lett., in press.

QMC6 **9:45 am**

Spectral evolution of pulse recovery in a femtosecond cavity-dumped Kerr-lens mode-locked oscillator

G. Chériaux, M. Joffre, F. Salin, J. P. Chambaret, *Laboratory of Applied Optics, ENSTA, Centre de l'Yvette, F-91120 Palaiseau, France*

The development of ultrashort-pulse oscillators by using Kerr–lens mode locking (KLM) has allowed the generation of pulses as short as 10 fs with pulse energies of a few nanojoules. To increase the available pulse energy by an order of magnitude without the use of separate amplification systems, cavity dumping of pulses can be used.[1,2] Cavity dumping exploits an ultrafast Bragg diffraction cell to switch out a single pulse from the oscillator with high efficiency. Because this represents a large perturbation to the KLM process, the investigation of the recovery of the oscillator immediately after the switchout gives a good example of a solitonlike evolution out of equilibrium. This can yield insight into the physics of the oscillator itself.

Here we report on the spectral evolution of a sub-50-fs hard-aperture KLM Ti:sapphire laser, after the operation of the Bragg cell. Experiments were undertaken by using a Ti:sapphire laser incorporating a 4-mm-long crystal and a fused-silica prism pair for dispersion control. Transform-limited pulses with full-width-at-half-maximum (FWHM) bandwidths between 20 and 100 nm can be achieved with an extracted energy of approximately 50 nJ for pulse durations between 15 and 45 fs. The dumping repetition rate can be varied between 400 Hz and 800 kHz with an amplitude stability of 3%.

Cavity dumping introduces a large perturbation into the pulse train, which then recovers to a steady state on a time scale of 10 μs (Fig. 1). We record the pulse spectrum as a function of time as the laser returns to the steady state. For the measurements, the laser output is taken through a 3% output coupler and the spectrum is monitored by using a gated intensified optical multichannel analyzer (OMA). The time delay between the 10-ns electric pulse used to gate the OMA and the rf pulse to the Bragg cell is varied, so that we can measure the spectrum after each round trip in the cavity.

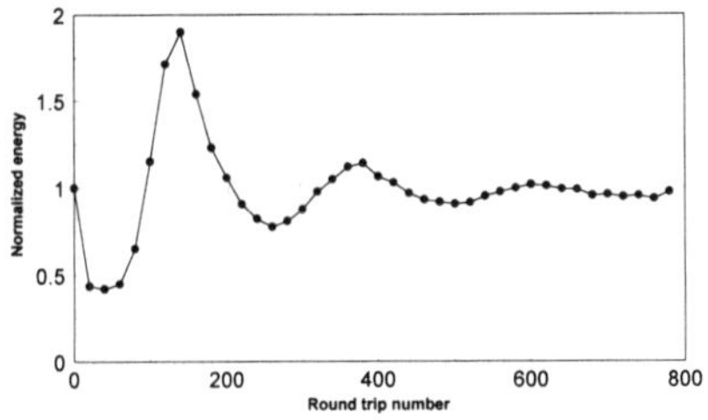

QMC6 Fig. 1. Evolution of the pulse energy after dumping. The energy is normalized to the steady state. The x axis is the number of round trips in the cavity, and the zero corresponds to six round trips before dumping. The time between two round trips is 12.3 ns.

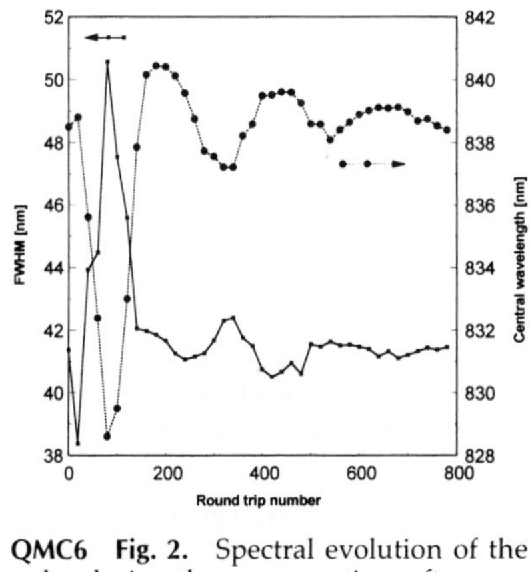

QMC6 Fig. 2. Spectral evolution of the pulse during the recovery time after dumping. The solid curve represents the FWHM of the spectrum, and the dotted curve represents the central wavelength.

During the pulse recovery we observe a variation in the FWHM of the spectrum between 39 and 51 nm (Fig. 2). This variation coincides with a variation in the central wavelength of the oscillator (Fig. 2). However, these two quantities vary in quadrature with the energy during the recovery time (Fig. 1).

The oscillations on the pulse train are well understood, considering the gain recovery and the population inversion in the gain medium. The behavior of the spectrum can be explained in terms of a balancing of oscillator gain, self-phase modulation, and intracavity residual high-order dispersion (third and fourth orders) and is similar to the recovery of a strongly perturbated soliton.

1. M. Ramaswamy, M. Ulman, J. Paye, and J. G. Fujimoto, Opt. Lett. **18**, 1822 (1993).
2. M. S. Pschenichnikov, W. P. de Boeij, and D. A. Wiersma, Opt. Lett. **19**, 572 (1994).

JMA Tutorial — 9:00 am, Hyatt E/F

Status of Short Pulse Optical Power Supplies

Ursula Keller, *ETH Zurich, Switzerland,* *Presider*

JMA1 (Tutorial) — **9:00 am**

Status of short pulse optical power supplies

Wayne H. Knox, *AT&T Bell Laboratories, Crawfords Corner Road, Holmdel, New Jersey*

Ultrafast optical technology is beginning to evolve from a laboratory curiosity into a real-world technology, with the promise of many novel applications in electronics, medicine, process control, imaging and the biomedical area. Although many interesting suggestions have appeared for such applications, there is still a strong need for development of compact, reliable and affordable sources of ultrashort optical pulses within a wide range of operation parameters in wavelength, pulsewidth, power, etc. In this tutorial, we discuss the status of short-pulse sources in the form of diode-pumped solid-state, fiber, and semiconductor lasers and try to outline the general requirements for such sources.

QMD — 10:30 am, Hyatt A

Novel Optical Effects

Alexander Luis Gaeta, *Cornell University, Presider*

QMD1 (Invited) — **10:30 am**

Parametric light scattering

Martti Kauranen, Thierry Verbiest, André Persoons, *University of Leuven, Celestijnenlaan 200 D, B-3001 Heverlee, Belgium*

Even-order nonlinear optical processes are forbidden, in the electric-dipole approximation, in centrosymmetric material systems such as an isotropic solution of achiral molecules. However, orientational fluctuations of noncentrosymmetric molecules can break locally the macroscopic centrosymmetry of such solutions and make incoherent nonlinear scattering processes possible. Second-order elastic and inelastic scattering at the second-harmonic frequency of the fundamental laser beam was first observed in 1965 from several materials.[1] More recently, elastic scattering at the exact second-harmonic frequency (hyper-Rayleigh scattering) has gained popularity as a routine technique to determine the first hyperpolarizability of organic molecules in solution.[2]

We present the first experimental results on a second-order scattering process in which the two incident photons are at different frequencies and sum-frequency radiation is detected.[3] In our first experiments, we have verified the fundamental scaling laws of this parametric light-scattering process as the various experimental parameters are varied. For these measurements, *para*-nitroaniline in methanol was used as the standard nonlinear molecule. The parametric light-scattering signal was found to exhibit the expected linear dependence on the intensity of each of the incident beams and on the number density of dissolved molecules. Similar measurements were repeated for *para*-nitroanisol to determine its hyperpolarizability with respect to that of *para*-nitroaniline. The results of these measurements are in perfect agreement with those obtained by using other techniques and show that, analogous to hyper-Rayleigh scattering, parametric light-scattering is well suited to determine the first hyperpolarizability of organic molecules.

We have also investigated the polarization dependence of parametric light scattering by varying the polarization of one of the incident beams and detecting unpolarized scattered radiation. We have used this technique to determine the ratio between the two dominant components of the first hyperpolarizability tensor for *para*-nitroaniline. The possibility of using unpolarized scattered radiation is a distinct advantage over hyper-Rayleigh scattering, which requires the use of depolarized scattered radiation to determine values of individual tensor components.

We also discuss the possibilities of parametric light scattering in increasing the number of components of the hyperpolarizability tensor that can be experimentally determined. This number can be significantly increased by utilizing the independent polarizations of the two incident beams and by detecting depolarized radiation at the sum frequency. However, the maximum number of independent measurements is shown to be fundamentally limited by the number of rotational invariants of the product of two components of the hyperpolarizability tensor.

1. R. W. Terhune, P. D. Maker, and C. M. Savage, Phys. Rev. Lett. **14**, 681 (1965).
2. K. Clays and A. Persoons, Phys. Rev. Lett. **66**, 2980 (1991).
3. T. Verbiest, M. Kauranen, and A. Persoons, J. Chem. Phys. **101**, 1745 (1994).

QMD2 **11:00 am**

Cascading without phase matching

Ch. Bosshard, R. Spreiter, M. Zgonik, P. Günter, *Nonlinear Optics Laboratory, Institute of Quantum Electronics, ETH Hönggerberg, CH-8093 Zürich, Switzerland*

Large, nonresonant optical-Kerr-like nonlinearities are a basic requirement for all-optical switching and related applications.[1] In addition to the nonlinearity that is proportional to $\chi^{(3)}$, there is another important contribution that occurs only in noncentrosymmetric materials. This contribution induces a nonlinear phase shift in nearly-phase-matched second-harmonic-generation interactions and other parametric processes.[2,3] Such effects were first observed in high-power pulsed second-harmonic-generation experiments. Recently the potential for all-optical switching was realized and large nonlinear phase shifts due to cascading were observed for the first time in potassium titanyl phosphate (KTP) waveguides.[3] Unexpectedly large nonlinear-refractive-index changes observed in fibers of 4-(*N,N*-dimethylamino)-3-acetamidonitrobenzene (DAN) could also be explained by cascaded nonlinearities.[4]

We now demonstrate that there exist other combined processes of second-order nonlinear optical effects that give rise to a large effective nonlinear refractive index n_2 without the need for phase matching. We discuss and compare the different possible contributions to the nonlinear refractive index n_2 in noncentrosymmetric media.

Experimental results of Z-scan measurements that verify our theory by using noncentrosymmetric $KNbO_3$ (point group mm2) and centrosymmetric $KTaO_3$ (point group m3m) are presented. The pure $\chi^{(3)}$ contribution to n_2 of $KNbO_3$ can be determined from measurements on $KTaO_3$. In a first approximation we can assume that the contributions to n_2 arising from $\chi^{(3)}$ are the same for $KTaO_3$ and $KNbO_3$.

Our investigations showed that this new contribution to n_2 is of magnitude comparable to that of the well-known one, which is due to second-harmonic generation and difference-frequency mixing. The measured n_2 of $KNbO_3$ was more than three times that of $KTaO_3$. A further advantage is that there is no loss of the incident beam because we have no depletion of the fundamental as occurs in the case of the large nonlinear phase shift generated in nearly-phase-matched second-harmonic-generation interactions. It is also interesting that this contribution is always positive. Moreover, it can also be present for light propagation along the polar axis. For a general propagation direction in a noncentrosymmetric material, both cascaded effects are important for induced nonlinear phase shifts. It is, however, possible to find special propagation directions for which one contribution is dominant, as is shown for $KNbO_3$.

1. G. I. Stegeman and A. Miller, in *Photonics in Switching, Vol. I: Background and Components*, J. E. Midwinter, ed. (Academic, Boston, 1993), p. 81.
2. J.-M. R. Thomas and J.-P. E. Taran, Opt. Commun. **4**, 329 (1972).
3. M. L. Sundheimer, Ch. Bosshard, E. W. Van Stryland, G. I. Stegeman, and J. D. Bierlein, Opt. Lett. **18**, 1397 (1993).
4. D. Y. Kim, W. E. Torruellas, J. Kang, Ch. Bosshard, G. I. Stegeman, P. Vidakovic, J. Zyss, W. E. Moerner, R. Twieg, and G. Bjorklund, Opt. Lett. **19**, 868 (1994).

QMD3 (Invited) **11:15 am**

Focused one-cycle optical pulses

R. W. Hellwarth, *Departments of Physics and Electrical Engineering, University of Southern California, Los Angeles, California 90089-0484*

As progressively shorter optical pulses are being achieved in the laboratory, the question arises: what are the shortest possible pulses and what are their properties? We will describe several families of focused pulse solutions of Maxwell's equations in vacuum that are either one optical cycle long or one-and-a-half cycles long. Each family of solutions has two adjustable parameters: (1) the optical wavelength and (2) the pulse diameter at focus. These parameters have essentially the same relation to the far-field diffraction angle of the pulse as in the case of a monochromatic gaussian beam. Four of the families of short-pulse solutions we have found constitute a small subset of the four-parameter pulse solutions discovered by Ziolkowski,[1] who called them "modified power spectrum" pulses. Two of the families have as their only nonzero electric and magnetic vectors components the real and imaginary parts, respectively, of the following complex expressions (in cylindrical ρ, θ, z coordinates and in S.I. units):

$$E_\theta = -i\sqrt{\frac{\mu_0}{\varepsilon_0}}\,\rho(q_1 + q_2 - 2ict)/D, \qquad (1a)$$

$$H_\rho = i\rho(q_2 - q_1 - 2iz)/D, \qquad (1b)$$

$$H_z = -[\rho^2 - (q_1 + i\tau)(q_2 - i\sigma)]/D, \qquad (1c)$$

where $D \equiv [\rho^2 + (q_1 + i\tau)(q_2 - i\sigma)]^3/4f_0$, $\tau \equiv z - ct$, and $\sigma \equiv z + ct$. We assume the amplitude f_0 to be real, and $q_2 \geq q_1 > 0$. These fields describe z-directed pulses having transverse electric (TE) vector. The pulse energy is contained in a doughnut encircling the z axis at $z = ct$. The radius of the doughnut decreases at negative times to a minimum radius $\sim \sqrt{q_1 q_2}$ at the focal point $z = 0$ (at $t = 0$), and then diverges afterward to form a diffracting outgoing doughnut-shaped pulse. Normal "weak focusing" occurs for $q_1 << q_2$. An exotic "strong-focusing" limit occurs at $q_1 = q_2$, where the doughnut actually implodes from the sides. Pulse behavior is easily visualized by viewing computer plots of (1). Two additional "TM" families of solutions are the dual of (1) with electric vector $\mathbf{F} = \sqrt{\mu_0/\varepsilon_0}\,\mathbf{H}$ and transverse magnetic field $\mathbf{G} = -\sqrt{\varepsilon_0/\mu_0}\,\vec{E}$. An outgoing (TE or TM) weakly-focused pulse described by the imaginary part of (1) has a field on axis at positive z ($>>q_1$) that crosses zero once. Therefore, we call this pair of solutions "one-cycle" pulses. The corresponding field of the real part of (1) crosses zero twice, and we call this pair of solutions "1½ cycle" pulses.

Each of this quartet of solutions has pulse energy $\pi^2\mu_0 f_0^2(q_1 + q_2)/(4q_1^3 q_2^3)$ Joules.[2] Each has the same energy spectrum, whose dependence on angular frequency ω, and on q_1 and q_2 is contained in the factor $\omega^4(a^{-2}\cosh a - a^{-3}\sinh a)\exp[-2\omega(q_1 + q_2)/c]$, where $a \equiv (q_2 - q_1)\omega/c$.[2] This is a simple bell-shaped curve that peaks at a frequency ω_0 near $2\pi c/q_1$ and decays exponentially with ω at higher frequencies.

The energy gained by an electron of arbitrary initial condition, interacting with a focused one-cycle pulse, has not yet been studied. However, the following simple case illustrates the ability of a weakly-focused doughnut pulse to accelerate an electron. Suppose an electron has a trajectory $z = ct$ and $\rho = 0$, and interacts with the dual F''_z of the one-cycle pulse represented by the imaginary part of (1c) from $t = -\infty$ to $t = 0$. The work done on this electron is $ec\int_{-\infty}^{t} ds F''_z(0, 0, cs, s)$, which is found directly to be $4e\pi^{-1}\varepsilon_0^{-1/2}U^{1/2}q^{-1/2}$.[2] Here U is the pulse energy and $q \equiv q_1 + q_1^2/q_2$ is very nearly $2\pi c/\omega_0$. Numerically, this says that a one Joule pulse of one micron (nominal) wavelength can give about 1 GeV of energy to an electron.[2]

Two families of similar one-cycle focused-pancake solutions will also be described. We review a number of unresolved questions surrounding all focused one-cycle pulse solutions.

1. R. W. Ziolkowski, Phys. Rev. A **39**, 2005 (1989).
2. R. W. Hellwarth and Pascale Nouchi (unpublished).

QMD4 **11:45 am**

Modulation-induced multitransparency and high-order frequency mixing in a two-level atomic system

A. E. Kaplan, E. Hudis, *Department of Electrical and Computer Engineering, Johns Hopkins University, Baltimore, Maryland 21218*

Electromagnetically induced transparency (EIT) of three-level atoms in a strong laser pulse[1] is attracting growing interest. We demonstrate here the feasibility of a modulation-induced inhibition of dynamics (MID), self-transparency, and very-high-order frequency mixing (HFM) in a two-level system (TLS) subjected to a periodically modulated laser field. In contrast to self-induced transparency (SIT) in 2π solitons characterized by the dramatic dynamics of population,[2] in MID the TLS dynamics (i.e., the oscillations of both the population and the polarization) are almost fully suppressed. For a sinusoidal modulation, the conditions for MID coincide with those for the resonances occurring if the frequency of modulation Ω_M is close to a subharmonic of the averaged Rabi frequency $\overline{\Omega}_R$.[3] It has also been shown[4] that similar resonances in TLS may occur even when the system is driven by a cw (nonmodulated) field strong enough to superdress TLS by having Ω_R exceed the TLS unperturbed frequency ω_0. We showed that at these resonances, in addition to the suppressed population difference δ, the TLS driven polarization p vanishes too, resulting, together with vanishing δ, in the collapse of Rabi sphere. Even when the driving spectral components are strongly off resonance, the field-induced Rabi splitting brings TLS at the MID points into exact resonance with the field. MID results in a self-transparency for the entire spectrum of the incident field (which could be very broad), allowing its propagation without dispersion suppressed now over the spectrum.

The MID resonances occur only at certain ratios $M = (\Omega_R)_{max}/\Omega_M$ (usually around $M \approx \pi n$, where n is an integer). However, outside of these narrow resonances, one can expect even more interesting effects whereby the periodic modulation induces dramatic TLS dynamics that can give rise to HFM. We found a parallel between HFM and recently discovered high-harmonic generation (HHG) in rare gases featuring a spectral plateau with a certain cutoff. One of the simplest models of HHG is an overdriven TLS, allowing an analytical solution[4] for the cutoff. We showed that a TLS-generated HFM spectrum, $\omega_n = \omega \pm (2n + 1)\,\Omega_M$, $n = 0, 1, 2, 3, \ldots$, features a similar plateau with a cutoff point $\Omega_{cut}/\Omega_M \approx (4 + M^2)^{1/2}$, similarly to Ref. 4. Unlike HHG, during the propagation HFM demonstrates large conversion efficiency, with the fundamental components almost vanishing at some points and their energy transversed into higher spectral components. This process gives rise to a train of solitonlike pulses with the energy redistributed to produce short spikes of high peak intensity, where the individual length of each is given by $\tau_M \approx \pi/[\Omega_M(4 + M^2)^{1/2}]$. (If $\lambda_0 = 1$ μm, $\Omega_M/\omega_0 = 1.5\%$, $M \approx 10$, and $\tau_M \approx 100$ fs.) Our numerical simulations showed the universal effect of full revival of the field envelope and its spectrum after certain propagation distances determined by the boundary conditions. HFM can be used for coherent generation of new frequencies (e.g., for nonlinear spectroscopy) and of short pulses of high intensity. The MID and HFM effects can be observed in a configuration similar to SIT (when a continuously modulated wave is used instead of a short pulse) or subharmonic-resonance experiments (when we use a long propagation path and observe the field transformation).

This work is supported by the U.S. Air Force Office of Scientific Research.

1. A. Imamoḡlu and S. E. Harris, Opt. Lett. **14**, 1344 (1989); K.-J. Boller, A. Imamoḡlu, and S. E. Harris, Phys. Rev. Lett. **66**, 2593 (1991).
2. S. L. McCall and E. L. Hahn, Phys. Rev. Lett. **18**, 908 (1967).
3. B. Blind, P. R. Fontana, and P. Thomann, J. Phys. B **13**, 2717 (1980); G. S. Agarwal *et al.*, J. Opt. Soc. Am. B **8**, 1163 (1991); W. M. Ruyten, Phys. Rev. A **42**, 4226, 4246 (1992); G. I. Toptygina and E. E. Fradkin, Sov. Phys. JETP **70**, 428 (1990); P. Thomann, J. Phys. B **13**, 1111 (1980); S. Chamakjian, K. Koch, and C. R. Stroud, Jr., J. Opt. Soc. Am. B **5**, 2015 (1988).
4. A. E. Kaplan and P. L. Shkolnikov, Phys. Rev. A **49**, 1275 (1994).

QMD5 **12:00 m**

No-carrier high-intensity subfemtosecond solitons

A. E. Kaplan, P. L. Shkolnikov, *Department of Electrical and Computer Engineering, Johns Hopkins University, Baltimore, Maryland 21218*

We propose two ways of generating high-intensity subfemtosecond ($\sim 10^{-17}$-s) pulses and solitons of very high repetition rate. One approach[1] is based on multifrequency cascade Raman stimulated scattering (CSRS). We predict that a pump-laser wave with frequency ω_L and many cascade-excited Stokes and anti-Stokes components with frequencies $\omega_j = \omega_L + j\omega_0$, $j = \pm 1, \pm 2, \pm 3, \ldots$, can be mode locked to each other through a fast full-swing 2π nutation of population at the Raman transition with frequency $\omega_0 << \omega_L$. Similar to bright–bright 2π solitons in CSRS with two[2] and three[3] components, these solitons have a new, very simple Lorentzian intensity profile. Because of the engagement of many mode-locked components, however, their total electromagnetic field in the time domain consists of a train of ultrashort, of the order of the pump cycle $2\pi/\omega_L$ or even shorter, pulses separated by the interval $2\pi/\omega_0$. The major feature of the proposed effect is that all the frequency components of the new soliton are so-called bright solitons, in contrast to the well-known bright–dark soliton combination in SRS.[4] The high-order CSRS required to observe the proposed effect was first observed experimentally in Ref. 5 and later in many other experiments, with the total number of components as high as 10–15.

The set of the Maxwell equations for the field components interacting through the Raman transition and the generalized Bloch equations[6] were solved analytically.[1] In general, they give rise to a rich family of solutions: solitary waves, stationary traveling fronts, bright and dark solitons, periodic stationary waves, etc. Of special interest to us are the so-called bright solitons, in particular, 2π solitons reminiscent of self-transparency in the one-photon process,[7] although now they have a very simple Lorentzian profile. To demonstrate the formation of subfemtosecond pulses, we choose a realistic example of hydrogen (with $\lambda_0 \approx 2.4$ μm) pumped by the third harmonics of a Ti:sapphire laser ($\lambda_L \approx 0.28$ μm), with 12 participating components whose intensities peak in the middle of the CSRS spectrum and fall off parabolically toward both of its ends. We found that the total mode-locked field consists of the train of subfemtosecond pulses, each ~0.2 fs long, separated by ~8 fs. With typical laser parameters one can achieve a peak intensity of 10^{14}–10^{15} W/cm^2.

Another process capable of producing very short, potentially subfemtosecond pulses is soliton formation in partially ionized gases. In particular, the presence of a very small fraction of free electrons (less than 1% of the neutral-atom density) in the air is enough for that medium to support solitons through Kerr-type nonlinear refraction of neutrals. At the same time, because this nonlinearity is very small, nonlinear Schrödinger equation (NLSE) predicts those solitons to be very short. Moreover, our numerical solution of NLSE with Kerr nonlinearities demonstrates that pumping intensities $\sim 10^{14}$ W/cm^2 are enough for the formation of a subfemtosecond soliton within a few centimeters of propagation length. Such short pulses, however, are well outside the applicability limits of the assumption of the steady-state nonlinear response of the air and the validity range of NLSE itself. To demonstrate the feasibility of subfemtosecond solitons in partially ionized air, we solved full (nonenvelope) Maxwell and full (nonperturbative) Bloch equations for a pulse propagating in the medium of two-level atoms and found solutions that correspond to $\sim 10^{-17}$-s and shorter solitonlike pulses with shock-wave-like fronts and intensities of 10^{13}–10^{15} W/cm^2.

This work is supported by the U.S. Air Force Office of Scientific Research.

1. A. E. Kaplan, Phys. Rev. Lett. **73**, 1243 (1994).
2. T. M. Makhviladze, M. E. Sarychev, and L. A. Shelepin, Sov. Phys. JETP **42**, 255 (1975).
3. A. E. Kaplan, P. L. Shkolnikov, and B. A. Akanaev, Opt. Lett. **19**, 445 (1994).
4. K. Druhl, R. G. Wenzel, and J. L. Carlsten, Phys. Rev. Lett. **51**, 1171 (1983).
5. R. W. Terhune and P. D. Maker, Phys. Rev. A **137**, 801 (1965).
6. V. S. Butylkin, A. E. Kaplan, Yu. G. Khronopulo, and E. I. Yakubovich, *Resonant Nonlinear Interactions of Light with Matter* (Springer, New York, 1989) (Russian edition: Nauka, Moscow, 1977).
7. S. L. McCall and E. L. Hahn, Phys. Rev. Lett. **18**, 908 (1967).

QMD6 **12:15 pm**

Inversionless superradiance in two-component media

A. V. Andreev, P. V. Polevoy, *Department of Physics, M. V. Lomonosov Moscow State University, Moscow 119899, Russia*

The phenomenon of superradiance (SR) is usually treated as a collective spontaneous emission of a multiatomic system consisting of the atoms of the same type. We consider here the SR of the multiatomic media consisting of two species of atoms, a and b, that have the same frequencies $\omega_a = \omega_b = \omega$ and different dipole matrix elements $d_a < d_b$ of the resonant transitions. The radiative lifetime T_1 is determined by

$$\frac{1}{T_1} = \frac{4}{3}\frac{\omega^3|d^2|}{\hbar c^3}.$$

Therefore, $T_1^{(a)} > T_1^{(b)}$ and we shall call the a atoms the slow atoms and the b atoms the fast atoms. We consider the case in which the slow atoms are excited by the external pump to the upper resonant level while the fast atoms are initially in the ground state.[1]

The results of numerical calculations show that the pulse of coherent radiation can be emitted even when the concentration of the resonantly absorbing component is higher than that of the amplifying component. Figure 1 shows the peak pulse intensity and the pulse width as a function of the concentration of the fast component. The concentration of the fast atoms is expressed in terms of the concentration of the slow ones, whereas the slow component concentration is taken to be equal to unity. The horizontal dashed lines show the peak pulse intensity and an SR pulse width for the monocomponent medium ($n_a = 1$, $n_b = 0$) under instantaneous pumping. It is seen from Fig. 1 that the adoption of the fast atoms, which are initially in the ground state, results in the increase of the peak pulse intensity and in a decrease of the SR pulse width.

The vertical dashed lines in Figs. 1(a) and 1(b) mark the medium with equal concentrations of the fast and slow atoms ($n_a = n_b = 1$). In the region $n_b > 1$ the concentration of fast atoms that are initially in the ground state is higher than the slow-component concentration. Therefore, the net population inversion is negative during the entire process of SR emission because both components are involved in the SR process. The emission process of such a medium can be called an inversionless SR. It should be noted [Fig. 1(c)] that in the process of inversionless SR the two-component medium can emit a more intense and shorter pulse than can the monocomponent medium of slow atoms under an ultrashort pumping pulse of the same energy. Another distinctive feature of the two-component SR is illustrated by Fig. 1(c), which shows the SR pulse profiles under different concentrations of the fast atoms. We can see that in this case the SR delay time drastically exceeds the pulse width. At the same time, the pulse width of the two-component SR is shorter than that of the monocomponent medium, and the delay time is higher than the delay time of the monocomponent SR. This is a very important feature of the two-component SR because it is a delay time that determines the duration of the pumping. The pumping process does not affect the SR pulse profile if the pumping-pulse temporal width is equal to or smaller than the SR delay time. This enables us to transform a broad pumping pulse of the lower intensity into an ultrashort SR pulse of high intensity. In our computer calculations we have found that the two-component SR pulse width can be three orders of magnitude smaller than the duration of pumping.

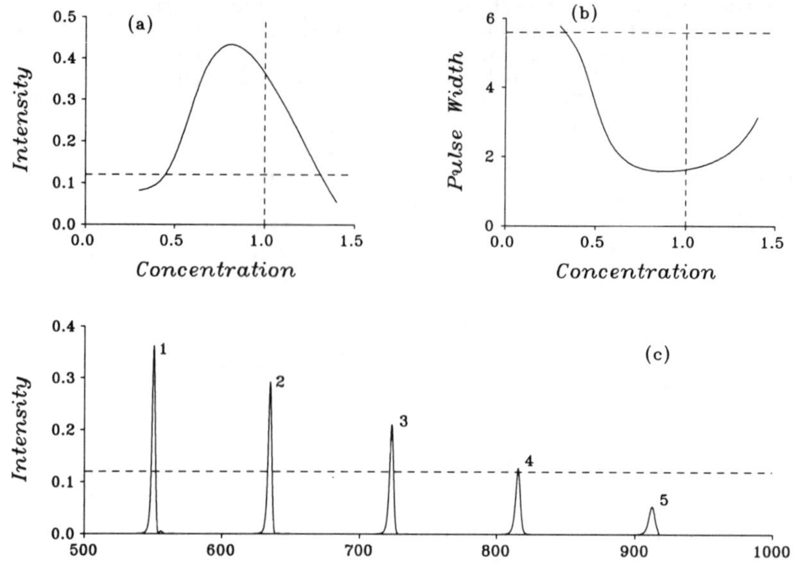

QMD6 Fig. 1. (a) Peak pulse intensity versus concentration of the fast component. The horizontal dashed line indicates the value of the peak intensity for the monocomponent medium of slow atoms. (b) Temporal pulse width versus concentration of the fast component. The horizontal dashed line is the pulse width for the monocomponent medium of slow atoms. (c) Intensity profiles for the inversionless superradiance for the following values of the concentration of the fast component n_b: 1.0 (curve 1), 1.1 (curve 2), 1.2 (curve 3), 1.3 (curve 4), and 1.4 (curve 5).

1. A. V. Andreev and P. V. Polevoy, Quantum Opt. **6**, 57 (1994).

QME **10:30 am**

Hyatt B

Squeezed Lasers and Photon Statistics

Michael Gordon Raymer, *University of Oregon, Presider*

QME1 (Invited) **10:30 am**

The microlaser: physics of a single atom interacting with a single mode of the radiation field

Michael S. Feld, Kyungwon An, *G. R. Harrison Spectroscopy Laboratory, Massachusetts Institute of Technology, Cambridge, Massachusetts 02139*

We describe a fundamental laser system, the microlaser, in which a single atom interacts with a single mode of the radiation field.[1] In the experiments a beam of ^{138}Ba atoms inverted on the $^1S_0 \rightarrow {}^3P_1$ transition at $\lambda = 791$ nm traverses an ultrahigh Q supercavity resonator. The excited atoms interact with the cavity one by one and emit photons into the resonator. Because of the long cavity storage time, a large number of photons can build up. As the average number of atoms in the cavity mode, $\bar{N}$, increases from 0.1 to 1, the average number of photons in the cavity mode, $\bar{\nu}$, increases from 0.1 to 12. Anomalously large signals are observed when $\bar{N}$ approaches unity and $\bar{\nu}$ is much larger than one. This behavior results from the standing-wave nature of the cavity mode, in combination with the saturation effect occurring at large cavity photon number. Spontaneous emission is negligible in the microlaser system. Therefore, unlike conventional lasers, it plays no role and amplification occurs via quantized Rabi oscillations,[2,3] which are the underlying manifestation of the stimulated emission process. The large energy and momentum associated with optical photons opens the possibility of studying new aspects of cavity QED, including the generation of photon-number trapped states and the study of entanglement of internal and mechanical degrees of freedom. This presentation shall review the current status of our microlaser research.

1. Kyungwon An, James J. Childs, Ramachandra R. Dasari, and Michael S. Feld, Phys. Rev. Lett. **73**, 3375 (1994).
2. E. T. Jaynes and F. W. Cummings, Proc. IEEE **51**, 89 (1963).
3. P. Filipowicz, J. Javanaine, and P. Meystre, Phys. Rev. A **34**, 3077 (1986).

QME2 **11:00 am**

Attosecond-duration antibunched electrons and squeezed photons

S. D. Vartak, N. M. Lawandy,* *Department of Physics, Brown University, Providence, Rhode Island 02912*

The development of ultrashort laser pulses has resulted in photoelectron pulses in the hundred-femtosecond regime for use in time-resolved structural studies in a variety of chemical processes.[1] More recently, rf-bunched electrons have been compressed with magnetic fields to produce 50-fs electron pulses[2] with mega-electron-volt energies. In this work we show how, by using an electron-acceleration process driven by optical rectification in a crystal, one can produce electron pulses with subfemtosecond durations and how these electrons can be used to produce significant amounts of attosecond-duration number-state-squeezed transition radiation from thin metallic targets. The method is straightforward and can be implemented by using currently available laser technology.

Because the temporal duration of an electron bunch is given by its spatial extent divided by its mean velocity, a straightforward way of decreasing the duration would be to accelerate electrons to a higher velocity. Unfortunately, conventional high-voltage plate accelerators first act on the front of the electron bunch and then on the back, resulting in a spatially longer electron bunch that is moving faster. To produce temporally compressed electron pulses, a field must act simultaneously on the entire photoelectron packet and be able to compensate for space-charge effects. These requirements may be satisfied by the use of dc electric fields generated by the nonlinear process of optical rectification.[3] The presence of a dc polarization parallel to a given crystal surface will, in turn, result in a dc electric field outside the crystal. Using the published values for $Sr_{0.3}Ba_{0.7}Nb_2O_6$ operated at one tenth of the rated damage threshold (~1 GW/cm^2), we expect fields of the order of 10^7 V/m outside the crystal face. Furthermore, the spatial profile of the accelerating optical pulse can be chosen to produce a field gradient that compensates for the space-charge field acting to spread out the pulse. The optical-rectification accelerator geometry is shown in Fig. 1. Figure 2 shows the electron-pulse duration at a target plate as a function of the relative delay between the initial optical pulse creating the photoelectrons and the peak of the accelerating pulse.

The generated attosecond electron pulses can be used to produce bursts of electromagnetic radiation by means of bremsstrahlung and transition radiation processes. The latter process presents an opportunity to experimentally infer the duration of the final electron pulse as well as to produce coherent attosecond electromagnetic pulses. Figures 3(a) and 3(b) show the theoretically calculated emission spectra for both the bremsstrahlung and transition-radiation processes for the parameters given in the simulations presented in Fig. 2. The results of Fig. 3(b) show that attosecond optical pulses can be created with total-wavelength integrated energies of the order of 5×10^{-15} J. Finally, it is expected that because of space-charge correlation that antibunches the electrons in the pulse striking the target, the emitted transition radiation should exhibit significant number-state squeezing.[4] A Fano factor for the electrons approaching due to the increased electron–electron interaction obtainable by spatial compression of the initial photoelectron pulse is expected. This antibunching will, in turn, be directly transferred to the transition-radiation spectrum.

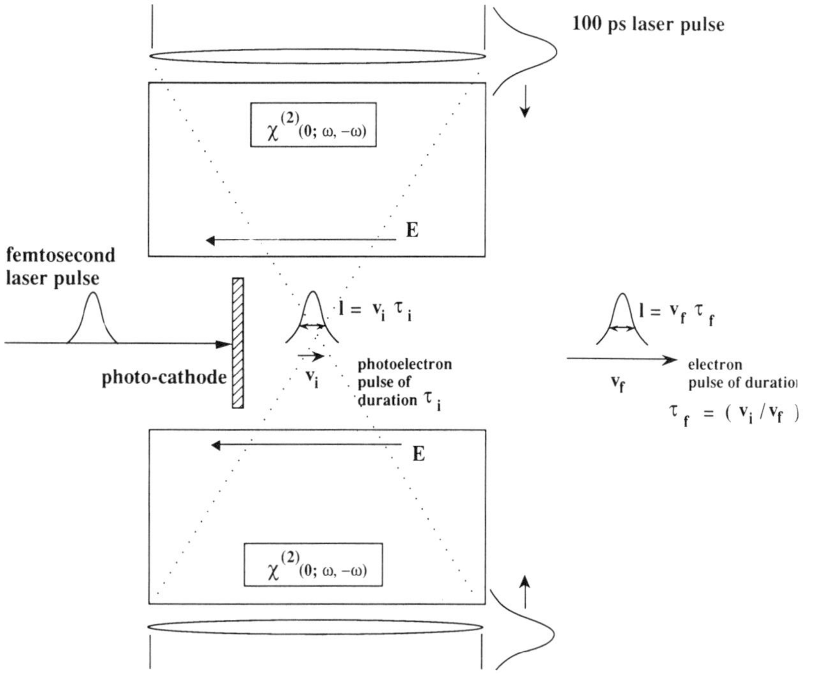

QME2 Fig. 1. Optical-rectification accelerator geometry. The use of two crystals allows for the partial cancellation of forces not in the direction of electron propagation.

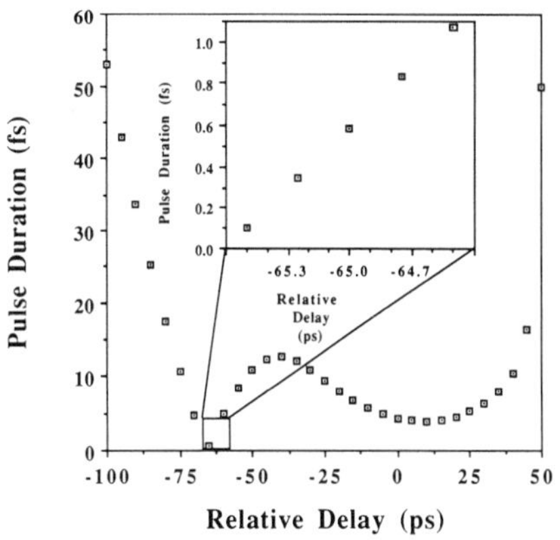

QME2 Fig. 2. Electron-pulse duration versus the relative delay between the initial photoelectron-pulse generation and the optical rectification pulse.

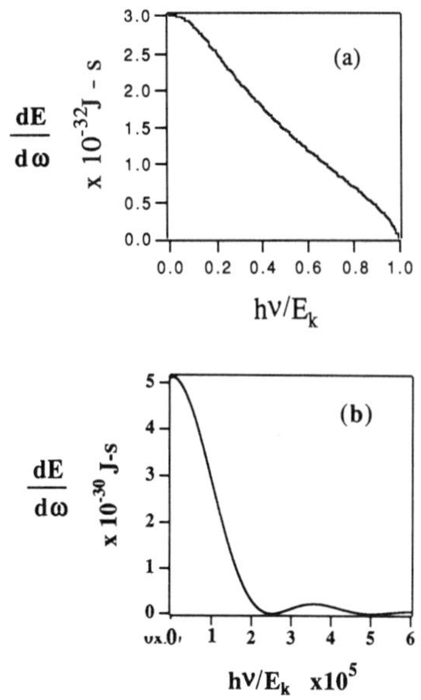

QME2 Fig. 3. Target radiation spectra in terms of the final photoelectron energy E_k from nonrelativistic 500-as pulses for (a) bremsstrahlung and (b) transition radiation.

**Also with Division of Engineering, Brown University, Providence, Rhode Island 02912*

1. J. C. Williamson and A. H. Zewail, J. Phys. Chem. **98**, 2766–2781 (1994).
2. P. Kung, H. Lihn, H. Weidemann, and D. Bocek, Phys. Rev. Lett. **73**, 967 (1994).
3. N. M. Lawandy and S. D. Vartak, in *International Quantum Electronics Conference*, 1994 OSA Technical Digest Series, Vol. 9 (Optical Society of America, Washington, D.C., 1994), p. 169.
4. M. C. Teich and B. E. A. Saleh, in *Progress in Optics*, E. Wolf, ed. (North-Holland, Amsterdam, 1988), Vol. 26, p. 1.

QME3 **11:15 am**

Change and uncertainty in quantum optics

J. D. Franson, *Applied Physics Laboratory, Johns Hopkins University, Laurel, Maryland 20723*

The Heisenberg uncertainty principle gives a lower bound on the product of the uncertainties in two observable quantities A and B:

$$(\Delta A)^2 (\Delta B)^2 \geq \frac{1}{4} (\langle i[A, B]\rangle)^2.$$

In many cases of interest, however, the expectation value of the commutator is zero even though A and B do not commute. The Heisenberg uncertainty principle then places no restriction on the uncertainties of interest.

It is possible to derive a somewhat different uncertainty principle that, instead, relates the uncertainty in a single observable Q to the change in its expectation value. Consider a closed system for which the Hamiltonian is constant. It can then be shown that there is no eigenstate $\psi_q(t)$ of operator Q such that

$$Q\psi_q(t) = q(t)\psi_q(t)$$

if the eigenvalue $q(t)$ changes in time. It follows from this that if there was initially no uncertainty in Q, then any change in its expectation value necessarily implies that Q will become uncertain.

This can be put in the form of a simple inequality. Let ΔQ be the usual uncertainty in Q defined by

$$(\Delta Q)^2 = \langle (Q - \langle Q\rangle)^2\rangle,$$

and define the change $\delta\langle Q\rangle$ in its expectation value as

$$\delta\langle Q\rangle = \langle\psi(t_f)|Q|\psi(t_f)\rangle - \langle\psi(t_i)|Q|\psi(t_i)\rangle,$$

where t_i and t_f are the initial and final times, respectively. Once again, if Q initially had no uncertainty, then for some time t between t_i and t_f the uncertainty in Q must satisfy

$$\Delta Q \geq |\delta\langle Q\rangle|.$$

In other words, the uncertainty in Q must become greater than or equal to the absolute value of the change in its expectation value.

Several examples are given of situations in quantum optics in which the expectation value of the commutator of the relevant operators vanishes so that the Heisenberg uncertainty principle places no restriction on the uncertainties of interest. In these examples the simple uncertainty principle outlined above continues to apply and is more restrictive than the Heisenberg uncertainty principle.

QME4 (Invited) **11:30 am**

Amplitude squeezed light generated by semiconductor quantum well lasers

D. C. Kilper, M. J. Freeman, J. Erland, D. G. Steel, R. Craig,* D. R. Scifres,* *Randall Laboratory of Physics, University of Michigan, Ann Arbor, Michigan 48109*

Amplitude squeezed light is characterized by a steady photon stream that, over certain time intervals, exhibits fluctuations in the photon population that are below the standard quantum limit (SQL, variance equal to the mean photon number). This nonclassical light may satisfy the number-phase minimum uncertainty relation if the optical phase noise is sufficiently low. The benefits of squeezed light in general are both practical, associated with the reduction in laser noise, and fundamental, due to the nonclassical features.

If a laser is driven far above threshold, such that the relaxation oscillations are suppressed and spontaneous emission is negligible, the statistics of the pump will be reflected in the output statistics of the laser intensity. Yamamoto *et al.*[1] first suggested that a semiconductor laser, when driven by a constant electrical pump current, may exhibit amplitude squeezed light. Although the noise associated with carrier injection into the lasing junction may be large, the use of a source resistance in the laser drive circuit that is much larger than the differential resistance of the lasing junction (high-impedance pump-noise suppression) to establish a constant current input to the laser imposes by conservation a constant photon output flux.

While the early experimental work is in qualitative agreement with the high-impedance pump-noise suppression model,[2] the difficulty in achieving significant amplitude squeezing at room temperature indicated the presence of other noise sources as well as the need to extend the theoretical description. Indeed, further experiments at low temperature showed improved noise reduction due to the stabilization of competing longitudinal side modes and reduction of the threshold current. The latter change allows for pump levels far above threshold, a limit believed necessary in order to suppress relaxation oscillations and observe significant squeezing. The use of cryogenic temperatures, however, is an experimental complication that introduces additional optical losses and hinders potentially simple application. Experiments by our lab demonstrate that the noise generated by near-threshold longitudinal side modes and amplitude fluctuations in the weak orthogonal polarization are two major problems that obstruct the observation of squeezing at room temperatures. In addition, the threshold current of most commercially available transverse junction stripe lasers is too large to allow for pumping beyond 2–3 times threshold.

These difficulties are overcome with the use of low threshold current quantum well lasers in dispersive feedback or injection-locking configurations[3]; for which suppression of the longitudinal side modes by 45–40 dB facilitates the observation of squeezing at room temperature (further suppression of the side modes did not affect the squeezing). Reduction of the noise due to the orthogonal polarization fluctuations is achieved by including a polarizer at the laser's output. The efficiency of optical elements has been observed to affect the squeezing level in external cavity configurations, but the major limitation on squeezing for injection-locked lasers remains an open question.

An important development for the application of this radiation is the demonstration that high frequency modulation of the laser drive current results in stable high frequency optical modulation without loss of amplitude squeezing. Measurements show that squeezing is maintained both outside the modulation frequency and at the modulation frequency.[4]

This research is supported by AFOSR. J. E. acknowledges support from the Danish Research Academy.

Spectra Diode Labs, San Jose, California 95135

1. Y. Yamamoto, S. Machida, and O. Nilsson, Phys. Rev. A **34**, 4025 (1986).
2. S. Machida, Y. Yamamoto, and Y. Itaya, Phys. Rev. Lett. **58**, 1000 (1987).
3. H. Wang, M. J. Freeman, and D. G. Steel, Phys. Rev. Lett. **71**, 3951 (1993); M. J. Freeman, et al., Opt. Lett. **18**, 2141 (1993).
4. M. J. Freeman, et al., accepted in Opt. Lett.

QME5 **12:00 m**

Ultrafast time-resolved photon-number statistics

M. Munroe, D. Boggavarapu, M. Anderson, M. G. Raymer, *Department of Physics, University of Oregon, Eugene, Oregon 97403*

Optical homodyne tomography[1] (OHT) has previously been used to determine the quantum state of fields,[2] and thus the photon-number statistics[3] of these fields in the 300-ps regime. We have demonstrated the use of OHT to measure the time-resolved photon-number statistics of a 5-ns pulsed field with a sampling time of 130 fs and time resolution of 60 ps. The ultrafast sampling allows us to measure the statistics of very-broadband fields, such as that of a diode laser below threshold.

We use an argon-laser-pumped Ti:sapphire laser in combination with a chirped-pulse regenerative amplifier to generate ultrashort local-oscillator (LO) pulses (330 fs) at a wavelength of 830 nm. Our signal is from a Sharp LT015 laser diode with a wavelength of 830 nm and a pulse width of 5 ns. The laser diode drive pulse is triggered by the LO through an adjustable electronic delay having 60-ps jitter, which is the time-resolution limit in this experiment.

Figure 1 shows the average photon number $\langle n\rangle$ during the 130-fs sampling time of the LO and the second-order coherence $g^{(2)}(0) = (\langle n^2\rangle - \langle n\rangle)/\langle n\rangle^2$ versus the delay time between the LO and the beginning of

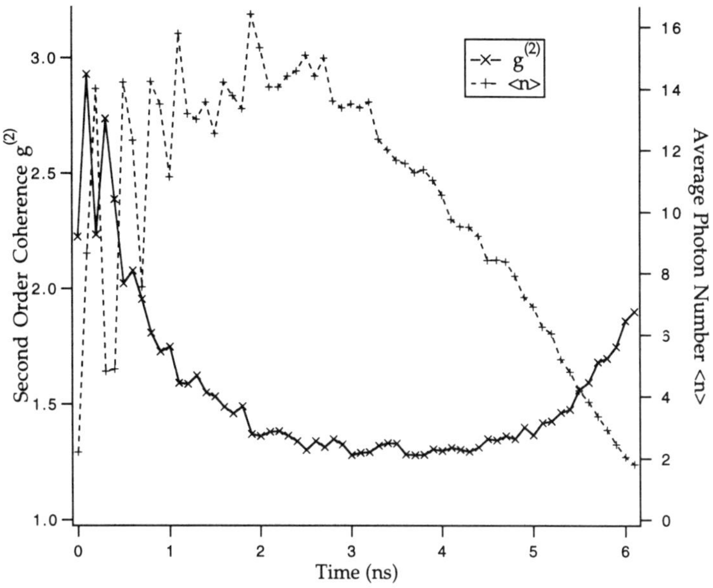

QME5 Fig. 1. Measured average photon number $\langle n\rangle$ and second-order coherence $g^{(2)}$ versus time delay between the LO and start of the signal pulse.

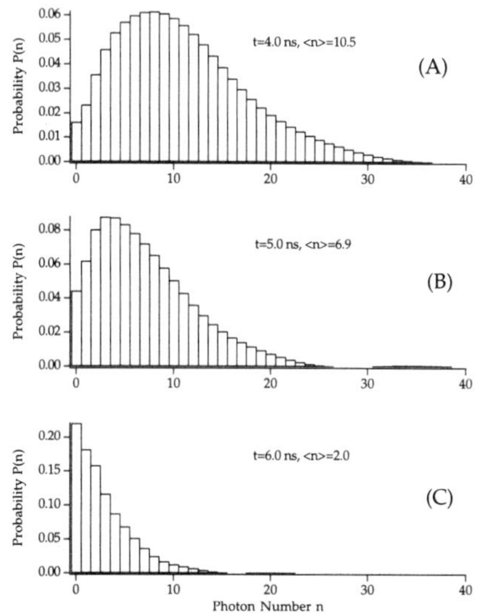

QME5 Fig. 2. Photon-number probability distributions $P(n)$ at three different times t in the pulse.

the signal pulse. Relaxation oscillations with a period of 300 ps are observed in the average photon number $\langle n\rangle$ as the laser first turns on. The value of $g^{(2)}$ also oscillates but is 180° out of phase with $\langle n\rangle$. We tentatively attribute this behavior to the laser oscillating from above threshold to below threshold in the initial turn on. The large values of $g^{(2)}$ during the relaxation oscillations is caused by the 60-ps jitter in our timing resolution, which is large compared to the width of the relaxation-oscillation spikes, thus adding noise to our measurement. As the relaxation oscillations damp out, the laser reaches a quasi-steady state, accompanied by a decrease in $g^{(2)}$ to a minimum of approximately 1.3, compared to a value of 1.0 for an ideal laser. The average photon number then decreases as the laser turns off. The second-order coherence increases in this region as the output light becomes thermal (ideally $g^{(2)} = 2$).

Figure 2 shows the photon-number probability distributions at three different times in the pulse, at 4, 5, and 6 ns after the laser diode turns on. The probability distributions in Fig. 2 show a transition from Poissonian-like statistics to thermal-like statistics as the laser drops below threshold.

In conclusion, the use of OHT with ultrashort LO pulses to measure the time-resolved photon-number statistics of a laser diode has been demonstrated. The new method offers promise for time resolving photon statistics of ultrafast optical sources such as molecular and semiconductor systems.

1. K. Vogel and H. Risken, Phys. Rev. A **40**, 2847 (1989).
2. D. T. Smithey, M. Beck, M. Belsley, and M. G. Raymer, Phys. Rev. Lett. **69**, 2650 (1992).
3. D. T. Smithey, M. Beck, J. Cooper, and M. G. Raymer, Phys. Rev. A **48**, 3159 (1993).

QME6 **12:15 pm**

Thresholdless microlasers?

U. Mohideen, R. E. Slusher, P. R. Rice,* L. Pedrotti,** *AT&T Bell Laboratories, Murray Hill, New Jersey 07974*

As the volume of a microlaser approaches a cubic wavelength in the material and the resonator is constructed so that a large fraction β of the spontaneous emission from the gain region is emitted into the lasing mode, the threshold for lasing, as measured by a sharp increase in the light emitted from the lasing mode as a function of pump power, begins to vanish. At β = 1 the laser appears to be thresholdless because the input/output light curve is a straight line with no threshold kink.[1] However, it has been argued[2] that in this limit there are still properties of the cavity light output that distinguish a threshold when there is of the order of one photon in the cavity averaged over time. One good indicator of the threshold phenomena is the level of amplitude fluctuations in the light output. There are large amplitude fluctuations in a low-β laser near threshold that indicate a transition to the coherent state of the light field. Theoretical models[3] show that in some cases these amplitude fluctuations nearly disappear in the limit of β, resulting in a very nearly thresholdless laser. We address this question of thresholdless lasing by studying a series of microdisk lasers[4] for which β varies between 0.04 and 0.2. Similar studies[5] have been made by using vertical-cavity lasers with a lower β value.

Experimentally, we use photon counting to measure the ratio of the variance of the amplitude fluctuations relative to the mean light level as a function of pump power and β value. The β value is varied from 0.04 to 0.2 by varying the diameter of the GaAs/AlGaAs microdisk lasers[4] used in these experiments from 5 to 2 μm. A peak in the ratio of the variance to the mean as a function of pump power is found near the threshold measured from the light input/output curve. The ratio of the FWHM pump-power width of this feature to the threshold pump power increases as the β value increases. This is the behavior expected from the simple models, indicating that the ability to define a distinct threshold is decreasing as β approaches unity. However, the peak amplitude of the fluctuations is measured to be much stronger than expected from a simple four-level model in which the system is pumped from the ground state to an excited level, rapidly decays to a lower excited level, and lases to an unoccupied state above the ground state. In the semiconductor system this four-level model is a reasonable approximation, except that the lower laser level is partially occupied, resulting in strong absorption for the unpumped system.

Our numerical models are in good agreement with the experiments only if we include the ground-state absorption for the semiconductor system. Our experiments extend to a reasonably large β value for which there is still a clear threshold in both the ratio of the variance to the mean and the input/output light curve. The extrapolation of our models to the β = 1 limit shows that a clear threshold persists for the case in which there is absorption at the laser transition wavelength; however, the threshold nearly vanishes when this absorption is removed.

**Department of Physics, Miami University, Oxford, Ohio 45056*

***Department of Physics, University of Dayton, Dayton, Ohio 45409*

1. F. DeMartini and G. R. Jascobovitz, Phys. Rev. Lett. **60**, 1711 (1988).
2. Y. Yamamoto and R. E. Slusher, Phys. Today **46**, 66 (1993).
3. P. R. Rice and H. R. Carmichael, Phys. Rev. A, in press.
4. U. Mohideen, W. S. Hobson, S. J. Pearton, F. Ren, and R. E. Slusher, Appl. Phys. Lett. **64**, 1914 (1994).
5. R. Jin, D. Boggavarapu, M. Sargent, P. Meystre, G. Kitrova, and H. Gibbs, Phys. Rev. A **49**, 4038 (1994).

QMF 10:30 am
Hyatt C/D

Laser Spectroscopy of Molecules and Dielectrics

Joan Frye, *Howard University, Presider*

QMF1 10:30 am

Saturation spectroscopy of molecular overtones for laser frequency standards in the visible and the near-visible domains

Long-Sheng Ma, Pierre Dubé, Peter Jungner, Jun Ye, John L. Hall, *Joint Institute for Laboratory Astrophysics, University of Colorado and National Institute of Standards and Technology, Boulder, Colorado 80309-0440*

Molecular vibration–rotation lines are interesting for laser-frequency stabilization because of their narrow widths and rich spectra. Now, with tunable diode lasers, it is of interest[1,2] to examine the overtone bands of simple molecules. An important advance was made recently by de Labachelerie, Nakagawa, and Ohtsu,[3] who demonstrated the possibility of intracavity saturated absorption spectroscopy on an acetylene-overtone transition at 1.54 μm. Here we push this concept further by using higher overtones to provide many transitions in the visible region while providing potential access to infrared widths in the subkilohertz domain. From these first experiments we report pressure-broadening measurements on the saturated absorption peak in the $P(11)$ line of the $\nu 1 + 3\nu 3$ band of C_2H_2 (acetylene) at 790.7 nm.

Detection of weak transitions, such as these molecular overtones, can be enhanced by locating the molecules in a high-finesse optical cavity, leading to an increased path length for absorption. The effective number of transits is $2F/\pi$, leading to a contrast in transmitted light of $C = \delta I/I = \alpha L \cdot 2F/\pi$, where αL is the absorptivity–length product of the intracavity gas. The internal power buildup offered by the cavity is fully sufficient to saturate even $\Delta v = 4$ transitions, as in our experiments here with acetylene. A spectroscopic absorption at the parts-per-million level thus becomes a potential reference line.

The molecular response must be distinguished from the cavity response. The laser will typically be locked to this cavity, using reflected light in the Pound–Drever system, with the optical sidebands lying outside the cavity resonance width. One detection possibility is to dither the cavity, and with it the laser frequency, leading to a first-derivative signal obtained by lock-in detection of the transmitted light. Figure 1 shows saturated absorption spectra obtained in this way. The estimated one-way absorption in the cavity is 13 parts in 10^6 at 30-mTorr pressure, which maps into a 2200-parts-in-10^6 change in cavity power. The Lorentzian saturation peak is about 25 parts in 10^6. A cavity peak-to-peak modulation of 1.2 MHz leads to a broadening and a Wahlquist line shape,[4] which was used in the fits. Our vastly-sub-Doppler spectra yield a preliminary pressure-broadening rate ~30 kHz/mTorr, about 3 times that known[1] in the Voigt limit. The line width extrapolated to zero pressure is ~510 kHz FWHM, to be compared with a transit FWHM of ~422 kHz and a natural width of <1 kHz. The laser contributes <40 kHz of the width. The present cavity (R = 60 cm, L = 33 cm) gives $2w_0$ = 520 μm and a finesse of 270. A 160-mW input gives 60-mW transmission, corresponding to a circulating power of 8.4 W. If the finesse is increased 40-fold, the present saturation ($S \sim 1$) can be maintained with a power as low as 4 mW, which would lower the noise and greatly increase the signal.

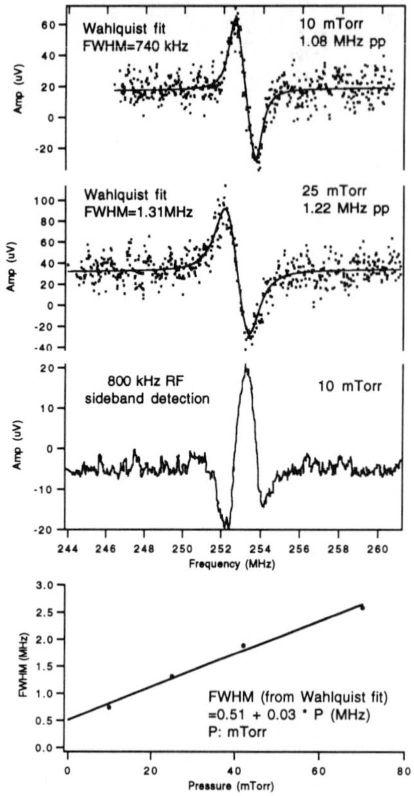

QMF1 Fig. 1. Pressure broadening of saturated absorption resonances of acetylene fourth-overtone band at 790 nm. Top two curves in top panel show signal obtained with synchronous detection at 700-Hz dither frequency. FM is 1.2 MHz peak to peak, the time constant is 0.4 s, and the sweep rate is 0.1 MHz/s. Bottom curve in top panel shows improved signal-to-noise ratio obtained when lock-in detects the 800-kHz dispersion signal. Bottom panel shows pressure-broadening slope for these preliminary data, ~30 kHz/mTorr, suggesting the enhanced role of velocity-changing collisions.

Data of improved signal-to-noise ratio were taken with 800-kHz FM sidebands, which (just) lie within the cavity linewidth; see Fig. 1c. These FM dispersion signals showed line-shape changes with pressure that may be associated with the saturation-dependent dynamic phase shifts observed previously[5] in Rb. An attractive system will match the sideband frequencies to adjacent cavity fsr resonances, leading to reduced AM conversion of laser residual FM noise. Reflection-mode topology lowers the light power on the detector and hence the noise in heterodyne detection.

1. F. S. Pavone, F. Marin, M. Inguscio, K. Ernst, and G. Di Lonardo, Appl. Opt. **32**, 259 (1993).
2. A. Lucchesini, I. Longo, C. Gabbanini, S. Gozzini, and L. Moi, Appl. Opt. **32**, 5211 (1993).
3. M. de Labachelerie, K. Nakagawa, and M. Ohtsu, Opt. Lett. **19**, 840 (1994).
4. H. Wahlquist, J. Chem. Phys. **35**, 1708 (1961).
5. H.-R. Xia, J. I. Cirac, S. Swartz, B. Kohler, D. S. Elliott, J. L. Hall, and P. Zoller, J. Opt. Soc. Am. B **11**, 721 (1994).

QMF2 10:45 am

Precise measurement of *EF* $^1\Sigma_g^+$ levels in H_2, D_2, and HD

N. Melikechi, S. Gangopadhyay, J.-C. Meiners, E. E. Eyler, *Department of Physics and Astronomy, University of Delaware, Newark, Delaware 19716*

We have completed a measurement of several *EF* $^1\Sigma_g^+$ levels in H_2 and its isotopomers with an absolute accuracy better than 1 part in 10^8. The $EF \leftarrow X$ interval is measured by Doppler-free two-photon absorption with counterpropagating beams at 202 nm. To obtain high peak power with bandwidths that approach the Fourier-transform limit, the output of a single-mode cw ring dye laser operating near 606 nm is pulse amplified in a three-stage dye amplifier pumped by a seeded Nd:YAG laser. The required 202-nm light is generated by frequency tripling the 606-nm pulses. Transitions to the *EF* states are detected by resonant multiphoton ionization induced by the absorption of a third photon at 202 nm.

Pulsed amplification can significantly perturb the optical phase, because the gain and the index of refraction of the amplifying medium are strongly time dependent.[1,2] These phase distortions cause time-dependent frequency shifts that can degrade the bandwidth and introduce significant systematic errors in frequency measurements. In particular, they were the dominant source of uncertainty in our previous measurement of *EF*-state energies.[3] Using an optical heterodyne technique, we measure the phase evolution of the amplified pulses, allowing us to correct for these perturbations and thereby to predict the spectral line shape of the transitions. The dashed and solid curves in Fig. 1 show the measured and simulated line shapes, respectively, of the $Q(0)$ transition of the $EF \leftarrow X$ (0–0) in D_2.

The new values for the energies of the *EF* $^1\Sigma_g^+$ states, accurate to better than 0.0009 cm^{-1}, are in disagreement with our previous results by about 0.008 cm^{-1}. This disagreement of up to 2.5 standard deviations arises because the size of the frequency-chirp corrections is larger than we had thought possible. To our knowledge, the

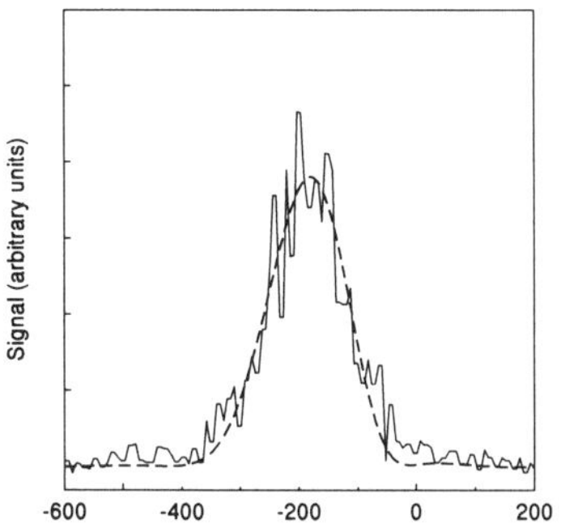

QMF2 Fig. 1. Measured (solid curve) and predicted (dashed curve) two-photon line shape of the $Q(0)$ transition in the (0–0) band of the $EF \leftarrow X$ in D_2. The frequency indicated on the x axis refers to the full VUV interval.

fractional accuracy is the best so far obtained for any measurement using pulsed lasers. The attainment of these high fractional accuracies presented a technical challenge. We have carefully designed both the dye amplifier chain to minimize frequency chirping and the optical lay-out so that the residual first-order Doppler shifts from angular misalignment can be compensated by the curvature of the Gaussian wave fronts.

This experiment, along with a future measurement of high molecular Rydberg states, will allow a determination of the ionization potential of molecular hydrogen, serving as one of the most sensitive tests of quantum electrodynamics in molecular systems.

This work was supported by the National Science Foundation.

1. S. Gangopadhyay, N. Melikechi, and E. E. Eyler, J. Opt. Soc. Am. B **1**, 231 (1994).
2. N. Melikechi, S. Gangopadhyay, and E. E. Eyler, J. Opt. Soc. Am. B, in press.
3. J. M. Gilligan and E. E. Eyler, Phys. Rev. A **46**, 3676 (1992).
4. A. V. Smith and M. S. Bowers, J. Opt. Soc. Am. B, in press.

QMF3 **11:00 am**

Fluorescence lifetimes of the alkoxy radicals

Xinming Zhu, Carron Sandifor, Mohammed M. Kamal, Prabhakar Misra, *Laser Spectroscopy Laboratory, Department of Physics and Astronomy, Howard University, Washington, D.C. 20059*

The alkoxy (RO; R = CH_3, C_2H_5, i-C_3H_7) radicals are important as oxidation intermediates in the combustion of hydrocarbons and in air pollution.[1] Alkyl nitrite (RONO) precursors (of typically 1–2% concentration) were transported by high-pressure helium (12–14 atm) and were supersonically expanded into a vacuum chamber through a commercial pulsed nozzle. The chamber pressure was $\sim 10^{-4}$ Torr with the pulsed valve in operation. UV radiation (typically 50-mJ pulses at 248 nm) was used to photodissociate RONO to generate the RO radicals *in situ* in the jet expansion.[2]

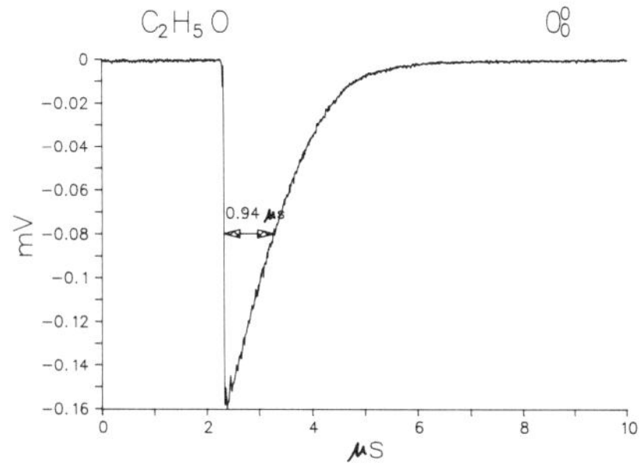

QMF3 Fig. 1. Time-resolved emission recorded with a digital oscilloscope after laser excitation of the 0^0_0 band of ethoxy at 342.68 nm.

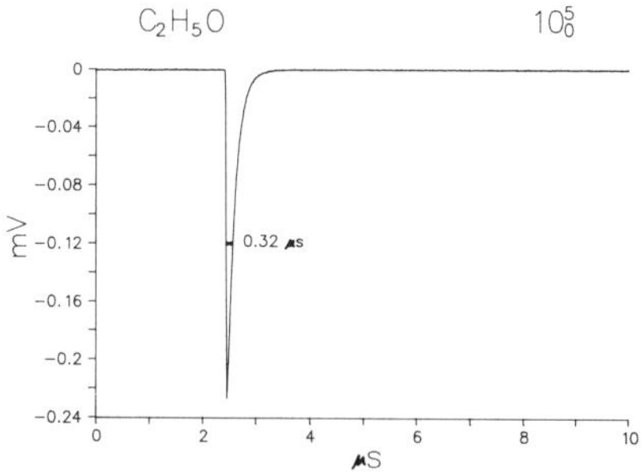

QMF3 Fig. 2. Fluorescence decay of ethoxy after laser excitation of the 10^5_0 vibronic transition at 311.15 nm.

Tunable radiation from an Nd:YAG-pumped dye laser, which was frequency doubled to generate near-UV wavelengths, was used for exciting the ground-state radicals to the first excited electronic state. The probe UV-pulse energy was typically 100 μJ. Fluorescence emission after excitation was detected with a photomultiplier tube, and the corresponding decay curve was displayed on the screen of a digital oscilloscope. Lifetime measurements involving the CO stretch mode for the excited vibrational levels v' = 0–5 of CH_3O, C_2H_5O, and i-C_3H_7O spanned the range 0.15–3 μs. Methoxy (CH_3O) had the longest-lasting fluorescence, and isopropoxy (i-C_3H_7O) had the shortest, whereas ethoxy (C_2H_5O) had an intermediate lifetime. For example, for two-quanta excitation of the CO stretch, the lifetimes were 1.55 μs for CH_3O and 0.62 μs for i-C_3H_7O. For C_2H_5O, Fig. 1 shows the fluorescence-decay curve excited on the **A–X** (or **B–X**) 0^0_0 transition, yielding a lifetime of 0.94 μs, and Fig. 2 depicts a trace of the time-resolved emission for the 10^5_0 vibronic band,[3] indicating a considerably reduced lifetime of 0.32 μs for the CO-stretch excitation. These observations are consistent with significant nonradiative decay routes being available in the excited states of the alkoxy radicals.

This work was supported by the U.S. Environmental Protection Agency's Office of Exploratory Research (R819720-01-0), NASA Lewis Research Center (NAG3-1677), and the Center for the Study of Terrestrial and Extraterrestrial Atmospheres (NASA NAGW-2950).

1. T. Ebata, H. Yanagishita, K. Obi, and I. Tanaka, Chem. Phys. **69**, 27 (1982).
2. P. Misra, X. Zhu, C.-Y. Hsueh, and J. B. Halpern, Chem. Phys. **178**, 377 (1993).
3. T. Shimanouchi, *Tables of Molecular Vibrational Frequencies*, Consolidated Vol. I, 1972, NSRDS-NBS 39 (U.S. Department of Commerce, Washington, D.C.

QMF4 **11:15 am**

Laser-induced fluorescence spectroscopy of the ethoxy radical in a supersonic jet expansion

Xinming Zhu, Mohammed M. Kamal, Prabhakar Misra, *Laser Spectroscopy Laboratory, Department of Physics and Astronomy, Howard University, Washington, D.C. 20059*

The ethoxy (C_2H_5O) radical is an important chemical intermediate in combustion and atmospheric reactions.[1] It belongs to the C_s point group and has 18 normal vibrational frequencies.[2] Four vibrational frequencies have been reported earlier,[3] but only one—the CO stretch—has been assigned definitively. The C_2H_5O radical was generated *in situ* by photolyzing ethyl nitrite (C_2H_5ONO) with K_rF (at 248 nm) excimer-laser pulses. The nitrite precursor (at 1% concentration) was transported by helium under high pressure (12–14 atm) and was seeded into the vacuum chamber in the form of a supersonic molecular beam through a commercial pulsed nozzle. We have recorded laser-excitation spectra of jet-cooled C_2H_5O with 0.15-cm^{-1} resolution, and we have obtained dispersed fluorescence spectra with the aid of a 0.275-m monochromator (equipped with a 300-grooves/mm grating of 1-nm resolution). A frequency-doubled Nd:YAG-pumped dye laser with a nominal linewidth of 0.07 cm^{-1} served as the probe beam. Optogalvanic transitions excited by using a commercial F_e–N_e hollow-cathode discharge lamp were used to calibrate the wavelength of the dye laser. The radicals were excited downstream in the jet expansion, and the delay between the photolysis and excitation beams was typically 8 μs.

Laser excitation spectra of the **A–X** (or **B–X**) electronic system[1] of C_2H_5O were obtained in the spectra region 310–350 nm with rotationally resolved resolution. In addition to the main CO stretch progression shown in Fig. 1, there were seven other

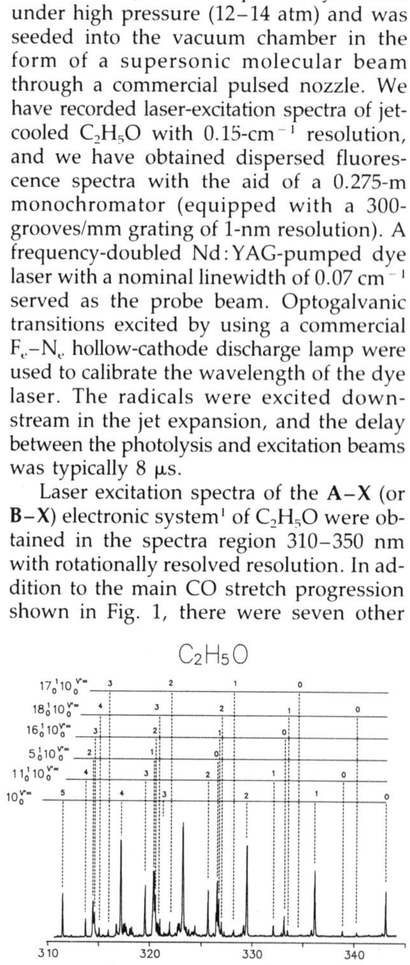

QMF4 Fig. 1. Laser-excitation spectrum of the ethoxy radical obtained in a supersonic jet expansion showing progressions involving the various vibrational modes ν_5, ν_{10}, ν_{11}, ν_{16}, ν_{17}, and ν_{18}.

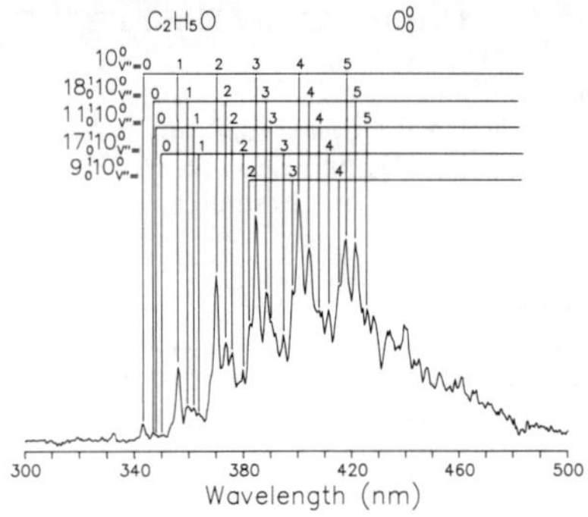

QMF4 Fig. 2. Dispersed fluorescence spectrum of the ethoxy radical obtained when the 0^0_0 band was pumped. A 0.275-m-focal-length monochromator equipped with a 300-grooves/mm grating was used to record the spectrum; it provided a resolution of 1 nm.

distinct excited-state frequencies that could be identified, and these have been assigned[2] as follows: CO stretch (ν'_{10} = 605 cm^{-1}), CC stretch (ν'_9 = 895 cm^{-1}), CCO deform (ν'_{11} = 365 cm^{-1}), CH_3 *d*-deform (ν'_5 = 1472 cm^{-1}), CH_3 *s*-deform (ν'_6 = 1459 cm^{-1}), CH_3 rock (ν'_8 or ν'_{16} = 871 cm^{-1}), CH_2 rock (ν'_{17} = 719 cm^{-1}), and torsion (ν'_{18} = 245 cm^{-1}).

Dispersed fluorescence spectra of C_2H_5O were recorded with an optical-multichannel-analyzer (OMA) system, and the wavelength calibration was carried out by using mercury lines. Seventeen bands excited in the near-UV provided strong emission and were dispersed with the OMA. Figure 2 shows a typical dispersed spectrum obtained by pumping the 0^0_0 band. To the best of our knowledge, 15 of the 17 bands are presented for the first time, together with detailed analysis of the structure of the wavelength-resolved emission spectra and the accompanying vibrational assignments.

This work was supported by the U.S. Environmental Protection Agency's Office of Exploratory Research (R819720-01-0), NASA Lewis Research Center (NAG3-1677), and the Center for the Study of Terrestrial and Extraterrestrial Atmospheres (NASA NAGW-2950).

1. X. Q. Tan, J. M. Williamson, S. C. Foster, and T. A. Miller, J. Phys. Chem. **97**, 9311 (1993).
2. T. Shimanouchi, *Tables of Molecular Vibrational Frequencies*, Consolidated Vol. I, 1972, NSRDS-NBS 39 (U.S. Department of Commerce, Washington, D.C.).
3. G. Inoue, M. Okuda, and H. Akimoto, J. Chem. Phys. **75**, 2060 (1981).

QMF5 **11:30 am**

High-resolution spectroscopy of diatomic lithium

K. Urbanski, A. Yiannopoulou, S. Antonova, A. M. Lyyra, Li Li,*, *Department of Physics, Temple University, Philadelphia, Pennsylvania 19122*

We use various high-resolution laser-spectroscopy techniques to observe details in the spectrum of $^7Li_2$ that are otherwise unresolved. Perturbation-facilitated optical–optical double-resonance (PFOODR) spectroscopy was used to access states in the triplet manifold. Optical–optical double-resonance (OODR) spectroscopy was used to detect Λ doubling. High-lying levels in singlet states were observed by using all-optical triple-resonance (AOTR) spectroscopy. Ionization detection greatly facilitated the observation of high-lying levels.

Molecular constants and RKR potential curves were obtained for the $2^3\Sigma_g^+$, $3^3\Sigma_g^+$, and $4^3\Sigma_g^+$ states. Our experimental T_e and R_e for the $2^3\Sigma_g^+$ state are 27297.45(16) cm^{-1} and 3.0797(18) Å, respectively, and for the $3^3\Sigma_g^+$ state they are 31043.93(53) cm^{-1} and 3.0378(19) Å, respectively. Hyperfine splitting for both states has been resolved. Both states follow a Hund's coupling case ($b_{\beta S}$) hyperfine coupling scheme. The experimental Fermi contact parameter b_F is approximately 96.0 ± 2 MHz for the $2^3\Sigma_g^+$ state and 95.6 ± 3 MHz for the $3^3\Sigma_g^+$ state. These values are in good agreement with the previously obtained value of 98.6 ± 4 MHz. One level of the $4^3\Sigma_g^+$ state has also been observed. Its hyperfine structure has been resolved and characterized with Hund's coupling case ($b_{\beta S}$) as well.

Both *e* and *f* parity levels of the $G^1\Pi_g$ state have been observed. This enabled us to calculate a Λ-doubling parameter of 4.7220(49) × 10^{-4}. This is significantly less than the value extracted from previous experimental work.

High-lying levels in the $A^1\Sigma_u^+$, $F^1\Sigma_g^+$, and $G^1\Pi_g$ states have been observed. Detecting ions rather than collision-induced side fluorescence improved the signal-to-noise ratio by a factor of 20 in some cases. From these data we have calculated molecular constants for these states. We are also in the process of applying a new near-dissociation analysis that should lead to improved RKR potential curves.

**Department of Modern Applied Physics, Tsinghua University, Beijing 100084, China*

QMF6 **11:45 am**

Large-dynamic-range spectroscopy of molecular relaxations

R. A. Cheville,* R. D. Averitt, N. J. Halas, *Department of Electrical Engineering and Rice Quantum Institute, Rice University, P.O. Box 1892, Houston, Texas 77251*

Over the last decade, time-resolved optical probes with subpicosecond temporal resolution have been used to directly measure fast relaxation processes in a wide variety of material systems. However, these techniques are impractical for measurements of relaxations occurring over several orders of magnitude in time. Relaxation processes that occur over many orders of magnitude in time are found in a wide variety of physical systems, including the excited-state dynamics of simple molecular systems and condensed matter systems, such as amorphous semiconductors and conductive polymers.

We have developed an experimental technique[1] that permits the measurement of relaxation processes with subpicosecond temporal resolution over a large temporal dynamic range of 10^7, from ~100 fs to >1 μs. The technique is applied to the measurement of relaxation in organic-dye solutions to determine both the singlet and the much-longer triplet lifetimes in the same experiment.

The relaxation can be determined with subpicosecond resolution on short time scales (up to several nanoseconds) by mechanically varying the delay between the excitation and probe pulses. Slower relaxation processes are then mapped out in time by incrementally increasing the pulse repetition period external to the laser cavity by using a fast acousto-optic modulator. Using an unamplified CPM source and differential lock-in detection, we have measured signals with as few as one pulse in 1000 switched to the experiment, corresponding to ~11 μs between pulses. Because this is an all-optical method, timing jitter is negligible on all time scales.

This technique is used to determine the complete S_1–T_1–S_0 relaxation dynamics of photoexcited organic-dye molecules in solution. In systems with small fluorescence quantum yield and fast intersystem crossing, the long-lived triplet state dominates relaxation dynamics. An example is the biological stain Azure A. Fast time-dependent processes are shown in Fig. 1(a), in-

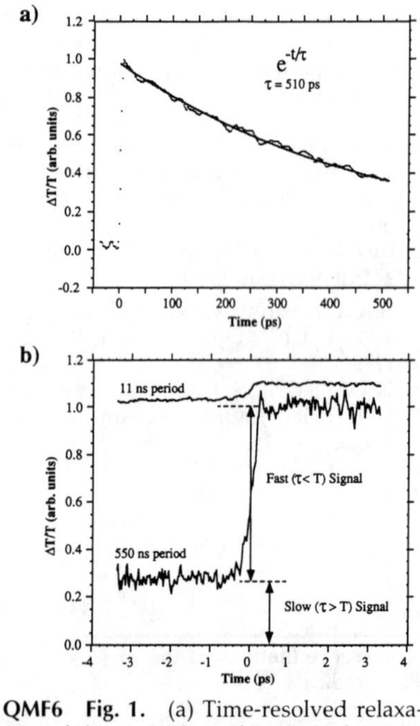

QMF6 Fig. 1. (a) Time-resolved relaxation of the excited singlet state of Azure A in ethylene glycol. The solid curve is a fit of exponential decay with τ = 510 ps. (b) Relative magnitude of the fast and slow time-varying signals for 11-ns (upper trace) and 550-ns (lower trace) laser repetition periods. Both traces have been normalized so that the peak value of the signal is unity, and the upper signal is offset by 0.1 for clarity. The magnitude of the time-dependent signals for both the 11-ns and the 550-ns repetition period is equal to within the noise level of the data. For an 11-ns repetition period the slow signal is ~15 times that of the time-dependent signal. For the 550-ns period the slow signal is reduced to half the magnitude of the time-dependent portion of the signal.

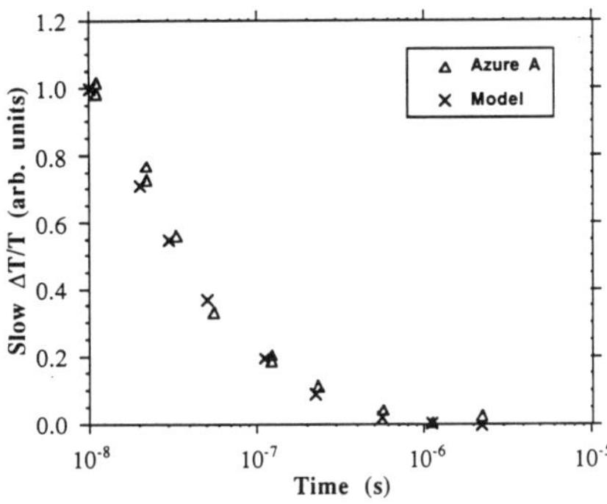

QMF6 Fig. 2. Magnitude of slow relaxation process in Azure A as a function of laser repetition period (triangles). A numerical simulation of the two-level system (crosses) is also shown. The decay corresponds to a triplet-state lifetime of 400 ns.

dicating depopulation of the excited singlet state in 510 ps. The fast decay signal occurs on top of a large, delay-line-position-insensitive bleaching signal due to ground-state depopulation, which is a function of the time between successive pulses. As the repetition period is increased, the magnitude of this slow signal decreases relative to that of the faster signal, as shown in Fig. 1(b). The triplet-state lifetime is determined by measuring the magnitude of the delay-line-position-insensitive portion of the photobleaching signal as a function of laser repetition period, as shown in Fig. 2. The results of a numerical model using a triplet-state lifetime of 400 ns are also shown in Fig. 2, with excellent agreement.

We have developed a technique that permits measurement of the relaxation dynamics of processes occurring on widely different time scales. The technique is illustrated by determining the complete relaxation dynamics, i.e., both the triplet- and singlet-state lifetimes, for the biological stain Azure A. To our knowledge this is the first time relaxation rates have been reported for Azure A.

*Current address: Department of Electrical Engineering and Center for Laser Research, Oklahoma State University, Stillwater, Oklahoma 74078

1. R. A. Cheville, R. D. Averitt, and N. J. Halas, Opt. Commun. **110**, 327 (1994).

QMF7 **12:00 m**

Chaotic light: Q spoiling and directional emission in deformed-dielectric-cavity resonators

A. Douglas Stone, J. U. Nöckel, A. Mekis, *Department of Applied Physics, Yale University, P.O. Box 208284, New Haven, Connecticut 06520-8284*

Cylindrical or spherical dielectric-cavity resonators in which high-Q modes are created by total internal reflection of electromagnetic waves circulating around the perimeter are commonly used as components in lasers and in sensitive detectors. The spectral properties of such resonators are crucial to the functioning of microdisk lasers[1] and to the nonlinear optics of liquid droplets.[2] An ideal resonator of this type consists of a dielectric body of perfect axial or spherical symmetry that totally internally reflects the light traveling in whispering-gallery modes almost tangential to the boundary. In an ideal resonator the light in these modes will escape only by evanescent leakage; thus they have very high Q ($>10^8$) and emit light isotropically (which can be a disadvantage in certain applications.[2] In this work we study the effect on these high-Q modes of substantial deformations from rotational symmetry (1%–20%) by using a ray-optics model and concepts from classical nonlinear dynamics.[3,4] Because the model is based on ray optics, it describes cavities whose typical dimensions are large compared to the wavelength and the smooth deformations.

The model describes the shape of the dielectric by a distortion parameter ε that may be, e.g., the strength of a quadrupolar deformation. When $\varepsilon = 0$, the shape is spherical (or cylindrical), and as ε increases, it becomes possible (in principle) for light rays that were trapped by total internal reflection to escape after many bounces as their angle of incidence falls below the critical angle. However, the Kolmogorov–Arnold–Moser theorem of Hamiltonian mechanics[5] implies that for sufficiently small deformations, the excursions of the rays away from their initial angle are bounded, so that typical whispering-gallery modes will not diffuse in angle far enough to escape and Q remains high. When the deformation is increased, a chaotic region in phase space will extend from the initial region to the region in which refractive escape is possible, opening up a classical escape channel. We can then calculate the mean escape rate τ^{-1}, and hence $\omega Q^{-1} = \tau^{-1}$, for the relevant set of rays. The basic results are the following: (1) Q remains quite high for deformations up to a threshold ε_c, which is typically 2%–10% of the radius, and it may be calculated if the type of deformation is known. (2) for $\varepsilon > \varepsilon_c$, Q decreases as a power law, $Q^{-1} = (\varepsilon - \varepsilon_c)^\alpha$, where $2.5 < \alpha < 3.2$. (3) Above ε_c, the emission becomes highly directional with maximal intensity in directions that may be calculated from the model. (4) The directional emission is even sharper for a quadrupolar deformation of the sphere than for a quadrupolar deformation of the cylinder.

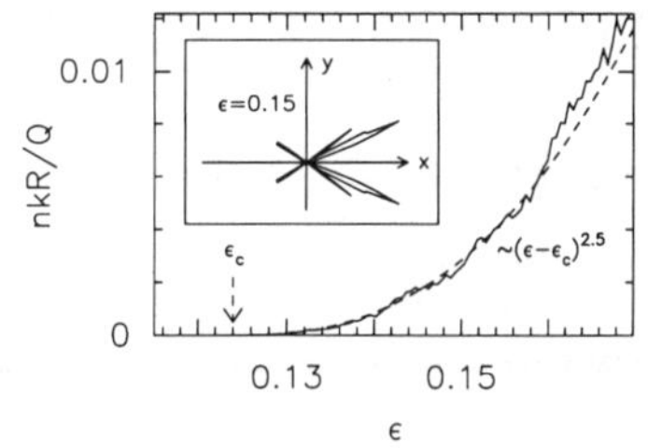

QMF7 Fig. 1. Q^{-1} versus deformation ε of the cylinder obtained from the average escape rate (see text). The unperturbed circle has radius R, k is the vacuum wave number, and n is the refractive index of the dielectric. Escaping rays are started in a strip with $0.66 \le \sin\theta \le 0.71$. A least-squares fit to the data gives a power law $nkR/Q = 23.7\ (\varepsilon - \varepsilon_c)^\alpha$, with $\varepsilon_c = 0.124$ and $\alpha = 2.5$. Inset: polar plot of the escape probability versus angle in the far field for $\varepsilon = 0.15$, showing the strong directionality of emission.

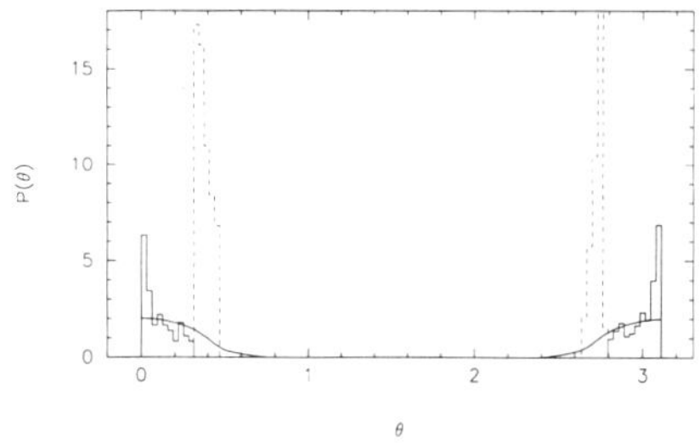

QMF7 Fig. 2. Polar angle distribution of the escape probability from a spheroid with a prolate quadrupolar deformation of ≈9%. The smooth curve is the result of the chaotic-mixing model and the histograms indicate directionality for a polar orbit (solid lines) and an orbit tilted ≈20° from the vertical (dashed lines).

Results (1) and (2) are illustrated in Fig. 1; results (3) and (4) are illustrated in the inset in Fig. 1 and in Fig. 2. The emission is most directional at threshold $\varepsilon = \varepsilon_c$ and tends to be concentrated in the directions tangential to the regions of minimal radius of curvature. A model that assumes completely chaotic mixing in phase space before escape locates correctly the regions of maximal escape but underestimates substantially the degree of directionality (see Fig. 2).

1. S. L. McCall *et al.*, Appl. Phys. Lett. **60**, 289 (1992).
2. S. C. Hill and R. E. Benner, in *Optical Effects Associated with Small Particles*, P. W. Barber and R. K. Chang, eds. (World Scientific, Singapore, 1988).
3. J. U. Nöckel, A. D. Stone, and R. K. Chang, Opt. Lett. **19**, 1693 (1994).
4. A. Mekis, J. U. Nöckel, and A. D. Stone, unpublished.
5. L. E. Reichl, *The Transition to Chaos* (Springer-Verlag, New York, 1992).

QMF8 **12:15 pm**

Spatially selective excitation of dielectric photonic atom resonances with an optical fiber

A. Serpengüzel, S. Arnold, G. Griffel,* *Microparticle Photophysics Laboratory, Polytechnic University, 6 Metrotech Center, Brooklyn, New York 11201*

Recently, μ-spheres have enjoyed a great deal of attention from the optical-spectroscopy community. Although a variety of linear and nonlinear optical processes have been observed in μ-droplets,[1] actual photonic devices require the use of solid μ-spheres. The first of these applications is the room-temperature hole-burning memory.[2] Addressing a solid three-dimensional photonic device composed of μ-spheres requires modification of the excitation geometry. One can no longer use a plane wave, which would give rise to multiple scattering. Instead, an optical-fiber coupler (OFC) can be used for addressing.

Compared with the plane wave, an OFC

has also the additional benefit of selectively and more efficiently coupling power to the μ-sphere morphology-dependent resonances (MDRs); one can excite an MDR with reduced scattering. Moreover, this on-resonant pump geometry, together with the cavity QED effects, will greatly reduce the threshold intensity required for a variety of nonlinear optical effects, such as stimulated Raman scattering, stimulated Brillouin scattering, wave mixing, and harmonic generation.

In this paper we report the coupling between an OFC and a μ-sphere. We have also observed that multiple μ-sphere MDRs can be separately excited on an OFC. This is particularly important for photonic applications, because the MDRs need to be addressed separately for a high spatial memory density.

Figure 1 shows our experimental setup. A μ-sphere covered with water is placed on the OFC. The scattering spectrum, obtained through a polarizer, is displayed in Fig. 2. The observed TE resonances have linewidths comparable to those of the laser (0.025 nm) and an on-resonance/off-resonance intensity ratio $>10^4$. Most of the features of this spectrum can be explained through the electromagnetic interaction of the μ-sphere with an off-axis Gaussian beam and by treating the μ-sphere as a photonic atom.

Figure 3 is a spectrum calculated by using such a Gaussian-beam model, developed by Lock,[3] in the generalized Lorentz–Mie formalism. The calculation bears a good relationship to the experimental data of Fig. 2. There is little off-resonance scattering, in agreement with the basic idea of localization,[4] which associates a light ray having a wave vector k and an angular momentum $(n + 1/2)h/2\pi$ with an impact parameter r_n; i.e., $r_n(h/2\pi)k = (n + 1/2)(h/2\pi)$. The res-

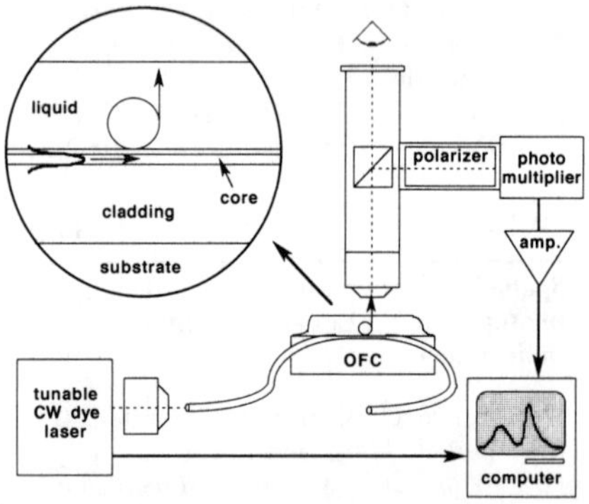

QMF8 Fig. 1. Experimental setup.

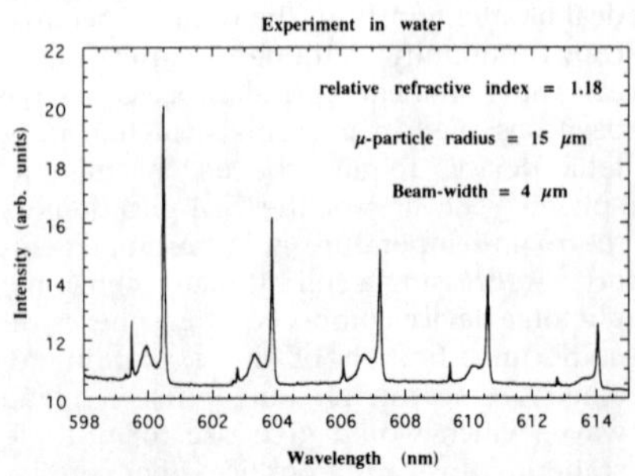

QMF8 Fig. 2. Experimental elastic-scattering spectrum for TE resonances.

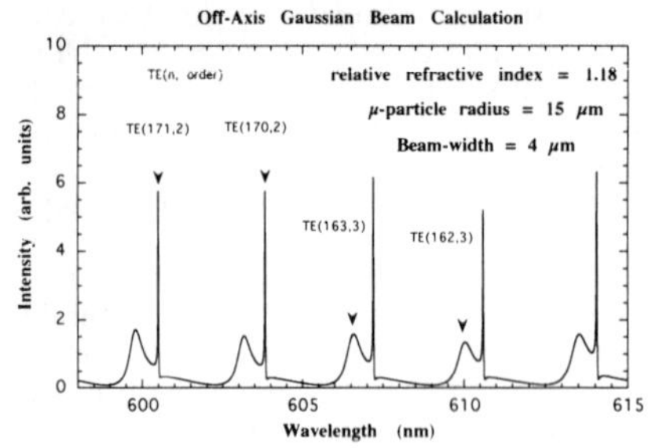

QMF8 Fig. 3. Calculated Gaussian-beam elastic-scattering spectrum for TE resonances.

onances appear to correspond to first-, second-, and third-order MDRs with calculated quality factors of approximately 10^7, 10^5, and 10^3, respectively. The experimental first-order MDRs are broadened because of the convolution with the laser linewidth.

The μ-sphere–OFC system has potential for future photonic applications. It can be used as (1) a building block for a photonic room-temperature hole-burning memory, (2) a piezoelectrically controlled microspectrum analyzer, (3) a test bed for external Gaussian-beam excitation of linear and nonlinear processes, or (4) a means for spatially selective photonic addressing.

We would like to acknowledge support from U.S. Air Force Office of Scientific Research (grant F49620-94-0195).

Department of Electrical Engineering, Polytechnic University, 6 Microtech Center, Brooklyn, New York 11201

1. S. C. Hill and R. E. Benner, in *Optical Effects Associated with Small Particles*, P. W. Barber and R. K. Chang, eds. (World Scientific, Singapore, 1988).
2. S. Arnold, C. T. Liu, W. B. Whitten, and J. M. Ramsey, Opt. Lett. **16**, 420 (1991).
3. G. Gouesbet and J. A. Lock, J. Opt. Soc. Am. A **11**, 2516 (1994).
4. H. C. van de Hulst, *Light Scattering by Small Particles* (Dover, New York, 1981), p. 208.

JMB Tutorial — 10:30 am
Hyatt E/F

Direct and Indirect Laser Fusion

Mark A. Henesian, *Lawrence Livermore National Laboratory, Presider*

JMB1 (Tutorial) — **10:30 am**

Direct and indirect laser fusion

Robert L. McCrory, *University of Rochester, Rochester, New York*

The pursuit of achieving controlled thermonuclear fusion has occupied many scientists since the potential benefits of the process were recognized and serious work began on it over 45 years ago. The two primary approaches to controlled fusion are magnetic confinement fusion and inertial confinement fusion (ICF). Inertial fusion uses high intensity lasers or particle beams to implode a capsule containing thermonuclear fuel. High implosion velocities are required ($>3 \times 10^7$ cm/sec) so that, on fraction of the fuel (the spark plug) is sufficient to ignite the spark and initiate propagating burn through the main fuel mass. There are two principal ICF approaches to using lasers to implode a fuel capsule: direct and indirect drive. This paper presents a tutorial discussion of the physics of both direct- and indirect-drive laser fusion and illustrates the physical principles by drawing principally on data from the OMEGA laser at the Laboratory for Laser Energetics at the University of Rochester for direct drive and from the NOVA laser at the Lawrence Livermore National Laboratory for indirect drive. Due to the recent declassification of many of the concepts and data from indirect drive, it is now possible to present a unified view of ICF research. The proposed National Ignition Facility design (1.8 MJ and 500 TW using 351 nm light) promises many opportunities for researchers in ICF, basic laser–matter interaction physics and quantum electronics.

JMC — 2:00 pm
Rooms 307–310

CLEO/QELS Plenary and Awards Session

Thomas F. Deutsch, John M. Goldsmith, *CLEO '95 General Chairs*, Tony F. Heinz, George I. Stegeman, *QELS '95 General Chairs*

2:00 pm

Introductory Remarks

2:05 pm

Awards Ceremony

Presentation of the 1995 Quantum Electronics Award of the Lasers and Electro-Optics Society of the Institute of Electrical and Electronics Engineers.

Presentation of the Charles Hard Townes Award of the Optical Society of America.

Presentation of the Distinguished Service Certificate of the APS/Laser Science Topical Group.

Presentation of the OSA fellow certificates.

Presentation of the IEEE/LEOS fellow certificates.

Presentation of the APS fellow certificates.

JMC1 **2:30 pm**

The renaissance in microscopy

E. Betzig, *NSOM Enterprises, 17 Webster Dr., Berkeley Heights, New Jersey 07922*

The last twenty years have seen explosive growth in the development and application of microscopic imaging techniques. This has not been by accident: advances in semiconductor technology, information storage, medical research, and the biological sciences have demanded more varied tools for characterization at progressively smaller length scales. It behooves researchers in these areas to be aware of the available microscopic methods.

These methods can be categorized at several different levels. First, there are those instruments which achieve resolution through spatial point sampling (the vast majority) and those which rely on frequency domain sampling (MRI). The former are often used for sample manipulation and modification (lithography, optical trapping) as well as imaging, while the latter may lead to molecular resolution in three or more dimensions. Spatial sampling systems are further divided into lens based instruments (conventional optical, x-ray, electron microscopes) and scanned probe techniques (STM, AFM, NSOM). Sub-optical resolution in the former case demands short wavelength irradiation but also yields tremendous depth of focus. Scanned probe instruments, on the other hand, offer high resolution under varied environmental conditions, but with extremely limited depth of focus and vexing image interpretation problems.

In short, although every available form of microscopy offers its own set of advantages and limitations, additional capabilities for existing techniques and entirely new methodologies are being introduced continually.

JMC2 **4:00 pm**

"Plastic" optoelectronics with conducting polymers

Alan J. Heeger, *UC–Santa Barbara*

Conjugated polymers have emerged as a novel class of materials with the optical and electronic properties of semiconductors in combination with the processing advantages and mechanical properties of polymers. Optical quality thin films of such materials can be readily fabricated by spin-casting from common solvents. Such thin films have been demonstrated to exhibit potentially important optoelectronic phenomena and to serve as the active elements in a growing number of "plastic" optoelectronic devices; including, for example, the following:

- polymer light-emitting diodes,
- polymer photodiodes and photodetectors,
- polymer photovoltaic cells and
- polymer opto-couplers.

In each case the device performance is comparable to or better than that obtainable with inorganic materials, but with specific advantages which include large area, ease of fabrication, and flexible mechanical properties etc.

In addition, the nonlinear optical properties of polymers have been utilized to successfully demonstrate optical image processing with optical correlators in which a conjugated (semiconducting) polymer serves as the active nonlinear optical (NLO) medium. For example, a four-wave mixing optical correlator was built with an air-stable conjugated polymer as the NLO medium. The continuously updatable processor correlates two 5000 pixel images in less than 160 femtoseconds. Related experiments have utilized a photorefractive polymer as the active NLO medium in a matched-filter four-wave-mixing optical correlator.

Progress in each of these areas will be reviewed and summarized.

JMC3 **5:00 pm**

Real world applications of laser spectroscopy

Sune Svanberg, *Department of Physics, Lund Institute of Technology, P.O. Box 118, S-221 00 Lund, Sweden*

Laser spectroscopy provides extreme spectral and temporal resolution and allows ultrasensitive monitoring. A host of techniques have been developed with major impact on basic as well as more applied sciences.[1,2] While the traditional realm of laser spectroscopy has been precision atomic and molecular physics, the techniques are now widely used in chemistry, both for sensitive analysis and for studies of ultrafast dynamics.[3] Laser spectroscopy methods have found many applications in energy, environmental and medical research. Combustion monitoring[4] and the study of other reactive media can be assessed with non-intrusive techniques. Atmospheric pollutants are effectively monitored by long-path absorption and laser radar methods.[5] Frequency-modulation spectroscopy using diode lasers is able to detect optical absorptions amounting to 1 part in 10^7. Water pollutants and vegetation[6] status can be remotely assessed by laser-induced fluorescence. Medical laser spectroscopy includes early cancer detection and atherosclerotic plaque monitoring.[7] Photodynamic tumor therapy is an example of laser-induced chemistry applied to the human body with a potential of having a major impact on cancer management. Light scattering spectroscopy and gated viewing techniques are being developed to recognize objects in the presence of heavy multiple scattering giving a hope for mammography without the use of ionizing radiation.[8]

1. W. Demtröder, *Laser Spectroscopy*, Springer Ser. Chem. Phys., Vol. 5 (Springer-Verlag, Berlin, Heidelberg, 1995); 2nd ed, in press.
2. S. Svanberg, *Atomic and Molecular Spectroscopy—Basic Aspects and Practical Applications* (Springer-Verlag, Berlin, Heidelberg, 1993).
3. A. H. Zewail, *Femtochemistry*, Vols. 1 and 2 (World Scientific, Singapore, Hong Kong, London 1994).
4. A. C. Eckbreth, *Laser Diagnostics for Combustion Temperature and Species* (Abacus Press, Turnbridge Wells, 1987).
5. S. Svanberg, "Differential absorption lidar (dial)," Chap. 3 in *Air Monitoring by Spectroscopic Techniques*, M. W. Sigrist, ed. (Wiley, New York, 1994).
6. S. Svanberg, "Fluorescence lidar monitoring of vegetation status," Physica Scripta, in press.
7. S. Svanberg, "Tissue diagnostics using lasers," Chap. 7 in *Lasers in Medicine*, G. Pettit and R. W. Waynant, eds. (Plenum, New York, in press).
8. Müller *et al.*, eds., *Medical Optical Tomography*, Vol. IS11 (SPIE, Bellingham, Washington, 1993).

JTuA 8:00 am
Room 308

Soliton Lasers

C. Randy Giles, *AT&T Bell Laboratories*, *Presider*

JTuA1 (Invited) 8:00 am

Soliton fiber lasers

James R. Taylor, *Femtosecond Optics Group, Physics Department, Imperial College, Prince Consort Road, London SW7 2BZ, U.K.*

Rare-earth-doped fibers, both oxide and fluoride, exhibit broad gain bandwidths capable of supporting sub 100 femtosecond pulsewidths. To date, the shortest pulses have been generated with Nd doped silicate fibers exploiting "soliton-like" shaping,[1] analogous to solid-state laser systems, where ultrashort pulse generation using such techniques have been particularly successful. The operating wavelength of Er-based lasers permits "true" soliton operation and many techniques and systems have been developed to investigate ultrashort pulse generation, including the widely investigated "figure-of eight" system[2] and loop lasers exploiting nonlinear polarization rotation as the pulse formation mechanism.[3]

Generally soliton fiber lasers utilizing active modulation techniques generate longer pulses but exhibit significantly greater stability than their passively-mode-locked and shaped femtosecond counterparts. Stability is achieved through operating in the path average soliton regime. On the generation of femtosecond pulses, it becomes difficult to remain within the flexible limits of this regime, with the pulses exhibiting spectral and temporal instabilities.[4] The relatively long fiber lengths involved also permit the observation of many additional nonlinear phenomena, which limit the pulse applicability.

Birefringence, either inherent or induced, can lead to interaction leading to spectral sideband splitting, which has been observed. Birefringence effects can also permit spectrally differing, but equivalent switching conditions in the lasers leading to independent dual wavelength operation and cross phase modulation. In addition, birefringence can lead to the formation of very high repetition rate pulse train bunches, with the number of pulses in the trains limited by the available powers. Self organization has also been observed in several higher power systems where the characteristic initially random but periodic outputs self organize in time into periodic pulse trains, which have been observed from the fundamental to greater than the four hundredth harmonic of the fiber laser. This effect has been attributed to electrostriction.

Clearly, the geometry of the present fiber soliton lasers prohibit ultrahigh pulse rate generation; the use of solitons also prohibit high power extraction and the uncontrolled spectral and temporal outputs make the sources impractical for many applications. Some of these aspects can be overcome to good effect by using nonsoliton sources[5] and other stabilization schemes can be used. In the future it is likely that in core chirped Bragg reflectors will play a role in the further simplification of compact ultrashort pulse fiber sources and initial results on this as well as the aspects described above will be presented.

1. M. H. Ober, M. Hofer, and M. E. Fermann, Opt. Lett. **18**, 367 (1993).
2. I. N. Duling II, Opt. Lett. **16**, 539 (1991).
3. V. J. Matsas, T. P. Newton, D. J. Richardson, and D. N. Payne, Electron. Lett. **28**, 2185 (1992.
4. S. M. J. Kelly, Electron Lett. **28**, 806 (1992).
5. K. Tamura, E. P. Ippen, H. A. Haus, and L. E. Nelson, Opt. Lett. **18**, 1080 (1993).

JTuA2 8:30 am

Synchronous and asynchronous dual-wavelength operation of a femtosecond neodymium-fiber laser

M. H. Ober, G. A. Reider, G. Sucha,* M. E. Fermann,* *Technische Universität Wien, Abteilung Quantenelektronik und Lasertechnik, Gusshausstr. 27/359/9, A-1040 Vienna, Austria*

Dual-wavelength operation of passively mode-locked solid-state lasers has recently gained considerable attention. The work has focused on Ti:sapphire lasers[1,2] and Er^{3+} fiber lasers.[3,4] Lack of controllability in these systems stems from homogeneous broadening of Ti:sapphire (leading to gain-competition) and from problems with polarization fluctuations in the Er^{3+} fiber laser cavity.[3,4] In addition the Er^{3+} fiber laser cavities did not allow for an independent tuning of the repetition rate of the two wavelengths. This asynchronous operation is not applicable to the Ti:sapphire systems unless two regions are pumped separately.

In our paper we show that the operation range and control of passively mode-locked dual-wavelength sources may be greatly expanded by resorting to singlemode Nd^{3+}-doped silica fiber as the gain medium. Intra-cavity dispersion compensation prisms are also used as an efficient means of separation of the two emission wavelengths. The Nd fiber laser generates nearly bandwidth-limited femtosecond pulses simultaneously at 1.060/1.100 or 1.057/1.077 μm. The experimental setup is shown in Fig. 1. To achieve mode-locking, we applied Kerr-type mode-locked (KTM)[5] and/or carrier-type mode-locking (CTM)[5] using multiple quantum well saturable absorbers. The mode-locked output powers were in the range from 2.5 to 15 mW. One cavity had a fixed repetition rate of 65 MHz, whereas the repetition rate of the other cavity was adjustable around that center frequency by mounting the end mirror on a translation stage. When applying KTM and CTM in the two cavity arms, a minimum repetition rate difference frequency Δf of 500 Hz was necessary to maintain the dual-wavelength mode-locked operation. By applying CTM

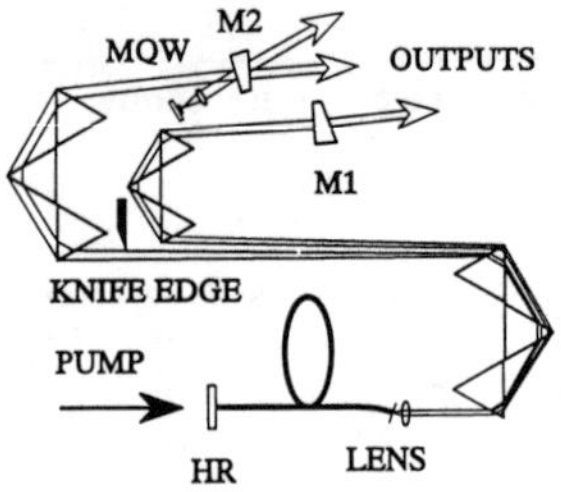

JTuA2 Fig. 1. Cavity design for the dual-wavelength asynchronous repetition rate Nd fiber laser oscillator.

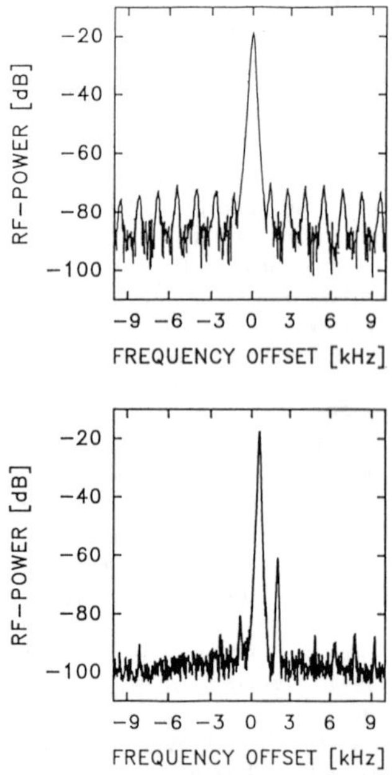

JTuA2 Fig. 2. RF spectra of the blue (top) and the red (bottom) pulse trains. The center frequency difference between the two repetition rates is 1.4 kHz.

for both wavelengths, the minimum Δf could be reduced to 300 Hz. Additionally, using cross polarizations for the two wavelengths, allowed synchronous operation of the two lines.

We verified that pulse interactions are negligible for highly asynchronous operation of this dual wavelength fiber laser by investigating the RF-spectra for the mode-locked pulse trains, as shown in Fig. 2. For both, CTM and KTM mode-locked wavelengths of any RF spectral sidebands were indeed 50 dB down from the carrier frequency. Bandwidth-limited femtosecond pulses were obtained when applying either CTM or KTM as shown in Fig. 3.

In conclusion, we have demonstrated controllable, synchronous, and asynchronous dual-wavelength operation of a passively mode-locked laser system in a single gain medium for the first time. By using Nd^{3+}-doped fiber as the gain medium, bandwidth-limited femtosecond pulses are generated at 1.060/1.100 or 1.057/1.077 μm, respectively. The difference between the repetition rates may be varied from 0 Hz to several MHz. This completely novel laser system can therefore serve as a very attractive candidate for high-resolution pump-probe experiments.

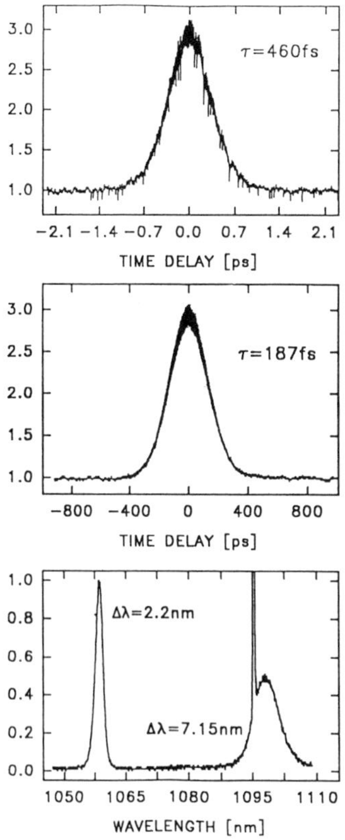

JTuA2 Fig. 3. Autocorrelation traces and output spectra of the mode-locked pulse trains of the dual-wavelength laser.

We are indebted to Ursula Keller and Heng Chiu for providing the MQW. We acknowledge discussions with Arnold Schmidt as well as partial support from the FWF P10515-PHY.

*IMRA America, Inc., 1044 Woodridge Ave., Ann Arbor, Michigan 48105

1. J. M. Evans, D. E. Spence, D. Burns, and W. Sibbet, Opt. Lett. **18**, 1074 (1993).
2. D. R. Dykaar and S. B. Darack, Opt. Lett. **18**, 634 (1993).
3. M. L. Dennis and I. N. Duling III, Electron. Lett. **29**, 409 (1993).
4. D. U. Noske, M. J. Guy, K. Rottwitt, R. Kashyap, and J. R. Taylor, Opt. Commun. **108**, 497 (1994).
5. M. H. Ober, M. Hofer, U. Keller, and T. H. Chiu, Opt. Lett. **18**, 1533 (1993).

JTuA3 **8:45 am**

High-power picosecond soliton fiber lasers using chirped fiber Bragg gratings

M. E. Fermann, K. Sugden,* I. Bennion,*
IMRA America, Inc., 1044 Woodridge Ave., Ann Arbor, Michigan 48105

Applications in telecommunication and instrumentation currently drive a vigorous research effort towards the construction of tunable high-power sources of psec pulses.[1] Particularly, passively-mode-locked fiber lasers appear very attractive for providing these functions; however a problem has been that passively-mode-locked fiber lasers typically tend to produce pulses near the bandwidth-limit of the gain medium, giving rise to pulses significantly shorter than 1 psec.[2]

Here we demonstrate that high-power psec fiber pulse sources are readily constructed by controlling the pulse width of a Kerr-mode-locked soliton fiber laser with chirped fiber Bragg gratings.[3] In a first demonstration wavelength-tunable 4 psec pulses with energies up to 12 nJ are generated at 1.55 μm in erbium fiber lasers. These are by far the highest pulse energies generated directly from fiber oscillators to date.

Chirped fiber gratings provide an excessive amount of negative dispersion inside a Kerr-mode-locked fiber laser cavity, which increases the pulse width, while keeping the nonlinearity of the cavity approximately constant. As a result a great increase in pulse energies compared to conventional soliton fiber lasers is obtained.

The experimental set up for an erbium fiber laser is shown in Fig. 1. We employed two chirped fiber gratings[3] with a bandwidth of 13 nm centered around 1.555 μm and a maximum reflectivity near 100%. Each grating had a length of about 5 mm and each grating provided a negative group-velocity dispersion (GVD) at 1.551 μm of around 3.5 psec.[2] The active fiber was doped with 1000 ppm Er^{3+}. Its numerical aperture was 0.15 and the core diameter was 6.6 μm. The active fiber length was 1.15 m and an additional length of 5.2 m of standard telecom fiber was also part of the cavity. The fiber laser was pumped via two wavelength-division multiplexing couplers with a total launched power up to 800 mW from two master oscillator power amplifier laser diodes operating at 980 nm. An intra-cavity polarizer was used to select a single polarization in the cavity and was also employed as an adjustable output coupler.

Passive mode locking was initiated by applying a small perturbation to the undoped fiber and was sustained by nonlinear polarization evolution,[4] where a minimum launched power of about 230 mW was required. By varying the pump power and adjusting the polarization state inside the laser, single pulse operation at fixed repetition rates of 15.3 MHz was obtained for average mode locked output powers between 46 and 185 mW (corresponding to pulse energies between 3 and 12 nJ).

Typical autocorrelation traces for 12, 6, and 3 nJ pulses and at 1.551 μm are shown in Fig. 2 and the corresponding spectra are shown in Fig. 3. The pulse widths were about 4.0 psec assuming a $sech^2$ shape and the corresponding time-bandwidth products rose from 0.52 to 0.67 with an increase in pulse energy.

Because of the reflection cut-off of the grating, spectral side bands were only generated on the red side of the pulse spectrum. Only at the highest pulse energy is a significant pedestal component observable on the autocorrelation trace. To obtain such high pulse energies we used an output coupling up to 95% and inserted most of the undoped fiber on the right hand side of the cavity, which greatly reduces the nonlinearity of the cavity. Thus, even for a 12 nJ pulse we only estimate a peak nonlinear phase delay of about 2 π inside the laser. Due to the large bandwidths of the chirped fiber gratings a wavelength tuning range of 12 nm was possible. Further optimization should allow the generation of pulses with energies up to 100 nJ.

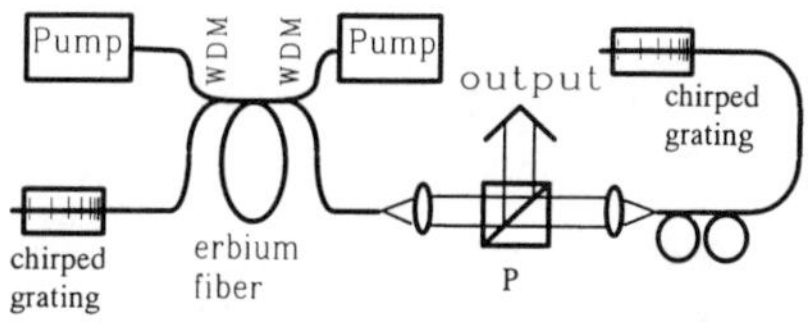

JTuA3 Fig. 1. Passively mode-locked fiber laser incorporating two chirped fiber gratings for pulse width control.

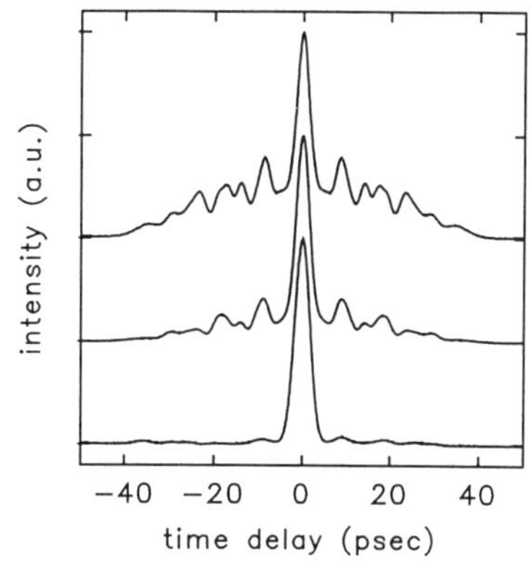

JTuA3 Fig. 2. Autocorrelations of 4.0 psec pulses with pulse energies of 12, 6, and 3 nJ (top to bottom).

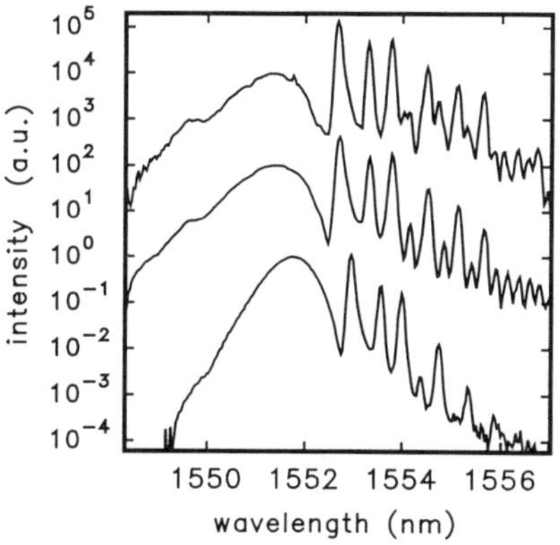

JTuA3 Fig. 3. Pulse spectra of 4.0 psec pulses with pulse energies of 12, 6, and 3 nJ (top to bottom).

Aston University, Aston Triangle Birmingham B4 7ET, U.K.

1. B. C. Barnett *et al.*, in *Conference on Lasers and Electro-Optics, CLEO 94*, 1994 OSA Technical Digest Series, Vol. 8 (Optical Society of America, Washington, D.C.), p. 52.
2. M. E. Fermann *et al.*, Opt. Lett. **16**, 244 (1991).
3. K. Sugden *et al.*, Electron. Lett. **30**, 440 (1994).
4. M. Hofer *et al.*, Opt. Lett. **16**, 502 (1991).

JTuA4 **9:00 am**

Passive harmonic mode locking of soliton fiber lasers

S. Gray, A. B. Grudinin, D. N. Payne, *Optoelectronics Research Centre, Southampton University, Southampton S017 1BJ, U.K.*

Passively-mode-locked fiber soliton lasers (FSL) are an attractive optical short pulse source for laboratory and telecommunications applications. The energy quantization effect caused by the soliton regime of operation of FSLs results in excellent stability of the duration and energy of the individual pulses, but also leads to pulse repetition rate instabilities, which for many applications is unacceptable. Recently, we have experimentally demonstrated that, under certain conditions, stable passive harmonic mode-locking occurs in a ring laser configuration,[1] and time jitter as low as 600 fs can be obtained at a repetition rate as high as 463 MHz in a fully passive ring configuration.[2]

In this paper we present the further investigation of this effect and show that a passive self-stabilization effect driven by transverse acoustic waves due to electrostriction becomes much stronger when the laser repetition rate is close to an acoustic eigen-frequency or its k-th harmonic.

The physical idea lying behind this phenomenon is quite simple: each pulse excites a weak acoustic wave that causes a change of the refractive index ($\sim 10^{-12}$ for 1 ps pulses). When several pulses exist inside the laser cavity (each generating an acoustic wave), after several round-trips all of them are trapped by the acoustic field of their predecessors, which results in passive phase modulation. If σ_T is the time jitter of the harmonically-mode-locked laser then it can be shown[2] that $\sigma_T \sim (\delta n)^{-1/2}$, where δn is the refractive index change induced by the soliton stream.

To find the value of the refractive index change induced by N (>> 1) pulses traveling inside a FSL, we have applied the theory developed in Ref. 3. The solution of the wave equation can be treated as a solution for a damped harmonic oscillator with an external force. Thus, one can expect strong enhancement of the refractive index perturbation (and consequently low jitter) at the repetition rates coinciding with acoustic eigen-frequencies. Our computer simulations have revealed that for stable operation the repetition rate should be slightly detuned from the corresponding resonant frequency.

Figure 1 shows the calculated temporal dependencies of refractive index change for 467 MHz repetition rate and damping time of 10 μsec. One can see that the refractive index perturbation increases $\sim 10^3$ times and its temporal variation looks like that produced by a conventional phase modulator. Note also, that refractive index enhancement occurs not only at resonant frequencies but also at its harmonics, which makes its spectral dependence rather complicated.

Knowing the acousto-induced refractive index change and using the expression for the time jitter, derived in Ref. 2, we can obtain the dependence of σ_T on repetition

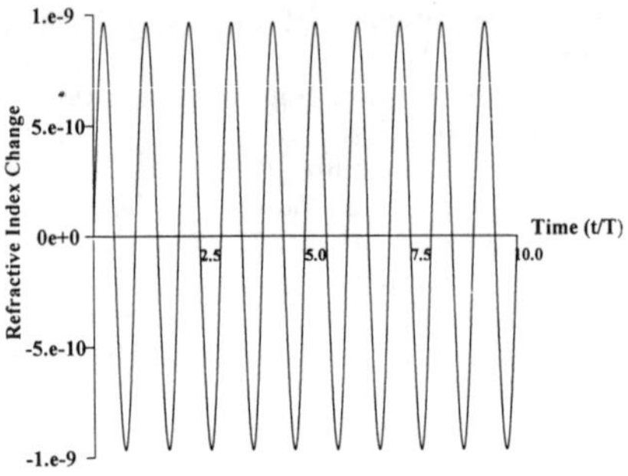

JTuA4 Fig. 1. Temporal dependence of the refractive index change when the laser repetition rate is equal to the 10-th acoustic eigen-frequency of the laser fiber. Refractive index change due to action of a single soliton is about 10^{-12}.

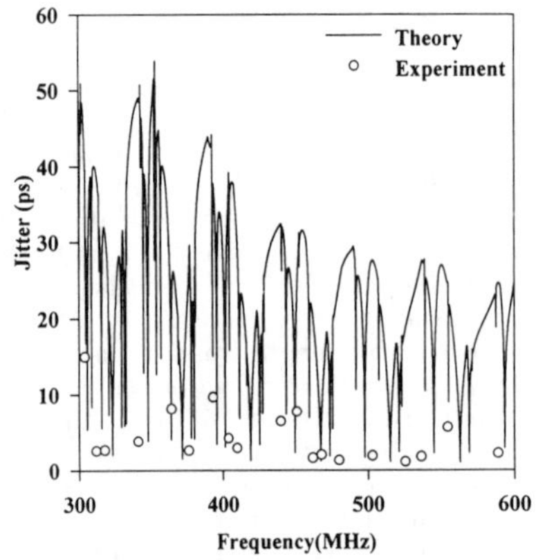

JTuA4 Fig. 2. Time jitter as a function of repetition frequency. Solid curve—theory, circles—experiment.

frequency. The solid curve in Fig. 2 represents the calculated time jitter and circles are experimental data from Ref. 2, which fit reasonably well with theoretical results.

In conclusion, we have shown that substantial enhancement of the refractive index perturbation and consequently lower jitter occur when the repetition rate of a harmonic passively-mode-locked laser is near an acoustic eigen-frequency of the FSL fiber.

1. A. B. Grudinin, D. N. Richardson, and D. N. Payne, Electron. Lett. **29**, 1860 (1993).
2. S. Gray, A. B. Grudinin, W. H. Loh, and D. N. Payne, Opt. Lett., accepted for publication.
3. A. N. Pilipetskii, A. V. Luchnikov, and A. M. Prokhorov, Sov. Lightwave Commun. **3**, 29 (1993).

JTuA5 **9:15 am**

Acoustic effect in passively-mode-locked fiber ring lasers

A. N. Pilipetskii, E. A. Golovchenko, C. R. Menyuk, *Department of Electrical Engineering, University of Maryland, Baltimore, Maryland 21228-5398*

Passively-mode-locked fiber ring lasers are an attractive source of picosecond and subpicosecond solitonlike pulses. In these lasers the basic dynamical balance that leads to passive mode locking of the individual pulses is due to a slow-saturable gain provided by erbium-doped optical amplifiers, a fast-saturable absorption provided by nonlinear polarization rotation and polarization-selective elements, and a frequency filter that is provided by frequency-dependent cavity resonances combined with the saturable absorber. These elements lead to no interpulse interactions, so that when the laser cavity contains many pulses, one finds that they are free to move about with respect to one another, leading in most cases to rich and complex dynamics.[1] However, recent experiments show that in some cases the pulses are all regularly spaced, indicating the presense of an interpulse interaction.[2] Not only is this result physically interesting, but it also has implications for potential applications that require regular pulse spacing.

We show theoretically that the interpulse interaction is due to acoustic waves stimulated by the pulses themselves. Physically, the acoustic waves are created by electrostriction due to the large transverse electric-field gradient in single-mode fibers. Each mode-locked pulse leaves in its train a perturbation of the refractive index due to the acoustic wave. This perturbation, which lasts about 1 ns and repeats every 20–100 ns, is due to reflections from the fiber cladding, so that mode-locked pulses affect those that come after far more than 100 ns.[3]

We investigate the acoustic interaction of pulses in the laser cavity, assuming that *N* pulses with equal amplitudes are circulating in the cavity, that the shape of these pulses in maintained by the laser itself, and that the pulses interact with each other only through the acoustic effect. We simulate the sequence of pulses by merely using their temporal positions and deviations from the central laser frequency, and we calculate the evolution of pulse trains inside the cavity for various initial conditions, various numbers of pulses inside the cavity, and various cavity lengths.

Our simulations indicate that the most probable evolution of an arbitrary pulse train in the laser cavity is that after some circulation time in the cavity the interaction force traps the pulses in bunches. The gen-

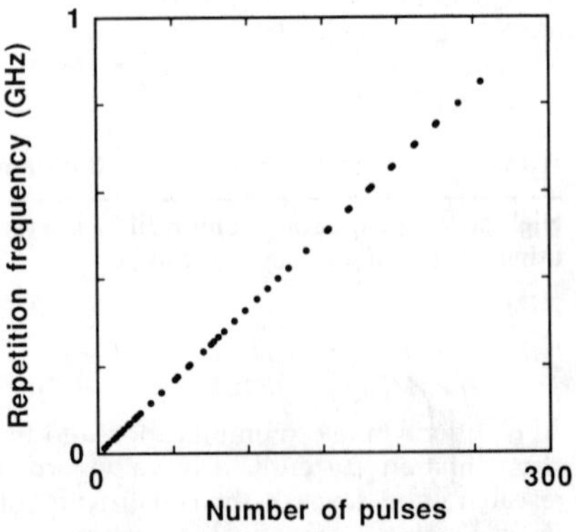

JTuA5 Fig. 1. Frequency dependence as a function of number of pulses *N* of stable harmonically-mode-locked operation. Dots are shown only at *N* values at which the operation is stable and regular pulse spacing is observed.

eration of such bunches has been experimentally observed.[1,2]

We derive an analytical expression to determine whether the acoustic interaction between the mode-locked pulses stabilizes or perturbs a periodic train of pulses inside the laser cavity. A typical result obtained from this analysis is shown in Fig. 1 for a 60-m-long cavity. Dots give the number of pulses N inside the cavity for which the periodic pulse train is stable. For other values of N the pulse train is unstable. Figure 1 is consistent with the experimentally obtained data of Ref. 2.

The work was supported by the Advanced Research Projects Agency through the U.S. Air Force Office of Scientific Research and by the National Science Foundation.

1. R. P. Davey, N. Langford, and A. I. Ferguson, Electron. Lett. **27**, 1257 (1991).
2. A. B. Grudinin, D. J. Richardson, and D. N. Payne, Electron. Lett. **29**, 1860 (1993); M. J. Guy, P. U. Noske, A. Boskovic, and J. R. Taylor, Opt. Lett. **19**, 828 (1994).
3. E. M. Dianov, A. V. Luchnikov, A. N. Pilipetskii, and A. N. Starodumov, Opt. Lett. **15**, 314 (1990); E. A. Golovchenko, and A. N. Pilipetskii, J. Lightwave Technol. **15**, 314 (1994).

JTuA6 **9:30 am**

Maximum filling factor in high-repetition-rate soliton lasers

M. Romagnoli, P. Franco,* M. Midrio,* F. Fontana,** *Fondazione Ugo Bordoni, va B. Castiglione, 59, 00143 Rome, Italy*

In view of the future application in long-distance and/or high-capacity fiber-based transmission systems, soliton sources seem to play a key role. The major drawback of the periodically amplified transmission systems is given by the amplifier spacing. This drawback leads to a lower limit in the generated pulsewidth. As a consequence of this the maximum repetition rate achievable for a given pulsewidth has to be found.

In this work, by means of soliton perturbation theory and numerical simulation, we investigate the problem of the soliton interactions in a fiber source operating in the harmonical regime and we find the minimum allowed soliton spacing that ensures both the soliton stability and low noise.

Our study includes the analysis of actively or passively-mode-locked structures and of the more recent scheme based on the sliding frequency soliton generation.[1]

For all the types of sources under consideration, we demonstrate that the maximum filling factor of the generated soliton train is 22%. In the specific case of mode-locked sources an improved form of the train stabilization along with further reduction of the fluctuation of the soliton train parameters is provided by the insertion in the cavity of a low finesse etalon whose free spectral range matches the soliton repetition rate.[2] The use of the low finesse etalon permits the control of both the phase relationship between solitons and the pulse to pulse timing jitter.

Pulse propagation in a ring laser with bandwidth-limited amplification and amplitude modulation may be described,[3] in the usual dimensionless soliton units, by the averaged perturbed nonlinear Schrodinger equation (NLS)

$$U_Z - \frac{i}{2} U_{TT} - i|U|^2U = Q(U) + I(U) \tag{1}$$

where T and Z are the time and propagation coordinates normalized respectively to the soliton l/e width, τ, and to the dispersion length $L_D = \tau^2/\beta_2$. The quantities Q and I accounts for the perturbing elements inside the cavity and the soliton-soliton interaction, respectively. The expression describing the perturbing elements reads

$$Q(U) = \delta U + \beta U + \frac{\alpha_M}{2} \Omega_M^2 T^2 + \gamma |U|^2 U$$
$$+ i\alpha T U + e^{-i\sigma} U(T + \Xi_0)$$

where $\delta = G - \Gamma - R$ is the excess gain that accounts for the distributed values of the amplifier gain, cavity loss and etalon reflectivity. The coefficient $\beta = 2/(B^2L)$ measures the distributed loss introduced by a filter of bandwidth B, α_M and Ω_M are the modulation depth and frequency. The coefficient γ rules the nonlinear gain term. The rate of sliding $\alpha = \Omega_S/L$ is defined as the frequency shift experienced by the light per dispersion length. The quantity Ξ_0 is the time delay corresponding to the inverse of the free spectral range of the etalon and the phase σ accounts for the detuning of the etalon transmission from the central frequency of the passband filter.

Soliton interaction occurs through the overlap of the tails of the preceding pulse and successive pulse. In our model, depicted in Fig. 1, we assume that the fiber loop contains only two closely spaced solitons, labeled "A" and "B". The interaction term in Eq. 1 takes the following form

$$I_{A,B} = 2i|U_{A,B}|^2 U_{B,A} + iU_{A,B}^2 U_{B,A}^*$$

where

$$U_{A,B}(Z, T) = \eta_{A,B} \operatorname{sech}[\eta_{A,B}(T - \xi_{A,B})]$$
$$\cdot e^{i[\kappa_{A,B}(T - \xi_{A,B}) + 1/2(\eta_{A,B}^2 - \kappa_{A,B}^2) + \theta_{A,B}]}.$$

Soliton perturbation theory yields the result that the soliton pair contained in the

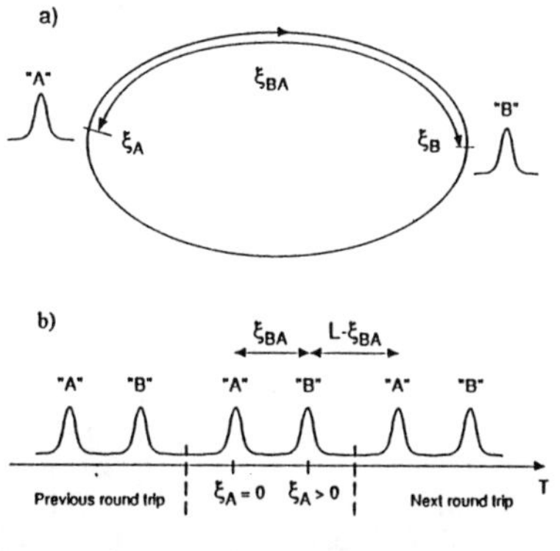

JTuA6 Fig. 1. Fiber loop containing the pair of solitons labeled "A" and "B", a) closed loop, and b) the equivalent open loop.

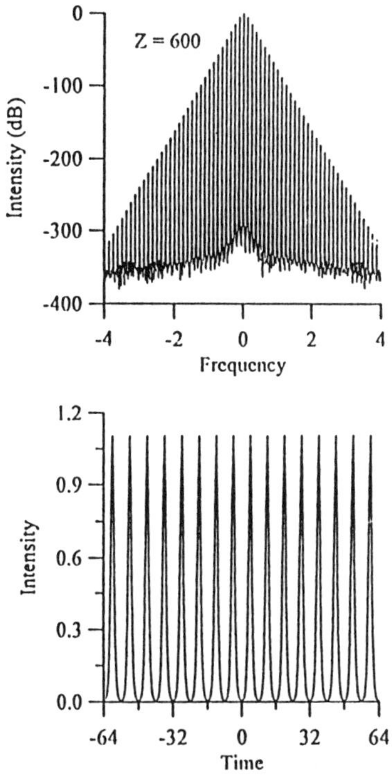

JTuA6 Fig. 2. Spectral and temporal evelope of an in-phase soliton train propagating in the passively-mode-locked laser including the etalon. The parameters are: $\beta = 0.15$, $\gamma = 0.06$, $R = 0.2$, $\sigma = 0$, $\Delta(0) = 6$, $\alpha = \alpha_M = \Omega_M = 0$. The final separation ($Z = 600$) is $\Delta = 8$.

loop described in Fig. 1 shows a stable steady state with fixed values of the soliton parameters. In both mode-locking schemes, the stability is obtained when the pair is out of phase. In the presence of the etalon the in-phase pair is stable and, taking into account quantum noise (amplified spontaneous emission) the fluctuation of the soliton-pair parameters approaches neglible values even for a low value of the etalon reflectivity. For the SFS case the phase relationship of the pair linearly evolves with distance, and this leads a substantial reduction of the interaction.

The numerical simulation was performed with the split-step spectral method. The fast Fourier transform (FFT) algorithm has been exploited for the simulation of an infinite train of solitons. The example of spectrum and time profile reported in Fig. 2 refers to the soliton train circulating in a passively-mode-locked source, including a low finesse etalon ($R = 0.2$) at a propagation distance $Z = 600$. Through this kind of numerical simulations we verified that a soliton train is stable in all considered schemes of soliton generation and we always found the same minimum soliton separation that is $\Delta_{min} = |\xi_i - \xi_{i-1}| = 8$ that corresponds to a filling factor equal to 0.22.

*Dipartimento di Elettronica e Informatica, Università di Padova, via Gradenigo 6/A, 35131 Padova, Italy
**Pirelli Cavi s.p.a., viale Sarca 202, 20146 Milano, Italy

1. G. T. Harvey and L. F. Mollenauer, Opt. Lett. **18**, 107 (1993).

2. F. Fontana, L. Bossalini, P. Franco, M. Midrio, M. Romagnoli, and S. Wabnitz, Electron. Lett. **30**, 321 (1994).
3. F. Fontana, P. Franco, M. Midrio, and M. Romagnoli, European Conference on Optical Communications (ECOC'94) (Florence, Italy, Sept. 25–29, 1994) p. 303.

JTuA7 9:45 am

Pulse self-ordering in soliton fiber lasers

H. A. Haus, E. P. Ippen, W. S. Wong, F. I. Khatri, K. R. Tamura,* *Department of Electrical Engineering and Computer Science, Research Laboratory of Electronics, Massachusetts Institute of Technology, Cambridge, Massachusetts 02139*

Erbium-doped fiber ring lasers pumped far above threshold often maintain for seconds on end evenly spaced trains of pulses of roughly equal energy.[1] The experimental setup we use to investigate this phenomenon is a simple soliton laser, similar to the setup shown in Ref. 2. Polarization additive pulse mode-locking (APM) is achieved with the bulk wave plates. A quartz filter is used for tuning. The loop length is 11.9 m, corresponding to a fundamental repetition rate of 16.8 MHz, and the total net dispersion is -0.16 ps^2. The erbium fiber (donated by AT&T) is 1.1-m long and has an estimated dispersion of $+0.075$ ps^2. The rest of the laser is constructed of 10.3 m of Corning SMF-28 fiber and a 0.5-m segment of Corning Flexcor 1060 fiber comprising the wavelength-division multiplexing coupler. The output pulse width is 0.64 ps and the spectral width is 7.6 nm, giving a time bandwidth product of 0.61.

Figure 1 shows an uneven arrangement of solitons produced by the laser. Figure 2 shows an even arrangement of solitons, corresponding to a repetition rate of 505 MHz. The rf spectrum measurement for this case is shown in Fig. 3 and shows the 30th harmonic at 505 MHz. The subharmonics at multiples of the fundamental repetition rate are suppressed by over 45 dB in between the first few harmonics of the 505 MHz pulse train. The question then arises as to what mechanism can produce such ordering of pulses, since there are no modulators or timing devices present in the system. In this study, we examine the interaction between solitons and continuum as a mechanism for pulse self-ordering.

Continuum generation by solitons has been a recent topic of study. Generation of

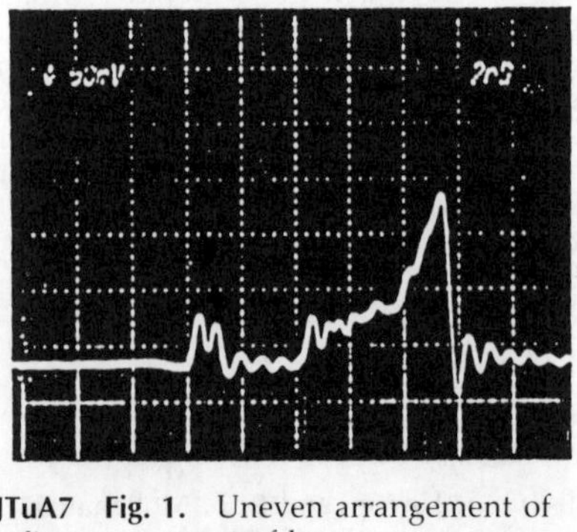

JTuA7 Fig. 1. Uneven arrangement of solitons at output of laser.

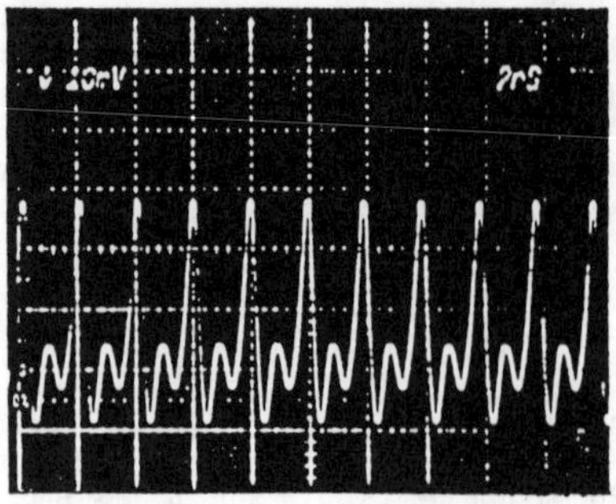

JTuA7 Fig. 2. Even arrangement of solitons at output of laser.

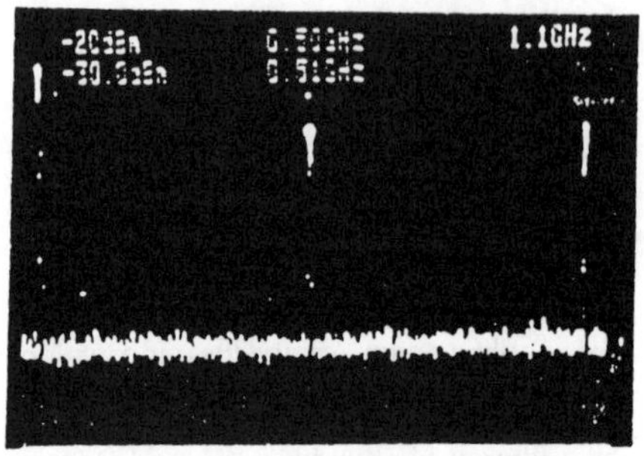

JTuA7 Fig. 3. rf spectrum of even arrangement of solitons.

resonant continuum has been studied in detail.[3,4] This continuum appears as nonuniformly spaced spectral sidebands whose spacing is dependent on the periodicity of the laser (or transmission system). In addition to this resonant contribution to the continuum, there is also a nonresonant component. This continuum is caused by deviations of the pulse away from a path-averaged soliton.[5,6] Gordon has developed analytic solutions to the resonant and nonresonant components of continuum.[7]

This continuum has been studied as it affects the pulse which generates it. However, to date no one has studied how continuum generated from one soliton affects a neighboring soliton. We have recently started investigating this problem analytically and numerically. In particular, we examine the problem using perturbation theory.[8] We derive equations of motion for the frequency and timing of a soliton, which is perturbed by continuum shed from its neighbor. To formulate the equations of motion, we use expressions for the continuous waves taken from Gordon's paper.[7] We find that the resulting first order differential equation for the timing of the perturbed soliton is similar to that of a particle in a viscous system which undergoes a repulsive force (away from the continuum generating soliton). Numerical simulation results confirm our theoretical predictions.

**NTT Access Network Systems Laboratory, Tokai, Ibaraki 319-11, Japan*

1. A. B. Grudinin, D. J. Richardson, and D. N. Payne, Electron Lett. **29**, 514.
2. K. Tamura, C. R. Doerr, H. A. Haus, and E. P. Ippen, Photon. Lett. **6**, 697 (1994).
3. N. Pandit, D. U. Noske, S. M. J. Kelly, and J. R. Taylor, Electron Lett. **28**, 455 (1992).
4. N. J. Smith, K. J. Blow, and I. Andonovic, J. Lightwave Technol. **10**, 1329 (1992).
5. L. F. Mollenauer, S. G. Evangelides, Jr., and H. A. Haus, J. Lightwave Technol. **9**, 194 (1991).
6. A. Hasegawa and Y. Kodama, Phys. Rev. Lett. **66**, 161 (1991).
7. J. P. Gordon, J. Opt. Soc. Am. *B*, **9**, 91 (1992).
8. H. A. Haus and A. Mecozzi, J. Quant. Electron **29**, 983 (1993).

QTuA 8:00 am
Hyatt A

Nonlinear Optical Effects

John F. Reintjes, *U.S. Naval Research Laboratory, Presider*

QTuA1 (Invited) 8:00 am

Coherent and population contributions in the photo-induced Faraday rotation

R. Frey, C. Buss, S. Hugonnard-Bruyère, J. Frey, C. Flytzanis, *Laboratoire d'Optique Quantique du Centre National de la Recherche Scientifique, Ecole Polytechnique - 91128 Palaiseau cédex, France*

Photo-induced Faraday rotation (PIFR) is a new nonlinear magneto-optical process of great interest both for fundamental reasons and also for applications. The choice of the appropriate materials is very crucial; it has been shown that this nonlinear process can be very efficient in semimagnetic semiconductors (SMSC) such as $Cd_{1-x}Mn_xTe$. The strong exchange interaction in these compounds between the spins of delocalized band electrons and those of the localized Mn^{++} ions, which provokes giant linear Faraday rotations,[1] is also the origin of the photo-induced Faraday rotations[2] observed there.

The enhanced effect of this spin exchange interaction is mediated through different mechanisms, namely the magnetic, population, and coherent processes and their relative impact has been analyzed in a series of recent studies.[3]

The magnetic contribution occurs because of heat production consequent to the absorption that leads to a magnetization decrease in the material. Due to the temperature dependence of the heat capacity this component is only significant at temperatures smaller than 20K; it has slow response and lifetimes, which are fixed by the spin-phonon relaxation of the Mn^{++} spins and the heat diffusion respectively.

The population contribution has its origin in the onset of saturation of the refractive indices consequent to the photogeneration of free carriers in conduction and valence bands through one or two-photon transitions. This contribution rises with the laser pulse characteristics and its lifetime is governed by the free carrier relaxation.

The coherent contribution is related to

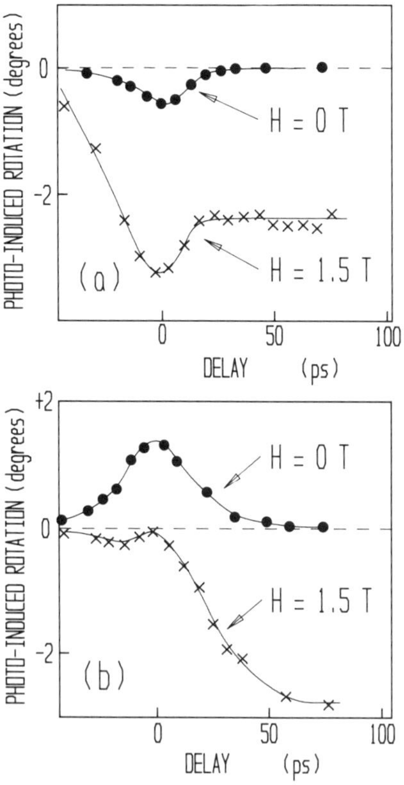

QTuA1 Fig. 1. Photo-induced rotation as a function of the delay between pump and probe pulses for parallel (a) and crossed (b) polarizations.

the optical Kerr effect in the presence of a static magnetic field through the nonlinear polarization

$$\underline{P}^{(4)}_{NL}(\omega) = \underline{\underline{\underline{\underline{\chi}}}}^{(4)}(\omega, -\omega, \omega, 0)\underline{E}(\omega)\underline{E}^*(\omega)\underline{E}(\omega)\underline{H}$$

where $\underline{E}(\omega)$ and $\underline{H}$ are the laser electric field and static magnetic field (applied along the propagation direction of the laser beam) respectively. This contribution arises through the combined effects of the nonlinear Kerr polarization $\underline{P}^{(3)}_{NL}(\omega) = \underline{\underline{\underline{\chi}}}^{(3)}(\omega, -\omega, \omega)\cdot \underline{E}(\omega)\underline{E}^*(\omega)\underline{E}(\omega)$ and the faraday polarization $\underline{P}^{(2)}_{NL}(\omega) = \underline{\underline{\chi}}^{(2)}(\omega, 0)\underline{E}(\omega)\underline{H}$. It is related to motion of electrons (virtual transition) and, therefore, follows the temporal evolution of the laser pulse.

In this paper, we will concentrate on the initial temporal behavior of the polarization direction of a probe beam when a short duration high intensity laser pulse of arbitrary polarization state propagates through a SMSC sample. This pump and probe experiment is performed in a $Cd_{0.75}Mn_{0.25}Te$ sample with a 111 orientation at a temperature of 30K in order to eliminate the thermally induced magnetic contribution; the pump and probe pulses of 25 ps duration are delivered by a mode-locked neodymium YAG laser.

As an example, in Figs. 1(a) and 1(b) we show the temporal dependence of the photo-induced Faraday rotation angle θ_{NL} at H = 0 and 1.5 T for parallel and crossed pump and probe polarizations. As expected, in both cases a negative population contribution occurs only at H = 1.5 T and remains for time much longer than 100 ps. On the contrary the coherent contribution (which is higher at H = 1.5 T than without magnetic field) has a sign which depends on the polarization state configuration of pump and probe beams. We will discuss different aspects of this experiment together with those of photo-induced absorption and phase changes which also occur when a two-photon resonance is excited for the interband transition.

1. J. K. Furdyna, J. Appl. Phys. **64**, R29 (1988).
2. S. Hugonnard-Bruyère, J. Frey, R. Frey, and C. Flytzanis, Nonlinear Optics, **5**, 259 (1993).
3. R. Frey, S. Hugonnard-Bruyère, C. Buss, and C. Flytzanis, to appear in *Il Nuovo Cimento*.
4. J. Frey, R. Frey, C. Flytzanis, and R. Triboulet, J. Opt. Soc. Am B, **92**, 132 (1992).

QTuA2 **8:30 am**

Generation of narrow-band fields from stochastic broadband phase-correlated fields

J. A. Moon,* P. R. Battle, M. Bashkansky, J. Reintjes, *Code 5640, Laser Physics Branch, U.S. Naval Research Laboratory, Washington, D.C. 20375*

We spectrally resolved the output of the difference-frequency-mixing process between two broadband stochastic phase-replica fields, $E_1(t)$ and $E_2(t)$, generated in two separate experiments. In the nondepletion limit, this spectrum can be represented as[1]

$$\tilde{E}_f(\omega, \tau) \sim \int dt e^{-i\omega t} E_1(t) E_2^*(t-\tau), \quad (1)$$

where τ represents a time delay between $E_1(t)$ and $E_2(t)$.

The broadband stochastic field $E_1(t)$ (centered at ω_1) is a phase replica of a second broadband field $E_2(t)$ (centered at ω_2) if the phases of the fields are related by $\phi_1(\omega) = \phi_2(\omega + \omega_f) + \phi_f$, where ω_f and ϕ_f are constants and $\omega_1 + \omega_f = \omega_2$.

There are two distinct features in the spectrum represented by relation (1) if E_1 and E_2 are phase replicas. We assume that E_1 and E_2 have Gaussian spectral envelopes with $1/e$ widths of Δ_{bb}, so that when $\omega = \omega_f$, relation (1) reduces to

$$|\tilde{E}_f(\omega_f, \tau)|^2 \propto \frac{\pi}{2} \tilde{E}_{10}^2 \tilde{E}_{20}^2 \Delta_{bb}^2 \exp\left[-\left(\frac{\Delta_{bb}\tau}{2}\right)^2\right], \quad (2)$$

where $\tilde{E}_{10}$ is the magnitude of E_1's Fourier transform at the center frequency [similarly for $E_2(t)$]. When $\omega \neq \omega_f$, we obtain

$$|\tilde{E}_f(\omega, \tau)|^2 \propto \frac{\pi}{2} \tilde{E}_{10}^2 \tilde{E}_{20}^2 \Delta_{bb}\Delta \exp\left[-\left(\frac{\omega - \omega_f}{\Delta_{bb}}\right)^2\right], \quad (3)$$

where $\Delta \cong 2\pi/T$ and T is the pulse length of $E_1(t)$ and $E_2(t)$. Notice that at zero time delay, the feature at ω_f is higher than the broad background by a factor of Δ_{bb}/Δ and that the energies contained in the broadband and narrow-band features here are the same.

We generated the phase-replica fields by two different means. The general experimental setup is shown schematically in Fig. 1. In the first experiment, we sum-frequency-mixed a narrow-band (~2-cm^{-1}) field centered at 1.064 μm and a broadband (~160-cm^{-1}) field centered at 0.738 μm in a 2-cm KD*P crystal to produce a phase-replica broadband field at 0.436 μm. The broadband pulses at 0.738 μm were obtained by Stokes-shifting the output of a rhodamine 6G dye laser oscillator/amplifier pumped by a frequency-doubled Nd:YAG oscillator/amplifier in a 1-m-long H_2 cell at 320 psi (16.5 kTorr). The resulting spectra as a function of time delay, obtained from the output of mixing the two broadband fields in the manner indicated in Fig. 1 in a 7-mm-long beta-barium borate (BBO) crystal, are shown in Fig. 2, where a narrow spectral feature may be clearly seen.

The H_2 cell also produced anti-Stokes pulses centered at 0.458 μm and second-Stokes pulses centered at 1.065 μm. The spectral output of the difference-frequency-mixing process between the second- and anti-Stokes pulses is shown in Fig. 3, where again the narrow spectral feature is seen when the relative delay between the pulses is less than a coherence length. It is known[2,3] that Stokes and anti-Stokes shifted fields are phase replicas of the initial pump field. To our knowledge, the degree of correlation between the second-Stokes and anti-Stokes fields has not been measured previously. This technique is thus useful for measuring the degree of phase correlation between broadband stochastic fields with widely differing center frequencies.

This work was supported by the Office of Naval Research. J. A. Moon and P. R.

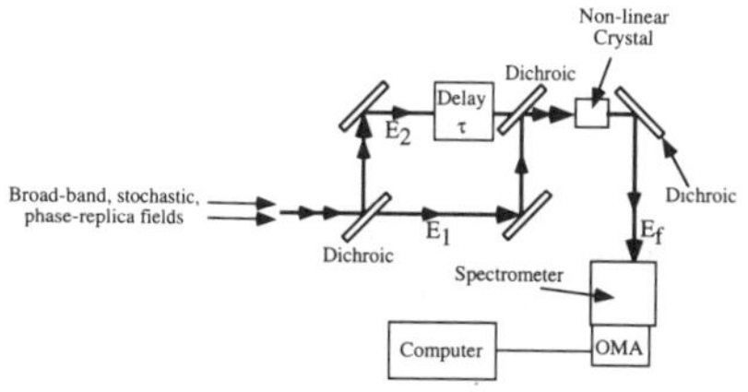

QTuA2 Fig. 1. Schematic diagram of experiment. Correlated fields generated from sum-frequency mixing or stimulated Raman scattering are time delayed relative to one another and are difference-frequency mixed.

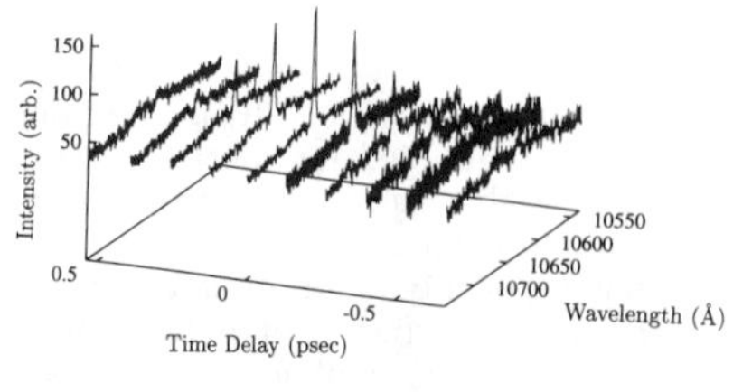

QTuA2 Fig. 2. Spectral output of the difference-frequency generator at various relative time delays between broadband fields initially generated in a sum-frequency process.

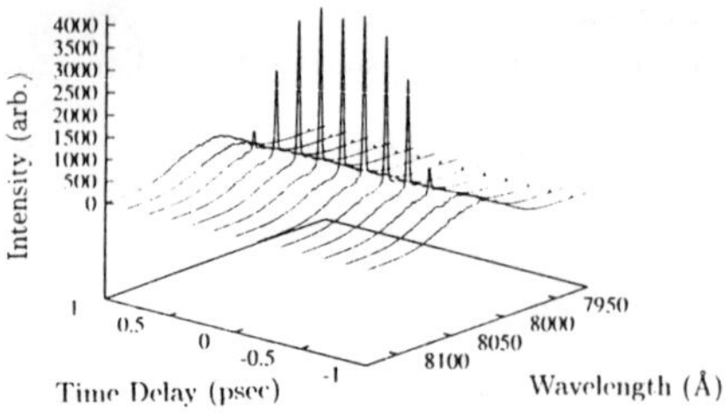

QTuA2 Fig. 3. Spectral output of the difference-frequency generator at various relative time delays between broadband second-Stokes and anti-Stokes fields initially generated in the same stimulated Raman process.

Battle acknowledge support from the National Research Council.

*Present address: *Code 5603.2, U.S. Naval Research Laboratory, Washington, D.C. 20375*

1. A. Yariv, *Quantum Electronics*, 3rd ed. (Wiley, New York, 1989), p. 408.
2. M. G. Raymer, J. Mistrowski, and J. L. Carlsten, Phys. Rev. A **19**, 2304 (1979).
3. M. Bashkansky and J. Reintjes, Opt. Commun. **83**, 103 (1991).

QTuA3 **8:45 am**

Raman depletion of an ultrashort light pulse during shock-wave creation

I. V. Aleksandrov, *S. I. Vavilov State Optical Institute, NITIOM Branch, Babushkina Street, 36-1, St. Petersburg, 193171, Russia*

When high-power picosecond light pulses traveled along hollow-core optical fibers filled by transparent neat organic liquids, envelope shock waves at the leading edge of these pulses were discovered.[1] It should be noted that optical shock waves are the spatial and temporal analogies of solitons. Such waves might be described by the set of differential equations identical to those for the self-focusing effect if one spatial coordinate in these equations is exchanged for a temporal one. Similarly to high-order optical solitons, the shock waves realize effective temporal compression for light pulses.

The practical observation of optical shock waves is very difficult. A new method of capillary filling by outgased liquids was developed in the experiments described. By using this method it became possible to propagate the high-power light pulses along the liquid core without fiber damage. A stable-shock-wave regime is realized when the strict balance of the nonlinearity and the dispersion of the liquid core is reached, as is needed for optical solitons. However, an additional and a very basic condition also must be fulfilled for shock-wave formation; namely, the values of the nonlinear relaxation constants must be chosen appropriately for the liquid investigated.

The modern theory of optical shock waves in isotropic nonlinear media operates according to the model, which deals with orientational nonlinearity as a possible mechanism for nonlinear response only.[2,3] Moreover, the loss of light-pulse energy during shock-wave formation is negligible.

In contrast to theory, the most interesting results were obtained when the liquids with a complex orientational and electronic nonlinearity were used. The influence of Raman depletion and envelope deformation on shock-wave formation are discussed in detail.

1. Z. V. Nesterova and I. V. Aleksandrov, Sov. Phys. JETP **88**, 96 (1985).
2. L. A. Ostrowskii, Sov. Phys. JETP **51**, 1189 (1966).
3. R. A. Fisher and W. K. Bishel, J. Appl. Phys. **46**, 4921 (1975).

QTuA4 **9:00 am**

Nonperturbative optimization of resonant BOXCARS signals

D. A. Coppeta, P. L. Kelley, P. J. Harshman,* V. Lee,* T. K. Gustafson,* J.-P. E. Taran,** *Electro-Optics Technology Center, Tufts University, Medford, Massachusetts 02155*

Coherent anti-Stokes Raman scattering (CARS) is an important tool in the study of dynamical processes in combustion, aerodynamics, fluid flow, and plasma physics. Frequently, molecular densities are low and available pump and Stokes intensities are limited. To compensate, resonant CARS is often used. To enhance separation of the anti-Stokes signal from the pump and Stokes signals, a folded geometry called BOXCARS is often employed (Fig. 1). In resonant BOXCARS, the pump and Stokes frequencies are chosen so that one-photon and/or three-photon resonances enhance the anti-Stokes signal. These multiple resonances can also be used to simplify spectra in molecules with a complex energy-level structure, such as NO_2. However, large field strengths near resonances can cause populations to change and/or levels to shift and mix (the optical Stark effect). This limits the resonance enhancement and causes deviations from linearity to occur. We report here a systematic study of the effect of these processes and give the conditions for obtaining optimum CARS signals.

We have developed a method for obtaining analytic results for complex nonlinear optical processes. The Liouville equation for the density matrix is recast so that the external field terms that couple density-matrix elements are second order in the field. The approach has two important effects. First, in the case for which parity or appropriate field restrictions apply, there is a reduction of the number of coupled equations that must be solved by at least a factor of 2. Second, this reformulation naturally leads to radiative renormalization terms that are directly related to single-photon population saturation, Stark shifts, and Rabi splittings.[1] This renormalization is equivalent to a Dyson equation analysis. In the four-level BOXCARS case studied here, there are strong fields coupling three of the level pairs and we are led to carry out two further renormalizations. These additional renormalizations give rise to results containing rate constants for one- two-, and three-photon processes.

We assume a BOXCARS geometry. We take the dephasing rates to be the same between all states and denote them by Γ. We take the Rabi Frequency of the Stokes field to be fixed at $\Omega_S = 0.1\ \Gamma$.

In Fig. 2, we assume the pump frequency $\omega_P = \omega_{P1} = \omega_{P2}$, is resonant with the 3–4 transition but is non-resonant with the 1–2 transition. We plot the anti-Stokes source polarization, $|\rho_{41}|$, as a function of the detuning of the Stokes field frequency, ω_S, from the Raman resonance at $\omega_P - \omega_{31}$ and the strength of the pump Rabi frequency, Ω_{P2}. As a function of Ω_{P2}, there is a single Lorentzian which grows rapidly until Ω_{P2} is comparable to Γ. The resonance then exhibits Autler-Townes splitting as the Rabi frequency increases.

In Fig. 3 we assume a system where the splitting of the upper and lower states are equal. We also assume that the polarization of the pumps can be chosen in such a way that Ω_{P2} does not couple level 1 to 2 and Ω_{P1} only weakly couples level 1 to 2 for all values of Ω_{P1} considered. We fix the Stokes field resonant with the 2–3 transition. The anti-Stokes source polarization $|\rho_{41}|$ is plotted as a function of detuning of the pumps from their respective one photon resonances and as a function of Rabi frequency Ω_{P2}. For $\Omega_{P2} << \Gamma$, rapid growth of the resonance occurs. As the Rabi frequency increases beyond Γ, we note a splitting of the original Lorentzian into three distinct lines whose strength fall off as $1/\Omega_{P2}$.

We also present results that incorporate Doppler broadening and anti-Stokes absorption, and we will compare the present results with the homologous problem of electromagnetic transparency.[2]

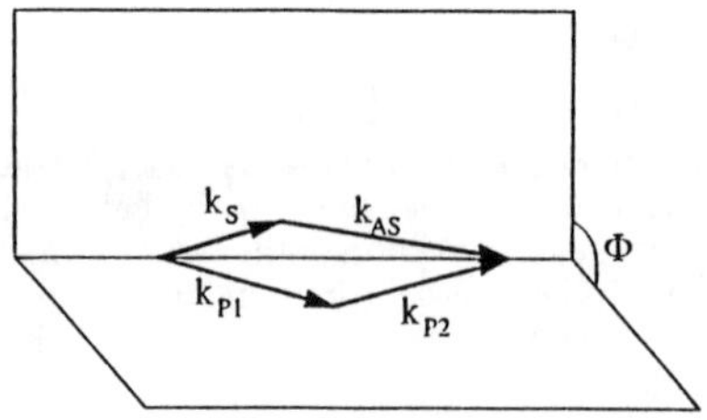

QTuA4 Fig. 1a. Wave vector diagram of folded BOXCARS geometry. Note that the pump wavevectors are contained in a plane approximately orthogonal to the plane containing both the Stokes and anti-Stokes wavevectors.

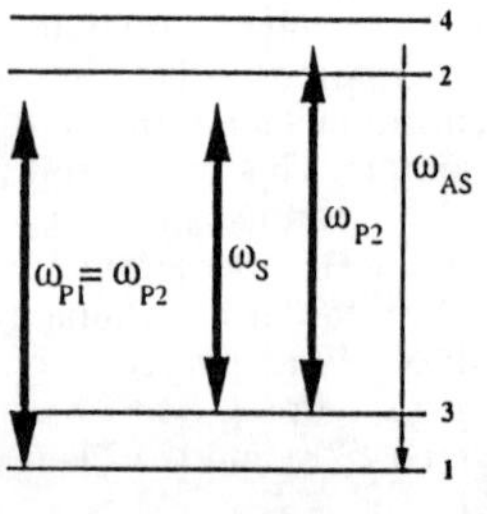

QTuA4 Fig. 1b. Energy-level diagram for resonant BOXCARS. ω_{P1} denotes the weakly coupled pump, ω_{P2} denotes the strongly coupled pump, ω_S denotes the Stokes field, and ω_{AS} denotes the anti-Stokes wave.

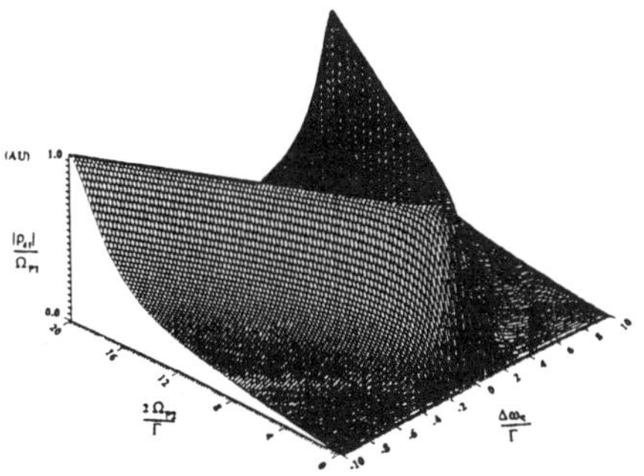

QTuA4 Fig. 2 Anti-Stokes source polarization, $|\rho_{41}|$, as a function of pump Rabi frequency and Stokes frequency detuning, $\Delta\omega_S = \omega_S - \omega_{31}$, normalized to the uniform dephasing rate Γ. Pump detuning from ω_{12} is taken to be over 1000 times larger than the largest Ω_{P1} and the Stokes Rabi frequency $\Omega_S = 0.1\ \Gamma$. Autler-Townes splitting is evident for large Ω_{P2}.

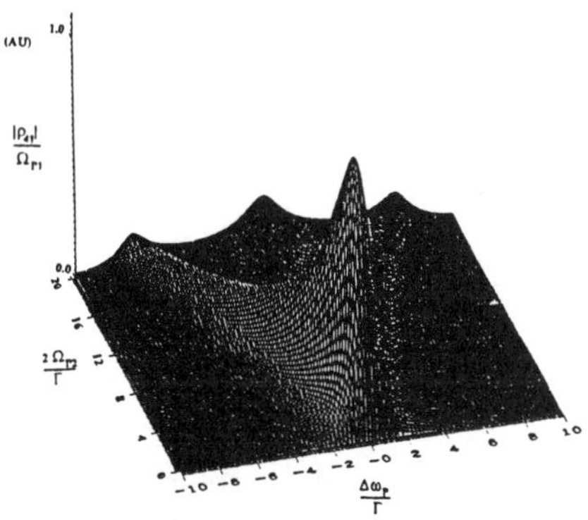

QTuA4 Fig. 3 Anti-Stokes source polarization $|\rho_{41}|$ as a function of pump frequency detuning, $\Delta\omega_P = \omega_P - \omega_{21}$ and pump Rabi frequency Ω_{P2}. The Stokes-field is resonant wit the 2–3 transition and Rabi frequency $\Omega_S = 0.1\ \Gamma$. Triple splitting of the line occurs for large Ω_{P2}.

*Department of Electrical Engineering and Computer Sciences and Electronics Research Laboratory, University of California at Berkeley, Berkeley, California 94720
**National Office of Aerospace Studies and Research, BP 72, 92322 Chatillon Cedex, France

1. O. Blum, P. J. Harshman, T. K. Gustafson, and P. L. Kelley, Phys. Rev. **47**, 5165 (1993).
2. P. J. Harshman, O. Blum, T. K. Gustafson, and P. L. Kelley, Opt. Lett. **18**, 1706 (1993).

QTuA5 **9:15 am**

High-efficiency optical phase conjugation in potassium vapor

M. Y. Lanzerotti,* R. W. Schirmer,* A. L. Gaeta, *School of Applied and Engineering Physics, Cornell University, Ithaca, New York 14853*

High-efficiency phase-conjugate mirrors (PCMs) can be constructed by using degenerate four-wave mixing (DFWM) in atomic vapors.[1,2] The nonlinearity that can be achieved by using atomic vapors can be both large and relatively fast (<10 ns). These characteristics are desirable for the creation of wide-bandwidth PCMs that have high phase-conjugate reflectivities. Such PCMs can be useful for optical communications and image processing.[3]

We have constructed a high-efficiency PCM by using a frequency-stabilized Ti:S laser operating at 767 nm to perform DFWM in a potassium-vapor cell in which the interaction length is 2 mm. The probe beam and pump beams are all linearly polarized along the same direction. The focusing of each pump beam is chosen such that the confocal parameter is much larger than the 2-mm interaction region and results in a typical pump intensity of 2×10^3 W/cm^2 at the cell.

This PCM simultaneously exhibits both of the following desirable characteristics: high phase-conjugate reflectivity and large angular acceptance. We obtain phase-conjugate reflectivities as high as 600% while avoiding the effects of self-focusing and self-defocusing (i.e., the optical quality of the conjugate beam is preserved). As the pump–probe angle is increased from 25 to 70 mrad, we observe that the reflectivity decreases from 600% to 250%. We keep the pump–probe angle larger than 17 mrad to avoid the effects of forward four-wave mixing. To our knowledge, this value of the reflectivity is the highest that has been achieved in an atomic vapor by using a continuous-wave laser. The dependence of the phase-conjugate reflectivity on the power in each of the pump beams is shown in Fig. 1.

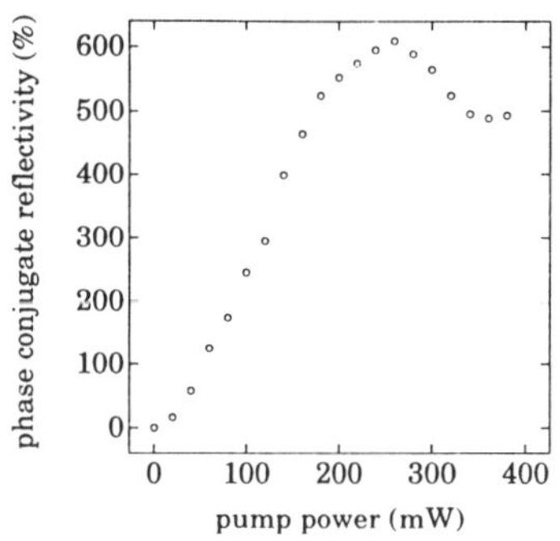

QTuA5 Fig. 1. Phase-conjugate reflectivity as a function of pump power, where the number density of potassium atoms is estimated to be 2×10^{14} cm^{-3}, the pump–probe angle is about 35 mrad, and the detuning below resonance is approximately 1 GHz.

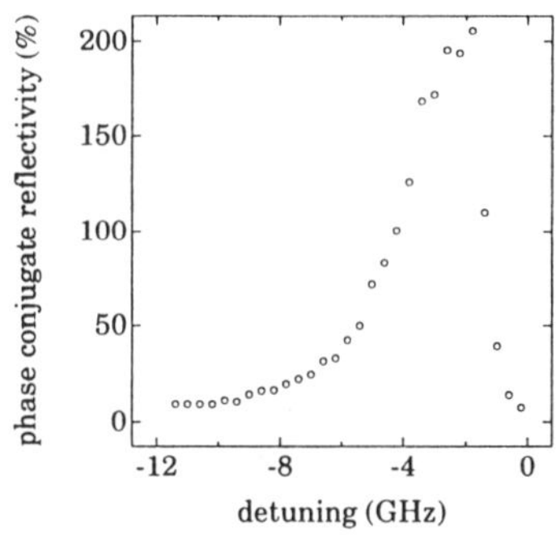

QTuA5 Fig. 2. Phase-conjugate reflectivity as a function of detuning from the D_2 resonance, where the number density of potassium atoms is estimated to be 2×10^{14} cm^{-3}, the pump–probe angle is about 30 mrad, and the pump power is 400 mW.

We also characterize the behavior of the reflectivity as a function of the laser detuning from resonance, the density of potassium atoms in the cell, and the frequency detuning between the probe beam and the pump beams. Figure 2 shows an example of the dependence of the reflectivity on the detuning of the laser from resonance. We find that the highest reflectivities are achieved when the laser is tuned below the D_2 resonance.

Department of Physics, Cornell University, Ithaca, New York 14853

1. J. R. R. Leite, P. Simoneau, D. Bloch, S. Le Boiteux, and M. Ducloy, Europhys. Lett. **2**, 747 (1986).
2. M. Vallet, M. Pinard, and G. Grynberg, Opt. Commun. **81**, 403 (1991).
3. B. Ai, D. S. Glassner, R. J. Knize, and J. P. Partanen, Appl. Phys. Lett. **64**, 951 (1994).

QTuA6 **9:30 am**

Microdroplet-cavity *Q* spoiling to enhance pumping of nonlinear processes

J. L. Cheung, J. M. Hartings, R. K. Chang, *Department of Applied Physics and Center for Laser Diagnostics, Yale University, New Haven, Connecticut 06520-8284*

Liquid microdroplets form optical cavities with very high *Q* factors (10^8). The high *Q* provides feedback for nonlinear processes such as stimulated Brillouin scattering (SBS) and stimulated Raman scattering (SRS).

When a droplet is illuminated by short pulses such that the laser bandwidth $\Gamma_L > \Gamma_{MDR}$ (the spectral width of the cavity resonance, which is also known as the morphology-dependent resonance, or MDR), the fraction of light coupled into the MDR is Γ_{MDR}/Γ_L. A larger fraction of the input light is thus coupled into lower-*Q* MDRs than into higher-*Q* MDRs. SRS is pumped more efficiently by light that is already on an MDR because of the longer path length and better spatial overlap between the pumping and feedback MDRs. For illumination by short pulses, reducing the *Q* of the pumping MDR while maintaining a high *Q* for the feedback MDR optimizes nonlinear processes such as SRS and SBS.

When an intense train of 100-ps pulses is focused at the equator of a microdroplet [see Fig. 1(a)], the short pulses create an electrostrictive perturbation to the droplet surface[1] such that the droplet *Q* is degraded. The effect of the pulses (separated by 13.2 ns) is cumulative, and it causes a pulse-to-pulse increase in the amount of light that is input coupled into MDRs. When the droplet is viewed from 180° to the input-laser direction (a geometry of interest in single-ended lidar studies), the input-coupled light leaks from the left rim of the droplet at F [see Figs. 1(b) and 1(d)]. The ratio of the light intensity from F to that of each input pulse is superimposed on the oscilloscope trace [Fig. 1(d)]. The intensity

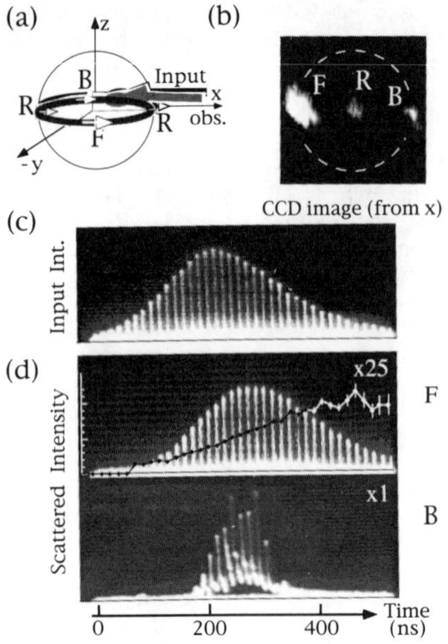

QTuA6 Fig. 1.

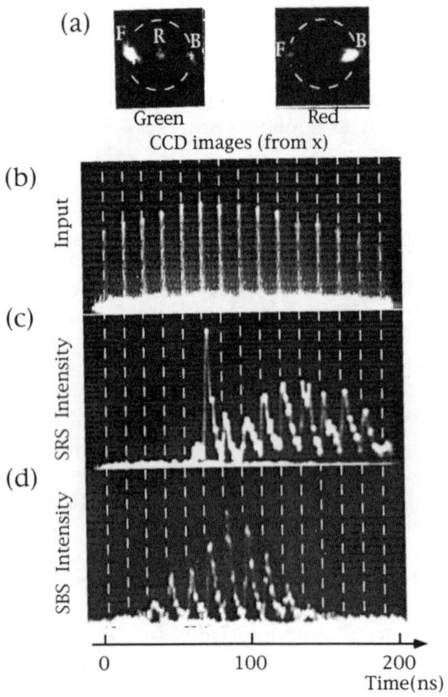

QTuA6 Fig. 2.

ratio increases from pulse to pulse as the Q is progressively degraded (to $\approx 10^5$) by the surface perturbation induced by the cumulative effect of the laser pulses. The intensity ratio continues to increase until the last few pulses, where the input intensity is low. This increased input coupling allows for more efficient pumping of the backward SBS [B in Fig. 1] and of the SRS [see Fig. 2].

The high-Q modes needed for feedback of SBS and SRS, however, can still exist, even though some MDRs within Γ_L are Q degraded. The degree of Q degradation of a particular MDR depends on the spatial overlap of the MDR mode volume with the localized electrostrictive surface perturbation. That high-Q MDRs can still exist is evidenced by the long retention time of the SRS light from B [Fig. 2(c)], indicative of Q $= 10^7$–10^8. Moreover, SRS light is retained within the droplet for longer than the 13.2-ns interpulse separation, acting as a seed for the next input pulse in the train and further enhancing the SRS process. Superimposed on the exponential decay of the SBS and SRS, we observed oscillations [see Figs. 2(c) and 2(d)], which we ascribe to mode beating between azimuthal modes.[2,3]

We gratefully acknowledge partial support of this research by the U.S. Army Research Office (grant DAAH 04-94-G-0031).

1. C. N. Ng, Ph.D. dissertation (Chinese University of Hong Kong, 1994).
2. G. Chen, R. K. Chang, S. C. Hill, and P. W. Barber, Opt. Lett. **16**, 1269 (1991).
3. H. M. Lai, P. T. Leung, K. Young, P. W. Barber, and S. C. Hill, Phys. Rev. A **41**, 5187 (1990).

QTuA7 **9:45 am**

Scattering of second-harmonic light from small spherical particles

Jordi Martorell, R. Vilaseca,* R. Corbalán, *Department of Physics, Autonomous University of Barcelona, 08193 Bellaterra, Barcelona, Spain*

Multiple scattering of classical waves in ordered and disordered structures is a growing field of interest. Among other relevant phenomena observed, cooperative scattering in three-dimensional periodic structures may lead to the appearance of photonic band gaps[1] and to the inhibition of spontaneous emission.[2] This range of phenomena may be expanded when nonlinear scattering effects are considered. Lawandy's group demonstrated a macroscopic orientation in sheared colloidal crystals and discussed its use for optical second-harmonic generation (SHG).[3]

In this paper we present experimental evidence of a new mechanism for SHG; it is attributed to the scattering of SH light from small spherical particles in the Rayleigh–Gans regime. SHG in the dipole approximation, a process forbidden in the bulk of centrosymmetric crystals, would be allowed in such materials by locally removing the inversion symmetry. As is well known, SH light may be reflected from the interface of two centrosymmetric media because of the lack of inversion symmetry of the surface layers. We report the experimental observation of SHG in a macroscopically centrosymmetric colloidal face-centered-cubic crystal composed of negatively charged small spherical particles of polystyrene suspended in water. An aromatic molecule with a strong nonlinear coefficient, such as malachite green, is adsorbed on the surface of each sphere. The intensity of the reflected SH light is shown in Fig. 1. Although the crystal composed of these coated spheres is centrosymmetric at a macroscopic level, this symmetry is broken at the surface of each sphere because of a preferred orientation of the dipole moment of the nonlinear molecule when it is adsorbed on a charged dielectric sphere. As is shown in a theoretical model developed, constructive interference of the light scattered at 2ω from different portions

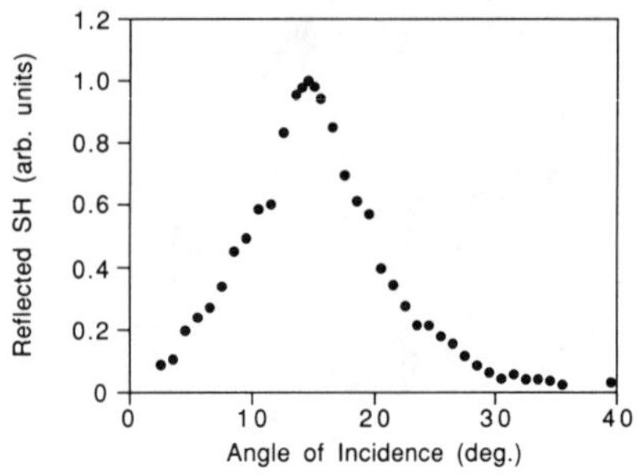

QTuA7 Fig. 1. Intensity of the second-harmonic light generated inside a single colloidal crystal, measured in a reflection geometry as a function of the angle of incidence. The wavelength of the incident laser pulse is 1064 nm. A single crystal is formed in a 1-mm cuvette from a concentrated aqueous suspension of 0.137-μm-diameter spheres. The spheres are coated in a 3×10^{-6} M solution of malachite green.

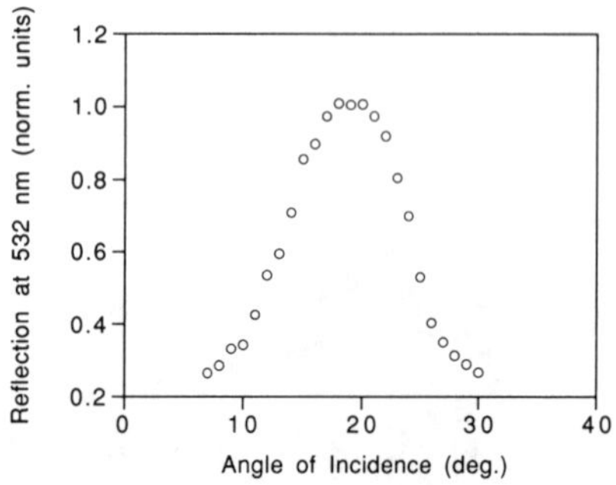

QTuA7 Fig. 2. Reflection of incident light at 532 nm as a function of the angle of incidence from the same colloidal crystal of Fig. 1. The change in index of refraction at the smaller-angle side of the forbidden gap is $\Delta n \sim -0.02$.

of the sphere's surface leads to a nonvanishing field when the radius of the sphere is comparable to the wavelength of the light. In other words, Rayleigh–Gans scattering is the mechanism responsible for the generation of light at 2ω. Contributions from the single spheres in a crystal plane are added coherently to form a plane-wave front propagating in the direction of the incident beam.

For a continuous growth of the SH field intensity, in addition to a nonlocal nonlinear coefficient $\chi^{(2)}$ it is necessary to phase-match the fundamental and SH beams. In a periodic material this mechanism is provided by the bending of the photon dispersion curve at the boundary of the forbidden zone.[4] The change of the effective index of refraction at the smaller angle side of the forbidden gap, determined from Fig. 2, is sufficient to overcome the normal dispersion of water. Then, maximum SHG will be observed at the left of the Bragg reflection notch at an angle of ~15°, as shown in Fig. 1.

Department of Physics and Nuclear Engineering, Polytechnic University of Catalunya, 08222 Terrassa, Spain

1. E. Yablonovitch, T. J. Gmitter, and K. M. Leung, Phys. Rev. Lett. **67**, 2295 (1991).

2. J. Martorell and N. M. Lawandy, Phys. Rev. Lett. **65**, 1877 (1990).
3. N. M. Lawandy, S. A. Johnston, and J. Martorell, Opt. Commun. **65**, 425 (1988).
4. N. Bloembergen and A. J. Sievers, Appl. Phys. Lett. **17**, 483 (1970).

QTuB 8:00 am
Hyatt B

Collisions in Traps

Randy G. Hulet, *Rice University, Presider*

QTuB1 (Invited) 8:00 am

Evaporative cooling of rubidium atoms in a time-averaged orbiting potential magnetic trap

M. H. Anderson, W. Petrich, J. R. Ensher, M. Mathews, E. A. Cornell,, *University of Colorado, Campus Box 440, Boulder, Colorado 80309-0440*

Evaporative cooling has been demonstrated to be a useful method for cooling atomic hydrogen.[1] In the case of alkali atoms, magneto-optical trapping and cooling with subsequent cooling in a polarization-gradient molasses can prepare dense samples of atoms with temperatures within ten photon recoils that can then be efficiently coupled into a purely magnetic trap.[2] Evaporative cooling can then be used to cool the sample to well below the photon recoil limit and, perhaps, down to the quantum regime and Bose-Einstein condensation (BEC).

In real experiments there are always loss processes that limit the available time for intratrap elastic collisions to rethermalize the sample after truncation of the energy distribution. Successful evaporative cooling requires the total loss rates to be small compared to the rethermalizing rate so that the density decrease due to loss is offset enough by the density increase resulting from evaporation to give an overall increase in the elastic collision rate. Our original experiments were carried out in a spherical quadrupole trapping potential because of its ability to augment the elastic collision rate by providing tight confinement. At the trap center the magnetic field vanishes and the trap suffers loss as a result of spin-flips of the atomic magnetic dipoles with respect to the direction of the local trapping field. The lifetime of the trap due to spin-flip loss can be shown to scale as $\tau = \alpha l^2$, where l is the half-width at half-maximum of the trapped atom cloud and α has the measured value for rubidium of 4×10^4 s/cm^2. We found that the spin-flip loss terminates evaporation in our quadrupole trap and limits the increase in phase-space density to about a factor of five.

To circumvent the problem of spin-flip loss, we have modified the spherical quadrupole trap by the addition of a rotating uniform magnetic field which causes the zero point of the field to orbit the trap center at the outer edge of the trapped atom cloud. The orbital frequency is chosen to be large compared to the natural frequencies of the resulting harmonic time-averaged orbiting potential (TOP). The resulting potential eliminates spin-flip loss and provides tighter harmonic confinement than can be obtained with static field traps constructed with similarly sized coils and similar power dissipation. In our first experiments with the new TOP trap we have observed an increase of phase space density of two orders magnitude even though the collision rate declined as evaporation proceeded. We also observe that the loss rate from the trap is independent of the temperature of the trapped atom sample leading to the expectation that the TOP trap will provide a stable trapping potential for the observation BEC. The current status of evaporative cooling experiments of rubidium in the TOP trap will be reported.

1. N. Masuhara, J. M. Doyle, J. C. Sandberg, D. Kleppner, T. J. Greytak, J. F. Hess, and G. P. Kochanski, Phys. Rev. Lett. **61**, 935 (1988).
2. C. Monroe, W. Swann, H. Robinson, and C. Wieman, Phys. Rev. Lett. **65**, 1571 (1990).

QTuB2 8:30 am

Collisions of doubly-spin-polarized ultracold rubidium atoms

J. R. Gardner, D. T. Robb, R. A. Cline, J. D. Miller, D. J. Heinzen, H. M. J. M. Boesten,* B. J. Verhaar,* *University of Texas, Austin, Texas 78712*

The rapidly developing techniques of laser cooling and trapping of neutral atoms make possible a variety of exciting experiments in fundamental atomic physics. These experiments include the search for a permanent electric-dipole moment of the electron, the development of very precise atomic frequency standards, and the study of quantum collective behavior, such as Bose–Einstein condensation. Progress in each of these experiments depends critically on understanding the long-range interactions between cold atoms. In the case of Bose–Einstein condensation in a dilute atomic gas, for example, a stable condensate can form only if the ground-state scattering length of the atoms is positive.

In this work we present new experimental data and a theoretical analysis of photoassociation spectra of rubidium at millikelvin temperatures. We are able to use our high-resolution photoassociation spectra to determine the ground-state scattering length of Rb_2, as well as the interaction coefficient C_6. Moreover, we show, for the first time to our knowledge, that the atomic ground-state interaction parameters can be determined from photoassociation spectra.

In our experiment we load approximately 10^4 atoms into a far-off-resonance optical-dipole force trap (FORT). The atoms are doubly spin-polarized by optical pumping into the ($F = 3$, $m_F = 3$) ground state, in which the nuclear spin and electronic spin are approximately parallel. This considerably simplifies the analysis, because the atoms interact only along the triplet potential.

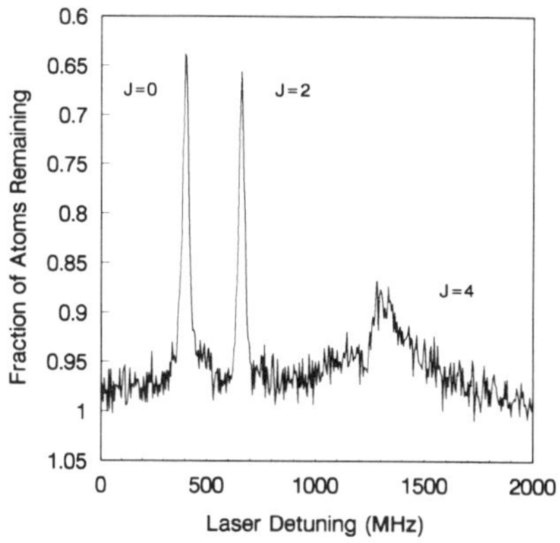

QTuB2 Fig. 1. Resolved rotational structure of one vibrational level of the 0_g^- electronic state. The absence of the odd J rotational levels is manifestation of the nuclear spin statistics.

We obtain high-resolution photoassociation spectra by recording the number of atoms remaining in the trap as a function of the frequency of an applied catalysis laser field. Because the atoms in the FORT are at a temperature of about 1 mK, the resulting rovibrational spectra show well-resolved lines. Figure 1 shows a typical photoassociation spectrum for a single vibrational level of the 0_g^- electronic state at 12 573.04 cm^{-1}.

To determine the Rb_2 ground-state parameters, we recorded data for a series of five vibrational levels of the 0_g^- electronic state that asymptotically connects to $5^2S_{1/2} + 5^2P_{1/2}$ separated-atom limit. Information on the ground-state-triplet potential is contained in the line intensities and shapes, both of which depend on the overlap of the excited-state wave function with the collisional ground-state wave function. To extract this information, we model these line shapes and intensities and find the ground-state potential parameters that provide the best fit to the data. These parameters are of crucial importance for the application of ultracold Rb atoms either to Bose–Einstein condensation or to the construction of very precise atomic clocks.

Our preliminary result for the triplet scattering length of ^{85}Rb is negative. Using mass-scaling arguments, we are able to predict that the triplet scattering length of ^{87}Rb is positive. Further experiments and analysis, currently in progress, will yield the parameters necessary to completely determine the interactions between both ^{85}Rb and ^{87}Rb atoms in any of their spin sublevels.

We gratefully acknowledge the support of the National Science Foundation and of the R. A. Welch Foundation.

* *Eindhoven University of Technology, Box 513, 5600MB Eindhoven, The Netherlands*

QTuB3 **8:45 am**

Spectroscopic determination of the *s*-wave scattering length of lithium

E. R. I. Abraham, W. L. McAlexander, C. A. Sackett, R. G. Hulet, *Department of Physics and Rice Quantum Institute, Rice University, Houston, Texas 77251-1892*

Recent advances in the trapping and cooling of spin-polarized atoms may lead to the observation of Bose–Einstein condensation (BEC) in a weakly interacting system. The formation of the condensate depends on the interaction between trapped atoms, which at low temperatures is described by a single parameter, the *s*-wave scattering length a. The scattering length is determined entirely by the interaction potential, which in the case of spin-polarized hydrogen or alkali atoms is the $^3\Sigma_u^+(a)$ potential. Because of the possibility of a zero-energy scattering resonance caused by a state very near the dissociation limit, the scattering length is sensitive to the binding energy of the least-bound state.

Recently, photoassociative spectroscopy of ultracold trapped atoms has proven to be a powerful spectroscopic tool for probing weakly bound, long-range molecular excited states.[1] We have extended this technique to probe ground-state molecular levels by means of a two-photon process (Fig. 1). We report the measurement of the binding energy of the least-bound state of the $^7Li_2$ $^3\Sigma_u^+(a)$ potential, and we extract a value for the triplet *s*-wave scattering length.[2]

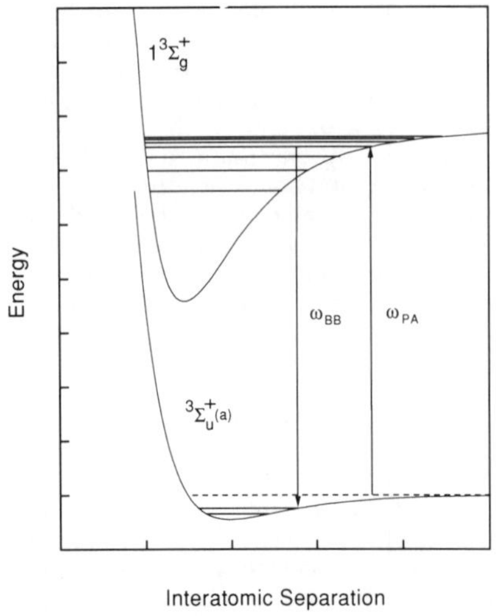

QTuB3 Fig. 1. Photoassociative spectroscopy of the ground-state triplet potential of $^7Li_2$. Two colliding ultracold atoms absorb a photon of frequency ω_{PA} that resonantly excites a high-lying vibrational level of the $1^3\Sigma_g^+$ potential, correlating asymptotically to $2S_{1/2} + 2P_{1/2}$ atoms. These atoms may be lost from the trap in a subsequent spontaneous-decay event, leading to a measurable reduction in trap-laser-induced fluorescence. A second laser of frequency ω_{BB} is tuned to resonance between a bound ground-state vibrational level and the bound excited-state vibrational level. This laser beam reduces the excitation into the excited state, leading to a recovery in the trap-laser-induced fluorescence signal.

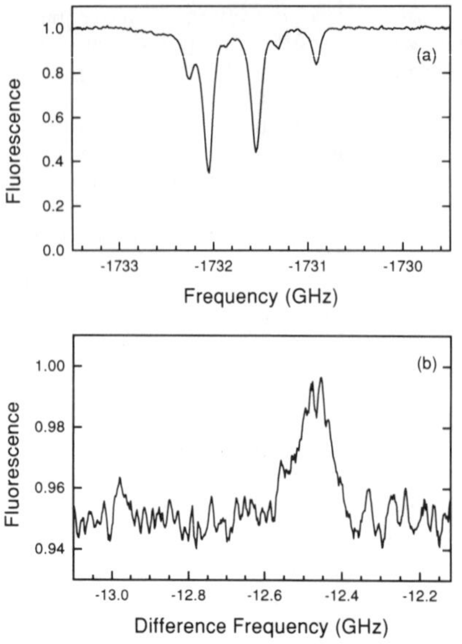

QTuB3 Fig. 2. (a) One-photon photoassociation spectrum of the $v = 64$ vibrational level of the $1^3\Sigma_g^+$ state of $^7Li_2$. The frequency is measured relative to the $2S_{1/2} \leftrightarrow 2P_{1/2}$ atomic transition frequency. The structure is due to the hyperfine interaction. (b) Two-photon spectrum of the $v = 10$ vibrational level of the $^3\Sigma_u^+(a)$ state. The baseline corresponds to the reduction of the fluorescence level due to the photoassociation laser beam. As ω_{BB} is tuned over the bound–bound transition frequency, the rate of photoassociative loss is reduced, thereby increasing the trap-laser-induced fluorescence. The fluorescence scales in both (a) and (b) are relative to the fluorescence level induced by the trap-laser beams in the absence of the photoassociation or bound–bound laser beams.

Figure 2(a) shows a high-resolution one-photon photoassociation spectrum of the $v = 64$ level of the $1^3\Sigma_g^+$ state of $^7Li_2$. The observed structure is due to the hyperfine interaction. Figure 2(b) shows the Raman spectrum of the least-bound ($v = 10$) state of the $^3\Sigma_u^+(a)$ potential, where the $v = 64$ state of the $1^3\Sigma_g^+$ is used as an intermediate resonance. To estimate the scattering length implied by the experimental results, we used a model potential to give a $v = 10$ eigenvalue, in agreement with the measured value. A numeric determination of the elastic-scattering phase shift from this potential gives the triplet *s*-wave scattering length for ^7Li to be $-26.8 \pm 1.4\ a_0$.

The sign of the scattering length has important implications for experiments seeking to observe BEC of ^7Li (Ref. 3) because of a prediction that BEC is not stable in systems with a negative a (Ref. 4).

We are grateful to K. Kobayashi of the NEC Corp. for providing diode lasers used for this experiment and to A. Dalgarno for providing the results of a calculation of the bound–bound matrix elements. This work is supported by the National Science Foundation and the Welch Foundation. W. I. McAlexander is a Fellow of the Fannie and John Hertz Foundation.

1. V. Bagnato, L. Marcassa, C. Tsao, Y. Wang, and J. Weiner, Phys. Rev. Lett. **70**, 3225 (1993); P. D. Lett, K. Helmerson, W. D. Phillips, L. P. Ratliff, S. L. Rolston, and M. E. Wagshul, Phys. Rev. Lett. **71**, 2200 (1993); L. P. Ratliff, M. E. Wagshul, P. D. Lett, S. L. Rolston, and W. D. Phillips, J. Chem. Phys. **101**, 2638 (1994); J. D. Miller, R. A. Cline, and D. J. Heinzen, Phys. Rev. Lett. **71**, 2204 (1993); R. A. Cline, J. D. Miller, and D. J. Heinzen, Phys. Rev. Lett. **73**, 632 (1994); W. I. McAlexander, E. R. I. Abraham, N. W. M. Ritchie, C. J. Williams, H. T. C. Stoof, and R. G. Hulet, submitted for publication.
2. E. R. I. Abraham, W. I. McAlexander, C. A. Sackett, and R. G. Hulet, submitted for publication.
3. J. J. Tollett, C. C. Bradley, C. A. Sackett, and R. G. Hulet, Phys. Rev. A, in press.
4. H. T. C. Stoof, Phys. Rev. A **49**, 3824 (1994).

QTuB4 **9:00 am**

Optical control of cold collisions

C. Orzel,* M. Hoogerland, U. Sterr, M. Walhout, S. L. Rolston, *Atomic Physics Division, PHY A167, National Institute of Standards and Technology, Gaithersburg, Maryland 20899*

Ultracold collisions between atoms offer a unique opportunity to exert control over collisional dynamics with laser radiation. The collisional energy can be much smaller than the light shifts created by the laser, and the collisional duration may be significantly longer than the excited-state lifetime.

Metastable xenon offers an ideal testing ground for such studies because it can be readily laser cooled and trapped in a magneto-optical trap[1] (MOT) and because the predominant collisional-loss mechanism is Penning ionization, with the production of an easily detected ion. We have investigated the effects of applying near-resonant laser light to modify the collision dynamics.[2] The rate constant for ionizing collisions between two $6s[3/2]_2$ metastable atoms has been measured for a number of different isotopes. This rate is then used as a benchmark to compare to the ionizing rates when a near-resonant control laser is applied (while the MOT beams are off). To the red of resonance, the collisional rate is enhanced, with a maximum near −20-MHz detuning for low intensity. This spectrum (Fig. 1) has been well characterized, especially close to resonance, and it suggests the need for the inclusion of relatively-long-lived molecular states to explain the results.

When the control laser is tuned to the blue of resonance, the ionization rate is found to drop below that with no light present. This is a form of optical shielding,[3] in which the atom pair is excited to a repulsive molecular potential and is prevented from reaching close range at which ionization occurs. We have been able to shield collisions by as much as a factor of 8 in this way.

If the MOT laser beams are left on, the

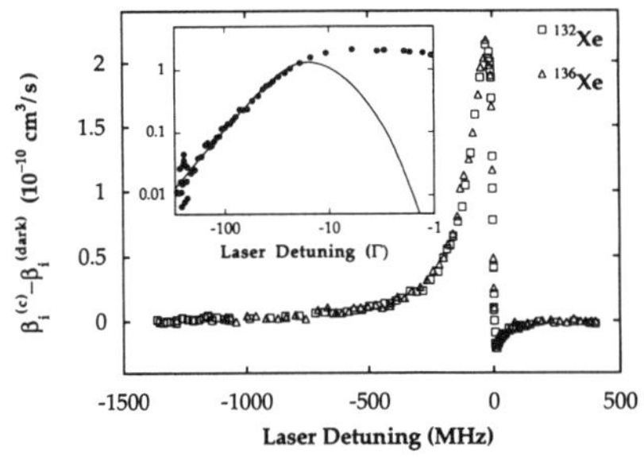

QTuB4 Fig. 1. Laser-induced-ionization rate coefficient versus control-laser detuning δ for two even isotopes of metastable xenon (intensity I = 1 mW/cm^2). The inset shows a log–log plot of the data and a scaled quantum-mechanical model of Cs trap loss (natural linewidth $\Gamma/2\pi$ = 5 MHz).

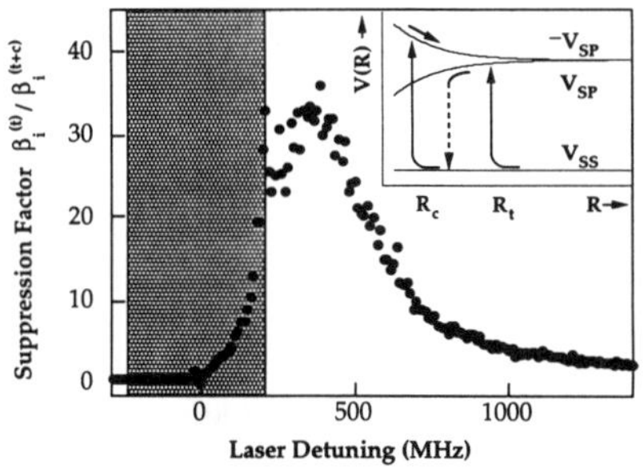

QTuB4 Fig. 2. Measured suppression factor (ratio of ion rates for MOT laser alone to MOT laser and control laser) versus detuning for ^{132}Xe (I = 1 W/cm^2). In the shaded region the trap is distorted by the control laser.

control laser has a much more significant effect, reducing the ionization rate by more than a factor of 35 (see Fig. 2). Using an absorptive imaging technique, we have measured the number and the density, and we find they both increase with the application of this control laser. The lifetime of the MOT is also substantially increased to the suppression of this loss channel.

Simple Landau–Zener theories have been applied to optical shielding,[3] and their predictions are discussed. The simplest theory predicts much larger shielding than we observed, so that some additional features, such as a distribution of molecular Rabi frequencies, must be included to reproduce the results.

*Department of Chemical Physics, University of Maryland, College Park, Maryland 20742

1. M. Walhout, H. J. L. Megens, A. Witte, and S. L. Rolston, Phys. Rev. A **48**, R879 (1993).
2. M. Walhout, U. Sterr, C. Orzel, M. Hoogerland, and S. L. Rolston, Phys. Lett., in press.
3. L. Marcassa *et al.*, Phys. Rev. Lett. **73**, 1911 (1994).
4. H. M. J. M. Boesten and B. J. Verhaar, Phys. Rev. A **49**, 4240 (1994).

QTuB5 9:15 am

Experimental investigation of optical suppression in cold collisions

L. Marcassa, R. Horowicz, S. Zilio, J. Weiner,* V. Bagnato, *University of São Paulo, IFSC, Caixa Postal 369, São Carlos, São Paulo 13.560, Brazil*

Cold collisions are a limiting process in obtaining high-density samples of cold atoms in traps. The suppression of inelastic collisional effects may be an alternative for the attainment of dense ultracold atomic samples. Suppression of the inelastic process works by preventing the colliding atoms from reaching small interatomic distances at which the inelastic process can take place. To obtain such a situation, one can excite the colliding partners to a repulsive state, forcing them to reverse their motion and fly apart from each other.

We have experimentally investigated the suppression of the photoassociative ionization (PAI) of sodium atoms held in a magneto-optical trap[1] by adding radiation from a blue-detuned laser in addition to the conventional frequencies existent in the trap. Variation on the suppressor laser allows us to obtain almost 100% suppression of the process. Figure 1 shows the path leading to the Na_2^+-ion product (solid line) and the suppression scheme (dashed line). In a normal situation[2] atom pairs on the (F = 2) + (F = 2) ground-state potential approach one another, absorbing a photon ω_p from a probe laser undergoing a transition to an attractive $-C_3/R^3$ potential. The atom pair accelerates, absorbing a second photon ω_2 that promotes the collision pair to the double-excited level. After entering the inner zone, the double-excited state becomes degenerate, with the ionized molecule producing Na_2^+. If this path is interrupted by a suppressor frequency ω_s that couples the incoming reactant flux to a repulsive level, the pair is prevented from approaching the autoionization zone, suppressing PAI.

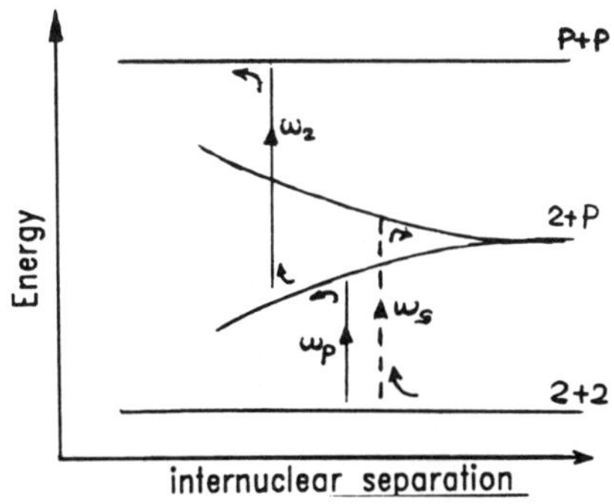

QTuB5 Fig. 1. Schematic representation of the normal PAI and the suppressed PAI.

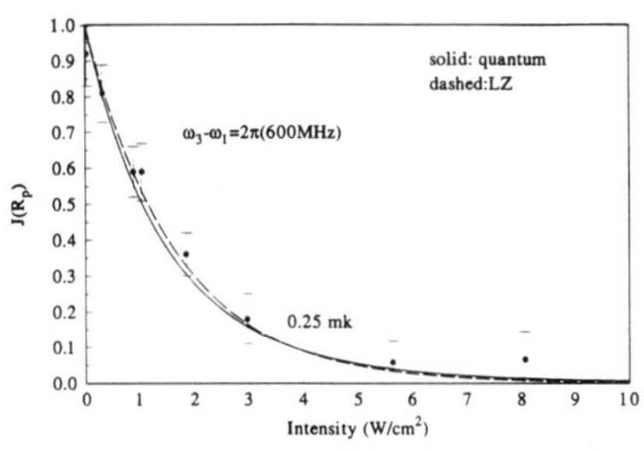

QTuB5 Fig. 2. Experimental results for the intensity dependence of the suppression and comparison with existing theories from Ref. 2. The frequency ω_1 represents the resonance with the transitions $3S_{1/2}$ (F = 2) to $3P_{3/2}$ (F = 3).

Figure 2 shows the observed suppression as a function of the suppressor laser intensity and its comparison with existing theory. In addition to suppression in PAI, we present results that confirm that excited-state-to-ground-state collision processes (such as radiative escapes and fine-structure change) are also suppressed, decreasing trap loss.

This work is supported by the Research Foundation of São Paulo, the National Research Council (Brazil), and the National Science Foundation (U.S.).

**University of Maryland at College Park*

1. L. Marcassa *et al.*, Phys. Rev. Lett. **73**, 1911 (1994).
2. K. A. Suominen *et al.*, submitted for publication.

QTuB6 9:30 am

Suppression of cold collisions at low temperatures

V. Sanchez-Villicana, S. D. Gensemer, C. D. Wallace,* T. P. Dinneen,** P. L. Gould, *Department of Physics, U-46, University of Connecticut, Storrs, Connecticut 06269-3046*

In a cold collision between a ground- and an excited-state atom, a key issue is the survival of the atomic excitation during the course of the collision. For small detunings from resonance, the excitation is created at long range (e.g., >100 nm). Once excited, the atom pair begins to accelerate on the attractive R^{-3} potential. For an inelastic-trap-loss collision to occur, the excited atom pair must reach short range (e.g., <5 nm). In the journey from long range to short range, there is a high probability that the excitation will decay, effectively turning off the collision. As the initial collision velocity is reduced, the collision time is lengthened and the probability for decay is enhanced. It is therefore predicted[1–4] that the collision rate is suppressed as the temperature is lowered. A key aspect of this prediction is localization of the initial excitation, as occurs in a Landau–Zener curve-crossing process.

We have observed this effect in ^{85}Rb by measuring the temperature dependence of the collisional loss rate from a magneto-optical trap.[5] The temperature is varied by changing the trap-laser intensity.[6] Because the overall collision rate between ground- and excited-state atoms increases with intensity, this expected linear dependence must be accounted for in the analysis. The collision rate, divided by intensity, is shown in Fig. 1. A strong suppression is observed for temperatures below ~70 μK. Alternative explanations for this suppression based on ground-state hyperfine-changing collisions, saturation, radiative-escape collisions, and localization of atoms in optical-wavelength-scale potentials have been

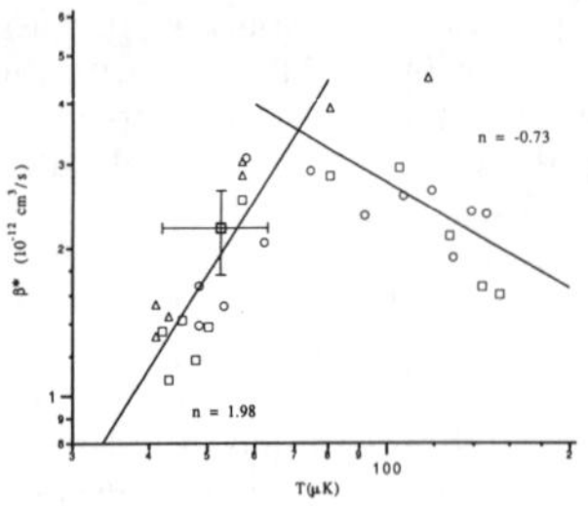

QTuB6 Fig. 1. Temperature dependence of the normalized collisional-trap-loss rate β* at a detuning $\Delta/2\pi = -5$ MHz. The solid curves are best-fit lines to data below and above 70 μK and have slopes as indicated.

considered and have been shown to be inconsistent with the data. The fact that trap-loss collisions are suppressed at low temperatures is a significant finding that may have important implications for obtaining high densities of cold atoms.

*Current address: *Department of Physics, Rice University, Houston, Texas 77251*

**Current address: *Lawrence Berkeley Laboratory, Berkeley, California 94720*

1. H. M. J. M. Boesten, B. J. Verhaar, and E. Tiesinga, Phys. Rev. A **48**, 1428 (1993).
2. P. S. Julienne, K.-A. Suominen, and Y. Band, Phys. Rev. A **49**, 3890 (1994).
3. K.-A. Suominen, M. J. Holland, K. Burnett, and P. S. Julienne, Phys. Rev. A **49**, 3897 (1994).
4. H. M. J. M. Boesten and B. J. Verhaar, Phys. Rev. A **49**, 4240 (1994).
5. C. D. Wallace, T. P. Dinneen, K. Y. N. Tan, T. T. Grove, P. L. Gould, Phys. Rev. Lett. **69**, 897 (1992).
6. C. D. Wallace, T. P. Dinneen, K. Y. N. Tan, A. Kumarakrishnan, P. L. Gould, and J. Javanainen, J. Opt. Soc. Am. B **11**, 703 (1994).

QTuB7 **9:45 am**

Permanent-magnet trap for cold atoms

C. A. Sackett, J. J. Tollett, C. C. Bradley, R. G. Hulet, *Department of Physics and Rice Quantum Institute, Rice University, Houston, Texas 77251-1892*

We have constructed a permanent-magnet trap that provides a tightly confining potential useful for producing dense ultracold atomic vapors.[1] This trap will be used to investigate quantum-statistical effects of ^{6}Li and ^{7}Li. Permanent magnets can produce field strengths and gradients comparable to those of superconducting-magnet traps but with reduced complexity and greater optical access. Our trap uses an arrangement of magnets in an Ioffe configuration,[2] which features a nonzero magnetic-field minimum. Atoms are trapped about this minimum through the interaction of their total magnetic moment with the trap's magnetic field. The nonzero minimum prevents trap-loss-producing spin-flip (Majorana) transitions.

As shown in Fig. 1, the trap consists of six axially magnetized, cylindrical NdFeB permanent magnets symmetrically positioned and aligned along three mutually orthogonal axes. The trap is estimated to be 7 mK deep for ^{7}Li and is essentially parabolic near the center, with an estimated oscillation frequency for ^{7}Li of 120 Hz. The trap is loaded from a laser-cooled atomic beam. In the trap, a fraction of the incident atoms is captured and cooled by optical molasses provided by three pairs of retroreflected laser beams.

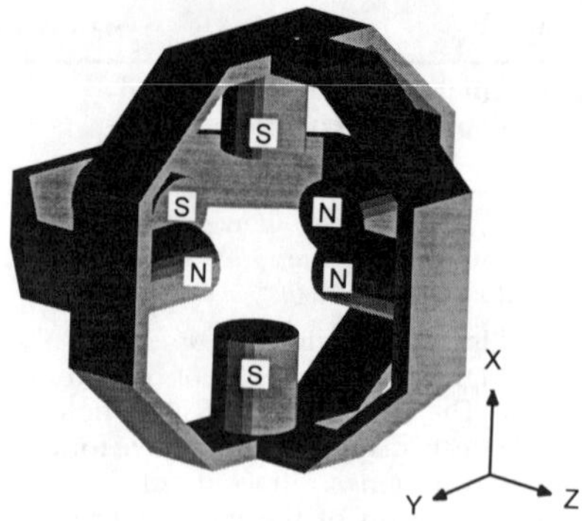

QTuB7 Fig. 1. Permanent-magnet trap. The magnets are supported by an iron yoke that recycles the magnetic flux. The letters indicate the inner-tip magnetization. The tip-to-tip spacing is 4.45 cm.

After the atoms are loaded, the trap beams can be turned off, leaving the atoms in the dark. The lifetime of atoms in the dark is 240 s, presumably limited by collisions with background gas. Fluorescence measurements indicate that 10^8 atoms are trapped with a temperature of 1.1 mK and a peak density of 3×10^9 cm^{-3}.

With modest improvements in temperature and loading we expect to achieve trapped-atom densities greater than 10^{10} cm^{-3}. The recently obtained[3] *s*-wave elastic ^{7}Li–^{7}Li scattering cross section of 2.5×10^{-13} cm^2 and the trap lifetime suggest that at these densities evaporative cooling should be possible by using radio-frequency or optical radiation to selectively remove the most energetic atoms. A density of 10^{13} cm^{-3} is needed to produce a degenerate quantum gas at a temperature of 1 μK, which would require 10^7 atoms for the present trap. Under these conditions, it may be possible to observe Bose–Einstein condensation of ^{7}Li or a degenerate Fermi gas in the case of ^{6}Li.

1. J. J. Tolleatt, C. C. Bradley, C. A. Sackett, and R. G. Hulet, Phys. Rev. A, in press.
2. Y. V. Gott, M. S. Ioffe, and V. G. Tel'kovskii, Nucl. Fusion, Suppl., Part 3, 1045 (1962); T. Bergeman, G. Erez, and H. J. Metcalf, Phys. Rev. A **35**, 1535 (1987); V. S. Bagnato, G. P. Lafyatis, A. G. Martin, E. L. Raab, R. N. Ahmad-Bitar, and D. E. Pritchard, Phys. Rev. Lett. **58**, 2194 (1987).
3. E. R. I. Abraham, W. I. McAlexander, C. A. Sackett, and R. G. Hulet, to be published; E. R. I. Abraham *et al.*, this conference.

QTuC **8:00 am**

Hyatt C/D

Harmonic Generation from Surfaces and Interfaces

William Rabinovich, *U.S. Naval Research Laboratory, Presider*

QTuC1 **8:00 am**

Second-harmonic generation from a self-assembled monolayer on a SiO_x/Si(111) surface

D. Hahn, H. Zhou, G. K. Wong, S. B. Roscoe,* T. J. Marks,* *Department of Physics and Materials Research Center, Northwestern University, Evanston, Illinois 60208*

Silicon owes its dominance as the electronic material of semiconductor technology to the high quality of its oxide layer. Understanding the nature of the SiO_x/Si interface and the SiO_x overlayer is quite important, and the problem has been investigated extensively by using a wide range of techniques, including AFM, STM, TEM, surface x-ray, and surface second-harmonic generation (SHG). An interesting, unresolved question raised by these studies is whether the SiO_x layer retains some degree of ordering or whether it is amorphous with an abrupt interface between it and the crystalline silicon surface. Earlier work seems to suggest that the latter is the case, with a transition region of about 5 Å. Recent evidence, however, indicates that the presence of crystalline SiO_x phases,[1-3] possibly tridymite[4,5] or cristobalite,[6] at or near the interface allows the long-range order of crystalline Si to decay gradually into an amorphous SiO_x overlayer. Because most interfacial probes are not capable of observing or interpreting disorder at the interface, a definitive answer to this question remains elusive. For example, the presence of small amounts of disorder can render high-resolution transmission-electron-microscopy images uninterpretable; hence TEM measurements cannot give an unambiguous answer.

We report experimental results that indicate that, unlike usual amorphous SiO_x on fused-quartz surfaces, the SiO_x overlayer on Si has a certain degree of ordering that is related to the structure of the underlying crystal surface. The experimental technique we used involves the chemical attachment of a chromophoric self-assembled monolayer (SAM) probe onto the native oxide of highly polished single-crystal Si (111) substrates. Detailed procedures for the sample preparation have been described elsewhere.[7] The oxide thickness was determined to be about 20 Å by ellipsometry. Surface SHG is used to study the surface ordering of the SAM.[8]

Figure 1 displays the *p*-polarized SH signal with *s*- and *p*-polarized fundamental input as a function of the sample in-plane rotation angle ϕ, where 0° corresponds to the $[11\bar{2}]$ direction of the Si(111) face. The SHG pattern shows a distinct threefold

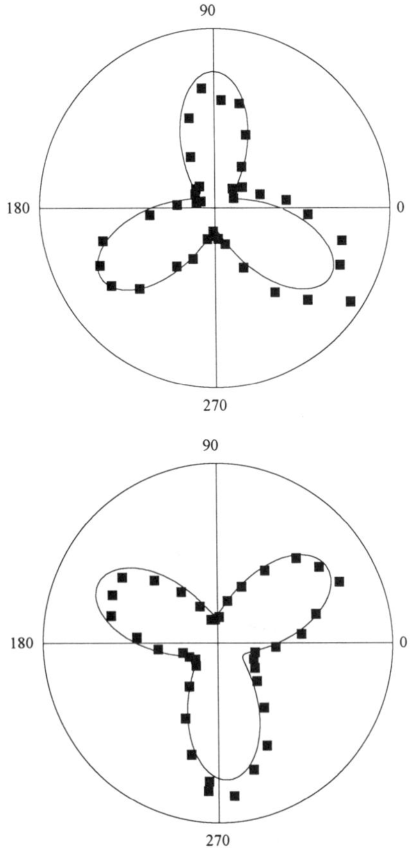

QTuC1 Fig. 1. SHG intensity (arbitrary units) versus sample rotation angle ϕ for two different fundamental input and second-harmonic output polarization combinations: (a) *s*-in/*p*-out and (b) *p*-in/*p*-out.

structure that is unmistakably due to the self-assembled monolayer. The experimental data can be fit to $\cos(3\phi) + c$ (solid curves in Fig. 1), which is compatible with a primitive hexagonal surface symmetry. Without the SAM, $SiO_x/Si(111)$ gives a very weak SHG signal with a sixfold symmetric pattern, reflecting the C_{3v} symmetry of the Si(111) crystal surface.[3] The preferred tilting of the chromophore on $SiO_x/Si(111)$ is presumably due to some degree of ordered structure on the surface of native oxide. Taken together, our results indicate that the native oxide layer on Si(111) retains some degree of the ordering of the underlying crystalline Si surface, even at distances of approximately 20 Å from the $SiO_x/Si(111)$ interface.

This work was supported by the National Science Foundation–MRL program through the Materials Research Center of Northwestern University (grant DMR 9120521) and by U.S. Air Force Office of Scientific Research contract 93-1-0114.

* *Department of Chemistry and Materials Research Center, Northwestern University, Evanston, Illinois 60208*

1. T. A. Rabedeau, I. M. Tidswell, P. S. Pershan, J. Bevk, and B. S. Freer, Appl. Phys. Lett. **59**, 706 (1991).
2. T. A. Rabedeau, I. M. Tidswell, P. S. Pershan, J. Bevk, and B. S. Freer, Appl. Phys. Lett. **59**, 3422 (1991).
3. G. Renaud, P. H. Fuoss, A. Ourmazd, J. Bevk, B. S. Freer, and P. O. Hahn, Appl. Phys. Lett. **58**, 1044 (1991).
4. A. Ourmazd, D. W. Taylor, J. A. Rentschler, and J. Bevk, Phys. Rev. Lett. **59**, 213 (1987).
5. A. Ourmazd, J. A. Rentschler, and J. Bevk, Appl. Phys. Lett. **53**, 743 (1988).
6. P. H. Fuoss, L. J. Norton, S. Brennan, and A. Fischer-Colbrie, Phys. Rev. Lett. **60**, 600 (1988).
7. S. Yitzchaik, A. K. Kakkar, S. B. Roscoe, Y. Orihashi, T. J. Marks, W. Lin, and G. K. Wong, Mol. Cryst. Liq. Cryst. **240**, 9 (1994).
8. See, for example, Y. R. Shen, Annu. Rev. Phys. Chem. **40**, 327 (1989); Y. R. Shen, Nature (London) **337**, 519 (1989).

QTuC2 8:15 am

In-situ monitoring of Si(001) during epitaxial growth by optical second-harmonic generation of femtosecond pulses

J. I. Dadap,* N. M. Russell, X. F. Hu,* J. G. Ekerdt, M. C. Downer,* *Science and Technology Center, University of Texas, Austin, Texas 78712*

New capabilities in interface nonlinear optics with femtosecond laser pulses have opened new possibilities for *in situ*, real-time nondestructive diagnostics of the growth of semiconductor films and heterostructures. Nonlinear optical probes such as SHG, because of interface specificity, are highly sensitive to growth kinetics and chemistry at the submonolayer level and to buried-interface properties. However, slow data acquisition and strong substrate heating have traditionally hampered its effectiveness as a real-time growth monitor. With SHG by state-of-the-art femtosecond excitation, real-time growth monitoring with negligible heating (<20 K) becomes possible.

The experiments were performed in a UHV environment (base pressure ~ 1 × 10^{-9} Torr). We used ~120-fs, 815-nm Ti:sapphire laser pulses focused onto a Si sample at 55° incidence at a laser fluence <0.5 mJ/cm² by using a *p*-in/*p*-out configuration, where *p* is parallel to Si ⟨100⟩. We monitored the SHG during the key atomic-layer-epitaxy[1] steps. Fig. 1(a) shows adsorption of a 2 × 10^{-5} Torr gas mixture of 4% Si_2H_6 (disilane)/He balance on Si(001)-2 × 1 at 425 K. Figure 1(b) shows the desorption of hydrogen as the surface is heated

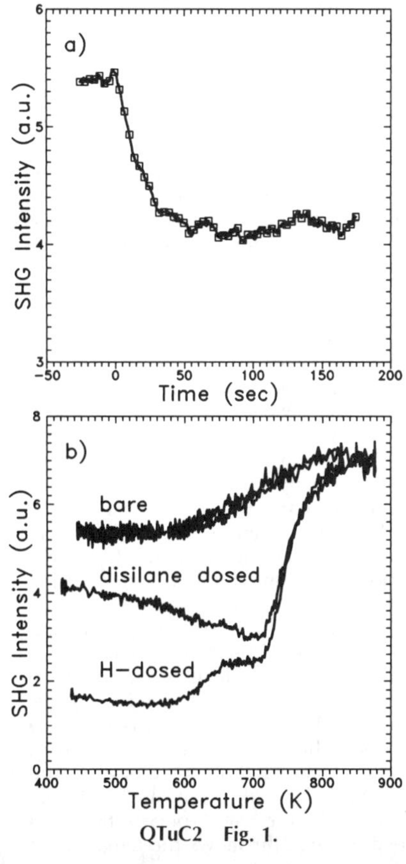

QTuC2 Fig. 1.

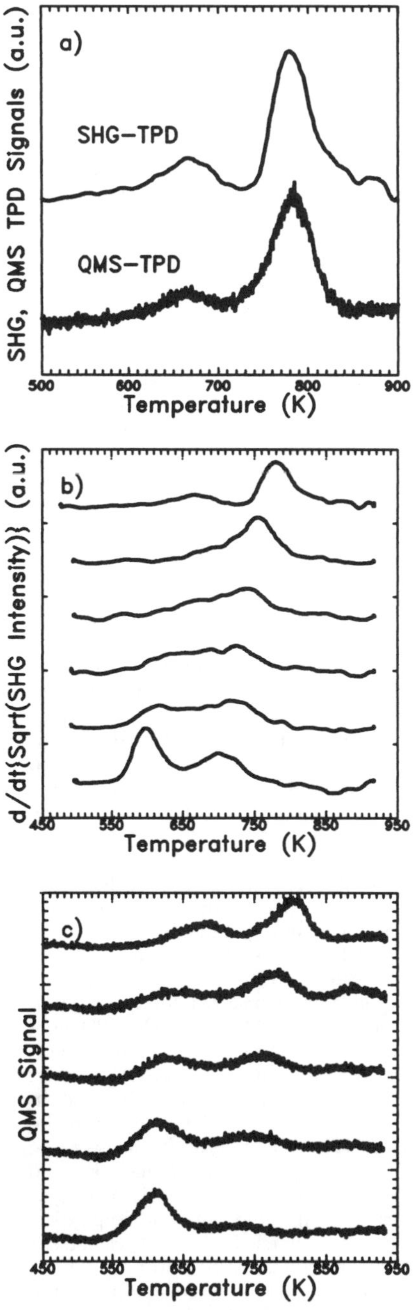

QTuC2 Fig. 2.

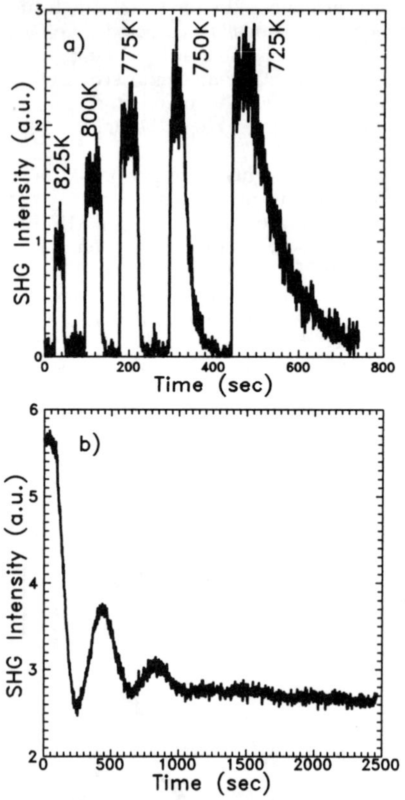

QTuC2 Fig. 3.

to ~900 K at the rate of 2 K/s. For comparison, the signals from the bare surface and an H-atom-dosed surface are shown. The heating after disilane dosing liberates the hydrogen, leaving 0.4 ML of silicon on the surface.

The nonlinear susceptibility of an adsorbate/substrate system is approximately proportional to the coverage of the adsorbate.[2] Hence the time derivative of the square root of the signal during the desorption yields the desorption rate. The resulting desorption rate of the H-saturated surface is compared directly to a temperature-programmed desorption (TPD) spectrum of H from Si at 2 K/s by using quadrupole mass spectrometry (QMS), as shown in Fig. 2(a), with excellent agreement. Notice the two-step desorption feature corresponding to β_2 (dihydride) and β_1 (monohydride) desorption observed in TPD.[3] To extend this technique further, five ALE cycles were performed by using Ge_2H_6, with each dose resulting in the deposition of 0.4 to 0.5 ML of Ge, such that a total of 2 ML of Ge were deposited on Si. The hydrogen desorption was monitored during each cycle as shown in Fig. 2(b). This SHG-TPD measurement is also in good agreement with the QMS-TPD measurements of Asami *et al.*,[1] which used a 5-K/s heating rate. In general, the relative phases between the dihydride and monohydride states can be different; hence one should take these into account to obtain the correct H coverage.

Figure 3(a) shows pressure-jump experiments, which used 4% Si_2H_6/He pulses at 715 nm and 5×10^{-4} Torr at several temperatures. At this wavelength the H/Si signal is stronger than the bare-substrate signal, unlike at the previously used wavelength of 815 nm. When the gas mixture is introduced, the *p*-in/*p*-out signals increase, depending on the steady-state H concentration. When the valve is closed, the SHG signals decrease with a characteristic decay rate. Notice the flat response of bare Si at these temperature ranges.

During UHV-CVD growth, a 20-Å $Si_{0.9}Ge_{0.1}$-on-Si(001) substrate is heated to a temperature of 925 K; then the disilane/He gas mixture is introduced into the chamber at 5×10^{-4} Torr. Using a 815-nm pump, we observe an oscillatory behavior in the *p*-in/*p*-out signal as the disilane mixture is introduced at $t = 0$ [Fig. 3(b)]. After about 1000 s, the signal is almost fully damped. This oscillatory behavior is attributed to the interference between the growth interface and the initial $Si_{0.9}Ge_{0.1}$ interface nonlinear sources. The damped behavior of the signal is due to the ~500-Å escape depth of the SH at 415 nm.

*Department of Physics, University of Texas, Austin, Texas 78712

1. S. Asami, N. M. Russell, A. Mahajan, P. A. Steiner IV, D. J. Bonser, J. Fretwell, S. Banerjee, A. Tasch, J. M. White, and J. G. Ekerdt, Appl. Surf. Sci., in press.
2. H. W. K. Tom, C. M. Mate, X. D. Zhu, J. E. Crowell, T. F. Heinz, G. A. Somorjai, and Y. R. Shen, Phys. Rev. Lett. **52**, 348 (1984).
3. K. Sinniah, M. Sherman, L. Lewis, W. Weinberg, J. Yates, and K. Janda, J. Chem. Phys. **92**, 5700 (1990).

QTuC3 **8:30 am**

Two-photon resonant-enhanced second-harmonic generation by femtosecond pulses from Si(001)-2 × 1 and H–Si(001) surfaces

J. I. Dadap,* X. F. Hu,* N. M. Russell, J. G. Ekerdt, M. C. Downer,* Science and Technology Center, University of Texas, Austin, Texas 78712

Daum *et al.*[1] have demonstrated a strong two-photon-resonance band at 3.3 eV in the $\chi^{(2)}_{zzz}$ susceptibility component of the surface second-harmonic (SH) spectrum of Si(001)-2 × 1, due to direct electronic transitions in the strained near-surface region. By using 120-fs Ti:sapphire-laser pulses to maximize surface SH efficiency while minimizing surface heating, our experiments (1) measure the thermal red shift of this two-photon $\chi^{(2)}_{zzz}$ resonance and agree with the calculated thermal shrinkage rate of the bulk Λ_3–Λ_1 transition; (2) measure the temperature dependence of the weaker $\chi^{(2)}_{zxx}$ and $\chi^{(2)}_{xxz}$ components, which do *not* track the Λ_3–Λ_1 resonance and must therefore be attributed to other temperature-dependent surface mechanisms; (3) measure a blue shift of the $\chi^{(2)}_{zzz}$ resonance on hydrogen termination and explain it by a coupled anharmonic Si–Si/Si–H oscillator model.

Pulses with wavelengths ranging from 700 nm to 1 μm and polarized along the sample ⟨100⟩ direction are focused onto a radiatively heated Si sample in ultrahigh vacuum. Figure 1(a) shows typical *p*- and *s*-polarized SHG outputs of bare Si(001) obtained while the input polarization was varied. Figure 1(b) shows the temperature

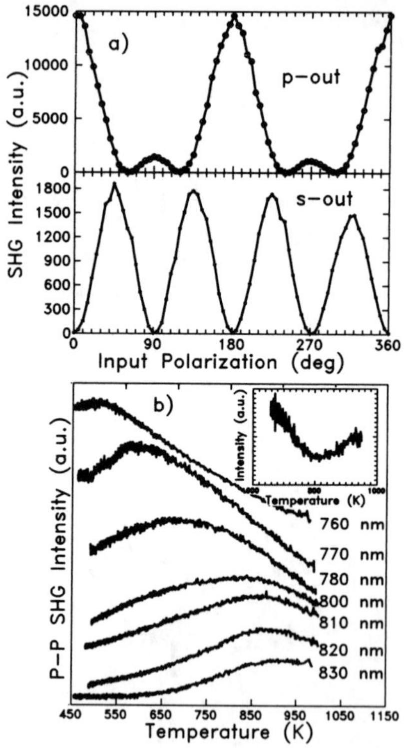

QTuC3 Fig. 1. (a) *p*-out and *s*-out SHG signals as a function of input polarization; e.g., input polarizations of = 0°, 90° mean *p* and *s* inputs, respectively. (b) *p*-in/*p*-out signals for several wavelengths as a function of temperature.

dependence of *p*-in/*p*-out signals for various wavelengths. The SHG peaks shift toward higher temperature with increasing wavelength. This corresponds to the thermal-band-gap shrinkage coming into resonance with the two-photon energy of the fundamental beam. By measuring the temperature of the peak SHG signal at different wavelengths, we obtained a band-gap shrinkage rate of ~6×10^{-4} eV/K in the temperature range 500–1000 K, in agreement with the bulk Si E_1 (Λ_3–Λ_1) transition-temperature-dependence calculation by Lautenschlager *et al.*[2]

The SH output field is given by $E_{pp} = f_1 \chi^{(2)}_{zzz} + f_2 \chi^{(2)}_{zxx} + f_3 \chi^{(2)}_{xxz}$, where f_1, f_2, and f_3 are the associated Fresnel factors. Measurements of I_{sp} (*s*-in/*p*-out) and I_{qs} signals (*q* means 45° relative to the incidence plane), both approximately 0.1 I_{pp}, isolate $\chi^{(2)}_{zxx}$ and $\chi^{(2)}_{xxz}$, respectively. They do not show any resonance feature and generally decrease as temperature increases. Thus only $\chi^{(2)}_{zzz}$ tracks the two-photon resonance. As an example, the inset in Fig. 1(b) shows I_{qs} at 780 nm, where it initially decreases and then increases. Similarly, I_{sp} decreases strongly with increasing temperature for 780–820 nm, showing no two-photon resonant enhancement. Electron–photon interactions may strongly affect $\chi^{(2)}_{zxx}$ at higher temperatures, as previously suggested by Höfer *et al.*[3]

Figure 2(a) shows I_{pp} signals from H-saturated Si from 720- and 750-nm input pulses as the temperature is increased at the rate of 2 K/s. The H/Si signals are superimposed on the bare Si signal. Notice that below the H desorption temperature

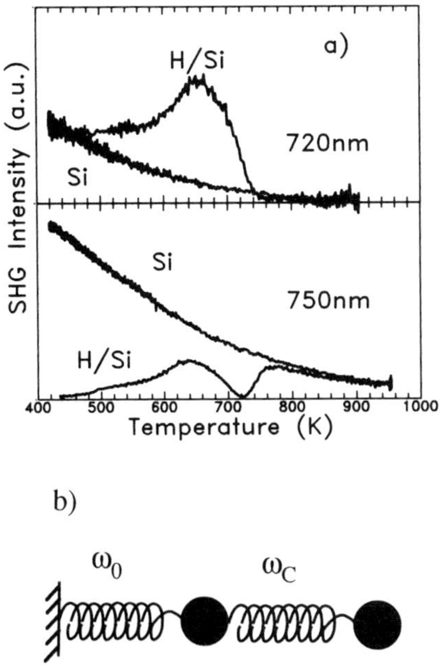

QTuC3 Fig. 2. (a) p-in/p-out signals at 720 and 750 nm for H-terminated and bare Si(001). (b) Schematic diagram of coupled-anharmonic-oscillator model.

(~800 K) the Si–H signal is stronger than the bare Si level at 720 nm but is weaker at 750 nm. Similar measurements at other wavelengths show that H/Si signals are stronger than bare Si signals above the bare Si two-photon resonance and are weaker below it; i.e., the resonance blue shifts. This is explained qualitatively by a classical coupled anharmonic oscillator model [Fig. 2(b)] in which oscillator frequencies ω_0 and ω_c represent electronic Si–Si and Si–H bond resonances, respectively. Compared to the uncoupled ω_0 oscillator, the calculated two-photon resonance of the coupled system is blue shifted by

$$\Delta\omega = \frac{\omega_0}{2}\left\{\frac{1}{\sqrt{2}}[(1 + 2\eta^2) + \sqrt{(1 + 2\eta^2)^2 - 4\eta^2}]^{1/2} - 1\right\},$$

where $\eta = \omega_0/\omega_c$.

**Department of Physics, University of Texas, Austin, Texas 78712*

1. W. Daum, H.-J. Krause, U. Reichel, and H. Ibach, Phys. Rev. Lett. **71**, 1234 (1993).
2. P. Lautenschlager, P. B. Allen, and M. Cardona, Phys. Rev. B **31**, 2163 (1985).
3. U. Höfer, L. Li, E. H. Ratzlaff, and T. F. Heinz, in *Quantum Electronics and Laser Science*, 1993 Technical Digest Series, Vol. 12 (Optical Society of America, Washington, D.C., 1993), p. 141.

QTuC4 **8:45 am**

Rotational anisotropy in third-harmonic generation from Si

Thomas Tsang, *Brookhaven National Laboratory, Upton, New York 11973*

The structural symmetry of Si surfaces has been probed by using second-harmonic generation (SHG) in reflection.[1] Subsequent experiments extended the technique to study detailed surface spectroscopy.[2] By using third-harmonic generation (THG) in reflection, ion-implanted Si has been studied, but with limited signal levels.[3] Because third-harmonic generation in reflection is dipole-allowed in the bulk, it was speculated that THG would unambiguously probe the structural characteristic of the bulk. In this experiment we demonstrated this technique by using Si as an example, and we contend that THG is more favorable than SHG, particularly when a femtosecond mode-locked Ti:sapphire laser is used.

The experiment is conducted by using a kilohertz-repetition-rate Ti:sapphire regeneratively amplified system. However, a self-mode-locked Ti:sapphire laser oscillator actually suffices to unfold the salient feature of the THG in reflection from Si. The femtosecond (250 fs for the amplifier and 100 fs for the oscillator) laser pulses are focused to a spot size of 4×10^{-3} cm^2 at a 17° angle of incidence on the Si wafers in air. After use of appropriate filters, the harmonic photons are collected by a monochromator and are processed with electronics. It is believed that because of the much larger excitation volume in the bulk than on the surface and because of the nature of the high peak power from a femtosecond laser source, the normally-much-weaker THG process dominates over SHG in reflection not only for Si but also for many other materials. Figure 1 shows the detected harmonic photons as the laser power is increased. THG clearly accentuates SHG at a remarkably low laser fluence of <1 GW/cm^2, which is accessible even by most femtosecond laser oscillators. By using the much stronger signal of the THG, the rotational anisotropy of Si(100) and Si(111) was investigated with a p-polarized fundamental input and p- and s-polarized THG output, $p(\omega) - p(3\omega)$ and $p(\omega) - s(3\omega)$, respectively. Figures 2 and 3 show the normalized signal for Si(100) and Si(111) as a function of the azimuth angle. For comparison, the corresponding SHG is also depicted. The fourfold and threefold symmetries of the Si(100) and Si(111) structures are clearly revealed, with a significantly better signal level in the THG than in the SHG. Although the symmetry properties are presented in both THG and SHG spectra, the p-in and p-out behaviors are markedly different. We address this issue by comparing the THG results to a theory. The easily accessible THG signal may provide a better diagnostic tool on the bulk than does the surface-sensitive SHG technique.

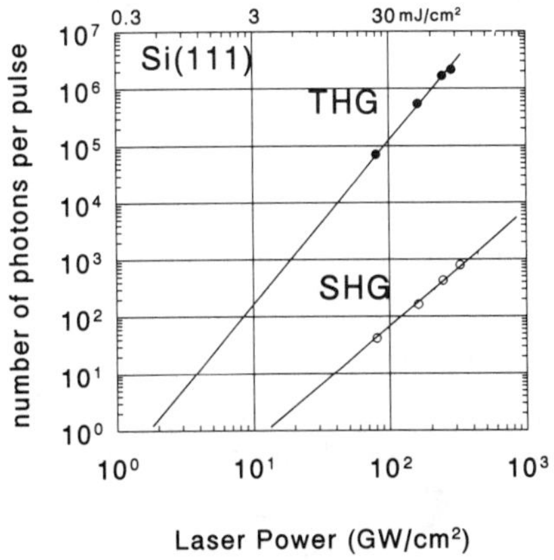

QTuC4 Fig. 1. Number of second- and third-harmonic photons generated on reflection from an Si(111) wafer.

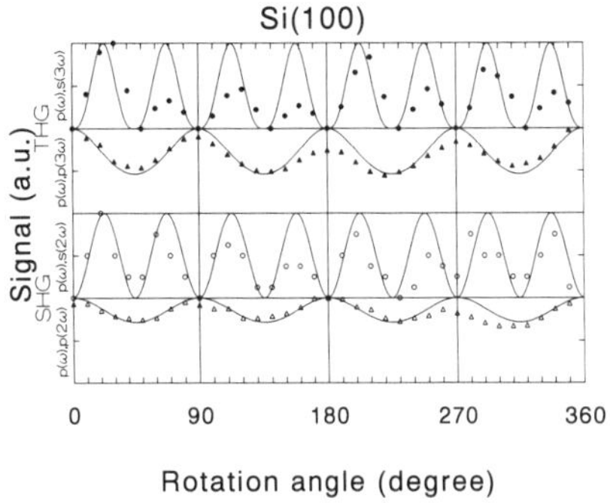

QTuC4 Fig. 2. Anisotropy response in the second and third harmonics from an Si(100) wafer.

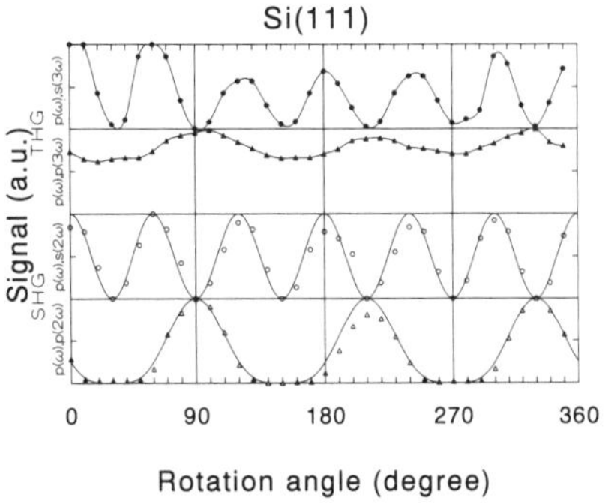

QTuC4 Fig. 3. Anisotropy response in the second and third harmonics from an Si(111) wafer.

1. H. W. Tom, T. F. Heinz, and Y. R. Shen, Phys. Rev. Lett. **51**, 1983 (1983).
2. Y. R. Shen, Nature **337**, 519 (1989).
3. D. J. Moss, H. M. van Driel, and J. E. Sipe, Appl. Phys. Lett. **48**, 1150 (1986); J. E. Sipe, D. J. Moss, and H. M. van Driel, Phys. Rev. B **35**, 1129 (1987).

QTuC5 **9:00 am**

Static-field–enhanced second-harmonic generation from epitaxial gallium nitride

Dennis K. Wickenden, Joseph A. Miragliotta, *Johns Hopkins University, Applied Physics Laboratory, Johns Hopkins Road, Laurel, Maryland 20723*

Second-harmonic (SH) generation has been used to probe the effects of a static electric (dc) field on the optical nonlinearities in epitaxial GaN, a wide-band-gap III–V semiconductor. Investigations were performed at SH photon energies near the absorption edge (E_{gap} = 3.4 eV) with a dc-field amplitude in the GaN sample of the order of 10^4 to 10^5 V/cm. Under these conditions, the reflected SH intensity exhibited a significant increase relative to the nonlinear response at the zero-static-field condition at SH photon energies above the band gap. These nonlinear results, coupled with complementary linear electroreflectance data, suggest a $\chi^{(3)}$ contribution to the nonlinear polarization at 2ω as the primary enhancement mechanism.

Experiments were performed in an aqueous electrochemical environment in which dc-field variations in the GaN film were potentiostatically controlled by means of the applied potential difference between the semiconductor sample and reference electrode. Surface dc-field amplitudes and sample dopant concentrations were determined from ac capacitance-applied potential measurements that confirmed the existence of an applied-potential-dependent Schottky barrier at the semiconductor/electrolyte interface. For the SH measurements, at 10-Hz pulsed dye laser (680–760 nm) was used as the fundamental generating laser source ω, which was p-polarized and was incident at an angle of 60° at the GaN surface. The specularly reflected SH signal was detected with a photomultiplier tube, was analyzed with a gated boxcar integrator, and was referenced to the SH signal generated in a nonlinear standard material (quartz).

The results in Fig. 1 illustrate the effects of applied potential, i.e., dc field, on the SH reflected intensity from the GaN/liquid interface. The scans are plotted versus SH wavelength, and all applied potentials are with respect to a saturated calomel electrode. The lower spectrum, recorded at an applied potential of −0.69 V, represents the flat-band, or zero-dc-field, limit. Note that there is very little structure on the scan at SH wavelengths less than the band-gap position [E_{gap} ~ 3.4 eV (365 nm)]. When the sample is reverse biased to a potential 0.05 V positive of this value (surface dc field ~ 20 kV/cm), a new feature emerges in the nonlinear response just above the band-edge position. The intensity of this peak increases linearly with the applied potential with a slight broadening in linewidth. In addition to this primary feature, a higher-energy, weaker peak in the SH spectra displays a behavior that is analogous to that of the lower-energy peak.

The behavior observed in the SH data is, in fact, similar to dc-field-dependent spectral variations that are commonly observed in linear electroreflectance measurements of direct-band-gap semiconductors. In the linear-reflectance case, the dc field modulates the dielectric function [$\varepsilon(2\omega)$] of the semiconductor near the absorption edge and generates above-band-gap oscillations in both linear-reflectance and absorption spectra. We have performed linear-electroreflectance investigations on the GaN sample of this study and have determined the dc-field-induced variations in $\varepsilon(2\omega)$ to be less than 0.02 for energies above the band-edge position. Attempts to numerically simulate the data in Fig. 1 with a dielectric variation of this amplitude show that dielectric variation cannot be responsible for the significant increases in the SH intensity. Therefore, we believe the SH enhancement mechanism is a $\chi^{(3)}$ contribution term that generates an additional source term at 2ω through the coupling of two optical fields at frequency ω and one static field. This nonlinearity would manifest itself as a reflected SH intensity that scales quadratically with applied static field (linearly with applied potential), such as the results shown in Fig. 1. Attempts are currently in progress to determine the dispersion profile of the $\chi^{(3)}$ susceptibility under these conditions.

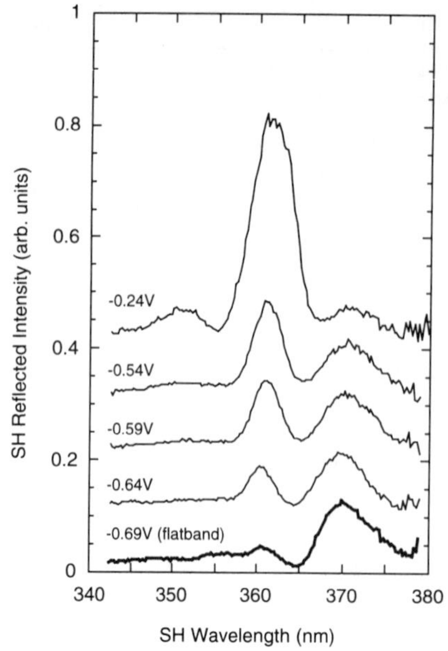

QTuC5 Fig. 1. SH spectra from a GaN film under potentiostatic control. Applied potentials are with respect to a saturated calomel electrode. Spectrum at −0.69 V represents the flat-band, or zero-static-field, result. Potentials positive of this value create reverse-bias conditions, i.e., large static fields in the space-charge region of the semiconductor. The absorption edge is located at ~365 nm.

QTuC6 **9:15 am**

Second-order susceptibility of the CH bond and its effect on determination of interfacial molecular orientation

M. S. Yeganeh, S. R. Hatch, S. M. Dougal, R. S. Polizzotti, R. Rabinowitz, *Exxon Research and Engineering Co., Route 22 East, Annandale, New Jersey 08801*

Liquid/solid and vapor/solid interfaces play critical roles in a wide range of commercially important processes such as corrosion, lubrication, and catalysis. Despite the importance of these processes there still does not exist a complete microscopic understanding of their chemistry and physics. Recently, surface sensitive probes such as sum-frequency generation (SFG) and second-harmonic generation (SHG) have been developed, which can significantly enhance our knowledge and characterization of such interfaces.

Previously, infrared-visible SFG was used to determine the interfacial molecular orientation of the CH_3CN/ZrO_2[1] and octadecyltrichlorosilane/silica[2] systems. In these experiments the vibrational SFG intensities generated by the CH_3 group were measured as a function of IR photon energy for two polarization configurations. The information obtained from the spectra was used to infer the molecular orientation at the interface. The average molecular orientation not only depends on the experimental value of $\sqrt{I_{ssp}/I_{sps}}$ but also depends on the ratio of the CH bond hyperpolarizability in (⊥, ⊥, ∥) to (∥, ∥, ∥), r_{CH}. Here I_{ssp} is the SFG intensity with SFG, visible and IR beams in s, p and p-polarization configurations, respectively. The value of r_{CH} was assumed to be zero in the previous work. In this report we present the first measurement of r_{CH} in the CH_3CN/Al_2O_3 system. Our results show that the CH hyperpolarizability ratio is not zero. As a result, using the assumed value of zero leads to a large error in the determination of the interfacial molecular orientation.

We have measured the interfacial vibrational spectra of CH_3CN/Al_2O_3 in a total-internal-reflection geometry using a tunable infrared and a visible (532 nm) laser beam. In this arrangement one face of a 60 × 60 × 60 Al_2O_3 prism was exposed to liquid acetonitrile at room temperature.

The vibrational SFG spectra of the interfacial acetonitrile molecules in ssp and sps polarization are shown in Fig. 1. The solid lines, depicted in Fig. 1, are the theoretical fits using a standard model consisting of a constant background and two Lorentzian resonances. The fitting parameters establish a numerical relationship between the r_{CH} and the tile angle, θ. In addition, we have determined the tilt angle, θ, using the vibrational SFG spectra of the C≡N stretching mode, shown in Fig. 2. As

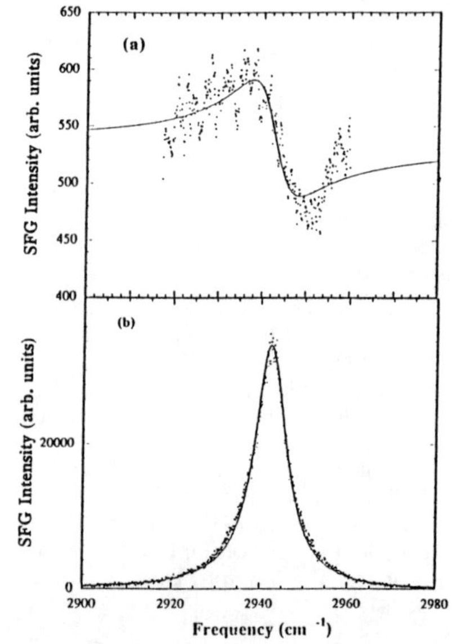

QTuC6 Fig. 1. SFG spectra of the CH_3 stretching region of CH_3CN/Al_2O_3 system in (a) sps- and (b) ssp-polarization configuration. The feature at 2943 cm^{-1} is assigned to the symmetric stretch of the CH_3 group.

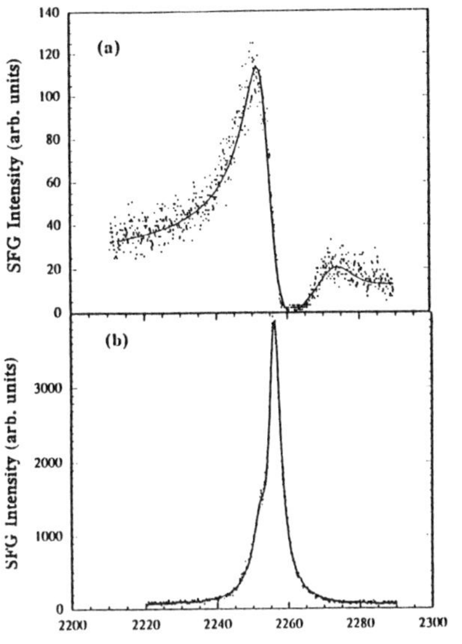

QTuC6 Fig. 2. SFG spectra of the CN stretching region of CH_3CN/Al_2O_3 system in (a) sps- and (b) ssp-polarization configuration. The resonance frequency at 2256 cm^{-1} is generated by the symmetric stretch of the C≡N bond and the feature at 2253 cm^{-1} is assigned to acetonitrile dimers at the interface.

in the case of CH_3 group, one can write a similar relationship between the tilt angle, the hyperpolarizability ratio of the CN bond, r_{CN}, and experimental ratio of $\sqrt{I_{ssp}/I_{sps}}$ at ~2555 cm^{-1}. The value of r_{CN}, in this case, can be deduced from Raman depolarization measurements. Using the fitting parameters and the known value of r_{CN}, we have determined that the tilt angle is 15° ± 1°. This value of the tilt angle leads to a CH hyperpolarizability ratio of, r_{CH} = 0.35 ± 0.02. The large value of r_{CH} in the CH_3CN/Al_2O_3 system may result from the strong permanent dipole moment of the acetonitrile molecule and may not be representative of other systems. The effect of the molecular environment, surrounding the CH_3 group, on the value of r_{CH} is under investigation in our laboratory.

1. S. R. Hatch, R. S. Polizzotti, S. Dougal, and P. Rabinowitz, Chem. Phys. Lett. **196**, 97 (1992).
2. P. Guyot-Sionnest, R. Superfine, J. H. Hunt, and Y. R. Shen, Chem. Phys. Lett. **144**, 1 (1988).

QTuC7 9:30 am

Second-harmonic generation with electrically controllable phases by using nemantic-liquid-crystal cells

R. Stolle, G. Marowsky, *Göttingen Laser-Laboratory, Hans-Adolf-Krebs-Weg 1, 37077 Göttingen, Germany*

For second-harmonic generation (SHG), the phase difference ΔΦ between the fundamental and the harmonic is determined by the phase of the underlying complex-value second-order susceptibility $\chi^{(2)}$. Hence, by controlling the phase of $\chi^{(2)}$, one should be able to control ΔΦ. This has been demonstrated by using nematic-liquid-crystal (NLC) cells with an externally applied dc field of alternating polarity. Their effective susceptibility $\chi^{(2)}_{eff}$ consists of an electrically controllable bulk contribution $\chi^{(2)}_{lc}$ proportional to the applied voltage U and a voltage-independent background $\chi^{(2)}_{const}$. This background contribution is due to the adjacent cell windows, which were IOT-coated and carried a thin rubbed polyimide layer for molecular alignment. We demonstrated the electrically controllable phase using the second-harmonic-interference technique, originally introduced by Bloembergen and Pershan.[1] Two types of NLC cells were studied: pure NLC cells and an NLC cell with an embedded guest of high optical nonlinearity.

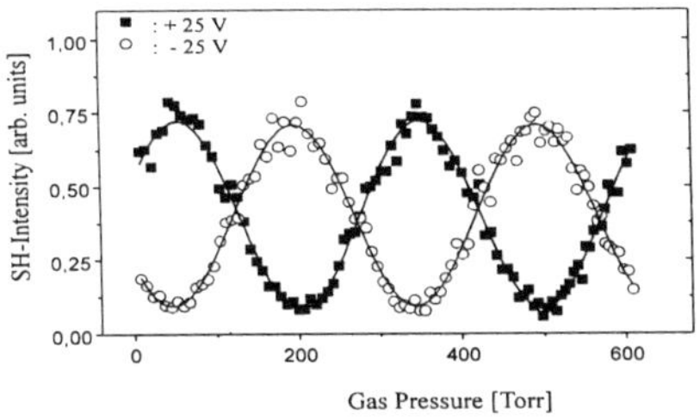

QTuC7 Fig. 1. Dependence of total SH signal on variation of the pressure in the gas cell for an applied voltage ±25 V.

Figure 1 shows as an example the oscillations of the total second-harmonic intensity as a function of variation of the gas pressure in a drift cell introduced between two interfering SH sources. The details of this approach for measuring the (relative) phases of the SH signal are reported in Refs. 2 and 3. The interference pattern in Fig. 1 clearly shows the sign reversal of the induced nonlinear polarization of the nematic mixture after a change in polarity of the externally applied electric field. This effect should, in principle, permit the realization of a hybrid electro-optical device based on second-harmonic interference.

1. N. Bloembergen and P. S. Pershan, Phys. Rev. **128**, 606 (1962).
2. E. Schwarzberg and G. Berkovic, Nonlinear Opt. **5**, 62 (1993).
3. E. Scharzberg, G. Berkovic, and G. Marowsky, "Nonlinear interferometry and phase measurements for surface second harmonic generation in a dispersive geometry," in press.

QTuC8 9:45 am

Interface states studied by second-harmonic generation from metal/GaAs junctions

J. Qi, W. Angerer, M. S. Yaganeh, A. G. Yodh, W. M. Theis,* *Department of Physics, University of Pennsylvania, Philadelphia, Pennsylvania 19104*

The identification of interface states and their role in Fermi-level pinning has attracted great attention.[1] By comparison to traditional noncontact methodologies, second-harmonic generation provides the intrinsic interface sensitivity and long penetration depth necessary for studies of buried interfaces.[2]

We have used second-harmonic and sum-frequency generation to spectroscopically study interface states at metal/GaAs(001) junctions with different bulk doping and interface preparation. The metal overlayers were Au and As. The Fermi level in n-type metal/GaAs(001) systems has been found to be pinned at 0.5–0.75 eV above the valance-band maximum.[1] This pinning is believed to be a result of defect states or metal-induced states. Our second-harmonic-generation (SHG) spectra reveal a strong resonance state within this energy range on Au/GaAs (n-type) and As/GaAs (n-type) samples. We have found that these spectra are quite sensitive to interface preparation and bulk doping.

The samples consist of 80-Å Au and As films grown by molecular-beam epitaxy (MBE) on an n-type and p-type MBE-grown GaAs(001) substrate. Tunable infrared second-harmonic-generating pulses were derived from a 10-ns Nd:YAG-pumped optical parametric oscillator (OPO) and amplifier system based on KTP. The bulk GaAs(001) has a zinc-blende structure and carries a single nonzero second-order susceptibility $\chi^{(2)}_{123}$, whose contribution to the output SH radiation is highly anisotropic. By appropriate adjustment of the input/output light polarizations and the sample orientation,[2] we have measured the spectral properties of predominantly $\chi^{(2)}_{333}$ and $\chi^{(2)}_{123}$ of this system for output SHG photons in the range 1.36–1.56 eV. By mixing OPO output photons with 1.06-μm Nd:YAG pulses, we have also obtained sum-frequency-generation (SFG) spectra in the range 0.68–0.78 eV for one-photon energy. We anticipate that $\chi^{(2)}_{333}$ preferentially probes states at the metal–semiconductor interface.[2]

In Fig. 1 we show the measured interface SHG ($\chi^{(2)}_{333}$) spectra on Au/GaAs(001) with As-rich-interface and As-capped GaAs(001) samples. Both types of interface spectra have strong resonance peaks with one-photon energy of ~0.72 eV. Separate SFG experiments confirmed that the resonance peak is a one-photon resonance. This resonance is consistent with suggested in-

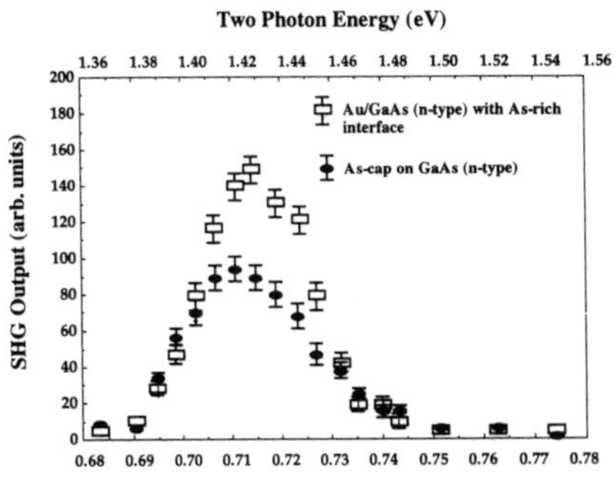

QTuC8 Fig. 1. Normalized SHG intensity from the interface ($\chi^{(2)}_{333}$) as a function of one-photon (ω) and two-photon (2ω) energies. The signals were detected in the p-in/p-out polarization configuration with the crystalline [100] axis parallel to the plane of incidence. A strong resonance peak was observed around 0.72 eV for both Au/GaAs (n-type) and As/GaAs (n-type) samples.

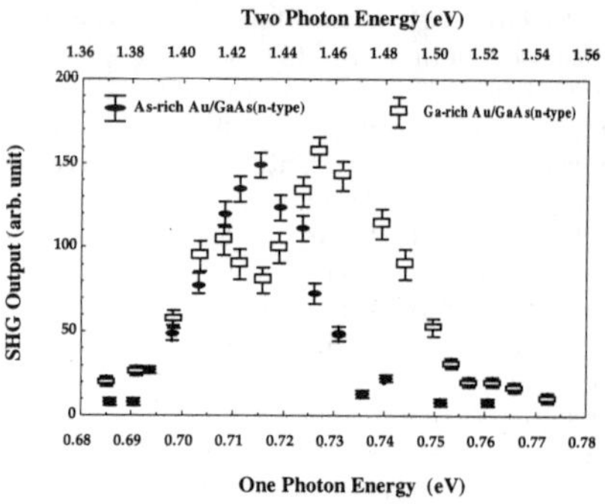

QTuC8 Fig. 2. Normalized SHG intensity from the interface $\chi^{(2)}_{333}$ as a function of one-photon (ω) and two-photon (2ω) energies. A shifted and perhaps multiple resonance was observed for a Ga-rich interface Au/GaAs (n-type) sample in comparison to an As-rich interface Au/GaAs (n-type) sample.

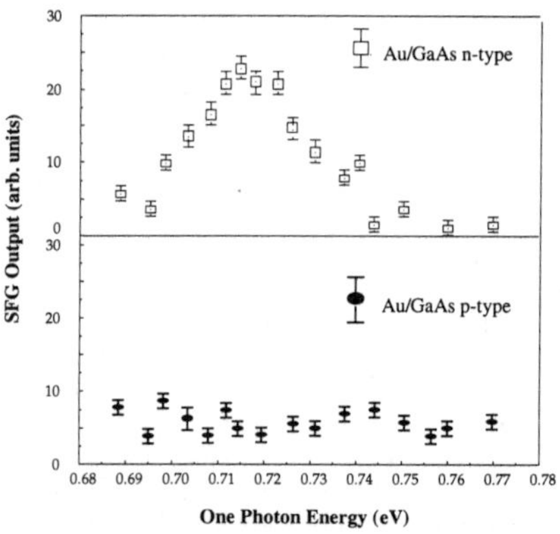

QTuC8 Fig. 3. SFG intensity from the interface $\chi^{(2)}_{333}$ as a function of the one-photon (OPO output photon) energy. The SFG experiments were performed by mixing the infrared OPO output photon with 1.06-μm Nd:YAG pulses. No resonance peak was observed for a Au/GaAs (p-type) sample in comparison to an Au/GaAs (n-type) sample.

terfacial midgap states.[1] The resonance was not observed in the bulk $\chi^{(2)}_{123}$ SH spectra at these same energies. Figure 2 demonstrates that the resonance is sensitive to interface preparation. In particular, we exhibit the resonance in n-type Au/GaAs systems with As- and Ga-rich interfaces, and we see that the Ga-rich interface appears to have multiple resonance structures. Finally, in Fig. 3 we show that SFG measurements in p-type Au/GaAs do not exhibit a resonance.

At present we believe that these one-photon resonances arise from the transitions starting in the midgap interface states and ending at the conduction-band edge. This is consistent with the observation of different Fermi-level pinning and, therefore, different occupied midgap states in p- and n-type metal/GaAs systems.[1] Our SHG spectra also suggest that this midgap state is related to interface As and Ga concentrations.

Naval Air Warfare Center, China Lake, California 93555

1. W. E. Spicer, Z. Lilient-Weber, E. Weber, N. Newman, T. Kendelewicz, P. Cao, C. McCants, P. Mahowald, K. Miyano, and I. Lindau, J. Vac. Sci. Technol. **B6**, 1245 (1988); R. E. Viturro, J. L. Shaw, C. Mailhiot, L. J. Brillson, N. Tache, J. McKinley, G. Margaritondo, J. M. Woodall, P. D. Kirchner, G. D. Pettit, and S. L. Wright, Appl. Phys. Lett. **52**, 2052 (1988).
2. T. F. Heinz, F. J. Himpsel, E. Palange, and E. Burstein, Phys. Rev. Lett. **63**, 644 (1989); M. S. Yeganeh, J. Qi, A. G. Yodh, and M. C. Tamargo, Phys. Rev. Lett. **68**, 3761 (1992); G. Lupke, D. J. Bottomley, and M. H. van Driel, Phys. Rev. B **47**, 10389 (1993).

JTuB Tutorial — 8:00 am

Hyatt E/F

Quantum Well Heterostructure Lasers

Daniel Chemla, *University of California, Berkeley, Presider*

JTuB1 (Tutorial) — **8:00 am**

Quantum well heterostructure lasers

James C. Coleman, *University of Illinois, Urbana*

Arguably the most significant advance in the 30-year history of the semiconductor laser was the introduction of the quantum well heterostructure. The attractive features of the physics of the quantum size effect are the basis for the incorporation of these structures in nearly all modern semiconductor laser diodes. Quantum well lasers have lower threshold current density, higher gain, higher efficiency, and in some cases an increased range of available emission wavelengths. Small variations, such as the incorporation of a modest amount of elastically accommodated strain, can result in dramatic changes in the operation of these devices. Although some of the main aspects have been known for many years, quantum well lasers are still the subject of much current research and development. In this tutorial, we describe the basic operating principles of quantum well heterostructure lasers, and outline recent advances in the understanding and application of these remarkable devices.

JTuC Tutorial — 9:00 am

Hyatt E/F

Quantum Teleportation, Key Distribution, and Practical Quantum Cryptography

Malvin C. Teich, *Columbia University, Presider*

JTuC1 (Tutorial) — **9:00 am**

Quantum teleportation, key distribution, and practical quantum cryptography

Paul R. Tapster, *Royal Signals and Radar Est. Research, U.K.*

I will describe the principles behind quantum cryptography and key shring systems, which can be either based on sources of pairs of entangled particles (typically photons), or based on the preparation and measurement of a single particle in one of two incompatible basis states. The first experimental demonstration of these techniques was based on polarization states and was limited to communications over distances of a few meters. More recent experiments use interferometric measurements on long wavelength photons transmitted through standard communications optical fibers and have demonstrated quantum cryptography at distances up to 10 kilometers. The information transmitted via quantum cryptography is completely secure—no eavesdropper can acquire any of the information without being detected. However, the complexity of the equipment required, and the relatively low rate at which data can be transmitted, means that the most likely application of these techniques is in the generation and distribution of keys for use with conventional cryptography systems. The sources of entangled pairs of particles can also in principle be used to demonstrate quantum teleportation, in which an unknown quantum state is destroyed at one place, and recreated at a remote location. Unfortunately, it is not yet possible to perform this experiment in practice, but I will discuss the fundamental quantum mechanical implications of this interesting thought experiment.

QTuD 10:30 am Hyatt A

Spatial Solitons II

Barry Luther-Davies, *Australian National University, Presider*

QTuD1 (Invited) 10:30 am

Physics of dark solitons

Yuri S. Kivshar, *Optical Sciences Centre, Australian National University, ACT 0200 Canberra, Australia*

The main objective of this talk is to give a simple introduction into the physics and properties of "dark" solitons and related localized modes, which occur in different problems of nonlinear optics. Unlike conventional (bright) solitons, dark solitons are spatially localized modes (or pulses) that exist on a background wave with nonvanishing asymptoics, the latter being usually modulationally stable. In the other words, a 2D dark soliton is a phase kink that connects two plane waves with the same amplitudes but different phases. In practice, such solitons are created of a background beam (or pulse) of finite extent but much longer than the soliton itself.

From the physical point of view dark solitons can be viewed as the reflectionless modes of the optical waveguide they induce whereas bright spatial solitons are the bound modes.[1] In the case of two transverse degrees of freedom such solitons may be observed as dark soliton stripes or grids[2] with the properties similar to those of 2D dark solitons. Dark solitons of circular symmetry (optical vortex solitons)[3] have been predicted and shown to be stable, and they have been already observed experimentally in self-defocusing materials. On the other hand, dark soliton stripes are unstable to transverse long-wavelength modulations.

Along with the topics which were previously reviewed in the literature (see, e.g., the review paper,[4] the present talk also mentions very recent results including polarization domain walls, optical vortex solitons, ring dark solitons,[5] dark solitons in media with $\chi^{(2)}$ nonlinear susceptibility, etc.

In particular, in dispersive materials with quadratic nonlinearities dark solitons may exist only for the opposite dispersion of the fundamental and second harmonics. For a small phase mismatch of the wavevectors of two harmonics such solitons are shown to possess nonmonotonous (oscillating) tails, which make the interaction between the neighboring solitons nontrivial: For example, such solitons may form stable bound states of two (or more) dark solitons.

A link between a variety of optical problems and corresponding problems of the theory of a weakly compressible fluid and nonideal Bose gas will be mentioned as well.

1. A. W. Snyder, D. J. Mitchell, and B. Luther-Davies, J. Opt. Soc. Am B **10**, 2341 (1993).
2. G. A. Swartzlander, Jr., D. R. Andersen, J. J. Regan, H. Yin, and A. E. Kaplan, Phys. Rev. Lett. **66**, 1583 (1991).
3. A. W. Snyder, L. Poladian, and D. J. Mitchell, Opt. Lett. **17**, 789 (1992); G. A. Swartzlander, Jr. and C. T. Law, Phys. Rev. Lett. **69**, 2503 (1992).
4. Yu. S. Kivshar, IEEE J. Quantum Electron. **29**, 250–264 (1993).
5. Yu. S. Kivshar and X. Yang, Phys. Rev. E **50**, R40 (1994).

QTuD2 (Invited) 11:00 am

Photorefractive dark solitons as waveguides

Matthew E. Morin, Galen C. Duree, Jr., Gregory J. Salamo, Mordechai Segev,* Bruno Crosignani,** Paolo Di Porto,** Amnon Yariv,† Edward J. Sharp,‡ Ratnakar R. Neurgaonkar,§ Department of Physics, University of Arkansas, Fayetteville, Arkansas 72701

Recently, we reported on the prediction[1,2] and observation[3] of a new type of spatial soliton. For this new spatial soliton, compensation for diffraction is accomplished through the photorefractive effect. In this case, a Gaussian beam incident on a photorefractive crystal is trapped at a diameter of about 30 μm and propagates through the crystal while maintaining its spatial shape and defying the rules of diffraction. We have referred to our observation as a photorefractive bright soliton.

Physically, this phenomenon can be understood by picturing diffraction of a Gaussian beam as if each Fourier component of the beam propagates independently. Because each component has a different k-vector projection along the propagation direction, the relative phase between Fourier components changes as a function of propagation distance z. Consequently, the sum of the Fourier components produces a different waveform at each z along the propagation direction. Thus, diffraction can be compensated if a material-index distribution can produce a k-vector distribution in which a Fourier component has the same magnitude projection along the propagation direction, thus ensuring a constant-phase relationship between Fourier components. For photorefractive solitons, the induced index change needed to produce the proper k-vector distribution arises from the gratings formed between all pairs of Fourier components through two-beam photorefractive coupling.

Our observations of bright photorefractive solitons have led to more recent observations of dark planar and vortex photorefractive solitons.[4,5] Here we report on the experimental observation of the use of a dark planar soliton to act as a waveguide for a second laser beam, which is injected into the dark notch of the planar soliton. To produce the planar dark soliton, we used the output of an argon-ion laser (Fig. 1) with a thin glass slide inserted into half of the beam to produce a wave front at the SBN photorefractive crystal that had the proper π phase jump at its center. Figure 2 shows the waveform at the output crystal face with and without dark-soliton formation. The external field required to trap the dark soliton was −400 V/cm along the c axis of the crystal. The width (FWHM) of the dark notch was 33 μm at the entrance face of the crystal. At the exit face, the dark notch was 49 μm without the applied voltage. With the applied voltage, the dark soliton formed about halfway through the crystal, with a width of 10 μm. Once the planar dark soliton was formed, the argon beam was blocked and a He–Ne beam was focused and injected into the notch. As shown in Fig. 3, the He–Ne Gaussian beam diameter was 22 μm at the entrance face of the crystal. The He–Ne beam diameter at the exit face was 105 μm without the use of a dark planar soliton and was 18 μm with the use of the dark planar soliton. Alignment of the He–Ne beam was found

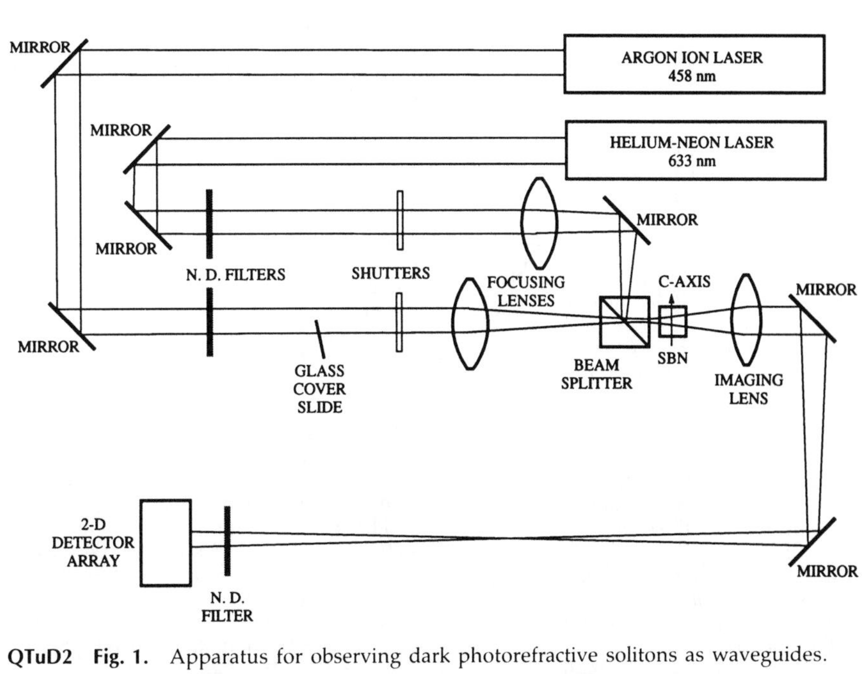

QTuD2 Fig. 1. Apparatus for observing dark photorefractive solitons as waveguides.

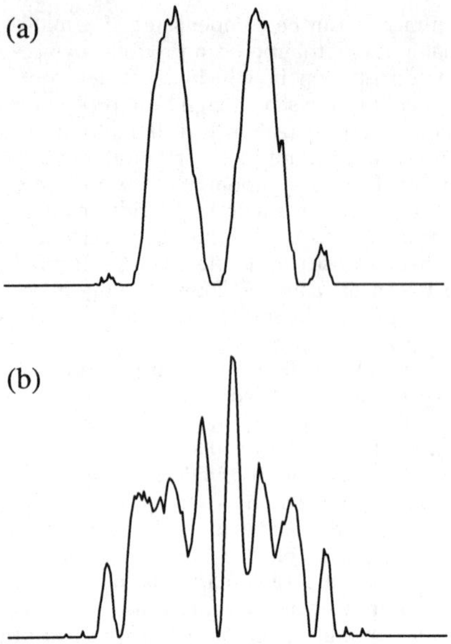

QTuD2 Fig. 2. Output waveform showing argon horizontal beam profile with dark notch (a) diffracted without applied voltage and (b) trapped with applied voltage.

QTuD2 Fig. 3. (a) Input horizontal profile of He–Ne beam to be injected into the dark notch of the soliton. Output horizontal profile of He–Ne beam (b) without and (c) with soliton-waveguide formation by argon laser beam.

to be critical, much like the coupling of a laser beam into an optical fiber.

*Department of Electrical Engineering, Princeton University, Princeton, New Jersey 09544
**Department of Physics, University of Aquila, 67010 L'Aquila, Italy
†California Institute of Technology, 128-95, Pasadena, California 91125
‡U.S. Army Research Laboratory, Fort Belvoir, Virginia 22060
§Rockwell International Science Center, Thousand Oaks, California 91360

1. M. Segev, B. Crosignani, A. Yariv, and B. Fischer, Phys. Rev. Lett. **68**, 923 (1992).
2. B. Crosignani, M. Segev, D. Engin, P. DiPorto, A. Yariv, and G. Salamo, J. Opt. Soc. Am. B **10**, 446 (1993).
3. G. Duree, J. L. Schultz, G. Salamo, M. Segev, A. Yariv, B. Crosignani, P. DiPorto, E. Sharp, and R. R. Neurgaonkar, Phys. Rev. Lett. **71**, 533 (1993).
4. G. Duree, M. Morin, G. Salamo, M. Segev, A. Yariv, B. Crosignani, P. DiPorto, and E. Sharp, submitted to Phys. Rev. Lett.
5. G. Duree, M. Morin, G. Salamo, M. Segev, A. Yariv, B. Crosignani, P. DiPorto, and E. Sharp, in *OSA 1994 Annual Meeting*, 1994 Technical Digest Series (Optical Society of America, Washington, D.C., 1994).

QTuD3 **11:30 am**

Dark photovoltaic spatial solitons

M. Taya, M. C. Bashaw, M. M. Fejer, M. Segev,* G. C. Valley,** B. Crosignani,† *Center for Nonlinear Optical Materials, Stanford University, Stanford, California 94305*

Spatial solitons, which propagate without variation in their transverse intensity profile, occur when nonlinear self-defocusing (focusing) exactly compensates the diffraction of dark (bright) amplitude profiles. Kerr-like media support spatial solitons,[1] typically requiring large powers (greater than watts and up to megawatts) and displaying widths that depend on intensity. Photorefractive media under the influence of an external electric field support two types of solitons. Quasi-steady-state, nonlocal photorefractive solitons occur in a transient (but long-lived) regime after the index perturbation forms but before background conductivity screens the external field[2]; they are intensity independent, permitting stable self-trapping in two dimensions. Photorefractive-screening solitons occur at steady state from nonuniform screening of an external electric field.[3,4]

We report here the experimental observation of photovoltaic spatial solitons and show that they display a number of properties predicted theoretically. At steady state the internal photovoltaic currents are balanced by drift resulting from an accumulated internal electric field; no external field is required. The index perturbation is due to the Pockels effect and can support dark or bright solitons.[5] We present experimental results on dark planar photovoltaic solitons (trapped in one transverse dimension) in a crystal of $LiNbO_3$, and we show that the associated index perturbation can guide a second beam of light.

Planar, scalar photovoltaic solitons result from an optically local index perturbation of the form $\delta n(x) = a(I/I_{dark})/(1 + I/I_{dark})$, where $I = I(x)$ is the intensity, I_{dark} is the intensity required to achieve a photoconductivity equal to the dark conductivity (dark irradiance), and $a = n_0^3 r E_p/2$ (E_p is the photovoltaic field parameter, r is the effective Pockels coefficient, and n_0 is the background index). This perturbation is mathematically equivalent to saturable Kerr nonlinearities. We have solved the wave equation for this nonlinearity for dark and bright solitons, and we have shown that observable soliton widths can occur at two different background intensities. Photovoltaic solitons are physically distinct from Kerr solitons because a is derived from the photovoltaic and Pockels tensors. Therefore, photovoltaic solitons strongly depend on the direction of propagation, the beam polarization, and the intensity gradient. Typically, the dark irradiance is of the order of microwatts, and we operate in the saturated regime.

To excite a dark soliton we generate a dark notch by imaging a phase step due to a glass plate put across a beam of 488-nm light onto a $LiNbO_3$ sample. The beam propagates perpendicular to the c axis, and the amplitude gradiant occurs essentially along the c axis. The input power of the 488-nm beam is 20 mW, and the average intensity is ~10 mW/cm². The dark notch is imaged onto the front surface of the crystal with a width of ~20 μm. Immediately after the beam is turned on, no index perturbation exists and the notch at the exit face diffracts significantly [Fig. 1(a)]. At steadystate, soliton behavior results in the balancing of diffraction by self-defocusing [Fig. 1(b)].

In the presence of the index perturba-

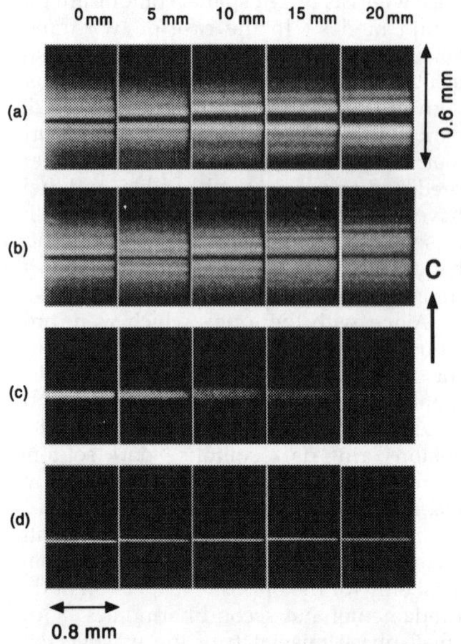

QTuD3 Fig. 1. Properties of a soliton launched into a (1 cm × 1 cm × 5 cm) sample of $LiNbO_3$. A mosaic of the side view of the profile of the 488-nm beam with a phase step is shown for (a) the initial condition before soliton formation and (b) steady state after soliton formation. A corresponding mosaic of the side view of the profile of a separate 514-nm beam is shown (c) before soliton formation and (d) as guided by the index perturbation left behind by the 488-nm soliton. The images are obtained at 5-mm intervals and show a width of 800 μm.

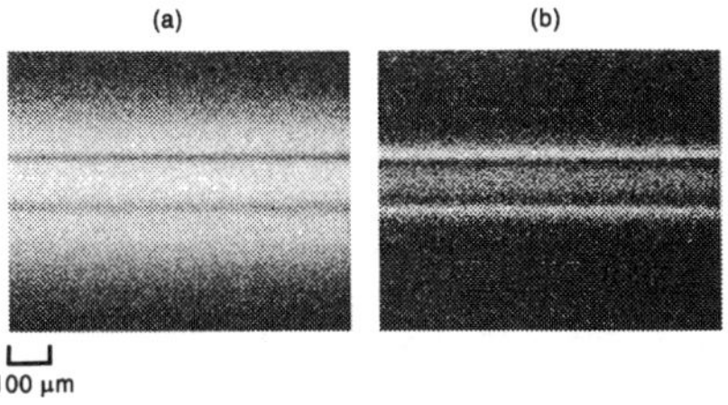

QTuD3 Fig. 2. Side view of two dark-soliton profiles in a (1 cm × 1 cm × 5 cm) sample of $LiNbO_3$, generated by a dark mask with no initial phase change. The images are obtained at a propagation distance of 2 cm, showing (a) a pair of dark solitons of 488-nm light and (b) a pair of guided 514-nm beams originating from one focused beam.

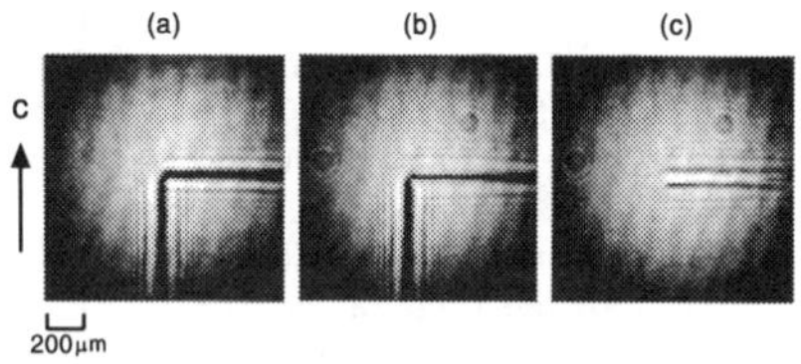

QTuD3 Fig. 3. Tensorial behavior of dark photovoltaic solitons illustrated in a (1 cm × 1 cm × 1 cm) sample of $LiNbO_3$. Shown are (a) the initial beam profile when the corner of the cover slip is in the beam path, (b) the beam profile at steady state, and (c) the beam profile immediately after removal of the cover slip at steady state.

tion left behind by a soliton at 488 nm, a second beam of light at 514 nm that would otherwise diffract [Fig. 1(c)] is guided and confined to the region of the dark notch [Fig. 1(d)]. Substituting an opaque strip for the phase step results in the propagation of pairs of solitons [Fig. 2(a)], whose associated index perturbations can be used to generate fan-in and fan-out configurations, such as Y and higher-order junctions [Fig. 2(b)]. The tensorial nature of photovoltaic solitons favors their formation when the index gradient is along the *c* axis, as indicated by the corner phase step in Fig. 3.

In summary, we have observed dark photovolotaic spatial solitons in $LiNbO_3$ and have described their distinct tensorial properties. We have further demonstrated their waveguiding properties in bulk media.

**Department of Electrical Engineering, Princeton University, Princeton, New Jersey 08544*
***Hughes Research Laboratories, Malibu, California 90265*
†Department of Physics, University of Aquila, L'Aquila, Italy, and Ugo Bordoni Foundation, Rome, Italy

1. R. Y. Chiao, E. Garmire, and C. H. Townes, Phys. Rev. Lett. **13**, 479 (1964); V. E. Zakharov and A. B. Shabat, Sov. Phys. JETP **34**, 62 (1972); A. Barthelemy, S. Maneuf, and C. Froehly, Opt. Commun. **55**, 201 (1985); J. S. Aitchinson, A. M. Weiner, Y. Silberberg, M. K. Oliver, J. L. Jackel, D. E. Leaird, E. M. Vogel, and P. W. Smith, Opt. Lett. **15**, 471 (1990).
2. M. Segev, B. Crosignani, A. Yariv, and B. Fischer, Phys. Rev. Lett. **68**, 923 (1992); B. Crosignani, M. Segev, D. Engin, P. DiPorto, A. Yariv, and G. Salamo, J. Opt. Soc. Am. B **10**, 446 (1993); G. Duree, J. L. Shultz, G. Salamo, M. Segev, A. Yariv, B. Crosignani, P. DiPorto, E. Sharp, and R. R. Neurgaonkar, Phys. Rev. Lett. **71**, 533 (1993); M. Segev, A. Yariv, B. Crosignani, P. DiPorto, G. Duree, G. Salamo, and E. Sharp, Opt. Lett. **19**, 1195, 1296 (1994).
3. M. Segev, G. C. Valley, B. Crosignani, P. DiPorto, and A. Yariv, "Steady-state spatial screening solitons in photorefractive materials with external applied field," Phys. Rev. Lett., in press.
4. Self-focusing related to soliton formation has been observed in bismuth titanium oxide by M. D. Iturbe Castillio, P. A. Marques Aguilar, J. J. Sanches-Montdragon, S. Stepanov, and V. Vysloukh, Appl. Phys. Lett. **64**, 408 (1994).
5. G. C. Valley, M. Segev, B. Crosignani, A. Yariv, M. M. Fejer, and M. C. Bashaw, "Dark and bright photovoltaic spatial solitons," Phys. Rev. A, in press.

QTuD4 **11:45 am**

Wave guiding by optical-vortex-soliton arrays

C. T. Law, G. A. Swartzlander, Jr.,*
Department of Electrical Engineering and Computer Science, University of Wisconsin, Milwaukee, Wisconsin 53201

Formation of the optical-vortex soliton (OVS) has been demonstrated in a self-defocusing Kerr nonlinear medium.[1] The OVS has a cylindrically symmetric dark core surrounded by a bright background. This nonlinear wave is a soliton because its dark core remains stable and stationary after a long propagation distance. The OVS also has the distinctive feature of a vortex characterized by a separable function of the azimuthal coordinate ϕ: $\exp(iM\phi)$, where $M = \pm 1, \pm 2, \ldots$ is the so-called topological charge. The OVS has promising applications in the area of light–light control. We address here two issues related to the use of OVS arrays in this area, the effects of background modulation and of cross coupling between OVS's. In a self-defocusing Kerr nonlinear medium, the refractive index around an OVS varies an $n(x, y) = n_0 - |n_2|I_\infty f(x, y)$, where $f(x, y)$ is the intensity profile of the OVS, (x, y) are transverse coordinates with the origin as the center of the OVS, I_∞ is the intensity of the background, n_0 is the linear refractive index, and n_2 is the nonlinear refractive index. Because the center of the vortex-soliton core has zero intensity, the refractive index in this region is equal to n_0. The surrounding bright area has a smaller refractive index, $n_0 - |n_2|I_\infty$. Therefore, the vortex filament can serve as a graded-index fiber waveguide for a probe beam of different frequency[2] or different polarization.[3] By perturbing the background intensity with another optical beam, one can provide modulation and amplification to the guided wave.[2] The change in coupling coefficient of the OVS core, the light-induced wave guiding, is caused by the decrease in core size with the increasing background intensity. We numerically simulate the propagation of guided beams in an OVS array, which consists of one guided beam for each OVS. We focus on the cases with all OVS's having topological charges $M = \pm 1$, where the neighboring OVS's can have either the same charge or opposite charges. We calculate wave guiding by OVS's when background intensity is modulated in the middle of the propagation path. If the separation between OVS's is large, this situation is effectively the same as the coupling efficiency of an OVS being modulated by another optical beam functioning as a controlling signal. When the modulation intensity is very high and the core size becomes very small, a large portion of the guided wave can be shaved away. A similar effect on the background intensity can occur when several OVS's are closely packed. However, the interactions among OVS's become significant in this case. Because OVS's obey the principles of hydrodynamics, they can cause translational and rotational motions in adjacent OVS's. If these motions are strong enough, the guided beam can be decoupled from the OVS. These and other results demonstrating the wave guiding in optical vortex soliton arrays are presented.

This work was supported by the U.S. Naval Research Laboratory and the Pittsburgh Supercomputer Center.

**Department of Physics, Worcester Polytechnic Institute, Worcester, Massachusetts 01609-2280*

1. G. A. Swartzlander, Jr., and C. T. Law, Phys. Rev. Lett. **69**, 2503 (1992); R. Y. Chiao, I. H. Deutsch, J. C. Garrison, and E. M. Wright, in *Serge Akhmanov: A Memorial Volume*, H. Walther, ed. (Adam Hilger, Bristol, U.K., 1992); A. W. Snyder, I. Poladian, and D. J. Mitchell, Opt. Lett. **17**, 789 (1992).
2. G. A. Swartzlander, Jr., D. L. Drugan, N. Hallak, M. O. Freeman, and C. T. Law, "Optical transistor effect using in optical vortex soliton," Laser Phys., in press.
3. C. T. Law and G. A. Swartzlander, Jr., "Polarized optical vortex solitons: instabilities and dynamics in Kerr nonlinear media," Chaos, Solitons, and Fractals, in press.

QTuE 10:30 am Hyatt B

Optical Properties of Trapped Atoms

Luis Adolfo Orozco, *State University, New York, Presider*

QTuE1 10:30 am

Optical lattices with metastable xenon

M. Walhout, C. Orzel,* U. Sterr, S. L. Rolston, *Atomic Physics Division, PHY A167, National Institute of Standards and Technology, Gaithersburg, Maryland 20899*

Optical lattices are stable periodic arrays of optical potentials that are capable of trapping and localizing atoms.[1] We have created a three-dimensional optical lattice for metastable xenon by using four laser beams directed along $+z \pm x$ and $-z \pm y$. This creates a tetragonal lattice with a periodicity of $\lambda/\sqrt{2}$ along x and y and creates half that along z. A weak probe is directed along z, and the absorption and gain near the lattice laser frequency are studied. Figure 1 shows a typical spectrum. Clearly visible are Raman sidebands at ± 110 kHz, corresponding to stimulated Raman transitions between vibrational levels of the optical potentials. A narrow central feature with a width of less than 1 kHz is also present. The mechanism for the production of such a narrow central feature is as yet unknown. The dependence of this feature has been studied, and we offer speculation as to possible mechanisms. There are intriguing differences between these results in metastable xenon and results in cesium,[2] which should be an almost identical system.

We have applied a magnetic field and have measured the effect on the spectrum. We find that the addition of a small field along z can greatly modify the absorption of oppositely circularly polarized probes, suggesting nearly complete spin polarization of the trapped atoms. The metastable system has an additional feature, which is the possibility of the application of an additional quench laser tuned to a transition that depopulates the metastable state. With this laser we can sensitively probe[3] the state of polarization, and we use it as a polarization filter as well.

Metastable xenon is an ideal system for studying collisions, because the predominant loss mechanism is the production of ions through Penning ionization. An optical lattice offers an interesting system in which collisions can be used as a probe of the localization. One can argue that if all the atoms are trapped in potential wells, they will have no opportunity to collide; hence the collision rate in a lattice will be greatly suppressed over that of correspondingly cold, but free, atoms. On the other hand, if two atoms fall into a single well, they will almost certainly collide, producing an effective collisional cross section that is roughly the size of the well ($\sim\lambda^2$), much larger than that for free atoms, suggesting an enhanced cross section. In this case the collisions become dominated by the random walk of the atoms from well to well and are a measure of the spatial diffusion in an optical lattice. Such effects of lattice dynamics on collisions will be investigated.

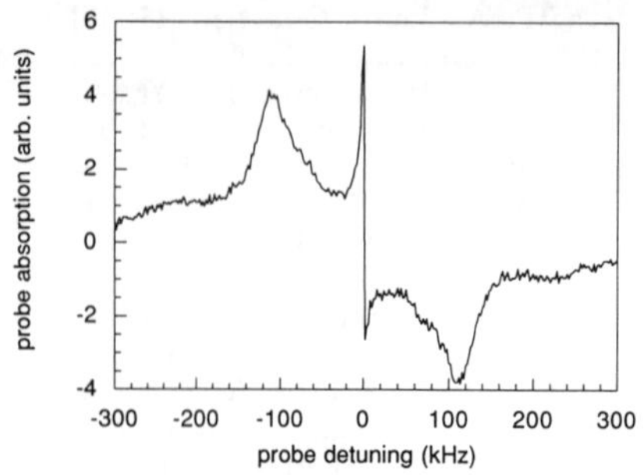

QTuE1 Fig. 1. Probe absorption spectrum for a metastable xenon optical lattice. The probe propagates along z and is polarized parallel to the two lattice beams propagating in the same plane. The frequency is referenced to the lattice laser frequency, which is detuned -10Γ from resonance.

*Department of Chemical Physics, University of Maryland, College Park, Maryland 20742

1. G. Grynberg *et al.*, Phys. Rev. Lett. **70**, 2249 (1993); A. Hemmerich *et al.*, Europhys. Lett. **22**, 89 (1993); P. S. Jessen *et al.*, Phys. Rev. Lett. **69**, 49 (1992).
2. P. Verkerk *et al.*, Europhys. Lett. **26**, 171 (1994).
3. M. Walhout *et al.*, Phys. Rev. Lett. **72**, 2843 (1994).

QTuE2 10:45 am

The spectrum of light scattered by atoms in optical lattices

G. Birkl, M. Gatzke, A. Kastberg,* P. S. Jessen,** W. D. Phillips, S. L. Rolston, I. H. Deutsch, R. J. Spreeuw,† *National Institute of Standards and Technology, Phys A167, Gaithersburg, Maryland 20899*

By intersecting four linearly polarized laser beams, we produce a 3D optical lattice: a standing wave pattern of laser polarization which induces a spatially periodic array of light shifts for near-resonant atoms. The optical potential associated with these light shifts constitutes a lattice of nearly harmonic potential wells in which atoms may become trapped.

In our experiment, we trap and precool cesium atoms in a magneto-optical trap. After turning off the magnetic field, we allow the atoms to equilibrate in the optical lattice where they become trapped. We detect the spectrum of fluorescence using an optical heterodyne technique[1] in which we Fourier transform the best signal between the scattered light and a local oscillator (see Fig. 1). Optical lattices have been studied in other groups by stimulated Raman spectroscopy with a probe beam.[2,3]

The spectrum is characteristic of a source which is executing oscillating motion as it radiates, exhibiting a narrow elastic scattering peak (the carrier) and sidebands which are offset by the oscillation frequency for atoms trapped in the potential wells. Quantum mechanically, the sidebands arise from spontaneous Raman transitions between different vibrational states of the optical potential. Assuming a Boltzmann distribution for the population of different vibrational states, we can infer a temperature for atoms in the lattice. The decreasing probability for atoms to occupy higher energy states gives rise to an asymmetry in the strength of the sidebands. From their ratio we can directly determine the temperature. By suddenly shutting off the optical lattice lasers and observing the time-of-flight distribution of the atoms as they fall through a probe beam several centimeters below, we can measure their momentum distribution, and independently determine the temperature.

Figure 1 shows a set of three heterodyne spectra taken under identical experimental conditions but for different atom densities. For high densities (Fig. 1a and 1b) the ratio of the strengths of the sidebands is artificially increased by the effect of wave mixing between the fluorescence photons and the lattice beams. This effect was observed in probe spectroscopy[2,3] before and gives rise to an amplification of the lower frequency sideband and a reduction of the higher frequency sideband. Using this sideband ratio would lead to a value for the temperature much too low. For low atomic densities (Fig. 1c) the real ratio of the sideband amplitudes is revealed. In the case of Fig. 1c the temperature derived from the fluorescence spectrum in $T_{Het} = 15.8^{+12.7}_{-5.1}$

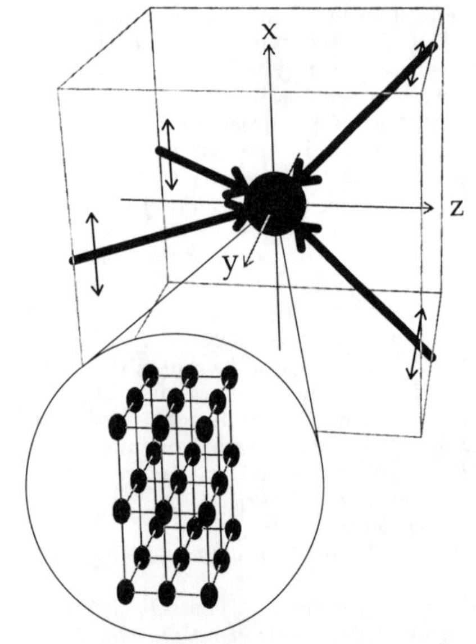

QTuE2 Fig. 1. Fluorescence spectra of atoms trapped in a 3D optical lattice for different atom densities. The density is decreasing from a to c. Only for the lowest density can the unperturbed shape of the fluorescence spectrum be observed. For higher densities gain on the lower frequency sideband and absorption on the higher frequency sideband distort the shape of the spectrum.

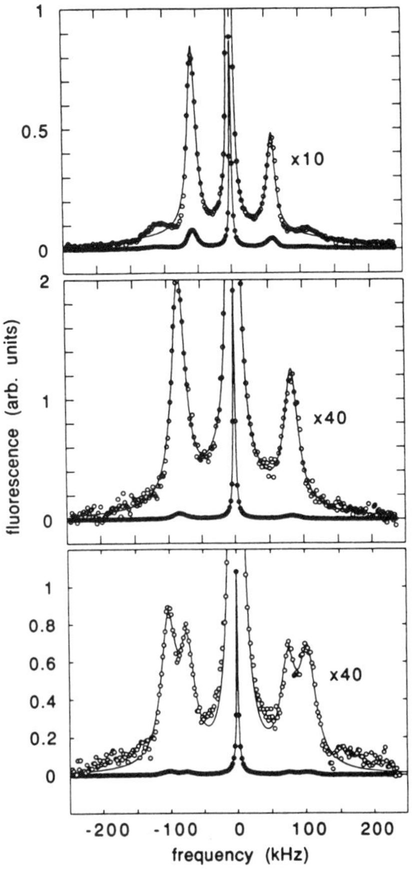

QTuE2 Fig. 2. Oscillation frequency of atoms trapped in a 3D optical lattice along a specific direction of observation. The vibrational frequency is only determined by the light shift independent of the specific values for intensity and detuning.

μK. The corresponding measurement of the momentum distribution gives a temperature of 13.8 ± 0.2 μK.

In addition to the temperature the spectrum of scattered light directly gives the oscillation frequency of atoms in the optical potential wells. It only depends on the value of the light shift the atoms experience in the lattice. Figure 2 shows the oscillation frequency of Cs atoms in a 3D optical lattice as a function of light shift. The data were obtained for a series of different intensities and detunings from atomic resonance.

**Stockholm University, Atomic Physics, Fescativägen 24, S-10405 Stockholm, Sweden*
***Optical Sciences Center, University of Arizona, Tuscon, Arizona 85721*
†Fakultät für Physik, Universität Konstanz, D-7750 Konstanz, Germany

1. P. S. Jessen, *et al.*, Phys. Rev. Lett. **69**, 49 (1992).
2. G. Grynberg, *et al.*, Phys. Rev. Lett. **70**, 2249 (1992).
3. A. Hemmerich, *et al.*, Europhys. Lett. **22**, 89 (1993).

QTuE3 **11:00 am**

Bragg scattering in optical lattices

I. H. Deutsch, R. J. C. Spreeuw,* M. Gatzke, G. Birkl, S. L. Rolston, W. D. Phillips, *National Institute of Standards and Technology, Physics A167, Gaithersburg, Maryland 20899*

Laser-cooled atoms can be trapped at periodic spatial positions by light-shift potential wells created by the polarization gradients and by interference of multiple laser beams.[1] These optical lattices constitute a novel form of matter in which atoms are trapped at sites separated on the scale of the optical wavelength. At currently achieved densities, atoms occupy very few of the available sites. However, the strict periodicity in the trapping position should result in long-range order in the atomic density correlation function.

This long-range spatial order will dramatically modify the propagation of a weak probe laser beam in the lattice, as compared with the propagation in a disordered gas. In particular, coherent interference of scattered waves in directions other than the incident direction can lead to substantial Bragg scattering, in analogy to x-ray dif-

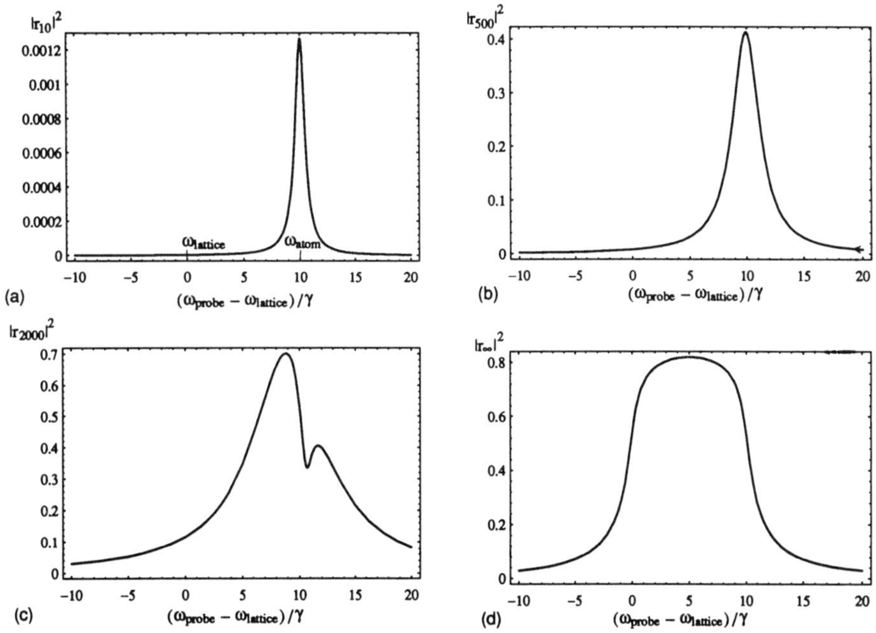

QTuE3 Fig. 1. Absolute square of the reflection coefficient as a function of $(\omega_{probe} - \omega_{lattice})/\gamma$ plotted for a lattice of Cs atoms with a mean density of 10^{11} cm^{-3} trapped by laser beams detuned to the red of resonance by $\delta_{lattice} = -10\gamma$ and for various numbers of atomic planes: (a) $n = 10$, (b) $n = 500$, (c) $n = 2000$, (d) $n = \infty$.

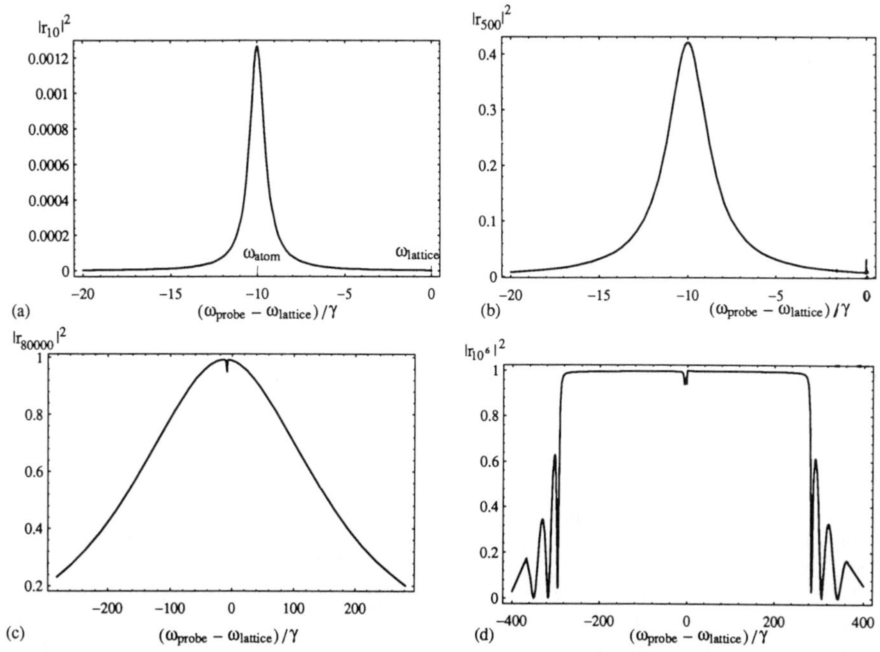

QTuE3 Fig. 2. Reflection coefficient for a Cs optical lattice with parameters given in Fig. 1 but with a blue-lattice laser detuning $\delta_{lattice} = +10\ \gamma$: (a) $n = 10$, (b) $n = 500$, (c) $n = 8 \times 10^4$, (d) $n = 10^6$.

fraction by solid-state crystals. This effect is highly polarization dependent, and the beam attenuation no longer follows Beers law, as in a disordered atomic gas. The polarization dependence of the scattering can be described by a tensor reciprocal lattice space, which gives information not only about energy and momentum conservation of the incident and scattered waves, but about angular momentum as well.

For a small total number of trapped atoms, multiple scattering is negligible, as in the kinetic theory of Bragg scattering, and the reflection as a function of probe frequency is a simple Lorentzian, following the atomic response function. For sufficient densities, multiple scattering becomes important, as in dynamic Bragg scattering, and the reflection is large over a wide band of probe frequencies, corresponding to a photonic band gap.

We first consider the interaction of the atoms with the laser beams that define the optical lattice. For a red-detuned lattice, atoms are localized at the lattice field antinodes; the presence of the atoms modifies the wavelength, and thus the lattice spacing, in a manner different from a bulk index of refraction. For blue-detuning, the atoms lie at the nodes of the field, which are spaced according to the vacuum wavelength. In both cases the self-consistent solution gives rise to a photonic band gap with one edge exactly at the trapping-laser frequency. The small difference between the trapping positions of the atoms for a red or blue lattice dramatically alters the photon-dispersion relation and the resulting band gap. For red lattices, the band gap extends between the trapping frequency and the atomic resonance (Fig. 1). For blue lattices the stop band extends symmetrically about resonance, typically for hundreds of atomic linewidths, except for the range between the trapping frequency and atomic resonance, which represents allowed propagation (Fig. 2). For red lattices with a mean density of 10^{11} cm^{-3}, 55% reflection of a weak probe should be observable. Experiments to observe this reflection are in progress.

*Present address: *Faculty of Physics, University of Konstanz, D-7750 Konstanz, Germany*

1. P. Verkerk, *et al.*, Europhys. Lett. **26**, 171 (1994); A. Hemmerich *et al.*, Europhys. Lett. **22**, 89 (1993); P. S. Jessen *et al.*, Phys. Rev. Lett. **69**, 49 (1992).

QTuE4 **11:15 am**

Quantized motion of atoms in a magneto-optical potential under coherent population trapping in a high-intensity laser field

N. P. Konopleva, A. V. Taichenachev, A. M. Tumaikin, V. I. Yudin, *Novosibirsk State University, 2 Pirogova Street, 630090 Novosibirsk, Russia*

Recently, different research teams have observed the transitions of laser-cooled atoms between vibrational levels in optical potential wells in the lin ⊥ lin laser configuration.[1,2] The case of $F_g = F \leftrightarrow F_e = F + 1$ transitions, in which F_g, F_e are angular moments of the ground and the excited states, respectively, has been examined. The physical mechanism of this phenomenon originates from the quantized motion of the atomic center of mass in periodic potential wells created by nonuniform light-induced shifts of Zeeman ground-state sublevels.

In the present work the quantized motion of atoms in magneto-optical potential wells is investigated in the case of the $F_g = 1 \leftrightarrow F_e = 1$ transition, when coherent population trapping (CPT) takes place. We suppose that atoms have been previously cooled to low temperatures, for example to $T \sim 10$ μK, as was done in three-dimensional molasses in Refs. 1 and 2. We represent the geometry in which the laser lin ⊥ lin configuration is combined with a uniform magnetic field directed along the laser beams (Z axis).

It is well known that for the $F_g = 1 \leftrightarrow F_e = 1$ transition the interaction of atoms and the laser lin ⊥ lin configuration does not lead to the creation of potential wells.[3] However, nonuniform laser polarization induces a spatially modulated magnetic momentum of the medium.

The turning on the magnetic field causes two effects: (1) As a result of the interaction of the periodic magnetic momentum with the magnetic field, a magneto-optical potential is formed, and (2) the dark (CPT) state in which atoms do not interact with the laser field is destroyed. However, when the Rabi frequency V is much greater than the Zeeman sublevel splitting Ω ($V >> \Omega$), slowly moving atoms with $kv << V$ remain predominantly in the CPT state and the relative population of the excited state is much less than unity. Then the change in the atomic energy arising from photon scattering is rather small. For this reason one can consider the atomic ensemble to be a conservative system. We show that in this case at stage $\gamma^{-1} < t < \tau$ the translational motion of cold atoms is described by the Schrödinger equation with potential

$$U = -\hbar\Omega \cos(2kz).$$

Note that the potential depth is determined only by the value of the magnetic field.

When the temperature of previously cooled atoms is much less than the depth of potential wells, $k_BT << \hbar\Omega$, they execute a periodic motion near the wells' bottoms. Using the harmonic approximation from the Shrödinger equation, we calculate an oscillation frequency

$$\Omega_{osc} = 2\sqrt{2\varepsilon_r\Omega/\hbar},$$

where ε_r is the one-photon recoil energy. Atoms are localized in potential wells within the range

$$\tilde{z} = \lambda/(4\pi)[2\varepsilon_r/(\hbar\Omega)]^{1/4},$$

which is much smaller than the wavelength λ. The dumping time of these oscillations τ connected with the finite lifetime of the CPT state is

$$\tau = \gamma^{-1}(V/\Omega)^2\sqrt{\hbar\Omega/2\varepsilon_r}.$$

Thus there are two reasons for the narrowing of vibrational levels. Apart from the well-known Lamb–Dicke effect, additional narrowing arises from the CPT effect in the high-intensity laser field (factor $(V/\Omega)^2 >> 1$).

In conclusion, note that because of the CPT effect, atom–atom resonance photon exchange is strongly reduced. Thus, the scheme under consideration is of interest for creating high-density atomic traps.

1. P. Verkerk, B. Lounis, C. Salomon, C. Cohen-Tannoudji, J.-Y. Courtois, and G. Grynberg, Phys. Rev. Lett. **68**, 366 (1992).
2. P. S. Jessen, C. Gerz, P. D. Lett, W. D. Phillips, S. L. Rolston, R. J. C. Spreeuw, and C. I. Westbrook, Phys. Rev. Lett. **48**, 49 (1992).
3. A. Aspect, E. Arimondo, R. Kaiser, N. Vansteenkiste, and C. Cohen-Tannoudji, J. Opt. Soc. Am. B **6**, 2112 (1989).

QTuE5 **11:30 am**

Anomalous diffraction by cold atoms

F. Strauch, D. Haubrich, H. Schadwinkel, D. Meschede,* *Institute for Quantum Optics, University of Hannover, Welfengarten 1, D-30167 Hannover, Germany*

Magneto-optical traps (MOTs) provide samples of typically 10^7 atoms that are optically thick near the trapping laser transition. As a consequence, both strong absorption of a weak probe laser beam sent through the trap and dispersion occur near an atomic resonance line. The light field experiences a phase shift due to atomic dispersion in the trapping volume, and it interferes with the transmitted light field, leading to a diffraction pattern. Dispersive phenomena, which are most prominent off resonance, may provide a method for probing trapped atoms without exchanging photon momenta, and hence heating, as in conventional absorption spectroscopy.

We use standard laser-cooling techniques to operate a cesium MOT with circularly polarized light.[1] The trap has a full width at half maximum of 1 mm, much smaller than the probe laser beam with a $1/e^2$ diameter of 5 mm. Its intensity is kept below 1 mW/cm^2 to minimize the influence on trapped-atom dynamics. On a screen 3 m away from the MOT and perpendicular to the probe-beam axis, the diffraction patterns are observed by means of a CCD camera. The image in Fig. 1 was obtained by taking the difference between the intensity profiles with and without trapped atoms. The system of fringes, which resembles the pattern of a circular aperture, is visible only by narrowing the probe laser to about a 100-kHz linewidth. To the red side of the cooling frequency a bright center of the diffraction pattern is observed, and to the blue side the center is dark. We can currently detect diffraction patterns within a range of $\pm 4\gamma$ (γ is the natural linewidth).

For a more systematic investigation of the diffraction pattern we have studied the variation of the central part as a function of probe-laser detuning with the trap size as a parameter by using an iris diaphragm of 1-mm diameter to define the detection area in front of the detector. The trap size is changed from 1.0 to 0.4 mm by slowly and continuously lowering the trapping-

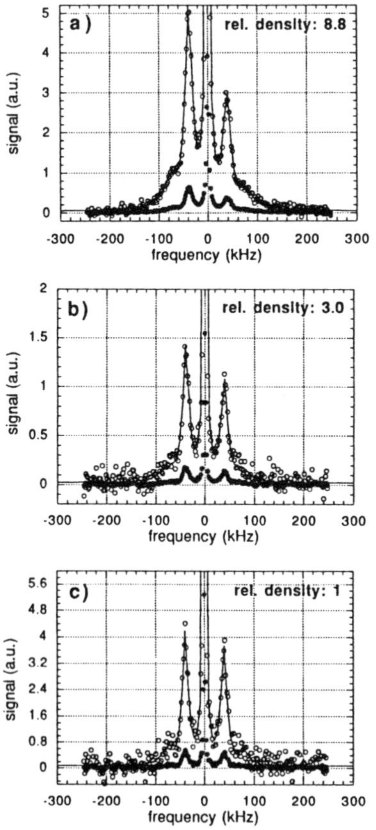

QTuE5 Fig. 1. Difference image of diffraction pattern and corresponding intensity cross profile.

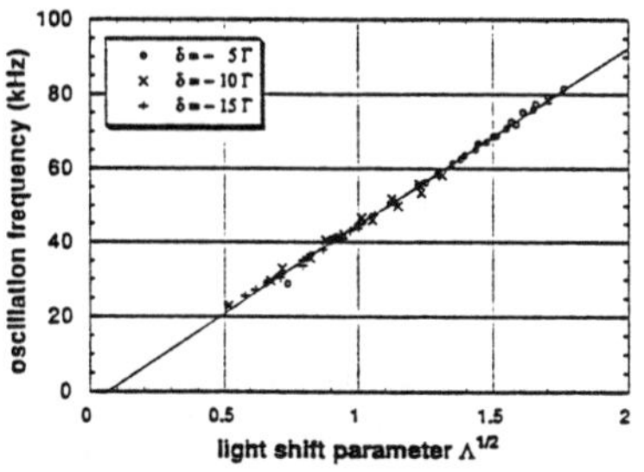

QTuE5 Fig. 2. Top: Total trapping laser intensity after the Pockels cell. Center: Piezo voltage for probe-laser frequency scan (rate, 24 MHz/V). Bottom: Central intensity of diffraction pattern.

laser intensity from 280 to 12 mW/cm^2 by means of a Pockels cell. At the same time, the probe-laser frequency is swept across the $F = 4 \rightarrow F' = 4$ transition at a rate of 4.8 MHz/ms. A spectrum is registered from a photodiode 6.5 m away from the MOT (Fig. 2). Starting on the right-hand side, a change of the power-broadened resonance curve shape from dispersive to absorptive to dispersive with opposite sign is observed.

For a simplified theoretical treatment of this behavior, we have replaced the Gaussian density distribution of cold atoms by a homogeneous dielectric sphere. This problem was treated by van de Hulst[2] for refractive indices near unity. In this case the light rays experience little bending on entering the scattering volume. The theory is qualitatively in accordance with the experimental results when we consider the fact that in the vicinity of a resonance both absorptive and dispersive contributions have to be taken into account.

*New address: *Institute for Applied Physics, Rheinische Friedrich-Wilhelms-Universität Bonn, Wegelerstrasse 8, D-53115 Bonn, Germany*

1. A. Höpe, D. Haubrich, H. Schadwinkel, F. Strauch, and D. Meschede, Europhys. Lett. **28**, 7 (1994).
2. H. C. van de Hulst, *Light Scattering by Small Particles* (Dover, New York, 1981).

QTuE6 **11:45 am**

Exposure of the internal spatial structure of resonance-cooled gas by means of optical susceptibility measurements

G. I. Surdutovich, A. V. Ghiner,*
V. S. Bagnato, *São Carlos Institute of Physics/USP, C.P. 369, CEP 13560-970, São Carlos, SP, Brazil*

What can we learn about the internal microstructure of an optical medium from macroscopic optical measurements? As is known, the microscopic symmetric properties of crystals manifest themselves in an anisotropy of the refractive index. Whereas the internal structure of the radiators exhibits itself by means of the multipolar expansion, neither the radiators' size nor the lattice-grain size ever enters into formulas for the refractive index n of the medium. Such an approach usually is well grounded because of the smallness of these sizes compared to the wavelength λ. According to the classical Lorentz–Lorenz (LL) formula, optical properties of the medium depend merely on the product of the density N and the polarizability α of an isolated radiator. This holds true for dense, $N\lambda^3 >> 1$, media only. Incorporated into the LL formula, local-field effects raise from interparticle interactions, and in this sense they reveal the very fact of the medium's discreteness, but not its magnitude. For a resonant gas, local-field corrections may be essential, even for a rarefied medium, when $N\lambda^3 \leq 1$, and medium discreteness in wavelength units becomes evident.

For calculation of such a gas susceptibility we applied the generalized method of integral equations[1] (MIE) adjusted to this specific case. By use of this technique we calculated the refractive index n, assuming $n - 1 << 1$, for a random medium under the condition $b << N^{-1/3}$:

$$n - 1 = 2\pi N\alpha(1 - N\alpha\beta' + 2N\alpha''\beta'')^{-1}, \tag{1}$$

where

$$\beta' = \pi\left\{1 + \left[\frac{4}{(kb)^2} - 1\right]\cos 2kb - 2\left[\frac{1}{(kb)^2} - 1\right]\frac{\sin 2kb}{kb}\right\},$$

$$\beta'' = \pi\left\{2kb - \frac{2}{kb} - \frac{2}{(kb)^3} + 2\left[\frac{1}{(kb)^2} - 1\right]\frac{\cos 2kb}{kb} + \left[1 - \frac{4}{(kb)^2}\right]\sin 2kb\right\}. \tag{2}$$

For small $kb << 1$ we have

$$\beta' = -\frac{4\pi}{3}\left[1 - \frac{11}{10}(kb)^2\right],$$

$$\beta'' = \frac{8\pi}{9}(kb)^3.$$

In the same long-wavelength limit, $kb << 1$, but for a liquidlike medium, $b \leq N^{-1/3}$, we obtain an essentially different result:

$$\beta' = -\frac{4\pi}{3}\left[1 - \frac{143}{160}(kb)^2\right],$$

$$\beta'' = \frac{\pi}{9}(kb)^3. \tag{3}$$

This distinction between gas and liquid with the same discrete parameter kb is connected to the difference between their correlation functions: the angular dependences are identical (isotropic), whereas the radial functions are different. Indeed, a gas radial function becomes constant at large distances, whereas for liquid it oscillates with period b, as is evident from the limiting case of a closely packed liquid. Thus, allowing for the rarity of the medium leads to a novel qualitative effect, the difference between the optical properties of two random media with the same discrete parameter kb. There is one more principal problem relating to rarefied media. It seems evident that in the limiting case of such media, an isolated radiator, radiation damping must be taken into account. Then the question of why radiation damping is not taken into account for a dense medium for $N\lambda^3 >> 1$ arises. The answer is that in this case the radiation damping of any radiator is significant, but for each of the radiators it is totally cancelled by the collective influence of all the rest of the radiators, so that the contribution of radiation damping to the polarizability α of an individual radiator is equal to zero. MIE, together with energy-conservation-law considerations, allowed us to calculate the radiation-damping contribution in the general case and to carry out the corresponding renormalization of the polarizability of an isolated atom.

**Federal University of Ceara, C.P. 6030, CEP 64450-970, Fortaleza, Brazil*

1. A. V. Ghiner and G. I. Surdutovich, Phys. Rev. A **49**, 1313 (1994).
2. A. V. Ghiner and G. I. Surdutovich, Laser Phys. **4**, 564 (1994).

Tuesday, May 23

QTuF 10:30 am
Hyatt C/D

Spatial Quantum Effects

Prem Kumar, *Northwestern University, Presider*

QTuF1 (Invited) 10:30 am

Quantum images and wave-particle aspects in nonlinear optical patterns

L. A. Lugiato, A. Gatti, G. Grynberg,* *Dipartimento di Fisica dell'Universitá di Milano, Via Celoria 16, 20133 Milano, Italy*

Optical systems are macroscopic, but they are able to display relevant quantum effects, even at room temperatures. This feature makes optics special with respect to other disciplines (e.g. hydrodynamics and nonlinear chemical reactions), which are more traditional in the study of pattern formation phenomena. Recent years have witnessed an increasing interest in the analysis of spatial aspects of squeezing, and of quantum effects in nonlinear optical patterns.[1-4]

Under appropriate parametric conditions, the signal field emitted by a degenerate optical parametric oscillator (OPO) above threshold has a stripe structure in the near field, while the far-field configuration corresponds to two separate and spatially localized signal beams.[5] The difference between the photon numbers of the signal beams exhibits squeezing, i.e. subshot noise fluctuations, with complete suppression of zero-frequency fluctuations. As a consequence of the uncertainty principle, in the phase difference between the two signal beams shows large fluctuations; suggesting the possibility that the near-field pattern is washed out by quantum fluctuations.[3]

In this paper we perform a systematic investigation of this problem by eliminating from the model the unphysical assumption that the system has a perfect translational symmetry. The symmetry breaking is introduced in a very simple and straightforward way and, as a consequence of this step, the phase difference no longer exhibits a purely dissipative motion because it is damped.

We calculate analytically the spectra of the quantum fluctuations for the photon number difference and for the phase difference with an arbitrary level of symmetry breaking. We can, therefore, determine the parametric conditions under which there is complementarity between the presence of squeezing in the far field and a presence of a stripe pattern in the near field. We argue that this complementarity is linked to the fact that the near field shows the wave aspect of the system, whereas the far field exhibits the particle aspects.

When the near field pattern becomes invisible, one has a "quantum image," which can be observed by measuring the spatial correlation function of the electric field. This is linked with the idea[6] of encoding informations in the space correlation function instead of the intensity distribution. The same phenomenology occurs also in a nonlinear cavity filled with a Kerr medium; in this case the behavior of the system is governed by four-wave mixing, which plays the same role as parametric downconversion in the OPO.

In Ref. 4 we showed that the spatial structure of squeezed states can be described by using a local oscillator field with an arbitrary spatial configuration. In this paper we show that the same result can be obtained by measuring the space-time correlation function of the electric field.

By applying these concepts to the degenerate OPO below threshold, we calculate analytically the spatial correlation function of the signal field. This function shows a modulation with a wavelength identical to that of the stripe pattern that appears above the OPO threshold; hence the onset of a spatial structure is heralded by the behavior of the correlation function. The correlation length diverges when the critical point, i.e. the threshold for signal generation, is approached (see Fig. 1).

Other examples of fields generated purely by quantum noise are well known (e.g. the case of the laser below threshold). However, our results show for the first time, to our knowledge, that, under appropriate conditions, a quantum noise-generated field (precisely the squeezed vacuum field emitted by the degenerate OPO below threshold) exhibits a modulated spatial structure. This corresponds again to a quantum image because the intensity distribution is uniform, and the structure appears only in the space correlation function.

**Laboratoire Kastler Brossel, Université P. et M. Curie, Case 74, 75252 Paris Cedex 05, France*

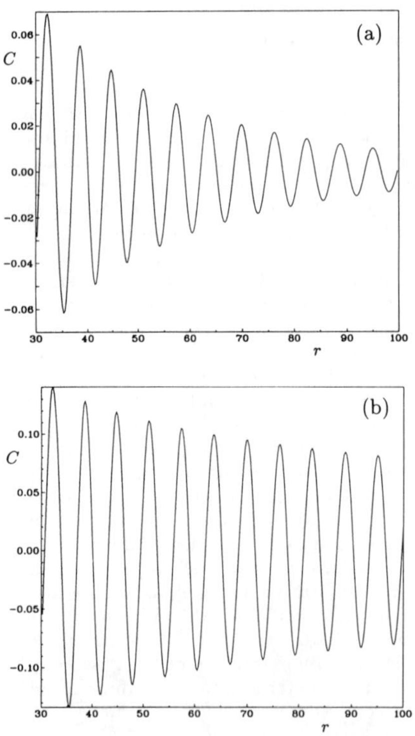

QTuF1 Fig. 1. The two-point correlation function of the signal field is graphed as a function of the normalized distance between the points. (a) The correlation decays exponentially; (b) when the threshold for signal generation is approached, the exponential damping vanishes because the correlation length diverges. The modulation of the correlation function is identical to that of the stripe pattern which appears above threshold.

1. M. I. Kolobov and I. V. Sokolov, Sov. Phys. JETP **69**, 1097 (1989); A. La Porta and R. E. Slusher, Phys. Rev. A **44**, 2013 (1991).
2. L. A. Lugiato and F. Castelli, Phys. Rev. Lett. **68**, 3284 (1992).
3. G. Grynberg and L. A. Lugiato, Opt. Comm. **101**, 69 (1993).
4. L. A. Lugiato and A. Gatti, Phys. Rev. Lett. **70**, 3868 (1993).
5. G. L. Oppo, M. Brambilla, and L. A. Lugiato, Phys. Rev. A **49**, 2028 (1994).
6. P. L. Knight, private communication.

QTuF2 11:00 am

Noiseless amplification of optical images

Mikhail I. Kolobov, Luigi A. Lugiato, *Department of Physics, University of Milan, I-20133 Milan, Italy*

It is well known from the literature that phase-insensitive amplifiers introduce at least 3 dB of extra noise in the output, whereas phase-sensitive amplifiers may preserve the input signal-to-noise ratio (SNR) and in this sense are noiseless. Although noiseless amplification in the temporal domain has been discussed in several papers,[1,2] its spatial counterpart has been neglected. Here we provide the first step toward the noiseless amplification of optical images.

We consider the parametric image amplifier shown in Fig. 1. A transparency with a faint image to be amplified is placed in the object plane P_1. The lens L_1 projects this transparency onto the input plane P_2 of a ring-cavity optical parametric amplifier (OPA). The amplified image from the output plane P_3 is projected onto the image plane P_4 by the lens L_4. All the optical intervals in the scheme are chosen to be equal to the focal length of the lenses L_1 and L_2.

We suppose that the amplified image is registered by a dense array of small photodetectors (pixels). Each pixel has a finite area S_d and quantum efficiency η, and the photocurrent from each pixel is collected during the observation time T_d. The quantity of interest to us is the number of photoelectrons $N_l(\boldsymbol{\rho}, t)$ registered by the pixel centered at the point $\boldsymbol{\rho}$ in the image plane in the time window from $t - T_d/2$ to $t + T_d/2$:

$$N_l(\boldsymbol{\rho}, t) = \int_{S_d} d\boldsymbol{\rho}' \int_{T_d} dt'\, i(\boldsymbol{\rho}', t'), \qquad (1)$$

where $i(\boldsymbol{\rho}, t)$ is the surface photocurrent density in photoelectrons per square centimeter per second. We consider the mean number $\langle N_l(\boldsymbol{\rho}, t)\rangle$ of registered electrons as the amplified signal of our scheme; its variance, $\sigma_l^2(\boldsymbol{\rho}, t)$, characterizes the noise properties of the amplified image, and the SNR of the amplified image is $\mathrm{SNR}_l = \langle N_l(\boldsymbol{\rho}, t)\rangle/\sigma_l(\boldsymbol{\rho}, t)$.

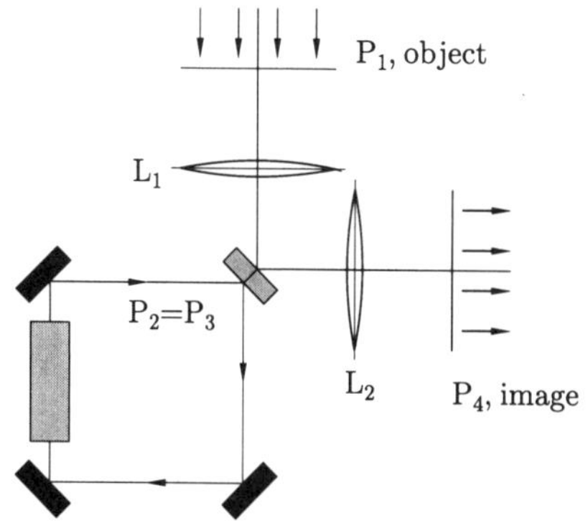

QTuF2 Fig. 1. Scheme of a parametric image amplifier.

The equivalent quantities $\langle N_O(\boldsymbol{\rho}, t)\rangle$, $\sigma_O^2(\boldsymbol{\rho}, t)$, and SNR_O in the object plane describe the signal and noise of the input image. To characterize quantitatively the amount of extra noise added by the amplifier to the input image, we introduce the noise transfer coefficient T as the ratio between the SNR in the object and the image lanes[2]:

$$T = SNR_O/SNR_I. \tag{2}$$

Because the SNR cannot be improved by the amplification process, this transfer coefficient must always be larger than unity, $T \geq 1$. We refer to the case of $T = 1$ as noiseless amplification.

To calculate the mean value of the photocurrent density and its variance, we have to express the field operator $e(\boldsymbol{\rho}, t)$ in the image plane through the corresponding field operator $a(\boldsymbol{\rho}, t)$ in the object plane and then to use the formulas of quantum photodetection theory.[3]

The behavior of the field operator $b(\boldsymbol{\rho}, t)$ inside the cavity is described by the following equation:

$$\frac{\partial}{\partial t} b(\boldsymbol{\rho}\ t) - i\frac{c}{2k}\nabla_T^2 b(\boldsymbol{\rho}, t) = -(\kappa + i\Delta)b(\boldsymbol{\rho}, t) + \sigma b^\dagger(\boldsymbol{\rho}, t) + \sqrt{2\kappa}\, b_{\text{in}}(\boldsymbol{\rho}, t), \tag{3}$$

where κ is the cavity decay constant, $k = 2\pi/\lambda$, where λ is the wavelength; Δ is the detuning parameter between the cavity mode and the signal; and σ is the constant of parametric interaction. The symbol ∇_T^2 denotes the transverse Laplacian with respect to the coordinate $\boldsymbol{\rho}$ in the transverse plane.

We assume that the field in the object plane is coherent with a complex amplitude $s(\boldsymbol{\rho})$ modulated in the transverse direction but that it is stationary in time. We obtain the following results for $\langle N_I(\boldsymbol{\rho}, t)\rangle$ and $\sigma_I^2(\boldsymbol{\rho}, t)$:

$$\langle N_I(\boldsymbol{\rho}, t)\rangle = \eta S_d T_d s^2(\boldsymbol{\rho}) G^2(\boldsymbol{\rho}),$$
$$G^2(\boldsymbol{\rho}) = \cos^2\phi(\boldsymbol{\rho})\exp[2r(\boldsymbol{\rho})] + \sin^2\phi(\boldsymbol{\rho})\exp[-2r(\boldsymbol{\rho})], \tag{4}$$

$$\sigma_I^2(\boldsymbol{\rho}, t) = \langle N_I(\boldsymbol{\rho}, t)\rangle \cdot (1 - \eta + \eta\{\cos^2\theta(\boldsymbol{\rho})\exp[2r(\boldsymbol{\rho})] + \sin^2\theta(\boldsymbol{\rho})\exp[-2r(\boldsymbol{\rho})]\}), \tag{5}$$

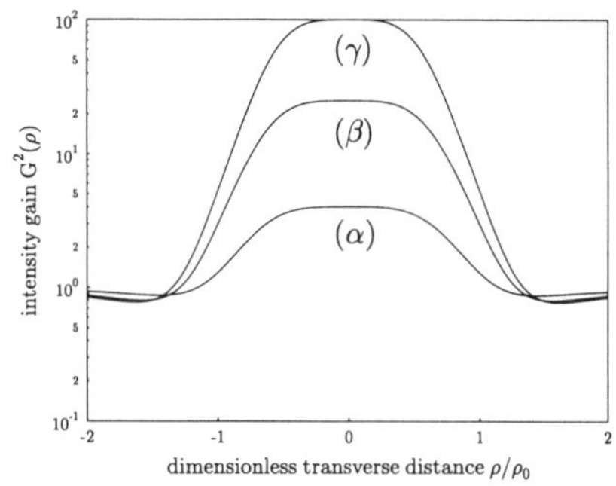

QTuF2 Fig. 2. Intensity gain $G^2(\boldsymbol{\rho})$ as a function of dimensionless transverse distance ρ/ρ_0 (see Ref. 4 for definition of ρ_0). Curves α, β, and γ correspond to the maximum gain of $G^2_{\max} = 4$, 25, and 100, respectively.

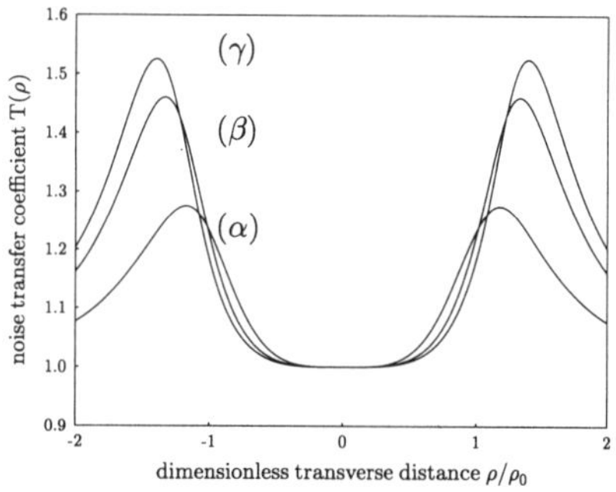

QTuF2 Fig. 3. Noise transfer coefficient $T(\boldsymbol{\rho})$ as a function of dimensionless transverse distance ρ/ρ_0 for $\eta = 1$. Curves α, β, and γ are for the same maximum gains as in Fig. 2.

where the orientation angles $\phi(\boldsymbol{\rho})$, $\theta(\boldsymbol{\rho})$ and the squeezing parameter $r(\boldsymbol{\rho})$ are functions of κ, Δ, and σ from Eq. (3). Their explicit expressions can be found in Ref. 4. In Eqs. (4) and (5) we discard the contributions due to spontaneous parametric scattering, which gives the inherent noise of the amplifier. This can be done when the input signal $s(\boldsymbol{\rho})$ is much larger than this noise. This condition gives an ultimate lower limit for the signal $s(\boldsymbol{\rho})$ that can be noiselessly amplified by our system (see Ref. 4).

The intensity gain $G^2(\boldsymbol{\rho})$ from Eq. (4) and the noise transfer coefficient $T(\boldsymbol{\rho})$ defined in Eq. (2) for $\eta = 1$ are shown in Figs. 2 and 3, respectively. Amplification conditions are chosen in such a way that the maximum amplification takes place at $\boldsymbol{\rho} = 0$. From Fig. 2 one can see that there is a circular region in the image plane where the gain $G^2(\boldsymbol{\rho})$ stays almost constant. Outside of this resolution area it falls sharply to unity. As follows from Fig. 3, the noise transfer coefficient in this region is equal to unity, thus realizing noiseless amplification within the resolution area of the amplifier.

1. C. M. Caves, Phys. Rev. D **26**, 1817 (1982).
2. I. E. Protsenko, L. A. Lugiato, and C. Fabre, Phys. Rev. A **50**, 1627 (1994), and references therein.
3. R. J. Glauber, Phys. Rev. **130**, 2529 (1963).
4. M. I. Kolobov and L. A. Lugiato, "Noiseless amplification of optical images," submitted to Phys. Rev. A.

QTuF3 **11:15 am**

Two-photon "ghost" interference–diffraction

Y. H. Shih, A. V. Sergienko, D. V. Strekalov, D. N. Klyshko, *Department of Physics, University of Maryland, Baltimore, Maryland 21228*

A pair of orthogonally polarized signal and idler photons, generated in spontaneous parametric downconversion (SPDC), are directed to different directions for coincidence measurement by two distant photon-counting detectors. A Young's double slit (or a single slit) is inserted into the path of the signal beam. Surprisingly, an interference (or diffraction) pattern is observed in the coincidences when the detector is scanned in the idler beam. This effect is even more striking when one considers that there is not any first-order interference–diffraction pattern behind the slits! This experiment demonstrated a typical Einstein–Podolsky–Rosen (EPR) phenomenon.[1]

The experimental setup is shown in Fig. 1. The 351.1-nm line of an argon-ion laser is used to pump type-II phase-matching BBO to produce pairs of orthogonally polarized signal (*e*-ray) and idler (*o*-ray) photons. The signal and idler beams are sent in different directions by a polarization-beam-splitting Thompson prism (BS). The

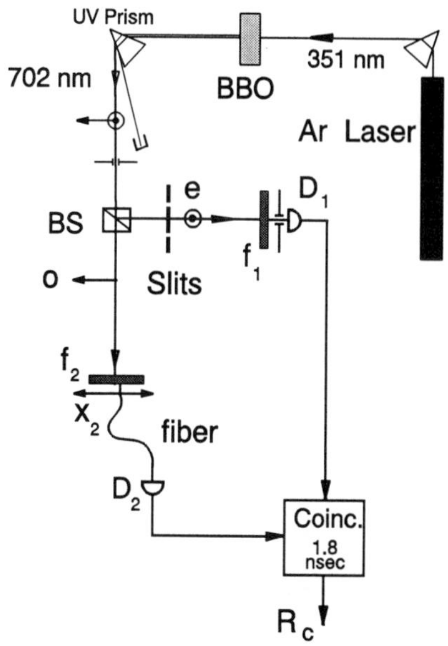

QTuF3 Fig. 1. Schematic setup for the two-photon ghost interference–diffraction experiment. The signal (*e*-ray) and idler (*o*-ray) photon pair is generated in the nonlinear crystal BBO. The ultraviolet pump beam is separated from the down-conversion beams by a UV-grade fused-silica dispersion prism. BS is a beam-splitting Thompson prism for splitting the signal and idler beams into different directions. f_1 and f_2 are spectral filters with 702.2-nm center wavelength and 10-nm bandwidth.

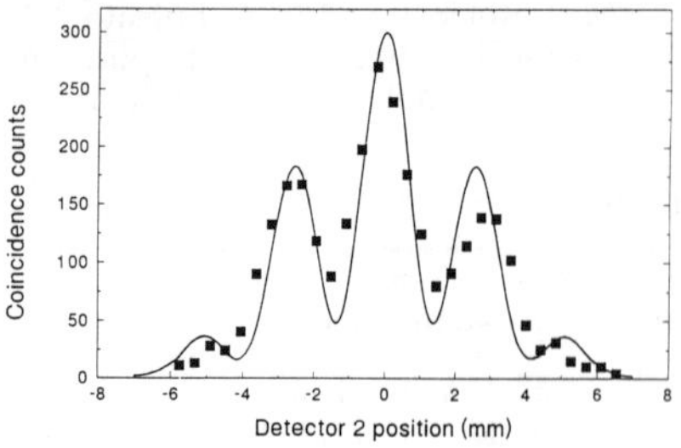

QTuF3 Fig. 2. Typically observed interference–diffraction pattern: the dependence of the coincidences (per 400 s) on the position of the optical fiber tip of detector D_2, which counts the idler photons, while the signal photons pass through a double slit with $a = 0.15$ mm and $d = 0.47$ mm. The solid curve is calculated from relation (1), corrected for the finite size of the detectors and the pump profile. If D_1 is moved to an unsymmetrical point, which results in unequal distances to the slits, the interference–diffraction pattern is observed to be simply shifted from the current symmetrical position.

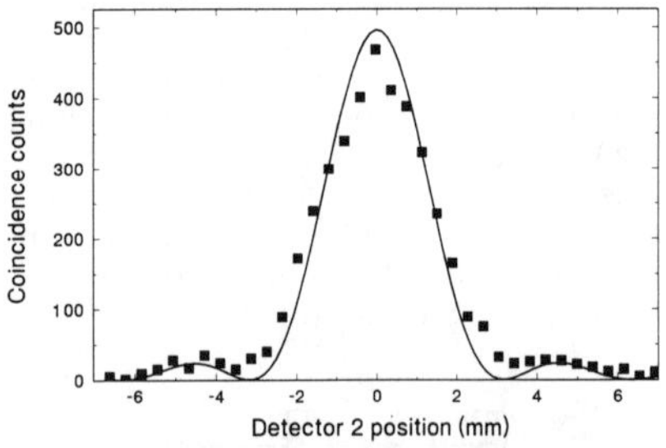

QTuF3 Fig. 3. Two-photon diffraction: coincidence counts (per 400 s) versus the idler-photon-counting detector position. A single slit of width $a = 0.4$ mm is in the signal beam. The solid curve is the sinc function of relation (1).

signal photon passes through a double-slit (or single-slit) aperture and then travels about 1 m to be counted by a pointlike photon-counting detector D_1 (0.5 mm in diameter) that is fixed on the axis of the signal beam. The idler photon travels a distance about 1.2 m from the BS to the input tip of the optical fiber, whose output end is coupled with another photon-counting detector D_2. The horizontal transverse coordinate x_2 of the fiber input tip is scanned by an encoder driver.

Figure 2 shows a typically observed double-slit interference–diffraction pattern. The Young's double slit has a slit width $a = 0.15$ mm and slit distance $d = 0.47$ mm. The interference period is measured to be 2.7 ± 0.2 mm, and the half-width of the envelope is estimated to be 8 mm. By curve fitting, we conclude that the observation is a standard Young's interference pattern, i.e., a sine function oscillation with a sinc function envelope:

$$R_c(x_2) \propto \mathrm{sinc}^2(x_2 \pi a/\lambda z_2)\cos^2(x_2 \pi d/\lambda z_2). \quad (1)$$

The remarkable feature here is that z_2 is the distance from the slit's plane, which is in the signal beam, back through the BS to the BBO crystal and then along the idler beam to the scanning fiber tip of detector D_2.

Furthermore, the single-detector counting rates are both observed to be constant when detector D_1 or D_2 is scanned. It is interesting to see the absence of the first-order interference–diffraction structure in the single counting rate of D_1, which is behind the double slit. The blurring out of the first-order interference fringes is due to the considerably large angular propagation uncertainty of a single SPDC photon.

Figure 3 shows a single-slit two-photon diffraction pattern. The slit has a slit width $a = 0.44$ mm. The solid curve is the sinc function of relation (1).

It is the essentially quantum nature of the entangled two-photon state in SPDC that allows these striking phenomena. The propagation direction of each single photon that belongs to a pair has a considerably large angular uncertainty. However, if one of them is detected in a certain direction, its conjugate twin must have been propagating in a defined certain direction. The above experiment demonstrated this EPR phenomenon.[1]

1. A. Einstein, B. Podolsky, and N. Rosen, Phys. Rev. **47**, 777 (1935).

QTuF4 **11:30 am**

Two-photon ghost image and quantum cropto-fax

Y. H. Shih, T. B. Pittman, D. V. Strekalov, A. V. Sergienko, *Department of Physics, University of Maryland, Baltimore, Maryland 21228*

The reported experiment is a two-photon optical imaging experiment[1] in which a spontaneous-parametric-downconversion light beam, which consists of pairs of orthogonally polarized signal and idler photons, is split into two diverging beams by a polarization beam splitter, so that coincidence detections may be performed between two distant photon-counting detectors. An aperture (mask) placed in front of one of the detectors is illuminated by the signal beam through a convex lens. Surprisingly, an image of this aperture is observed in the coincidence counting rate when the other detector is scanned in the transverse plane of the idler beam, even though both detectors' single counting rates remain constant.

The schematic diagram is shown in Fig. 1. The 351.1-nm line of an argon-ion laser is used to pump a type-II phase-matching BBO crystal to generate pairs of orthogonally polarized signal (*e*-ray) and idler (*o*-ray) photons. The signal and idler beams are sent in different directions by a Thompson prism. The signal beam passes through a convex lens with a 400-mm focal length and illuminates an aperture (mask), for example, the letters "UMBC." Behind the aperture is the photon-counting-detector package D_1. The idler beam is met by the input tip of a 0.5-mm-diameter multimode optical fiber, whose output end is coupled with another photon-counting detector D_2. The input tip of the fiber is scanned in the transverse plane. The output pulses of each detector are sent to a coincidence-counting circuit with a 1.8-ns acceptance window.

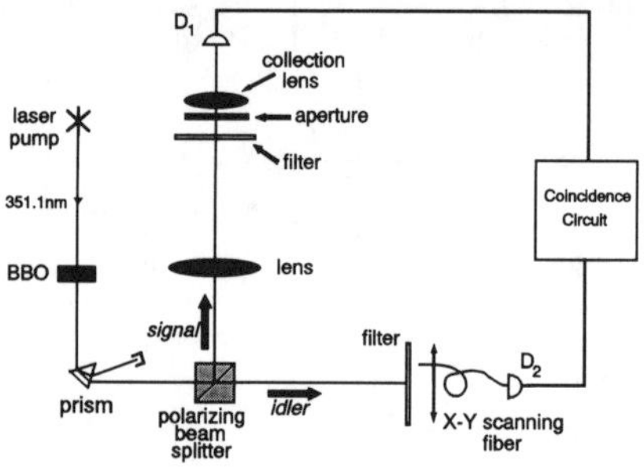

QTuF4 Fig. 1. Schematic setup for the two-photon ghost-image experiment. The signal (*e*-ray) and idler (*o*-ray) photons are generated in the nonlinear crystal BBO. The ultraviolet pump beam is separated from the downconversion beams by a UV-grade fused-silica dispersion prism. BS is a beam-splitting Thompson prism for splitting the signal and idler beams into different directions. f_1 and f_2 are spectral filters with 702.2-nm center wavelength and 83-nm bandwidth.

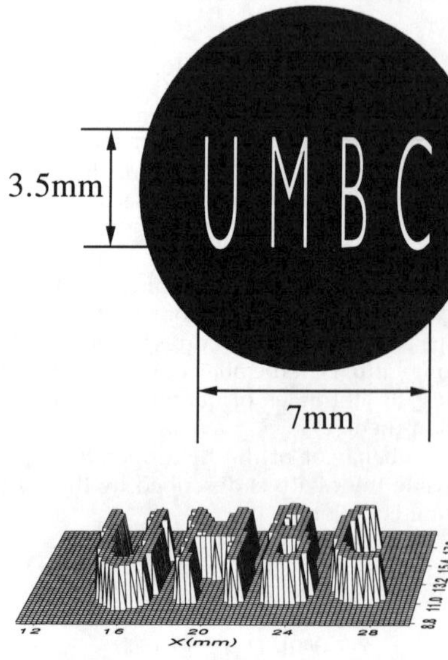

QTuF4 Fig. 2. (a) Reproduction of the actual aperture "UMBC" placed in the signal beam. Note that the size of the letters is of the order of that of standard text. (b) Image of "UMBC": coincidence counts as a function of the fiber tip's transverse plane coordinates. The scanning step size is 0.25 mm. The data shown represent a slice at the half-maximum value, with no image enhancement.

By recording the coincidence counts as a function of the fiber tips' transverse coordinates, we see the image of the aperture, as is shown in Fig. 2. It is interesting that although the size of the "UMBC" aperture inserted in the signal beam is only about 3.5 mm × 7 mm, the observed image measures 7 mm × 14 mm. Despite the completely different physical situation, the remarkable feature here is that the relationship between the focal length of the lens f, the aperture's optical distance from the lens S, and the image's optical distance from the lens (from the lens back through the beam splitter to the BBO and then along the idler beam to the image) S' satisfies the Gaussian thin-lens equation

$$\frac{1}{S} + \frac{1}{S'} = \frac{1}{f}. \qquad (1)$$

In this experiment we chose $S = 600$ mm, and the twice-magnified clear image was found when the fiber tip was in the plane with $S' = 1200$ mm.

In the sense of recent work performed on quantum cryptography,[2] the above image-transfer experiment demonstrates that the possibilities of a quantum crypto-fax are, in principle, quite real. Because the information is essentially carried in coincidences, an eavesdropper between detectors would require information from two communication channels. Therefore, even though the random pulse electric signal from detector D_1 is quite available, the use of indivisible light quanta in the other channel would foil any spying attempts.

1. For related theory, see D. N. Klyshko, Sov. Phys. Usp. **31**, 74 (1988).
2. A. K. Ekert, Phys. Rev. Lett. **67**, 661 (1991).

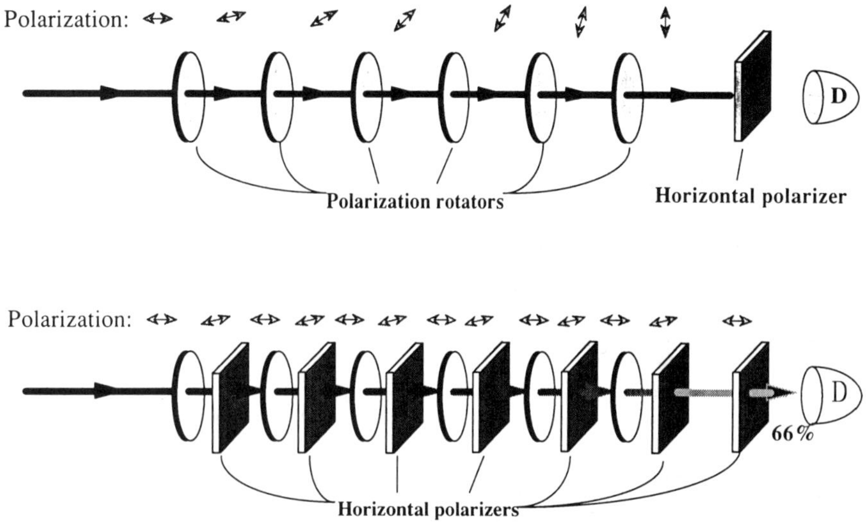

QTuF5 Fig. 2. Simple optical version of the quantum Zeno effect. (a) A series of polarization rotators are used to rotate the polarization of the input photon from horizontal to vertical. No light is seen at the final detector. (b) When a series of horizontal polarizers are interspersed between the rotators, the light is at every stage projected back onto a state of horizontal polarization, resulting in light at the final detector.

QTuF5 **11:45 am**

Experimental realization of interaction-free measurements

Harald Weinfurter, Paul Kwiat, Thomas Herzog, Anton Zeilinger, Mark Kasevich,* *Institute for Experimental Physics, University of Innsbruck, 6020 Innsbruck, Austria*

Recently, it has been pointed out that it is possible to make a quantum-mechanical interaction-free measurement (IFM), in which the existence of an object in a given region of space may be determined seemingly without interacting with it.[1,2] In the embodiment first proposed by Elitzur and Vaidman,[3] a standard Michelson interferometer is aligned so that incident single photons (obtained in our experiment by using parametric downconversion; see Fig. 1) must exit by the same port in which they entered, in the absence of any object within the interferometer. The presence of an object (e.g., the mirror leading to detector D_B) in one of the arms completely changes the possible outcomes because the constructive interference is interrupted. Now it is possible that the photon will exit from the *other* exit port, to detector D_{Ifm}. By detecting the photon in this case, one can claim with certainty that an object was within the interferometer, even though the photon could not have interacted with it. If the beam splitter has a very low reflectivity, then the fraction of IFMs approaches 1/2.

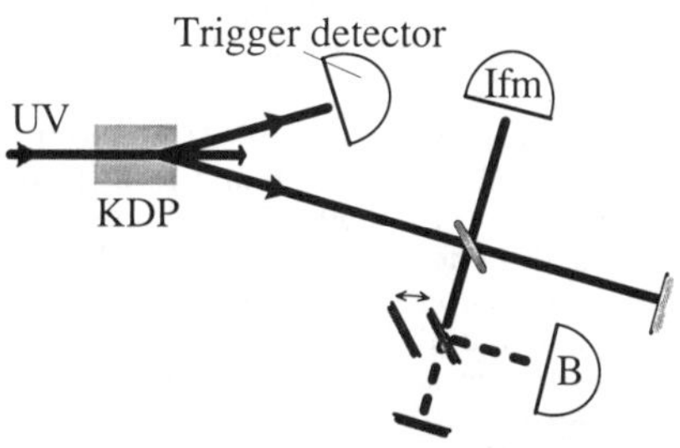

QTuF5 Fig. 1. (a) Downconversion implementation of the interaction-free measurement proposal of Elitzur and Vaidman. The Michelson interferometer was adjusted to have a minimum count rate at D_{Ifm} in the absence of the mirror directing photons to D_B. When the beamsplitter reflectivity was reduced, it was possible for nearly half of the measurements to be interaction free.

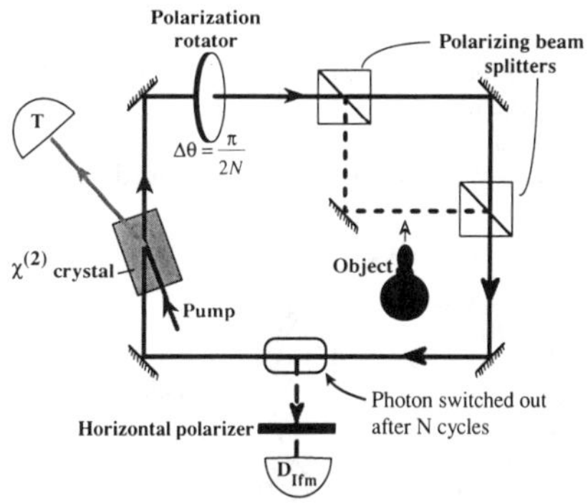

QTuF5 Fig. 3. One method of observing greater than 50% interaction-free measurement. The light makes N cycles before being removed, and its polarization is measured. Without the object, the polarization will with certainty be vertical, but with the object, it will be horizontal.

We have discovered that by applying an optical analog of the quantum Zeno phenomenon[4] one can, in fact, make the probability of an interaction-free measurement arbitrarily close to unity. In Fig. 2(a) the polarization of a photon undergoes an evolution from horizontal to vertical by passing through several polarization rotators. The presence of interspersed horizontal polarizers [Fig. 2(b)] inhibits this evolution. As the number of stages becomes large, the probability that the photon is absorbed becomes vanishingly small. However, the output polarization (still horizontal) reveals the presence of the polarizers.

Figure 3 indicates one method of detecting *any* nontransmitting object by using an interferometer with polarizing beam splitters. At each of N cycles the polarization is rotated by $\pi/2N$ by a half-wave plate, so that, in the absence of an object, the final polarization is vertical. With the object present in the vertical-polarization arm of the interferometer, at each cycle the *nonabsorption* of the photon projects the wave function into the other interferometer arm and the polarization stays horizontal. When the number of cycles is ≥ 4, the 50% limit of Elitzur–Vaidman is exceeded. As the number becomes large, the fraction of IFMs approaches unity.

Curiously, the new technique permits the use of a classical pulse of interrogating light, which substantially simplified our first experiment. When it is combined with an object in a superposition state, it is possible to create a superposition of macroscopic states of light, or even macroscopic entangled states.

*Department of Applied Physics, Stanford University, Palo Alto, California 94305

1. P. G. Kwiat, H. Weinfurter, T. Herzog, A. Zeilinger, and M. A. Kasevich, submitted to Phys. Rev. Lett.
2. C. Bennett, Nature **371**, 479 (1994).
3. A. Elitzur and L. Vaidman, Found. Phys. **23**, 987 (1993).
4. A. Peres, Am. J. Phys. **48**, 931 (1980).

JTuD Tutorial

10:30 am
Room E/F

Epitaxial Growth and Optoelectronics

Alan R. Kost, *Hughes Research Laboratories, Presider*

JTuD1 (Tutorial) **10:30 am**

Epitaxial growth and optoelectronics

Won T. Tsang, *AT&T Bell Laboratories*

This tutorial gives an overview in both epitaxial growth technology and optoelectronic device technology. The fundamentals, unique capabilites, current status, and differences among metalorganic vapor phase epitaxy, molecular beam epitaxy and chemical beam epitaxy will be given. Emphasis will be made on more recent developments such as selective-area-growth for lateral bandgap engineering, and in-situ chemical beam etching/epitaxy with atomic layer control. For the device technology, which is made possible by the powerful growth technique, an overview of the principal of operation, device characteristics and system applications of quantum well lasers in optical fiber communications will be given. Emphasis will be made on recent progress in various advanced laser structures, the strained-layer QW single-frequency lasers, the integrated electro-absorption modulated lasers, the fiber Braff reflector soliton lasers, the 1.48 μm and 0.98 μm pump lasers for optical fiber amplifiers, and the gain-coupled DFB lasers.

JTuE

10:30 am
Room 308

Joint Session on Soliton Lasers and Propagation

Martin Fermann, *IRMA America, Inc., Presider*

JTuE1 **10:30 am**

Colliding pulse mode-locked lasers using Er-doped fiber and semiconductor saturable absorber

Hong Lin, David K. Donald,
Kok-Wai Chang, Steven A. Newton,
Hewlett Packard Laboratories, 3500 Deer Creek Road, Palo Alto, California 94303

A mode-locked Er-fiber laser using a semiconductor saturable absorber[1,2,3] has advantages of self-starting and insensitivity to fiber polarization variation, when compared with either a Figure-8 laser[4] or a nonlinear polarization rotation laser.[5] We present in this paper a new configuration: a colliding pulse mode-locked (CPM) fiber laser with an optimized ion-implanted semiconductor saturable absorber. The laser produced near transform-limited 0.42 ps pulses and showed less polarization dependence than unidirectional ring lasers using the same components.

The scheme of the CPM fiber laser is shown in Fig. 1. The CPM configuration is simpler than a unidirectional ring laser because the isolator in the cavity is not needed. The Er-doped fiber had a length of 8 meters and an Er-doping of 300 ppm. Two 0.22 pitch GRIN lenses were used to image the fiber cores to a 3-μm spot on the saturable absorber. The output coupler had a 17% coupling ratio. The laser operated at its fundamental repetition rate of 9.3 MHz with 20 mW of 980-nm pump power.

Figure 2 shows the measurement results at output A. The FWHM of the autocorrelation trace was 0.65 ps, which corresponds to a 0.42 ps pulsewidth for a sech^2 pulse. The spectral width of 6.19 nm gave the time-bandwidth product of 0.33, which is very close to the transform-limited value of 0.32. The typical output power was 298 μW at port A and 19.4 μW at port B at its fundamental repetition rate.

Previously reported semiconductor saturable absorbers[1,2,3] used for mode-locked Er-fiber lasers have relied on high defect density crystal growth to achieve a short carrier lifetime. However this characteristic is difficult to control and often inadequate for generating subpicosecond pulses. Our approach is to use ion implantation to reduce the carrier lifetime. This relaxes the requirement of crystal growth and results in a very fine control of the carrier lifetime. The saturable absorber used here had a 2-μm thick InGaAsP active layer grown by MOCVD at normal temperature. The bandgap was 1.57 μm and the absorption at 1.55 μm was about 3 dB. The backside of the substrate was polished and both sides of the sample were AR coated.The optimal H^+ implantation was found to be 1.5×10^{15} cm^{-2} with an energy of 260 keV. Figure 2 shows the results of the output pulses. The sample was very uniform and the laser performance did not depend on the "sweet" spots. Figure 3 shows the results when a non-optimal ion-implantation was used. Detailed analysis about the behavior of this laser for different ion implantation dosages will be presented.

We compared the CPM configuration with a unidirectional ring laser in which a polarization independent isolator is inserted at point C in Fig. 1. The CPM configuration produced pulsewidths about 20% to 40% shorter than the comparable unidirectional ring laser. We also found that the CPM laser could tolerate significantly more fiber polarization variation than a unidirectional ring laser. The laser was operated in a turn-key fashion and no polarization adjustment was needed for several months once the laser was set to a correct polarization. We will also present data on the amplitude and phase noise characterization of our lasers.

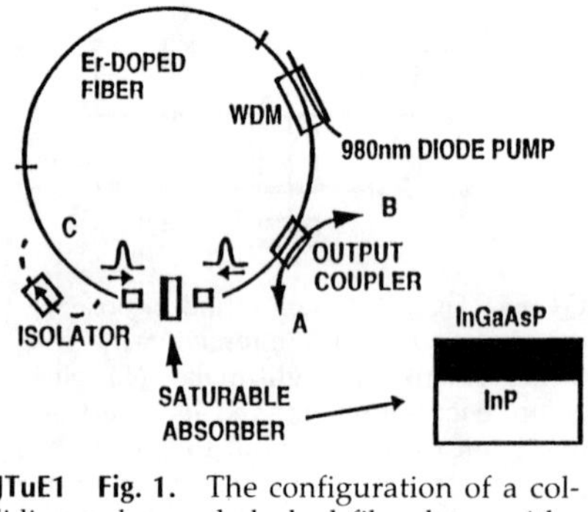

JTuE1 Fig. 1. The configuration of a colliding pulse mode-locked fiber laser with semiconductor saturable absorber. The isolator is used for the comparable unidirectional ring laser.

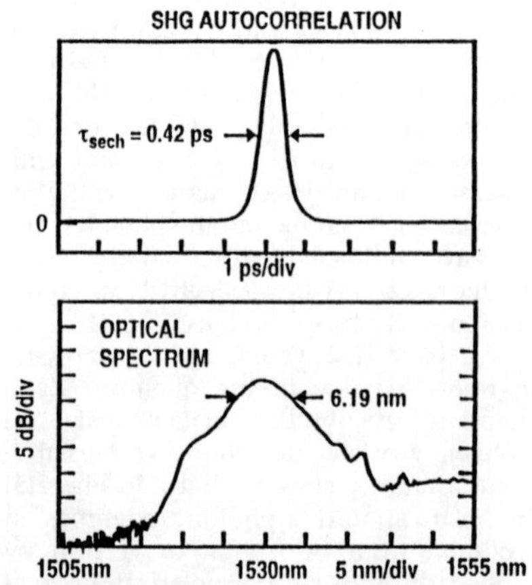

JTuE1 Fig. 2. The autocorrelation trace and optical spectrum of the CPM fiber laser using optimized saturable absorber.

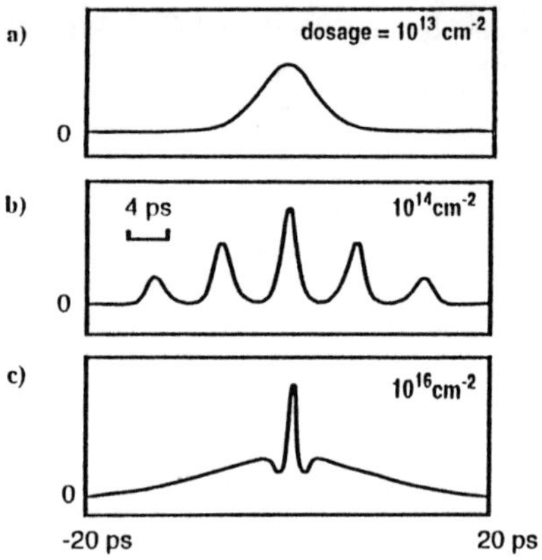

JTuE1 Fig. 3. The autocorrelation traces for non-optimal saturable absorbers. a) and b) implantation dosages below optimal value. c) implantation dosages above optimal value.

1. M. Zirngibl, L. W. Stulz, J. Stone, J. Hugi, D. DiGiovanni, and P. B. Hansen. Electron. Lett. **27**, 1734 (1991).
2. K. V. Reddy, W. Riha, A. Grudinin, D. Richardson, J. Townsend, D. Payne, and M. N. Islam, in *Optical Fiber Communication Conference*, 1993, paper PDP17.
3. E. De Souza, C. Soccolich, W. Pleibel, R. Stolen, J. Simpson, and D. DiGiovanni, Electron. Lett. **29**, 447 (1993).
4. Irl N. Duling III, Opt. Lett. **16**, 539 (1991).
5. M. Nakazawa, E. Yoshida, and Y. Kimura, Electron. Lett. **29**, 1327 (1993).

JTuE2 10:45 am

Noise characterization of EDFLs mode locked by semiconductor saturable absorbers

L. Rahman, B. C. Barnett, M. N. Islam, H. Lin,* R. Fortenberry,* K. W. Chang,* S. A. Newton,* *Department of Electrical Engineering and Computer Science, University of Michigan, 1301 Beal Ave., Ann Arbor, Michigan 48109-2122*

We compare the noise performance of a 980-nm laser diode (LD)-pumped ring erbium-doped fiber laser (EDFL) with a Fabry-Perot (FP) EDFL that is pumped by a diode-pumped ND:YLF laser. For frequencies greater than 200 Hz, the measured rms timing jitter is 3.2 ps for the FP EDFL and 1.8 ps for the ring EDFL. We find that the phase noise below 200 Hz in the FP EDFL is primarily due to mode-hopping in the Nd:YLF pump laser. The performance of the ring EDFL is competitive with passively mode-locked semiconductor lasers[1] and should be appropriate for transmission, switching and sampling applications.

We measured the phase noise using the phase detector method,[2] which demodulates the carrier so that the noise can be measured on the resulting baseband signal. The experimental configurations of the FP and the ring EDFL's are shown in Fig. 1. The FP EDFL, pumped by 415 mW from a LD-pumped cw Nd:YLF laser (1050 nm), produced 4 ps, 0.6 nJ pulses at a 35 MHz repetition rate. The ring EDFL, pumped by 15 mW from a single-mode cw LD (980 nm), produced 0.5 ps, 9 pJ pulses at a 11 MHz repetition rate. Both EDFL's are passively mode-locked using semiconductor saturable absorbers.[3]

The single-sideband phase noise L(f) for the two EDFL's is shown in Fig. 2(a) while the integrated jitter is included in Fig. 2(b). The rms timing jitter for the FP EDFL calculated beyond 200 Hz is 3.2 ps. However, when the integration begins at 10 Hz and includes the peaks near 30 Hz and 140 Hz, the rms timing jitter increases to 38 ps.The low frequency peaks are absent in the phase noise of the LD-pumped ring EDFL. The timing jitter increases from 1.8 ps for f > 200 Hz to only 1.9 ps for f > 10 Hz. We attribute the relatively high jitter at low frequencies in the FP EDFL to random mode-hopping between five longitudinal modes in the Nd:YLF (observed with an optical spectrum analyzer). We tested and ruled out other possible noise sources such as mechanical vibrations, thermal drift, and electrically induced fluctuations. We also checked the FP EDFL frequency stability and used bandwidth limiting elements, isolating the possible noise sources to the Nd:YLF.

The low frequency phase noise of the FP EDFL is due to fluctuations in the absorbed pump power as the Nd:YLF mode-hops rather than to pump noise. We show the relative intensity noise (RIN) of the pump before and after the codoped fiber in Fig. 3. The RIN of the 1050 nm exiting the fiber exhibits a peak near 130 Hz, corresponding to the 130 Hz peak observed in Fig. 2(a). Absorbed pump power fluctuations can lead to refractive index changes in the codoped

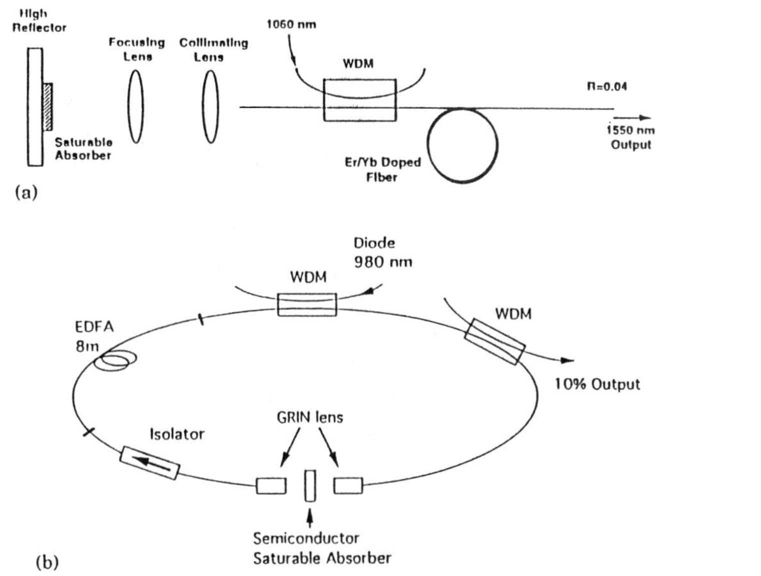

JTuE2 Fig. 1. Experimental configurations of the (a) high-power Fabry-Perot and the (b) ring EDFL's.

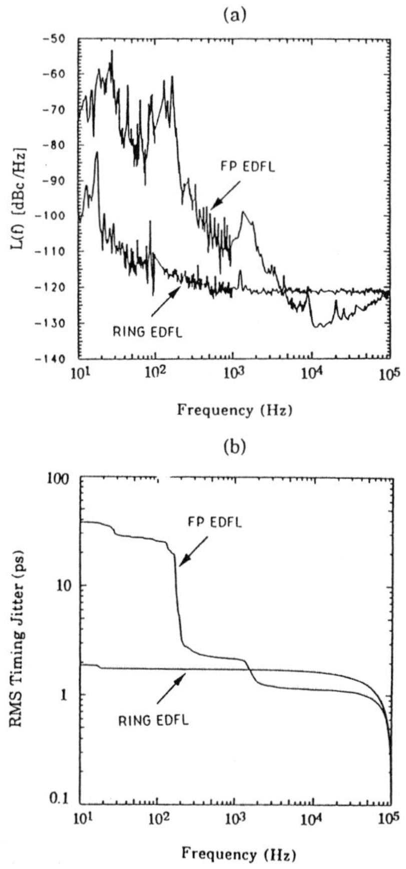

JTuE2 Fig. 2. Single sideband phase noise spectral densities L(f) (a) and integrated timing jitters (b) of the FP and the ring EDFLs.

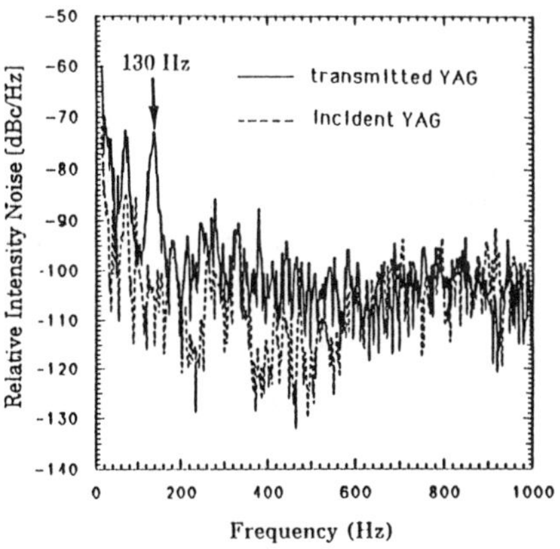

JTuE3 Fig. 3. Relative intensity noise (normalized to a 1 Hz bandwidth) for the transmitted (solid line) and incident (broken line) pump power to the Er/Yb fiber in the high-power FP EDFL.

fiber adding to the timing jitter. The low frequency noise may be lessened by changing the Nd:YLF laser cavity to reduce spatial hole burning effects, flattening the absorption spectrum of the codoped fiber, or by active stabilization of the FP EDFL. The 980 nm semiconductor LD pump for the ring EDFL operates in a single longitudinal mode, thus avoiding the low frequency noise.

In summary, we measure the timing jitter of a Nd:YLF pumped FP EDFL (3.2 ps, 200 Hz < f < 100 kHz) and a semiconductor LD-pumped ring EDFL (1.9 ps, 10 Hz < f < 100 kHz). We provide evidence that the low frequency noise of the FP EDFL may result from the frequency-dependent absorption of the Nd:YLF pump as it mode-hops.

This work is supported in part by ARPA.

**Hewlett-Packard Laboratories, 3500 Deer Creek Road, MS 26M10, Palo Alto, California 94304-1392*

1. S. Sanders, *et al.*, App. Phys. Lett. **59**, 1275 (1991).
2. K. Feher, *Telecommunications, measurements, analysis and instrumentation* (Prentice Hall 1987)
3. B. C. Barnett, *et al.*, High power erbium-doped fiber laser mode-locked by a semiconductor saturable absorber," submitted to Optics Letters.

JTuE3 **11:00 am**

Experimental study of dispersion in fiber-optic ring lasers

D. Prasad, R. Fortenberry, R. Zanoni, *Oklahoma State University, 413 NRC, Center for Laser Research, Stillwater, Oklahoma 74078*

Passively mode-locked fiber lasers have attracted attention because these lasers are of interest as compact sources of ultrashort optical pulses. The minimum pulsewidth of the laser is determined by the gain bandwidth of the erbium amplifier and dispersion in the laser cavity.[1] We have studied the minimum pulsewidth for various laser cavity configurations, and have found regions of operation where the minimum pulsewidth is not limited by linear dispersion in the laser cavity.

To the best of our knowledge, passively mode-locked ring lasers have been operated in the regime where the length of the laser is comparable to the soliton period. In this paper we report results from recent experimental studies of the ring laser in the regime of operation where the length of the laser is in the range of 2 to 12 times the soliton length. Several different ring laser configurations were studied, and it was observed that the relative position of the delay line in the cavity plays an important role in determining the pulsewidth of the laser.

A schematic of the ring laser is shown in Fig. 1. An erbium-doped fiber amplifer 1.4-m long with a doping concentration of 1000 ppm (D = 13 ps/nm/km) is used as the gain medium. The laser is constructed using standard telecommunications fiber Corning SMF-28 (D = 17 ps/nm/km). In the first set of experiments the length of coil A was fixed at 10 m, and the length of coil B was varied from 100 m to 10 m. In this case, the observed pulsewidth closely matched the predicted pulsewidth limited by linear dispersion as illustrated in Fig. 2 (open circles). The theoretical curve shows the expected pulsewidth based on linear dispersion using the result from Dennis *et al.*[1]

$$\tau_{min} = \sqrt{L\beta_2}$$

As the length of coil B was increased the laser became progressively unstable, and it

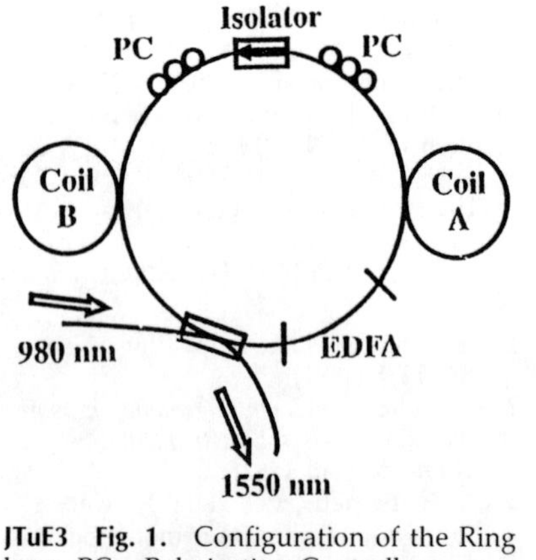

JTuE3 Fig. 1. Configuration of the Ring laser. PC—Polarization Controller, EDFA—Erbium-Doped Fiber Amplifer.

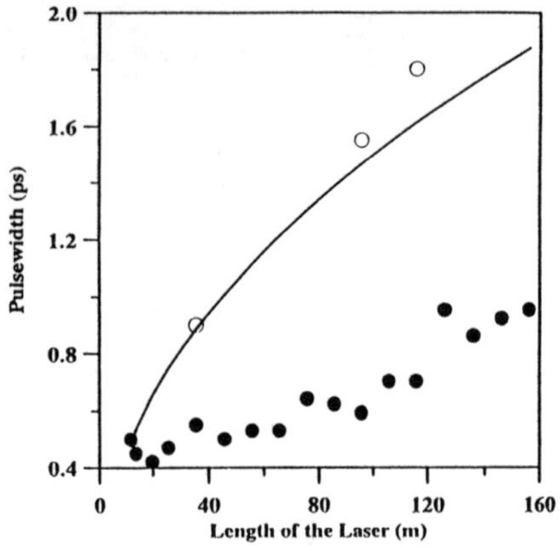

JTuE3 Fig. 2. Plot of the expected pulsewidth and the observed pulsewidth. Solid circles correspond to configuration where the length of coil B is fixed at 2 m and the length of coil A is varied from 150 to 10 m. Open circles correspond to configuration where the length of coil A is fixed at 10 m and the length of coil B is varied from 100 m to 10 m.

was difficult to maintain stable mode-locking with a single pulse circulating in the cavity. However, when the length of coil A is changed in steps of 10 m from an initial length of 150 m while keeping the length of coil B fixed at 2 m (see Fig. 1) we find a very different result. Figure 2 shows a plot of the shortest pulsewidth obtained as a function of the length of the laser cavity (solid circles). These pulses are not limited by the linear dispersion in the laser cavity.

This new regime of operation where the pulsewidth is not limited by linear dispersion can be understood very simply in terms of the intensity of the pulse. Immediately after the isolator, the pulse has low intensity, and therefore its pulsewidth is determined by linear dispersion. Immediately after the amplifier, high intensity pulses are limited by nonlinear dispersion as well.

1. Michael L. Dennis, and Irl Duling III, Experimental Study of Sideband Generation in Femtosecond Fiber Lasers, IEEE J. Quant. Elect. **30**, 1469 (1994).

JTuE4 **11:15 am**

Solitary pulse shaping and optical turbulence in a nonlinear resonator

G. Steinmeyer, F. M. Mitschke,* *Institute for Quantum Optics, University of Hannover, Welfengarten 1, 30167 Hannover, Germany*

Optical cavities containing a Kerr-type nonlinearity have been predicted to show period doubling and chaos.[1] Numerous experimental approaches have confirmed the occurrence of periodic and irregular behavior in such a system. Most of these experiments used a single-mode fiber to suppress transverse effects and a mode-locked laser to induce sufficient self-phase modulation inside the resonator. To date, no experiments have been able to prove the deterministic chaotic origin of these irregularities. In this paper we present a detailed analysis of the irregular behavior of the synchronously pumped passive fiber ring resonator. We will provide, for what we believe to be the first time, evidence of the deterministic character of the irregularities based on time-series analysis. Our analysis reveals a dependence of the dynamics of the system on fiber dispersion and the duration of the pump. When we compare measurements at normal dispersion, close to the zero-dispersion wavelength of the fiber, and in the soliton regime we find striking differences in the dynamical behavior of the system. From our experimental data we identify regimes of low-dimensional chaos that are in close correspondence to the predictions of Ref. 1, but we also find regimes of fully developed spatiotemporal turbulence.

To explain the significant differences in the dynamical behavior, we carried out numerical simulations. These simulations confirm the occurrence of a pulse-fragmentation process. Moreover, the results clearly indicate differences in the pulse-shaping mechanisms encountered in the experiments. In particular, we identify solitary pulse shaping in the regime of anomalous dispersion.

Our numerical results are in excellent agreement with an experimental examination of dynamical changes in the pulse structure. In these experiments we measure the time constant of the fragmented substructures in the pulse. We compare the coherence time of the pulses with a model based on the numerical investigations.

The pulse-shaping process in the device presented here is in direct analogy to the widely studied formation of transverse structure in optical systems. It has also direct implications for instabilities in mode-locked lasers. Because of the conceptional simplicity of the nonlinear fiber ring resonator, we can more easily achieve an understanding of chaotic dynamics in our system; such understanding of the passive fiber resonator will also be helpful in those related fields of research.

*Institute for Applied Physics, University of Münster, Münster, Germany

1. K. Ikeda *et al.*, Phys. Rev. Lett. **45**, 709 (1980).

JTuE5 **11:30 am**

Optical solitons due to competing nonlinearities

Alexander V. Buryak, Yuri S. Kivshar, Stefano Trillo,* *Optical Sciences Centre, Australian National University, Canberra, ACT 0200, Australia*

Recently, it was demonstrated experimentally that nonlinearity-induced phase shifts may be achieved in $\chi^{(2)}$ materials by means of cascaded nonlinearities.[1] This opens new perspectives for analyzing well-known nonlinear effects in cubic materials, such as self-focusing and (spatial or temporal) optical solitons, that exist solely because of $\chi^{(2)}$. As shown in Ref. 2, bright solitons can exist in $\chi^{(2)}$ materials in the form of two-wave symbiotic (i.e., strongly coupled) localized modes of the fundamental and the second-harmonic fields. The explicit analytical forms[2] of such solitons are only a particular case of a family of localized solutions of an effective dynamical system.[3] However, in $\chi^{(2)}$ materials there always exist next-order (i.e., cubic nonlinearities, which might become very important and even strongly compete with quadratic nonlinearities (see, e.g., Ref. 4). In this paper we discuss the effects of the competing nonlinearities on the existence and stability properties of bright optical solitons. Using analytical and numerical (shooting) methods, we find families of the simplest localized soliton solutions and their bifurcations to more complicated ones. Some of these solutions resemble the familiar solitons known for either pure $\chi^{(2)}$ or $\chi^{(3)}$ media, but we also find other solitons that reflect the effect of the competing nonlinearities. We consider the coupled equations obeyed by the envelope amplitudes E_1 and E_2 of the first ($\omega_2 = \omega$) and second ($\omega_2 = 2\omega$) harmonics in a $\chi^{(2)} + \chi^{(3)}$ medium,

$$i\frac{\partial E_1}{\partial z} + i\delta_1\frac{\partial E_1}{\partial \xi} + \gamma_1\frac{\partial^2 E_1}{\partial \xi^2} + \chi_2 E_1^* E_2 e^{i\Delta k z} + \chi_3(|E_1|^2 + \sigma|E_2|^2)E_1 = 0,$$

$$i\frac{\partial E_2}{\partial z} + i\delta_2\frac{\partial E_2}{\partial \xi} + \gamma_2\frac{\partial^2 E_2}{\partial \xi^2} + \chi_2 E_1^2 e^{-i\Delta k z} + 2\chi_3(|E_2|^2 + \sigma|E_1|^2)E_2 = 0 \qquad (1)$$

where z is the propagation distance and ξ represents time in the temporal case (γ_j and δ_j, $j = 1, 2$, are dispersions and group velocities, respectively) or a transverse coordinate for spatial solitons ($\gamma_j + 1/2k_j$). Here $\Delta k \equiv k_2 - 2k_1$ is the wave-vector mismatch, χ_2 and χ_3 are the coefficients of quadratic and cubic nonlinearities, and σ is the cross-phase modulation.

To investigate the propagation of stationary waveforms, we apply the exact transformation $E_1 = [\kappa/(2\sigma\chi_2^2)^{1/2}]\, w \exp(i\beta z + i\Omega\xi)$ and $E_2 = (\kappa/\chi_1)\, v \exp(i\beta_2 z + 2i\Omega\xi)$, where β and $\beta_2 \equiv 2\beta - \Delta k$ are the nonlinearity-induced wave-vector shifts, $\rho \equiv |\gamma_1/\gamma_2|$, $\kappa = \beta + \delta_1\Omega + \gamma_1\Omega^2$, and $\Omega = (\delta_1 - \delta_2)/2(2\gamma_2 - \gamma_1)$. The following reduced equations for real w and v describe parametric interactions in dispersive/diffractive $\chi^{(2)} + \chi^{(3)}$ media:

$$i\frac{\partial w}{\partial \zeta} + r\frac{\partial^2 w}{\partial \tau^2} - w + w^* v + \chi\left(\frac{1}{2\rho}|w|^2 + \sigma|v|^2\right) w = 0,$$

$$i\rho\frac{\partial v}{\partial \zeta} + s\frac{\partial^2 v}{\partial \tau^2} - \alpha v + \frac{1}{2}w^2 + 2\rho\chi\left(|v|^2 + \frac{\sigma}{2\rho}|w|^2\right) v = 0 \qquad (2)$$

where $\zeta = \kappa z$, $\tau = (|\kappa|/|\gamma_1|)^{1/2}(\xi - vz)$, $v = (2\gamma_2\delta_1 - \gamma_1\delta_2)/(2\gamma_2 - \gamma_1)$, $r = \mathrm{sign}(\kappa\gamma_1)$, $s = \mathrm{sign}(\kappa\gamma_2)$, and $\alpha = (\beta_2 + 2\delta_2\Omega + 4\gamma_2\Omega^2)\rho/\kappa$ and where $\chi \equiv \chi_3/\chi_2^2\kappa$ accounts for the effect of competing nonlinearities.

All bright or dark solitons are separatrix solutions of the rather complicated system of Eqs. (2) with $\partial/\partial\zeta = 0$. We have found an exact solution of these equations that exists for a special relation between the system parameters. Stable propagation of this two-wave soliton is shown in Fig. 1. To analyze the existence of more general types of bright solitons), we focus on the case $\Omega = 0$, $\Delta k = 0$ (perfect matching and, as a result, $\rho = 2$ for spatial solitons), $\sigma = 2$ (scalar process), and $r = s = +1$ (as for spatial solitons, or anomalous dispersion for pulses). For this realistic choice of the system parameters, we find the bright solitons as the separatrix trajectories followed by an effective particle that initially sits at the origin ($w = v = 0$) in a potential well $U(w, v)$, which is parametrized by the only (positive or negative) parameter χ. We have found the main types of these trajectories (i.e., lowest-order solitons) and bifurcations. In particular, for $\chi > \chi_{cr} \approx 8.76$ the solutions are solitons of the second harmonics that exist solely because of $\chi^{(3)}$ nonlinearity. For smaller χ's a nontrivial two-wave soliton appears as a bifurcation of the trivial $\chi^{(3)}$ soliton. Such two-wave solitons exist for $\chi > -0.0616$, i.e., even for defocusing nonlinearities for which standard $\chi^{(3)}$ solitons are forbidden, and they strongly compete (i.e., the energies are comparable) with the $\chi^{(3)}$ solitons (still existing for $0 < \chi < \chi_{cr}$). We also discuss the stability diagram for all these (and higher-order) solitons and demonstrate the effects resulting from their propagation and collisions.

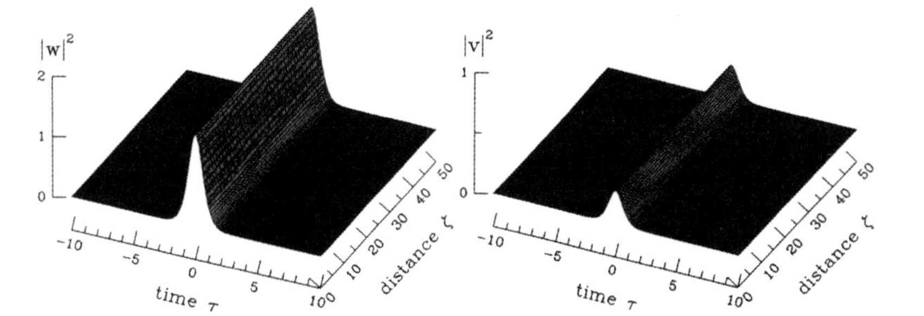

JTuE5 Fig. 1. Propagation of the two-wave bright soliton in the presence of both quadratic and cubic nonlinearities at $\chi = 1$ and $\sigma = 2$.

In conclusion, we have shown that bright solitons exist in dispersive media because of interplay of quadratic and cubic nonlinearities, and we have found novel classes of stable two-wave solitons.

Ugo Bordoni Foundation, Via B. Castiglione 59, 00142 Rome, Italy

1. R. DeSalvo, D. J. Hagen, M. Schiek-Bahae, G. Stegeman, and E. W. Van Stryland, Opt. Lett. **17**, 28 (1992).
2. Y. N. Karamzin and A. P. Sukhorukov, JETP Lett. **20**, 339 (1974); R. Schiek, J. Opt. Soc. Am. B **12**, 1848 (1993); K. Hayata and M. Koshiba, Phys. Rev. Lett. **71**, 3275 (1993); M. J. Werner and P. Drummond, J. Opt. Soc. Am. B **12**, 2390 (1993).
3. A. V. Buryak and Yu. S. Kivshar, Opt. Lett. **19** (October 14, 1994).
4. S. Trillo and S. Wabnitz, Opt. Lett. **17**, 157 (1992).

JTuE6 **11:45 am**

Bright and dark solitons in dispersive $\chi^{(2)}$ materials

Alexander V. Buryak, Yuri S. Kivshar, *Optical Sciences Centre, Australian National University, 0200 Canberra, ACT, Australia*

Recently, it was shown experimentally[1] that nonlinearities of $\chi^{(2)}$ materials may create a nonlinearity-induced phase shift due to cascaded effects. Here we consider parametric interactions between the fundamental and second harmonics in a dispersive (or diffractive) $\chi^{(2)}$ medium and demonstrate that a variety of nonlinear effects related to cubic materials, such as modulational instabilities, self-focusing, and (spatial or temporal) optical solitons, can exist solely because of nonlinear parametric interactions. We present novel families of stable bright and dark two-wave solitons in a purely quadratic dispersive nonlinear medium.

Consider interaction of the first ($\omega_1 = \omega$) and second ($\omega_2 - 2\omega$) harmonics in a $\chi^{(2)}$ medium described by the envelopes E_1 and E_2,

$$i\frac{\partial E_1}{\partial z} + i\delta_1\frac{\partial E_1}{\partial \xi} + \gamma_1\frac{\partial^2 E_1}{\partial \xi^2} + \chi_1 E_1^* E_2 e^{i\Delta k z} = 0,$$

$$i\frac{\partial E_2}{\partial z} + i\delta_2\frac{\partial E_2}{\partial \xi} + \gamma_2\frac{\partial^2 E_2}{\partial \xi^2} + \chi_2 E_1^2 e^{-i\Delta k z} = 0 \qquad (1)$$

where $\chi_1 \equiv (4\pi\omega^2/k_1c^2)\chi^{(2)}(\omega; 2\omega, -\omega)$ and $\chi_2 \equiv (8\pi\omega^2/k_2c^2)\chi^{(2)}(2\omega; \omega, \omega)$, z is the propagation distance, and $\Delta k \equiv (2k_1 - k_2)$ is the wave-vector mismatch between the harmonics. Equations (1) can be used for both spatial and temporal soliton propagation.

We are interested in stationary propagation and apply the following exact transformation: $E_1 = (\kappa/\sqrt{2\sigma\chi_1\chi_2})\, w \exp(i\beta_1 z + i\Omega\xi)$ and $E_2 = (\kappa/\chi_1)\, v \exp(i\beta_2 z + 2i\Omega\xi)$, where β_1 and $\beta_2 \equiv 2\beta_1 - \Delta k$ are the nonlinearity-induced shifts of the propagation constant, $\sigma = |\gamma_1|/|\gamma_2|$, $\kappa = \beta_1 + \delta_1\Omega + \gamma_1\Omega^2$, and $\Omega = (\delta_1 - \delta_2)/2(2\gamma_2 - \gamma_1)$. Now equations for w and v take the form

$$i\frac{\partial w}{\partial \zeta} + r\frac{\partial^2 w}{\partial \tau^2} - w + w^* v = 0,$$
$$i\sigma\frac{\partial v}{\partial \zeta} + s\frac{\partial^2 v}{\partial \tau^2} - \alpha v + \frac{1}{2}w^2 = 0, \qquad (2)$$

where $\zeta = \kappa z$, $\tau = (|\kappa|/|\gamma_1|)^{1/2}(\xi - vz)$, $v = (2\gamma_2\delta_1 - \gamma_1\delta_2)/(2\gamma_2 - \gamma_1)$, $r = \mathrm{sign}(\kappa\gamma_1)$, $s = \mathrm{sign}(\kappa\gamma_2)$, and $\alpha = (\beta_2 + 2\delta_2\Omega + 4\gamma_2\Omega^2)\sigma/\kappa$.

For large α we obtain $v \approx w^2/2\alpha$ and Eqs. (2) can be formally reduced to a single NLS equation for w that possesses bright ($r = +1$) or dark ($r = -1$) solitons. More general stationary solutions are described

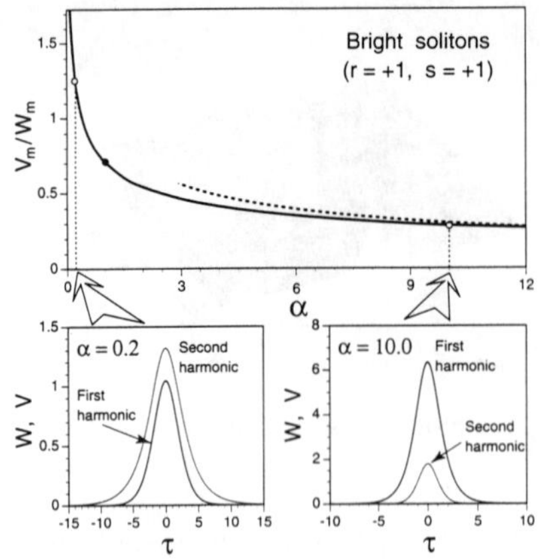

JTuE6 Fig. 1. Characteristic profiles of two-wave bright solitons versus α. The filled circle corresponds to the exact solution, and the dashed curve, $v_m/w_m = 1/\sqrt{\alpha}$, corresponds to the asymptotic solution described by an effective NLS equation.

by system (2) for the real envelopes $w(\tau)$ and $v(\tau)$ without derivatives in ζ. This stationary problem may be considered as a mechanical problem with Hamiltonian $H = rp_w^2/2 + sp_v^2/2 + U(w, v)$, where $p_w = w'$, $p_v = v'$, and $U(w, v) = 1/2(w^2v - \alpha v^2 - w^2)$.

To find separatrix solutions corresponding to solitons, we analyze Eqs. (2) for real $w(\tau)$ and $v(\tau)$ by means of the numerical shooting technique. In the case $r = s = +1$ for $\alpha > 0$ we have found a family of stable, bright one-soliton solutions (see also Ref. 2. The examples of such solitons are shown in Fig. 1. Note that this family of bright solitons includes, as a particular case at $\alpha = 1$, an exact solution.[3,4]

In the case $r = s = -1$ (for $\alpha > 0$) we have found a discrete set of unstable, dark-soliton solutions of a complicated form. These solutions do not form a family and exist only at some particular values of the parameter α. In the case $r = 1$, $s = -1$ (for $\alpha > 0$) there exists a discrete set of unstable, bright solitons. These solutions also do not form a family. One of these unstable solutions (at $\alpha = 1$) is known in an explicit analytical form.[5]

Finally, in the case $r = -1$, $s = 1$ we have found a family of dark solitons. Solitons of this family exist for every positive value of α. However, the modulational-instability analysis predicts that at $r = -1$ and $s = +1$ the cw background becomes unstable provided that $\alpha < 2$. Additionally, for $\alpha < 8$, dark solitons have nonmonotonic tails that result in local minima for an effective interaction potential of interacting solitons. As a result, we show that dark solitons with nonmonotonic tails attract each other and can form bound states. These solitons are shown to exist for every positive value $\alpha < 8$, and the solitons of different families can be classified by the number of single dark solitons forming these bound states and also by the relative distance between the neighboring solitons.

1. R. DeSalvo *et al.*, Opt. Lett. **17**, 28 (1992).
2. A. V. Buryak and Yu. S. Kivshar, Opt. Lett. **19** (October 15, 1994).
3. Y. N. Karamzin and A. P. Sukhorukov, JETP Lett. **20**, 339 (1974).
4. K. Hayata and M. Koshiba, Phys. Rev. Lett. **71**, 3275 (1993).
5. M. J. Werner and P. D. Drummond, J. Opt. Soc. Am. B **10**, 2390 (1993).

QTuG 1:00 pm

2nd and 3rd Level of the Convention Ctr.

Poster Session

Physics of Coherent Light Sources

QTuG1

Stimulated emission in a dye solution with scatterers

Daozhong Zhang, Bingying Cheng, Wei Hu, Zhaolin Li, Junhui Yang, *Optical Physics Laboratory, Institute of Physics, Chinese Academy of Sciences, Beijing 100080, China*

Recently, Lawandy *et al.* have observed laser action in a laser-pumped high-concentration dye solution with the addition of scattering particles. A quite different result has been found also: when a low-concentration pure-dye solution is excited, superradiance may appear, and with the addition of scatterers the threshold of superradiance is enhanced. Here we report an experimental study on the relation between the bandwidth of the emission and the gain length by using two kinds of gain area: point and line. For point excitation the laser beam was focused to a small point with a diameter less than 2 mm. In the other case the laser beam was focused by a cylindrical lens. Then the gain length was increased by an order of magnitude over that of the point excitation. We have tried two kinds of commercial TiO_2, particles with an average diameter of 1 μm and powdered TiO_2, and we have used homemade $SrTiO_3$ with particles with diameters of several micrometers.

For point excitation, a high dye concentration of 5×10^{-3} mol/L, when the density of TiO_2 particles was around 10^9 cm^{-3} (corresponding to 20 mg of TiO_2 in 10 mL of dye solution) and the incident energy was more than 1 mJ, a strong isotropic emission centered at 616 nm with a bandwidth of 7 nm was detected, which was much narrower than the bandwidth of normal fluorescence. As the density of TiO_2 particles increased to 4×10^9 cm^{-3}, the shape of the emission spectrum was almost unchanged. However, the intensity of the emission increased dramatically if the laser power was increased from 1 to 25 mJ.

When 5 mg of TiO_2 powder was put into 2 mL of dye solution and the laser output was similar, we also observed narrow-bandwidth emission from such a colloid. The intensity distributions for particles and powder were nearly the same.

In addition, $SrTiO_3$ particles were used. Under a lower concentration of dye (<3.5 $\times 10^{-4}$ mol/L) and low density of $SrTiO_3$, only normal fluorescence was observed. When we put 10 mg of $SrTiO_3$ into 4 ml of Rh-640 methanol solution of 5×10^{-3} mol/L (the average density of $SrTiO_3$ here was about 10^9 cm^{-3} also), a strong emission with a bandwidth of 7 nm appeared, which is comparable to that for TiO_2. When the pump energy was increased from 12 to 25 mJ, the emission intensity went up by a factor of 8. The situation was much the same as that when TiO_2 particles were applied.

At a lower dye concentration of 5×10^{-4} mol/L we observed only the ordinary fluorescences in both cases, with and without scatterers.

For line excitation, for a dilute pure Rh-640 dye solution of 5×10^{-4} mol/L an intense directional emission with a bandwidth of 4 nm and a divergence of 0.1 rad occurred along the pumping line when the incident energy was more than 1 mJ. However, when TiO_2 particles of 10^9 cm^{-3} were added to the solution, the bandwidth of the emission became much wider, and its intensity decreased by nearly 2 orders of magnitude.

We also put a dispersive grating at one end of the cuvette, which contained 3.5×10^{-3} mol/L dye solution and a small amount of TiO_2 particles. When the grating was adjusted properly, a bright spot could be readily seen. In the spectrum there were two peaks. The broader one coincided with the peak that appeared without the grating. The much narrower one was caused by the grating, and its wavelength varied with the diffraction angle of the grating. This phenomenon implied that a population inversion of dye molecules in the solution may be present.

For the pure dye solution (>3×10^{-3} mol/L), under line excitation the emission was always of narrow bandwidth, regardless of the density of the scatterers, as long as the pumping energy was more than 1 mJ.

In conclusion, our experimental results show that a narrow-bandwidth emission from a laser-pumped dye solution with added scatterers can be expected if the gain length is long enough; it is due to multiple scattering of photons with scatterers, so that the stimulated process dominates. The shape and size of added scatterers should not be crucial.

QTuG2

New repumping scheme for a sodium magneto-optical trap

A. M. L. Oien, I. T. McKinnie, P. J. Manson, N. J. Mulgan, W. J. Sandle, D. M. Warrington, *Department of Physics, University of Otago, Dunedin, New Zealand*

The role of repumping in the magneto-optical trap (MOT) has often been seen as a necessary evil, requiring the introduction of either an electro-optical modulator or extra laser.[1,2] However, repumping can be exploited to improve the performance of the MOT,[3] or to study further the behavior of such a trap. We report results of a modified MOT using for repumping the D_1 transitions $3S_{1/2}$ $F = 1$ to $3P_{1/2}$ $F = 1$ and $F = 2$. Because these wavelengths are well displaced from the usual (D_2) repump wavelengths, the direct repump influence on the trapping and cooling laser interaction should be significantly reduced. We also expect further possibilities for cold-atom spectroscopy within the MOT. The possibility of a dark trap with this repumping transition has been suggested.[3]

The cooling/trapping and repump beams are provided by separate cw ring dye lasers. The trap laser, slightly red-detuned from $3S_{1/2}$ $F = 2$ to $3P_{3/2}$ $F = 3$, is split in the normal configuration into three beams that intersect orthogonally in a sodium-vapor cell. The repumping laser is mixed into only one dimension of the trapping laser beams in order to simplify the repump polarization in the trap environment. This polarization is the same σ^+/σ^- configuration as the trapping beam pair it joins.

The temperature is obtained by applying an oscillating magnetic field to the trap and measuring the phase lag of the oscillation in trap position.[4] An investigation of the behavior of this modified MOT is presented with the temperature and density of the trap characterized, and the dependence of trap qualities on detuning and intensity is discussed.

1. E. Raab, M. Prentiss, A. Cable, S. Chu, and D. Pritchard, Phys. Rev. Lett. **59**, 2631 (1987).
2. C. Monroe, W. Swann, H. Robinson, and C. Wieman, Phys. Rev. Lett. **65**, 1571 (1990).
3. W. Ketterle, K. Davis, M. Joffe, A. Martin, and D. Pritchard, Phys. Rev. Lett. **70**, 2253 (1993).
4. P. Kohns, P. Buch, W. Suptitz, C. Csambal, and W. Ertmer, Europhys. Lett. **22**, 517 (1993).

QTuG3

Polarization bistability in strained 1.3-μm InGaAsP/InP ridge-waveguide lasers: theory and experiment

A. Klehr, G. Berger, R. Müller, M. Voss, *Max Born Institute for Nonlinear Optics and Ultrafast Spectroscopy, Rudower Chaussee 6, 12489 Berlin, Germany*

The polarization of light emitted from semiconductor lasers can be strongly influenced by introduction of stress in the active layer. Especially, strained InGaAsP lasers were found to exhibit dominant TM-mode emission, TE/TM-mode coexistence as well as TE/TM-mode switching, and polarization bistability.[1-3] The latter effect is of interest for device applications such as optical switches and memory for computing and signal processing.

In this paper we present a detailed numerical analysis of the physical mechanisms underlying TE/TM bistability in 1.3-μm ridge-waveguide InGaAsP/InP bulk lasers at room temperature. The theoretical model is based on the paraxial wave equations for TE- and TM-polarized modes and the diffusion equation for the carrier distribution. Calculated light power–current (P–I) characteristics, as well as lateral near- and far-field patterns, are compared with measurements.

Figure 1 shows calculated and measured P–I characteristics, which indicate polarization bistability within a hysteresis of width $\Delta I_H \sim 2$ mA. Lasing starts with TE-mode emission and switches abruptly from TE to TM at switching current I_{S1}. Above this current value the laser exhibits TM

Tuesday Posters

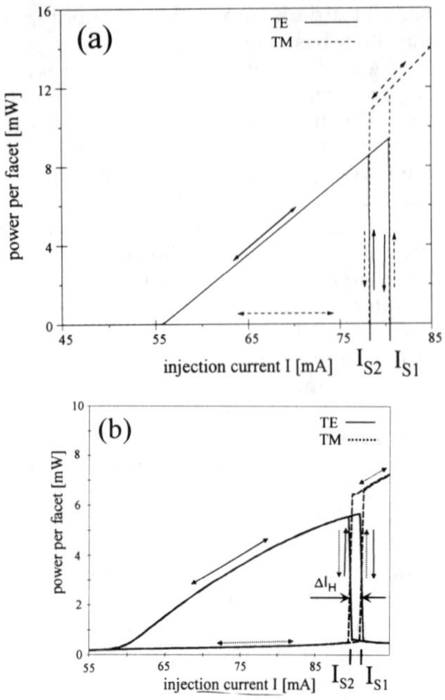

QTuG3 Fig. 1. (a) Calculated and (b) measured polarization-resolved light power versus current for a polarization-bistable InGaAsP/InP ridge-waveguide laser.

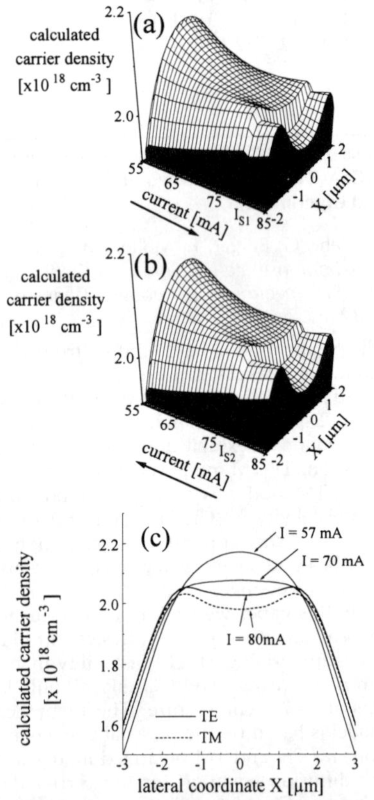

QTuG3 Fig. 2. Calculated lateral profiles of carrier concentration near a laser facet for (a) increasing and (b) decreasing injection current within a specified current interval of 55 mA to about 85 mA and (c) for selected injection currents, showing a significant difference between the depths of the spatial holes for TE and TM lasing in the bistable region (I = 80 mA).

emission. Reducing the current the polarization switches back to TE at I_{S2}.

Figure 2 shows calculated lateral profiles of carrier concentration below the ridge for increasing and decreasing injection current near a laser facet. At higher current, a spatial hole is burnt in the carrier distribution at the center of the light beam $X = 0$, which becomes much more pronounced above I_{S1}, after the transition from TE to TM lasing. Figure 2(c) allows a direct comparison of the spatial holes burnt in the carrier distribution corresponding to TE and TM lasing in the bistable region at I = 80 mA.

Our combined experimental and theoretical investigations indicate that the observed TE/TM bistability is due to the interplay of three different effects: (1) Tensile stress promotes the TM gain strongly enough to compete with the TE mode. (2) Improved TM waveguiding due to an enhancement of the effective refractive index near the beam axis caused by carrier depletion with increasing current leads to the onset of TM lasing and TE/TM switching. (3) The TE/TM transition is accompanied by an abrupt increase of spatial hole burning in the lateral carrier distribution. Because of this nonlinear effect, a lower current is needed to switch the laser back to TE, giving rise to a hysteresis loop in the P–I characteristics and to TE/TM polarization bistability. Consequences of these results for the TE/TM switching dynamics are discussed.

1. Y. C. Chen and J. M. Liu, Opt. Quantum Electron. **19**, S93 (1987).
2. C. S. Adams and D. T. Cassidy, J. Appl. Phys. **64**, 6631 (1988).
3. A. Klehr, A. Bärwolff, R. Müller, M. Voss, J. Sacher, W. Elsässer, and E. O. Göbel, Electron. Lett. **27**, 1680 (1991).

QTuG4

Coherent pulse amplification in a two-component medium

A. V. Andreev, P. V. Polevoy, *Department of Physics, M. V. Lomonosov Moscow State University, Moscow 119899, Russia*

We report here the results of computer simulations of coherent pulse amplification in a two-component medium. The two-component medium consists of two species of resonant atoms with the same frequency of the resonant transition and different values of the dipole matrix elements of transition $d_a < d_b$. It is well known that the Rabi frequency is proportional to the dipole matrix element of transition ($\Omega \sim d$). Here $\Omega_a < \Omega_b$ for $d_a < d_b$, and we shall call the a atoms slow atoms and the b atoms fast atoms. Initially the slow atoms are in the excited state and the fast atoms are in the ground state. In the case of pure coherent interaction of the incident pulse with both components, the dynamics of the Bloch angle $\theta(x, t) = \sqrt{\beta_a} \int_{-\infty}^{t} a(x, t)\, dt$ are determined by the equation

$$\frac{\partial^2\theta}{\partial t^2} + \frac{\partial^2\theta}{\partial x \partial t} = R\beta_a \sin\theta - r\sqrt{\beta_a\beta_b}\,\sin\left(\sqrt{\frac{\beta_b}{\beta_a}}\,\theta\right), \qquad (1)$$

where $R(r)$ is the concentration of the slow (fast) atoms and the parameters $\beta_{a,b}$ are

$$\beta_{a,b} = \frac{2\pi\omega|d_{a,b}|^2}{\hbar}\frac{N_a}{V}\tau^2. \qquad (2)$$

The signs of the two terms in the right-hand side of Eq. (1) indicate the initial state of the slow and fast atoms. It follows from Eq. (1) that in the case of an incident pulse of small area, the pulse area $\Sigma(x) = \theta(x, t = \infty)$ will increase if the right-hand side of Eq. (1) is positive. Therefore, the threshold condition of coherent amplification for the weak pulses is

$$R\beta_a > r\beta_b, \qquad r_{th} = \frac{\beta_a}{\beta_b}R. \qquad (3)$$

If the condition $r < r_{th}$ holds, we have above-threshold amplification. In the opposite case the regime of subthreshold amplification occurs. In the latter case the area of the propagating pulse increases only if the incident pulse area is above the critical area Σ^*. For the case $\beta_b = 4\beta_a$ the critical pulse area Σ^* is

$$\Sigma^* = \arccos\left(\frac{R}{4r}\right). \qquad (4)$$

In the case $\Sigma_{in} < \Sigma^*$ the pulse area tends to zero. Therefore such a pulse is absorbed by the two-component medium.

Let us consider the case $\beta_b = 4\beta_a$. Computer simulations show that in the process of amplification the pulse area tends to π. Hence we obtain the π-pulse with respect to the slow atoms and two π-pulses with respect to the fast ones. In spite of the fact that the peak pulse area remains constant, the pulse profile varies with the distance traveled. Figure 1 shows the results of computer simulations for the dependency of the peak pulse intensity on the distance traveled in the two-component medium for various values of the fast-atom concentrations $r_1 < r_2 < r_3 < r_4 < r^* = R/(4 \cos \Sigma_0)$. It is seen that the amplification increases as the fast-component concentration approaches the critical value $r^* = R/(4 \cos \Sigma_0)$. However, when $r > r^*$, which means that $\Sigma_{in} < \Sigma^*$, the incident pulse is absorbed by the two-component medium. It should be noted that the spatiotemporal dynamics of pulse amplification in a two-component medium are substantially different from those for amplification in a monocomponent medium. Figure 2 shows the peak pulse

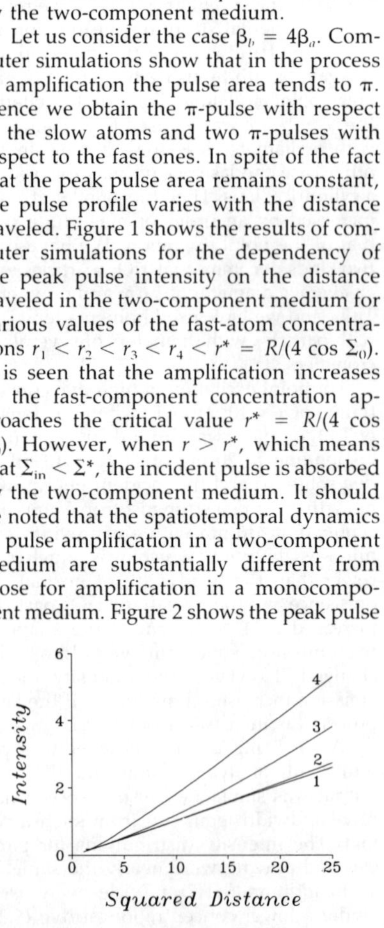

QTuG4 Fig. 1. Pulse intensity as a function of the distance traveled in a two-component medium with the following values of the concentration of the fast component r: 0.25 (curve 1), 0.35 (curve 2), 0.4 (curve 3), and 0.45 (curve 4).

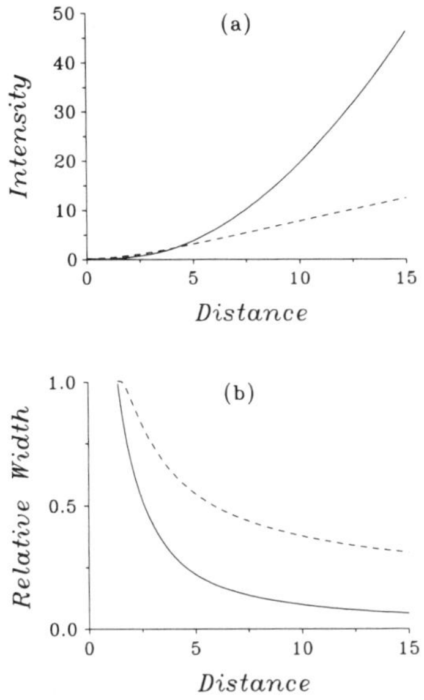

QTuG4 Fig. 2. (a) Intensity and (b) pulse width as a function of the distance traveled in two-component (solid curves) and monocomponent (dashed curves) media.

intensity and the pulse width as a function of the distance traveled for two-component (solid curve) and monocomponent (dashed curve) media. The two-component subthreshold medium and the monocomponent medium with the same value of R and with $r = 0$ are compared here. It is seen that the dependencies for the two-component and monocomponent media differ substantially. In the two-component medium the rise of the peak pulse intensity with distance is higher than that in the monocomponent medium. As in the two-component medium, the pulse width decreases inversely proportional to the distance traveled. In the monocomponent medium an incident pulse of small area is transformed into a soliton and the pulse temporal width at the large distances tends to a constant, which is a soliton width.

1. A. V. Andreev and P. V. Polevoy, ZETP, number 11 (1994).

QTuG5

Theoretical modeling of bimode erbium-doped fiber lasers

François Sanchez, Pascal Besnard, Guy Michel Stephan, *ENSSAT, Laboratory of Optronics, National Center for Scientific Research (EP0001), 6 rue de Kérampont, B.P. 447, 22305 Lannion Cedex, France*

Multimode lasers are a fascinating field of investigation because of their various and complex dynamics. Among them, antiphase, or winner-takes-all, dynamics have recently received great attention.[1] In this context, erbium-doped fiber lasers are of particular interest because of their large variety of dynamical behaviors: the cw or self-pulsing regime[2] when operating around 1.55 μm and nT-periodic orbits and chaotic motion when operating simultaneously at 1.55 and 1.536 μm.[3]

A simple modeling of the dynamics has been developed[2] in the case of single-mode oscillation. The model was based on the existence of isolated erbium ions (concentration $1x-2x$) and ion pairs (concentration x) (or clusters) that act as a saturable absorber. The dynamics depend on both the ion-pair concentration and the pumping rate[2]: for low x, the laser is cw; for intermediate x values, the laser is self-pulsing from pumping rates ranging from the threshold to a higher value beyond which a cw operation is attained; for high x, the laser is self-pulsing in a large range of pumping rates. More recently, a quasi-periodic route to chaos has been reported for dual-wavelength operation.[3] The typical scenario, for $x = 7.5\%$, is cw → T-periodic → $2T$-periodic → $3T$-periodic → chaos → T-periodic, for decreasing pumping rates r.

In this paper we present a theoretical modeling of bimode erbium-doped fiber lasers.[4] The model is based on the well-known two-mode laser model, to which we have added two equations for the population dynamics of ion pairs. The numerical results are similar to the experimental data for $x = 10\%$. Figure 1 gives the bifurcation diagram that represents the evolution of the extrema of the intensity versus the pumping ratio r. Antiphase dynamics between the two modes are observed in transient regimes and in the asymptotic regime for the nT-periodic solutions such that $n \geq 2$. The chaotic regime takes place between the $3T$- and the T-periodic orbits. The first return map given in Fig. 2(a) shows a two-dimensional distribution, suggesting that its strange attractor has a higher dimension than that of the single-mode chaos, for which the first return map is close to a one-dimensional curve. Figure 2(b) shows a representation of the strange attractor in the phase space. Note that the attractor is distributed three dimensionally. The calculated correlation dimension is about 2.8.

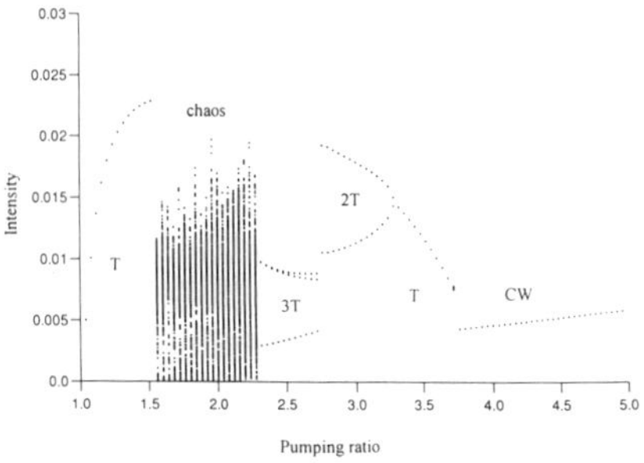

QTuG5 Fig. 1. Bifurcation diagram for $x = 10\%$.

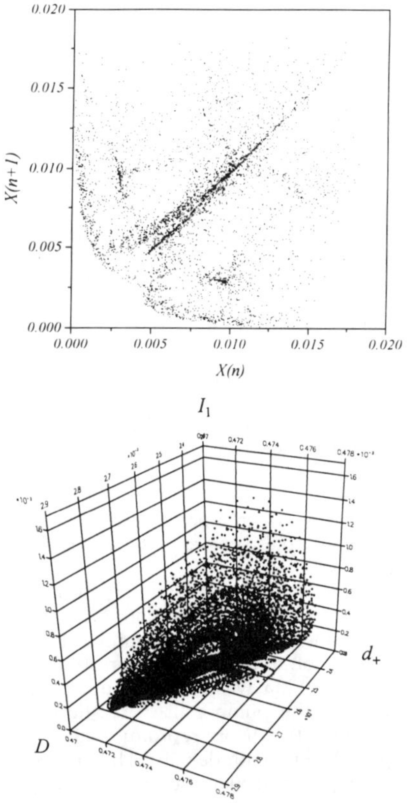

QTuG5 Fig. 2. (a) First return map in the chaotic regime; (b) portrait of the strange attractor.

1. K. Wiesenfeld, C. Bracikowski, G. James, and R. Roy, Phys. Rev. Lett. **65**, 1749 (1991).
2. F. Sanchez, P. LeBoudec, P. L. François, and G. Stephan, Phys. Rev. A **48**, 2220 (1993).
3. F. Sanchez, M. LeFlohic, G. Stephan, P. LeBoudec, and P. L. François, "Quasi-periodic route to chaos in erbium-doped fibre laser," IEEE J. Quantum Electron, in press.
4. F. Sanchez, M. LeFlohic, P. Besnard, P. L. François, and G. M. Stephan, "Erbium laser model for two-wavelength operation," submitted to J. Phys. III (Paris).

QTuG6

Density dependence of free-carrier scattering

David Snoke, *Department of Physics and Astronomy, University of Pittsburgh, 3941 O'Hara Street, Pittsburgh, Pennsylvania 15218*

In the past few years, a number of experiments have probed the rate of scattering of free carriers in semiconductors, primarily free conduction-band electrons in GaAs, as a function of electron density.[1–6] The beauty of these experiments is that although the absolute rate of scattering at a given density is difficult to determine, the relative change of the carrier–carrier scattering rate as density varies can be measured extremely accurately. Therefore, these results provide stringent tests of the theory of scattering of charged particles.

Simulations of the evolution of a nonequilibrium electron gas have been performed by numerically integrating the Boltzmann equation for the electron energy distribution.[5,7,8] The matrix element for charged-particle scattering that enters the Boltzmann equation, $M_{121'2'} = 4\pi e^2/\varepsilon(K, \omega)VK^2$, must be treated with care. Most numerical studies of carrier–carrier scattering have used the quasi-static approximation,

$$\frac{\varepsilon(K, \omega)}{\varepsilon_\infty} \approx \frac{\varepsilon(K, 0)}{\varepsilon_\infty} = 1 + \frac{q_D^2}{K^2}, \qquad (1)$$

with Debye screening given by

$$q_D^2 = -\frac{4\pi e^2}{\varepsilon_\infty} \sum_{\mathbf{k}} \frac{\partial f(E(\mathbf{k}))}{\partial E} \qquad (2)$$

This quasi-static approximation is actually the $\omega \rightarrow 0$ and long-wavelength limit of the more general random-phase approximation (RPA) for the dielectric function, which is valid for scattering energies much less than the plasmon energy of the gas. It turns out, however, that typical energies exchanged in the thermalization of an electron gas at low density are comparable to the plasmon energy.

In the full RPA calculation for an isotropic gas, ε has the same form as Eq. (1) but q_D is replaced by q_{RPA}, which is a function of ($\hbar\omega/K$). Figure 1(a) shows a calculation of the screening parameter q_{RPA} for a test particle with kinetic energy E, for a Maxwellian energy distribution of the particles. Figure 1(a) shows that for most of the particles the effect of the RPA correction is to reduce the screening. A small fraction of the particles at high energy see a negative value of q^2_{RPA}, which corresponds to plasmon emission. This is consistent with the result of Pines and Bohm,[9] that only electrons with energy well above the average will emit plasmons.

Figure 1(b) shows the same calculation for a Gaussian energy distribution of the particles, which is a typical early nonthermal distribution generated by a laser. In this case, plasmon emission is prohibited for essentially all of the particles. The screening is slightly greater in this case than it is for a thermal distribution with the same average energy. An important result of these calculations, however, is that the RPA corrections do not change the density dependence of the scattering rates.

A typical solution of the Boltzmann equation simulating the experiment of Kash[4] is shown in Fig. 2. In these experiments a

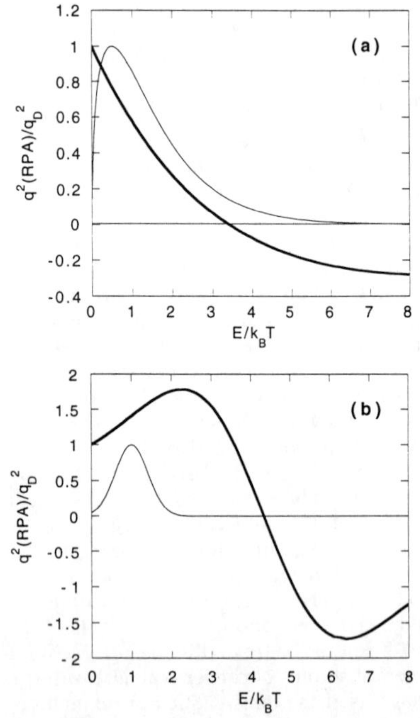

QTuG6 Fig. 1. Screening wave vector q_{RPA} relative to the quasi-static screening wave vector $q_D \equiv q_{RPA}$ (0) (heavy line) for (a) the Maxwell-Boltzmann distribution (light line) and (b) the Gaussian distribution (light line).

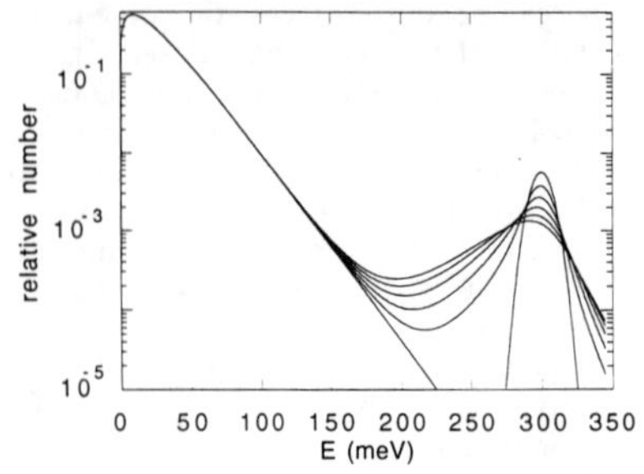

QTuG6 Fig. 2. Boltzmann-equation solution of a model of the Kash pulse–probe experiments.[4] The probe is assumed to be created instantaneously at $t = 0$ in a Gaussian distribution at $E = 300$ meV in the presence of a Maxwell–Boltzmann distribution with density 2×10^{16} cm^{-3}. The curves correspond consecutively to the distribution at $t = 0$, 0.35, 0.71, 1.09, 1.47, and 1.82 ps.

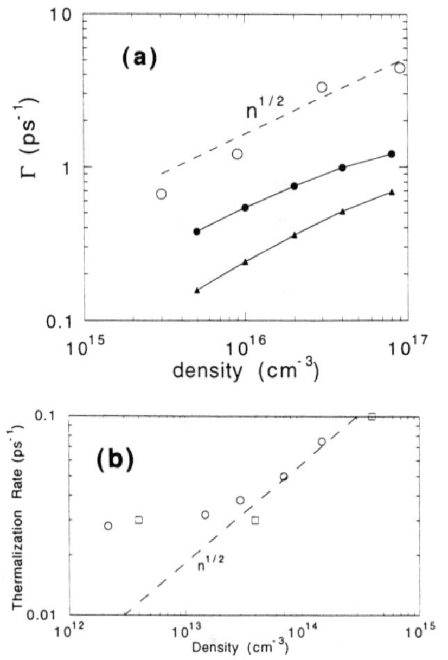

QTuG6 Fig. 3. (a) Summary of the energy-loss rate found by Kash[4] (open circles) and calculated from the model shown in Fig. 2, as a function of density: filled circles, initial probe energy width 10 meV (corresponding to the experiment of Kash); filled triangles, initial probe energy width 20 meV. (b) Summary of the thermalization-rate data of Refs. 5 (open squares) and 6 (open circles).

cold thermal distribution is created by an intense laser pulse preceding the probe pulse by 50 ps.

Figure 3(a) shows a summary of the instantaneous energy-relaxation rate, compared to the energy-loss rate found by Kash,[4] as a function of the electron density in the low-energy Maxwellian distribution generated by the pump pulse. As shown in this figure, the theory reproduces the scaling of the rate with density. The factor-of-2 discrepancy in the absolute rates can be attributed to experimental uncertainties and to corrections from the RPA-screening and electron–hole-scattering calculations.

Figure 3(b) shows a summary of the

QTuG6 Table 1. Density dependence of rates for electron–electron scattering.

Density	Dephasing	Energy Relaxation	Experimental References
Low			
3D	const.	$n^{1/2}$	4–6
2D	const.	n	
High			
3D	$n^{1/3}$	$n^{2/3}$	1
2D	$n^{1/2}$	n	2, 4

thermalization rates measured in Refs. 5 and 6. The thermalization rate scales approximately as $n^{1/2}$ at the highest densities, similar to the results of Kash and consistent with this theory.

The screening length cannot be less than the average interparticle distance. The quasi-static formula (2) predicts a screening length decreasing as $n^{-1/2}$, however, whereas the interparticle separation decreases only as $n^{-1/3}$. At some point, the screening length will begin to follow the interparticle spacing, decreasing as $n^{-1/3}$. Similar considerations can be made for momentum relaxation in two dimensions. Therefore, a high-density, nondegenerate regime occurs in both two and three dimensions, in which the relaxation rates scale differently with density. Table 1 summarizes the density and temperature dependence of the various rates found in this work, in two and three dimensions in both the low-density regime and the screening-limited (high-density) regime. The electron gas is assumed nondegenerate in all cases. A notable result is the deduction of a constant total scattering rate at low density, which gives the density dependence of quantum dephasing. This is consistent with calculations that find finite dephasing, even in the collisionless regime.[10] Also, the results of Kash in two-dimensional quantum wells,[4] which showed a thermalization rate proportional to n, are found to be consistent with the results of Bigot *et al.*,[2] which found a dephasing rate proportional to $n^{1/2}$ in the same density range.

1. P. C. Becker *et al.*, Phys. Rev. Lett. **61**, 1647 (1988).
2. J.-Y. Bigot *et al.*, Phys. Rev. Lett. **67**, 636 (1991).
3. J. A. Kash, Phys. Rev. B **40**, 3455 (1989).
4. J. A. Kash, Phys. Rev. B **48**, 18336 (1993).
5. D. W. Snoke, W. W. Rühle, Y.-C. Lu, and E. Bauser, Phys. Rev. Lett. **68**, 990 (1992); Phys. Rev. B **45**, 10979 (1992).
6. R. M. Hannak and W. W. Rühle, Proc. Soc. Photo-Opt. Instrum. Eng. **2142**.
7. D. W. Snoke, Phys. Rev. B **47**, 13346 (1993).
8. D. W. Snoke, Phys. Rev. B **50** (1994).
9. D. Pines and D. Bohm, Phys. Rev. **85**, 338 (1952).
10. E.g., E. M. Lifshitz and L. P. Piaevskii, *Physical Kinetics* (Pergamon, Oxford, 1981).

QTuG7

Robust operation of a dual-core fiber ring laser

Yun Je Oh, J. W. Haus, S. L. Doty,* R. L. Fork,** *Department of Physics, Rensselaer Polytechnic Institute, Troy, New York 12180-3590*

Recently, Winful and Walton[1] proposed a dual-core mode-locked fiber laser. This approach with dual-core fiber is particularly attractive because the second core provides a saturable-absorber action that not only strongly reduces pulse duration, but also efficiently removes energy shed by the pulse. To the best of our knowledge, no one has published an experimental realization of this laser. Our investigations extend the work of Winful and Walton and identify an experimentally realizable configuration. In particular, our simulations predict pulses that are robust against distortions common to single-mode fiber lasers. Unlike the shortest pulses generated in typical single-mode fiber lasers, these pulses closely approximate lowest-order solitons, have minimal chirp, and show virtually no pedestal.

The dual core in our example forms a unidirectional ring geometry in which a portion of one of the cores is active, i.e., erbium-doped, whereas the other core is entirely passive. The active core section forms a cavity with an output coupler that retains 90% of the pulse energy in the cavity. The other core is passive, and any energy transferred from the active core to the passive core is lost after one round trip. The laser length is approximately half a beat length for the dual-core fiber. We set the gain at 4 dB.

Our numerical simulations examine the intrapulse Raman scattering, the amplifier bandwidth, the gain dispersion, the self-steepening, the third-order dispersion, the Raman self-frequency shift, and the robustness of the laser against variations of the cavity length. The number of FFT points is 4096, and the temporal window size is approximately 30 in normalized time units. We compared our results with a broad segment of the literature and have tested two separate programs against each other for consistency.

Stimulated Raman scattering does lower the central frequency of the pulse. The restriction imposed by the gain bandwidth, however, eventually halts the Raman frequency shift because as the pulse moves to lower frequencies, it experiences smaller gain. The spectrum undergoes significant modification during the initial shaping, but it eventually settles into a smooth and broad shape, yielding a pulse duration of about 100 fs. The pulses are close to the lowest-order transform-limited soliton in temporal and spectral shape for the given fiber.

The third-order dispersion causes an additional small chirp on the pulse. Normal values of the self-steepening parameter do not change the results of our simulations. Changing the cavity length from half a beat length to about a quarter of a beat length makes no appreciable change in the pulse characteristics. The dual-core mode-locking mechanism appears to be robust, and we find no pedestal on the pulses for any of our examples.

The experimental conditions require relatively large core separations, long fiber lengths, and minimization of the influence of bending effects. The simulation and other work to date, however, indicate that the conditions of interest are experimentally accessible. Our experimental studies regarding realization of the predicted behavior are encouraging.

* *Department of Science and Math, GMI Management and Engineering Institute, Flint, Michigan 48504-4898*

** *Department of Electrical and Computer Engineering, University of Alabama in Huntsville, Huntsville, Alabama 35898*

1. H. G. Winful and D. T. Walton, Opt. Commun. **17**, 1688 (1992).

QTuG8

Vector solitons in birefringent optical fibers

V. V. Bryksin, M. P. Petrov, R. V. Kiyan, *Physico-Technical Institute, Russian Academy of Sciences, 194021, Polytechnicheskaya Street, 26, St. Petersburg, Russia*

Two basic approximations are used to describe propagation of optical solitons through birefringent optical fibers. In the case of strong birefringence, optical pulses polarized along the slow-x and fast-y axes travel with different group velocities and interact with each other only on collision.[1] For weak birefringence, a vector soliton having both x and y components, but propagating with the same group velocity, can be formed.[2]

Vector solitons are described by the nonlinear Shrödinger equations:

$$-i\frac{\partial E_x}{\partial z} - \frac{\alpha}{2}\frac{\partial^2 E_x}{\partial \tau^2} - \kappa E_x - i\Delta\beta_1\frac{\partial E_x}{\partial \tau} = \gamma\left[\left(|E_x|^2 + \frac{2}{3}|E_y|^2\right)E_x + \frac{1}{3}E_y^2E_x^*\right]$$

$$-i\frac{\partial E_y}{\partial z} - \frac{\alpha}{2}\frac{\partial^2 E_y}{\partial \tau^2} + \kappa E_y + i\Delta\beta_1\frac{\partial E_y}{\partial \tau} = \gamma\left[\left(|E_y|^2 + \frac{2}{3}|E_x|^2\right)E_y + \frac{1}{3}E_x^2E_y^*\right]. \quad (1)$$

Here, E_x, E_y are the complex field amplitudes; $\kappa = \pi(n_x - n_y)/\lambda$, where λ is the light wavelength and $n_x - n_y$ is the difference between the refractive indices of radiation polarized along the slow and fast axes, respectively; $\Delta\beta_1 = (\beta_{1x} - \beta_{1y})/2 = \kappa\lambda/(2\pi c)$, where $(\beta_{1x})^{-1}$, $(\beta_{1y})^{-1}$ are the group velocities of pulses polarized along the slow and fast axes, respectively, and c is the velocity of light in vacuum; $\alpha = -\beta_2$, where β_2 is the dispersion of group velocities ($\beta_2 < 0$); $\tau = t - \beta_1 z$, where $\beta_1 = (\beta_{1x} + \beta_{1y})/2$; $\gamma = 2\pi n_2/(\lambda A_{eff})$, where A_{eff} is the effective cross sectional area of the mode in fiber and $n_2 = 3.2 \times 10^{-20}$ m²/W.

We substitute $E_{x,y} = A_{x,y}\exp(\pm i\kappa z)$ into Eqs. (1). For strong birefringence, the terms $A_y^2A_x^*\exp(4i\kappa z)$ and $A_x^2A_y^*\exp(-4i\kappa z)$ are typically omitted because they oscillate quickly along z.[3]

Christodoulides and Joseph[2] considered the case of weak birefringence and found one of the particular solutions of Eqs. (1) corresponding to a vector soliton.

Here we give a more general solution of Eqs. (1) and show that a difference in the carrier frequencies of orthogonally polarized components of vector solitons can lead to the formation of a periodic sequence of vector solitons.

Solution of Eqs. (1) is sought in the form

$$E_{x,y}(z,\tau) = A_{x,y}(\tau)\exp[i(q + \Omega\Delta\beta_1/2)z \pm (-i\Omega\tau)], \quad (2)$$

where q is a propagation constant that depends on the optical pulse energy and $\Omega = \Delta\beta_1/\alpha$. The plus and minus signs correspond to the x and y components, respectively.

Then, instead of Eqs. (1) we have

$$(q - \kappa)A_x - \frac{\alpha}{2}\frac{d^2A_x}{d\tau^2} = \gamma\left[\left(|A_x|^2 + \frac{2}{3}|A_y|^2\right)A_x + \frac{1}{3}A_y^2A_x^*\exp(+4i\Omega\tau)\right]$$

$$(q - \kappa)A_y - \frac{\alpha}{2}\frac{d^2A_y}{d\tau^2} = \gamma\left[\left(|A_y|^2 + \frac{2}{3}|A_x|^2\right)A_y + \frac{1}{3}A_x^2A_y^*\exp(-4i\Omega\tau)\right]. \quad (3)$$

Solution in the form of Eqs. (2) corresponds to a pulse having two components, E_x and E_y, with the same group velocity. This is achieved by shifting their carrier frequencies by $\pm\Omega$ from the central frequency. If we assume that the pulse duration is $\tau_p << 1/4\Omega$, $\exp(\pm 4i\Omega\tau)$ changes only slightly during time τ_p and can be replaced by $\exp(\pm 4i\Omega\tau_0)$, where τ_0 is the center of gravity of the pulse. After this, by substituting $A_{x,y} = C_{x,y}\exp(\mp i\Omega\tau_0)$, Eqs. (3) can be reduced to a set of two equations for motion of a classical particle in a two-dimensional potential for real variables $C_{x,y}$ (τ). The solution of these equations is

$$C_x = 2\sqrt{\kappa/\gamma}[\eta_- \mathrm{sh}(\eta_+\tau - \delta)]/[\eta_+\mathrm{ch}(\eta_+\tau - \delta)\mathrm{ch}(\eta_-\tau - \delta) - \eta_-\mathrm{sh}(\eta_+\tau - \delta)\mathrm{sh}(\eta_-\tau - \delta)]$$

$$C_y = 2\sqrt{\kappa/\gamma}[\eta_+ \mathrm{ch}(\eta_-\tau - \delta)]/[\eta_+\mathrm{ch}(\eta_+\tau - \delta)\mathrm{ch}(\eta_-\tau - \delta) - \eta_-\mathrm{sh}(\eta_+\tau - \delta)\mathrm{sh}(\eta_-\tau - \delta)] \quad (4)$$

where δ is an arbitrary parameter and $\eta_\pm^2 = 2(q \pm \kappa)/\alpha$. Equations (4) correspond to a bright vector soliton. The soliton energy is independent of δ and is

$$W = \int_{-\infty}^{+\infty} (C_x^2 + C_y^2)\, d\tau = \frac{\sqrt{8\kappa\alpha}}{\gamma} \left[\left(\frac{q}{\kappa} + 1\right)^{1/2} + \left(\frac{q}{\kappa} - 1\right)^{1/2} \right]. \quad (5)$$

Roughly speaking, solution (4) is a superposition of two pulses with pulse width $\tau_{\perp} = \sqrt{\alpha/2(q \pm \kappa)}$.

Note that Eqs. (3) allow periodic solutions $A_{x,y}(\tau) = A_{x,y}(\tau + \pi m/2\Omega)$, where m is an integer. If the condition $4\Omega\tau_p << 1$ is met, this solution can be formed on the basis of a periodic sequence of solitons (4):

$$\mathbf{A}(\tau) = \sum_{-\infty}^{\infty} \mathbf{A}_0 \left(\tau + \frac{\pi m}{2\Omega}\right), \quad (6)$$

where $\mathbf{A}_0$ is the one-soliton solution of Eqs. (4). It is assumed here that the overlap between solitons with different values of m is negligibly small. The soliton repetition period is $\Delta\tau = \pi/2\Omega = \pi|\beta_2|/2\Delta\beta_1$.

1. M. N. Islam, C. D. Poole, and J. P. Gordon, Opt. Lett. **14**, 1011 (1989).
2. D. N. Christodoulides and R. I. Joseph, Opt. Lett. **13**, 53 (1988).
3. C. R. Menyuk, Opt. Lett. **12**, 614 (1989).

QTuG9

Excess noise in one- and two-dimensional geometries

G. H. C. New, M. A. Rippin, *Laser Optics and Spectroscopy Group, Department of Physics, Imperial College, London SW7 2BZ, U.K.*

In an important 1979 paper,[1] Petermann pointed out that in certain laser-resonator geometries a typical mode may exhibit more noise than expected on the basis of the standard quantum-mechanical one-photon-per-mode prescription. It was subsequently shown by other authors (e.g., Ref. 2) that this arises because the self-reproducing field patterns that are commonly referred to as laser "modes" are not true modes of the radiation field. Rather than obeying the familiar condition of power orthogonality, these satisfy the more general condition of biorthogonality with respect to a set of adjoint modes that correspond to the self-reproducing patterns traveling in the opposite direction.[3]

In unstable resonator geometries, the Petermann *K*-factor (or excess-noise factor) can be very large, possibly as high as 10^4 (Ref. 4). In a paper currently in press[5] we discuss the physical origin of this phenomenon. For the confocal unstable slit resonator example of Ref. 4 we relate the large *K*-values to another key feature of non-power-orthogonal situations, namely, that optimum excitation of a particular mode is achieved by matching to the mode in a *time-reversed* sense.[2,3] The injected-wave excitation factor (IWEF), which measures the numerical advantage of time-reversed over normal excitation can be shown to be identical to the ENF. The physical interpretation of large *K*-factors is that for several (typically 10–15) transits after time reversal, the wave energy is confined near the cavity axis and so suffers minimal round-trip loss. By contrast, the self-reproducing field in our example typically suffers 50% loss per transit, and so its time-reversed counterpart has a relative energy advantage of 2^{10}–2^{15}, figures of the same order as the *K*-factors.

In the present paper we extend this work to two transverse dimensions. An intuitive argument based on the physics of injected wave excitation suggests that in a two-dimensional geometry the *K*-factor should be roughly the *square* of its corresponding value in one dimension; this would clearly imply enormous excess-noise factors of around 10^8! Although this would appear to be realized in a rectangular cavity, in a cylindrical geometry the detailed mode calculations presented here indicate that the *K*-factors are closer to those that apply in one dimension. This occurs because the cylindrical modes are concentrated along the resonator axis and therefore suffer less loss per round trip than do their one-dimensional counterparts. The relative energetic advantage immediately after time reversal is therefore reduced.

Detailed calculations of excess-noise factors for cylindrical modes are presented. Emphasis is placed on the physical interpretation and the significance of the results. The implications for noise levels in microcavity lasers are considered.

1. K. Petermann, IEEE J. Quantum Electron. **QE-15**, 566 (1979).
2. H. A. Haus and S. Kawakami, IEEE J. Quantum Electron. **QE-21**, 63 (1985).
3. A. E. Siegman, Phys. Rev. A **39**, 1253, 1264 (1989).
4. M. A. Lauder and G. H. C. New, Opt. Commun. **67**, 343 (1988).
5. G. H. C. New, J. Mod. Opt., in press.
6. W. H. Southwell, J. Opt. Soc. Am. A **3**, 1885 (1986).

QTuG10

Linewidth reduction in diode lasers with negative electrical feedback

Sungman Lee, Lloyd W. Hillman, *Department of Physics and Center for Applied Optics, University of Alabama in Huntsville, Huntsville, Alabama 35899*

Many applications, such as coherent optical measurement and coherent optical communication, rely on the narrow linewidth of diode lasers. Several techniques, such as negative electrical feedback, optical feedback, a combination of optical and electrical feedback, and amplitude or phase squeezing,[1–3] have been used to achieve narrow linewidths. Among these techniques, negative electrical feedback is effective in preserving the small cavity and mechanical stability of a solitary diode laser. Unlike optical feedback and injection locking, in which both AM and FM noise are reduced simultaneously, negative electrical feedback is associated with an increase in intensity noise accompanying a reduction in frequency noise. The theoretical relationship between frequency noise and intensity noise has been obtained for negative electrical feedback by using heterodyne detection.[5] On the basis of single-mode rate equations, we present a theoretical model for a diode laser under negative electrical feedback using a Fabry–Perot cavity as a frequency discriminator. Experimentally, we also analyze the intensity- and frequency-noise spectrum of the diode laser under negative electrical feedback.

The calculated frequency- and intensity-noise spectral densities show strong dependence on both the gain and the delay time of the feedback circuit. Figure 1 shows the calculated frequency-noise spectrum $S_\varphi(f)$ versus frequency f, from 10^2 to 10^{10} Hz. For a 20-dB feedback gain, the frequency-noise spectrum decreases by about 40 dB for frequencies less than 10 MHz. The rate of frequency-noise reduction decreases at higher frequencies because of the phase delay of the feedback loop. Because of intrinsic coupling, the frequency-noise reduction causes an increase in intensity noise, as shown in Fig. 2. The larger the feedback gain is, the larger is the value of the intensity-noise spectrum. This is due to the spurious intensity modulation of the electrical feedback. Increased delay time in the feedback circuit reduces the bandwidth of the frequency-noise reduction and enhances noise for frequencies greater than the laser linewidth.

Experimentally, we achieved frequency-noise reduction by using negative electrical

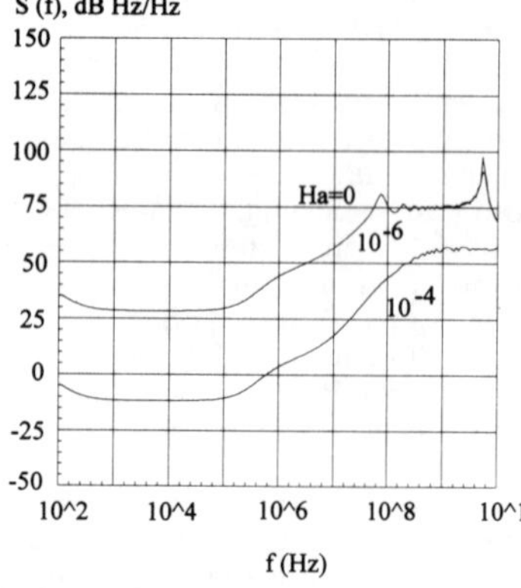

QTuG10 Fig. 1. Dependence of the feedback gain on the frequency-noise spectrum.

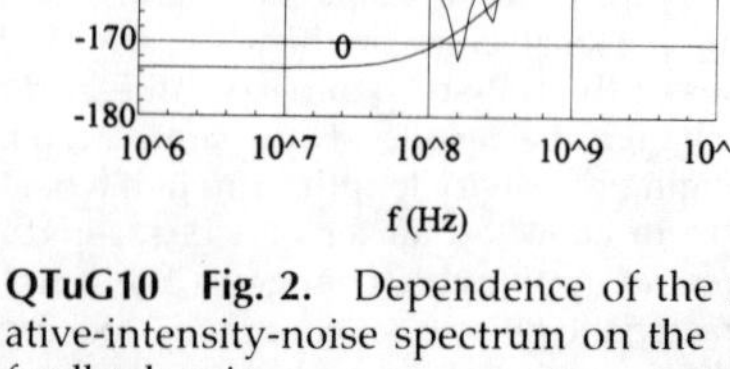

QTuG10 Fig. 2. Dependence of the relative-intensity-noise spectrum on the feedback gain.

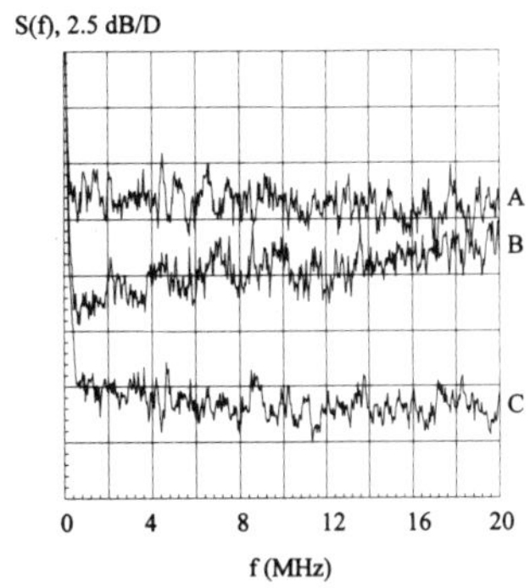

QTuG10 **Fig. 3.** Frequency-noise spectrum measured free running (curve A), with electrical feedback (curve B), and with no incident light (curve C).

feedback. In diode lasers low-frequency noise (less than 20 MHz) dominates the central part of the laser's linewidth. Figure 3 shows the typical frequency-noise spectrum measured between 0 to 20 MHz. Signal A is for the free-running condition, and signal B is for negative electrical feedback. Signal C is the thermal noise of the detector. The frequency-noise spectrum is reduced by half for a loop gain of 20 dB in the lower-frequency range. The frequency-noise spectrum measured in the range from 0 to 1 MHz shows that $1/f$ noise dominates the spectrum for frequencies less than 100 kHz. Low-level residual mechanical vibrations and higher harmonics from an ac electric source are two main sources of the peak. This work also involves measurement of the laser's linewidth reduction by using a homemade delayed self-heterodyne interferometer.

1. S. Machida and Y. Yamamoto, Phys. Rev. A **60**, 792 (1988).
2. C.-H. Shin and M. Ohtsu, Opt. Lett. **15**, 1455 (1990).
3. M. Ohtsu, M. Murada, and M. Kourogi, IEEE J. Quantum Electron. **QE-26**, 231 (1990).
4. L. Li, IEEE J. Quantum Electron. **27**, 174 (1991).

QTuG11

Dual-frequency modulated Nd:YAG laser

G. W. Baxter, J. M. Dawes, D. S. Knowles, P. Dekker, *Centre for Lasers and Applications, School of Mathematics, Physics, Computing, and Electronics, Macquarie University, North Ryde 2109, NSW, Australia*

Nd:YAG and other lasers operated unpolarized exhibit orthogonally polarized modes, which occur with a difference frequency that is a function of the intracavity birefringence.[1,2] These modes are typically elliptically polarized, with an approximate ellipticity of 20:1. Because the modes experience the same laser-cavity optics, they are very stable with respect to each other (of the order of 1 in 10^{11} frequency stability[1]). Control of the difference frequency is possible by stressing the Nd:YAG crystal to vary the birefringence[3] or by adding another birefringent element to the cavity. If the laser is operated in one polarization, varying the birefringence can be used to tune the operating frequency.[4]

Using an end-pumped diode-pumped Nd:YAG laser, we have demonstrated tuning of the difference frequency over a gigahertz range by applying a dc voltage to an intracavity lithium niobate crystal. The response, which is linear and reproducible, is shown in Fig. 1. Also shown is the theoretically expected difference frequency as a function of lithium niobate voltage, with a constant correction for residual cavity birefringence added. The two laser modes are polarized mutually orthogonally, with the axis of polarization determined by the orientation of the lithium niobate crystal. Each mode has a polarization ellipticity of approximately 25:1. The output power is distributed equally between the two modes. The laser operates in a single transverse mode, and its axial modes are restricted by an uncoated étalon (at high pump power).

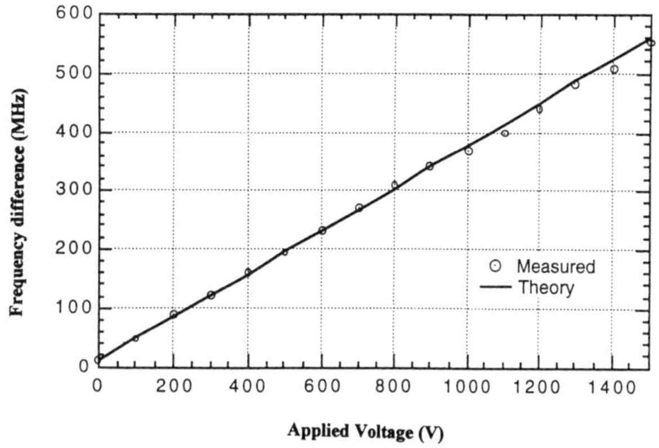

QTuG11 **Fig. 1.** Frequency difference as a function of the voltage applied to the lithium niobate crystal, showing theoretical and measured results.

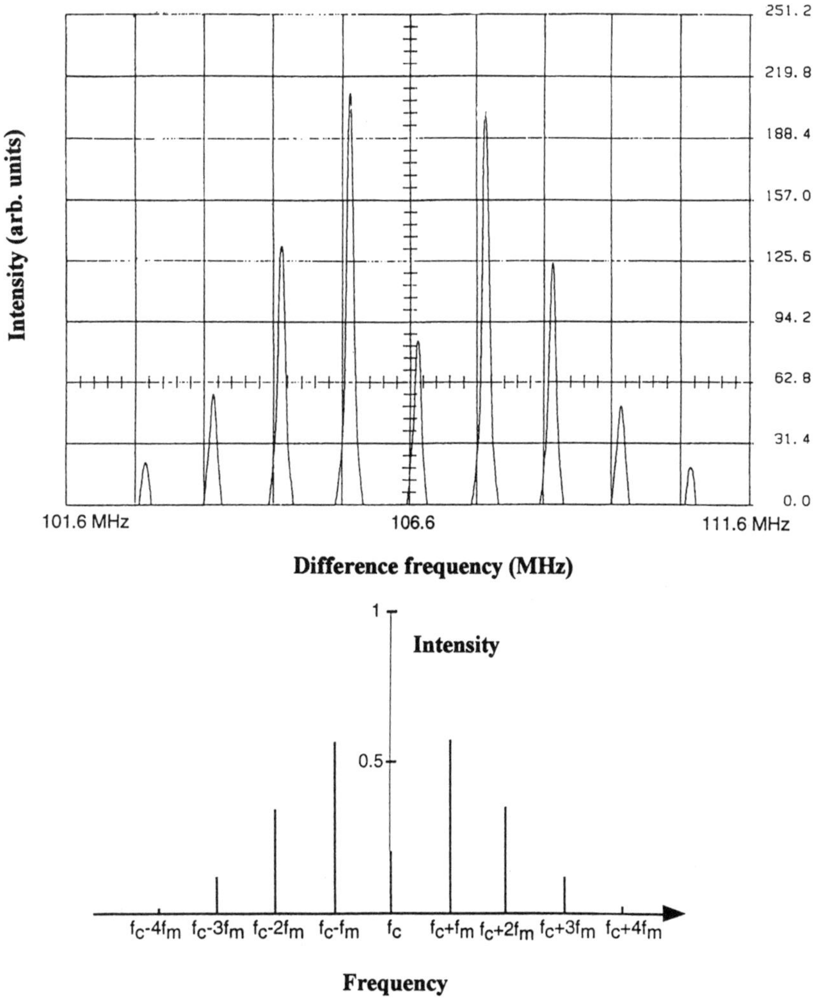

QTuG11 **Fig. 2.** Frequency-modulated laser difference frequency for a modulation index of 2.0 and a modulation frequency of 1 MHz. The cavity longitudinal mode spacing is 1 GHz.

The width of the difference-frequency beat is less than 10 kHz, giving an indication of the instantaneous relative stability of the two optical modes. This behavior occurs for a Nd:YAG laser operating at 1.06 μm or at 1.32 μm. The difference frequency can be locked to a specific value by a feedback circuit controlling the electro-optic crystal voltage.

It is also possible to frequency modulate the difference frequency, giving one laser mode with varying frequency and one mode with an orthogonally polarized fixed frequency from the same laser oscillator. An rf spectrum of the difference frequency under conditions in which the modulation index is 2.0 and the drive frequency is 1 MHz is shown in Fig. 2. This measurement is equivalent to a heterodyne measurement in which the local-oscillator reference frequency is obtained from the same source as the modulated frequency. The standard FM laser-frequency spectrum is shown for comparison.

1. S. P. Bush, P. F. Mead, and C. C. Davis, in *1991 Lasers and Electro-Optics Society Conference Digest* (Institute of Electrical and Electronics Engineers, New York, 1991), paper ELT 4.3.
2. C. He and D. K. Killinger, Opt. Lett. **19**, 396 (1994).
3. A. Owyoung and P. Esherick, Opt. Lett. **12**, 999 (1987).
4. J. J. Zayhowski, P. A. Schultz, C. Dill, and S. R. Henion, IEEE Photon. Technol. Lett. **5**, 1153 (1993).

QTuG12

Critical fluctuations in quantum parametric oscillation

Paul Kinsler, Peter D. Drummond, *Department of Physics, University of Queensland, 4072 Brisbane, Australia*

Optical parametric oscillators are some of the most interesting and well-characterized devices in nonlinear quantum optics. Novel discoveries made with them include demonstrations of large amounts of squeezing,[1] significant quantum intensity correlations,[2] and a quadrature correlation measurement that provided the first experimental demonstration of the original EPR paradox.[3] Practical applications include their use as highly efficient and tunable frequency converters.[4] Despite these numerous investigations, very little work has been done on what is perhaps the most intriguing property of these devices. They provide an example of a critical-point phase transition that is both far from thermal equilibrium and completely dominated by quantum effects.

In the case of critical-point phase transitions at thermal equilibrium, we generally start with a solution to the quantum density matrix, which is assumed to be a canonical ensemble. This simple procedure is not possible in a nonequilibrium phase transition, and so we cannot expect that the usual behavior found in thermal-equilibrium phase transitions will generalize to systems like those of the optical parametric oscillator. At the same time, information about the critical point is needed, because this is the limit in which a crude linear analysis would predict the strongest quantum correlations and noise reductions. A difference has already emerged between quantum-nonequilibrium and thermal-equilibrium systems—in thermal equilibrium, all fluctuations tend to increase at the critical point.[5]

In summary, in this driven, open quantum system there is no known solution to the equilibrium density matrix. For this reason these results cannot be obtained by using standard canonical ensemble methods. Instead, we use quantum stochastic calculations,[6] in which an asymptotic (small-noise) expansion in appropriately scaled critical variables is employed. The calculations are compared to time-dependent quantum stochastic simulations of the fully nonlinear coupled mode equations. We find that the critical squeezing variance in the output spectrum will scale as $N^{-1/2}$ for a pump-photon flux of N, and we verify this by using "infodynamic" or information-theoretic arguments.

Finally, we point out that an unusual feature of quantum parametric oscillators is that the triple quadrature correlations cannot be correctly predicted semiclassically. Not only are these significant at the critical point, but, more generally, they could provide a route to new tests of measurement theory in a semimascroscopic limit.

1. L. A. Wu, H. J. Kimble, J. Hall, and H. Wu, Phys. Rev. Lett. **57**, 2520 (1986).
2. A. Heidmann, R. J. Horowicz, S. Reynaud, E. Giacobino, C. Fabre, and G. Camy, Phys. Rev. Lett. **59**, 2555 (1987).
3. Z. Y. Ou, S. F. Pereira, and H. J. Kimble, Phys. Rev. Lett. **68**, 3663 (1992); note that this EPR experiment is different from the Bell inequality experiments of A. Aspect, P. Grangier, and G. Roger, Phys. Rev. Lett. **47**, 460 (1981); **49**, 91 (1982).
4. See, for example, Special Issue on Parametric Oscillation and Amplification, J. Opt. Soc. Am. B. **10** (September 1993).
5. H. Haken, Synergetics (Springer-Verlag, Berlin, 1977).
6. P. Kinsler and P. D. Drummond, Phys. Rev. A **43**, 6194 (1991).

QTuG13

Nonspreading wave packets in cavity quantum electrodynamics

S. M. Chumakov, A. B. Klimov,* C. Saavedra,** *IIMAS, National Autonomous University of Mexico, A.P. 139-B, 62191 Cuernavaca, Morelos, Mexico*

A single quantized field mode interacting simultaneously with many resonant two-level atoms and a Kerr medium is described by the Hamiltonian ($\hbar = 1$):

$$\hat{H} = \omega(\hat{n} + \hat{S}_z) + \chi\hat{n}^2 + \Delta\hat{S}_z + g(\hat{a}\hat{S}_+ + \hat{a}^\dagger\hat{S}_-). \quad (1)$$

Here $\hat{a}^\dagger$, $\hat{a}$, and $\hat{n} = \hat{a}^\dagger\hat{a}$ are photon operators and $\hat{S}_z$ and $\hat{S}_\pm$ are collective atomic oper-

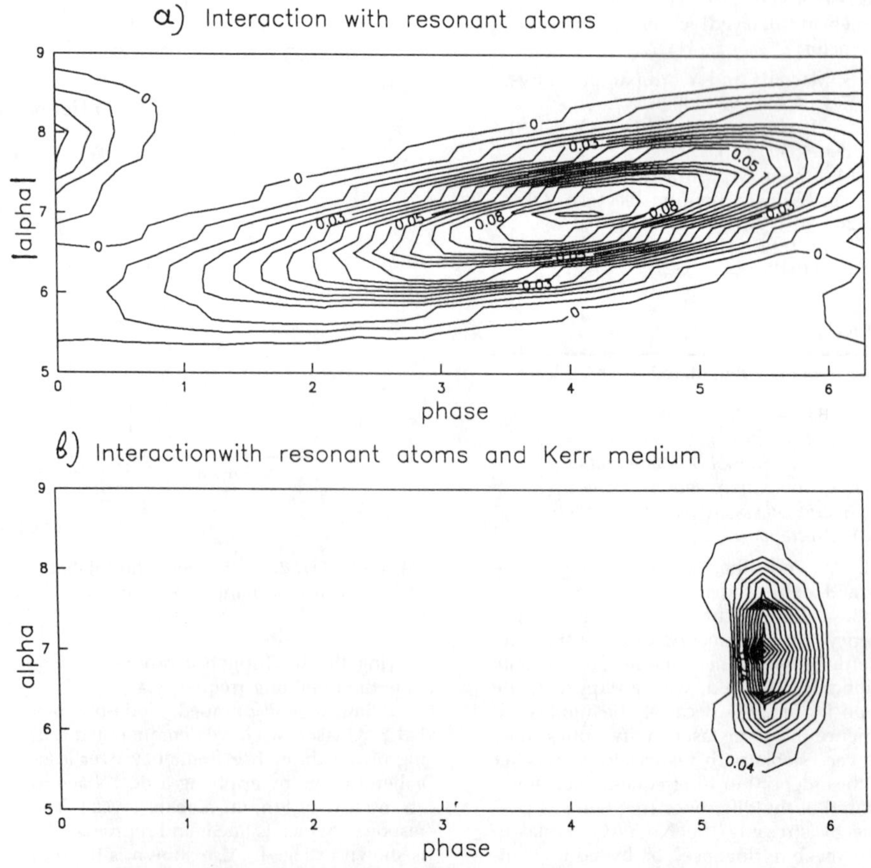

QTuG13 Fig. 1. $Q(\alpha)$ level plots in the $\phi - |\alpha|$ plane. The horizontal axis denotes ϕ, and the vertical axis denotes $|\alpha|$, where $\alpha \equiv |\alpha|e^{i\phi}$.

ators. For simplicity we assume that the atomic coupling constant g does not depend on the atomic position and consider only symmetrical atomic states. The Kerr coupling constant χ is proportional to the third-order nonlinear susceptibility.

In the classical-field limit the field operators are replaced by c-numbers, $\hat{a} \rightarrow \alpha \equiv \sqrt{\bar{n}}e^{i\phi}$, and $\hat{a}^{\dagger} \rightarrow \alpha^*$. We define the semiclassical Hamiltonian as

$$\hat{H}_{cl}(\phi) = e^{i\phi}\hat{S}_+ + e^{-i\phi}\hat{S}_-.$$

The semiclassical atomic eigenstates are

$$\hat{H}_{cl}(\phi)|p, \phi\rangle_{at} = \lambda_p|p, \phi\rangle_{at},$$

$$\lambda_p = A - 2p, \quad p = 0, 0, \ldots, A$$

(A is the number of atoms). They are known from standard angular momentum theory.

To compensate the dynamical Stark shift introduced by the Kerr medium, we choose the detuning value as $\Delta = 2\chi\bar{n}$. The addition of a Kerr medium preserves one remarkable property of the resonant interaction. Let the initial field state be the coherent state with a large photon number, $\bar{n} >> A$, and zero phase, $\phi = 0$, and let the initial atomic state be a semiclassical eigenstate. Then the field and atomic subsystems remain nearly in pure states, in spite of their interaction. The wave function is approximately factorized, and the evolution of the field and atomic subsystems can be described by the field and atomic effective Hamiltonians. The field evolves as

$$|\Phi(t)\rangle_f = e^{-it\hat{H}_f}|\alpha\rangle_f,$$

$$\hat{H}_f = \psi\hat{n}^2 + g\lambda_p\sqrt{\hat{n} + 1/2}.$$

$\hat{H}_f$ depends on the initial semiclassical eigenvalue λ_p and the initial photon number $\bar{n}$. (The factorization holds for times $gt < \bar{n}$. It has been previously discovered for the pure resonant interaction.)[1,2]

The Kerr term introduces a phase shift that grows with the field intensity. The opposite is the case for the resonant interaction term. It gives the possibility of compensating the phase spreading. This happens if the second derivative of the Hamiltonian over the photon number vanishes, i.e., $d^2\hat{H}_f/d^2n = 2\chi - g/4\bar{n}^{3/2} = 0$. We obtain the condition of least dispersion

$$\chi = \frac{g\lambda_p}{8\bar{n}^{3/2}}. \qquad (2)$$

For given values of Kerr and atomic constants and the initial atomic parameter λ_p, this condition determines the field intensity for which the phase dispersion disappears.

We show the field quasi probability $Q(\alpha)$ for time $t = 150$ in Fig. 1. The field evolves from the initial coherent state (a) interacting with resonant atoms only, $g = 1$, $\chi = 0$, and (b) with the addition of a Kerr medium, $g = 1$, $\chi = g\lambda_p/8\bar{n}^{3/2}$. $A = 2$, $\bar{n} = 49$, and the atomic state evolves from the semiclassical eigenstate $|p = 0\rangle$ with $\lambda_0 = 2$. Fig. 1(b) shows the strong suppression of phase spreading. Our diagrams are produced by computer diagonalization of the full Hamiltonian (1). Hence they provide independent tests of our theory, showing full agreement with it.

*Department of Physics, University of Guadalajara, Guadalajara, Mexico
**Department of Physics, University of Concepción, Concepción, Mexico

1. J. Gea-Banacloche, Phys. Rev. A **44**, 5913 (1991); **50**, 2040 (1994).
2. S. M. Chumakov, A. B. Klimov, and J. J. Sanchez-Mondragon, Phys. Rev. A **49**, 4972 (1994).

QTuG14

Quantum measurement of force with an error less than the standard quantum limit

S. P. Vyatchanin, E. A. Zubova, *Department of Physics, Moscow State University, Moscow 119899, Russia*

Under continuous coordinate measurement, the back-action noise of the displacement transducer is responsible for the limit of sensitivity known as the standard quantum limit (SQL): for a force $F = F_S \sin(\omega_F t)$ acting on a free mass m during time t_f that is a multiple of $2\pi/\omega_F$, SQL is equal to $F_{SQL} \simeq (m\hbar\omega_F^2/t_f)^{1/2}$. We show that for the simplest optical transducer (Fig. 1) the signal force can be measured with an error less than SQL, i.e., $\Delta F_S << F_{SQL}$, if one measures not the phase but the quadrature amplitude B of the reflected wave.

Let the incident wave be coherent so that the fluctuations of amplitude and phase are not correlated. On a phase diagram these fluctuations are described by a circle (Fig. 2). The light-pressure force (its fluctuations are responsible for back-action perturbation) is proportional to the square of the amplitude, and as the amplitude of the incident wave increases, so do the light pressure and the displacement of the mirror, leading to an increase in the phase shift of the reflected wave. On the phase diagram points A and C for the incident wave transform to points A' and C', respectively, for the reflected wave. Amplitude–phase correlation also appears. This means that the reflected wave is in the squeezed state. One can see that the light-pressure mechanism (in fact, the ponderomotive type of $\chi^{(3)}$ susceptibility) is causing back action that leads to squeezing of the reflected wave.

The action of the minimal signal force F to be registered has to change the phase of the reflected wave by $\Delta\varphi_{coh}$, but only if one measures quadrature amplitude $B(\vartheta)$. We should not measure the phase, because back action strongly disturbs the phase and the uncertainty of phase is much greater than $\Delta\varphi_{coh}$. The two states of the reflected wave with and without the action of the signal force are shown by thick and thin curves, respectively, in Fig. 2. Registration of $B(\vartheta)$ allows us not to see the back-action perturbation.

The balanced homodyne detector (Fig. 1) is used to measure the quadrature amplitude B: the differential photocurrent $J_-(t) \sim B(\vartheta, t)$, where the angle ϑ is the phase of the local-oscillator field. Measurement of $B(\vartheta, t)$ at the constant angle ϑ does not allow us to detect the special squeezing caused by the prolonged action of the probe mass on the light-pressure force fluctuations. Instead, one should measure (in the spectral domain) the value $\int d\Omega[b(\omega_0 + \Omega)e^{-i\theta} + b^+(\omega_0 - \Omega)e^{i\theta}]$, where angle θ formally depends on the spectral frequency $\theta = \theta(\Omega)$ (b and b^+ are the annihilation and creation operators, respectively, describing fluctuations in the reflected wave, and ω_0 is the mean frequency of the incident wave). For this measurement we propose modulation of the phase ϑ of the local oscillator during the time of measurement.

The linear theory of optimal filtering gives the optimal filtering function $f(t)$ that maximizes the signal-to-noise ratio. We find that one can observe considerable suppression of the noise of the sufficient statistic $J = \int_{-\infty}^{\infty} f(t)J_-(t)\, dt$ (J is the output of the filter) when

$$g_C(\Omega) = \frac{-4\omega_0\delta_R(\omega_M^2 - \Omega^2)}{(\omega_M^2 - \Omega^2)^2 + 4\delta_R^2\Omega^2}g_S(\Omega), \qquad (1)$$

where $g_S(t) = f(t)\sin\vartheta(t)$, $g_C(t) = f(t)\cos\vartheta(t)$, $g_S(\Omega)$ and $g_C(\Omega)$ are the Fourier transforms of $g_S(t)$ and $g_C(t)$, respectively, ω_M is the frequency of the mechanical oscillator, $\delta_R = 2W/(mc^2)$ is the coefficient of radiative friction (radiative friction results from the dependence of the light-pressure force on the velocity of the probe mass), W is the mean power of the incident wave, and c is the velocity of light. For example, in the free-mass approximation ($\omega_M = 0$), Eq. (1)

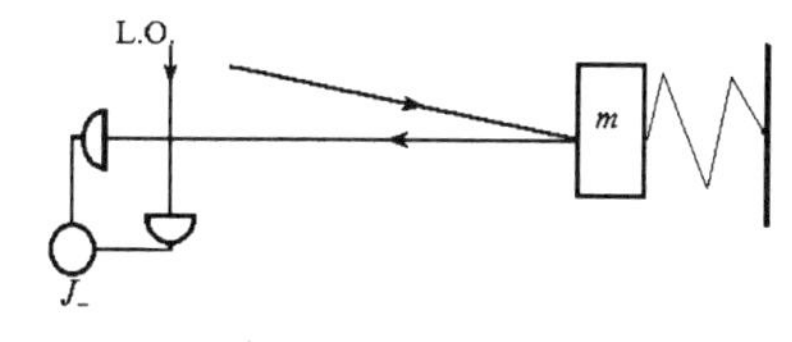

QTuG14 Fig. 1. Optical transducers of mechanical displacement. A plane electromagnetic wave falls normal to the surface of the mirror. The action of the signal force changes coordinate X of the mirror. This causes a phase change in the reflected wave; this phase change is measured by a balanced homodyne scheme.

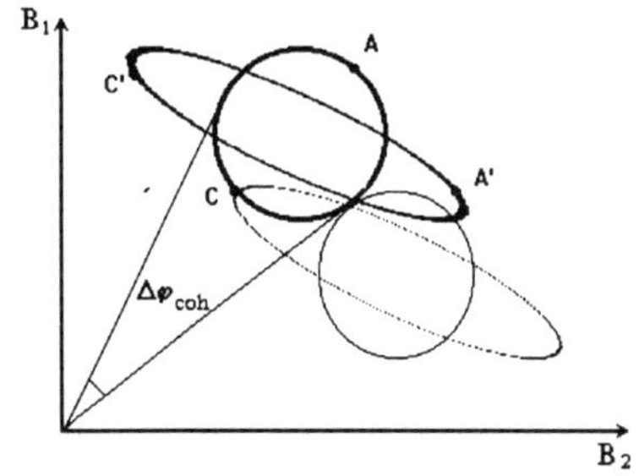

QTuG14 Fig. 2. Phase diagram for the input and output waves of the transducers. If the incident wave is coherent, its fluctuations are described by a circle at the distance $\sqrt{n}$ from the origin of the coordinates; i.e., the uncertainties in quadrature amplitudes, $\langle\Delta^2B_1\rangle$ and $\langle\Delta^2B_2\rangle$, are equal. The ponderomotive nonlinearity (light pressure) produces phase–amplitude correlation in the reflected wave. This produces squeezing of the quadrature amplitude's Fourier transform $B(\Omega, \vartheta)$; i.e., the circle turns into an ellipse. Points A and C for the incident wave transform to points A' and C', respectively, for the reflected wave.

corresponds to the quadrature dependence of $\vartheta(t) = \vartheta_0 + \alpha t^2$ (ϑ_0 and α are constants) for the function $f(t) = \exp(-t^2/t_0^2)$ (which is not optimal).

Under this condition, the minimum error for the force measurement is $F_{min} \simeq \xi F_{SQL}(\omega_F/\omega_0)^{1/4} << F_{SQL}$ (ξ is a numerical factor of order unity), which is achieved at optimum pump power. It is worth emphasizing here that there is no contradiction between this result and the Heizenberg uncertainty principle because full information about the coordinate cannot be extracted from a measurement of quadrature amplitude.

Also, this procedure with phase modulation of the local oscillator can be used to observe squeezing with spectral-frequency dependence [for example, this occurs when light passes through a nonlinear medium with a dispersion of $\chi^{(3)} = \chi^{(3)}(\Omega)$].

QTuG15

Cw-laser-induced transparency in Rb atomic systems

Jun Lin, Yifu Zhu, *Department of Physics, Florida International University, Miami, Florida 33199*

Electromagnetically induced transparency relies on quantum coherence and correlations in atomic systems induced by coherent radiation among various atomic-state configurations[1] and can be traced back to the earlier observation of coherent population trapping.[2] Because of the destructive interference between two absorption paths, the atoms can be rendered transparent to a probe laser tuned to the center of the resonant transition. By using pulsed-laser coupling, electromagnetically induced transparency in the time-dependent regime has been observed in a number of experiments.[3,4] Recently, cw-laser-induced transparency has been studied in a three-level ladder-type atomic system.[5]

We present the results of our experimental work exploring cw-laser-induced transparency in both three-level Λ-type and V-type Rb atomic systems. We used a cw single-frequency Ti:sapphire laser as the coherent pump laser and an external-cavity-stabilized diode laser as the probe laser. The experiment was carried out in a Rb-vapor cell operating at room temperature. The Rb Λ-type system is formed by the atomic transitions $5S_{1/2}$ ($F = 2$) $\rightarrow 5P_{3/2} \rightarrow 5S_{1/2}$ ($F = 3$). The coherent pump laser is coupled to the Rb atomic transition $5S_{1/2}$ ($F = 2$) $\rightarrow 5P_{3/2}$ [$5S_{1/2}$ ($F = 3$) $\rightarrow 5P_{3/2}$] and generates a pair of dressed states. The probe laser is tuned near the Rb atomic transition $5S_{1/2}$ ($F = 3$) $\rightarrow 5P_{3/2}$ [$5S_{1/2}$ ($F = 3$) $\rightarrow 5P_{3/2}$] separated from the pump transition by about 3 HGz. During the experiment, the probe transmission spectrum is recorded while the frequency of the coherent pump laser is locked to the pump transition. The probe transparency is observed at the center of the atomic transition $5S_{1/2}$ ($F = 3$) $\rightarrow 5P_{3/2}$ [$5S_{1/2}$ ($F = 3$) $\rightarrow 5P_{3/2}$]. The three-level V-type Rb system is formed by tuning the strong pump laser to the Rb transition. $5S_{1/2}$ ($F = 3$) $\rightarrow 5P_{1/2}$ while the probe laser is coupled to the Rb transition $5S_{1/2}$ ($F = 3$) $\rightarrow 5P_{3/2}$. A free-running diode laser tuned to the Rb atomic transition $5S_{1/2}$ ($F = 2$) $\rightarrow 5P_{3/2}$ is added in the experiment involving the Rb V system to eliminate the optical-pumping effect. When the Rabi frequency of the pump laser is greater than the Doppler width (about 500 MHz), the probe transparency is observed at the center of the atomic transition $5S_{1/2}$ ($F = 3$) $\rightarrow 5P_{3/2}$.

Laser-induced atomic coherence can lead to lasing without population inversion.[6–8] Three-level Λ and V atomic systems have been proposed as model systems to study lasing without population inversion.[9,10] We are planning to extend the experiment to study lasing without inversion in the Rb Λ and V systems described here. When additional incoherent pumping provided by a free-running diode laser is applied to the appropriate Rb transition, the atomic population in the Rb excited states can be manipulated and the atomic coherence responsible for the electromagnetically induced transparency can be tailored to induce probe-laser amplification without population inversion.

1. S. E. Harris, J. E. Field, and A. Imamoglu, Phys. Rev. Lett. **64**, 1107 (1990).
2. H. R. Gray, R. M. Whitley, and C. R. Stroud, Jr., Opt. Lett. **3**, 218 (1978).
3. K. J. Boller, A. Imamoglu, and S. E. Harris, Phys. Rev. Lett. **66**, 2593 (1991).
4. J. E. Field, K. H. Hahn, and S. E. Harris, Phys. Rev. Lett. **67**, 3062 (1991).
5. M. Xiao *et al.*, in *International Quantum Electronics Conference*, 1994 Technical Digest Series, Vol. 9 (Optical Society of America, Washington, D.C., 1994), p. 151.
6. S. E. Harris, Phys. Rev. Lett. **62**, 1033 (1989).
7. O. Kocharovskaya, Phys. Rep. **219**, 175 (1992).
8. M. O. Scully, Phys. Rep. **219**, 191 (1992).
9. A. Imamoglu, J. E. Field, and S. E. Harris, Phys. Rev. Lett. **66**, 1154 (1991).
10. Y. Zhu, Phys. Rev. A. **45**, R6149 (1992).

QTuG16

Coherence-induced lasing in simple atomic systems

Yifu Zhu, *Department of Physics, Florida International University, Miami, Florida 33199*

Laser-induced atomic coherence is a subject of much current interest.[1–3] Under suitable conditions, light amplification without population inversion can be induced by optical coherence among the coupled states driven by laser fields. Therefore, an inversionless laser oscillator may be realized by using coherently driven atomic systems as active media. It has been predicted that optical coherence and quantum interference may lead to interesting statistical properties, such as narrower intrinsic linewidths and amplitude squeezing, in inversionless lasers.[4–6]

Here we present an analysis of a laser model that consists of an ensemble of closed ladder-type three-level atoms incoherently pumped from the ground state to the upper excited state; a coherent pump field drives the transition between two excited states. Lasing occurs from the intermediate excited state to the ground state. The steady-state lasing for the resonant coherent pumping is discussed, and the threshold condition and the laser-intensity behavior above the threshold are analyzed. Our results show that, depending on the pumping and atomic-decay parameters, the three-level system may exhibit either lasing without population inversion or lasing with population inversion. In particular, under a sufficiently strong coherent pump, the atomic system may be made to start lasing from population inversion and then display a transition from lasing with population inversion to lasing without population inversion as the laser intensity increases. This situation corresponds to the highest lasing efficiency. Far above the threshold, the laser-intensity dependence on the atomic density is determined by the light-amplification mechanism. If lasing is initialized and maintained by population inversion, the laser intensity will depend linearly on the atomic density; if lasing is initialized by coherence, the laser intensity is independent of the atomic density and is proportional to the intensity of the coherent pump field; if lasing is initialized by population inversion and is then maintained by the atomic coherence, after displaying the transition from population inversion to noninversion the laser intensity will increase nonlinearly with the increasing atomic density and will asymptotically approach a linear dependence.

1. S. E. Harris, Phys. Rev. Lett. **62**, 1033 (1989).
2. O. Kocharovskaya, Phys. Rep. **219**, 175 (1992).
3. M. O. Scully, Phys. Rep. **219**, 191 (1992).
4. G. S. Agarwal, Phys. Rev. Lett. **67**, 980 (1991).
5. K. M. Gheri and D. F. Walls, Phys. Rev. Lett. **68**, 3428 (1992); K. M. Gheri and D. F. Walls, Phys. Rev. A **45**, 6675 (1992).
6. Y. Zhu and M. Xiao, Phys. Rev. A **48**, 3895 (1993).

QTuG17

CO_2 laser trap for cesium atoms

Tetsu Takekoshi, Jeng-Rong Yeh, R. J. Knize, *Department of Physics, University of Southern California, Los Angeles, California 90089-0484*

Neutral-atom traps are valuable tools for precision spectroscopy, the study of cold collisions, and quantum collective effects. In this paper we study an optical trap in which the laser frequency ω and the Rabi frequency are much smaller than the frequency of the first allowed electric-dipole resonance. The trap frequency is far below all the atomic electric-dipole resonances. In many respects this is essentially a quasi-electrostatic trap (QUEST). The proposed trap can be realized at the focus of a high-intensity CO_2 laser beam ($\lambda = 10.6$ μm).

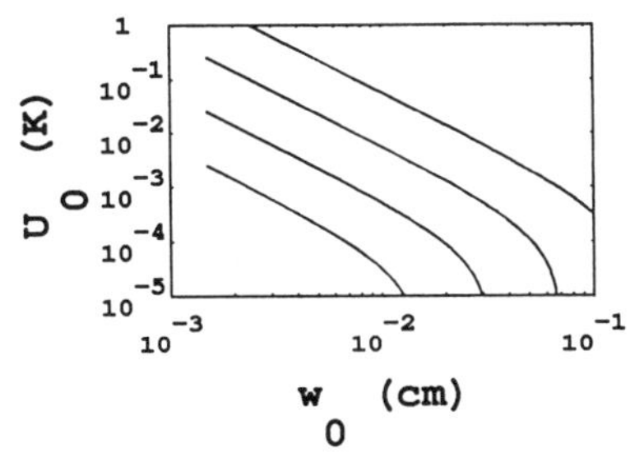

QTuG17 Fig. 1. Trap depth U_0/k_b (including gravity) versus waist radius w_0 for cesium atoms. The curves (from top to bottom) correspond to laser powers of 10 kW, 1 kW, 100 W, and 10 W.

This trap can confine all ground-state atoms of several atomic species simultaneously for long periods with small scattering and heating rates.

At the CO_2-laser focus the overall trapping potential for an alkali atom, including gravity, can be approximated by

$$U = -\frac{1}{2}\alpha_S|\vec{E}_{rms}|^2 + mgz,$$

where m is the mass of the atom, α_S is the polarizability, and E_{rms} is the rms value of the electric field. At CO_2-laser wavelengths, the Stark shift approaches the dc-Stark-shift value of $-\alpha_S E^2/2$. We have analyzed a CO_2-laser QUEST for alkali and hydrogen atoms. The trap is formed by focusing a horizontal TEM_{00} laser beam to a waist of radius w_0. Figure 1 shows the trap depth U_0 for cesium atoms as a function of power and waist radius. It is shown that for laser powers from 10 to 10^4 W, the trap depth varies from 10^{-5} to 1 K.

Our experiment demonstrates the QUEST by using a 25-W CO_2-laser beam focused to a waist radius of 50 μm. In this case the trap depth will be about 500 μK and the trap volume will be about 10^{-1} mm^3. The photon scattering lifetime of this particular QUEST is expected to be $\tau_{trap} \sim 4 \times 10^{10}$ s. Details of the experiment are discussed.

Because of the far-off-resonance nature of the trap, it is possible to trap different neutral atoms in a single CO_2-laser optical trap. If background collisions and three-body recombinations are excluded, the trap lifetime can be of the order of years. Two possible QUEST applications include a search for an atomic electric-dipole moment (EDM) and an attempt to produce Bose–Einstein condensation of a cold atomic vapor.

QTuG18

Suppression of atomic radiation in a cylindrical nanocavity

P. S. Davids, P. B. Lerner*
Materials Division, Los Alamos National Laboratory, Los Alamos, New Mexico 87545

An excited atom placed inside a cylindrical nanocavity cannot radiate if the frequency of emission is below the cutoff frequency for electromagnetic wave propagation in the cavity. However, we demonstrate that it can be de-excited through the emission of surface plasmons. The observable effect will be a substantial enhancement in the lifetime of dipole-allowed transitions. The recently discovered carbon nanotubules are explored as potential nanometer-scale cylindrical cavities, and the lifetimes of excited states of helium ($2p \rightarrow 1s$) in the carbon nanocavities are calculated as an example of this nonradiative decay mechanism. The variation of the transition rate as a function of the radial distance from the tube surface is studied by varying the initial radial quantum number of the center-of-mass wave function for an atom confined within the nanotube. The results of the calculation are used to explore the possible use of concentric carbon nanotube structures as TEM waveguides.

**Theoretical Division, Los Alamos National Laboratory, Los Alamos, New Mexico 87545*

QTuG19

Modifications to the Planck blackbody-radiation formula for a microcavity: thermal corrections to the cavity-induced Lamb shift

Jonathan P. Dowling, Asim O. Barut,* *U.S. Army Missile Command, AMSMI-RD-WS-ST, Redstone Arsenal, Alabama 35898-5428*

The Planck blackbody law is usually derived in the large-cavity limit in which the density of electromagnetic modes is treated in the free-space-continuum limit. However, in a microcavity the normal mode structure is manifestly different from that in free space, and hence the form of the blackbody spectrum can be radically altered. We calculate the modified blackbody spectrum in the simple example of a cavity consisting of two parallel mirrors separated by a short distance. Specifically, we compute the cavity-modified blackbody distribution and Stefan–Boltzmann law.

We relate this result to recent experiments by H. Walther's group in Garching that measure the cavity-induced Lamb shift. In this experiment cavity modifications to the vacuum Lamb shift in Rydberg atoms are measured at a nonzero temperature of around 100 K. We find that resonant thermal cavity photons provide the dominant effect. Our result is in agreement with experiment.

**Department of Physics, University of Colorado, Boulder, Colorado*

QTuG20

Spectrum of quasi energies for a two-level atom driven by a modulated laser field

M. Z. Smirnov, *Department of Linear and Angular Measurements, D. I. Mendeleyev Institute for Metrology, 19 Moskovsky Prospect, St. Petersburg 198005, Russia*

The nonlinear dynamics of a two-level system driven by a modulated optical field show various resonance phenomena that have been extensively studied both theoretically and experimentally (see, for instance, Refs. 1–4 and references therein). To interpret these phenomena one must analyze the quasi-level structure of the driven atom.[5,6]

The existing asymptotic techniques are applicable provided that one can point out dimensionless parameters of the problem that are many times as small (or as large) as unity. Therefore, these techniques do not cover the important region of parameters for which the Rabi frequencies, the modulation frequency, and the detuning are of the same order of magnitude.

This paper presents both the numerical and analytical considerations of the quasi-level structure in bichromatic and pulse-trained driving fields. The analytical considerations take advantage of the étalon-equation method and yield approximate analytical expressions for the quasi energies. These expressions are valid even though there are no small or large parameters in the problem.

When a two-level atom is driven by a modulated laser field, each energy level splits into two equidistant sequences of quasi levels. In particular, the quasi energies near the atomic ground state are

$$\hbar\omega'\alpha_n = \hbar\omega'(-\lambda_1 + 2n),$$

$$\hbar\omega'\alpha'_m = \hbar\omega'(-\lambda_2 + 2m + 1),$$

where ω' is the modulation frequency, m and n are integers, and $\hbar$ is Planck's constant. The Floquet indices, λ_1 and λ_2, are given by

$$\lambda_{1,2} = \frac{1}{2}\delta \pm \frac{1}{2\pi}\arccos \cdot \left\{\left[1 - \frac{a^2\delta^2}{2\sigma_0^2}\sin^2(2\sigma_0)\right]\cos^2(\delta b) - \cos(4\sigma_0)\sin^2(\delta b) - \frac{a\delta}{2\sigma_0} \cdot \sin(4\sigma_0)\sin(2\delta b)\right\}, \quad (1)$$

where δ is the detuning of the optical carrier from the atomic transition scaled by the modulation frequency ω', $\sigma = |dE/(\hbar\omega')|$ is the saturation parameter, $\sigma_0 = [\sigma^2 + (a\delta/2)^2]^{1/2}$ is the generalized Rabi frequency scaled by ω', $a \approx 1.1423$ is the solution of the equation $a^2 + 3a\cos(a) - 3\sin(a) = 0$ (for the bichromatic driving field), and $b = \pi/2 - a$. Figure 1 shows a comparison of the numerical results for the Floquet indices (solid curve) with the analytical ones (dashed curve) for $\sigma = 1.5$. Figure 2 is the plot of the quasi-level structure versus detuning, illustrating the quasi-level anticrossing effect (for $\sigma = 1.2$).

Suppose the driving field to be a sequence of very short pulses described by the following modulation function:

$$g(\tau) = 2\sum_{k=-\infty}^{+\infty}(-1)^k\delta(\tau - k\pi), \quad (2)$$

where $\tau = \omega't$ is the dimensionless time, ω' is the repetition frequency, and $\delta(\cdot)$ is the Dirac delta function. Then, the expression for the Floquet indices takes the form

$$\lambda_{1,2} = \frac{1}{2}\delta \pm \frac{1}{2\pi}\arccos\left[\cos^2\left(\frac{\pi}{2}\delta\right) - \cos(4\sigma)\sin^2\left(\frac{\pi}{2}\delta\right)\right]. \quad (3)$$

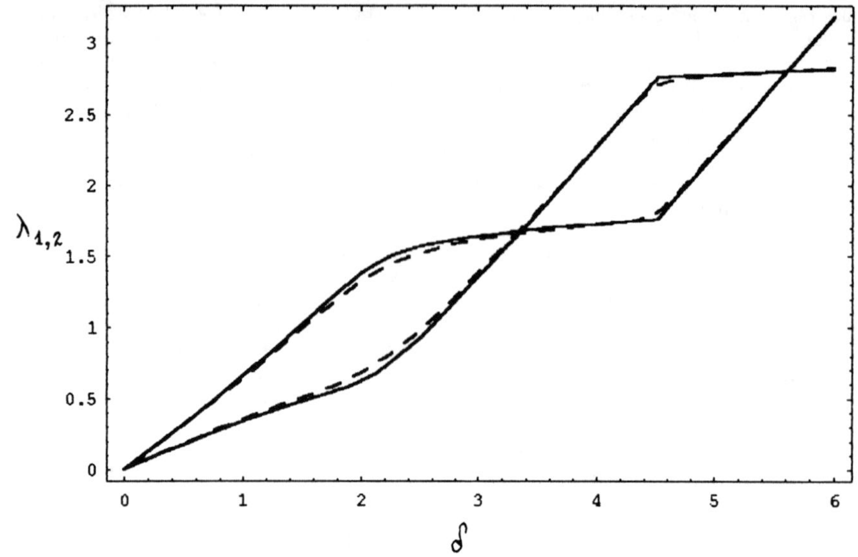

QTuG20 **Fig. 1.** Comparison of the numerical and analytical results for λ_1 and λ_2 (σ = 1.5, bichromatic driving field).

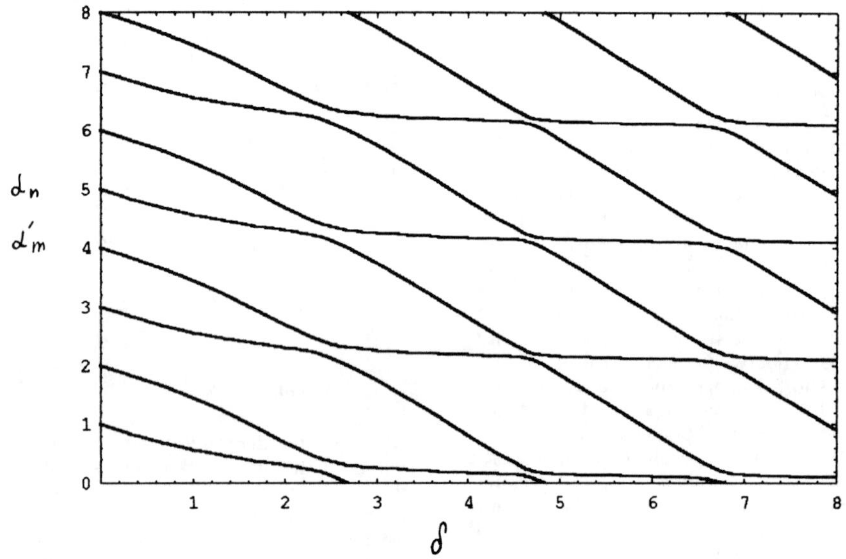

QTuG20 **Fig. 2.** Two equidistant sequences of quasi levels versus detuning (σ = 1.2, bichromatic driving field).

When $\delta \to \infty$, Eq. (1) yields $\lambda_1 = 0$ and $\lambda_2 = \delta$, which corresponds to the absence of interaction between the atom and the field. Simultaneously, Eq. (2) has no limit in the above case. To recognize this distinction one can note that a driving field with a modulation function given by Eq. (2) has an infinitely large spectral width. Therefore, increasing detuning δ does not cause the atom–field interaction to disappear.

1. Z. Zhu, Q. Wu, A. Lezama, D. J. Gauthier, and T. W. Mossberg, Phys. Rev. A **41**, 6547 (1990).
2. Z. Ficek and H. S. Freedhoff, Phys. Rev. A **48**, 3092 (1993).
3. M. Z. Smirnov, J. Opt. Soc. Am. B **9**, 2171 (1992).
4. M. Z. Smirnov, J. Opt. Soc. Am. B **11**, 109 (1994).
5. M. Z. Smirnov, in *International Quantum Electronics Conference*, 1994 Technical Digest Series, Vol. 9 (Optical Society of America, Washington, D.C., 1994), p. 48.
6. T. Yabuzaki, S. Nakayama, Y. Murakami, and T. Ogawa, Phys. Rev. A **10**, 1955 (1974).

QTuG21

Measurements of quantum noise in optical phase conjugation

M. Y. Lanzerotti,* R. W. Schirmer,* A. L. Gaeta, *School of Applied and Engineering Physics, Cornell University, Ithaca, New York 14853*

The light generated by a phase-conjugate mirror (PCM) is predicted to exhibit excess quantum noise that is inherent to the phase-conjugation process.[1] It is important to characterize the fundamental noise limit of a PCM, because this limit determines the smallest signal that can be phase conju-

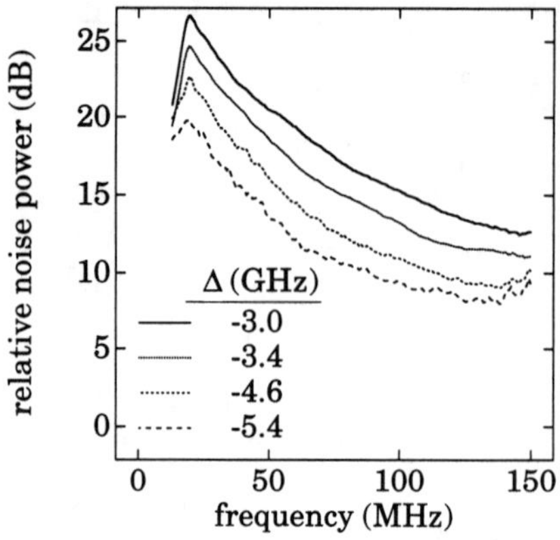

QTuG21 **Fig. 1.** Relative-noise-power spectra of the phase-conjugate beam for laser detunings of −3.0, −3.4, −4.6, and −5.4 GHz, which correspond to reflectivities of 355%, 280%, 80%, and 40%, respectively.

gated. This issue is important for applications such as optical communications and information processing. Furthermore, the noise properties of the PCM are important for characterizing the quantum-statistical properties of phase-conjugate resonators and phase-conjugate lasers.

Through use of degenerate four-wave mixing in a potassium-vapor cell, we create a PCM that is capable of exhibiting phase-conjugate reflectivities greater than 600%. We measure the noise spectrum of the photocurrent generated by the phase-conjugate beam incident on a fast (150-MHz-bandwidth) photodiode by using a high-gain, low-noise electronic amplifier and an electronic spectrum analyzer. The noise spectrum of the phase-conjugate beam is normalized to the noise spectrum of a shot-noise-limited beam having a power level equal to that of the phase-conjugate beam. We find that for a PCM having a reflectivity of 350%, the noise power of the phase-conjugate beam is far above the shot-noise limit within the bandwidth of the photodiode and is greater than 25 dB for frequencies near 20 MHz. The frequency at which the peak noise levels occur is similar to that observed in four-wave-mixing squeezing experiments[2] in sodium vapor. Figure 1 is a plot of the relative-noise-power spectra of a phase-conjugate beam for various values of the detuning of the laser below the D_2 resonance. As the laser frequency is tuned away from resonance, the relative noise power is decreased.

We present additional results that characterize the excess noise added by the phase-conjugation process as a function of the detuning of the laser, the density of potassium atoms, and the relative frequency detuning between the probe beam and the pump beams.

Department of Physics, Cornell University, Ithaca, New York 14853

1. A. L. Gaeta and R. W. Boyd, Phys. Rev. Lett. **60**, 2618 (1988); M. Y. Lanzerotti, A. L. Gaeta, and R. W. Boyd, to be published.
2. R. E. Slusher and B. Yurke, *Frontiers in Quantum Optics*, E. R. Pike and S. Sarkar, eds. (Adam Hilger, Boston, 1986), p. 1.

QTuG22

Quantum correlation between semiconductor-laser photon-flux noise and pump noise near the threshold

A. S. Trifonov, P. A. Usachev, *A. F. Ioffe Physical-Technical Institute, Russian Academy of Sciences, 26, Polytechnicheskaya Street, 194021 St. Petersburg, Russia*

In the present paper we describe investigations of quantum correlation between photon-flux noise and pump noise in a semiconductor laser near the threshold. The laser was a separate-confinement-heterojunction quantum-well (SCH QW) semiconductor laser at $\lambda = 0.8$ μm, similar to that described in Ref. 1. The two-channel-correlation technique with a delay line was used for measurements of quantum correlations. As shown in Ref. 2, the coefficient of correlation between laser photon flux noise and the junction voltage noise is

$$C = -\sqrt{\frac{2}{p + 2}}, \qquad (1)$$

where $p = I/I_{th} - 1$ is the pump parameter and I_{th} is the threshold current.

If there are any optical losses η between laser and photodiode, this expression can be modified to

$$C = -\sqrt{\frac{2}{p + 2 + \eta p^2/(1 - \eta)}}. \qquad (2)$$

As is clear from Eq. (1), near the threshold, correlation is perfect and negative. The correlation coefficient decreases with increasing pump current and optical losses.

Figure 1 shows our experimental setup. The delay line was a 30-m-long coaxial cable with a 0.12-μs delay. This corresponds to an 8-MHz interval between the noise signals' beat maximums. The frequency range was 10–110 MHz. A p-i-n photodiode with 80% quantum efficiency was used as a detector. The amplification of each channel was estimated to be around 35 dB.

Figure 2 shows the noise spectra for the junction voltage (curve 1), the photon flux (curve 2), and the difference between the junction voltage noise and the delayed photon flux noise (curve 3). The correlation coefficient C obtained from Fig. 2 was 1.06 ± 0.10. This means that the correlation is perfect near the threshold. A neutral light filter placed between the laser and photodiode decreases photon-flux noise, but within the measurement accuracy the correlation between junction-voltage noise and photon-flux noise remains in perfect accordance with Eq. (2).

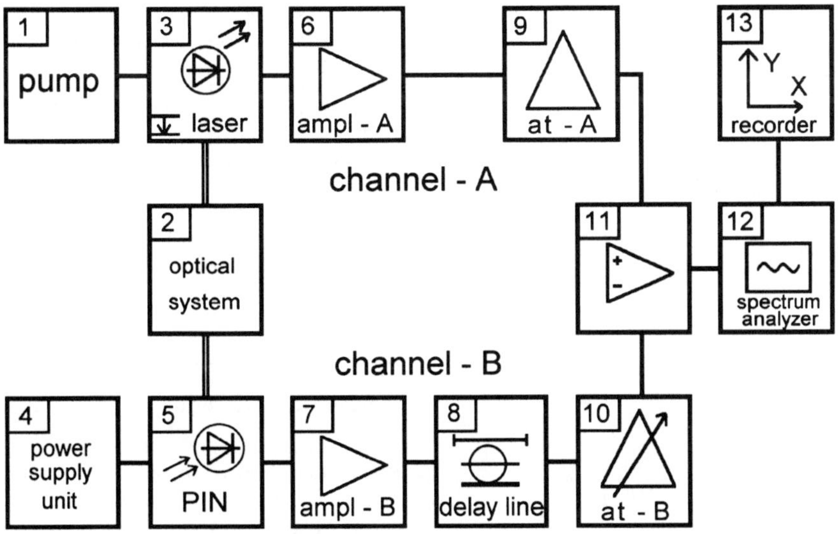

QTuG22 Fig. 1. Experimental setup.

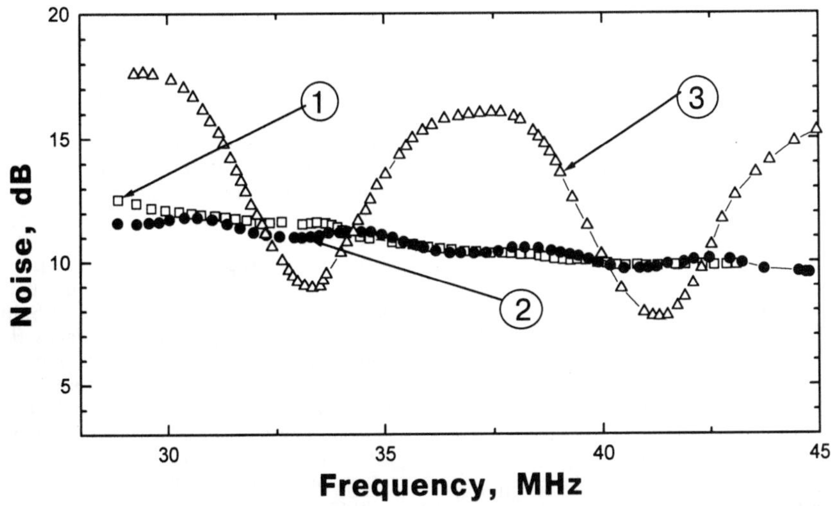

QTuG22 Fig. 2. Quantum correlation between the photon flux and the junction voltage.

1. D. Z. Garbuzov, I. E. Berishev, Yu. V. Il'in, D. N. Il'inskaya, A. V. Ovchinnikov, N. A. Pikhtin, and I. S. Tarasov, J. Appl. Phys. **72**, 319 (1992).
2. Y. Yamamoto, W. Richardson, and S. Machida, in *Quantum Measurements in Optics*, P. Tombesi and D. F. Walls, eds. (Plenum, New York, 1992), p. 65.

QTuG23

Visibility control by stimulated downconversion

P. H. Souto Ribeiro, S. Pádua, J. C. Machado da Silva, G. A. Barbosa, *Department of Physics, Federal University of Minas Gerais, CP 702, CEP 30161-970, Belo Horizonte, Minas Gerais, Brazil*

In parametric-downconversion luminescence, one photon incident in a nonlinear crystal is spontaneously converted in two simultaneous photons (signal and idler beams). It was shown recently that Young's double-slit interference patterns can be detected by performing coincidence measurements between the signal and the idler.[1] The degree of visibility obtained with a double slit placed at the signal beam can be nonlocally controlled through the idler beam.

In this work it is shown that the degree of visibility of the interference fringes produced by a signal beam transmitted through a double slit can also be controlled by aligning with the idler beam an auxiliary laser of the same wavelength and by varying its power. The degree of coherence of the source is varied directly by varying the inducing-laser intensity without requiring any measurements to be performed on the idler beam.

An argon-ion laser beam of wavelength 3511 Å (~100 mW) is incident on a $LiIO_3$ crystal to produce parametric-downconversion luminescence. A 3-mW He–Ne laser, aligned with the 6328-Å downconverted beam, stimulates the 7887-Å (signal)/6328-Å (idler) conjugated-pair emission (see Fig. 1).

The double slit is positioned 8 cm from the crystal in the path of the signal beam. The interference-pattern distribution is measured with a photomultiplier (PMT) mounted on a Z-axis translation stage. A 100-Å-bandwith interference filter placed at the PMT entrance assures that the detected light is almost monochromatic.

The interference patterns shown in Fig. 2 were obtained by varying the inducing He–Ne laser intensity I_s. The slits were placed at a distance from the light source such that the coherence area for the spontaneous emitted light is smaller than the distance between the slits.[2] Thus the visibility for $I_s = 0$ is nearly zero. We see clearly that the increase of I_s produces interference patterns with increasing visibilities. In this way we can control the spatial coherence of the signal beam by varying the intensity of the laser beam aligned with the idler

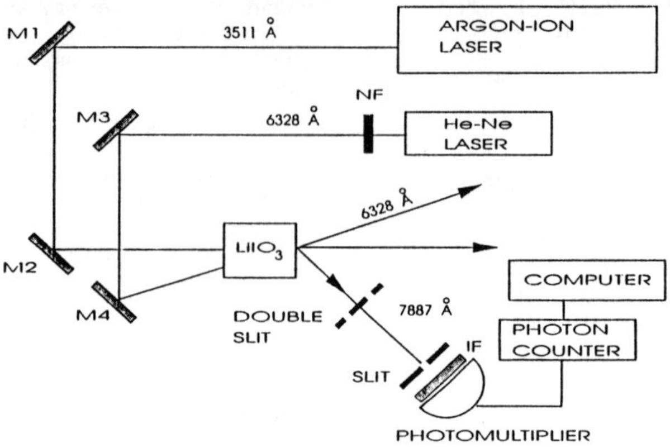

QTuG23 Fig. 1. Schematic diagram of the experimental setup for Young's double-slit experiment: M_1, M_2, M_3, M_4 are mirrors, IF is an interference filter, and NF is a neutral-density filter.

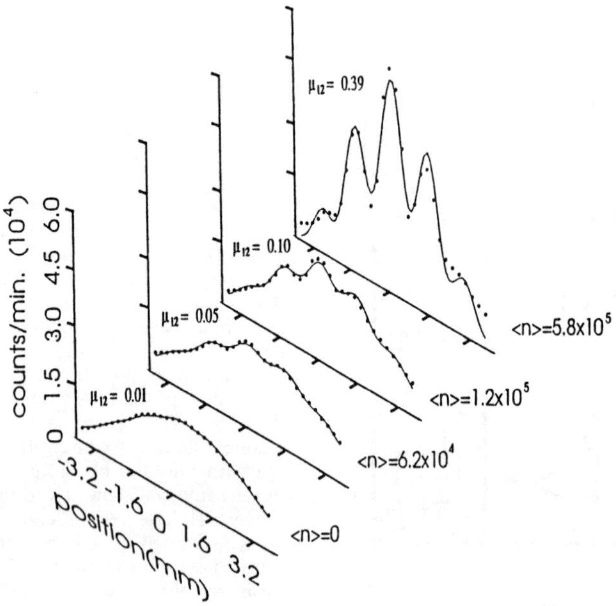

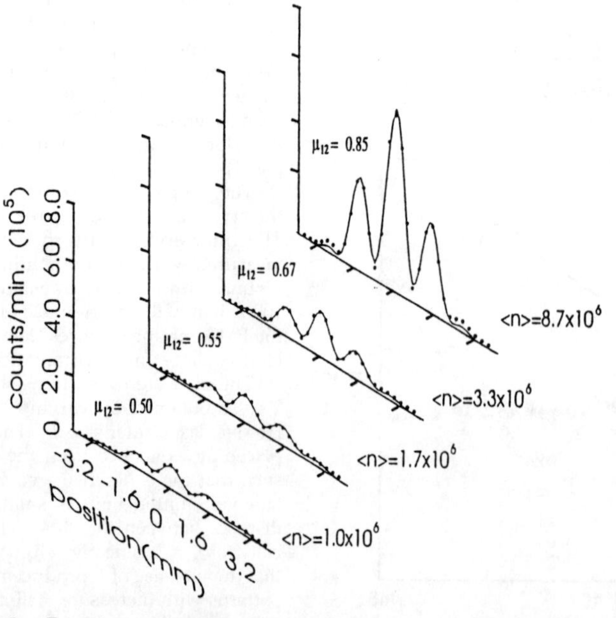

QTuG23 Fig. 2. Experimental points (dots) showing the intensity patterns as a function of the detector position. Curves are fits to the data, and errors bars are the same size as the dots.

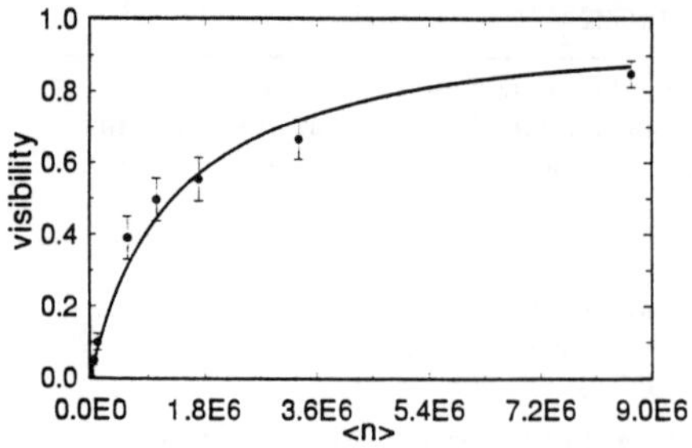

QTuG23 Fig. 3. Experimental results (dots) and theory (curve) for the visibility of Young's fringes as a function of the mean photon number of the inducing laser.

beam. The visibilities are obtained from the interference patterns by a fit to the standard intensity-distribution expression for a Young's double-slit experiment.

We have also compared theory and measurements. A simple theoretical description was developed and is presented in detail. The measured visibilities are compared with the theory in Fig. 3, showing a reasonable agreement.

The experimental data indicate a change in the downconversion-radiation-field statistics from thermallike to laserlike as the transition from spontaneous to stimulated regime occurs.

This work was supported by the National Research Council (Brazil) and by FINEP.

1. P. H. S. Ribeiro, S. Pádua, J. C. Machado da Silva, and G. A. Barbosa, Phys. Rev. A **49**, 4176 (1994).
2. P. H. S. Ribeiro, C. H. Monken, and G. A. Barbosa, Appl. Opt. **33**, 352 (1994).

QTuG24

Ultrasensitive measurement with amplitude-squeezed light

D. C. Kilper, J. Erland,* D. G. Steel, *Randall Laboratory of Physics, University of Michigan, Ann Arbor, Michigan 48109*

Squeezed light provides the opportunity for enhanced sensitivity in spectroscopy and for the measurement of nonclassical optical interactions, as shown in recent experiments.[1] Because photon-number fluctuations are easily measured by photosensitive devices, methods have been developed to use this information to control the amplitude fluctuations in a laser.[2] Significant noise reduction [up to 20 dB (Ref. 3)] below the standard quantum limit (SQL) has been observed in the amplitude fluctuations of light when the photocurrent is fed back to the laser source. Application of feedback-mediated squeezed-light generation has been prevented by the lack of a practical light-extraction technique that preserves the nonclassical statistics.[2-4] Benefits are not realized within the feedback loop because a signal applied to the beam within the loop induces a response in the feedback circuit that reduces the signal magnitude, resulting in no improvement in signal-to-noise ratio (SNR).

Here, we experimentally demonstrate a

technique that allows recovery of the full amplitude of the optical signal within the feedback loop and achieves an improved SNR, with sensitivity 5.5 dB below the SQL (limited by the frequency response of the feedback electronics). Figure 1 shows the setup for measurement of the acoustic response of a flint-glass acousto-optic crystal. The optical beam from the injection-locked semiconductor laser is amplitude modulated by the acousto-optic crystal at 9.25 MHz. When the photodetected feedback signal was applied, the laser noise was reduced from the SQL to 5.5 dB below it, at the relevant frequencies, and was detected in a balanced configuration.[5]

Although feedback reduces the amplitude noise of the light by squeezing the quantum-mechanical fluctuations of the quadrature in phase with the mean field, the amplitude of a classical amplitude-modulated signal is reduced in the feedback loop by modulation of the laser intensity out of phase (negative feedback) with the optical signal.[4] Through standard rf/microwave subtraction techniques, however, this out-of-phase modulation signal may be removed (without introducing noise) from the electrical-feedback input to the laser, as we recently demonstrated.[6] The 9.25-MHz (recovery) signal from the signal generator is applied at point A in Fig. 1 to remove the out-of-phase modulation signal. As shown in Fig. 2, without the presence of the out-of-phase modulation signal (i.e., with recovery) in the feedback loop, the magnitude of the optical signal within

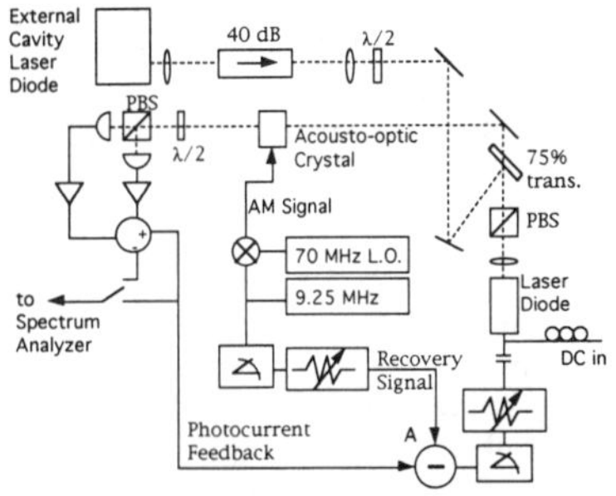

QTuG24 Fig. 1. Experiment setup. The out-of-phase modulation signal is removed from the photocurrent by subtraction of the recovery signal at point A.

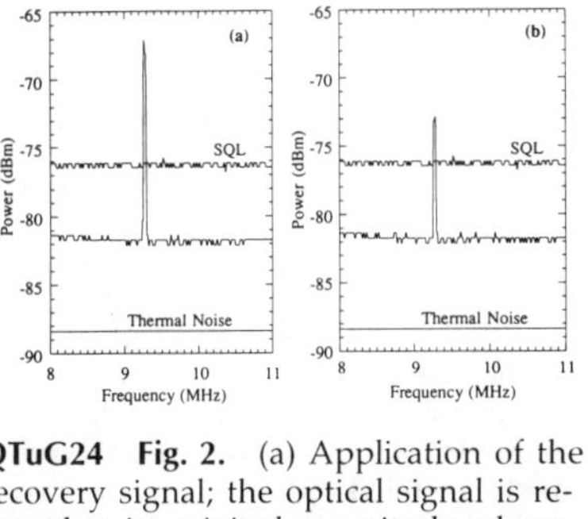

QTuG24 Fig. 2. (a) Application of the recovery signal; the optical signal is restored to its original magnitude, above the 5.5-dB squeezed-noise level. (b) Without the recovery signal; the optical signal is reduced the same as the noise (5.5 dB). The SQL and amplifier thermal noise levels are also shown; 30-kHz RBW, 300-Hz VF.

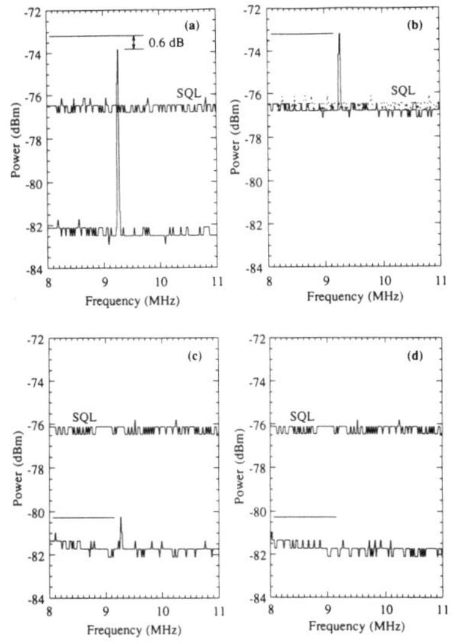

QTuG24 Fig. 3. Sensitivity of technique: the optical-signal amplitude (a) with feedback and the recovery signal applied is equal to (b) the optical signal without feedback. The 0.6-dB-larger spectral amplitude without feedback is due to the extra 5.5 dB of noise that is added to the total signal amplitude integrated over the 30-kHz RBW when the feedback is removed; with the recovery signal applied (c), an optical signal that is otherwise unobservable below the noise (d) because of the reduction in the feedback loop is observed 4.0 dB below the SQL; 30-kHz RBW, 300-Hz VF.

the loop is fully recovered above the noise background.

When the photocurrent with the recovery signal applied is fed back to the laser, the observed optical-signal amplitude becomes equal to the unreduced signal amplitude, i.e., that observed when no feedback or other external signal is applied to the laser [Figs. 3(a) and 3(b)]. The difference in peak spectral power of 0.6 dB between Fig. 3(b) and Fig. 3(a) is due to the additional 5.5 dB of noise that is added to the total integrated signal power at the modulation frequency when the feedback is removed. In Fig. 3(d), when no recovery signal is applied, the optical signal is reduced by the feedback to below the noise. On the other hand, when the recovery signal is applied [Fig. 3(c)], the optical-signal amplitude is restored and can be observed above the noise (but still 4.0 dB below the SQL).

Feedback-mediated amplitude-squeezed light has an advantage over open-loop squeezing techniques because the noise reduction occurs at the detectors and is not degraded by losses in the optical train—squeezing is not limited by the losses in the sample or by the detector quantum efficiencies.

This research is supported by the U.S. Air Force Office of Scientific Research. J. Erland acknowledges support from the Danish Research Academy.

* *Permanent address: Department of Physics, University of Odense, DK-5230 Odense M, Denmark*

1. Grangier *et al.*, Phys. Rev. Lett. **59**, 2153 (1987); E. S. Polzik, J. Carri, and H. J. Kimble, Phys. Rev. Lett. **68**, 3020 (1992).
2. S. Machida and Y. Yamamoto, Opt. Commun. **57**, 290 (1986); J. G. Rarity, P. R. Tapster, and E. Jakeman, Opt. Commun. **62**, 201 (1987).
3. Youn *et al.*, J. Opt. Soc. Am. B **11**, 102 (1994).
4. H. A. Haus and Y. Yamamoto, Phys. Rev. A **34**, 270 (1986); Y. Yamamoto, N. Imoto, and S. Machida, Phys. Rev. A **33**, 3243 (1986); Shapiro *et al.*, J. Opt. Soc. Am. B **4**, 1604 (1987).
5. M. J. Freeman *et al.*, Opt. Lett. **18**, 2141 (1993).
6. M. J. Freeman *et al.*, Opt. Lett., in press.

QTuG25

Homodyne correlation-measurement schemes with weak local oscillators

W. Vogel, *Department of Physics, University of Rostock, Universitätsplatz 3, Postfach 999, D-18051 Rostock, Germany*

Recently, the use of a homodyne cross-correlation scheme (HCCS) with a strong local oscillator has been proposed for the detection of squeezed light.[1] Moreover, a homodyne intensity-correlation scheme (HICS) with a weak local oscillator has been found to be appropriate for detecting squeezing and anomalous moments in resonance fluorescence.[2] Correlation methods of this type are useful when efficient detection of the light field is impossible. In this paper a detailed analysis of the two kinds of homodyne correlation schemes (see Figs. 1 and 2, respectively, for HICS and the HCCS) is given for a weak local oscillator, and the problems arising from its classical noise (see Ref. 1) are resolved.

Let us denote the (stationary) correlation functions measured in the HICS and the HCCS by $G^{(2,2)}(\tau)$ and $\mathcal{G}^{(2,2)}(\tau)$, respectively. Moreover, we introduce $\Delta G^{(2,2)} = G^{(2,2)}(0) - G^{(2,2)}(\infty)$ and $\Delta\mathcal{G}^{(2,2)}$ accordingly. These functions may be decomposed with respect to the orders in the local oscillator field, labeled by the index i for the ith order.

In the zeroth order with respect to the local-oscillator field the normally ordered intensity fluctuations of the signal field are observed:

$$\langle :(\Delta \hat{I}_{si})^2: \rangle = \frac{\Delta G_0^{(2,2)}}{|\tilde{T}|^4} = \frac{\Delta \mathcal{G}_0^{(2,2)}}{|\tilde{T}|^2|\tilde{R}|^2}, \quad (1)$$

where $|\tilde{T}|$ and $|\tilde{R}|$ are the amplitudes of the transmission and the reflection of the beam splitters, respectively. For negative values of this quantity, the signal field shows sub-Poissonian statistics. The squeezing effect is derived from the second-order term:

$$\langle :(\Delta \hat{E}_{si\phi})^2: \rangle = \frac{\Delta G_2^{(2,2)}}{|\tilde{T}|^2|\tilde{R}|^2\tilde{E}_{lo}^2} = -\frac{\Delta \mathcal{G}_2^{(2,2)}}{|\tilde{T}|^2|\tilde{R}|^2\tilde{E}_{lo}^2}, \quad (2)$$

where $\tilde{E}_{lo}$ is the amplitude of the local-oscillator field and $\hat{E}_{si\phi}$ is the slowly varying operator of the signal-field strength. Neg-

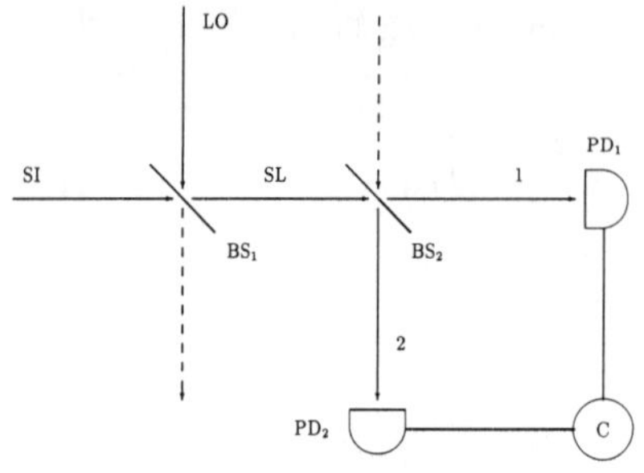

QTuG25 Fig. 1. Homodyne intensity-correlation scheme: SI, signal field; LO, local oscillator; BS, beam splitter; SL, superimposed light; PD, photodetector; C, correlator.

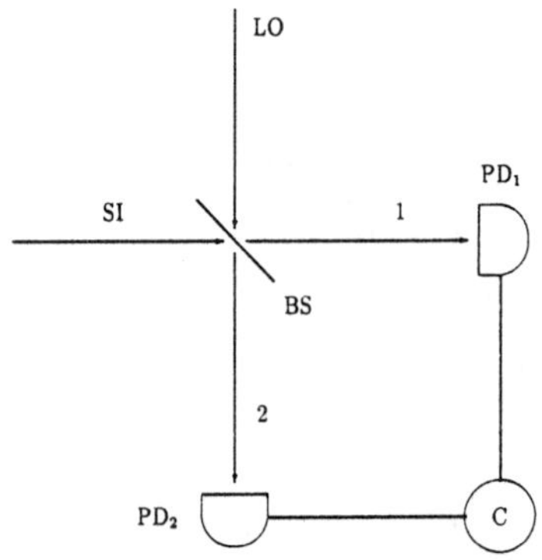

QTuG25 Fig. 2. Homodyne cross-correlation scheme.

ative values of the normally ordered signal-field variance indicate squeezing. In the first order in $\bar{E}_{lo}$ the (normally ordered) correlation of intensity and field-strength fluctuations can be observed:

$$\langle : \Delta\hat{E}_{si\phi}\Delta\hat{I}_{si} : \rangle = \frac{\Delta G_1^{(2,2)}}{2|\bar{T}|^3|\bar{R}|\bar{E}_{lo}} = \frac{\Delta \mathcal{G}_1^{(2,2)}}{|\bar{T}||\bar{R}|(|\bar{R}|^2 - |\bar{T}|^2)\bar{E}_{lo}}, \quad (3)$$

which represents the correlation of two noncommuting observables. In the HCCS the observation of this effect requires beam splitters differing from a 50%:50% partition.

Whereas in the limit of a strong local oscillator the squeezing effect dominates, all the terms are measured when the local oscillator is weak. Their effects are easily separated from each other, so that the correlation techniques are suited to simultaneously determining the statistics of two observables and their correlation.

Besides the loss of information, the strong local oscillator limit has the disadvantage that small noise of the local oscillator may essentially disturb the detection of squeezed light.[1] This problem is resolved by decreasing the strength of the local oscillator, leading to an attenuation of the disturbing noise effects.

Eventually, the possibility of spectral measurements must be considered. In addition to the well-known intensity-fluctuation spectrum and the squeezing spectrum, it becomes possible to observe the spectrum

$$S_{E/I}(\omega) = \frac{1}{2\pi}\int d\tau e^{i\omega\tau}\langle : \Delta\hat{E}_{si\phi}(\tau)\Delta\hat{I}_{si}(0) : \rangle \quad (4)$$

of the correlation of intensity and field-strength noise.

This work was supported by the German Research Association.

1. Z. Y. Ou, C. K. Hong, and L. Mandel, Phys. Rev. A **36**, 192 (1987).
2. W. Vogel, Phys. Rev. Lett. **67**, 2450 (1991).

QTuG26

Squeezed polarized light and quantum precision measurement of the Stokes parameters

A. P. Alodjants, A. S. Chirkin,* S. M. Arakelian, *Vladimir State Technical University, Vladimir 600026, Russia*

The squeezed polarized states in a cubic nonlinear anisotropic medium of Kerr-like nonlinearity are described by the Stokes parameters S_i, $i = 0, 1, 2, 3$:

$$S_{0,1} = b_x^+ b_x \pm b_y^+ b_y,$$

$$S_2 = b_x^+ b_y + b_y^+ b_x,$$

$$S_9 = i(b_y^+ b_x - b_x^+ b_y), \quad (1)$$

where $b_{x,y}$ are the components of annihilation operators and $b_{x,y}^+$ are the components of creation operators. We found that the fluctuations of the values representing some combinations of the S_i parameters should be less in the squeezed states than for coherent light. The main point of our study is to describe the procedure for observing squeezed polarized light in an actual experiment. This problem is closely connected with the possibility of nondemolition measurement of polarized characteristics and also with generation of quantum polarized stochasticity. We consider a range of questions for a special type of optical nonlinear fiber for which a strong energy exchange between two modes occurs in space and for which the intrinsic anisotropy of the system, resulting in the interaction of modes of different polarizations, is essential (screw fiber).

The Stokes parameters given by Eqs. (1) can be rewritten by coupling relations as follows (we display them, e.g., for the $S_{2,9}$ values only):

$$S_2 = |b_x|[\exp(-i\varphi_x)b_y + \exp(i\varphi_x)b_y^+] \equiv |b_x|Q_y,$$

$$S_9 = |b_x|i[\exp(i\varphi_x)b_y^+ - \exp(-i\varphi_x)b_y] \equiv |b_x|P_y, \quad (2)$$

where a classic strong-mode (pump-radiation) $b_x = |b_x|\exp(i\varphi_x)$, with φ_x the phase parameter, is assumed. $Q_y = \exp(-i\varphi_x)b_y + \exp(i\varphi_x)b_y^+$ is the amplitude component and $P_y = i[\exp(i\varphi_x)b_y^+ - \exp(-i\varphi_x)b_y]$ is the phase-quadrature component of the light with the y-directed linear polarization.

The procedure for measuring the quadratures Q_y, P_y is very well known, but the vital point for our consideration is that such a measurement should be carried out by a nondemolition technique, such as, e.g., back-action-evading measurements.

By using Eqs. (2) we can reduce our polarization problem to the standard problem of determining the fluctuations of the light quadrature. The final relations for fluctuations $\langle \Delta S_{2,9}^2(z) \rangle$ of the Stokes components $S_{2,9}$ for our spatially distributed system (the propagation coordinate is the z coordinate) can be written in general form as follows (if we assume that the b_x component is in a coherent state but that the b_y component is in a vacuum state at the input of the fiber $z = 0$):

$$\langle \Delta S_2^2(z) \rangle = |b_x|^2[1 + \psi \sin\theta \cos 2kz(\psi \cos 2kz \sin\theta - 2\cos\theta)],$$

$$\langle \Delta S_9^2(z) \rangle = |b_x|^2[1 + \psi \cos\theta \cos 2kz(\psi \cos 2kz \cos\theta + 2\sin\theta)], \quad (3)$$

where θ is the phase parameter arising from the propagation of the modes, ψ is the nonlinear phase retardation in anisotropy (i.e., induced birefringence), k is the wave vector for the pump radiation; some approximations have been made.

As we can see from Eqs. (3), the fluctuations are distributed over space, in fact, and the squeezed polarized light is determined by conditions $\langle \Delta S_{2,9}^2(z) \rangle < |\langle \Delta S_1(z) \rangle|^2$, which results in the necessary values of the parameters used (for coherent light in our case $\langle \Delta S_i^2(z) \rangle = |b_x|^2$, $i = 1, 2, 3$).

Our analysis shows that precision measurement, e.g., S_1 for input light, could be carried out under certain conditions that describe the coupling of the control parameters of the problem, both amplitude and polarization. Moreover, we need polarized squeezed light at input. An experimental setup for that consists of a nonlinear distributed-feedback system that realizes the energy exchange between various modes. Such a measurement procedure using the polarization parameters of light is also very promising cryptographically when the communication between two correspondents is carried out by varying amplitude parameters (in our case, the photon-number difference for two orthogonal linearly polarized components of the light) but when injection in the channel and nondemolition measurement is realized by varying the polarization characteristics.

The analysis of quantum polarized stochasticity and control of it have been accomplished by a procedure that uses the Stokes wave functions in the Schrödinger representation; this allows us to write the necessary equations in the form of the classical approach. So the problem is reduced to the analysis of very-well-known equations, e.g., the Duffing equation. But, in contrast to the classical case, we have the additional interference between stable and unstable states for the quantum description. The results obtained facilitate a simple physical interpretation for the Stokes parameters and also the description of an obvious process for their measurement.

To observe the above-described physical phenomena in an experiment, we have carried out numerical estimations for the two-channel system, i.e., for a special type of tunnelly coupled optical fiber. The procedure made use of laser-induced nonlinear phase retardation in the medium. The necessary intensities for achieving that are of the order of watts per square centimeter for a length of fiber of a few centimeters.

*Moscow State University, Moscow 117899, Russia

QTuH 2:30 pm
Hyatt A

Spatial Solitons II

Ian Alexander Walmsley, *University of Rochester, Presider*

QTuH1 (Invited) 2:30 pm

Two-dimensional solitary waves in a quadratic medium

William E. Torruellas, Zuo Wang, David J. Hagan, Eric W. VanStryland, George I. Stegeman, Lluis Torner,* Curtis R. Menyuk,** *Center for Research and Education in Optics and Lasers, University of Central Florida, 12424 Research Parkway, Orlando, Florida 32826*

Early on it was recognized that second-order processes could mimic the behavior of a third-order nonlinear response.[1] Indeed, to efficiently generate the third harmonic of a laser, two second-order processes are commonly cascaded, namely, second-order harmonic generation followed by sum-frequency generation. Similarly, a Kerr-type nonlinear effect has been observed recently in both bulk and waveguide geometries.[2-4] The question then arises: Are optical solitary waves possible in such media? Recent numerical experiments have shown that it is possible to generate, under the present state of the art, both one- and two-dimensional spatial solitary waves in optical media with quadratic responses.[5-7] Here, we investigate experimentally the latter case.

We explore the intriguing possibility of going beyond what an intrinsic third-order process can do. Because of the well-known collapse of a two-dimensional beam described by the nonlinear parabolic Schrödinger equation, the formation of stable two-dimensional solitons in Kerr-like media is prevented. However, under strong nonlinear coupling in a phase-matchable second-order crystal, two copropagating waves can form stable solitonlike beams.[6,7]

We show, under phase-matching conditions, the measurement of the waist of a beam that has propagated in a 1-cm-long KTP crystal; see Fig. 1. Self-focusing induced by cascading of second-harmonic generation and downconversion is present, even at low input fundamental intensities. However, beyond a certain input intensity the beam profile stabilizes into a two-dimensional cylindrically symmetric beam with a waist smaller than that measured in air; see Fig. 2.

Our experimental arrangement demonstrates, we believe for the first time, the formation of two-dimensionial solitary waves in a quadratic medium. A 1064-nm beam was focused at the entrance face of a 1-cm-long KTP crystal. The measured beam waist in air was 20.5 μm. Thus, the KTP crystal represented five diffraction lengths. We could, in fact, measure a stable beam whose waist at the output of the crystal measured 12.5 μm for input intensities exceeding 10 GW/cm^2; see Fig. 1. The formation of stable two-dimensional solitary waves under a variety of experimental conditions is presented and discussed.

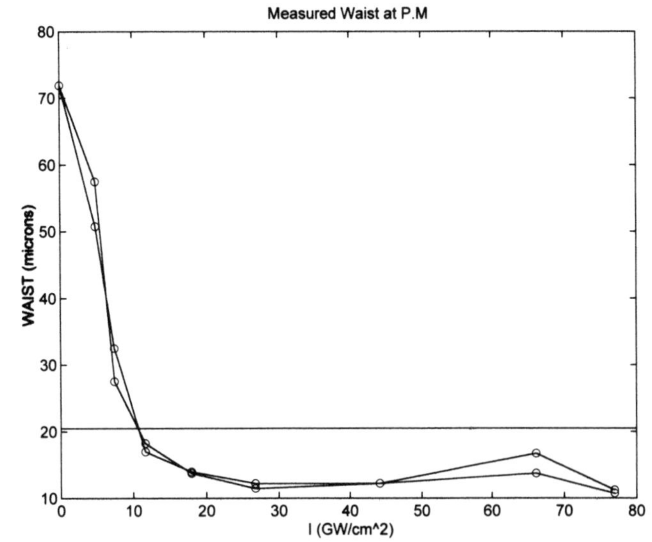

QTuH1 Fig. 1. Measured beam radius at the output of a 1-cm-long crystal under phase-matching conditions. The input beam was focused to 20.5 μm at the entrance face of the crystal. The crystal length corresponds to five diffraction lengths.

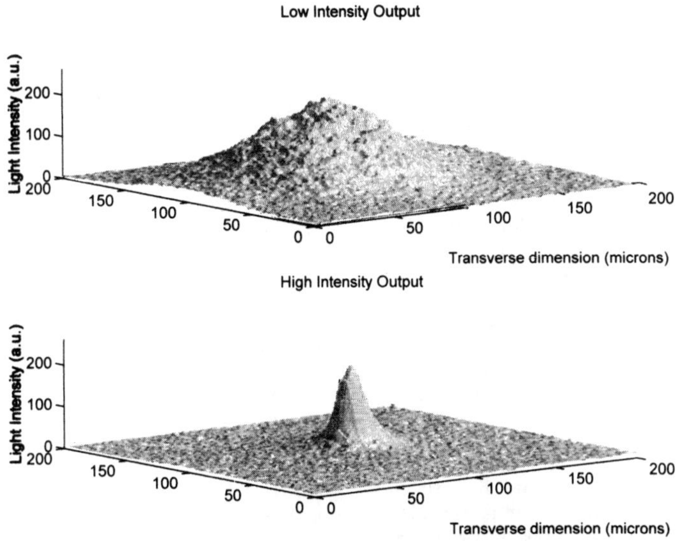

QTuH1 Fig. 2. Beam profile of the fundamental beam at the output of the KTP crystal for both low and high input intensities. A stable, cylindrically symmetric beam that defeats and compensates diffraction is definitively formed beyond 10 GW/cm^2.

* *Department of Signal Theory and Communications, Polytechnic University of Catalunya, P.O. Box 30002, 08080 Barcelona, Spain*
** *Department of Electrical Engineering, University of Maryland, Baltimore, Maryland 21228-5398*

1. C. Flytzanis and N. Bloembergen, Prog. Quantum Electron. **4**, 271 (1976).
2. R. Desalvo, D. J. Hagan, M. Sheik-Bahae, G. I. Stegeman, E. W. Van-Stryland, and H. Vanherzeele, Opt. Lett. **17**, 28 (1992).
3. M. L. Sundheimer, Ch. Bosshard, E. VanStryland, and G. I. Stegeman, Opt. Lett. **18**, 1397 (1993).
4. D. Y. Kim, W. E. Torruellas, J. Kang, C. Bosshard, G. I. Stegeman, P. Vidakovic, J. Zyss, W. E. Moerner, B. Twieg, and G. Bjorklund, Opt. Lett., in press.
5. R. Schiek, J. Opt. Soc. Am. B **10**, 1848 (1993).
6. K. Hayata and M. Koshiba, Phys. Rev. Lett. **71**, 3275 (1993).
7. L. Torner, C. Menyuk, W. E. Torruellas, G. I. Stegeman, Opt. Lett., in press.

Tuesday, May 23

QTuH2 **3:00 pm**

Two-dimensional spatial solitons with second-order nonlinearities

Lluis Torner, Curtis R. Menyuk,*
William E. Torruellas, George I. Stegeman,
Center for Research and Education in Optics and Lasers, University of Central Florida, 12424 Research Parkway, Orlando, Florida 32826

In addition to harmonic generation and frequency mixing, intense light beams propagating in $\chi^{(2)}$ nonlinear media exhibit a variety of self- and cross-phase-modulation effects that until recently were commonly thought to occur only in $\chi^{(3)}$ interactions. One important example is the propagation of solitonlike waves in which the fundamental and second-harmonic waves mutually focus and trap each other. The (1 + 1) trapping (i.e., one transverse dimension and one propagation dimension) yields both spatial and temporal solitons that can, in principle, occur in planar waveguides and fibers, respectively.[1,2] In this paper we report our investigations on the propagation of Gaussian beams in bulk $\chi^{(2)}$ media.[3] We show that, contrary to the conventional self-focusing case in bulk Kerr-like media, stable two-dimensional-solitonlike propagation is possible, in the paraxial, in phase-matchable materials with a quadratic nonlinear response.

We consider continuous-wave light beams traveling in a $\chi^{(2)}$ active bulk crystal in the slowly-varying-envelope approximation. First, we explored configurations with negligible linear walk-off between the fundamental and the second-harmonic waves. Figures 1(a) and 1(b) show the outcome for typical excitation conditions. The plots show the amplitude of the fundamental wave as a function of the propagation distance, measured in units of coherence lengths, for various values of the amplitude of the fundamental and the second-harmonic signals (A and B, respectively). In Fig. 1(a) the linear phase mismatch ($\Delta\kappa = 2\kappa_\omega - \kappa_{2\omega}$) is positive, whereas in Fig. 1(b) it is negative. Below a threshold input intensity, the beams spread. In contrast to this, when suitable signals are injected with sufficient intensity at the input face of the crystal, we observed that the beams evolve into a soliton. Solitonlike states form with both signs of the phase mismatch, but for identical excitation conditions the sign of $\Delta\kappa$ significantly affects the beam evolution. In particular, with negative phase mismatch when no second harmonic signal is initially supplied, the beams always spread. However, when a second-harmonic beam is colaunched with the fundamental one, stable propagation is observed. For comparison, we have also plotted the evolution of a beam governed by the nonlinear Schrödinger equation. Above the collapse threshold, formation of the characteristic spike is clearly visible. In a second numerical experiment, we examine the beam evolution in the presence of moderate linear walk-off. We set a walk-off length comparable to the diffraction length. As in the case for (1 + 1) trapping,[2] we found that the beams stick together and evolve into a soliton.

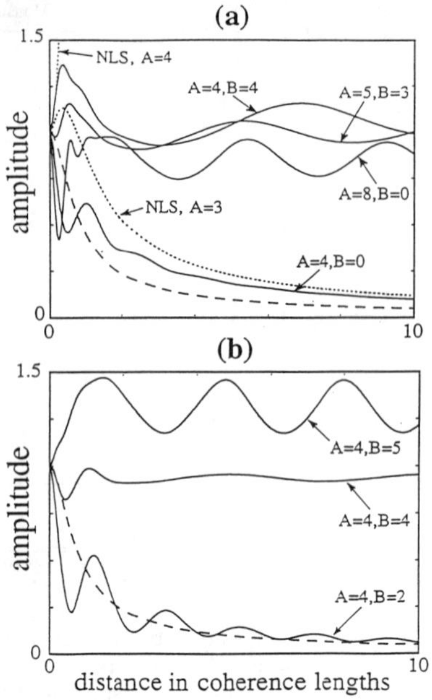

QTuH2 Fig. 1. Evolution of the amplitude of the fundamental beam. Dashed curves show the evolution of the beams in the absence of nonlinearity, and dotted curves show the evolution of a beam governed by the nonlinear Schrödinger equation. All values scaled to the input amplitudes. (a) $\beta = 3$; (b) $\beta = -3$. Here $\beta = \kappa_\omega \eta^2 \Delta\kappa$, where η is the beam width and $\Delta\kappa = 2\kappa_\omega - \kappa_{2\omega}$ is the linear phase mismatch.

The experimental observation of the mutual beam trapping requires large, nonresonant $\chi^{(2)}$ nonlinearities, which have to be accessible in configurations with negligible or moderate walk-off. A large damage threshold and long samples are also required. Such conditions can be fulfilled by ferroelectric materials such as lithium niobate ($LiNbO_3$) and potassium titanyl phosphate (KTP), although photorefractive effects could limit the usefulness of $LiNbO_3$. For the parameter values in Fig. 1 the phase-matchable nonlinear coefficients of these materials accessible through birefringence tuning lead to peak intensities in the range 0.1–1 GW/cm^2, which are accessible with common picosecond-pulsed lasers. However, organic materials, with large nonlinear coefficients, constitute a promising alternative.

In summary, we have investigated the formation of (2 + 1) solitonlike waves in materials with second-order nonlinearity. We have observed numerically, for what we believe to be the first time, that (2 + 1) solitonlike waves form at both signs of the phase mismatch and continue to exist in the presence of moderate linear walk-off between the fundamental and the second-harmonic waves. Provided that the required condition for the soliton formation can be fulfilled with long enough samples in ferroelectric and organic materials, we ultimately conclude that the solitons can be observed under currently available experimental conditions. Work in this direction is in process in our laboratory.

**Department of Electrical Engineering, University of Maryland, Baltimore, Maryland 21228-5398*

1. Q. Guo, Quantum Opt. **5,** 133 (1993); A. G. Kalocsai and J. W. Hause, Opt. Commun. **97,** 239 (1993); Phys. Rev. A **49,** 574 (1994); R. Schiek, J. Opt. Soc. Am. B **10,** 1848 (1993); M. J. Werner and P. D. Drummond, J. Opt. Soc. Am. B **10,** 2390 (1993); Opt. Lett. **19,** 613 (1994).
2. L. Torner, C. R. Menyuk, and G. I. Stegeman, Opt. Lett. **19,** 1615 (1994).
3. Yu. N. Karamzin and A. P. Sukhorukov, Sov. Phys. JETP **41,** 414 (1976); A. A. Kanashov and A. M. Rubenchik, Physica D **4,** 122 (1981); K. Hayata and M. Koshiba, Phys. Rev. Lett. **71,** 3275 (1993).

QTuH3 **3:15 pm**

Spatial solitons caused by cascaded second-order nonlinearity in a $LiNbO_3$ planar waveguide

Y. Baek, R. Schiek, G. I. Stegeman, *Center for Research and Education in Optics and Lasers, University of Central Florida, 12424 Research Parkway, Orlando Florida 32826*

Self-trapped optical solitons were proposed in the early days of nonlinear optics, and their demonstration based on the third-order nonlinearity has been reported in both the temporal and spatial regimes.[1-3] Recently it has been predicted that a cascaded second-order nonlinearity can produce effects similar to those of third-order nonlinearities with nonlinear phase shifts that are higher by orders of magnitude. In fact, large nonlinear phase shifts have already been reported.[4,5] Presented here is, to our knowledge, the first demonstration of the existence of one-dimensional spatial solitons due to the cascaded second-order nonlinearity.

Experiments were performed in a Ti diffused $LiNbO_3$ slab waveguide by using an Nd:YAG Q-switched mode-locked pulse laser at 1.32 μm. The sample is 47 mm long with its propagation direction along the *x* axis of a *y*-cut crystal. Second-harmonic generation was performed by coupling a fundamental TM mode to a second-harmonic TE mode through temperature tuning. We used TM_0-to-TE_1 conversion, for which the overlap integral has a reasonable value. Because of a nonuniform temperature profile along the waveguide, asymmetric SHG was measured around the phase-matching temperature, as reported previously.[5] The advantage of this effect is that in the region below the phase-matching temperature the phase shifts are accumulated, so that we can obtain large phase shift with only small depletion of the fundamental.

Spatial solitons were observed in the region in which the soliton formation condition $\chi^{(3:casc)} \cdot \beta_1 < 0$ is satisfied (Fig. 1). (β_1 is the dispersion of the propagation constant of the fundamental). By using a cylindrical telescope, an input beam with an elliptical profile with a 60–70-μm FWHM was coupled into the waveguide. Figure 2 shows the output-beam profiles. The beam profiles were measured by scanning the temperature. At 332.8°C, far away from the

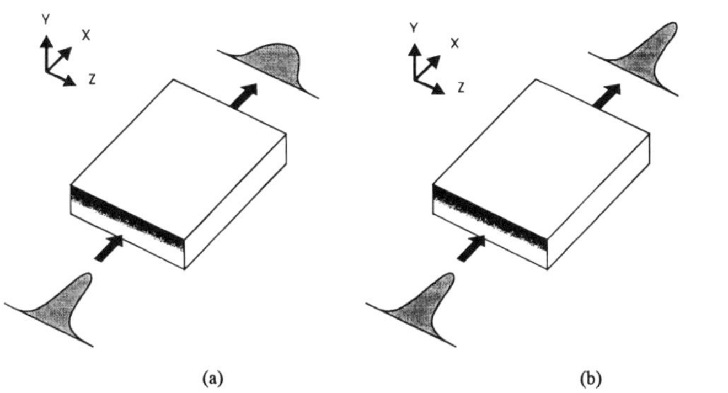

QTuH3 **Fig. 1.** Schematic illustrations for soliton propagation in a planar waveguide (a) without soliton formation and (b) with soliton formation.

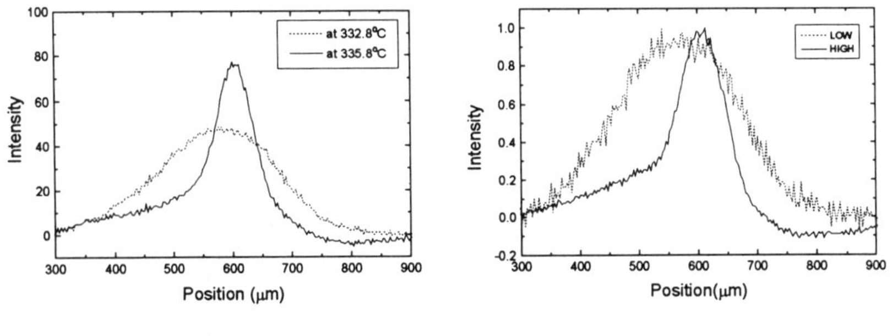

QTuH3 **Fig. 2.** Experimentally measured output-beam profiles obtained (a) by tuning the temperature and (b) by changing the input power (normalized); HIGH corresponds to 20 times the power of LOW.

phase-matching temperature, the output beam is diffracted to ~3 times that of the input beam. But as the temperature approaches the phase-matching temperature, the spatial width of the output beam decreases. At 335.8°C the output beam is approximately the same size as the input beam. We also measured the output-beam size at a fixed temperature to study the intensity dependence. As can be clearly seen in Fig. 2(b), the width of the output beam changes according to the input intensity. The peak input power of the laser is about 1–2 kW. The observed peak power was much lower than that required for third-order soliton generation.

In conclusion, we have reported what we believe to be the first observation of a one-dimensional spatial optical soliton based on the cascaded second-order nonlinearity.

1. L. F. Mollenauer, R. H. Stolen, and J. P. Gorden, Phys. Rev. Lett. **45**, 1095 (1980).
2. L. F. Mollenauer and K. Smith, Opt. Lett. **13**, 675 (1988).
3. J. S. Aitchison, Y. Silberberg, A. M. Weiner, D. E. Leaird, M. K. Oliver, J. L. Jackel, E. M. Vogel, and P. W. E. Smith, J. Opt. Soc. Am. B **8**, 1290 (1991).
4. M. L. Sundheimer, Ch. Bosshard, E. W. Van Stryland, G. I. Stegeman, and J. D. Bierlein, Opt. Lett. **18**, 1397 (1993).
5. R. Schiek, M. L. Sundheimer, D. Y. Kim, Y. Baek, and G. I. Stegeman, submitted to Opt. Lett.

QTuH4 **3:30 pm**

Interaction of a weak probe beam with a bright spatial soliton in an AlGaAs planar waveguide

J. U. Kang, G. I. Stegeman,
J. S. Aitchison,* *Center for Research and Education in Optics and Lasers, University of Central Florida, 12424 Research Parkway, Orlando, Florida 32826*

Solitary waves (solitons), which frequently occur in nature, exhibit a truly stable form of wave propagation. In a nonlinear material, optical beams that propagate with no temporal dispersion (temporal solitons) or spatial diffraction (spatial solitons) have been experimentally demonstrated recently.[1,2] In complete analogy to temporal solitons, spatial solitons can propagate long distances in a planar waveguide without any diffraction. In a planar waveguide, spatial solitons can be steered to guide incoming signals to different output ports.[3] In this paper we demonstrate the deflection and possible trapping of an orthogonally polarized weak signal by a spatial soliton when it is simultaneously launched at a given angle with respect to the signal so that the signal pulse crosses the path of the spatial soliton.

The experiment was performed with a NaCl color-center laser operating at a wavelength that corresponds to the low-loss telecommunication window of 1.55 μm. Additive-pulse mode locking (APM) the laser produces 660-fs pulses at 76 MHz. A 6-mm-long planar AlGaAs waveguide whose guiding region was made of $Al_{0.2}Ga_{0.8}As$ and whose cladding region was made of $Al_{0.4}Ga_{0.6}As$ was used for this experiment.

The strong pump (transverse electric field) and the weak signal (transverse magnetic field) were formed by using a polarizing beam splitter. These beams were elliptically shaped by a cylindrical telescope and were end-fire coupled into the planar waveguide by using a microscope objective. Approximately 25-μm ($1/e^2$ radius) beams were separated by about 5 μm from center to center at the input facet. The weak signals were launched at an angle so that in the absence of the pump, the signals emerged at the output displaced by 5 μm on the other side of the soliton formed by the strong pump.

Without the presence of the soliton, the weak signal propagates through a linear waveguide, diffracting to about 1.5 times the original width. However, when the soliton is simultaneously launched, the signal is deflected and is possibly trapped by the soliton through cross-phase modulation and emerges coincident with the soliton. The observed beam profile without and with the presence of the spatial soliton is shown in the upper and lower images in Fig. 2, corresponding respectively to the upper and lower cases in Fig. 1. In Fig. 2, note that the width of the signal beam decreased and that the center is shifted in the presence of the soliton. This shape is identical to the shape of the soliton beam (Fig. 3). The solid curve and the dotted curve represent the trapped signal and soliton beams, respectively. It is clear from this line scan that the centers and the shapes of the two beams are very similar to each other. When the time delay between the soliton and the signal were larger than their pulse width, the signal beam was not affected by the soliton.

In conclusion, we have observed a strong

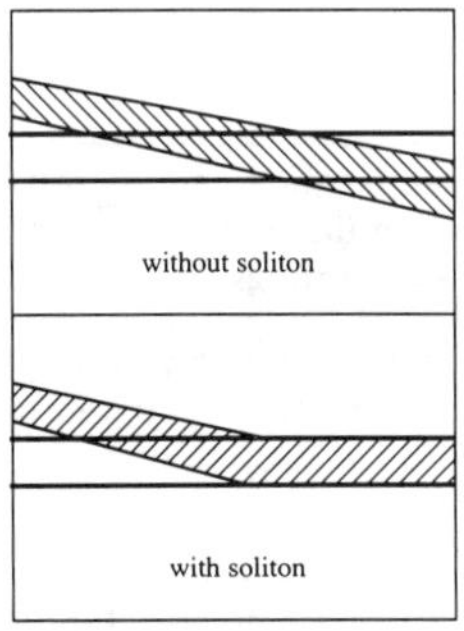

QTuH4 Fig. 1. Schematic diagram of the experimental geometry. The shaded area represents the path of a signal beam, and the unshaded area represents the path of the soliton beam.

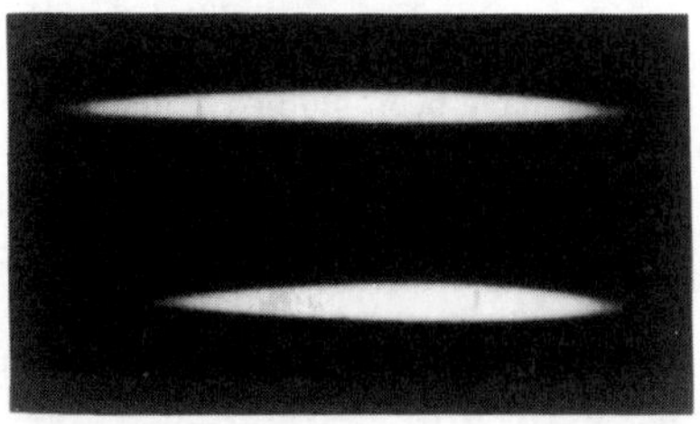

QTuH4 Fig. 2. Upper beam and lower beam showing the output signal beam without and with the soliton beam, respectively.

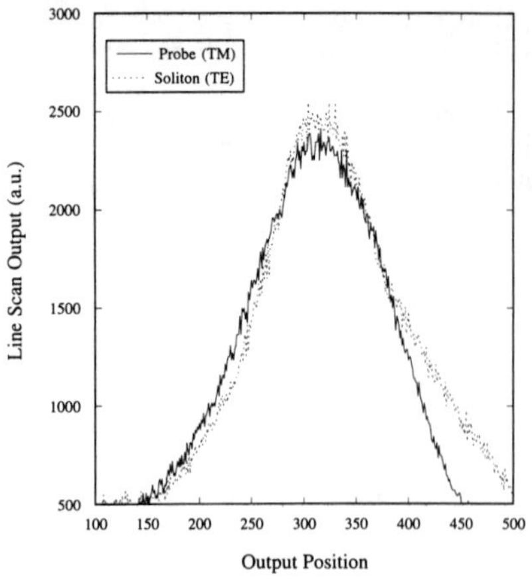

QTuH4 Fig. 3. Intensity profiles of a spatial soliton and signal beam with a soliton beam is present at the waveguide output. The solid curve and the dotted curve represent the signal beam and the soliton beam, respectively.

interaction between a bright spatial soliton and an orthogonally polarized weak probe beam. We believe larger spatial shift can be obtained with a longer sample.

Department of Electronics and Electrical Engineering, Glasgow University, Glasgow G12 8LT, U.K.

1. L. F. Mollenauer, R. H. Stolen, and J. P. Gorden, Phys. Rev. Lett. **45**, 1095 (1980).
2. J. S. Aitchison, A. M. Weiner, Y. Silberberg, M. K. Oliver, J. L. Jackel, D. E. Leaird, E. M. Vogel, and P. W. E. Smith, Opt. Lett. **15**, 471 (1990).
3. P. V. Mamyshev, A. Villeneuve, G. I. Stegeman, and J. S. Aitchison, Electron. Lett. **30**, 726 (1994).

QTuH5 **3:45 pm**

Solitonlike propagation of the grating envelope during the readout of photorefractive gratings

Muthu Jeganathan, Matthew C. Bashaw, Lambertus Hesselink, *Durand 358, MC 4035, Stanford University, Stanford, California 94305-4035*

Because of two wave mixing, the observed diffraction efficiency during the readout of photorefractive gratings frequently does not display exponential decay predicted by the band-transport model.[1] The interaction of the diffracted signal with the incident readout wave results in the formation of new gratings that can enhance or degrade the existing gratings. These self-diffraction effects are responsible for interesting dynamics, which have been studied numerically in the past.[2]

We present an elegant approach to determine transients by studying the evolution of grating envelopes (GE). For negligible absorption and planar gratings (no applied field and negligible photovoltaic effect), the materials equation (from the linearized solution to the band-transport model) and field equations (from coupled-wave theory) can be combined into a single partial differential equation for the rate of change of grating amplitude during coherent erasure. This grating-envelope-evolution equation (GEEE), which describes the erasure process, is governed by the damped sine-Gordon equation:

$$2\pi\Theta_{t'z'} + 2\pi\Theta_{z'} - \sin\pi\Theta = 0,$$

where $n(z', t') = 2\pi\Theta_{z'}(z', t')$ is the normalized index of refraction. Here, $z' = \gamma z$ and $t' = t/\tau$ are the normalized space and time coordinates, respectively, γ and τ are the photorefractive gain and response time, respectively, Θ represents the area of the GE, and the subscripts denote derivatives with respect to that variable. The diffraction efficiency is given by $\sin^2 \pi\Theta/2$. The numerical solution to the GEEE predicts alteration of the recorded grating into a hyperbolic-secant-like envelope, which then propagates through the medium. This supports the experimental observation that the readout time depends on γd, where d is the crystal length (see Fig. 1). As shown in Fig. 2, the solution to the GEEE for n differs only slightly from the hyperbolic secant function (which is the soliton to the sine-Gordon equation[4]). Like a soliton in Kerr media, the GE propagates without change in shape and the GE attains the sech-like function irrespective of the initial conditions. Because in photorefractive media the index change is due to a charge distribution, the propagating grating envelope represents a collection of light-assisted solitary charge waves.

Several extensions to the simple model described above have been made. With the inclusion of absorption, the GE is shown to slow down as it propagates, leading to a tradeoff between diffraction efficiency and the readout time. Numerical results for nonzero applied field, when phase coupling or fringe bending is significant, are presented. Furthermore, GE propagation is predicted in the reflection geometry, for which the GE approximates the hyperbolic-tangent function and behaves like a dark soliton. To provide a physical picture of the erasure process, motion pictures of the GE evolution and propagation are shown.

In summary, we present a comprehensive analysis of transients due to two-wave mixing in photorefractive media that predicts solitonlike propagation of the grating envelope during readout in the appropriate geometry.

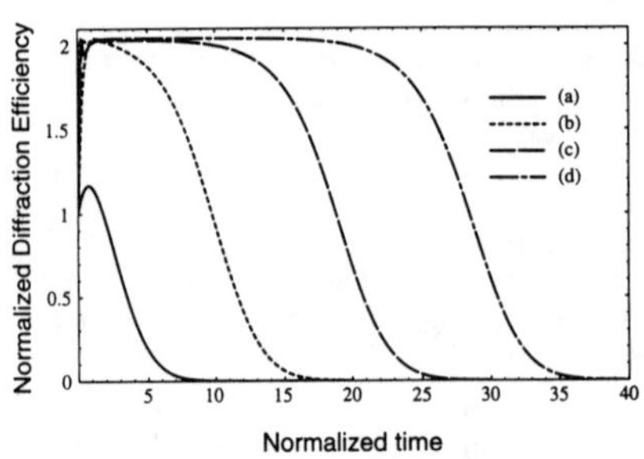

QTuH5 Fig. 1. Normalized diffraction efficiency as a function of normalized time (t') for (a) $\gamma d = 5$, (b) $\gamma d = 10$, (c) $\gamma d = 15$, and (d) $\gamma d = 20$. The grating was recorded with a beam ratio of 1.

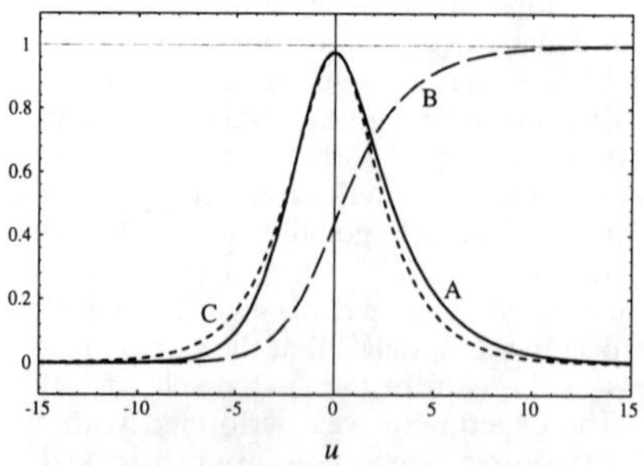

QTuH5 Fig. 2. Shape of the propagating grating envelope $n(u)$ (curve A) and plot of $\Theta(u)$, the integral of the GE (curve B), along with the function sech(u/2) (curve C) for comparison. Here, $u = \gamma z - ct/\tau$ is the coordinate in a moving reference frame (with velocity $c \simeq 0.5$) in which the GE is stationary.

1. See, for example, L. Hesselink and M. C. Bashaw, Opt. Quantum Electron. **25**, S611 (1993).
2. L. B. Au and L. Solymar, Appl. Phys. **49**, 339 (1989).
3. M. Jeganathan, M. C. Bashaw, and L. Hesselink, "Evolution and propagation of grating envelopes during erasure in bulk photorefractive media," submitted to J. Opt. Soc. Am. B.
4. G. L. Lamb, *Elements of Soliton Theory* (Wiley, New York, 1980).

QTuI — 2:30 pm, Hyatt B

Quantum Physics of Coherent Sources

George Harvey, *AT&T Bell Laboratories, Presider*

QTuI1 (Invited) **2:30 pm**

Squeezing in a Sagnac fiber loop

H. A. Haus, *Department of Electrical Engineering and Computer Science, Research Laboratory of Electronics, Massachusetts Institute of Technology, Cambridge, Massachusetts 02139*

Squeezing, using optical parametric amplification (i.e. a second-order nonlinearity, or $\chi^{(2)}$) has a long history. Squeezing with a third-order nonlinearity ($\chi^{(3)}$) was pioneered by the IBM group of Levenson and Shelby[1] using cw excitation. The use of pulses in a Sagnac loop proposed by Shirasaki and Haus[2] was followed up by experiments at IBM using solitons and at MIT using Gaussian pulses at 1.3 μm, at the zero dispersion wavelength of the fiber.

Squeezing changes the quantum noise associated with a mode of electromagnetic radiation. Squeezed-vacuum excitation of a mode has zero expectation value of the electric field, just like regular vacuum excitation, but the expectation value of photon number in squeezed vacuum is nonzero. The fluctuations of the electric field component in-phase with a reference signal are different from the fluctuations of the quadrature component. If the reference signal is used as the probe in an interferometric measurement in a Mach-Zehnder configuration, injection of squeezed vacuum of proper phase into the, normally, unexcited port of the Mach-Zehnder, results in a reduction of the detector noise below shot noise. This is the main motivation for the generation of squeezed radiation.

The use of pulses in squeezing experiments has two advantages: (a) it produces the requisite nonlinearity at a reduced power level and (b) leads to an increase of the stimulated Brillouin scattering threshold. It has the disadvantage that the noise produced by Guided Acoustic Wave Brillouin Scattering (GAWBS), first observed by Shelby *et al.*,[3] convolves into the (low-) frequency range of the measurement and may mask the noise reduction due to squeezing.

We shall describe the experiments of generation of squeezed vacuum using pulses in a Sagnac fiber loop. We have realized two schemes for the elimination of GAWBS. One scheme uses a pulse source of a repetition rate high enough (1 GHz) to avoid the convolution of GAWBS noise into the low-frequency measurement window. The other scheme uses two pump pulses separated by less than 1 nsec and reverses the phase of one of the two pulses upon detection. A noise reduction of 5.1 dB below the shot noise level has been accomplished. An experiment that resulted in a sub-shot-noise measurement of phase will be described.

Currently we are preparing modifications of the system to achieve greater noise reduction. This involves the proper use of positive dispersion for the shaping of the pump pulse, and by the use of shorter wavelengths for which detectors of higher quantum efficiency are available.

1. R. M. Shelby, M. D. Levenson, D. F. Walls, and A. Aspect, Generation of squeezed states of light with a fiber-optic ring interferometer, Phys. Rev. A **33**, 4008 (1986).
2. M. Shirasaki and H. A. Haus, Broadband squeezing with cotraveling waves, Conference on Nonlinear Guided-Wave Phenomenon, paper SA2, Optical Society of America Technical Digest Series **2**, 232 (1989).
3. R. M. Shelby, M. D. Levenson, and P. W. Bayer, Guided acoustic-wave Brillouin scattering, Phys. Rev. B **31**, 5244 (1985).

QTuI2 **3:00 pm**

Electromagnetically induced transparency in semiconductor quantum wells

Dong S. Lee, Kevin J. Malloy, Olga Blum,* John Klem,* *Center for High Technology Materials, University of New Mexico, Albuquerque, New Mexico 87131*

Because electromagnetically induced transparency[1,2] has been observed in atomic systems, analogous semiconductor-quantum-well configurations have been introduced.[3,4] Following our modulated reflectivity result for undoped multiple quantum wells in an asymmetric Fabry–Perot cavity,[5] we report reduced absorption by moderate intersubband coupling (10 kW/cm^2) at room temperature for what we believe to be the first time. This is direct evidence of the pure optical Stark effect, not electromagnetically induced transparency (EIT) as defined in Ref. 2. To achieve electromagnetically induced zero absorption through dressed states in semiconductor quantum wells, we propose an asymmetric quantum

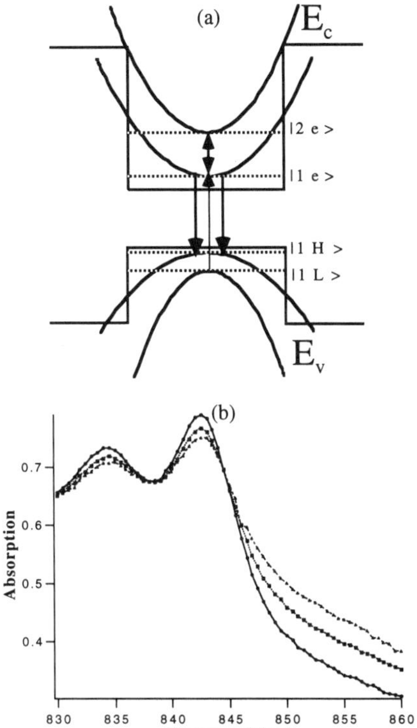

QTuI2 Fig. 1. (a) Single symmetric quantum well (8.5-nm GaAs/23.2-nm $Al_{0.7}Ga_{0.3}As$). (b) Absorption of the multiple quantum wells (65 pairs) on a ZnSe window, measured by controlling boxcar gate delay times to vary the coupling strength.

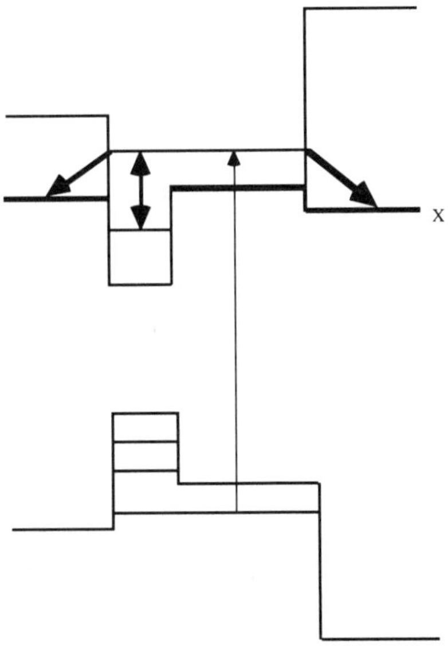

QTuI2 Fig. 2. Induced zero-absorption configuration in asymmetric semiconductor quantum wells ($Al_{0.5}Ga_{0.5}As$/5-nm $Al_{0.25}Ga_{0.75}As$/10-nm $Al_{0.4}Ga_{0.6}As$/AlAs).

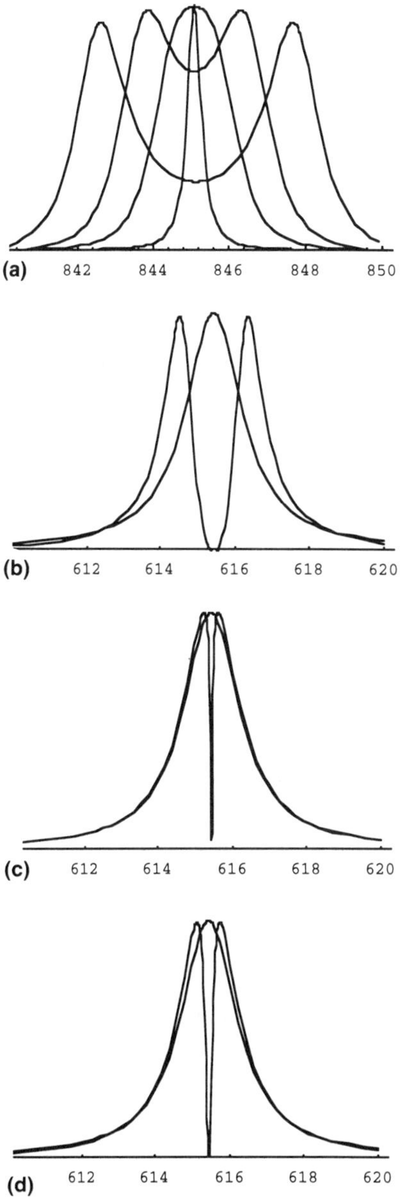

QTuI2 Fig. 3. (a) Excitonic absorption in a symmetric quantum well showing that only the optical Stark effect is present when the intersubband coupling strength is comparable to the subband relaxation; coupling strengths vary from 1 to 5 meV, and the absorption scales drop from 10 to 1. (b)–(d) Excitonic absorption in an asymmetric quantum well showing that (b) zero absorption is induced by quantum interference and a broadened Rabi splitting results from the optical Stark effect; the absorption scales are the same for the case without coupling and the case with 3-meV coupling; (c) a narrow zero-absorption dip results from quantum interference, but there is no Rabi splitting by the optical Stark effect; the absorption scales are the same for the case without coupling and the case with 0.3-meV coupling; (d) a zero-absorption dip is covered by quantum interference, and a weak optical Stark effect is present; the absorption scales are the same for the case without coupling and the case with 0.6-meV coupling.

well with Γ–X band mixing. Our calculations illustrate the significant difference between the pure optical Stark effects in undoped symmetric quantum wells and the pure quantum interference effects in asymmetric quantum wells.

Although it may be possible to observe EIT in light-hole excitonic transitions in symmetric quantum wells,[2] our direct absorption measurement gives evidence only for the optical Stark effect for both heavy- and light-hole excitonic transitions. The absorption results also imply that conduction-subband lifetimes are long enough to observe reduced absorption at room temperature. Figure 1 illustrates our symmetric single quantum well and the reduced absorption result obtained by using a properly polarized CO_2 laser for intersubband coupling. The other polarization introduces only thermal effects. Unlike reflectivity measurements, the absorption measurement removes multilayer interference and cavity effects and shows the presence of inhomogeneous broadening.

Because semiconductor subbands have long lifetimes, intersubband coupling easily creates the dressed states necessary for quantum interference resulting from a common one-way decay path. Figure 2 describes how band-gap engineering can reproduce the atomic system in semiconductors. The asymmetric-quantum-well configuration of Fig. 2 results in an $n = 1$ conduction subband with a long lifetime and an $n = 2$ conduction subband with a short lifetime and large oscillator strength. Therefore, most carriers decay to the X-band continuum. In Fig. 3 we show the calculated heavy-hole excitonic absorption for this configuration and for the symmetric multiple-quantum-well sample of Fig. 1, using the same formula that we derived from the one-electron density-matrix equation.[4] Figure 3(a) shows only Stark splitting for the symmetric quantum well, and Fig. 3(b) shows zero absorption (from quantum interference) and large splitting for the asymmetric quantum well. Figures 3(c) and 3(d) illustrate how quantum interference creates a narrow window with weak intersubband coupling in our new configuration.

Sandia National Laboratories, Albuquerque, New Mexico 87185

1. S. E. Harris, Phys. Rev. Lett. **62**, 1033 (1989).
2. J. E. Field, K. H. Hahn, and S. E. Harris, Phys. Rev. Lett. **67**, 3062 (1991).
3. K. Maschke, P. Thomas, and E. O. Gëbel, Phys. Rev. Lett. **67**, 2646 (1991).
4. D. S. Lee and K. J. Malloy, IEEE J. Quantum Electron. **30**, 85 (1994).
5. D. S. Lee, K. J. Malloy, O. Blum, and J. Klem, in *International Quantum Electronics Conference*, 1994 Technical Digest Series, Vol. 9 (Optical Society of America, Washington, D.C., 1994), postdeadline paper QPD24.

QTuI3 **3:15 pm**

Rosonance Raman scattering in Cr^{4+}-doped-forsterite laser crystal

Dana M. Calistru, W. B. Wang, V. Petricevic, R. R. Alfano, *Institute for Ultrafast Spectroscopy and Lasers, New York State Center of Advanced Technology for Ultrafast Photonic Materials and Applications, Department of Physics, City College and Graduate School of the City University of New York, 138th Street and Convent Avenue, New York, New York 10031*

The laser properties of solid-state tunable lasers are affected by nonradiative relaxation processes. To date, there is no theory to predict which vibronic (impurity-associated) and phonon (lattice) modes are involved in such processes or what the most important features that determine their coupling are. One step in understanding nonradiative processes is to study the phonon and vibronic modes, their interaction, and their dynamics.

We have investigated the Cr^{4+} modes in $Cr:Mg_2SiO_4$ by resonance Raman scattering at room temperature with an excitation wavelength of $\lambda_0 = 575$ nm, which corresponds to the Cr^{4+} dipole allowed transition ${}^3A_2 \rightarrow {}^3B_2({}^3T_1)$. (See Figs. 1–3.)

Cr^{4+} substitutes for Si in T_d molecular symmetry, and it has nine normal modes

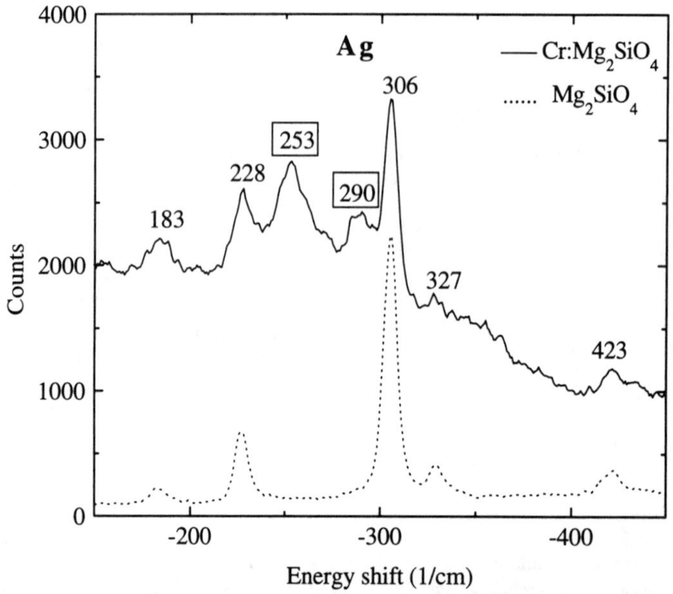

QTuI3 Fig. 1. Polarized Raman spectrum (dotted curve) for Mg_2SiO_4 and resonance Raman spectrum (solid curve) for $Cr^{4+}:Mg_2SiO_4$ at room temperature. The intensity of the lines is given in arbitrary units. The scattering configuration is $y(zz)x$, with A_g symmetry. The spectral range is 150–425 cm^{-1}. Resonance-enhanced modes are boxed.

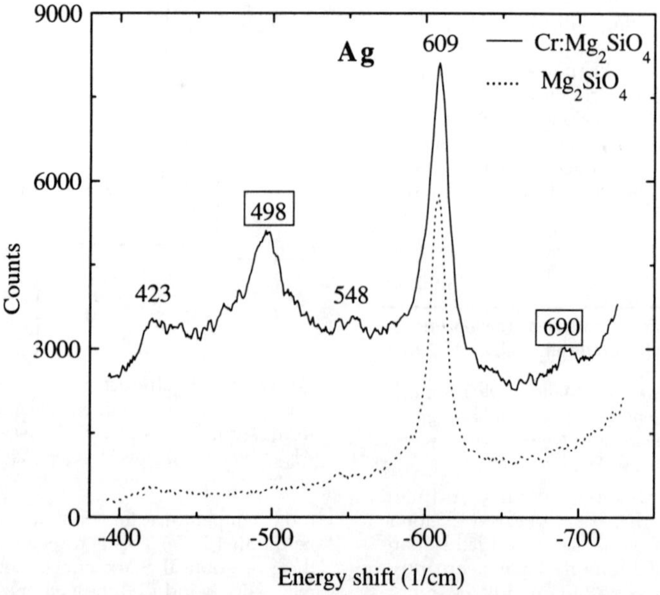

QTuI3 Fig. 2. Polarized Raman spectrum (dotted curve) for Mg_2SiO_4 and resonance Raman spectrum (solid curve) for $Cr^{4+}:Mg_2SiO_4$ at room temperature. The intensity of the lines is given in arbitrary units. The scattering configuration is $y(zz)x$, with A_g symmetry. The spectral range is 425–700 cm^{-1}. Resonance-enhanced modes are boxed.

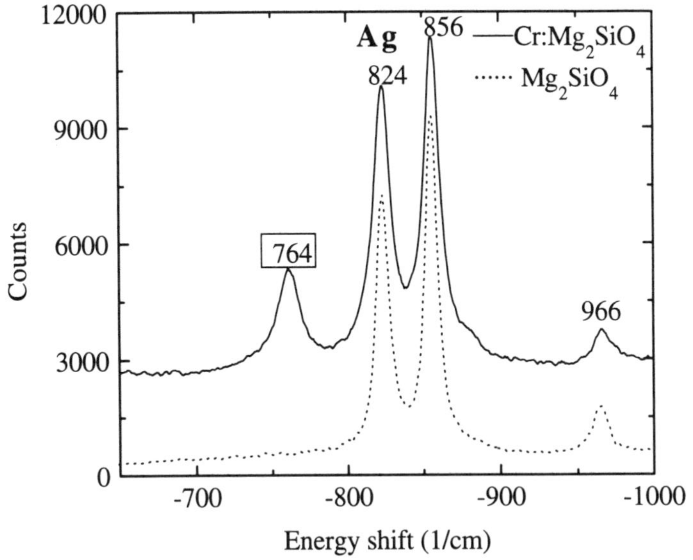

QTuI3 Fig. 3. Polarized Raman spectrum (dotted curve) for Mg_2SiO_4 and resonance Raman spectrum (solid curve) for Cr^{4+}:Mg_2SiO_4 at room temperature. The intensity of the lines is given in arbitrary units. The scattering configuration is $y(zz)x$, with A_g symmetry. The spectral range is 700–1000 cm^{-1}. Resonance-enhanced modes are boxed.

of vibration, $\Gamma = A_1 + E + 2T_2$, denoted by $\nu_1(A_1)$, $\nu_2(E)$, $\nu_3(T_2)$, and $\nu_4(T_2)$. Because of C_s site symmetry, the degeneracy is totally lifted. There are six totally symmetric modes, which were measured at 253, 290, 498, 690, and 764 cm^{-1}. There are three nontotally symmetric modes, which were measured at 480 and 740 cm^{-1}. By using group-theoretical methods it was determined that 764 cm^{-1} belongs to $\nu_1(A_1)$ and is the tetrahedral breathing mode, that 690 and 740 cm^{-1} belong to a $\nu_3(T_2)$, that 480 and 498 cm^{-1} belong to $\nu_2(E)$, and that 253 and 290 cm^{-1} belong to $\nu_4(T_2)$. The missing totally symmetric mode may be due to a small Stokes shift,[1] and the missing nontotally symmetric mode may be due to a weak vibronic coupling between $^3B_2$ and other electronic levels.[1] Estimates for the Si–O and Cr^{4+}–O elastic constants were made by using the Lagrange equations in two models in which only intratetrahedral interactions are considered. When the O–O elastic constant is assumed not to change when Cr^{4+} substitutes for Si, the effective elastic constant for $\nu_1(A_1)$ can be written in T_d and C_{3v} (effective symmetry when C_s distortions are taken into account). The determined Si–O and Cr^{4+}–O elastic constants (model dependent[2]; k given in mdyn/A) are k_{1Si-O} $(T_d) \in [3.12–3.77]$, k_{1Cr-O} $(T_d) \in [2.07–2.7]$ and k_{1Si-O} $(C_{3v}) \in [3.27–3.8]$, k_{1Cr-O} $(C_{3v}) \in [2.19–2.76]$. When no assumption is made regarding the O–O elastic constant k_2, the effective elastic constant for the four T_d modes and their mean measured values give k_1 = 3.27 mdyn/A, k_2 = 0.57 mdyn/A, $\nu_3(T_2)$ = 737 cm^{-1} (as compared to a mean measured value of 715 cm^{-1}) and $\nu_4(T_2)$ = 293 cm^{-1} (as compared to a mean measured value of 272 cm^{-1}). Comparison of k_1 with theoretical values for k_{1Si-O} leads to a decrease by ≈15% when Cr^{4+} substitutes for Si. The decrease in elastic constant shows that no localized modes exist in Cr^{4+}:Mg_2SiO_4.

For possible phonon-vibronic coupling between the host and the Cr^{4+} ion, from the energetic point of view the 253-cm^{-1} mode would be a good candidate for coupling with the 225-cm^{-1} phonon mode that was shown to be involved in nonradiative processes following photoexcitation in Cr^{4+}:Mg_2SiO_4.[3]

1. A. C. Albrecht, J. Chem. Phys. **43**, 5 (1961).
2. K. Iishi, Am. Mineral. **63**, 1198 (1978).
3. S. G. Demos, J. M. Buchert, and R. R. Alfano, Appl. Phys. Lett. **61**, 660 (1992).

QTuI4 **3:30 pm**

Intracavity injection lasing of a composite neodymium laser

Lei Ding, Li Yan, *Department of Electrical Engineering, University of Maryland, Baltimore, Maryland 21228*

Injection locking is very useful for obtaining high-power coherent laser signals. Usually, a weak external coherent laser beam derived from a stable master oscillator is injected into another high-power laser oscillator or amplifier. Both high-spectral-coherence and high-spatial-coherence laser beams of high powers can be obtained. In this paper we report, for the first time to our knowledge, the intracavity injection lasing of a composite laser that contains an Nd:YLF and an Nd:glass medium and the cooperative lasing behavior. Potentially, the coherence property of the homogeneously broadened Nd:YLF medium and the broadband property of the inhomogeneously broadened Nd:glass medium can be combined to produce (more easily) ultrashort laser pulses.

Figure 1 shows the schematic diagram of the composite neodymium laser. The five-mirror cavity contained two laser media: Nd:YLF and Nd:phosphate glass. The Nd:YLF was homogeneously broadened with its line center at 1.053 μm, and the Nd:glass was partially inhomogeneously broadened with its line center at 1.054 μm. A cw Ti:sapphire laser was used to pump the two gain media.

When one gain medium was pumped and the other was not pumped (by blocking the corresponding pump beam), we obtained the respective lasing behaviors for Nd:YLF and Nd:glass [Figs. 2(a) and 2(b)]. Nd:YLF lased at 1.053 μm with a narrow bandwidth of 2.7 Å (presumably due to spatial hole burning[1]). Nd:glass lased at ~12 Å away from the Nd:YLF line, as expected, with a broad bandwidth of ~9 Å. When both Nd:YLF and Nd glass were pumped simultaneously, we obtained lasing of the composite laser [Fig. 2(c)]. The Nd:glass lasing line at 1.054 μm was suppressed, and lasing occurred only at 1.053 μm with a bandwidth of 2.7 Å. Essentially, Nd:glass was locked to the lasing spectrum of Nd:YLF. This can be viewed as lasing of Nd:glass by the injection-stimulated emission from the Nd:YLF lasing signal. Under this injection-stimulated emission, the spectral components coincided with those of the Nd:YLF lasing signal, extracting most of the stored energy in the Nd:glass. But a more appropriate way to characterize this type of lasing is cooperative lasing, because even when pumping of Nd:YLF was attenuated to some level below its threshold (measured when Nd:glass was not pumped) while Nd:glass was pumped at a high level [same as that in Fig. 2(b)], the composite lasing was still similar to lasing of Nd:YLF alone. Only when pumping of Nd:YLF was attenuated by a factor of ~6 below its

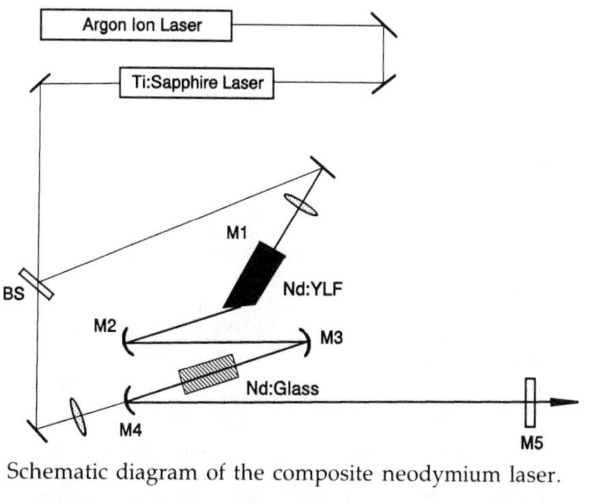

QTuI4 Fig. 1. Schematic diagram of the composite neodymium laser.

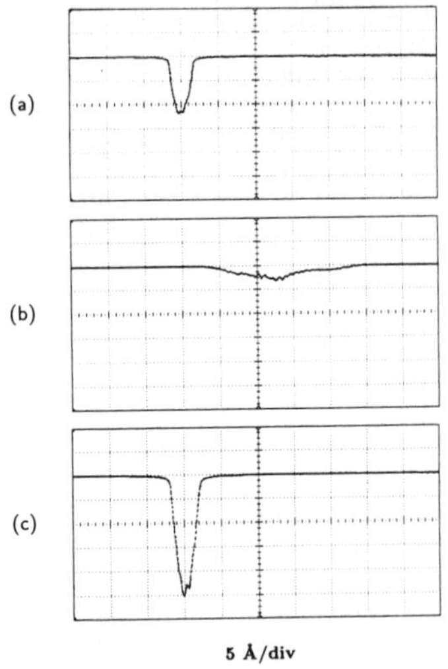

QTuI4 Fig. 2. Lasing spectra: (a) only Nd:YLF pumped, (b) only Nd:glass pumped; (c) both pumped.

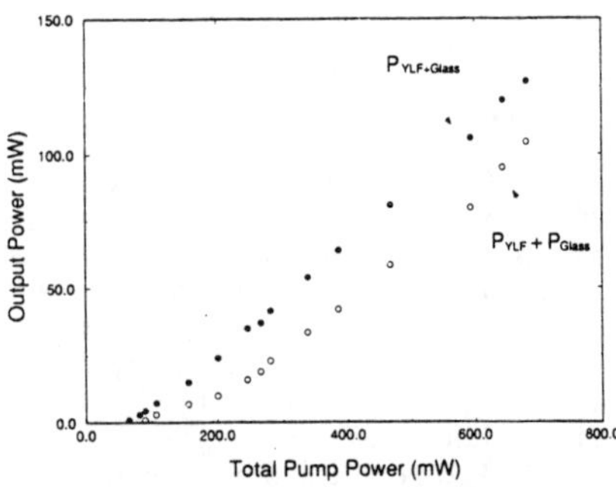

QTuI4 Fig. 3. Laser output powers for Nd:YLF and Nd glass pumped simultaneously (filled circles). Also shown is the simple sum of the individual lasing powers of Nd:YLF and Nd:glass measured when one gain medium was pumped and the other was not (open circles).

threshold value did lasing of the composite return to that of Nd:glass. We also found that the locking of pulling of the Nd:glass spectrum to that of Nd:YLF is determined by the relative ratio of the pumping level of Nd:YLF to that of Nd:glass but not by the absolute pumping levels. The lasing spectrum of the composite laser is determined by the broadening nature of the two gain media and by gain competition of the resonance modes.

Figure 3 shows the output powers of the composite laser when only Nd:YLF, only Nd:glass, and both Nd:YLF and Nd:glass were pumped. Under the injection-lasing condition, Nd:YLF and Nd:glass were not saturated independently. The cooperative saturation led to a lower threshold and a larger output power than the simple sum of the individual Nd:YLF and Nd:glass laser powers. We approximate the lasing power of Nd:glass as that from a homogeneously broadened laser. Results of calculation agreed with the observed output powers.

1. P. A. Schulz and S. R. Henion, Opt. Lett. **16**, 1502 (1991).

QTuI5 **3:45 pm**

Comparison of lasing mechanisms in ZnSe and GaAs

I. Galbraith, S. W. Koch,* *Department of Physics, Heriot-Watt University, Edinburgh EH14 4AS, U.K.*

In this paper we consider various mechanisms, both excitonic and free-carrier, that may give rise to lasing, and we compare the archetypal II–VI semiconductor, ZnSe, with the archetypal III–V semiconductor, GaAs. We calculate the gain spectrum for each lasing mechanism and thereby determine which process will dominate under given conditions in each material. Such a comparison can be expected to yield generic information about the differences between GaAs-based and ZnSe-based lasers.

We have calculated the gain for the following process: (1) exciton–electron recombination (ex–e), (2) exciton–hole recombination (ex–h), (3) exciton–exciton recombination (ex–ex), (4) exciton–LO-phonon recombination (ex–LO), and (5) simulated emission from an ionized electron–hole plasma (e–h). By calculating gain spectra and determining which process has the highest gain at a particular temperature, we are able to determine which process will dominate the lasing at threshold and hence the lasing wavelength.

In Fig. 1 we show the threshold density in ZnSe as a function of temperature for the various processes. At very low temperatures ex–ex recombination has the lowest threshold. Between 1 K and approximately 100 K ex–LO recombination has the lowest threshold. At still higher temperatures the ex–e process takes over. The ex–h process is not shown because its threshold density is always above the ex–e threshold because of the larger mass. Typical gain coefficients at threshold are $\approx 100\ cm^{-1}$, compared to a band-edge absorption peak of $\approx 10^4\ cm^{-1}$. Note that the lasing thresholds are substantially below the Mott density in ZnSe.

In Fig. 2 we show the threshold density in GaAs as a function of temperature for the various processes. Qualitatively, the curves are very similar, but the crucial fact

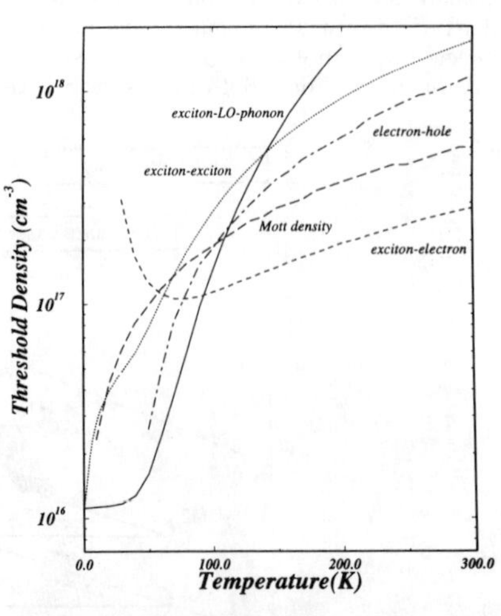

QTuI5 Fig. 1. Threshold excitation densities and Mott density as a function of temperature in ZnSe.

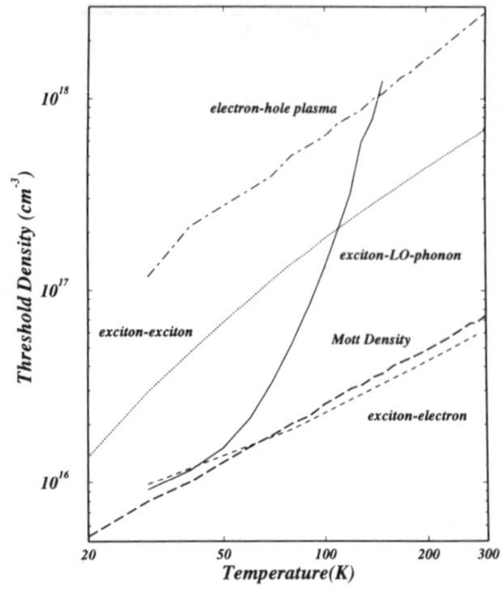

QTuI5 Fig. 2. Threshold excitation densities and Mott density as a function of temperature in GaAs.

is that all excitonic processes have thresholds at or above the Mott density. This means that they cannot proceed because there will be no excitons in the inversion at that density. Thus only the e–h process can contribute to the gain. This is in accord with many experimental findings in GaAs. Note that although the carrier density is above the Mott density, there can remain a peak in the nonlinear absorption spectrum because of the Coulomb enhancement of the continuum states.

We have shown from calculation of various gain mechanisms that the lowest threshold occurs below the Mott density in ZnSe, unlike in GaAs. This indicates that the lasing mechanism in ZnSe-based materials can be qualitatively different from that in GaAs-based lasers. This has implications for laser design in that for lowest-threshold operation the excitonic nature of the pumped plasma must be maintained. At higher pump intensities, at which the screening of the Coulomb interaction dominates, both materials will lase on the e–h mechanism.

**Department of Physics, Philipps University, Marburg, Germany*

QTuJ **2:30 pm**

Hyatt C/D

Near-Field Optical Microscopy

Daniel S. Chemla, *Lawrence Berkeley Laboratory, Presider*

QTuJ1 (Invited) **2:30 pm**

Ultrafast near-field optical probing

Jason B. Stark, Umar Mohideen,*
Richart E. Slusher, *AT&T Bell Laboratories, Murray Hill, New Jersey 07974*

Direct measurement of carrier dynamics, by using femtosecond light sources, has yielded a wealth of understanding within the field of condensed-matter physics.[1] Also, recently, near-field optical-probe techniques have permitted optical measurements with spatial resolution below the diffraction limit.[2] An understanding of many scientically and technologically important issues requires a knowledge of the local behavior of carriers within materials and microstructures. We report the development of the required instrumentation and what we believe to be the first measurement of carrier dynamics combining spatial resolution below the diffraction limit and temporal resolution below 60 fs.

The experimental instrumentation consists of a broadband mode-locked light source, a pair of acousto-optic modulators (AOMs), and a near-field optical microscope (Fig. 1). Infrared optical pulses are generated in a mode-locked Ti:sapphire laser, with a repetition rate of 82 MHz and average power of 400 mW. The spectral bandwidth is adjustable up to a maximum of 100 nm about the 800-nm center. The pulse width, from the 3-mm crystal, is below 15 fs. The output enters a dispersive delay line for compensation of the positive dispersion of the output coupler and subsequent optics. Within this delay line the beam is spectrally separated into pump and probe. The pump beam passes through a variable delay line, followed by an AOM operating at 3.1 MHz. It then enters the far-field microscope and is reflected downward from a beam splitter, through the objective, and into a 5-μm spot on the sample.

The probe beam enters a second dispersive delay line to precompensate the dispersion of the 1-m fiber probe tip. The probe beam is modulated at 3.0 MHz. Pulses enter the probe tip and emerge with a pulse width of 50 fs. The fiber probe tip is a single-mode optical fiber that is tapered and coated with aluminum, leaving a small aperture, typically 50–100 nm.[3] Light propagating through this aperture is coupled into the far field and is detected. The coupling efficiency is typically −50 to −60 dB, leaving ~1 nW radiating into the far field. The sample is held in the near field of the fiber tip by using a feedback technique,[4] stabilizing the tip–sample separation at 50 nm. The fiber tip is imaged into a multimode optical fiber and is delivered to a spectrometer for detection. In a nonlinear transmission measurement, the nonlinear response is recorded at the 100-kHz modulation sideband.

Samples of GaAs quantum wells were etched to produce 150-nm-thick disks, of varying diameters from 3 to 10 μm, comprising eight periods of 10-nm wells separated by 5-nm barriers. The room-temperature excitonic absorption edge is at 850 nm. Measurements indicate that the carrier recombination time is 500 ps. The pump beam was filtered before the sample, to a 10-nm bandwidth about 800 nm, and the probe was filtered after the sample to a 10-nm bandwidth about 820 nm. In this case the probe measures carriers well above the band edge, as they relax from the initial distribution injected by the pump. Figure 2 shows the nonlinear signal, measured by the probe, in two microdisks (upper right and lower left) of diameters 5 and 10 μm. The nonlinear optical response at 67-fs delay [Fig. 2(a)] shows the carriers relaxing after excitation by the leading edge of the pump. Within 133 fs [Fig. 2(b)] the nonlinear response is near its peak. After the initial excitation, measurements at 2, 20, and 67 ps [Figs. 2(c)–2(e)] show that carriers relax in a time of ~80 ps, indicating the rate of carrier cooling. Furthermore, carriers near the edges of the microdisks relax most quickly, leaving the nonlinear response peaked toward the center.

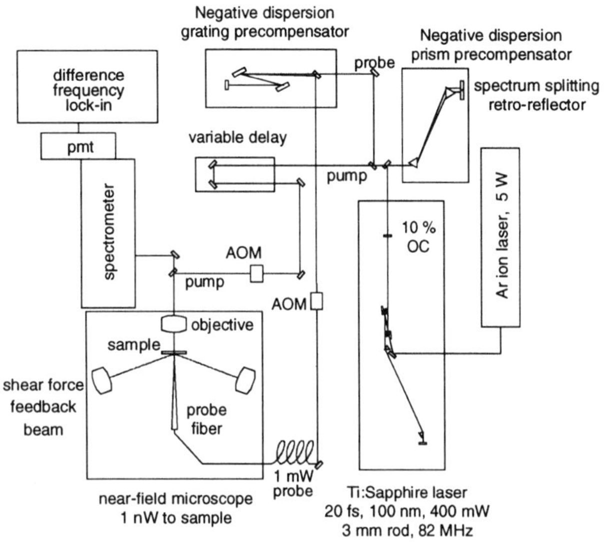

QTuJ1 Fig. 1. Ultrafast near-field probe. Ti:sapphire laser pulses are spectrally split into the pump and the probe. The probe is precompensated for dispersion, the pump is delayed, and the pump and probe are modulated by AOMs. The probe is coupled into the fiber tip, through the sample in the near field, and into the spectrometer. The pump is focused onto the sample, and the difference frequency is measured.

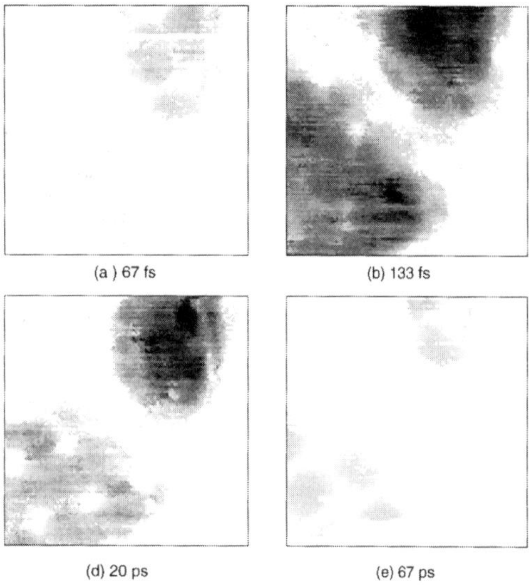

QTuJ1 Fig. 2. Time-resolved images (9 μm × 9 μm) of carrier relaxation in semiconductor microdisks for time delays (a) 67 fs, (b) 133 fs, (c) 2 ps (omitted), (d) 20 ps, and (e) 67 ps. Two disks are shown, upper right and lower left. Carriers are injected by an 800-nm pump and are measured with an 820-nm probe. Carriers injected uniformly (a) relax most quickly near edges (b)–(e).

We demonstrate, for the first time to our knowledge, submicrometer spatially resolved images of carrier dynamics, with a time resolution of 60 fs. Carriers injected uniformly into the microdisks cool in a time of 80 ps, most quickly near the edges. Studies continue to measure the details of this relaxation as a function of probe wavelength and with better spatial resolution. Similar studies are under way to study nonequilibrium transport in narrow-channel electronic devices and to study local carrier dynamics in operating semiconductor lasers. Measurements with high spatial and temporal resolution have now evolved from the stage of instrumentation development to begin producing results with important scientific and technological implications. The rapid development of this field presents many exciting prospects, in the immediate future, for the study of carrier dynamics in materials and microstructures.

*Department of Physics, University of California, Riverside, California 92521

1. Jagdeep Shah, ed., *Hot Carriers in Semiconductor Nanostructures* (Academic, San Diego, Calif., 1992).

2. E. Betzig and J. K. Trautman, Science **257**, 189 (1992).
3. E. Betzig, J. K. Trautman, P. D. Harris, J. S. Wiener, and R. L. Kostelak, Science **251**, 1468 (1991).
4. E. Betzig, P. L. Finn, and J. S. Wiener, Appl. Phys. Lett. **60**, 2484 (1992).

QTuJ2 **3:00 pm**

Near-field optical-beam-induced-current measurements for the investigation of quantum structures

W. D. Herzog, B. B. Goldberg, M. S. Ünlü, H. F. Ghaemi, D. Sun,* and E. Towe,*
Center for Photonics Research, Boston University, Boston, Massachusetts 02215

We report near-field optical-beam-induced-current (NOBIC) measurements on semiconductor-quantum-well (QW) structures. A subwavelength fiber tip is coupled with a tunable laser source and is scanned over a sample surface. The induced photocurrent reveals the compositional profile of quantum structures. Semiconductor QW structures were designed and were fabricated by molecular-beam epitaxy (MBE) to study the wavelength dependence and resolution capability of NOBIC. We demonstrate that the resolution of the probe depends strongly on the excitation wavelength because of the variation of the optical penetration depth.

Near-field scanning optical microscopy (NSOM) is a recent technique[1] in which a tapered optical-fiber probe of aperture size ~100 nm is controlled in close proximity ($z \sim 10$ nm) to a sample and is scanned over the surface. By using the tip as a localized radiation source, induced photocurrent can be measured.[2] Topological imaging and height control are achieved by shear-force techniques. A tunable Ar^+-pumped Ti:sapphire laser provides excitation at λ = 700–855 nm for NOBIC measurements.

We have designed and fabricated structures consisting of three sets of GaAs QW pairs surrounded by $Al_{0.35}Ga_{0.65}As$ barriers placed in the depletion region of a p-i-n diode. The spacing between the QWs is varied from 300 to 50 nm, and the pairs are separated by 500 nm [see Fig. 1(a)]. Mesa photodiodes were fabricated and were cleaved to reveal the vertical structure. Scanning the tip across this facet produces a photocurrent response used to generate NOBIC images. Figure 1 shows a NOBIC line scan at λ = 700 nm in the epitaxial-growth direction and its correlation to the layer structure. The QWs separated by 300 nm are clearly resolved. In this experiment we were unable to resolve more-closely-spaced QW pairs because of the relatively large tip diameter (~120 nm) and the lack of depletion caused by a high p-type background in the unintentially doped region. An additional limitation on the resolving power is due to the penetration of the optical field into the semiconductor.

To investigate the wavelength dependence of NOBIC, the excitation wavelength was varied from 700 to 855 nm; the results are shown in Fig. 2. The barriers ($Al_{0.35}Ga_{0.65}As$) are transparent for all wavelengths, and the absorption coefficient in the GaAs QW is estimated to vary from

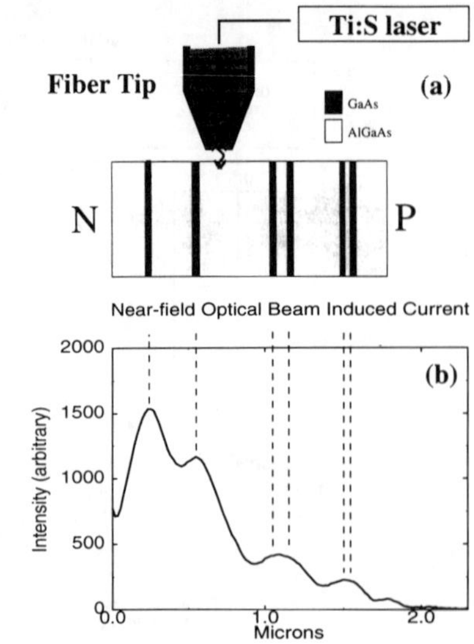

QTuJ2 **Fig. 1.** (a) Layer structure and the fiber tip and (b) NOBIC line scan. The GaAs QWs are 30 nm wide and are separated by distances of 300, 120, and 50 nm. The sample was reverse biased at 2 V.

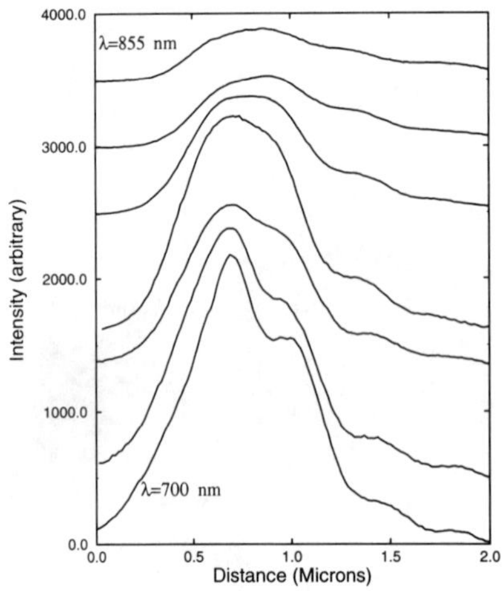

QTuJ2 **Fig. 2.** NOBIC line scans as a function of excitation wavelength. The wavelength varies from 700 nm (bottom) to 855 nm (top). Note the reduction in resolving power as the wavelength is increased.

4×10^4 to 1×10^4 cm^{-1} with increasing wavelength, corresponding to optical penetration depths of 250 nm to 1 μm. Increasing the excitation wavelength reduces the resolving power because of the increase in the penetration depth into the semi-infinite 2-D structure. Our low-temperature luminescence and reflectivity results on quantum wires do not have this limitation, and higher resolutions are achieved.

In conclusion, we demonstrate the importance of the excitation wavelength on NOBIC studies and other NSOM imaging techniques. The NOBIC technique allows correlation of layer composition directly with local optical properties. Further studies will include facet patterning by selective recessing of the AlGaAs barriers to directly correlate NOBIC with topography. Measurements as a function of wavelength and tip-to-surface distance will provide detail on the coupling constants of the electromagnetic field to the sample.

This research is supported by the National Science Foundation and the Office of Naval Research.

Department of Electrical Engineering, University of Virginia, Charlottesville, Virginia 22903

1. E. Betzig, J. K. Trautman, T. D. Harris, J. S. Weiner, and R. L. Kostelak, Science **251**, 1468 (1991).
2. J. W. P. Hsu, E. A. Fitzgerald, Y. H. Zie, and P. J. Silverman, Appl. Phys. Lett. **65**, 344 (1994).

QTuJ3 **3:15 pm**

Nanometric integrating tip: enhanced sensitivity of fluorescence detection in photon STM

T. Saiki, S. Mononobe, M. Ohtsu,*
Kanagawa Academy of Science and Technology, 3-2-1 Sakado, Takatsu-ku, Kawasaki, Kanagawa 213, Japan

Fluorescence detection with subwavelength resolution is important in biological applications[1] such as the observation of a membrane selectively labeled with dye molecules. Photon scanning tunneling microscopy (PSTM), which is also called scanning near-field optical microscopy (SNOM),[2] enables us to detect fluorescence with the resolution determined by the tip diameter at the apex of the fiber probe. In comparison with the illumination mode in PSTM, the most favorable quality of the collection mode is that probe–sample distance can be controlled by all-optical means without a feedback mechanism based on shear force, which may damage samples. It is, therefore, essential to estimate the resolution and the sensitivity of PSTM as fluorescence microscopy. Also, local detection of fluorescence with a fiber tip is an attractive subject in quantum optics. The approach of the tip to a fluorescent material in the near-field region will largely modify its radiation modes. This modification leads to a change of its radiation pattern and of its lifetime, which offers the possibility of highly effective collection of fluorescence. From these viewpoints, we here report the fluorescence imaging of dye-doped 110-nm microspheres, and we estimate the pickup efficiency, which demonstrates that the fiber tip acts as something like a nanometric integrating tip in collecting fluorescence.

Our experimental setup is shown in Fig. 1. As a sample, rhodamine-doped polystyrene microspheres of 110-nm diameter are dispersed on a silica substrate. The sharpened fiber probe with a 20-nm tip diameter is fabricated by chemical etching.[3] A part of the tip is coated with gold to eliminate background fluorescence [a practical coating scheme is shown in Fig. 3(a)]. Fluorescence from the rhodamine molecules is picked up by the fiber tip and is measured by a photon-counting method. Figure 2 shows a near-field fluorescence image of microspheres. A sample–probe separation of ~20 nm is kept during scanning. The image size of the spheres is estimated as ~180 nm full width at half maximum, beyond the diffraction limit.

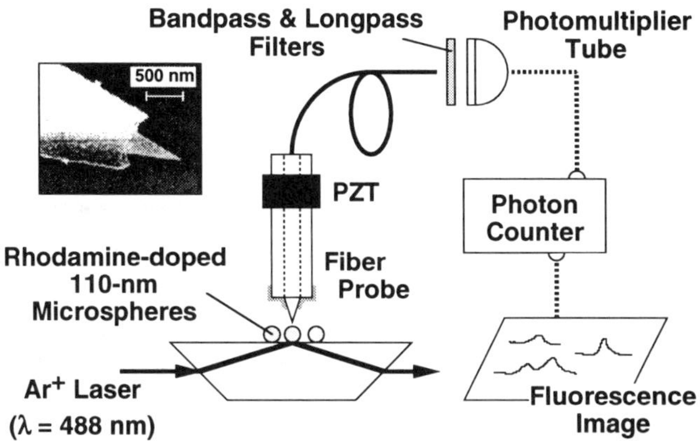

QTuJ3 Fig. 1. Experimental scheme of fluorescence detection with the PSTM system. Inset shows SEM image of a fiber tip.

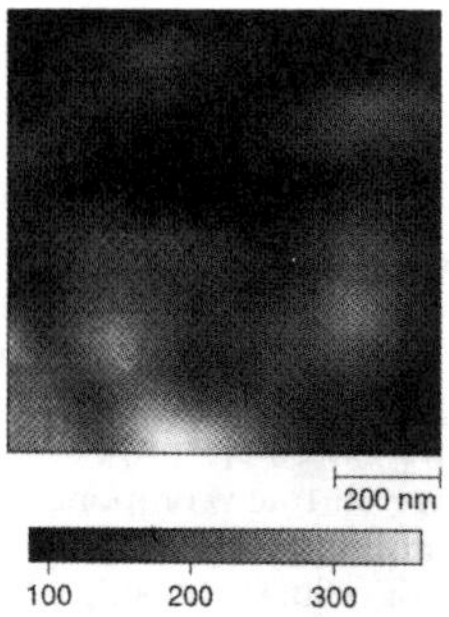

QTuJ3 Fig. 2. Near-field fluorescence image of dye-doped 110-nm microspheres.

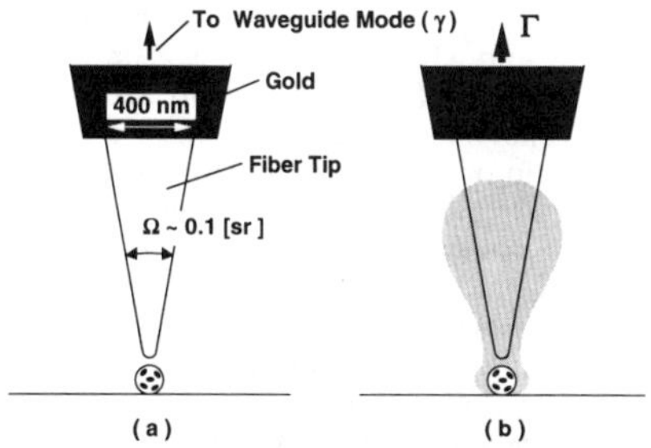

QTuJ3 Fig. 3. (a) Pickup efficiency determined by the solid angle at the aperture plane from the sample (Ω) and coupling coefficient to waveguide mode (γ). (b) Enhanced pickup efficiency (Γ) due to near-field coupling between the tip and the sample.

To clarify the mechanism of fluorescence detection with high resolution, we estimate the pickup efficiency Γ (defined as the ratio of fluorescence intensity finally coupled to the waveguide mode of the fiber probe to the total fluorescence intensity). (By applying the parameters of the dye molecules and detection systems, a Γ of 2×10^{-3} to 5×10^{-3} is obtained. If the radiation pattern of the dye molecules is isotropic, even when the probe approaches the sphere to a distance of less than 20 nm, the pickup efficiency is estimated as $(0.1/4\pi)\gamma$, where γ ($\ll 1$) is the coupling coefficient to the waveguide mode [Fig. 3(a)]. By comparing these two values, it is confirmed that the enhanced sensitivity of Γ is induced by the modification of radiation modes due to the strong coupling between the fiber tip and the sphere in the near-field region [Fig. 3(b)]. That is, the fiber probe works as a nanometric integrating tip in the collection of fluorescence.

*Also with *Interdisciplinary Graduate School of Science and Engineering, Tokyo Institute of Technology, 4259 Nagatsuta-cho, Midori-ku, Yokohama, Kanagawa 226, Japan*

1. R. C. Dunn, G. R. Holtom, L. Mets, and X. S. Xie, J. Chem. Phys. **98**, 3094 (1994).
2. E. Betzig and R. Chichester, Science **262**, 1422 (1993).
3. T. Pangaribuan, S. Jiang, and M. Ohtsu, Electron. Lett. **29**, 1978 (1993).

QTuJ4 **3:30 pm**

GaAs/AlGaAs quantum-dot near-field scanning optical spectroscopy

U. Mohideen, M. J. Yoo,* H. Hess,* W. S. Hobson,* F. Ren,* R. Kopf,* R. E. Slusher,* *Department of Physics, University of California at Riverside, Riverside, California 92521*

Quantum dots (QDs) with dimensions less than the optical wavelength and of the order of the Bohr radius of the exciton are expected to exhibit novel types of carrier–phonon and carrier–photon relaxation mechanisms[1–3] because of the radical change in the carrier, phonon, and photon density of states due to the three-dimensional confinement. Other predicted optical nonlinearities, such as a concentrated oscillator strength and superradiance,[4] might be very useful in the fabrication of optoelectronic devices. The near-field scanning optical microscope (NSOM) with a spatial resolution better than a tenth of an optical wavelength is ideally suited for studying individual nanostructures such as QDs. We have obtained images and spatially resolved spectra of 100-nm QDs of surface-passivated GaAs/AlGaAs quantum wells of varying thickness with a low-temperature NSOM.[5]

An MOCVD-grown multiple-quantum-well sample was used. The QDs were made by electron-beam lithography followed by ECR etching. The surfaces of the QDs were then passivated by sulfide treatment.[6] A low-temperature SiN_x deposition was carried out to encapsulate and stabilize the sulfide-treated surface. The emission in the dot region is one-sixth that from the planar-quantum-well region because of the residual surface recombination centers left after the passivation step. Emission corresponding to well-width fluctuations and defects has also been observed in the planar-quantum-well region of these MOCVD samples. Narrow emission spectra corresponding to single exciton states in individual QDs were observed. The integrated spectra from a 2.2-nm quantum well in the QD region showed a blue shift of 1 nm, which can be attributed to the 100-nm lateral confinement of the QDs. A continuous spectrum from spatially localized single-exciton states was also observed in some cases from the deeper-lying 4.5-nm quantum well in the QD region.

**AT&T Bell Laboratories, Murray Hill, New Jersey 07974*

1. U. Bockelmann and G. Bastard, Phys. Rev. B **42**, 8947 (1990).
2. U. Mohideen, R. E. Slusher, F. Jahnke, and S. Koch, Phys. Rev. Lett. **73** (1944).
3. P. D. Wang, C. M. Sototmayor Torres, H. Benisty, C. Weisbuch, and S. P. Beaumont, Appl. Plys. Lett. **61**, 946 (1992).
4. E. Hanamura, Phys. Rev. B **37**, 1273 (1988).
5. H. F. Hess, E. Betzig, T. D. Harris, L. N. Pfeiffer, and K. W. West, Science **264**, 1740 (1994).
6. U. Mohideen, W. S. Hobson, S. J. Pearton, F. Ren, and R. E. Slusher, Appl. Phys. Lett. **64**, 1911 (1994).

QTuJ5 **3:45 pm**

Equal pulse correlation in GaAs thin films with near-field microscopy

S. Smith, B. G. Orr, R. Kopelman, T. Norris, *Center for Ultrafast Optical Science, University of Michigan, Ann Arbor, Michigan 48109*

We are developing a femtosecond near-field scanning optical microscope (NSOM). Here

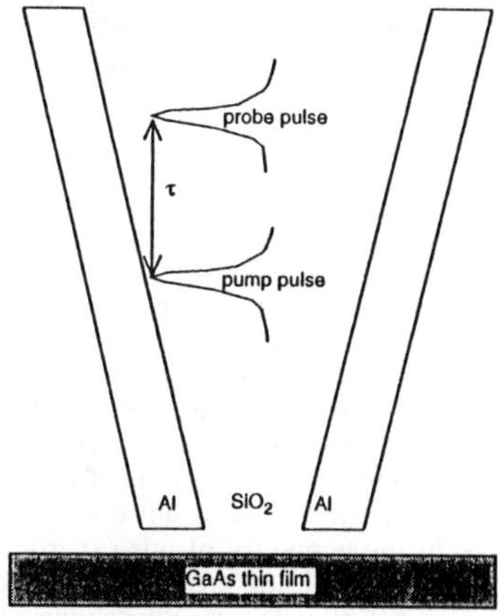

QTuJ5 Fig. 1. The tapered aluminum-glass-aluminum structure of the near-field tip forms a subwavelength aperture through which both the pump and probe are sent with variable time delay τ. If the near-field tip to sample separation is $\ll\lambda$, a highly localized pump-probe measurement can be made.

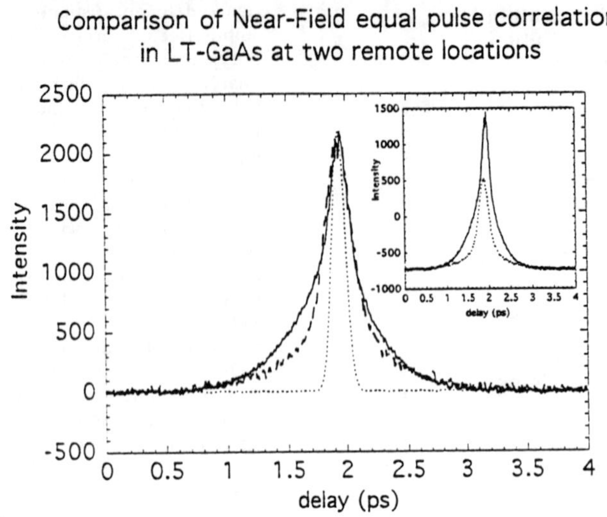

QTuJ5 Fig. 2. Near-field equal pulse correlation at two remote locations in a 1-μm thick LT-GaAs film (outer two traces). Inner trace is the autocorrelation of the pulse. Inset: equal pulse correlations before normalization.

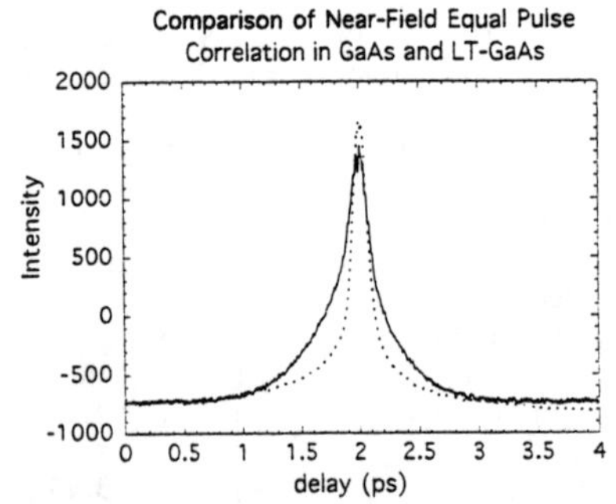

QTuJ5 Fig. 3. Near-field equal pulse correlation in GaAs and LT-GaAs. The GaAs correlation is the narrower of the two.

we present preliminary results from an equal pulse correlation[1] (EPC) measurement in GaAs thin films (1-μm thick) taken using our NSOM. These results deomonstrate the possibility of making highly localized, femtosecond carrier relaxation and/or carrier diffusion measurements in a semiconductor thin film.

Our NSOM uses the tapered fiber probe developed by Betzig *et al.*,[2] through which we send two equal intensity pulses as shown in Fig. 1. The well known saturation of the absorption in an optically thin semiconductor film leads to an increased transmission, which is symmetric about time zero. As discussed in the original paper by Tang *et al.*,[1] the signal can be interpreted as a convolution of the pulse autocorrelation and a double-sided decaying exponential, where the decay is attributed to hot carrier relaxation. In the geometry of our experiment, we expect an additional contribution due to carrier diffusion because the volume being sampled is much smaller than typical diffusion lengths in GaAs.

Figure 2 shows the EPC at two remote locations in a 1-μm thick low temperature-grown GaAs (LT-GaAs) film taken using our NSOM. The autocorrelation through an identical length fiber without an NSOM tip is also shown for comparison.[3] We observe variations in both relaxation time and signal amplitude. These variations can be related to the local density of states and the proximity of material defects and thus are a signature of the local environment on a submicron scale.

Figure 3 shows a comparison of the EPC taken with our NSOM in standard epitaxially grown GaAs and LT-GaAs 1-μm thick films. The GaAs EPC width is much narrower, even though the conduction band lifetime in LT-GaAs is known to be much shorter than in GaAs. One possible reason for this could be the effect of carrier diffusion away from the volume being probed. Our current understanding of LT-GaAs assumes a high density of defects and thus a decrease in the mobility and diffusion length. This in turn would show a slower relaxation component due to diffusion for the LT-GaAs, which is consistent with what is observed.

It should be noted that the aperture size in these measurements was undetermined; however, an estimate based on the output power observed compared to that of other NSOM tips of known size is 250 nm (0.3λ for the wavelength used). The obtainable resolution of NSOM in this application will be limited by signal-to-noise considerations, and has yet to be determined.

In light of these results, the possibility of producing 2-D images at varying time delays between the "pump" and "probe" pulses now exists. Such images of the nonlinear optical response of thin films may prove valuable in understanding the ultrafast optical response of materials like LT-GaAs. Our work now focuses on generating these images and determining the practical resolution limit of near-filed optics in this application.

This work was funded by: the National Science Foundation through the Center for Ultrafast Optical Science STC PHY 8920108 and NSF grant DMR-9111622, and by DOE grant DE-FG02-90ER61085.

1. C. L. Tang and D. J. Erskine, Phys. Rev. Lett. **51,** 844 (1983).
2. E. Betzig *et al.*, Science **251,** 1468 (1991).
3. We have previously shown the pulse width from a single-mode fiber without an NSOM tip is the same as a fiber with an NSOM tip. Paper CTuM5, IQEC 1994.

JTuF Tutorial **2:30 pm**

Hyatt E/F

Applications of Lasers in Bioanalytical Chemistry

David Rakestraw, *Sandia National Laboratories, Presider*

JTuF1 (Tutorial) **2:30 pm**

Applications of lasers in bioanalytical chemistry

Edward S. Yeung, *Iowa State University*

Lasers have grown from being an exotic device in the research laboratory to become an important part of practical bioanalytical instrumentation. Laser-excited fluorescence offers extremely high sensitivity for the detection of trace components in ultramicro environments. This has led to a new generation of cellular and subcellular probes for the biochemical and the clinical laboratory. Examples of upcoming technologies that are relevant to medicine and biotechnology are presented, including (1) high-speed DNA mapping and sequencing, where the analysis time is decreased by 1000× and where the accuracy is improved substantially; (2) the detection of various intracellular enzymes in blood cells, providing information about aging and carcinogenesis; (3) the characterization of individual enzyme molecules, showing how protein folding and refolding can lead to different activities among a population of otherwise identical molecules; (4) the measurement of zeptomole quantities of antigens, paving the way for early diagnosis of diseases; and (5) the monitoring of trace impurities in biopharmaceuticals, potentially leading to new standards for quality control in their production. New types of lasers optimized for these applications will therefore have significant market potential.

QTuK 4:30 pm
Hyatt A

Spatial Pattern Formation

Grover A. Swartzlander, Jr., *Worcester Polytechnic Institute, Presider*

QTuK1 4:30 pm

Drift instability in a rubidium cell with a feedback mirror

A. Petrossian, L. Dambly, G. Grynberg, *Kastler-Brossel Laboratory, Department of Physics, l'Ecole Normale Supérieure, 24 rue Lhomond, 75231 Paris, Cedex 05, France*

Spatiotemporal dynamics of extended systems is presently a very active field of research. If the static behavior of patterns can be considered as being well established, the situation is quite different for the dynamics. In fact, in most cases the first instability gives rise to a static pattern and the temporal dynamics appear only at higher intensities. A much simpler situation would occur if both the spatial and the temporal aspects of the instability appeared at the first threshold. We present an experimental study of such a situation for a laser beam transmitted through a rubidium cell with a tilted feedback mirror.

We consider a system consisting of a thin slab of nonlinear medium with a single feedback mirror.[1] When a plane wave is incident on such a system, an instability appears above threshold leading to modulation of the field amplitude or of the field polarization. In the case in which the system can be described with only one transverse coordinate, the pattern that naturally appears is the roll pattern, which is associated with a spatially periodic variation of some observable of the nonlinear medium. A stationary pattern varying as $\cos Kx$ is then predicted. The transverse wave vector K is proportional to $1/\sqrt{\lambda d}$, where d is the distance between the nonlinear medium and the mirror. When the incident beam and the normal of the mirror make an angle α, the incident beam and the beam reflected from the feedback mirror are shifted by an amount $2d\alpha$ in the nonlinear medium. As a result, there is a phase shift $\Phi_x = 2Kd\alpha \approx \alpha\sqrt{d/\lambda}$. This phase shift can be compensated by a drift of the pattern, leading to a wave varying as $\cos(Kx - \delta t)$ because for a time-dependent excitation, the steady-state solution exhibits another phase shift $\Phi_t \approx \delta/\gamma_t$ (where γ_t^{-1} is the atomic relaxation time) between the exciting wave and the atomic response. A self-consistent solution for the pattern is found when $\Phi_x = \Phi_t$ (Ref. 2). For this reason, the drift frequency δ is expected to vary as $\delta \approx \gamma_t\alpha\sqrt{d/\lambda}$.

The experimental setup consists of a 3-cm rubidium cell at a temperature of about 100°C, which is placed in front of a flat multidielectric mirror. The light source is a single-mode Ti:sapphire laser operating at 795 nm around the $5S_{1/2}$–$5P_{1/2}$ resonance line. The laser-beam cylindrical symmetry is broken by reducing the beam dimension along one transverse direction until a stable

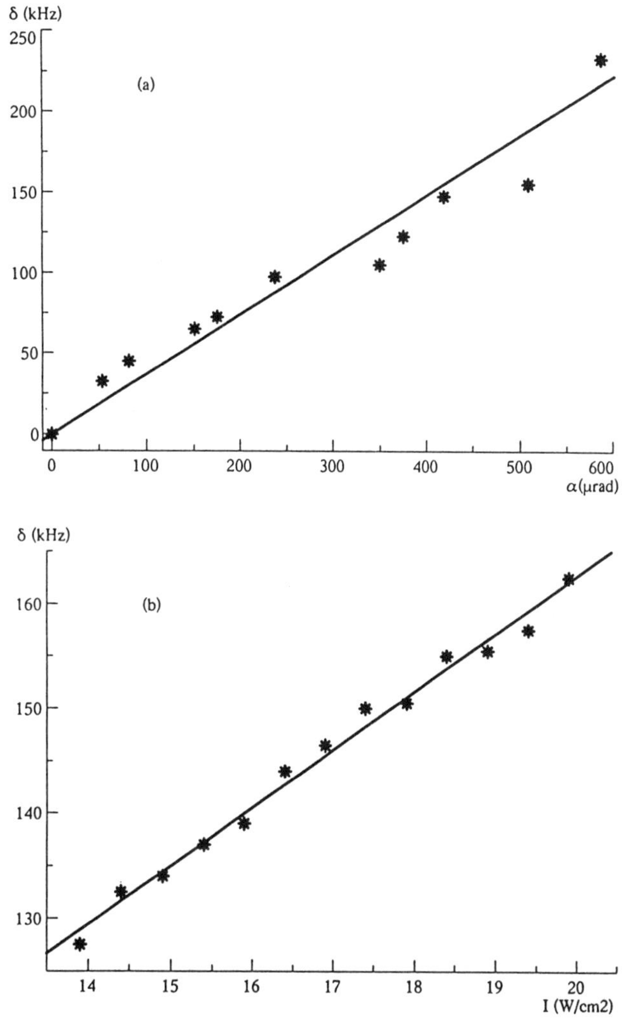

QTuK1 Fig. 1. Drift frequency δ (a) versus α for I = 18 W/cm² and (b) versus field intensity for α = 0.4 mrad. Both experiments were performed with d = 0.1 m.

double-spot far-field pattern is obtained. Such a pattern corresponds to rolls in the near field. A part of the instability component was sent on a photodiode. When $\alpha \neq 0$, the signal exhibits a periodic variation at frequency 2δ. The linear variation of δ versus α is shown in Fig. 1. The variation of δ versus the field intensity I is also linear (as expected from the origin of γ_t). In conclusion, the experimental results are in excellent agreement with predictions.

1. W. J. Firth, J. Mod. Opt. **37**, 151 (1990).
2. G. Grynberg, Opt. Commun. **109**, 483 (1994).

QTuK2 4:45 pm

Simulating the dynamics of laser-beam pattern formation in sodium vapor

S. Dangel, R. Holzner, *Institute of Physics, University of Zurich, Winterthurerstrasse 190, CH-8057 Zurich, Switzerland*

Optical pumping[1] of alkali vapors by circularly polarized low-intensity laser light causes a variety of astonishing nonlinear effects, such as the formation of slow-moving steep longitudinal intensity gradients (pumping-light fronts),[2] mutual extinction of circularly polarized laser beams (beam switching),[3] deflection of copropagating beams (beam bouncing),[4] highly structured polarization and intensity patterns,[5] asymmetry of absorption profiles,[6] and beam splitting.[7]

We present a theory and a numerical model for the 3-dimensional and time-dependent simulation of the propagation and pattern formation of polarized laser beams in sodium vapor. The beams are tuned closely to the D_1 line of sodium and pass the medium only once (no cavity). Our model is semiclassical; i.e., we use the wave equation for the classical field amplitudes, and we use the density-matrix formalism for the quantum-mechanical description of the atomic polarization. It can be shown that the slowly-varying-envelope approximation (SVEA) is still valid, even for dynamical processes.

In simulations of the steady state, the sodium hyperfine structure has always been neglected; i.e., the sodium D_1 transition has been treated as a $(J = 1/2)$-to-$(J = 1/2)$ system.

However, the $(J = 1/2)$-to-$(J = 1/2)$ model leads to predictions for time-dependent effects that are about an order of magnitude

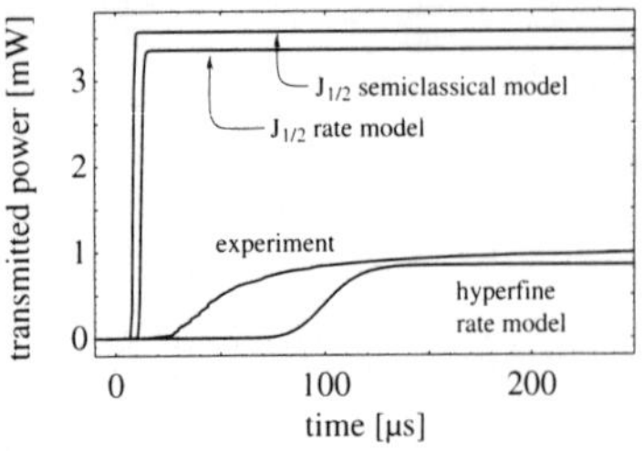

QTuK2 Fig. 1. Transmitted power as a function of time after one circularly polarized beam is switched on at $t = 0$. The ($J = 1/2$)-to-($J = 1/2$) rate model, ($J = 1/2$)-to-($J = 1/2$) semiclassical model, hyperfine rate model, and experiment are compared. No adjustable parameters were used, but the measured beam powers were corrected for cell-window losses.

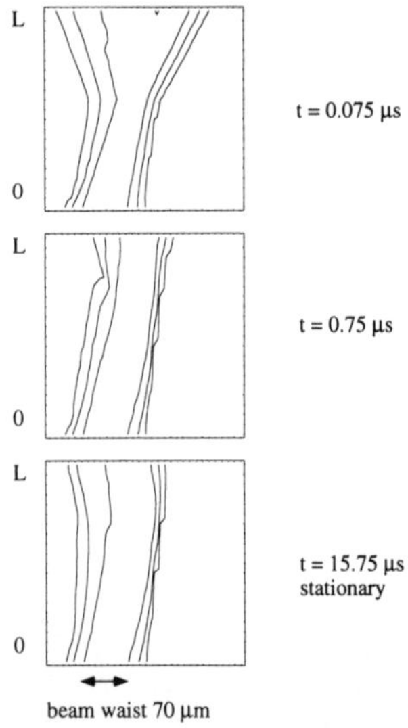

QTuK2 Fig. 2. Intensity profiles of beam bouncing seen perpendicular to the beam propagation at three different times after both beams have been switched on. The beams enter the medium from below and travel up. Only one beam polarization is shown; the other is symmetrical. The intensities have been normalized for each plane along the propagation so the innermost contour lines show the spot size. (a) Beams self-focus, (b) self-focusing has finished, and (c) stationary beam-bouncing pattern, typically after 10–100 μs.

faster than in the actual experiments. It has already been pointed out[8] that the reason for this is neglect of the nuclear spin effects. It has long been believed that these effects are too small to be of much importance. The energy of the nuclear spin interaction is indeed negligible, because in the experiments argon buffer gas for homogeneous line broadening larger than the hyperfine splitting is used, but the fact remains that in reality there are (for sodium, $I = 3/2$) eight lower and eight upper states to deal with. Therefore, most atoms can no longer be transferred directly into the pumped state; instead, several pump cycles are needed. This makes optical pumping in sodium much slower (as observed in the experiment) than in a hypothetical ($J = 1/2$)-to-($J = 1/2$) atom. Also, diffusion has a much stronger effect than before: whereas in the ($J = 1/2$)-to-($J = 1/2$) model half of the atoms entering the light beam through diffusion are already in the pumped state, now only one eighth of them are. Thus the same diffusion rate leads to much stronger absorption if the hyperfine structure is not neglected.

So, instead of a 4-by-4 matrix, the density matrix that has to be solved is 16 by 16. It describes the polarizations that couple to two complex wave equations for the two light polarizations.

In Fig. 1 a circularly polarized beam has been turned on at $t = 0$ and travels through sodium vapor. The power at the end of the medium is compred with experiment and with various models.

As an example for the 3-D dynamical simulation, we demonstrate the time dependence of beam bouncing in Fig. 2.

1. W. Happer, Rev. Mod. Phys. **44**, 169 (1972).
2. N. D. Bhaskar, M. Hou, B. Suleman, and W. Happer, Phys. Rev. Lett. **43**, 519 (1979).
3. R. Holzner, P. Eschle, C. Wigger, B. Röhricht, E. Brun, and A. W. McCord, in *Proc. International Quantum Electronics Conference*, Vienna, June 1992, p. 518.
4. R. Holzner, P. Eschle, A. W. McCord, and D. M. Warrington, Phys. Rev. Lett. **69**, 2192 (1992).
5. B. Röhricht, P. Eschle, S. Dangel, and R. Holzner, Z. Naturforsch. **48a**, 621 (1993).
6. B. Röhricht, P. Eschle, C. Wigger, S. Dangel, R. Holzner, and D. Suter, "Large frequency shifts of absorption profiles due to the combination of light-shift, optical pumping and magnetic fields in sodium vapor," Phys. Rev. A, in press.
7. R. Holzner, B. Röhricht, P. Eschle, S. Dangel, and E. Brun, in *Quantum Electronics and Laser Science*, 1993 Technical Digest Series, Vol. 12 (Optical Society of America, Washington, D.C., 1993), p. 207.
8. D. Suter, Phys. Rev. A **46**, 344 (1992).

QTuK3 **5:00 pm**

Observation of dynamical laser-light patterns in sodium vapor

R. Holzner, S. Dangel, B. Röhricht, H. Schmid, U. Rusch, R. Richard, P. Eschle, *Institute of Physics, University of Zurich, Winterthurerstrasse 190, CH-8057 Zurich, Switzerland*

We investigate experimentally several dynamical effects as they occur when one or two polarized resonant laser beams of a few milliwatts of power and about a 100-μm beam waist penetrate a glass cell filled with sodium vapor of $3 \times 10^{18}\ m^{-3}$ density and 250 Torr of argon buffer gas.

If a single circularly polarized laser beam is switched on, the temporal evolution of the penetrated light (Fig. 1) shows two characteristic time constants. During the time t_d the light front[1] passes through the cell before the full transmitted intensity is built up during the time t_r. These observed time constants are about an order of magnitude slower[2] than predicted by the commonly used ($J = 1/2$)-to-($J = 1/2$) model. To describe such dynamical behavior accurately, the hyperfine structure of the ground state has to be taken into account in spite of the fact that line broadening by the buffer gas is larger than the hyperfine splitting.

If the beam is slightly elliptically polarized and the frequency is detuned by about 10 GHz to the blue, ring patterns can be observed (Fig. 2). The temporal evolution has been measured with 1-μs resolution by a commercial CCD camera by using a stroboscopic technique. The modeling of such 3-dimensional pattern-formation dynamics is difficult, but the steady-state ring pattern can be modeled by a semiclassical theory.[3]

If left- and right-hand circularly polarized beams are superposed, one can observe beam switching, in which both beams extinguish each other while one beam alone penetrates the medium.[4] This provides an

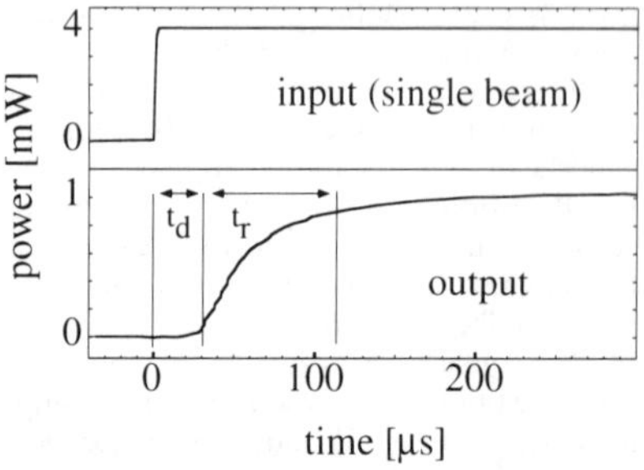

QTuK3 Fig. 1. Time evolution of transmitted power behind the 6-cm-long sodium-vapor cell (output) when a single circularly polarized laser beam has been switched on at $t = 0$ (input).

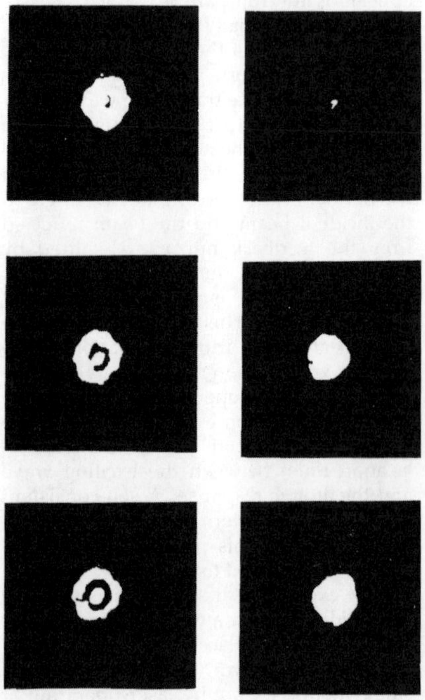

QTuK3 Fig. 2. Time evolution of a ring pattern captured in steps of 1 μs after a slightly elliptically polarized laser beam is switched on.

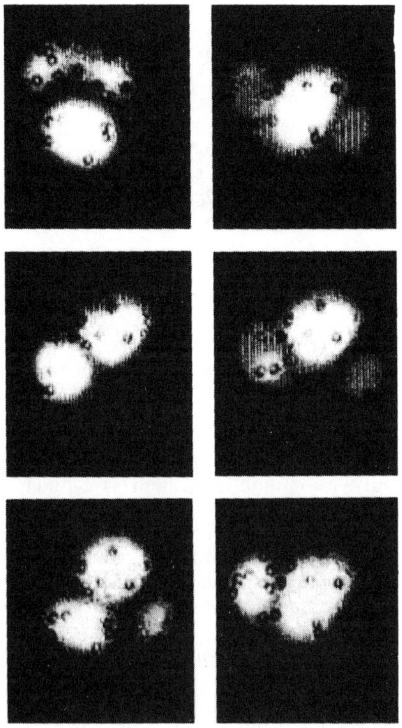

QTuK3 Fig. 3. Output beam patterns for slightly offset and tilted left- and right-hand circularly polarized input beams.

all-optical logical XOR gate for which the switching time scale is also defined by the optical pumping efficiency. Under the same conditions but with one beam slightly tilted to form an angle of a few millirads, beam bouncing, in which both beams mutually deflect each other, takes place.[5] Instead of one beam being tilted, it can be slightly offset by about 15 μm, which leads to beam splitting,[6] in which the initial almost-overlapping beams are well separated by about two beam waists at the end of the cell. With the same stroboscopic method as above, the temporal evolution of both beam bouncing and beam splitting was measured, and a time scale of a few microseconds was found, which is characteristic for the optical-pumping process involved.

A most interesting spatiotemporal pattern occurs when both beams are slightly tilted in one plane and are offset perpendicular to this plane. The output pattern jumps chaotically among a few patterns, as shown in Fig. 3, on a time scale of about 10 Hz, which is much slower than the optical pumping time or the time it takes for the light to penetrate the cell. We believe that the splitting and the bouncing processes compete with each other under the chosen critical conditions. We are working on a model for describing the full three-dimensional spatial pattern as well as the dynamics; this involves at least one fast and one slow time scale.

1. N. D. Bhaskar, M. Hou, B. Suleman, and W. Happer, Phys. Rev. Lett. **43**, 519 (1979).
2. D. Suter, Phys. Rev. A **46**, 344 (1992).
3. A. W. McCord and R. J. Ballagh, J. Opt. Soc. Am. B **7**, 73 (1990).
4. R. Holzner, P. Eschle, C. Wigger, B. Röhricht, E. Brun, and A. W. McCord, in *Proc. International Quantum Electronics Conference*, Vienna, June 1992, p. 518.
5. R. Holzner, P. Eschle, A. W. McCord, and D. M. Warrington, Phys. Rev. Lett. **69**, 2192 (1992).
6. R. Holzner, B. Röhricht, P. Eschle, S. Dangel, and E. Brun, in *Quantum Electronics and Laser Science*, 1993 Technical Digest Series, Vol. 12 (Optical Society of America, Washington, D.C., 1993), p. 207.

QTuK4 **5:15 pm**

Flowerlike-mode formation in a Nd:YAG laser with coupled feedback from a nonlinear gain grating

M. J. Damzen, R. P. M. Green, W. Hubbard, K. S. Syed, *Blackett Laboratory, Imperial College, London SW7 2BZ, U.K.*

Much recent activity has focused on spontaneous spatial-pattern formation in nonlinear optical systems with feedback,[1,2] with most of these systems involving the use of a Kerr-like nonlinearity. Recent work by Grynberg *et al.*[3] has shown spatial-pattern formation by using a saturable nonlinearity in rubidium vapor in which they observe high-order-mode formation of the Laguerre–Gaussian type. These modes have the appearance of a multipetaled flower and are different in spatial form from patterns normally observed in Kerr-like systems. In our work we show experimentally that flowerlike-pattern formation also occurs in a laser resonator in which the saturable gain nonlinearity of the laser amplifier induces gain-grating formation that acts as a nonlinear diffractive feedback mechanism.

In our experimental system, shown schematically in Fig. 1, a plane–plane cavity is formed by a back mirror with reflectivity 4% and an output mirror with reflectivity 10% placed about a flashlamp-pumped Nd:YAG amplifier G_1. Output from the back mirror is directed by a set of mirrors in a loop arrangement to intersect with the amplifier G_1 at a small angle ~10 mrad to the cavity axis. Gain gratings induced in the amplifier G_1 are able to diffract cavity flux back into this loop to produce feedback into the linear cavity. In our system, to enhance this feedback coupling we include a second Nd:YAG amplifier G_2 and a nonreciprocal transmission element (NRTE) into the loop section. The NRTE allows the transmission of the radiation in the forward loop direction to be different from the transmission in the backward direction.

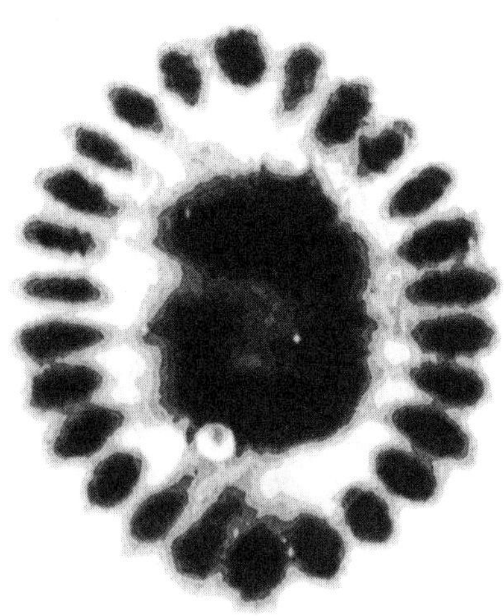

QTuK4 Fig. 2. Experimental spatial output showing flowerlike-mode formation from the Nd:YAG resonator with nonlinear feedback from the gain medium.

The spatial form of the output from the 10% mirror, shown in Fig. 2, consists of a flowerlike pattern with an outer region with a high-order even number of petals and a central region with a low-order mode. Depending on the alignment of the coupled cavity system, the number of petals was 24, 26, or 28, with form similar to the Laguerre–Gaussian modes $L_{0,12}$, $L_{0,13}$, and $L_{0,14}$, respectively, and the inner central mode with the approximate appearance of a Hermite–Gaussian mode TEM_{00}, TEM_{01}, or TEM_{11}. Under this interpretation, the pattern in Fig. 2 therefore approximates the case $L_{0,13}$ + TEM_{01}. It is noted that it was possible to achieve different inner-region modes with a fixed outer-mode pattern. In all cases the inner and outer regions were individually of a single mode type and were not themselves superpositions of modes.

In previous work on a saturable gain medium in a self-intersecting loop geometry[4] it was shown that gain gratings induced in the amplifier by spatial hole burning are capable of diffracting light and can lead to laser oscillation and self-pumped phase conjugation. The loop scheme as depicted

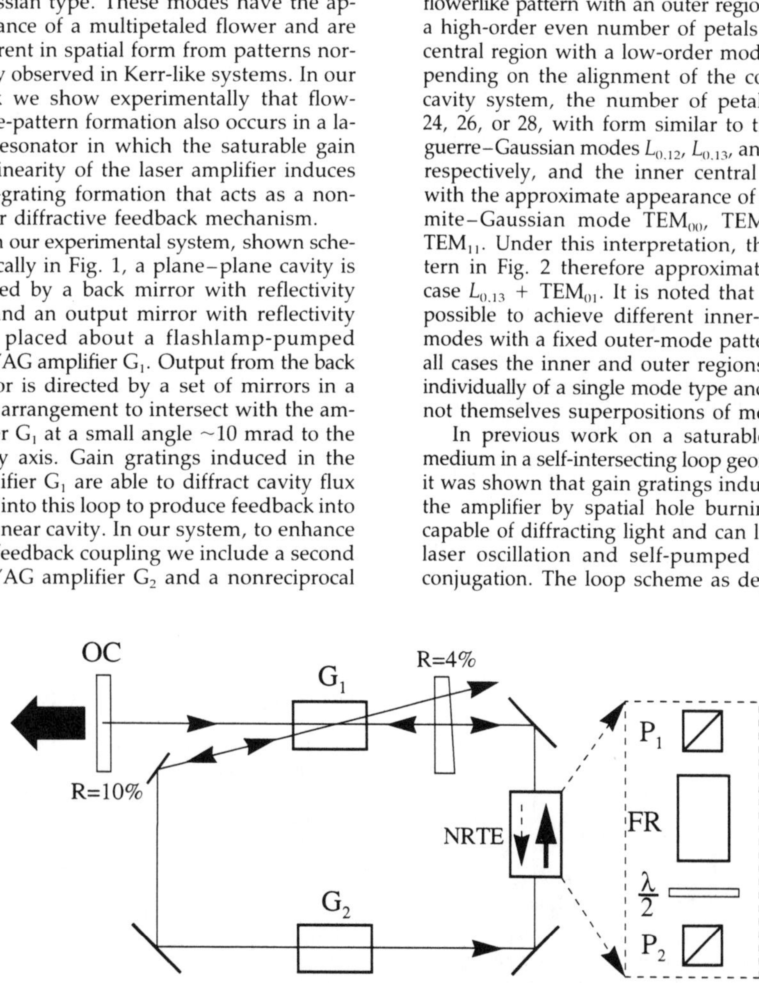

QTuK4 Fig. 1. Schematic diagram of experimental Nd:YAG resonator with nonlinear feedback coupling from the laser amplifier in an angularly intersecting loop geometry.

in Fig. 1 has a form similar to this arrangement, and this leads us to predict that induced-gain-grating formation is responsible for the feedback mechanism.

We present our interpretation of this pattern-forming process. We also show the variation of the output pattern as a function of the length of the linear-resonator section of the coupled nonlinear system.

1. N. B. Abraham and W. J. Firth, J. Opt. Soc. Am. B **7**, 948 (1990).
2. R. Macdonald and H. J. Eichler, Opt. Commun. **89**, 289 (1992).
3. G. Grynberg, A. Maitre, and A. Petrossian, Phys. Rev. Lett. **72**, 2379 (1994).
4. M. J. Damzen, R. P. M. Green, and G. J. Crofts, Opt. Commun. **110**, 152 (1994).

QTuK5 **5:30 pm**

Order–disorder phase transition and critical slowing down in photorefractive self-oscillators

Doruk Engin, Sergei Orlov, Amnon Yariv, Mordechai Segev,* George C. Valley,**
California Institute of Technology, 128-95, Pasadena, California 91125

Two of the characteristic properties of phase transitions are critical slowing down and large change in an order parameter in the vicinity of the critical point.[1] These effects are observed in ferromagnets, ferroelectric materials, and many other systems that change their actual structures (ordering) at a specific phase-transition point.[1]

We show that nonlinear self-oscillation[2] exhibits features of an order–disorder phase transition. We analyze one example, the photorefractive double-phase-conjugate mirror (DPCM), and show that it undergoes a phase transition at the critical point of self-oscillation. Both the nonlinear medium and the phases of the participating waves go from disorder below the critical point to order above it. The order parameter is associated with phase-conjugate reflectivity, and the correlation distance is associated with both the period of the nonlinear perturbation in the refractive index and the bandwidth (temporal and angular) of the scattered waves. Our analysis is general and can be extended to include all photorefractive self-oscillators and, with some modifications, other nonlinear self-oscillators.

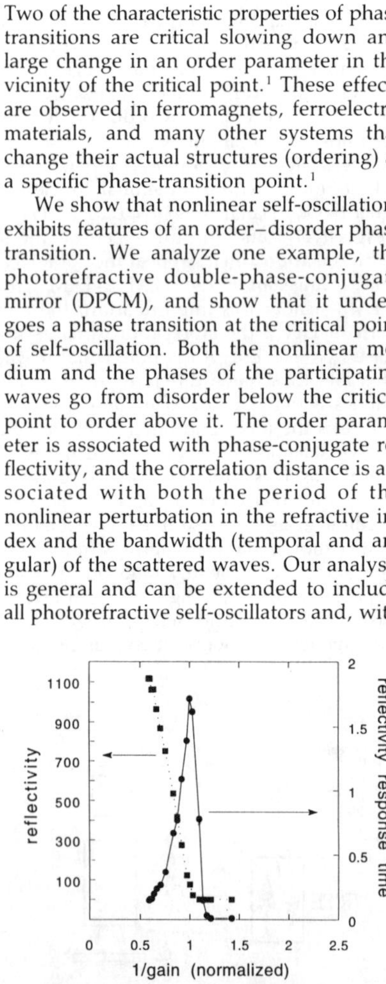

QTuK5 Fig. 1. Experimental results of the phase-conjugate reflectivity and time response as a function of gain for a single resolution image (1.26 line pairs/mm). The curves are guides to the eye.

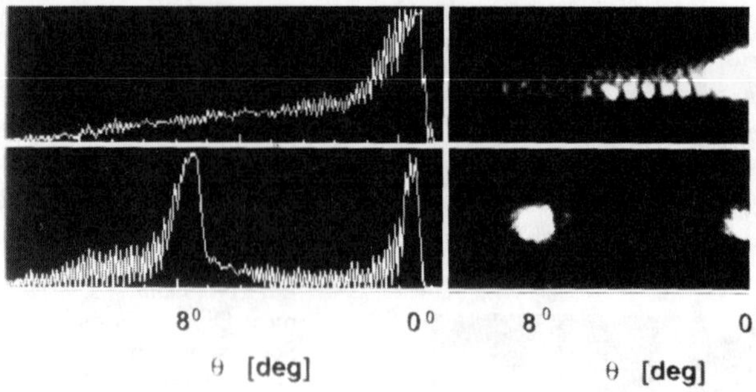

QTuK5 Fig. 2. Experimental results of the angular spectrum (far field) of one of the output beams below (upper section) and above (lower section) the transition point (threshold).

The time-dependent one-dimensional analysis treats the DPCM as a four-(plane)-wave mixing problem.[3] The steady-state solutions of the coupled wave equations for the simplest case of equal-intensity input beams yield

$$a = \tanh(-a\gamma L/2 + \gamma L\varepsilon), \qquad (1)$$

where a is the phase-conjugate amplitude reflectivity, γ is the coupling coefficient, L is the interaction length, and ε is a small level of phase-matched seed noise (of random phase). We identify the critical value $(\gamma L)_c = -2$ and observe that at that point a has a finite value even in the absence of seeding noise ($\varepsilon \to 0$). The transient solution gives the characteristic time to reach steady state $T = \tau/[\gamma L - (\gamma L)_c]$. These two properties indicate self-oscillation ($|a| > 0$ for $\varepsilon \to 0$) and critical slowing down [$T \to \infty$ for $\gamma L \to (\gamma L)_c$]. The similarity to other systems that undergo second-order phase transitions is apparent from the analogy to, for example, ferromagnetism,[1] where

$$M = \tanh[(MT_c + H)/T], \qquad (2)$$

where M is the magnetization, H is the external magnetic field, and T, T_c are the absolute and the Curie temperatures, respectively. Equation (2) (obtained from mean-field theory) minimizes the free-energy function, which in the DPCM case is $F = -(1/\gamma L)\log[2\cosh(a\gamma L/\gamma L_c)] + a^2/\gamma L_c$. The reflectivity a is analogous to the magnetization M and is the order parameter of the system. The physical property that becomes ordered at the phase transition is the index change $\Delta n(z)$, which assumes the form $\Delta n = \Delta n_0 \sin(2\pi z/\Lambda)$, where $\Lambda = 2\pi/K_0 = \lambda/(2\sin\theta/2)$ and θ is the angle between the two input beams. This corresponds to a narrowing in **K** space from a broadband to a near-monochromatic behavior. For temperatures $T > T_c$ in ferromagnetism, or for gain levels below $(\gamma L)_c$ for the DPCM, the system goes from order to disorder; that is, M and a vanish.

Our two-dimensional analysis,[4] which treats the DPCM as it evolves from random noise, goes beyond the results of mean-field theory. It considers other possible Fourier components of the gratings and is analogous to the Ginzburg–Landau[1] theory of spin fluctuations. In the one-dimensional model, the order parameter a represents the contribution of a single grating component K_0 (all others are neglected), and we examine its growth as a function of gain. In the two-dimensional case, one may look at the equivalent of spin correlations. In the DPCM, both the spectrum of the optical field and the spectrum of the refractive-index perturbation show correlation effects.

Experimental results shown in Figs. 1 and 2 illustrate the order–disorder phase transition. The conjugate reflectivity (normalized) and the response time as a function of gain are shown by the triangles and circles, respectively, in Fig. 1. The response time reaches its maximum value exactly at the critical point of the reflectivity curve $(\gamma L)_c = 2.06$. The ratio between the response times at the threshold and at the highest gain is about 10. To illustrate the change in the correlation distance when the DPCM goes through its threshold, we show in Fig. 2 experimental results (photographs, right column; profiles, left column) of the angular spectrum of one of the output beams below (upper) and above (lower) the transition point. The spectrum narrows to practically a single Fourier component, broadened only by the pictorial information borne on the input beam. The behavior of the reflectivity, the response time, and the angular spectrum at the threshold are conclusive experimental evidence of a second-order phase transition.

**Department of Electrical Engineering, Princeton University, Princeton, New Jersey 08544*

***Hughes Research Laboratories, 3011 Malibu Canyon Road, Malibu, California 90265*

1. L. D. Landau and E. M. Lifshitz, *Statistical Physics* (Pergamon, Oxford, 1958); and R. Kubo, *Statistical Mechanics* (North Holland, Amsterdam, 1965), p. 49, Chap. 5.
2. A. Yariv, *Quantum Electronics*, 3rd ed. (Wiley, New York, 1988), Chap. 17.
3. S. Weiss, S. Sternklar, and B. Fischer, Opt. Lett. **12**, 114 (1987).
4. M. Segev, D. Engin, A. Yariv, and G. C. Valley, Opt. Lett. **18**, 1828 (1993); D. Engin, M. Segev, S. Orlov, A. Yariv, and G. C. Valley, J. Opt. Soc. Am. B **11**, 1708 (1994).

QTuK6 5:45 pm

Correlated and decorrelated alternating patterns in a bidirectional photorefractive resonator

Z. Chen, D. McGee, N. B. Abraham, *Department of Physics, Bryn Mawr College, 101 North Merion Avenue, Bryn Mawr, Pennsylvania 19010-2899*

Experimental studies of the bidirectional emission of a photorefractive resonator with a $BaTiO_3$ crystal pumped by counterpropagating beams of 10:1 relative intensity show spontaneous pattern alternation in definite and repeatable sequences, as observed previously for unidirectional emission with a single pump beam.[1] The number of distinct patterns and the complexity of the dynamics depend on the Fresnel number.

When the Fresnel number is small, the pattern sequences consist of relatively long residency times with patterns closely equivalent to low-order Gauss–Laguerre or Gauss–Hermite modes (or zero intensity). As the Fresnel number is increased, the number of modes alternating in the sequence increases, the period for a complete sequence decreases, and the transitions from one pattern to another become more continuous. For low Fresnel numbers the counterpropagating patterns are highly correlated; they are essentially the same pattern in space and time, including the tracking of the intensity and the orientation of the patterns. For higher Fresnel numbers, as the dynamics become more rapid, the correlation between the counterpropagating emissions breaks down, with differences in the patterns or differences in the brightness of sections of geometrically similar patterns being the most notable changes. In addition, for high Fresnel numbers the pattern alternation no longer includes the zero-intensity option and the patterns have dynamics that may be termed "restlessness," with both radial and orientational oscillations.

Some intermediate patterns during switchings (as in Fig. 1) can be clearly interpreted to be the result of intermode beating, for example, between a TEM_{00} mode and a $(\text{Gauss–Hermite})_{01}$ mode of different frequencies (causing alternate blinking of two spots) or between a TEM_{00} mode and a $(\text{Gauss–Laguerre})_{01}$ mode (causing a rotating dark spot).[1-3] The interaction (rotation, approach, departure, and annihilation) of two dark spots is also observed as part of the intermediate transition between two patterns in an alternation sequence when there is competition among a few excited modes, as discussed recently for a related system.[4]

For increasing Fresnel numbers (as seen in Fig. 2), the patterns lose their relative-intensity correlation in different parts of the pattern while retaining the same basic geometry, particularly when a modal pattern such as $(\text{Gauss–Laguerre})_{04}$ is identifiable. For the highest Fresnel numbers shown in Fig. 2, the patterns lose both geometrical and relative-intensity correlation over most of their evolution.

These experimental results are in substantial agreement with recent theoretical predictions involving numerical solutions.[5]

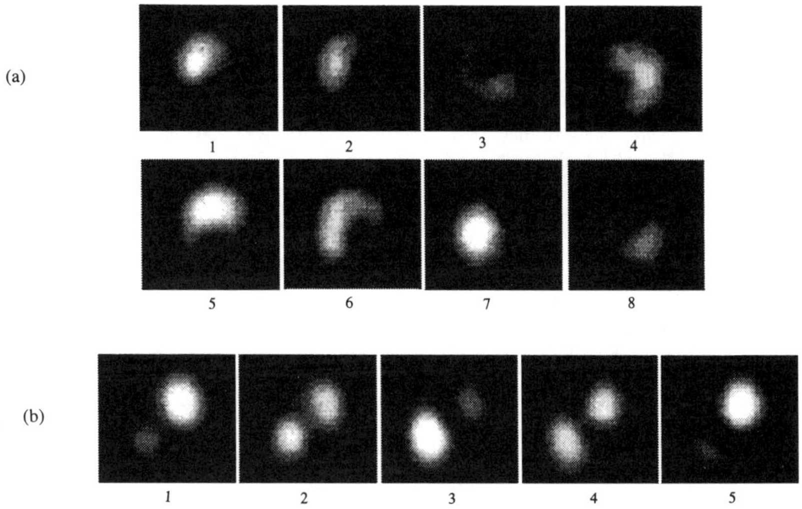

QTuK6 Fig. 1 Examples of (a) rotating dark spot and (b) blinking two-spot pattern. Sequence of snapshots is taken from one of the counterpropagating beams at $F = 2.0$. The time interval between snapshots is 5 s.

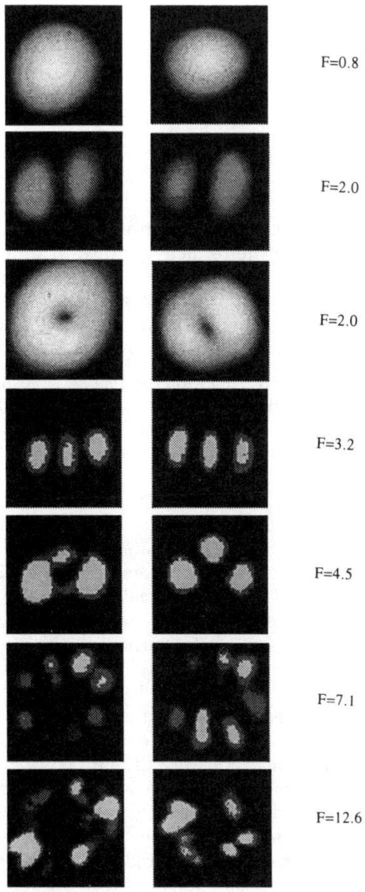

QTuK6 Fig. 2. Samples of the highest-order modal pattern seen in the pattern-alternation sequence that also involves lower-order modes and a period of zero intensity for the Fresnel number indicated. Patterns are viewed from opposite sides of the crystal, so that highly correlated patterns in the two beams (opposite columns) are those that are mirror images under reflection about a central vertical axis. This correlation is progressively lost for increasing F, as seen in $F = 4.5$, 7.1, and 12.6.

1. F. T. Arecchi, G. Giacomelli, P. L. Ramazza, and S. Residori, Phys. Rev. Lett. **65**, 2531 (1990); **67**, 3749 (1991).
2. D. Hennequin, L. Dambly, D. Dangoisse, and P. Glorieux, J. Opt. Soc. Am. B **11**, 676 (1994).
3. B. M. Jost and B. E. A. Saleh, private communication.
4. G. Indebetouw and D. R. Korwan, J. Mod. Opt. **41**, 941 (1994).
5. Z. Chen and N. B. Abraham, Appl. Phys. B, in press.

QTuL 4:30 pm

Hyatt B

Induced Transparency and Inversionless Lasers

Philip Robert Hemmer, *U.S. Air Force, Presider*

QTuL1 4:30 pm

Cross focusing in electromagnetically induced transparency

Richard R. Moseley, Sara Shepherd, David J. Fulton, Bruce D. Sinclair, Malcolm H. Dunn, *J. F. Allen Physics Research Laboratories, Department of Physics and Astronomy, University of St. Andrews, North Haugh, St. Andrews, Fife KY16 9SS, U.K.*

Recently, within an experiment to produce electromagnetically induced transparency (EIT),[1] we observed focusing and defocusing on the weak probe laser at separate frequency detunings within the transparency window.[2] The mechanism for this is that the radial intensity profile of the strong coupling laser generates a spatial refractive-index modulation that is experienced by the probe laser beam. We have dubbed this effect "electromagnetically induced focusing" (EIF) and believe it to be of importance in the whole field of quantum-interference experiments, including EIT, lasing without inversion,[3] and phaseonium.[4]

The atomic scheme employed in this study of EIF is the $5S_{1/2}$–$5P_{3/2}$–$5D_{5/2}$ cascade system in rubidium vapor; see Fig. 1. Two continuous-wave, single-frequency titanium sapphire lasers provide the probe and coupling radiation at 780 and 776 nm, respectively. The light from these sources is introduced in a counterpropagating geometry to the rubidium cell to allow observation of EIT with Rabi frequencies of less than the inhomogeneous linewidth of the probe transition. As the probe laser is tuned through the transparency window, strong changes in the transmitted beam size are observed; see Fig. 2. We stress that these changes are not due to any self-effect of the weak probe beam but are caused by the spatial conditions produced by the separate coupling laser beam. During the transparency window, strong changes in the probe-beam parameters may be produced with low attendant absorption of either beam. However, at exact double resonance the spatial conditions allow near-diffraction-limited propagation of the probe beam, a fact we have confirmed experimentally.

The spatial form of the lens may be inferred from calculations of the atomic susceptibility experienced by the probe laser at different radial positions across the beam. From these calculations the change of lensing conditions with the tuning of either laser frequency may be deduced. At some detuning points the lens profile approximates a Gaussian, causing either a positive or negative focal length lens, depending on the tuning; see Fig. 3(a). However, over some frequency regions the profile is shown to consist of both focusing and defocusing elements; see Fig. 3(b). Detailed calculations, using a matrix method,[5] of the focusing conditions produced by such a lens are reported and are compared with the experimental results. These calculations include the spatial variations in absorption due to the changing Rabi frequency induced by the coupling laser.

Further experimental characterization of EIF are presented. In particular, the effect of varying the initial relative laser-beam spot sizes and the lengths of the rubidium cell are reported. These results are compared with those of the theoretical model, and general conclusions are drawn.

Cross focusing will be an important factor in the design of any experiment in the quantum-interference area and may have applications in its own right as a smooth, highly controllable lens induced on one laser beam by another at a separate wavelength.

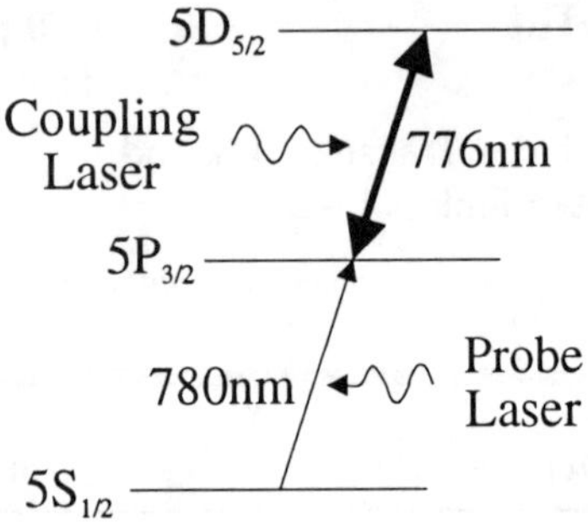

QTuL1 Fig. 1. Cascade scheme in rubidium.

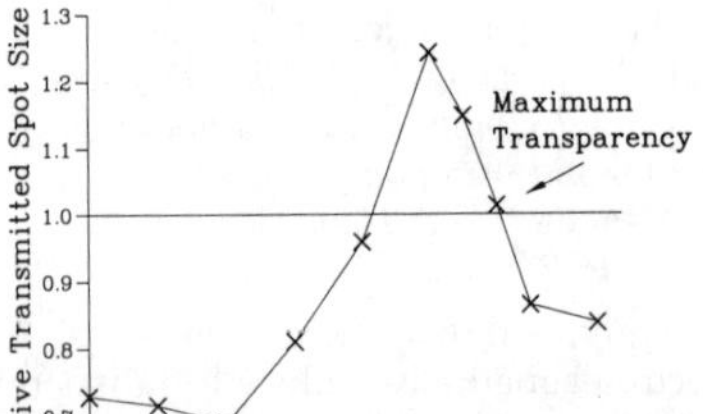

QTuL1 Fig. 2. Experimentally observed changes in the probe beam horizontal waist after the rubidium cell, normalized to the unperturbed waist, as the probe laser frequency is scanned through the EIT window induced in the ^{85}Rb, $5S_{1/2}$ (F = 3) → $5P_{3/2}$ transition.

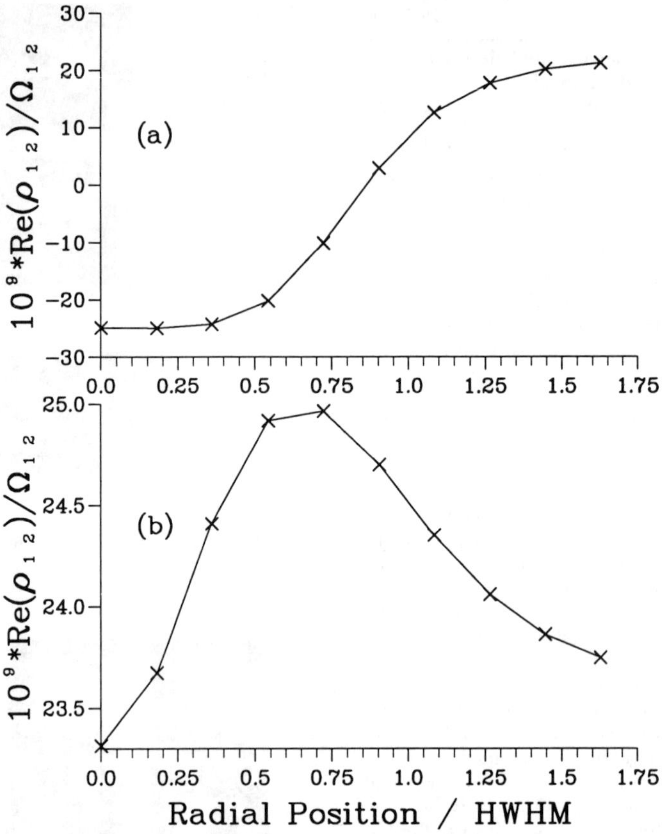

QTuL1 Fig. 3. Two sample theoretical radial variations in Re(ρ_{12}), which is directly proportional to the refractive index, at different probe frequency detuning points of (a) −6 MHz and (b) −14 MHz from line center (coupling-laser Rabi frequency, 20 MHz; homogeneous linewidth, 20 MHz). The radial position is normalized to the half-width at half-maximum of the Gaussian intensity profile of the coupling laser.

1. K.-J. Boller, A. Imamoğlu, and S. E. Harris, Phys. Rev. Lett. **66**, 2593 (1991).
2. R. R. Moseley, S. Shepherd, D. J. Fulton, B. D. Sinclair, and M. H. Dunn, Phys. Rev. Lett., in press.
3. O. Kocharovskaya, Phys. Rep. **219**, 175 (1992).
4. M. O. Scully, Phys. Rep. **219**, 191 (1992).
5. R. McDuff, Appl. Opt. **29**, 802 (1990).

QTuL2 **4:45 pm**

Electromagnetically induced transparency schemes within rubidium vapor

Sara Shepherd, David J. Fulton, Richard R. Moseley, Bruce D. Sinclair, Malcolm H. Dunn, *J. F. Allen Physics Research Laboratories, Department of Physics and Astronomy, University of St. Andrews, North Haugh, St. Andrews, Fife KY16 9SS, U.K.*

We present a detailed investigation of three contrasting schemes for the observation of electromagnetically induced transparency[1,2] (EIT) in rubidium vapor.

The first scheme is a cascade system that uses a coupling field on the $5P_{3/2}$–$5D_{5/2}$ transition and a probe field on the $5S_{1/2}$–$5P_{3/2}$ transition, as shown in the energy diagram in Fig. 1. This system is the most unambiguous in the observation of EIT because it is not affected by the competing effect of optical pumping. The other two schemes are a vee (V) system and a lambda (Λ) system, as shown by the energy diagrams in Figs. 2 and 3, respectively. In each of these systems the coupling field is resonant with the ground state; therefore at high pump powers the observation of EIT is complicated by optical pumping.

The schemes were investigated by using two single-frequency cw titanium sapphire lasers, either counter- or copropagating through a 10-cm-long cell of rubidium vapor. In the cascade scheme the coupling

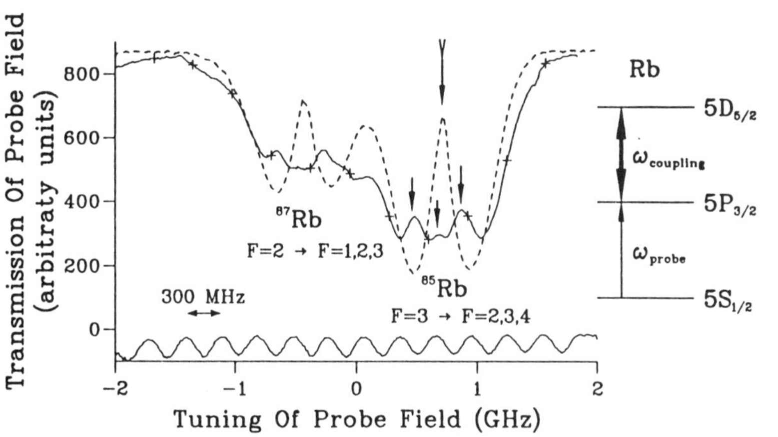

QTuL2 Fig. 1. Probe-field transmission for low (dashed curve) and high (solid curve) applied magnetic fields. A signal from a confocal interferometer is included as a frequency marker. The frequency position of the EIT features are shown by a wedge-tailed arrow for the low-magnetic-field regime and by a solid arrow for the high-magnetic-field regime. Throughout, the probe and coupling fields are linearly polarized in the same direction, this being perpendicular to the applied magnetic field such that the two-photon selection rules are $\Delta m_l = 0, \pm 2$, $\Delta m_s = 0$. A partial energy-level diagram for rubidium, showing the cascade scheme used for this experiment, is also shown.

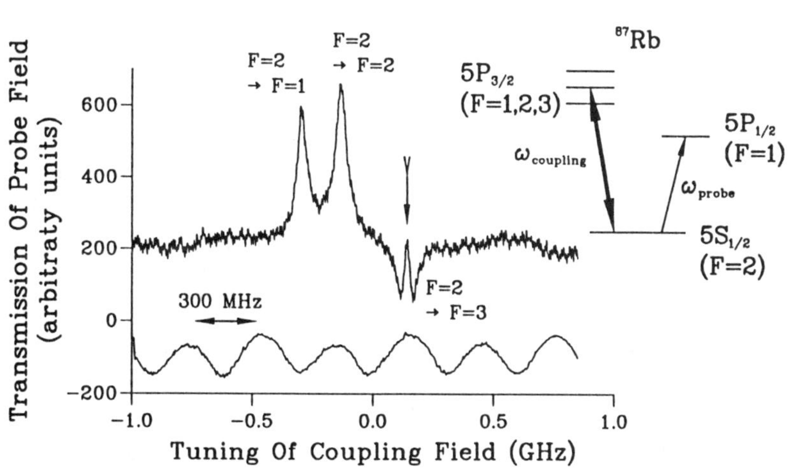

QTuL2 Fig. 2. Probe-field transmission from the ^{87}Rb $5S_{1/2}$–$5P_{1/2}$–$5P_{3/2}$ V-scheme experiment. A signal from a confocal interferometer is also included as a frequency marker. The probe and coupling fields are linearly polarized in the same direction. The frequency position of the EIT feature is shown by a wedge-tailed arrow, and a partial energy-level diagram of the rubidium V scheme is also included.

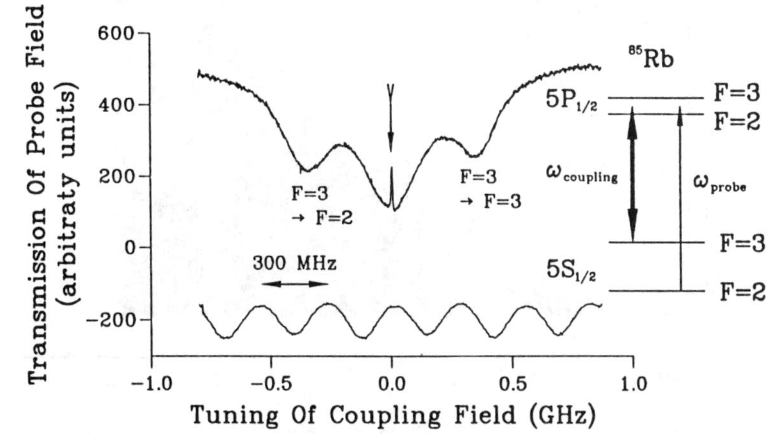

QTuL2 Fig. 3. Probe-field transmission from the ^{85}Rb $5S_{1/2}$–$5P_{1/2}$ Λ-scheme experiment. A confocal signal is also included as a frequency marker. The probe and coupling fields are linearly polarized with an orthogonal orientation. The frequency position of the EIT feature is shown by a wedge-tailed arrow, and a partial energy-level diagram of the rubidium Λ scheme is also included.

(pump) field is fixed in frequency and the transparency of the vapor is observed as the counterpropagating probe field is scanned through the two rubidium absorptions, as shown in Fig. 1.

Figures 2 and 3 show the changing absorption of a fixed resonant problem frequency as the copropagating pump field is scanned through the coupling transition. In each figure there are three optical pumping features. For the V scheme they correspond to the three hyperfine resonances as the coupling field is scanned over the $5S_{1/2}$ ($F = 2$)–$5P_{3/2}$ transition, with only the $F = 2$–$F = 3$ resonance producing the increased absorption on which the EIT can be observed.[3] For the Λ scheme the three features are now velocity-selected optical pumping resonances, of which only the middle one obeys the two-photon-resonance condition necessary for EIT. We present a more detailed analysis of both these schemes, with particular emphasis on the previously uninvestigated Λ scheme.

We have also studied the effect of Zeeman splitting on EIT by using the cascade scheme. The solid curve on Fig. 1 shows the transparency being split into three separate features by the application of a magnetic field. The number of features is determined by the polarization of the two lasers, and their frequency positions are determined by the magnetic sublevel splitting. Only those sublevels selected by two-photon resonance contribute to the transparency. Explanations of the two-photon-selection rules and the varying EIT window depths are provided.

The cascade scheme provides the best system for observing EIT because optical pumping dramatically affects EIT features within the V and Λ schemes. It was also found that EIT windows are reduced in depth while simultaneously undergoing frequency shifting and broadening by the magnetic field. These effects must be taken into consideration when experiments to study EIT and related topics are designed.

1. K. J. Boller, A. Imamoğlu, and S. E. Harris, Phys. Rev. Lett. **66**, 2593 (1991).
2. S. E. Harris, Phys. Rev. Lett. **72**, 52 (1994).
3. A. Weis, F. Sander, and S. Kanorsky, in *Technical Digest, 5th European Conference on Quantum Electronics*, 1994, p. 252.

QTuL3 **5:00 pm**

Nonlinear four-wave mixing using electromagnetically induced transparency in rubidium atoms

Yongqing Li, Min Xiao, *Department of Physics, University of Arkansas, Fayetteville, Arkansas 72701*

Electromagnetically induced transparency (EIT) may be used to enhance the net efficiency of nonlinear generation processes in multilevel atomic systems.[1,2] Stoicheff and co-workers experimentally demonstrated the possibility of obtaining higher conversion efficiency for second-harmanoic generation and four-wave mixing by using EIT in atomic hydrogen.[2] Recently, we demonstrated the EIT effect in lambda-type and cascade-type three-level systems of rubidium atoms using cw low-power diode lasers and a Doppler-free configuration.[3] This paper reports theoretical and experimental investigations of enhancement of nonlinear-optical processes by using the EIT effect in these three-level systems.

Our system consists of Rb^{87} atoms in a Doppler-broadened vapor cell. The two hyperfine levels $F = 1$ and $F = 2$, separated by 6.8 GHz, of the ground state $5S_{1/2}$ serve as the two lower states of the lambda system. The exited state $5P_{1/2}$, $F' = 2$ serves as the common upper state. When a strong diode-laser beam at frequency ω_2 is tuned to resonance with transition $5S_{1/2}$, $F = 2$–$5P_{1/2}$, $F' = 2$, a weak probe-laser beam at frequency ω_1, tuned on the other transition ($5S_{1/2}$, $F = 1$–$5P_{1/2}$, $F = 2$), sees 90% reduction in absorption [reduced linear susceptibility Im$\chi^{(1)}$], although almost all the population is transferred to the ground state of $5S_{1/2}$, $F = 1$ because of the optical-pumping effect. The nonlinear susceptibility $\chi^{(3)}$ is measured by nondegenerate four-wave mixing (NFWM). A sideband frequency ω_3 ($= \omega_2 + \Omega_0$) is created in the pumping beam by an acousto-optic modulator (AOM) with $\Omega_0 \sim 80$ MHz. As the probe laser at frequency ω_1 is tuned near resonance, a sideband frequency ω_4 ($= \omega_1 - \Omega_0$) in the probe beam is generated by nonlinear multiwave mixing. This NFWM signal is detected by beating the output of the probe beam with a strong reference beam at the probe frequency of a p-i-n photodiode, and the beating frequency at Ω_0 is measured by an rf spectrum analyzer. This hyterodyne detection is shot-noise limited and can be used to pick up a very small NFWM signal. The nonlinear susceptibility $\chi^{(3)}$ can be determined from this measurement.

Theoretical analysis shows that at the resonant frequency the linear susceptibility Im $\chi^{(1)}$ (absorption) is reduced because of atomic coherence and the nonlinear susceptibility $\chi^{(3)}$ is still resonantly enhanced to a limit value. Thus, the total conversion efficiency [$\sim\chi^{(3)}$/Im $\chi^{(1)}$] is resonantly enhanced. Another enhancement effect is the result of the pumping-induced ac-Stark effect when the sideband frequency Ω_0 is equal to half of the pumping Rabi frequency. The experimental measurement agrees with the theoretical prediction.

Studies of nonlinear-optics processes in a cascade-type three-level system in rubidium atoms ($5S_{1/2}$–$5P_{3/2}$–$5D_{5/2}$) are underway. We believe that these are useful applications of cw-induced transparency in nonlinear processes.

1. S. E. Harris, J. E. Field, and A. Imamoğlu, Phys. Rev. Lett. **64**, 1107 (1990).
2. K. Hakuta, L. Marmet, and B. P. Stoicheff, Phys. Rev. Lett. **66**, 59 (1991); G. Z. Zhang, K. Hakuta, and B. P. Stoicheff, Phys. Rev. Lett. **71**, 3099 (1993).
3. Y. Q. Li, S. Z. Jin, and M. Xiao, in *International Quantum Electronics Conference*, 1994 Technical Digest Series, Vol. 9 (Optical Society of America, Washington, D.C., 1994); J. Gea-Banacloche, Y. Q. Li, S. Z. Jin, and M. Xiao, Phys. Rev. A, in press.

QTuL4 5:15 pm

Cw lasing with and without inversion in a reversibly driven three-level-system

Peter B. Sellin, Geoffrey A. Wilson, Krishna K. Meduri, T. W. Mossberg, *Department of Physics, University of Oregon, Eugene, Oregon 97403*

Recently, it has been pointed out that laser action is possible under conditions that fall outside the range normally considered practicable. A substantial literature has developed concerning the creation of inversionless laser gain, i.e., gain in the absence of an inversion of population between the terminal levels of the lasing transition.[1] Several recent experiments have demonstrated transient inversionless amplification of light.[2] Here, we report on the ob-

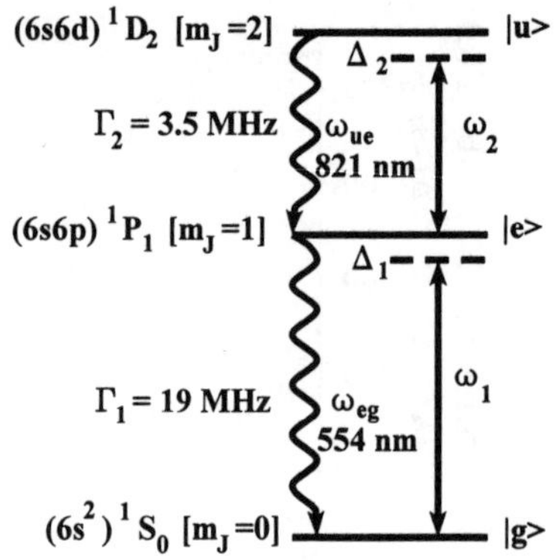

QTuL4 Fig. 1. Three-level cascade system in barium. Circular polarization selects transitions in which the magnetic quantum number m_J increases by one for an absorption.

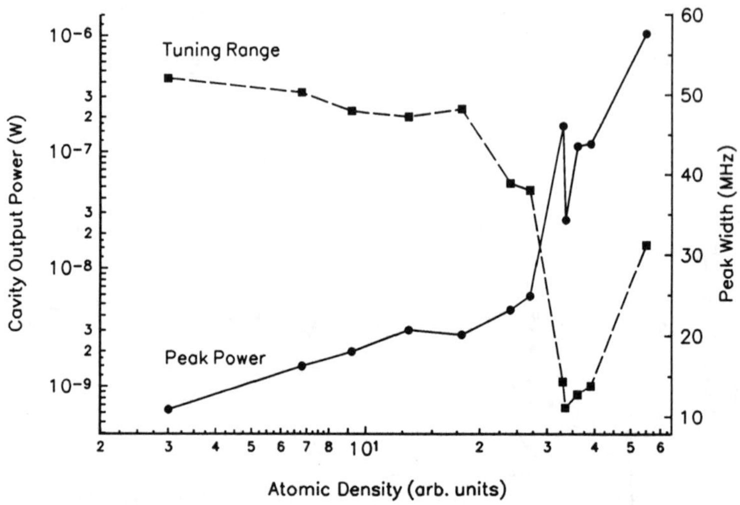

QTuL4 Fig. 2. Lasing-cavity output power and half-power cavity tuning range as a function of atomic-beam number density.

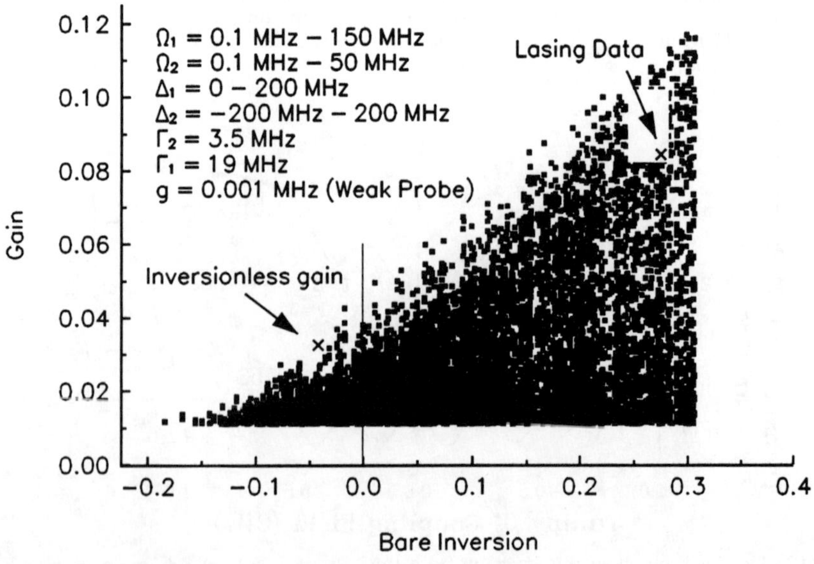

QTuL4 Fig. 3. Calculated gain versus atomic inversion. Each dot corresponds to a different set of control parameters (detunings, drive field strengths, etc.) accessible in our experiment. Positive gains observed to the left of zero on the horizontal axis correspond to inversionless gain.

servation of continuous lasing in a cascade-type three-level atom driven by two coherent fields, each nearly stepwise resonant with one of the two allowed transitions (see Fig. 1). Lasing is observed between the uppermost level u and the middle level e. This configuration is unusual because the upper level of the lasing transition has no traditional irreversible source of population pumping, such as spontaneous emission from a higher-lying atomic level. Furthermore, the lower laser level e experiences strong pumping from the ground level g by one of the excitation fields.

Our ongoing experiment uses atomic-beam barium atoms passing through a high-finesse optical-laser cavity. The use of barium and suitably polarized excitation beams allows us to realize a very clean atomic system (see Fig. 1) that can be subjected to quantitative theoretical analysis. Results of measurements of cavity output power as a function of atomic-beam density, obtained with the laser cavity tuned to the maximum-gain frequency, are shown in Fig. 2. The transition to superlinear growth of output power with density indicates the onset of lasing. Also shown is the half-power cavity tuning range. These observations are discussed.

It has been pointed out[3] that population inversions can arise in doubly driven cascade-type systems. We show by means of simulation (see Fig. 3) that gain in the cascade system does not require such inversions. Our measurements have already demonstrated inversionless gain. Current efforts focus on bringing observed lasing clearly into the inversionless regime, and these results are discussed. We also discuss the close relationship between the inversionless gain found in multilevel systems and the driven two-level-atom gain.[4] We also consider the experimental distinctions (or lack thereof) between gain with inversion and inversionless gain.

We thank the U.S. Army Research Office and the National Science Foundation for financial support.

1. For a review see O. Kocharovskaya, Phys. Rep. **219**, 175 (1992).
2. A Nottelmann, C. Peters, and W. Lange, Phys. Rev. Lett. **70**, 1783 (1993); W. E. van der Veer, R. J. J. van Diest, A. Donszelmann, and H. B. van Linden van den Heuvell, Phys. Rev. Lett. **70**, 3243 (1993); E. S. Fry, X. Li, D. Nikonov, G. G. Pabmabandu, M. O. Scully, A. V. Smith, F. K. Tittel, C. Wang, S. R. Wilkinson, and S.-Y. Zhu, Phys. Rev. Lett. **70**, 3235 (1993).
3. R. M. Whitley and C. R. Stroud, Jr., Phys. Rev. A **14**, 1498 (1976).
4. B. R. Mollow, Phys. Rev. A **5**, 2217 (1972); D. Grandclement, G. Grynberg, and M. Pinard, Phys. Rev. Lett. **59**, 40 (1987); G. Khitrova, J. F. Valley, and H. M. Gibbs, Phys. Rev. Lett. **60**, 1126 (1988); A. Lezama, Y. Zhu, M. Kanskar, and T. W. Mossberg, Phys. Rev. A **41**, 1576 (1990); D. J. Gauthier, Q. Wu, S. E. Morin, and T. W. Mossberg, Phys. Rev. Lett. **68**, 464 (1992).

QTuL5 5:30 pm

Lasing without population inversion in ruby by using low-frequency coupling

Yang Zhao, Cunkai Wu, M. K. Kim,*
Department of Electrical and Computer Engineering, Wayne State University, Detroit, Michigan 48202

We report experimental observations of field-induced transparency in a solid medium. Lasing without population inversion (LWI) and electromagnetically induced transparency (EMIT) can be realized when quantum coherence and interference are created in an atomic system. One requirement for quantum coherence is that the Rabi frequency of the coupling field must be larger than the linewidth of the transition. LWI and EMIT in solid media are complicated by the large linewidth of optical transitions. To create the coherence, large coupling power is needed, even at low temperature.

We have demonstrated a new scheme using low-frequency coupling for LWI and EMIT in solids, since the linewidth of transitions is proportional to the central frequency. We used microwaves to couple the lower two levels of a three-level system in ruby. The two closely spaced lower levels are the Zeeman split sublevels of the ground state $4A_2$ of ruby. The measured linewidth of the microwave transition is around 66 MHz, which is much smaller than that for optical transitions. When the angle between the constant magnetic field and C axis of ruby is 90°, the transition-matrix element is quite large. By changing the constant magnetic field intensity, exact resonant excitation can be established. The upper level of the lasing process is the Zeeman split sublevel $M = -1/2$ of the $2E$ state. The ruby crystal, with about a 0.05% chromium concentration, is placed at the center of a microwave cavity and is cooled by liquid helium. With a 2450-G magnetic field, the coupling microwave frequency is 16.3 GHz and the required power is around 4.6 kW/cm^2.

By tuning the probe frequency, we studied EMIT for both the lambda and ladder schemes. With the coupling microwave, we observed reduction in the absorption in both schemes. Under our experimental conditions the ladder scheme is better than the lambda scheme. For the ladder scheme, the reduction in absorption coefficient is more than 20% when the coherent coupling field is applied. This result indicates that it is possible to realize LWI in ruby if there is enough incoherent pump. Effects of detuning, coupling-field intensity, and incoherent pump rate on the performance of LWI in ruby, as well as the design of a cavity for LWI, are discussed.

* *Department of Physics and Astronomy, Wayne State University, Detroit, Michigan 48202*

QTuL6 5:45 pm

Optical-dipole noise

A. M. Bacon, H. Z. Zhao, L. J. Wang, J. E. Thomas, *Department of Physics, Duke University, Durham, North Carolina 27708-0305*

We measure phase-dependent optical noise in the radiation emitted by coherently driven, long-lived two-level atoms in an atomic beam that is resonant with a preparation field that the atoms traverse. The total noise measured in the experiments contains contributions from phase-dependent Bloch-vector projection noise, spontaneous-emission, noise, and phase-dependent Poisson noise due to the moving optical dipoles that traverse the preparation region. Noise spectra are measured by using a new homodyne detection scheme in which signals from two identically prepared atomic samples are subtracted as described below. The shape and magnitude of the measured noise spectra are found to be in good agreement with predictions using no free parameters.

The experiments (Fig. 1) make use of the 556-nm $^1S_0 \rightarrow {}^3P_1$ transition of ^{174}Yb, which forms a two-level system comprising the $J = 0$ and $J = 1$, $M = 1$ states, with a radiative lifetime of 875 ns. A 1-cm-wide supersonic Yb beam crosses a continuous laser preparation field that is focused with a cylindrical lens to a half width at a $1/e$ intensity of 0.1 mm horizontally and 1 mm vertically.

The preparation field is polarized in the x–z plane, where, because of an applied Zeeman field, only the x component is resonant with the atoms. A Babinet–Soleil compensator is used to vary the relative phase between the z and x fields. A projection polarizer with its axis $\hat{e}$ oriented at 45° to the z axis is placed after the atomic sample to mix the radiation field from the

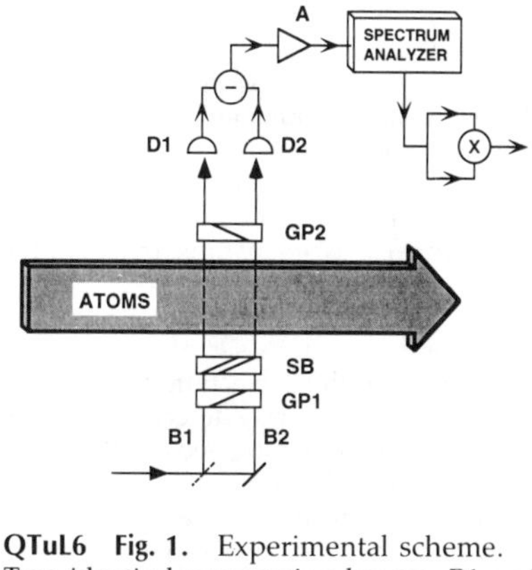

QTuL6 Fig. 1. Experimental scheme. Two identical preparation beams, B1 and B2, are polarized by GP1 in the x–z plane. A Babinet–Soleil compensator SB controls the relative phase between the x- and z-polarized field components. The total fields emerging from the sample are analyzed by polarizer GP2 with intensities measured by photodiodes D1 and D2 for which output currents are subtracted. Because the two preparation beams are generated with a beam splitter and interact with independent atoms, the technical noise in the difference signal subtracts and the quantum noise adds.

Tuesday, May 23

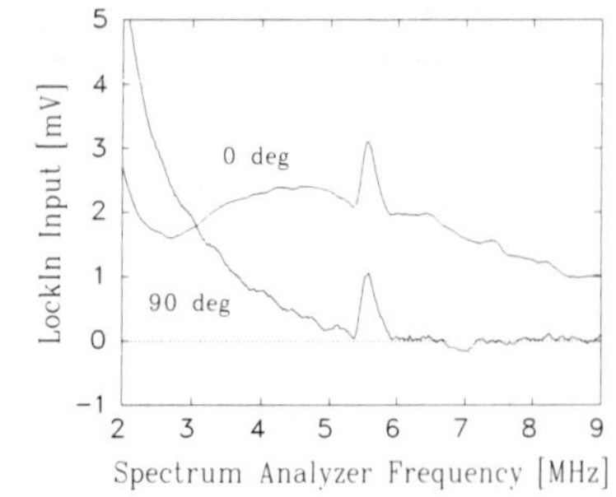

QTuL6 Fig. 2. Measured in-phase and out-of-phase optical-noise spectra with the shot noise subtracted to display the atom-noise contribution. The vertical axis is the lock-in input signal, which is proportional to the mean-square noise voltage. The data shown are divided by the corresponding phase-dependent local-oscillator intensities. Note that the small peak near 6 MHz is due to the ion laser.

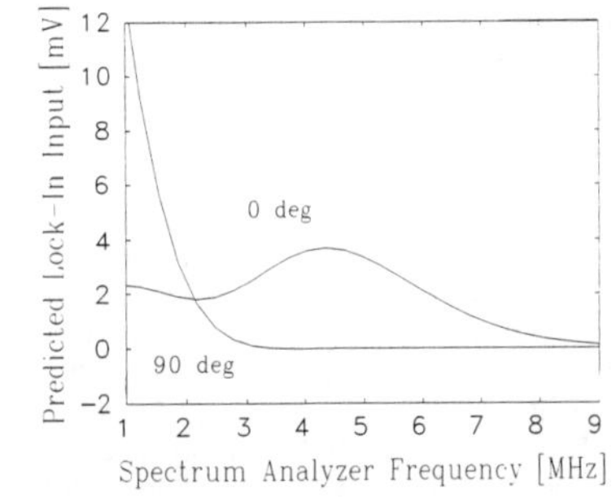

QTuL6 Fig. 3. Noise spectra calculated by using no free parameters. Atomic density is estimated from the measured phase-dependent absorption. The Rabi frequency and atom transit time are estimated from the measured preparation-laser-beam parameters.

atoms with the net local oscillator field along $\hat{e}$. Although the local oscillator intensity is phase dependent, this method assures that the relative phase between the driving field and the net local oscillator field is stable.

To eliminate technical noise in the quadrature signals, two separate, independent regions of the atomic beam are excited by identical preparation fields; see Fig. 1. The intensities of the two signal fields transmitted through the projection polarizer are each measured by diode detectors. The detector outputs are subtracted. In this way, the technical noise in the local-oscillator amplitude and in the nonvanishing quadrature signals is suppressed. However, the quantum noise in the signals adds because the independent regions are not quantum mechanically correlated. The subtracted detector output is fed to a low-noise transimpedance amplifier and then to an rf spectrum analyzer.

To measure the optical noise with high sensitivity and on a linear scale, the analog output of the spectrum analyzer is squared in real time by using a low-noise multiplier (Fig. 1) and fed to a lock-in amplifier, which subtracts the squared noise signals obtained with the laser fields on and off. As a calibration, the mean-square shot-noise voltage was measured and was found to vary linearly for laser powers from 3 mW down to 2 μW. The slope of the linear fit agrees with predictions to better than 10%.

The measured phase-dependent noise spectra $S(\omega)$ can be compared to those calculated for resonant, coherently driven two-level atoms that move across the optical-field regions. $S(\omega)$ contains both a shot-noise term due to the total transmitted power and atom-noise terms proportional to the local-oscillator power P_{LO}. Because the measured lock-in input voltage is proportional to the mean-square noise voltage $S(\omega)$, the atom-noise terms are obtained by subtracting the shot-noise contribution from the measured noise spectra and dividing by the (phase-dependent) local-oscillator power. This yields the experimental atom-noise spectra for the in-phase and out-of-phase quadratures, as shown in Fig. 2. The data are in good agreement with the calculated spectra; see Fig. 3.

QTuM 4:30 pm
Hyatt C/D

Quantum Dots

Arto V. Nurmikko, *Brown University, Presider*

QTuM1 (Invited) 4:30 pm

Optical properties of InAs/GaAs quantum dots

J.-Y. Marzin, J.-M. Gérard, A. Izraël, G. Bastard,* *France Telecom, CNET-PAB, Laboratoire de Bagneux, BP107, 92225 Bagneux cedex, France*

Among the various possible techniques to produce low-dimensional structures, their direct fabrication by epitaxy is very attractive, in particular in order to minimize the size fluctuations. We have in this spirit optimized the 3-D growth of InAs islands on GaAs by molecular beam epitaxy to produce small and regular quantum dots. These dots consist of small InAs pyramids, which are very regular in shape and size. When these InAs islands are overgrown by GaAs, they exhibit efficient photoluminescence. The typical dimensions are a height of 2 nm and a half base of 9 nm, with a relative standard deviation of 5%. Even if they are much smaller than those obtained for etched quantum dots, these fluctuations of size and thus of quantum confinement, broaden the emission line of a large number of dots. We could however observe the narrow (<0.15 meV) emission line of one individual quantum dot at low temperature (<100 K). This experimental result was achieved by reducing the total number of emitting InAs dots, through the study of small mesa structures, processed by standard electron beam lithography and etching techniques. The smaller mesas (200 × 200 nm^2) contain about 10 dots. The analysis of the statistics of the dots emission, compared to AFM measurements,[1] gives a new insight on their formation. Our calculations suggest that only the fundamental transition is observed. They also show that the energy separation between the lowest two electron (or hole) levels is larger than both kT at 300 K and the LO phonon energy. These easily fabricated single quantum dots are thus ideal objects for the experimental study of energy relaxation in these structures, which constitutes the key issue for their future use in devices.

* *Laboratoire de physique de la matière condensée, Ecole Normale Supérieure, 24 rue Lhomond, 75005 Paris, France*

1. J.-M. Moison, F. Houzay, F. Barthe, L. Leprince, E. André, and O. Vatel, Appl. Phys. Lett. **64**, 196 (1994).

QTuM2 (Invited) 5:00 pm

Persistent hole-burning spectroscopy in semiconductor quantum dots

Yasuaki Masumoto, *Institute of Physics, University of Tsukuba, Tsukuba, Ibaraki 305, Japan*

Nanometer-size semiconductor crystals abbreviated to "semiconductor nanocrystals" are known as zero-dimensional "quantum dots." Their optical properties have been characterized by the quantum confinement effect and the lowest excited states show blue shifts depending on their size. Quantum dots are sharply different from other low-dimensional quantum structures, such as quantum wells and quantum wires, because they are made of as small as 10^3~10^6 atoms. Considerable fraction of atoms face the surface or the interface of quantum dots in the surrounding materials, so that the energy of quantum dots is expected to be easily modified by the surroundings.

Nonlinear optical properties of semiconductor quantum dots have been extensively investigated. When the spectrally narrow light excites the ensemble of semiconductor quantum dots having the inhomogeneously broadened absorption band, a spectral hole is formed at the excitation photon energy in the absorption band. So far, inhomogeneous broadening is believed to come from only the size distribution and spectral hole burning observed in quantum dots has been believed to be transient. Nevertheless, persistent spectral hole burning (PSHB) phenomenon was observed generally in a number of quantum dots, CdSe, CdSSe, CuCl, CuBr and CuI, embedded in glass and crystal.[1–3] [See also Figs. 1 and 2.] In inhomogeneously broadened exciton absorption spectra of these quantum dots, the spectral hole and its associated structure are made by the narrow-band laser excitation and are conserved for more than several hours at 2 K.[3–6] Hole depth grew in proportion to logarithm of the burning fluence. Thermally annealing and light-induced hole-filling phenomena were observed. The hole burning takes place by the tunneling process through potential barriers with broadly distributed barrier height and thickness. Photoionization of dots and the carrier tunneling into traps in the surrounding host is the most probable origin of the PSHB phenomenon.

Application of the PSHB phenomenon to the size-selective optical spectroscopy of quantum dots is possible by "laser mark-

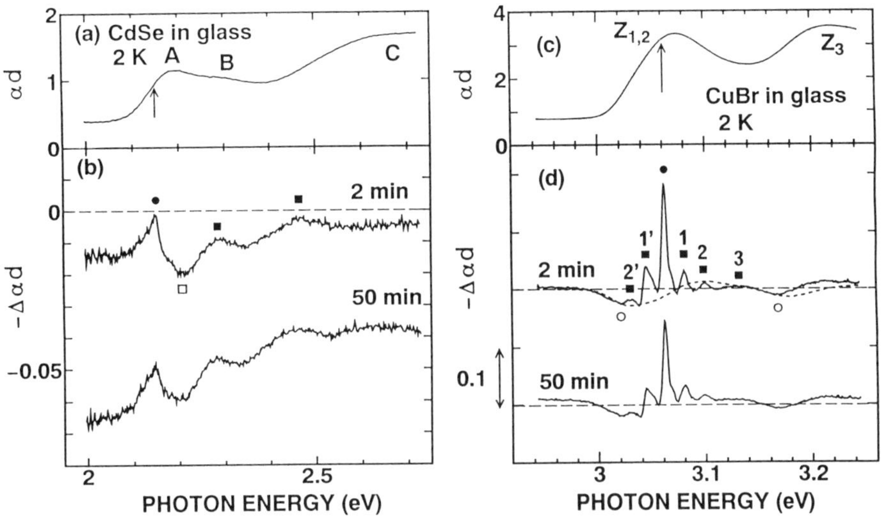

QTuM2 Fig. 1 Observation of the persistent spectral hole-burning phenomenon in CdSe and CuBr quantum dots embedded in glass at 2 K. The upper Figs. (a) and (c), show the absorption spectrum of the virgin samples, CdSe (R = 2.4 nm) and CuBr (R = 3.6 nm) in glass, respectively. The lower Figs. (b) and (d), show the spectral change observed at 2 min and 50 min after the burning laser exposure is stopped. The CdSe sample is excited by 1,800 shots of dye laser pulses with the photon energy of 2.148 eV, and energy density of 0.5 mJ/cm^2. The CuBr sample is excited by 9,000 shots of dye laser pulses with the photon energy of 3.060 eV, and energy density of 5 μJ/cm^2. The spectral hole and associated structure are preserved for more than several hours at 2 K. A short dashed line in Fig. (d) shows the derivative of the absorption spectrum (c).

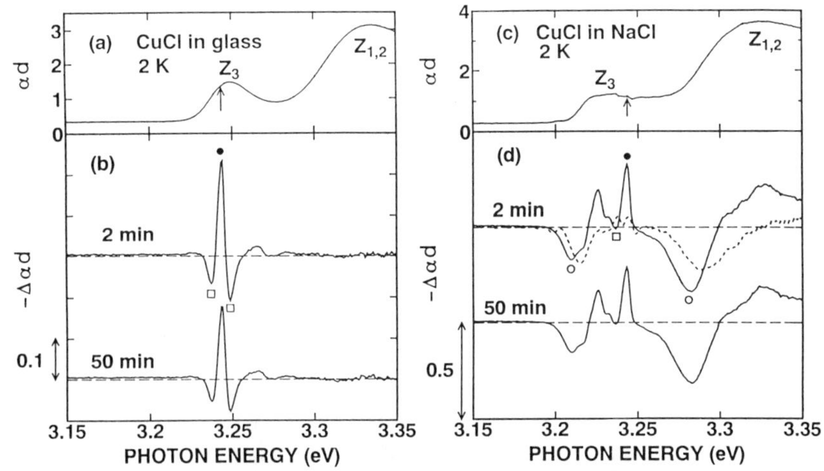

QTuM2 Fig. 2. Observation of the persistent spectral hole-burning phenomenon in CuCl quantum dots embedded in glass and NaCl. The upper Figs. (a) and (c), show the absorption spectrum of the virgin samples, CuCl (R = 2.5 nm) in glass and CuCl (R = 3.5 nm) in NaCl, respectively. The lower Figs. (b) and (d), show the spectral change observed at 2 min and 50 min after the burning laser exposure is stopped. The samples are excited by 1,800 shots of dye laser pulses with the photon energy of 3.245 eV, and energy density of 33 μJ/cm^2. The spectral hole and associated structure are preserved for more than several hours at 2 K. A short dashed line in Fig. (d) shows the derivative of the absorption spectrum (c).

ing'' certain size dots. For example, acoustic phonons confined in CuCl, CuBr, and CuI quantum dots are investigated precisely as a function of the radius of dots.[7] The first-order and second-order confined optical phonon modes are observed as the energy difference between the main hole and the satellite holes, 1, 1' and 2', in Fig. 1(d). Excited excitonic or electronic quantum states in dots are observed as satellite holes. Satellite holes denoted by 2 and 3 in Fig. 1(d) are excited quantum states of $Z_{1,2}$ excitons in CuBr dots. Satellite holes in Fig. 1(b) show the excited quantum states in CdSe dots. Precise size-selective spectroscopy of semiconductor quantum dots by laser marking under external perturbations, such as temperature, electric field, magnetic field, pressure, and uniaxial stress, is also helpful.

1. Y. Masumoto, L. G. Zimin, K. Naoe, S. Okamoto, and T. Arai, Mater. Sci. Eng. B **27**, L5 (1994).
2. K. Naoe, L. G. Zimin, and Y. Masumoto, Phys. Rev. B **50**, Dec./15 (1994).
3. Y. Masumoto, to be published in *Proc. 22nd Int. Conf. on Phys. Semicond.* (Vancouver, August, 1994).
4. Y. Masumoto, Lev G. Zimin, K. Naoe, S. Okamoto, T. Kawazoe, and T. Yamamoto, to be published in J. Lumin.
5. Y. Masumoto, to be published in Jap. J. Appl. Phys.
6. Y. Masumoto, S. Okamoto, T. Yamamoto, and T. Kawazoe, to be published in Phys. Stat. Solidi (b) **188**, No. 1 (1995).
7. S. Okamoto and Y. Masumoto, to be published in J. Lumin.

QTuM3 **5:30 pm**

Excitons in complex quantum nanostructures

Garnett W. Bryant, Paul S. Julienne, Yehuda B. Band,* *National Institute of Standards and Technology, B268 Physics, Gaithersburg, Maryland 20899*

As nanotechnology becomes more developed, the nanostructures being fabricated become more complex, with complicated geometries and strong coupling between structures. Wegscheider *et al.*[1] have shown that T-shaped quantum wires at intersections of quantum wells formed by cleaved edge overgrowth (see Fig. 1) exhibit stimulated emission; these nanostructures are the first to lase from the exciton state. More recently, Pfeiffer *et al.*[2] studied strongly coupled T-shaped wires formed by cleaved edge growth on double-well systems. Complex multilayer quantum-dot structures (see Fig. 1), called quantum-dot quantum wells, in analogy with quantum-well structures, are also now being fabricated.[3,4]

An accurate theory of the optical properties of complex nanostructures requires that one account for quantum confinement by nanostructures with complicated shapes, include effects of coupling between nanostructures, incorporate electron–hole correlation in complicated spatial potentials,

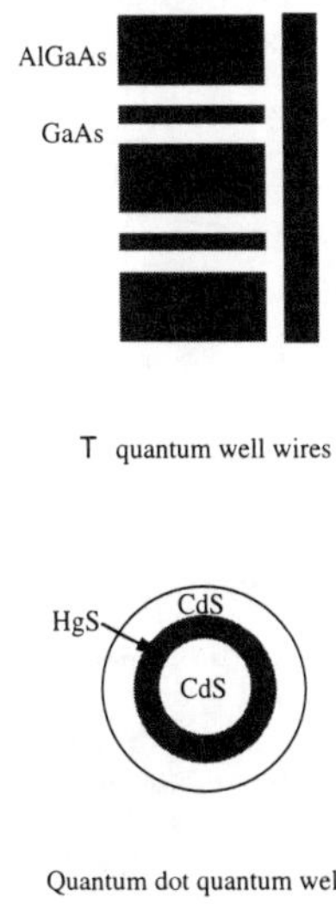

QTuM3 Fig. 1. Cross-sectional view of coupled T-shaped quantum-well wires and of a quantum-dot quantum well.

and treat electrons and holes with different masses such that scales for electron confinement, hole confinement, and pair correlation are different.

Different approaches are needed for different nanostructures. To solve the interacting electron–hole Schrödinger equation for excitons in T-shaped wires and coupled T-shaped wires, we use single-particle states, localized to a grid of points that covers the cross section of the wires, to define basis functions for the problem. To study quantum-dot quantum wells, we use states obtained by solving the single-particle radial Schrödinger equation for a radial quantum-well potential.[4,5] In both cases, we use these single-particle states to define electron–hole-pair states for large-scale configuration-interaction calculations that account for pair correlation in these complicated geometries.[6]

Electron energy levels calculated for the coupled pair of T-shaped wires studied experimentally[2] are shown in Fig. 2. The experimental T spacing is 7 nm. Our results show substantial level repulsion between the ground state and the first excited state for this spacing. Strong interwire coupling of exciton states is expected in this case. In coupled nanostructures, the relative strength of this coupling to the pair correlation determines the coherence of the exciton state in the coupled structure and its optical properties.[6] We present exciton energies and oscillator strengths in single and coupled T-shaped wires to explain the large red shift of the exciton energy observed for coupled wires[2] and to determine the coherence of excitons in coupled wires and how this affects their optical properties.

We also present energies and oscillator strengths for excitons in quantum-dot quantum wells. In a layered structure the electron and hole may be trapped in different regions. We determine how this trapping depends on the layer structure in quantum-dot quantum wells, how the trapping is affected by pair interaction, and how exciton energies, oscillator strengths, and coherence are affected by this trapping. The first quantum-dot quantum wells are now being made. We suggest interesting systems to be explored.

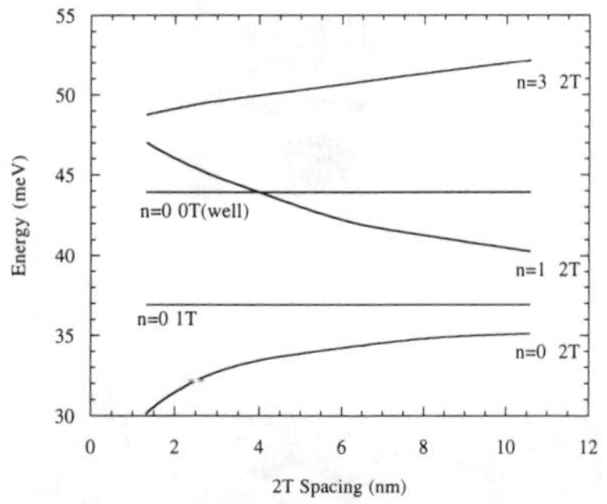

QTuM3 Fig. 2. Dependence of the energy for the ground state and the first two excited states of an electron in a coupled T nanostructure on the T separation. Ground-state energies for an isolated T and for a well with no T are shown for comparison. The wells that define the T are 7-nm-wide GaAs/AlGaAs wells.

*Departments of Chemistry and Physics, Ben-Gurion University, Beer Sheva, Israel

1. W. Wegscheider, L. N. Pfeiffer, M. M. Dignam, A. Pinczuk, K. W. West, S. L. McCall, and R. Hull, Phys. Rev. Lett. **71**, 4071 (1993).
2. L. Pfeiffer, W. Wegscheider, A. Pinczuk, and K. West, in *Proc. 7th International Conference on Superlattices, Microstructures, and Microdevices*, Banff, Alberta, Canada, August 1994.
3. A. Mews, A Eychmuller, M. Giersig, D. Schooss, and H. Weller, J. Phys. Chem. **98**, 934 (1994).
4. D. Schooss, A. Mews, A. Eychmuller, and H. Weller, Phys. Rev. B **49**, 17072 (1994).
5. J. W. Haus, H. S. Zhou, I. Honma, and H. Komiyama, Phys. Rev. B **47**, 1359 (1993).
6. G. W. Bryant, Phys. Rev. B **47**, 1683 (1993).

QTuM4 **5:45 pm**

Ultrafast gain dynamics in strongly confined CdSe quantum dots

H. Giessen, B. Fluegel, G. Mohs, N. Peyghambarian, Ulrike Woggon,* *Optical Sciences Center, University of Arizona, Tucson, Arizona 85721*

Strongly confined quantum dots are believed to exhibit excellent laser properties (e.g., low threshold) because of the zero-dimensional confinement of the oscillator strength in the quantized energy levels. However, a decreased carrier relaxation rate due to a phonon-bottleneck[1] effect could prevent efficient operation of a quantum-dot laser.

CdSe quantum dots with a radius of 25 Å are in the strong-quantum-confinement regime because their energy-state separation is large compared with the bulk exciton binding energy. By pumping the lowest optical transitions with a pump beam of 120-fs duration (up to 220 GW/cm^2), we have obtained optical gain.

Figure 1(a) shows the early gain-buildup dynamics. The three lowest-lying electronic transitions are simultaneously bleached. Figure 1(b) shows the gain-decay dynamics: the bleaching signal decays fastest at higher energies, whereas gain at the low-energy side can still be seen after 200 ps.

We explain the early bleaching dynamics by the creation of two-electron–hole-pair states, or biexcitons, by the strong pump beam. Because of their fermionic character, only two electrons can occupy the lowest electronic state. Therefore, the probe beam cannot create any further excitons, and the lowest-lying excitonic transitions are bleached simultaneously. The late bleaching dynamics are explained by the decay of a biexciton into an exciton and a photon. Therefore, a probe photon can be absorbed and it creates an excited biexciton, which causes the reduction of the bleaching signal at the high-energy side of the spectrum.[2]

Figure 2(a) shows an intensity-dependent absorption measurement at +5 ps. At the highest pump intensities the gain region stretches from below the band gap to the pump wavelength; it is about 100 nm wide. The gain arises from both exciton-to-ground-state and biexciton-to-exciton transitions. The low-energy gain is purely due to biexciton-ground-state-to-excited-exciton transitions. The agreement with a theory that solves the optical Bloch equations for a quantum-dot system and takes excitons and biexcitons into account is excellent.[3] Band-gap renormalization does not occur in our situation because two electrons with different spins do not show exchange interaction due to Pauli blocking.

Figure 2(b) shows the spectral-hole-burning dynamics in the gain region. At 5 ps after the gain is created, a second pump pulse of 160-fs duration is tuned into the gain region. It reduces the gain by stimulated emission, which causes an induced

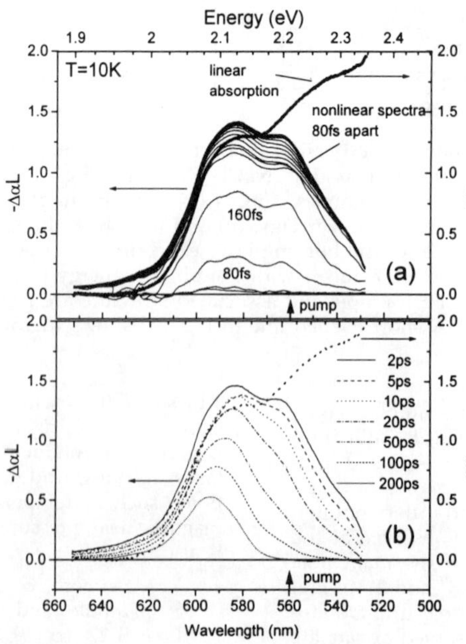

QTuM4 Fig. 1. (a) Gain-buildup dynamics of CdSe quantum dots for maximum pump intensity; (b) gain-decay dynamics.

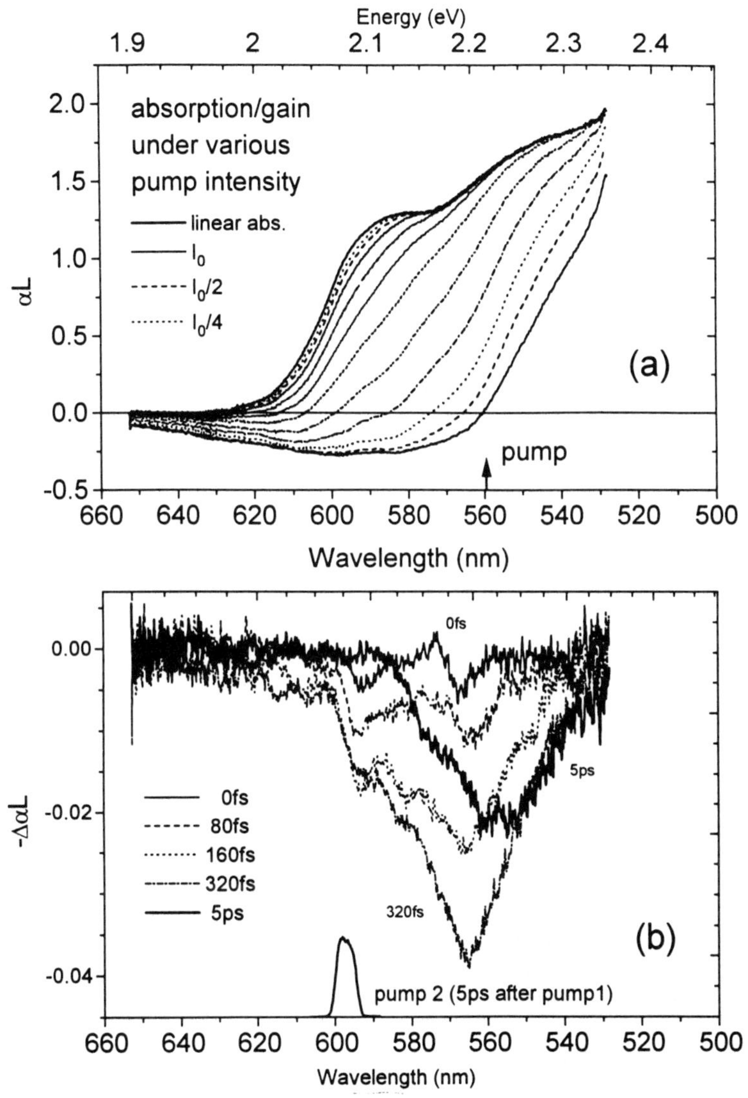

QTuM4 Fig. 2. (a) Absorption and gain at +5 ps under increasing pump power; (b) differential-transmission spectra for spectral hole burning in the gain region of the quantum dots.

absorption feature in the differential-transmission spectra. A spectral hole in the gain region around the pump pulse disappears after 320 fs. The induced absorption feature around 565 nm, which is attributed to biexciton-to-exciton decay, is reduced to half its peak amplitude after 5 ps. The time scales on which these spectral holes are replenished allow us to assume that the predicted phonon bottleneck does not prohibit the fast carrier relaxation in our system. This is important for a possible laser operation of quantum dots.

Support from the National Science Foundation, the U.S. Air Force Office of Scientific Research, and the Ballistic Missile Defense Organization is acknowledged.

**Institute for Applied Physics, University of Karlsruhe, Kaiserstrasse 12, D-76128, Karlsruhe, Germany*

1. H. Benisty *et al.*, Phys. Rev. B **44**, 10945 (1991).
2. Y. Z. Hu *et al.*, Phys. Rev. Lett. **64**, 1805 (1990).
3. Y. Z. Hu *et al.*, submitted for this conference.

QWA 8:00 am

Hyatt A

Nonlinear Optical Materials I

Peter Gunter, *ETH Honggerberg, Presider*

QWA1 8:00 am

Wavelength and polarization dependence of hole-burning efficiency in highly oriented J-aggregates

Takayoshi Kobayashi, Kazuhiko Misawa, Shinjiro Machida,* Kazuyuki Horie,* *Department of Physics, University of Tokyo, 7-3-1 Hongo, Bunkyo-ku, Tokyo 113, Japan*

A great deal of effort has been made to clarify the optical properties of one-dimensional excitons in J-aggregates of cyanine dyes.[1,2] However, the macroscopic sizes of the J-aggregates have not yet been established, because the samples of J-aggregates in previous studies have been limited to isotropically dispersed systems in solutions.[1] Recently, we developed a novel preparation method for highly oriented J-aggregates dispersed in a polyvinyl alcohol (PVA) film.[2–4] The samples show distinctly different dichroic spectra $A_\parallel$ and $A_\perp$ for the parallel and perpendicular polarizations, respectively, with respect to the orientation axis (Fig. 1).

In the present paper we describe what we believe to be the first measurement of the polarization dependence of the spectral hole-burning on the oriented J-aggregates of 1,1'-diethyl-2,2'-quinocyanine bromide [pseudoisocyanine bromide (PIC-Br)]. The solid-state oriented films prepared by the vertical spin coating enabled us to measure the polarization dependence of the spectra and the efficiencies of the hole burning for what we believe to be the first time.

To investigate the polarization dependence of the hole burning, we measured the quantum efficiency of the hole formation in the four different configurations with parallel and perpendicular polarizations of the laser and probe light to the orientation axis of the J-aggregates ($\eta_{\perp pr}^{\perp ex}$, $\eta_{\parallel pr}^{\perp ex}$, $\eta_{\perp pr}^{\parallel ex}$, and $\eta_{\parallel pr}^{\parallel ex}$).

Figure 2 shows the quantum efficiencies $\eta_{\perp pr}^{\perp ex}$ (open circles), $\eta_{\parallel pr}^{\perp ex}$ (open triangles), $\eta_{\perp pr}^{\parallel ex} \cdot A_\parallel/A_\perp$ (filled triangles), and $(\eta_{\parallel pr}^{\parallel ex} - \eta_{\perp pr}^{\parallel ex}) \cdot A_\parallel/(A_\parallel - A_\perp)$ (filled circles) obtained at the several burning wavelengths. The derivation of the above efficiencies is described in detail elsewhere.[5]

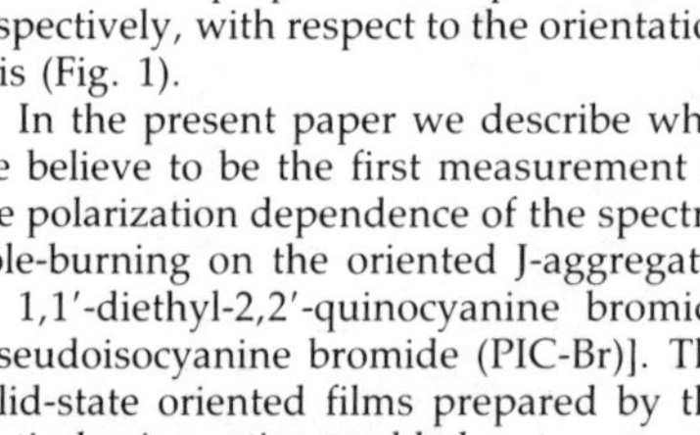

QWA1 Fig. 1. Absorption spectra of the oriented J-aggregates at 5 K with parallel and perpendicular polarization of incident light with respect to the orientation axis.

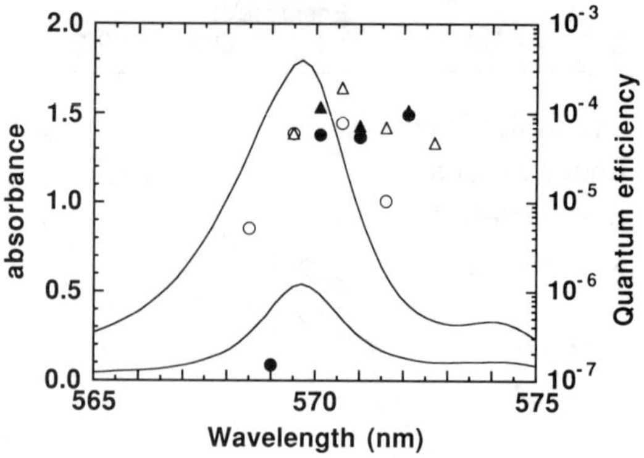

QWA1 Fig. 2. Wavelength dependence of the quantum efficiency. The polarizations of the excitation and probe light are both parallel (filled circles), parallel and perpendicular (filled triangles), perpendicular and parallel (open triangles), and both perpendicular (open circles).

The efficiencies $\eta_{\perp pr}^{\perp ex}$, $\eta_{\parallel pr}^{\perp ex}$, and $\eta_{\perp pr}^{\parallel ex} \cdot A_\parallel/A_\perp$ obtained in the different polarization configurations are nearly constant around 10^{-4} on the *lower-energy* side of the J-band. This result implies that the oriented film of J-aggregates contain an isotropic component, which we call a coherent mesoaggregate.[5]

The value of $(\eta_{\parallel pr}^{\parallel ex} - \eta_{\perp pr}^{\parallel ex}) \cdot A_\parallel/(A_\parallel - A_\perp)$ corresponds to the efficiency of oriented J-aggregate[5] η_{or}, in agreement with the value of the individual coherent aggregates.

At the higher-energy side of the parallel J-band, the quantum efficiency was extremely low, below the detection limit of the present measurement. The higher-energy side of the parallel J-band is due to the excitons with higher transition energy in an ensemble of the coherent aggregates, which we call a macroaggregate. From the low burning efficiency $\sim 10^{-7}$ observed at the higher-energy side of the J-band, an extremely large relaxation rate of $(10\ \text{fs})^{-1}$ is concluded to take place among the aligned aggregates in a macroaggregate. The ~100-times-enhanced oscillator strength due to the coherently delocalized excitation in a mesoaggregate and the close spacing among the mesoaggregates in a macroaggregate are found to be quantitatively consistent with such an extremely fast energy transfer.

In conclusion, the hole-burning efficiency of the mesoaggregates and macroaggregates at the lower-energy side of the J-band was obtained to be about 10^{-4} in all the polarization configurations. An extremely large relaxation rate of $(10\ \text{fs})^{-1}$ among the mesoaggregates in a macroaggregate is estimated from the very low (10^{-7}) efficiency at the higher-energy side of the J-band.

Department of Chemistry and Biotechnology, University of Tokyo, 7-3-1 Hongo, Bunkyo-ku, Tokyo 113, Japan

1. K. Minoshima, M. Taiji, K. Misawa, and T. Kobayashi, Chem. Phys. Lett. **218**, 67 (1994).
2. K. Minoshima, M. Taiji, A. Ueki, K. Miyano, and T. Kobayashi, Nonlinear Opt., in press.
3. K. Misawa, H. Ono, K. Minoshima, and T. Kobayashi, Appl. Phys. Lett. **63**, 577 (1993); J. Luminescence **60/61**, 812 (1994).
4. K. Misawa, K. Minoshima, H. Ono, and T. Kobayashi, Chem. Phys. Lett. **220**, 251 (1994).
5. K. Misawa and T. Kobayashi, Chem. Phys. Lett., in press; J. Luminescence, in press.

QWA2 8:15 am

Third-harmonic dispersion measurements in squaraines

M. H. Wu,* R. F. Shi, A. F. Garito, *Department of Physics, University of Pennsylvania, Philadelphia, Pennsylvania 19104*

Squaraines have attracted much research interest because of their large, nonresonant third-order optical susceptibilities $\gamma_{ijkl}(-\omega_4; \omega_1, \omega_2, \omega_3)$.[1,2] Evidence has accumulated that a two-level model for $\gamma_{ijkl}(-\omega_4; \omega_1, \omega_2, \omega_3)$ is not adequate even for molecules with negative $\gamma_{ijkl}(-\omega_4; \omega_1, \omega_2, \omega_3)$. Earlier we discussed the experimental observation of high-lying two-photon states that are critical to the negative $\gamma_{ijkl}(-\omega_4; \omega_1, \omega_2, \omega_3)$ in squaraines, achieved by using femtosecond pump–probe techniques.[3] We report here results from our recent full third-harmonic dispersion measurements on squaraine dyes using a tunable picosecond laser source based on optical parametric generation and amplification in KTP crystals pumped by a picosecond Nd:YAG laser.[4]

The linear optical spectrum of BSQ squaraine dye in chloroform (CH_2Cl_2) shown in Fig. 1 has a sharp major absorption peak at 1.92 eV (647 nm). The large transparency window is ideally suited for third-harmonic-generation (THG) dispersion measurements. Figure 2 shows the full dispersion results for the real part of the isotropically averaged $\langle\gamma(-3\omega, \omega, \omega, \omega)\rangle$ for BSQ dye in chloroform as a function of the fundamental input photon energy. The results clearly show that below all electronic resonances, $\text{Re}\langle\gamma(-3\omega, \omega, \omega, \omega)\rangle$ is negative, with a magnitude of -46×10^{-36} esu at $\hbar\omega = 0.60$ eV. This agrees with earlier dc-induced second-harmonic-generation results[1] and calculations.[2]

The solid curve in Fig. 2 is generated from a four-level model with the parameters illustrated in Table 1. The results clearly

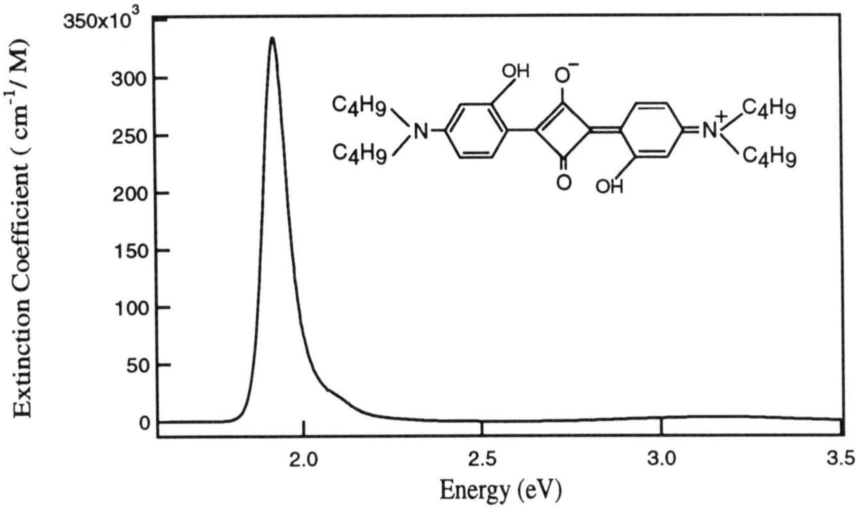

QWA2 Fig. 1. Ground-state linear absorption spectrum of BSQ dye in a chloroform solution. Inset: schematic molecular structure of BSQ. Typical squaraines have a strong absorption peak in the visible region with a molar extinction coefficient of the order of 10^5 mol^{-1}·liter·cm^{-1}, with the exact value and peak position depending on the detailed structure and solvent used.

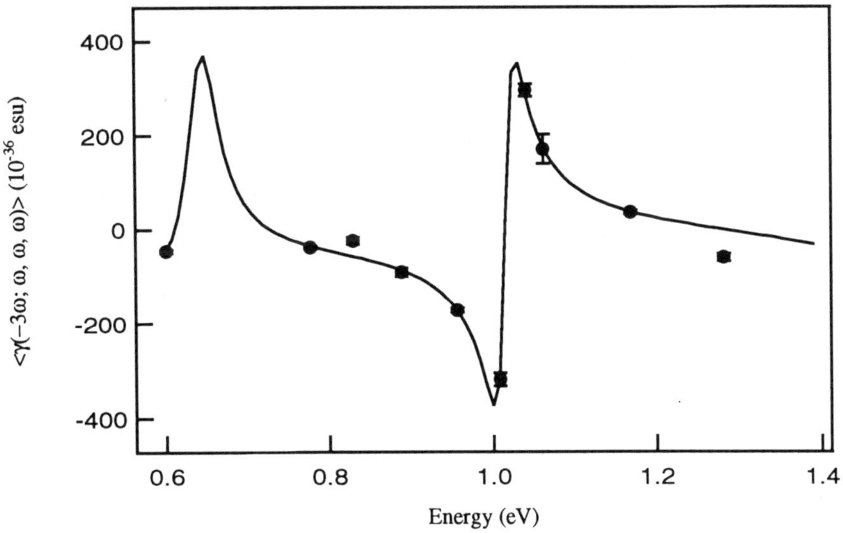

QWA2 Fig. 2. Dispersion results for the real part of $\langle\gamma(-3\omega, \omega, \omega, \omega)\rangle$ for BSQ dye in a chloroform solution as a function of the fundamental input photon energy. Points are from measurements, and the solid curve is from a four-level model with the parameters illustrated in Table 1. The gap in the experimental points is due to the three-photon resonance at $\hbar\omega = 0.64$ eV.

QWA2 Table 1. Transition moments μ between corresponding states, transition energies E between the excited and ground states, and width Γ used in the four-level model.

	$1^1A_g \rightarrow 1^1B_u$	$1^1B_u \rightarrow 2^1A_g$	$1^1B_u \rightarrow n^1A_g$
$\mu\,(D)$	12.4	2.05 ± 0.05	14.5 ± 0.1
$E\,(eV)$	1.92	2.03 ± 0.01	3.62 ± 0.03
$\Gamma\,(eV)$	0.067	0.026 ± 0.004	0.030 ± 0.015

demonstrate that the high-lying two-photon state (n^1A_g) is critical to understanding the features of the dispersion curve. The fact that $\mu_{1^1B_u,n^1A_g} = 14.5$ D, which is comparable to $\mu_{1^1B_u,1^1A_g} = 12.4$ D deduced from the linear absorption spectrum confirms both the earlier experimental observation[3] and the theoretical prediction[2] that the coupling strength between the 1^1A_g and 1^1B_u states is of the same order as that between the 1^1B_u and n^1A_g states. Therefore, a multilevel model, with both types of transition sequences, is necessary for accurate calculation of $\gamma_{ijkl}(-\omega_4; \omega_1, \omega_2, \omega_3)$. Interestingly, the sign change for $\mathrm{Re}\langle\gamma(-3\omega, \omega, \omega, \omega)\rangle$ at $\hbar\omega = 1.01$ eV indicates that the low-lying two-photon state 2^1A_g has a nonnegligible contribution to γ, which confirms the independent finding by other researchers.[5] Identification of the position of the low-lying two-photon state can be used to study the solvatochromism effect on two-photon states, which is important, considering the fact that squaraines exhibit a negative solvatochromism effect for one-photon states.

**Current address: Laboratory of Chemical and Biological Dynamics, University of Leuven, Celestijnenlaan 200D, B-3001 Leuven, Belgium*

1. C. W. Dirk, L.-T. Cheng, and M. G. Kuzyk, Int. J. Quantum Chem. **43**, 27 (1992).
2. Q. L. Zhou, R. F. Shi, O. Zamani-Khamiri, and A. F. Garito, Nonlinear Opt. **6**, 145 (1993).
3. Y. Z. Yu, R. F. Shi, A. F. Garito, and C. H. Grossman, Opt. Lett. **19**, 786 (1994).
4. M. H. Wu, Y. M. Cai, and A. F. Garito, in *Quantum Electronics and Laser Science*, 1993 Technical Digest Series, Vol. 12 (Optical Society of America, Washington, D.C., 1993), 59.
5. J. H. Andrews, J. D. V. Khaydarov, and K. D. Singer, Opt. Lett. **19**, 984 (1994).

QWA3 **8:30 am**

Molecular asymmetrization: a route to third-order-susceptibility enhancement

Vincent Ricci, William E. Torruellas, George I. Stegeman, Sandra Gilmour,* Seth Marder,** Mireille Blanchard-Desce,†
Center for Research and Education in Optics and Lasers, University of Central Florida, 12424 Research Parkway, Orlando, Florida 32826

The polyene *trans*-β-carotene has become a standard in the study of third-order nonlinear optical properties of organic compounds.[1] It shows one of the largest instantaneous off-resonant nonlinearities reported to date. The most common approach to enhancing this nonlinear response has been to increase the electron conjugation with longer backbones.[2] However, it has been recognized recently that removing the center of molecular symmetry represents another feasible route to increasing the third-order susceptibility without deleterious effects, such as smaller molecular densities, an increased molecular instability, and poor solubility.[3] This approach changes the relative contributions of the one- and two-photon resonances associated with states of mixed symmetry, enhancing the positive contributions to the nonlinearity. Here we report an enhancement of up to approximately a factor of 50, which is to our knowledge one of the largest enhancements reported to date. Indeed, extrapolations to, for example, a 40%-loaded side-chain polymer could produce a material with one of the largest third-order susceptibilities known.

We have investigated, by linear absorption and third-harmonic generation, a family of five carotenoids with a single acceptor attached to one end of *trans*-β-carotene, as

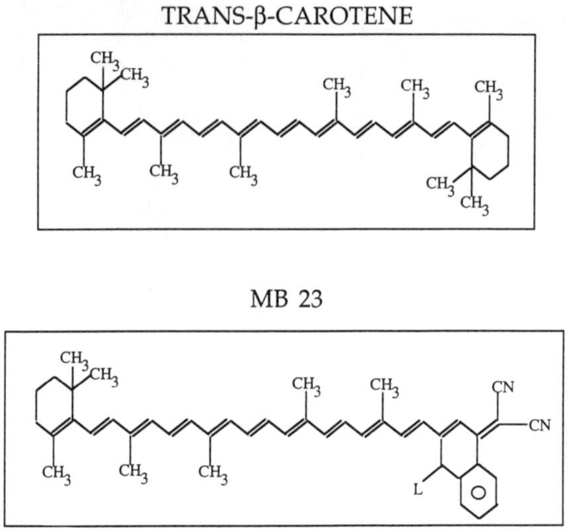

QWA3 Fig. 1. Molecular structure of *trans*-β-carotene and of MB 23.

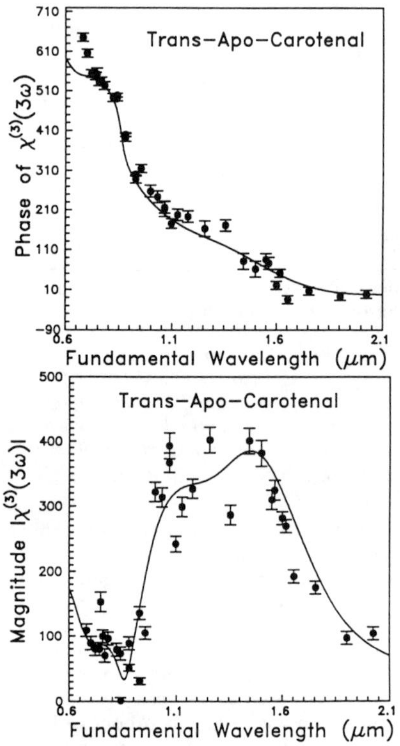

QWA3 Fig. 2. Experimental THG spectrum of *trans*-apocarotenal. The curve corresponds to the best fit to a four-level model.

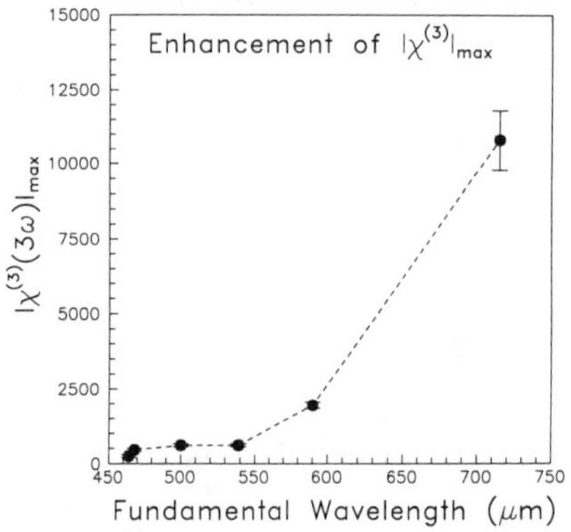

QWA3 Fig. 3. Enhancement of the third-order susceptibility versus the position of the peak of the charge-transfer band. The amplitude of the third-order response was taken at the peak of the THG spectra.

well as the *trans*-β-carotene itself. The structures of the two molecules representing the no-acceptor and strongest-acceptor cases, *trans*-β-carotene and MB 23, are shown in Fig. 1. The corresponding peak absorption wavelengths were 464 and 715 nm, respectively. An example of the third-harmonic-generation spectrum for an intermediate case, *trans*-β-apocarotenal, is shown in Fig. 2. Also shown is the best fit to a four-level model based on one ground state and three mixed-parity excited states. From such data, the peak THG response was obtained for each molecule and was plotted in Fig. 3 versus the wavelength of the absorption maximum associated with the charge-transfer band.

The samples measured had a 4% molar loading in the PMMA thin films. In the case of the molecule with the largest susceptibility, MB 23, increasing its present loading fraction to the 40% typically obtained in side chain polymers would result in a disordered polymer with one of the largest third-order nonlinear responses known to these authors,[4] 10^5 times that of fused silica. These results show that adding charge-transfer complexes to linear polyenes can lead to a very large enhancement in the third-order susceptibility.

**Jet Propulsion Laboratory, California Institute of Technology, 4800 Oak Grove Drive, Pasadena, California 91109*

***Jet Propulsion Laboratory, Beckman Institute and California Institute of Technology, 4800 Oak Grove Drive, Pasadena, California 91125*

†*Collège de France, 11, Place Marcelin Berthelot, 75005 Paris, France*

1. J. P. Hermann and J. Ducuing, J. Appl. Phys. **45**, 5100 (1974).
2. K. H. Haas, A. Ticktin, A. Esser, H. Fisch, J. Paust, and W. Schrof, J. Phys. Chem. **97**, 8675 (1993).
3. A. F. Garito, J. R. Heflin, K. Y. Wong, and O. Zamani-Khamiri, in *Organic Materials for Nonlinear Optics*, R. A. Hann and D. Bloor, eds., Vol. 69 of Royal Society of Chemistry Special Publications (Scolium International, Port Washington, N.Y., 1989), p. 16.
4. B. Lawrence, M. Cha, J. U. Kang, W. Torruellas, G. I. Stegeman, G. Baker, J. Meth, and S. Etemad, Electron. Lett. **30**, 447 (1994).

QWA4 **8:45 am**

Third-order optical nonlinearities in zwitterionic molecules

E. C. Smith, A. K. Kar, B. S. Wherrett, M. Szablewski,* G. Cross,* D. Bloor,*
Department of Physics, Heriot-Watt University, Edinburgh EH14 4AS, U.K.

Zwitterions are molecules with permanently separated positive and negative charges. DEMI-3CNQ and its derivatives are of interest for nonlinear-optical applications because of their large measured dipole moment (45 D) and charge-transfer excitation. Many nonlinear organic materials are based on a conjugated backbone structure that, when excited, moves the charge distribution down the molecule, giving it a larger dipole moment. It is thought that these zwitterionic molecules work in reverse, the charge being redistributed such that the dipole moment is lessened with excitation. With the zwitterion's large dipole moment and ease of alignment in the presence of an electric field, $\chi^{(2)}$ effects have been observed and a strong photorefractive phenomenon should be possible. DEMI has already been shown to be a good possible candidate for low-cost frequency-doubling applications.[1,2] We report the measurement of huge real (2.0×10^{-7}-esu) and imaginary (-4.7×10^{-8}-esu) components of $\chi^{(3)}$, leading to good figures of merit for optical-switching applications.

All the experiments were performed by using a 0.8-ps synchronously-mode-locked dye laser operating on or near DEMI's long-wavelength absorption peak at 712 nm. A subpicosecond continuum was also produced for dynamic spectroscopic studies.

Samples of 1%- and 2%-doped DEMI-3CNQ in PMMA were investigated by excite–probe and degenerate-four-wave-mixing (DFWM) techniques at a variety of incident pump energies and wavelengths (Fig. 1). All samples showed a range of relaxation times (<0.5, 0.6, 7, and 75 ps; see Fig. 2) and corresponding $\chi^{(3)}$ values, requiring a remodeling of the current theory to include multiple lifetimes. Subsequent excite–continuum–probe experiments (Fig. 3) showed that these relaxation times affected the three absorption bands of DEMI from 600 to 800 nm approximately equally. The figures of merit for switching applications[3] showed DEMI to be a suitable material for both nonlinear Fabry–Perot resonators and directional couplers, with a low required switching energy. Other derivatives of DEMI, designed to alter the location of the absorption band (Fig. 3), were then tested and were found to have comparable figures of merit. This will allow the manufacturing of materials to suit a particular wavelength rather than the cur-

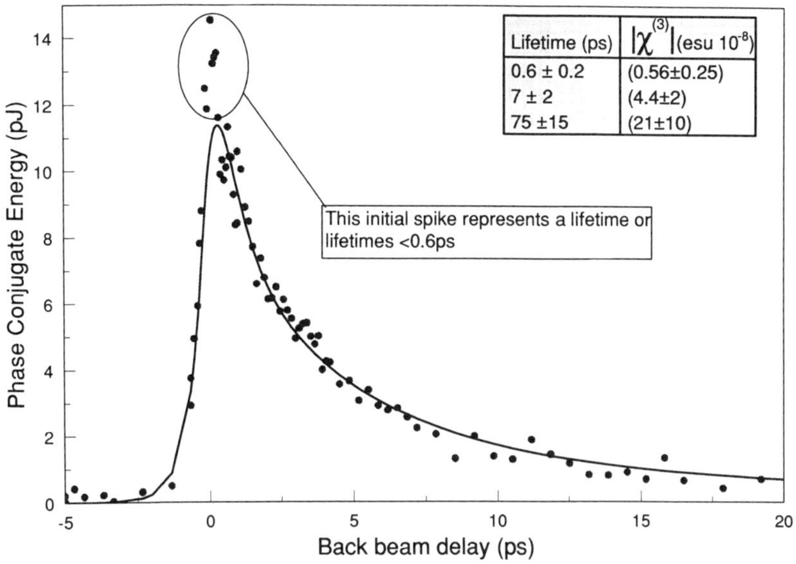

QWA4 Fig. 1. Absorption spectra of DEMI-doped PMMA.

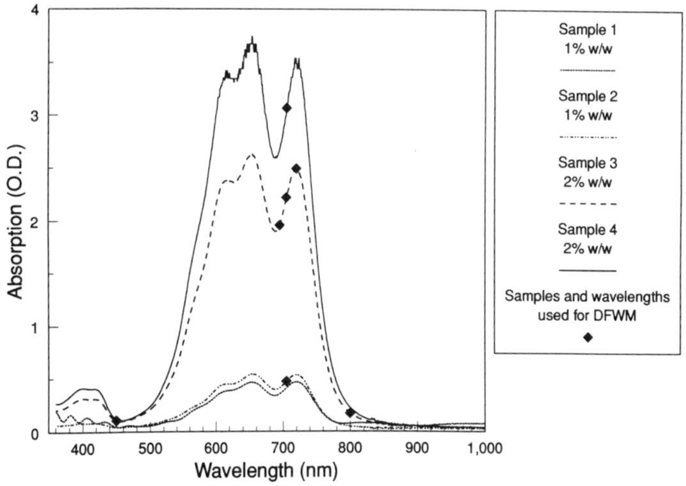

QWA4 Fig. 2. DFWM on DEMI sample showing >3 excitation lifetimes.

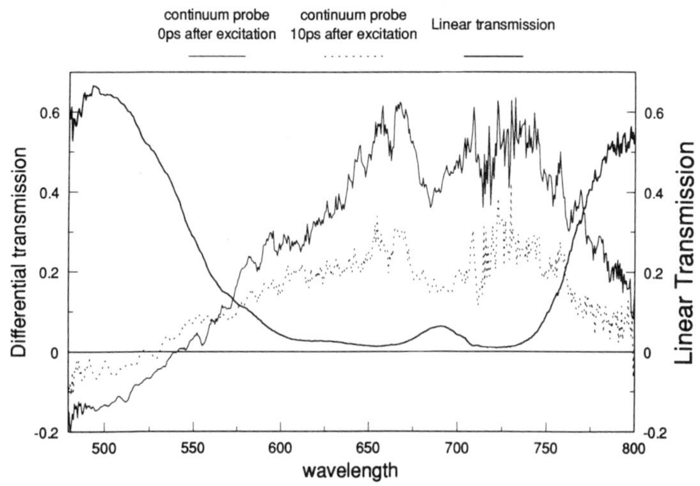

QWA4 Fig. 3. Selected excite–continuum-probe results.

rent reverse situation of finding wavelengths to suit a particular material.

In summary, a new class of material for third-order nonlinear-optical applications has been tested and has been found to be a good candidate. These materials showed promising figures of merit and had the advantage of wavelength tunability by chemical alteration of the molecule.

**Department of Physics, Applied Physics Group, Durham University, Durham DH1 3LE, U.K.*

1. G. H. Cross, D. Bloor, and M. Szablewski, "'Blue window' organics for frequency doubling devices," submitted to Nonlinear Opt.
2. M. Szablewski, J. Org. Chem **59**, 954 (1994).
3. B. S. Wherrett and D. C. Hutchings, in *Nonlinear Optics in Signal Processing*, R. W. Eason and A. Miller, eds. (Chapman and Hall, London, 1993), p. 145.

QWA5 **9:00 am**

Self-alignment supramolecule nonlinear optical thin film

H. Y. Zhang, X. H. He, Y. H. Shih, L. S. Choi, O. K. Kim, L. J. Buckley,*
Department of Physics, University of Maryland at Baltimore County, Baltimore, Maryland 21228

We report here a new type of self-aligned nonlinear organic material made from supramolecules. The supramolecule is formed from a guest trapped inside of a host by hydrophobic interaction. Hemicyanine dye, 4-[4-(dimethyamino)styryl]-1-alkypyridinium bromide, is chosen for the guest. The amylose is chosen for the host. It is a linear macromolecule and has a helix conformation in DMSO. The cavity of the helix is hydrophobic. The resulting supramolecule is shown in Fig. 1.

The film is made from the aqueous solution of the supramolecule on the glass substrate at room temperature without external electric-poling treatment. By studying the second-harmonic generation (SHG), we found the supramolecules are self-aligned in the resulting film. The SHG is pumped by the 1064-nm fundamental wavelength of a Spectra-Physics Q-switched Nd:YAG laser. The SHG maker fringes of the film with thickness 7.5 μm are shown in Figs. 2(a) and 2(b). In Fig. 2(a) the pump beam is incident from the air–film interface, and in Fig. 2(b) the pump is incident from the glass–film interface. Two symmetrical peaks were observed at ±52°. In the case in which the laser beam is incident from the air–film side [Fig. 2(a)], the measured SHG signal was generated in the aligned layer, which is close to the glass substrate near the output surface of the film. Because of the strong absorption at a wavelength of 532 nm ($\alpha = 0.42\ \mu m^{-1}$), the signals generated far from the output surface are absorbed by the film. Figure 2(b) shows the SHG maker fringes generated from the aligned layer close to the air–film interface. This SHG signal is weaker. However, the pattern is similar to that in Fig. 2(a). These are typical SHG maker fringes of an aligned film with strong absorption at the harmonic wavelength.[1] The result indicates that the whole film is aligned. Similar maker fringes are observed in films with variable thicknesses. The ratio of the SHG intensity in Fig. 2(a) to that in Fig. 2(b) varied with the thickness of the film, with a larger intensity ratio observed in thicker film. For extremely thin films (0.1 μm), the intensities of the SHG signals generated from the two sides (air–film and glass–film) are close to each other. The molecular orientations of the film are also studied by polarization measurements of SHG, which has been established in Refs. 2 and 3.

The $\chi^{(2)}_{33}$ of the film was measured to be 4.2×10^{-9} esu at 532 nm. The long-term stability of $\chi^{(2)}$ is reported in Figs. 3(a) and 3(b). Within the experimental error, $\chi^{(2)}$ remains a constant for 1100 h at room temperature. At 90°C, $\chi^{(2)}$ does not have significant changes for at least 100 h [Fig. 3(b)].

The mechanism of the self-alignment of the film is discussed in detail.

Authors at the U.S. Naval Research Laboratory gratefully acknowledge partial support from the Office of Naval Research.

**Code 6120, U.S. Naval Research Laboratory, Washington, D.C. 20375*

1. L. M. Hayden, G. F. Sauter, F. R. Ore, P. L. Pasillas, J. M. Hoover, G. A. Lindsay, and R. A. Henry, J. Appl. Phys. **68**, 456 (1990).
2. T. F. Heinz, H. W. K. Tom, and Y. R. Shen, Phys. Rev. A **28**, 1883 (1983).
3. K. Kajikawa, K. Kigata, H. Takezoe, and A. Fukuda, Mol. Cryst. Liq. Cryst. **182**, 91 (1990).

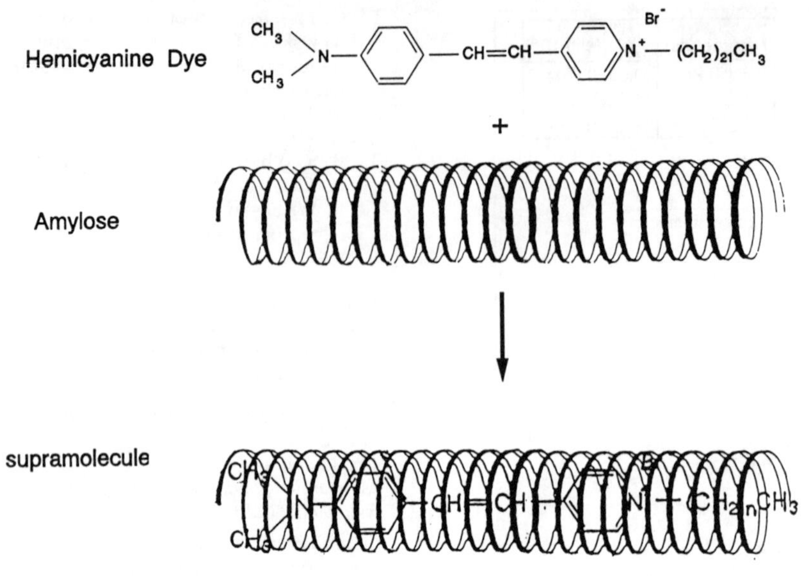

QWA5 Fig. 1. Formation of a supramolecule.

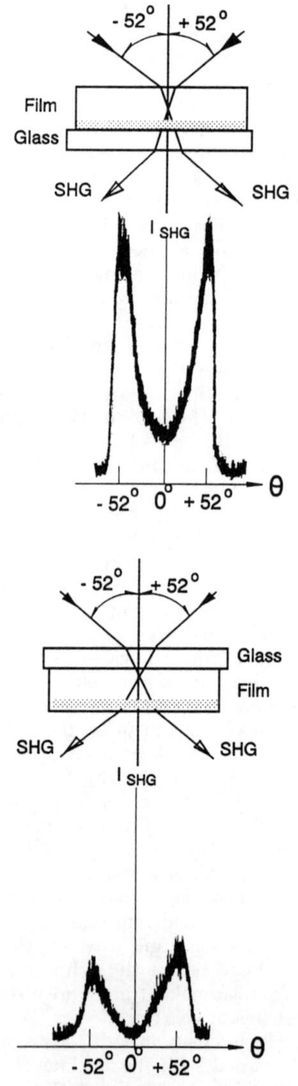

QWA5 Fig. 2. SHG maker fringes on the film, where pump beam is incident from (a) the air–film interface and (b) the glass–film interface. Intensity scale is the same for parts (a) and (b).

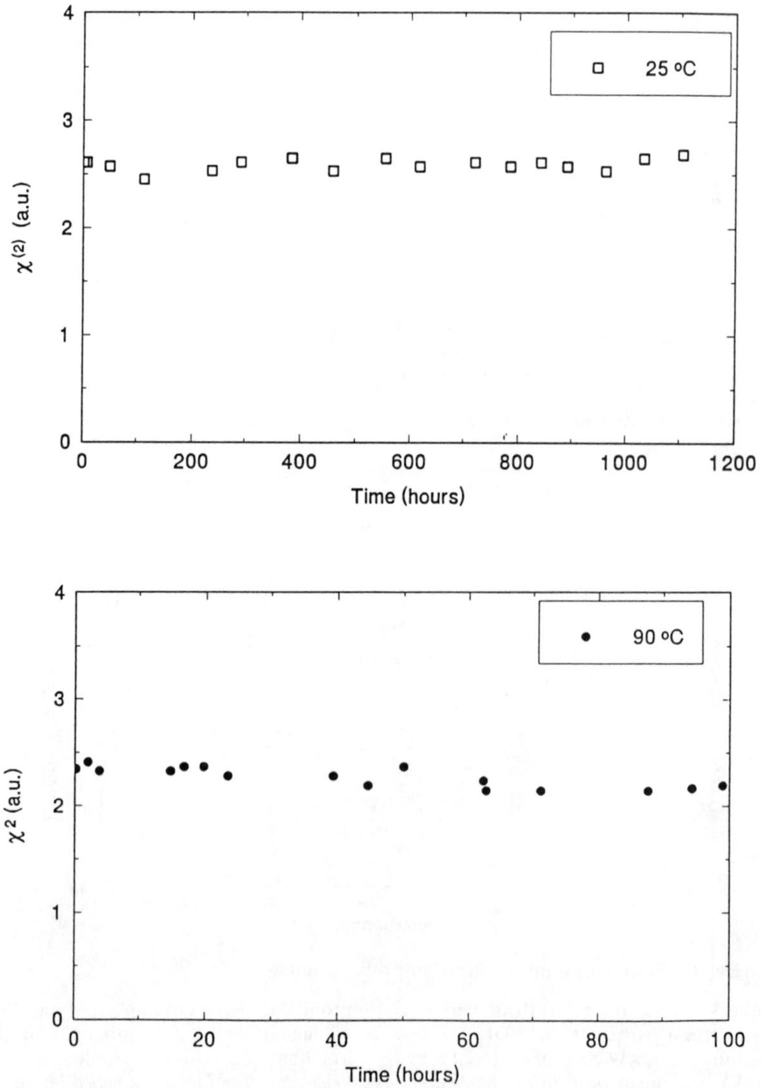

QWA5 Fig. 3. $\chi^{(2)}$ of the film versus time at (a) room temperature and (b) 90°C.

QWA6 **9:15 am**

Neodymium-doped potassium titanyl arsenate (Nd:KTA): a new nonlinear optical material

Narasimha S. Prasad, Allen R. Geiger,
LāSen, Inc., Suite 600, 300 North Telshor, Las Cruces, New Mexico 88001

KTA, with its broad transparency range extending up to 5.5 μm, has been demonstrated to be an attractive optical-parametric-oscillator (OPO) material[1] with a high damage threshold (>500 MW/cm^2). The development of Nd:KTA is a result of the efforts to develop new nonlinear materials that can exhibit optical parametric interaction as well as lasing action. We have developed an Nd:MgO:$LiNbO_3$-based OPO/laser (OPOL). This device uses diode pumps at 810 nm to produce lasing at 1085 nm, and this lasing acts as a pump to induce optical parametric interaction in a single crystalline medium.[2] It is anticipated that Nd:KTA will provide an extended tuning range with a damage threshold higher than that of Nd:MgO:$LiNbO_3$.

Nd:KTA was grown from a polycomponental flux melt instead of from a pure melt. The conventional Czochralsky technique was suitably modified to facilitate controlled pulling of the seed crystal from the flux composition. In this process, crystallization of 14 growth sectors around the seed allow crystal growth to occur. Layered controllable growth of a single growing facet has provided high-optical-quality crystals with appreciable volumes. Because the crystals are grown within a single growth sector, the grown crystals are found to contain fewer defects than are produced when other growth techniques are used. Neutron-activation analysis has yielded a neodymium number density of (2.5–3.0) × 10^{19} cm^3 for a weight level of 0.4% dopant concentration.

Figure 1 shows the absorption spectrum. A 0.003-wt. % Nd:KTA sample with dimensions of 2 × 2 × 2 mm was excited by an 810-nm laser diode. The resultant fluorescence observed in the vicinity of 1 μm had a fluorescent lifetime of 235 ± 5 μs at FWHM.

The intensity as a function of emission wavelength for 0.003-wt. % Nd:KTA is shown in Fig. 2. The crystal was excited at 585 nm, and the emission was recorded from 800 to 1600 nm. In this spectra, three overlapping curves are obtained for three different faces with respect to an arbitrary

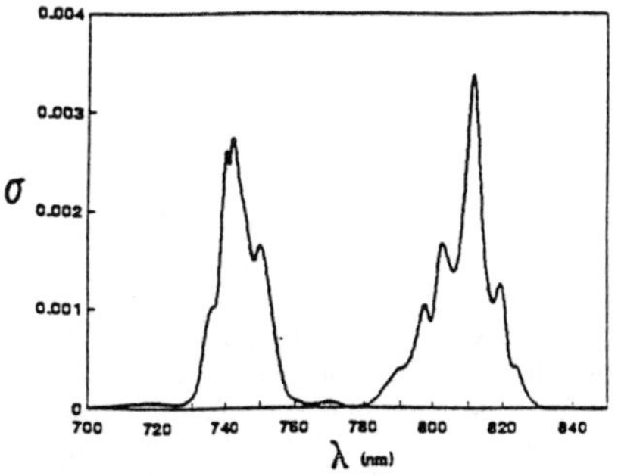

QWA6 Fig. 1. Absorption spectrum of Nd:KTA.

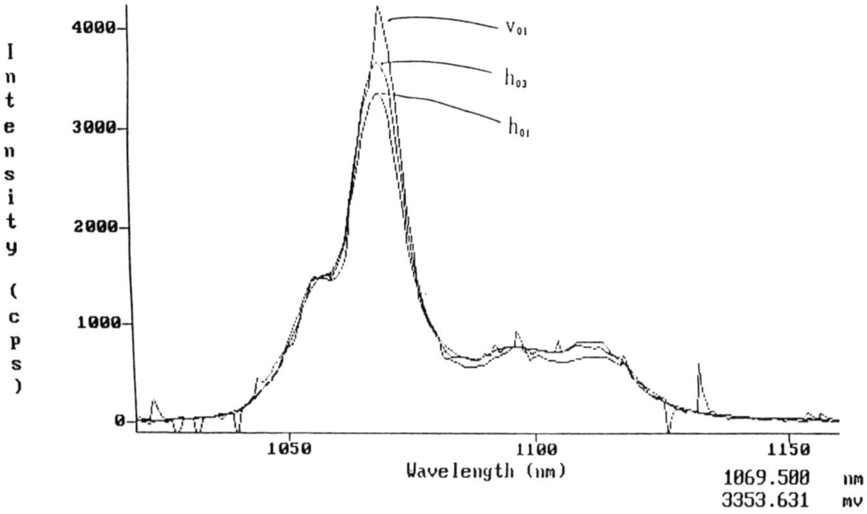

QWA6 Fig. 2. Emission spectrum of 0.003-wt. % Nd:KTA.

QWA6 Table 1. Stimulated-emission cross-section coefficients of Nd:KTA.

Band	Peak Wavelength	Branching Ratios 0.003 % wt	0.2 % wt	0.5 % wt	S.E.C Coefficient
$^4F_{3/2} \rightarrow {}^4F_{9/2}$	1075 nm	0.4	0.25	0.22	4×10^{-20} cm^2
$^4F_{3/2} \rightarrow {}^4F_{11/2}$	1070 nm	0.48	0.62	0.68	5.19×10^{-20} cm^2
$^4F_{3/2} \rightarrow {}^4F_{13/2}$	1100 nm	0.12	0.14	0.11	5.71×10^{-20} cm^2

reference scratch. The peak wavelength at 1069.50 nm corresponds to the $^4F_{3/2} \rightarrow {}^4F_{11/2}$ band. The other two bands of significance are $^4F_{3/2} \rightarrow {}^4F_{9/2}$ and $^4F_{3/2} \rightarrow {}^4F_{13/2}$. Similar experiments were carried out with 0.2-wt. % and 0.5-wt. % Nd:KTA crystals. Only in the case of 0.5-wt. % Nd:KTA was the fluorescence polarized. The branching ratios and the corresponding stimulated-emission cross-section coefficients are listed in Table 1. From preliminary investigations, it appears that two sites are available for Nd-ion substitution in the KTA matrix, and the substitution of an ND ion at a desired site can be controlled to reduce homogeneous broadening and to modify branching ratios to obtain better stimulated-emission cross-section values, which at present are comparable to those of Nd:glass materials.

1. W. R. Bosenberg *et al.*, in *Proc. Advanced Solid-State Lasers Conference*, Washington, D.C., 1993, 134.
2. A. R. Geiger and N. S. Prasad, in *Proc. Advanced Solid-State Lasers Conference*, Salt Lake City, Utah, 1994, 283.

QWA7 **9:30 am**

New development of nonlinear optical crystals for the ultraviolet region by molecular engineering

Chuangtian Chen, Baichang Wu,
Yebin Wang, Wenrong Zeng, Linhua Yu,
Fujian Institute of Research on the Structure of Matter, Chinese Academy of Sciences, Fuzhou, Fujian, 350002, China

Both theoretical estimates and recent experimental studies show that potassium fluoroboratoberyllate, $KBe_2BO_3F_2$ (KBBF), possesses excellent linear-optical (LO) and nonlinear-optical (NLO) properties. For example, its transparent range expands from 155 nm to 3700 nm, and it also has a moderate birefringence with Δn = 0.073 in the 589 nm–294.5-nm region. These two properties give the crystal a very wide phase-matching capacity. The curve of Fig. 1 indicates that the shortest wavelength of second-harmonic generation (SHG) of KBBF is at 184.7 nm with the phase-matching angle Θ_{pm} = 61. The crystal's SHG coefficient $2d_{36}(KDP) \leq d_{11} \leq 2.2d_{36}(KDP)$, which is large enough for deep UV harmonic generation. However, KBBF has a strong layer habit of growth and its decomposition temperature is low (T = 825°C), and these characteristics make it very difficult to grow in large enough sizes. It is obvious that growing larger KBBF crystals is a precondition for further measurements of phase-matching characteristics, in the entire spectral region, from the visible region to the UV, and also for its NLO application.

The abovesaid shortcoming of KBBF forced us to look for new compounds, which we hope not only retain the major favorable structural characteristics of KBBF [i.e., $(BO_3)^{3-}$ group keeps a coplanar structure with its three dangling bonds eliminated], but also can overcome the layer habit of KBBF. From a series of synthesis and structure studies, we have found a new boron–beryllium compound, $Sr_2Be_2B_2O_7$ (SBBO),[1] whose characteristics of crystal structure are ideal and whose LO and NLO properties are similar to those of KBBF. Also, the study of crystal growth shows that the strong layer habit of KBBF has been removed to a great extent by oxygen atoms bridged between two $(Be_3B_3O_6)$ layers. In addition, the density of NLO active groups (BO_3) per unit volume in SBBO is about twice that of KBBF, which makes the SHG coefficient of SBBO twice that of KBBF.

SBBO crystallizes in the space group *P*-6c2 (hexagonal system) with cell dimensions a = 4.683(3) A, c = 15.311(7) A, and z = 2. Figures 2(a) and 2(b) depict the crystal structure of SBBO. Its main structure features may be described as follows: (1) The nearly coplanar $(Be_3O_3O_6)$ ring is expanded by sharing every member of the ring with those of its nearest neighbors to form a two-dimensional layer. At the same time, all $(BO_3)^{3-}$ groups contained in the layer are parallel with the z axis (C_3 axis) of the crystal. This network structure eliminates the three dangling bonds of $(BO_3)^{3-}$, as in the case of KBBF [as shown in Fig. 2(a)]. (2) The whole structure of SBBO consists of pairs of $(Be_3B_3O_6)$ infinite-network layers. In each pair there are oxygen atoms (the layer-out ligand of beryllium atoms) bridged between the $(Be_3B_3O_6)$ layers, as shown in Figs. 2(a) and 2(b). It is reasonable to consider that this special structural feature is responsible for the layer habit of SBBO being weaker than that of KBBF, and it makes growing SBBO crystals somewhat easier than growing crystals of KBBF. Using the top-seeding high-temperature-flux method with the growth temperature of 900–1000°C, we have succeeded in growing SBBO single crystals in the size of 7 × 7 × 2 mm³ with good optical quality.

SBBO is transparent from 165 nm to 3870

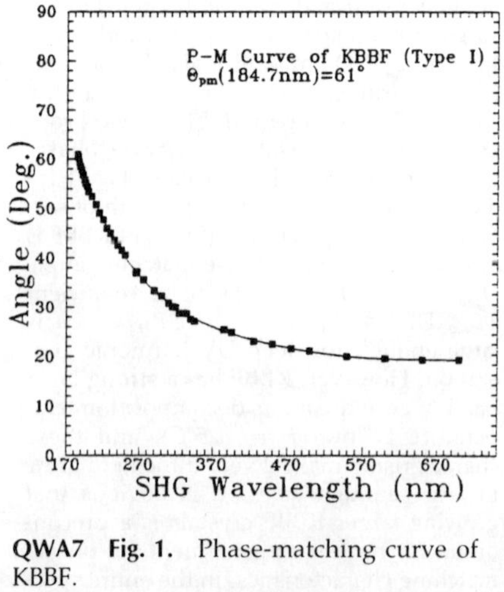

QWA7 **Fig. 1.** Phase-matching curve of KBBF.

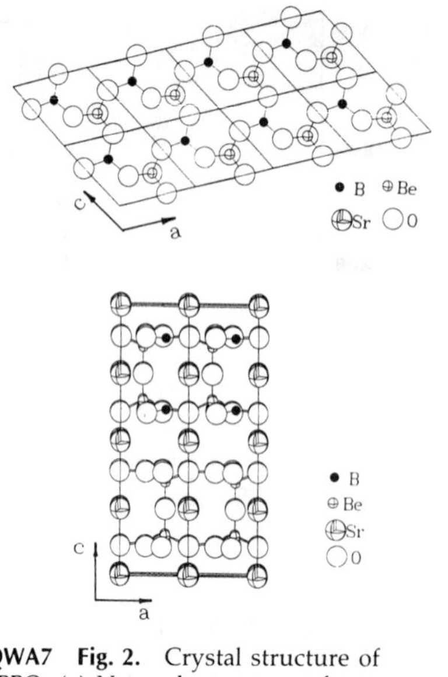

QWA7 **Fig. 2.** Crystal structure of SBBO. (a) Network structure of a $(Be_3B_3O_6)_\infty$ layer in SBBO in crystal. (b) $(Be_3B_3O_6)_\infty$ layer structure of SBBO crystal with bridge oxygen atoms bonded to Be atoms on the layers.

nm, which is in agreement with the theoretical prediction of the π–π^* transition of the $(BO_3)^{3-}$ group. According to the point group D_{3h}, we need to determine only the d_{22} coefficient, and the representation of the d_{eff} formula is $d_{eff} = -d_{22} \times \text{Cos}\ \theta \times \text{Sin}\ 3\theta$ (Type I). Using the phase-matching method and taking BBO as a standard, we preliminarily measured the d_{22} coefficient. The result is d_{22} = 0.8 ~ 1.0 × d_{22}(BBO), which is nearly equal to the theoretical prediction: d_{22}(SBBO) ≅ 2 × d_{11}(KBBF). Owing to the $(BO_3)^{3-}$ co-planar structure of SBBO that is similar to KBBF, the crystal also possesses a moderate birefringence (Δn – 0.063 at λ = 589 nm) and, therefore, has a very wide phase-matching range. A preliminary measurement of phase-matching characteristics shows that the crystal is useful, at least, for formation of SHG output from 200 nm to near 1800 nm, which is wider than that of BBO both in the UV and IR side. On the other hand, the crystal also possesses a lower angle sensitivity than does BBO, because of the relatively small birefringence. SBBO is also not deliquescent, and it has good mechanical properties for cutting, polishing, and coating.

In conclusion, by using molecular engineering, including theoretical evaluations of the microscopic second-order susceptibilities, the energy gap, and the anisotropy of linear susceptibilities of anionic groups in the boron-beryllium series, we have successfully found another new UV NLO crystal, SBBO, which possesses better LN and NLO properties in UV than does BBO. It is beyond doubt, therefore that the crystal has great potential for application in UV NLO devices.

1. C. T. Chen, Y. B. Wang, B. C. Wu, K. C. Wu, W. R. Zeng, and L. H. Yu, Nature **373**, 323 (1995).

QWA8 **9:45 am**

Cubic optical nonlinearity of bulk metals: probing by the incoherent specular inverse Faraday effect

P. J. Bennett, H. Loh, S. V. Popov, I. R. Shatwell, Y. P. Svirko, N. I. Zheludev, V. F. Kamalov,* E. Slobodchikov,*
Department of Physics, University of Southhampton, Southampton SO17 1BJ, U.K.

It is often anticipated that bulk metals do not yield any significant nonlinearity $\chi^{(3)}$ in the infrared region. For example, the nonlinearities $\chi^{(3)}(3\omega, \omega, \omega, \omega)$ responsible for the *coherent* process of third-harmonic generation are rather small. However, here we report the results of direct measurements of the cubic nonlinearities in bulk gold, aluminum, and nickel, which are responsible for the incoherent specular inverse Faraday effect (SIFE). We found that in the near-infrared region (at λ = 1260 nm, below the energy necessary for interband transitions), these incoherent nonlinearities are rather high (10^{-7}–10^{-8} esu) and very fast, with relaxation times shorter than 90 fs (see Fig. 1).

The reported characterization of bulk metals became possible as a result of the development of a new pump–probe polarization-sensitive technique based on the incoherent SIFE.[2] This effect is a nonlinear optical phenomenon in which the medium is stimulated by a strong circularly polarized pump wave of intensity I. This results in the alteration of the polarization of a probe wave reflected from the surface because of pump-induced circular dichroism (see Fig. 2). SIFE uniquely complements conventional techniques used for measurements of nonlinear optical properties of opaque materials, such as transient pump–probe reflectivity and reflective harmonic generation, because it provides information on those combinations of the $\chi^{(3)}$ tensor that are not accessible by either of these traditional methods. If $\varepsilon_1 + i\varepsilon_2 = |\varepsilon|e^{i\beta} = n^2$ is the complex dielectric tensor of a strongly absorbing medium, the incoherent

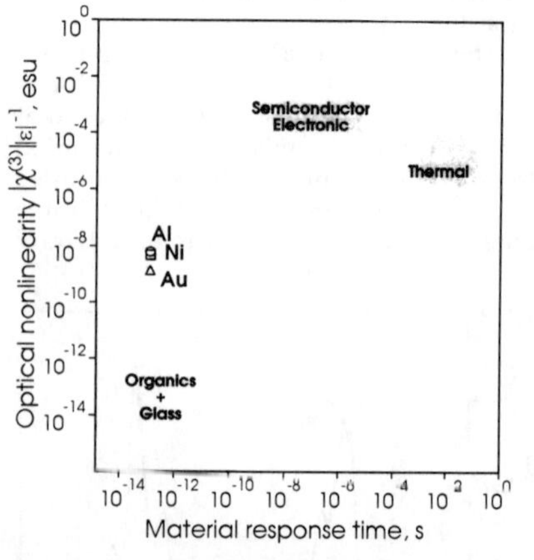

QWA8 **Fig. 1.** Typical values of the normalized cubic optical nonlinearity $|\chi^{(3)}||\varepsilon|^{-1}$ versus material response time in various classes of materials. Susceptibilities of bulk metals were measured in this work, whereas all other data were extracted from the survey by D. Cotter.[1]

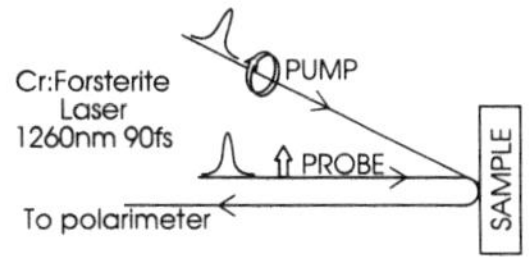

QWA8 Fig. 2. Experimental arrangement for measuring optical nonlinearities by using the specular inverse Faraday effect.

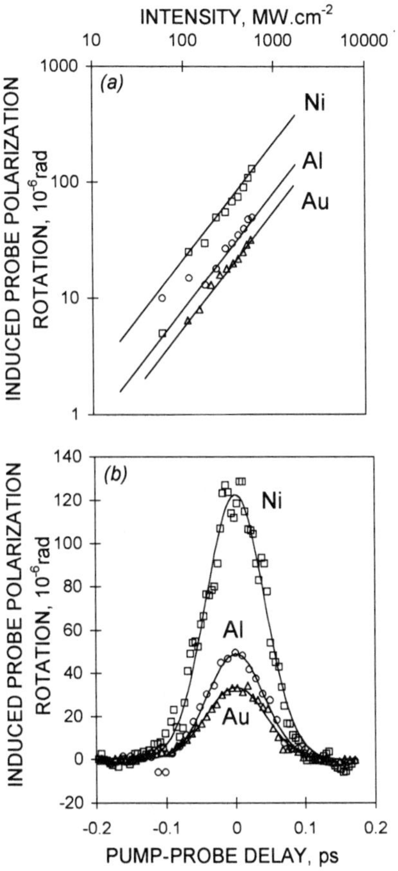

QWA8 Fig. 3. SIFE rotation in various metals (λ = 1260 nm, room temperature) versus (a) pump-pulse intensity (zero pump–probe delay) and (b) delay time between pump and probe pulses ($I \sim 600$ MW/cm²).

contribution to the pump-induced alteration of the probe-wave-polarization azimuth α_r in SIFE is

$$\delta\alpha_r = 32\pi I/c|1+n|^2 \mathrm{Im} \times \left[\frac{\chi^{(3)}_{xxyy}(\omega,\omega,\omega,-\omega) - \chi^{(3)}_{xyyx}(\omega,\omega,\omega,-\omega)}{n(1-n^2)}\right]. \quad (1)$$

Thus, measurement of $\delta\alpha_r$ gives the value of $\chi^{(3)}_{xxyy}(\omega, \omega, \omega, -\omega) - \chi^{(3)}_{xyyx}(\omega, \omega, \omega, -\omega)$.

We studied mirrorlike samples of gold, nickel, and aluminum deposited on glass substrate in vacuum. A femtosecond Cr:forsterite laser (λ = 1260 nm), pumped by a Nd^{3+}:YAG laser, was used as the optical source. The temporal resolution of the measurements was 90 fs. In all the materials the optical response was instantaneous within the time resolution available, and no transient dynamics were observed (see Fig. 3). For reference, some measurements were also taken by using a frequency-doubled, mode-locked Nd^{3+}:YAG laser (τ = 80 ps, λ = 532 nm). For a correct derivation of optical nonlinearities of metals, it is crucial to properly account for the dielectric factor $|1+n|^2 n(1-n^2) \sim |\varepsilon|^{2.5}$ in Eq. (1) (a similar correction appears in all reflective nonlinear interactions with metals). A huge value of the material dielectric constant in the infrared region ($|\varepsilon|^{2.5} = 3.8 \times 10^4$ for gold at 1260 nm) dramatically reduces the observable magnitudes of the reflective nonlinear phenomena, although the nonlinearity of the material might be very large. The values of $|\chi^{(3)}_{xxyy}(\omega, \omega, \omega, -\omega) - \chi^{(3)}_{xyyx}(\omega, \omega, \omega, -\omega)|$ at 1260 nm derived from our experimental data are (see also Fig. 1)

$$\text{Au: } |\chi^{(3)}_{xxyy}(\omega,\omega,\omega,-\omega) - \chi^{(3)}_{xyyx}(\omega,\omega,\omega,-\omega)| = 8.0 \times 10^{-8}\ \text{esu};$$

$$\text{Al: } |\chi^{(3)}_{xxyy}(\omega,\omega,\omega,-\omega) - \chi^{(3)}_{xyyx}(\omega,\omega,\omega,-\omega)| = 9.6 \times 10^{-7}\ \text{esu};$$

$$\text{Ni: } |\chi^{(3)}_{xxyy}(\omega,\omega,\omega,-\omega) - \chi^{(3)}_{xyyx}(\omega,\omega,\omega,-\omega)| = 1.9 \times 10^{-7}\ \text{esu}.$$

We believe that the observed effect in the infrared region is mainly due to imaginary intraband nonlinearities of free carriers in metals. The nonlinearity is attributed to the difference between the population distributions for the free electrons with spins parallel and antiparallel to the excitation beam direction, which is induced by absorption of the circularly polarized pump. We note that this is not a coherent process, and it changes only the values of $\varepsilon_{\perp}$; therefore, it does not contribute to the third-harmonic generation. The fast nonlinearity response time observed is associated with a free-electron spin-relaxation time, and it is compared with published data on electron-spin-resonance experiments. We discuss the roles of scattering by phonons and impurities and the interaction of carriers with the border in spin-flipping free-electron absorption leading to this nonlinearity. We also address the role of multipole contributions to the nonlinearity. We compare infrared results with our data obtained with picosecond pulses at 532 nm and discuss the role of surface states and transitions from *d* states to the free-electron states near the Fermi level for the nonlinearity in the visible range.

**Institute of Chemical Physics, 4 Kosigina Street, Moscow 117977, Russia*

1. D. Cotter, in *Ultrafast Phenomena V*, G. R. Fleming and A. E. Siegman, eds. (Springer-Verlag, New York, 1986), p. 274.
2. S. V. Popov, N. I. Zheludev, and Yu. P. Svirko, Opt. Lett. **19**, 13 (1994); Yu. P. Svirko and N. I. Zheludev, J. Opt. Soc. Am. B **11**, 1388 (1994); N. I. Zheludev, S. V. Popov, Yu. P. Svirko, A. Malinowski, W. Y. Liang, and C. T. Lin, Solid State Commun. **90**, 287 (1994); N. I. Zheludev, M. A. Brummel, R. T. Harley, A. Malinowski, S. V. Popov, D. E. Ashenford, and B. Lunn, Solid State Commun. **89**, 823 (1994).

QWB — 8:00 am, Hyatt B

Novel Quantum Effects

Harry Jeffrey Kimble, *California Institute of Technology, Presider*

QWB1 (Invited) — **8:00 am**

The Aharonov-Casher effect in an atomic system

Stephen M. Barnett, Erling Riis, Alastair G. Sinclair, Karin Sangster,* E. A. Hinds,* *Department of Physics and Applied Physics, University of Strathclyde, Glasgow G4 0NG, Scotland*

We describe a new experimental configuration suitable for observing the Aharonov-Casher phase in an atomic system. We have verified the effect to 2% and have demonstrated its independence of velocity and linearity in an electric field.

**Physics Department, Yale University, New Haven, Connecticut 06520*

QWB2 — **8:30 am**

New interpretation of the Aharonov–Bohm effect

J. D. Franson, *Applied Physics Laboratory, Johns Hopkins University, Laurel, Maryland 20723*

A single electron passing through an interferometer will produce a small path-dependent change in the electromagnetic field, which causes the state of the field to become entangled with the state of the electron, as illustrated in Fig. 1. In the usual analysis of the Aharonov–Bohm (AB) effect,[1] the magnetic field produced by a solenoid S is treated classically and the electron is treated quantum mechanically, which neglects any effects of the entanglement. One might, instead, consider the opposite approach, in which the phase shift along a specific path is calculated by quantizing the field but treating the electron as a classical charge distribution. This gives a phase shift ϕ_D in the state of the field of the form

$$|\psi_p\rangle = \exp(i\phi_D)\exp\left(-i\int \langle H\rangle\, dt/\hbar\right)|\psi_f\rangle,$$

where ϕ_D is given by

$$\frac{d\phi}{dt} = \frac{1}{2}\frac{\int \vec{j}(\vec{r},t)\cdot\langle\vec{A}(\vec{r})\rangle\, d^3\vec{r}}{\hbar} = -\frac{1}{2}\frac{\langle H'\rangle}{\hbar}.$$

As discussed previously, ϕ_D vanishes in the quasi-static limit of slowly varying currents and has therefore been referred to as the dynamic phase of the field.[2]

The only correct approach for the analysis of such an interferometer is to quantize both the field and the electron. The inclusion of the dynamic phase shift ϕ_D of the field gives a somewhat unexpected form for

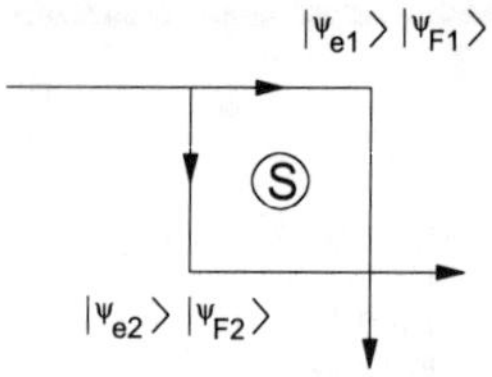

QWB2 Fig. 1. Entanglement of the electromagnetic field with an electron in an interferometer, which can modify the Aharonov–Bohm effect.

the overall phase shift:

$$\frac{d\phi_T}{dt} = \frac{1}{2}\sum_i \frac{\int \langle \vec{j}_e\rangle \cdot \langle \vec{A}_{Si}(\vec{r}_e)\rangle \, d^3\vec{r}_e}{\hbar} + \frac{1}{2}\sum_i \frac{\int \langle \vec{j}_{Si}\rangle \cdot \langle \vec{A}_e(\vec{r}_{Si})\rangle \, d^3\vec{r}_{Si}}{\hbar}.$$

Here index i labels the electrons in the solenoid, and $\langle A_{Si}\rangle$ and $\langle A_e\rangle$ are the fields produced by the electrons in the solenoid and in the interferometer, respectively. The form of this expression suggests a new interpretation of the AB effect in which half of the usual phase shift is associated with the test particle (electron) in the interferometer and the other half of the phase shift occurs in the source of the external field. When retardation effects are negligible, the two integrals in the above equation can be shown to be equal and the new interpretation reduces to the usual one. But when retardation effects are significant, the two interpretations are not equivalent and these effects can substantially modify the experimental results.

1. Y. Aharonov and D. Bohm, Phys. Rev. **115**, 485 (1959).
2. J. D. Franson, in *Proc. Conference on Fundamental Problems in Quantum Theory*, June 1994, in press; submitted to Phys. Rev. A.

QWB3 **8:45 am**

Atom interferometer with phase gratings

E. M. Rasel, M. K. Oberthaler, H. Batelaan, J. Schmiedmayer, A. Zeilinger, *Institute for Experimental Physics, Technickerstrasse 25/ IV, A-6020 Innsbruck, Austria*

The de Broglie-type interferometer for atoms has been realized, until now, by diffraction on very precisely machined absorption gratings or slits, dividing the beam into the various orders of diffraction. Our interferometer, however, does not use absorption gratings. Instead, it uses phase gratings and a small grating period as well as low absorption of a far-detuned standing light wave. We choose for separation and recombination of the beam the three-grating configuration. This setup is free of dispersion and thus accepts a source with a small longitudinal coherence and a rather broad beam divergence. The divergence is limited only by the diffraction angle, which is 30 μrad for the first order.

We used the metastable $[3p^5s]1s_5$ state of argon for interferometry, and the related brightness of our source is about 3.8×10^{15} counts·s^{-1}·sr^{-1}·cm^{-2}. The most probable speed of the beam is about 850 m/s, corresponding to a de Broglie wavelength of 0.12 Å for ^{40}Ar, with a velocity spread of ±30%. The beam is collimated to below one photon recoil.

The spacing between the standing light waves is roughly 0.25 m, which corresponds to a length of the complete interferometer of 0.5 m. The relative position of the mirrors defines the antinodes of the standing waves, and thus they have to be vertically adjusted to each other to better than 3×10^{-5} rad. The focus of the standing light waves has the form of a 30-mm-high light sheet with a waist of 50 μm. The transit time of the atomic wave is 60 ns, which is short enough for the light potential to be considered as a two-dimensional grating. In this regime, the so-called Raman–Nath or Kapitza–Dirac regime, the far-detuned standing light wave can be described as a phase grating with a sinusoidal form. In contrast to absorption gratings, the phase grating permits the strength of the diffraction orders to be varied. Consequently, the focus and the intensity of each of the three standing light waves could be separately optimized in our setup.

The total intensity of the atom beam oscillates by a transversal shift of, e.g., third-light-grating multiples of the grating period. Thus we scanned the interference pattern by moving the third mirror on a piezo-driven translation stage. The interference contrast of the first results is about 8%. The reasons for this contrast are the cross talk of the various diffraction orders, imperfect adjustment of slits and mirrors, and vibrations.

QWB4 (Invited) **9:00**

Atom optics in the transition from classical stability to chaos

C. Bharucha, F. L. Moore, J. C. Robinson, M. G. Raizen, *Department of Physics, University of Texas at Austin, Austin, Texas 78712*

Atomic momentum transfer in a modulated standing wave is studied experimentally. We observe quantum effects in a regime that goes from stability through a mixed phase space to full chaos as the modulation amplitude is varied.

QWB5 **9:30 am**

Nonlinear Jaynes–Cummings dynamics of trapped and laser-driven ions

W. Vogel, R. L. de Matos Filho, *Department of Physics, University of Rostock, Universitätsplatz 3, Postfach 999, D-18051 Rostock, Germany*

The Jaynes–Cummings model (JCM) is of great interest because it is an exactly solvable model for a quantum-physical interaction. Progress in micromaser experiments[1] demonstrates the practical relevance of this model. More recently, it has been found that the motion of a trapped and laser-driven ion in the Lamb–Dicke regime can be described under appropriate resonance conditions by the JCM.[2]

Whereas in the Lamb–Dicke limit an electronic transition of the ion can be accompanied by one-quantum transitions of the trap vibration, multiquantum effects occur when the Lamb–Dicke limit does not apply. The vibration-assisted interaction Hamiltonian between ion and laser reads as

$$\hat{H}_{\rm int} = \lambda E_l \hat{A}_{12} \hat{g}(\hat{x}) + \text{h.c.}, \qquad (1)$$

where λ is the (dipole or quadrupole) coupling-matrix element, E_l is the laser amplitude, $\hat{A}_{12}$ is the atomic flip operator, and $\hat{g}(\hat{x})$ is the (operator-valued) radiation mode function (e.g., $\exp(ik\hat{x})$ for a traveling wave) depending on the position operator $\hat{x}$ of the ion. The JCM Hamiltonian is obtained when a linear approximation of the mode function ($\hat{g}(\hat{x}) \propto \hat{x}$) applies together with a resonance approximation $\hat{x} \propto \hat{a}^\dagger$, where $\hat{a}^\dagger$ is the creation operator for a vibrational quantum.

With increasing Lamb–Dicke parameter, k-quantum-assisted transitions become more probable and may lead to new effects, such as second-sideband laser cooling.[3] In the resonance approximation the Hamiltonian of Eq. (1) can be simplified as

$$\hat{H}_{\rm int} = \lambda E_l \hat{f}(\hat{a}^\dagger \hat{a}) \hat{A}_{12} (\hat{a}^\dagger)^k + \text{h.c.}, \qquad (2)$$

which represents a nonlinearly coupled k-photon JCM Hamiltonian with nonlinear coupling $\lambda E_l \hat{f}(\hat{a}^\dagger \hat{a})$. The operator function $\hat{f}$ depends on the mode function $\hat{g}$ in Eq. (1).

The interaction problem described by the Hamiltonian of Eq. (2) is exactly solvable and gives rise to new physical properties. In Fig. 1 we show the electronic dynamics of the ion based on the nonlinear JCM (NJCM) and the exact solution (without resonance approximation) of the equations of motion in Ref. 3. Except for relaxation, it is seen that the NJCM describes the ionic dynamics very well.

In Fig. 2 we show the (vibration-assisted) Rabi frequencies Ω^{nn+1}, corresponding to the NJCM with $k = 1$, as a function of the vibrational quantum number. In the standard JCM we have $\Omega^{nn+1} \propto \sqrt{n+1}$ or $\Omega^{nn+2} \propto \sqrt{(n+1)(n+2)}$ in the two-quantum JCM (setting $\hat{f} = 1$ and $k = 2$ in Eq. (2)). In the latter case, the quasi-linear dependence of Ω^{nn+2} on n leads to sharp revivals.[4] Figure 3 reveals that the revivals of the one-quantum ($k = 1$) NJCM may exhibit features of the standard two-quantum JCM, provided that the initial state of the vibration is prepared close to a quasi-linear region of the nonlinear Rabi frequencies. On the other hand, the dynamics of the two-quantum NJCM may be close to those of the standard one-quantum JCM.

These examples demonstrate that the complex behavior of a trapped and laser-driven ion far from the Lamb–Dicke limit may show (exactly solvable) NJCM dynamics. This opens new possibilities for fundamental experiments in quantum physics.

This work was supported by the German Research Association.

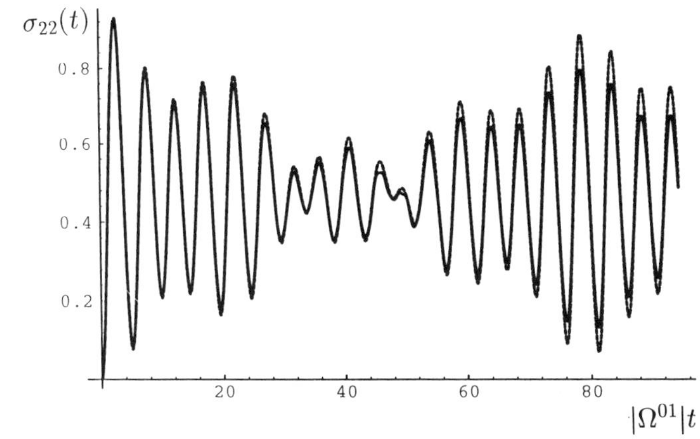

QWB5 Fig. 1. Excited-state occupation σ_{22} as a function of time derived from the equations of motion of the ion in Ref. 3 (solid curve) and from the NJCM ($k = 1$, dashed curve). The vibration is initially in a coherent state with a mean number of three quanta. Parameters: vibrational frequency $\nu = 5 \times 10^3\Gamma_1$, electronic Rabi frequency $\Omega_0 = 5 \times 10^2\Gamma_1$, Lamb–Dicke parameter $\eta = 0.5$, where Γ_1 is the electronic energy-relaxation rate.

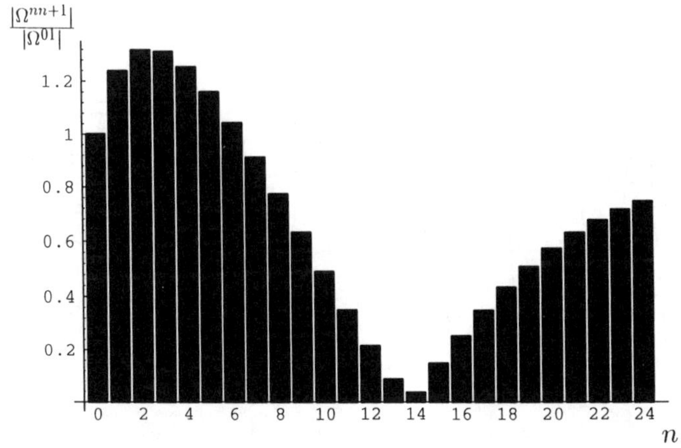

QWB5 Fig. 2. Nonlinear Rabi frequencies Ω^{nn+1} for $\eta = 0.5$.

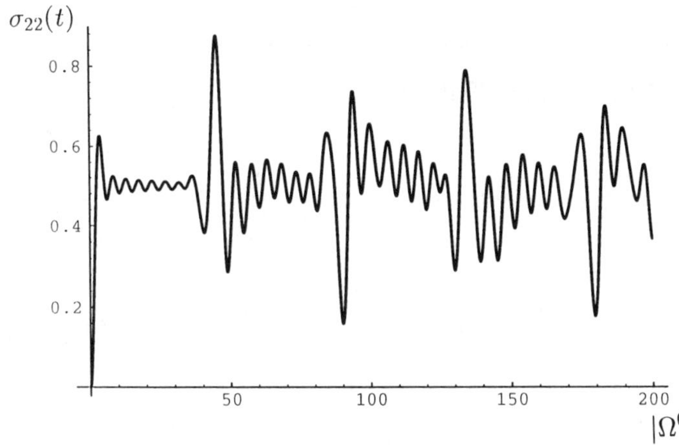

QWB5 Fig. 3. Excited-state occupation σ_{22} in the one-quantum NJCM. The vibration is initially in a coherent state, with the mean number of quanta being 10 and $\eta = 0.5$.

1. D. Meschede, H. Walther, and G. Müller, Phys. Rev. Lett. **54**, 551 (1985).
2. C. A. Blockley, D. F. Walls, and H. Risken, Europhys. Lett. **17**, 509 (1992).
3. R. L. de Matos Filho and W. Vogel, Phys. Rev. A **50**, R1988 (1994).
4. W. Vogel, D.-G. Welsch, and L. Leine, J. Opt. Soc. Am B **4**, 1633 (1987).

QWB6 **9:45 am**

Observation of large continuous-wave two-photon optical amplification

Hope M. Concannon, Daniel J. Gauthier, *Department of Physics, Duke University, P. O. Box 90305, Durham, North Carolina 27708*

Two-photon lasers have intrigued researchers for years because the dynamical behavior of this new class of highly nonlinear quantum oscillator is very different from that of normal one-photon lasers. Experimental tests of the numerous, often conflicting, predictions regarding their behavior have been hindered by the lack of suitable gain media until the recent development of the dressed-atom two-photon gain medium. Although work with the dressed-atom gain medium was remarkably successful, culminating in the first observation of two-photon optical lasing,[1] the small two-photon gain in this system (~0.1%) limits its practicality.

We have observed significantly larger two-photon gain in a new type of laser-driven two-photon optical amplifying medium based on the two-photon Raman-scattering process shown in Fig. 1. Most notably, it is relatively insensitive to broadening mechanisms, so that the gain feature remains spectrally resolved even in a vapor. To test our predictions, we perform pump–probe spectroscopy in a potassium vapor (length 7.5 cm, vapor pressure ~5.5 × 10^{-4} Torr). The pump laser (frequency ω_P), an 850-mW actively stabilized Ti:sapphire laser, is linearly polarized, has a ~1-mm beam spot size as it passes through the vapor cell, and is tuned ~800 MHz below the 770-nm $4S_{1/2}$ ($F = 1$) → $4P_{1/2}$ ($F = 1$) transition. The pump laser serves two purposes: (1) it creates the necessary inversion between the ground-state hyperfine levels (splitting $\Delta_{hfs} = 462$ MHz) by optically pumping atoms from the $F = 2 \rightarrow F = 1$ levels as they move into the pump beam, and (2) it acts as the pump field for the two-photon Raman process. The probe laser (frequency ω_R), a grating-stabilized diode laser, is polarized orthogonal to the pump laser, crosses the pump laser at an ~12 mrad angle, and has an ~500-μm beam spot size. Significantly smaller crossing angles (<6 mrad) could not be used because of the appearance of four-wave-mixing processes. We observe large (>5000%) amplification of the probe laser due to one-photon Raman scattering[2] when $\omega_R \approx \omega_P - \Delta_{hfs}$ and its power is low (< 10 μW). For

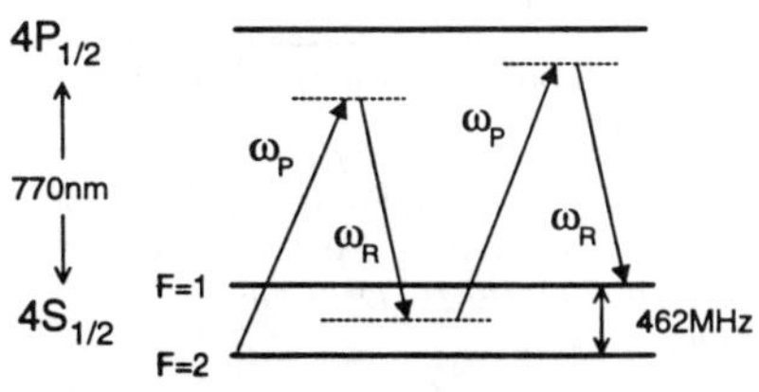

QWB6 Fig. 1. Two-photon Raman scattering as a mechanism for two-photon gain. Two-photon gain occurs at $\omega_R = \omega_P - \Delta_{hfs}/2$ when the pump laser (frequency ω_P) is tuned to the low-frequency side of the potassium resonance.

Wednesday, May 24

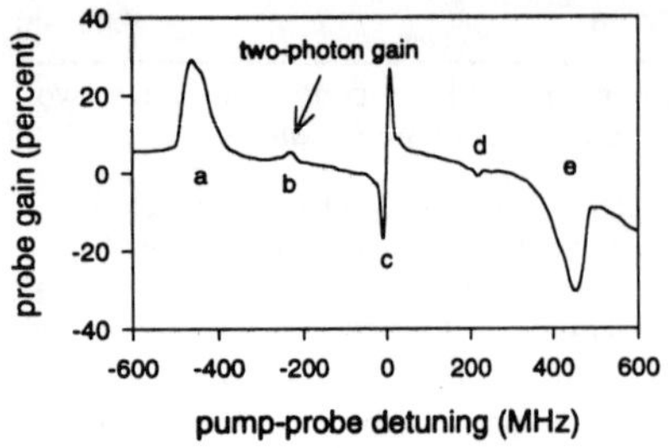

QWB6 Fig. 2. Gain experienced by a strong probe beam. The two-photon gain is ~3%. We identify the spectral features as (a) one-photon Raman gain at $\omega_R - \omega_P - \Delta_{hfs}$, (b) two-photon Raman gain $\omega_R = \omega_P - \Delta_{hfs}/2$, (c) stimulated Raleigh gain near ω_P, (d) two-photon Raman absorption at $\omega_R = \omega_P + \Delta_{hfs}/2$, and (e) one-photon Raman absorption at $\omega_R = \omega_P + \Delta_{hfs}$.

higher powers (~10 mW), the one-photon Raman gain decreases dramatically and a new gain feature appears when $\omega_R \approx \omega_P - \Delta_{hfs}/2$, as shown in Fig. 2. We attribute this probe-dependent feature to the two-photon Raman scattering process. We observe ~3% two-photon amplification, which should be sufficient to observe two-photon lasing. The left edge of the two-photon gain peak is broadened because of small one-photon Raman gain from ^{41}K (Δ_{hfs} = 255 MHz, natural abundance 6%), and the background amplification is due to broadened one-photon dressed-state gain. We will discuss our progress toward achieving two-photon Raman lasing and the ability of a buffer gas to suppress the background caused by the one-photon dressed-state gain process.

1. D. J. Gauthier, Q. Wu, S. E. Morin, and T. W. Mossberg, Phys. Rev. Lett. **68**, 464 (1992).
2. M. Poelker and P. Kumar, Opt. Lett. **17**, 399 (1992).

QWC 8:00 am
Hyatt C/D

Polaritons and Phonons

Jorn M. Hvam, *Technical University of Denmark, Presider*

QWC1 (Invited) 8:00 am

Coherent phonon polaritons as a probe of anharmonic phonons

H. J. Bakker, S. Hunsche, H. Kurz, *Institut für Halbleitertechnik II, RWTH Aachen, Sommerfeldstrasse 24, D-52056 Aachen, Germany*

We investigate the low-frequency dielectric response of the ferroelectrics $LiNbO_3$ and $LiTaO_3$ via the impulsive excitation and phase-sensitive detection of coherent phonon polaritons. Phonon polaritons are mixed light-polarization states that arise from the coupling of far-infrared light with phonon resonances and can be used to investigate these resonances. For ferroelectric crystals there are often more optical-phonon resonances observed than there are normal vibrations in the unit cell.[1,2] The additional resonances are usually explained from the anharmonic coupling of different optical phonons (difference bands, combination bands).[1,2] However, it was shown recently that the additional resonances can also result from transitions in the microscopic anharmonic potential of the phonon itself.[3,4]

We excite phonon polaritons using two intense femtosecond laser pulses. The wave vector of the polaritons is determined by the difference of the wave vectors of the two pulses and can be tuned by varying the angle between the beams. The polariton frequencies that correspond with this wave vector, according to the polariton dispersion, will be generated with the highest efficiency since the generation process is phase matched for these frequencies. The polaritons are detected phase-sensitively with a time-delayed third pulse.

We find that at most wave vectors only a single polariton frequency is generated. However, in several wave-vector intervals strong beating phenomena (polariton beats) are observed (Fig. 1), indicating that two or more different polariton frequencies are coherently excited at one wave vector. This observation indicates the presence of low-frequency phonon resonances that lead to avoided crossings in the polariton dispersion (Fig. 2). These low-frequency resonances have not been found previously and cannot be assigned to transitions between different normal lattice vibrations, because the observed frequencies do not fit the energy differences of the optical phonons of $LiNbO_3$ and $LiTaO_3$.[1,2]

For both $LiNbO_3$ and $LiTaO_3$ we find that the polariton dispersion and damping

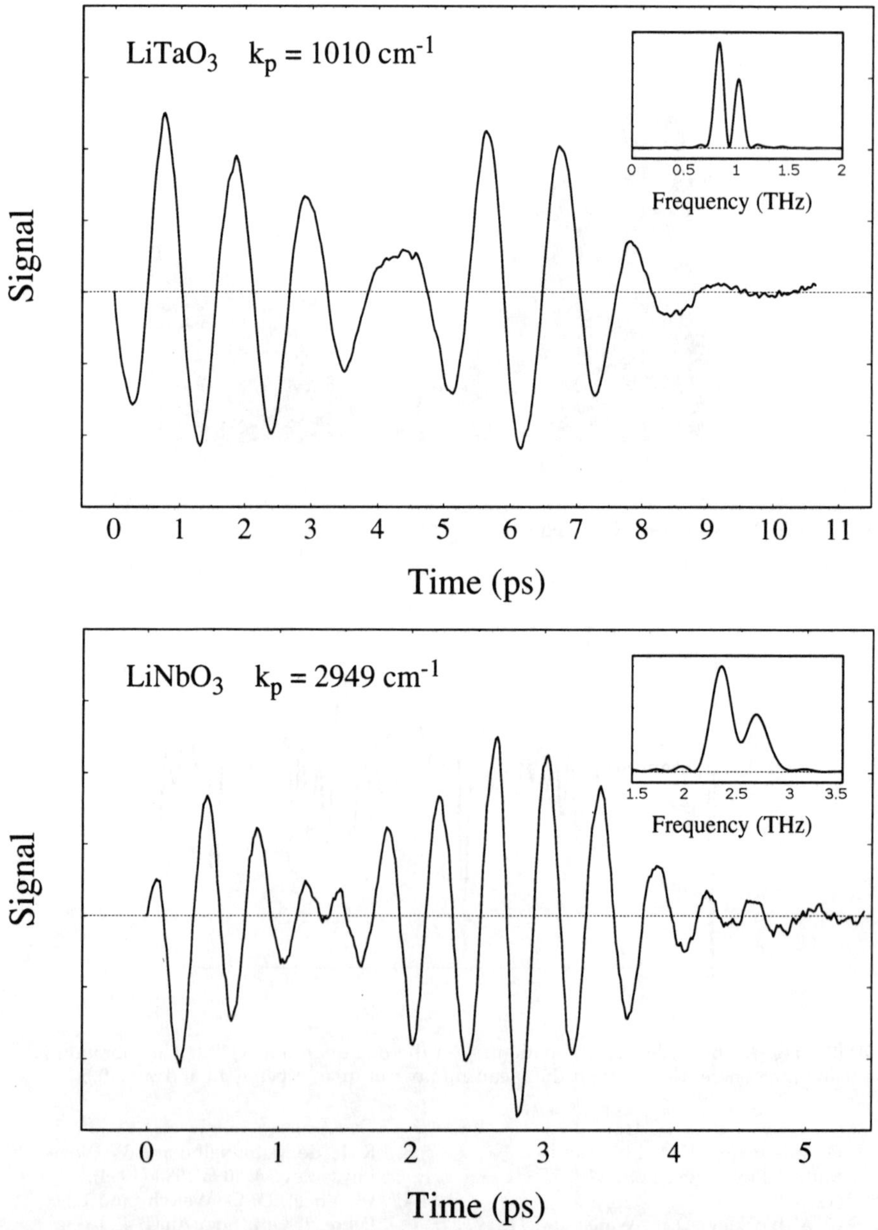

QWC1 Fig. 1. Time-resolved measurements of phonon polaritons in $LiTaO_3$ and $LiNbO_3$. In the inset the Fourier transform of the measurements are shown.

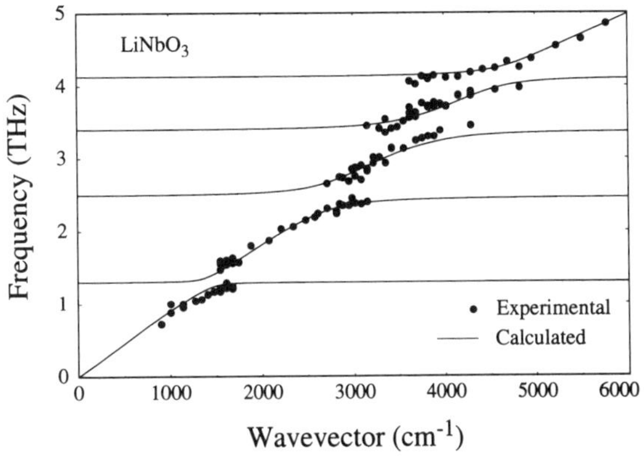

QWC1 Fig. 2. Phonon-polariton dispersion of $LiNbO_3$ at 300 K. The solid curves are calculated from the transitions between the vibrational levels in the potential presented in Fig. 3.

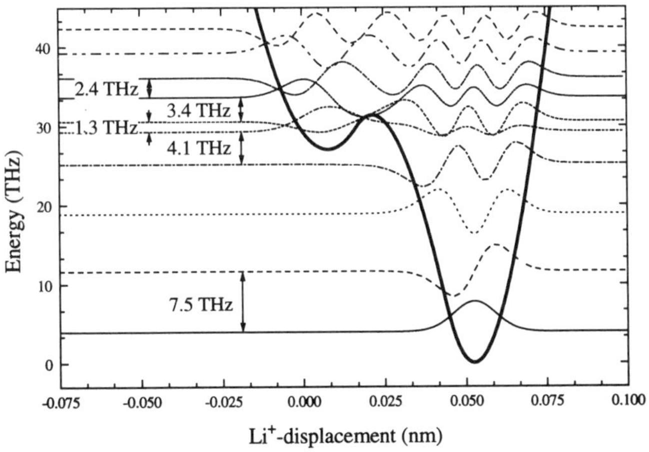

QWC1 Fig. 3. The two lowest wells of the triple-well potential and the 10 lowest-energy vibrational wavefunctions of the ferroelectric phonon of $LiNbO_3$ as a function of the displacement of the Li^+ ion along the optical axis.

can only be fitted with an extremely anharmonic vibrational potential for the ferroelectric phonon that consists of three wells (Fig. 3). The newly discovered low-frequency resonances can all be assigned to transitions between vibrational levels that are delocalized over the two lowest wells of the triple-well potential. The triple-well potential also provides an accurate description for the temperature dependence of the spontaneous polarization and the static dielectric constant of $LiNbO_3$ and $LiTaO_3$. Thereby a consistent description of the low-frequency dielectric response and the ferrolectric phase transition of these crystals is obtained.

1. W. D. Johnston, Jr. and I. P. Kaminov, Phys. Rev. **168**, 1045 (1968).
2. A. S. Barker, Jr. and R. Loudon, Phys. Rev. **158**, 433 (1967).
3. C. M. Foster, Z. Li, M. Grimsditch, S.-K. Chan, and D. J. Lam, Phys. Rev. B **48**, 10160 (1993).
4. H. J. Bakker, S. Hunsche, and H. Kurz, Phys. Rev. Lett. **69**, 2823 (1992); H. J. Bakker, S. Hunsche, and H. Kurz, Phys. Rev. B **50**, 914 (1993).

QWC2 **8:30 am**

Generation and observation of coherent surface optical phonons

Li Xu, Yu-Ming Chang, Harry W. K. Tom, *Department of Physics, University of California–Riverside, Riverside, California 92521*

We report what we believe to be the first observation of coherent surface optical phonons generated by impulsive stimulation and detected by time-resolving second-harmonic generation (SHG). Several groups have reported the observation of coherent bulk phonons in various media by using short optical pulses to impulsively excite the vibrational mode.[1] In contrast to linear reflectivity, SHG in reflection geometry is known to have decided surface sensitivity.[2] Therefore, the observation of coherent surface phonons by using time-resolved SHG is possible. Indeed, we observe an oscillation in time-resolved SHG at 8.36 THz from GaAs (100) excited with 30-fs, 855-nm pump pulses. Because this oscillation frequency is in the forbidden gap for the bulk LO-phonon–plasmon coupling mode and is in good agreement with the oscillation frequency of the Fuchs–Kliewer surface phonon of GaAs,[3] we assign it to the surface optical phonon.

Laser pump–probe experiments were performed on the surface of a GaAs (100) wafer that was n-doped with 3×10^{18} Si/cm^3. The sample was held in air and had a native oxide. The depletion field is estimated to be about 400 kV/cm. In general, the SH comes from the surface and the bulk. Figure 1 shows the orientation dependence of the SHG of the probe beam (p-polarized) from the sample. It is important to note that the SHG is very sensitive to the optical wavelength of the probe.

In Fig. 2 we show the time-resolved SHG results with various injected-carrier densities. The 855-nm s-polarized pump and the

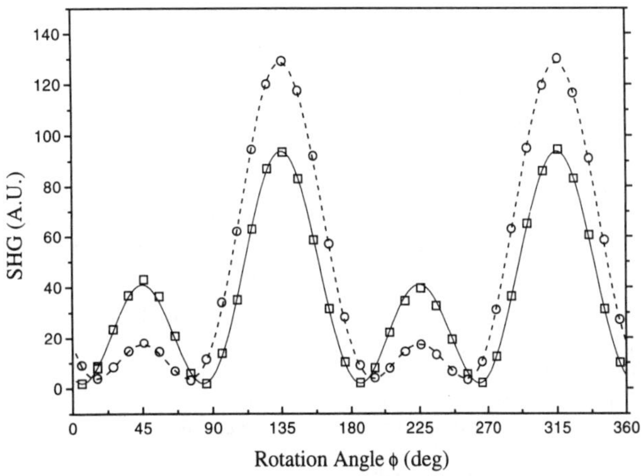

QWC2 Fig. 1. Orientation dependence of SHG (p-polarized) generated by the probe pulse (p-polarized). The dashed and solid curves correspond to SHG without and with the pump pulse, respectively. The rotation angle φ is defined as the angle between the polarization of the probe beam and the (010) crystal axis of GaAs. Although the bulk nominally has $\bar{4}3m$ symmetry, which gives fourfold symmetry of the orientation dependence of SHG, both the surface and the near-surface bulk have mm2 symmetry because the surface and the depletion field break symmetry normal to the surface. Therefore, twofold symmetry is observed and the pump influence on the SHG can be explained by the screening of the depletion field as shown.

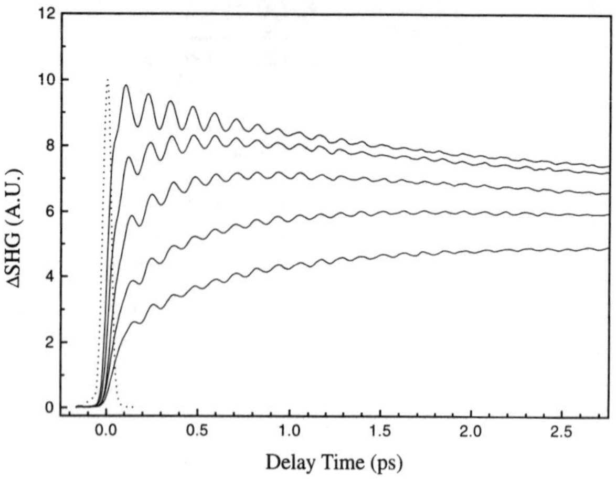

QWC2 Fig. 2. Change in SHG by the probe pulse from GaAs (100) versus time delay with respect to the pump pulse under various carrier injection levels. The zero of time delay is measured by the cross correlation of the pump and the probe on the sample (dotted curve). The change in SHG increases with the injected carrier density. The injected carrier densities are $3 \times 10^{18}/cm^3$, $1 \times 10^{18}/cm^3$, $6 \times 10^{17}/cm^3$, $3 \times 10^{17}/cm^3$, and $1.5 \times 10^{17}/cm^3$, respectively.

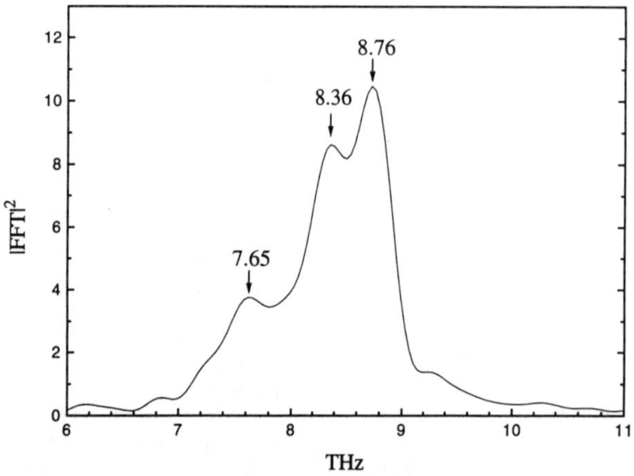

QWC2 Fig. 3. Fourier transform of time-resolved SHG under $3 \times 10^{18}/cm^3$ injected carrier density. Three different oscillation frequencies are clearly indicated.

p-polarized probe were incident at 45° from the surface normal. After the rapid increase of SHG due to the depletion-field screening, which returns the bulk symmetry to $\bar{4}3m$, there is a slow decrease and significant oscillations caused by the generation of coherent phonons. The Fourier transform of the oscillation part of the time-resolved SHG is shown in Fig. 3. It indicates that the oscillation has three frequencies: 7.65, 8.36, and 8.76 THz. The oscillation at 8.76 THz is assigned to the bulk Γ_{15}–LO-phonon mode, in agreement with earlier work[4] using linear reflectivity. The oscillation at 7.65 THz is tentatively assigned to the plasmon–LO-phonon coupling mode. The origin of this oscillation is not understood at this time. Here we assign the oscillation at 8.36 THz to the Fuchs–Kliewer surface optical phonon in the GaAs (100)–native-oxide interface. For the clean GaAs (100) surface in ultrahigh vacuum, the surface-phonon frequency was measured to be 8.7 ± 0.12 THz by electron-energy-loss spectroscopy.[3] The difference in frequency can be ascribed to the presence of the native oxide, which has a dielectric constant different from that of vacuum. By following Ref. 5, the frequency is predicted to be $\nu_s = \nu_{TO}\{[\varepsilon_{GaAs}(0) + \varepsilon_{ox}(\infty)]/[\varepsilon_{GaAs}(\infty) + \varepsilon_{ox}(\infty)]\}^{1/2}$, where ν_{TO} is the bulk TO phonon frequency, and the dielectric constants ε for GaAs and the native oxide are at dc and optical frequencies. We obtain $\nu_s = 8.36$ THz for $\varepsilon_{ox}(\infty) = 12.75$, which is a reasonable value.

In conclusion, we have observed a coherent surface phonon at the GaAs (100)–native-oxide interface with time-resolved surface SHG. This technique may be applied to a variety of buried interfaces and to surfaces of interest in surface chemistry.

1. W. Kutt, W. Albrecht, and H. Kurz, IEEE J. Quantum Electron. **QE-28**, 2434 (1992), and references therein.
2. C. Yamada and T. Kimura, Phys. Rev. Lett. **70**, 2344 (1993).
3. Z. J. Gray-Grychowski *et al.*, Surface Sci. **186**, 482 (1987).
4. G. C. Cho, W. Kutt, and H. Kurz, Phys. Rev. Lett. **65**, 764 (1990).
5. H. Ibach and D. L. Mills, *Electron Energy Loss Spectroscopy and Surface Vibrations* (Academic, New York 1982).

QWC3 **8:45 am**

Measurement of the resonant-polaron effect in the Reststrahlen band of GaAs:Si by using far-infrared two-photon excitation

P. C. M. Planken, P. C. van Son, J. N. Hovenier, T. O. Klaassen, W. Th. Wenckebach, P. W. Barmby,* J. L. Dunn,* C. A. Bates,* C. T. Foxon,*
Department of Applied Physics, Delft University of Technology, P.O. Box 5046, 2600 GA Delft, The Netherlands

In polar materials such as GaAs:Si shallow-donor transitions become strongly coupled to LO phonons when the transition energy approaches the LO-phonon energy. This resonant polaron effect manifests itself around the LO-phonon energy (296 cm^{-1}) as an avoided crossing of the electronic-transition energy versus magnetic field.[1,2] However, in the region between the TO and LO phonons, the Reststrahlen band, photoconductivity measurements are extremely difficult because of the very small penetration depth of light in this region.[2] In addition, the strongly varying dielectric function of the GaAs around the TO/LO-phonon frequencies gives rise to artifacts in the weak photoconductivity spectra.[2]

Here, we present the results of photoconductivity measurements of the resonant electron–phonon interaction in the middle of the Reststrahlen band obtained by using two-photon excitation with intense picosecond pulses with frequency around 143 cm^{-1} (70 μm). We use two photons, rather than a single photon, for the excitation of the resonant polaron to avoid the difficulties encountered in direct single-photon excitation in the Reststrahlen band mentioned above.

The sample is a 10-μm-thick Si-doped GaAs epitaxial layer on a 400-μm semi-insulating GaAs substrate grown by MBE at Philips Research Laboratory, Redhill, U.K. The electronic levels of the (Si) shallow donor can be tuned by the application of a magnetic field up to 10.3 T. Intense tunable picosecond pulses with a frequency of around 143 cm^{-1} from the Dutch Free-Electron Laser for Infrared EXperiments (FELIX) at the FOM Institute "Rijnhuizen" are weakly focused onto the sample, which is kept at 8 K. Electrons excited to the $3d^{+2}$ state through the electric-dipole allowed two-photon transition out of the $1s_0$ ground state decay to the conduction band and give rise to an increase in the conductivity. In Fig. 1 we plot a typical measurement of this photoconductivity versus excitation wavelength at a fixed magnetic field. The small peak around 284 cm^{-1} in the middle of the Reststrahlen band corresponds to the $3d^{+2}$ transition. It is superimposed on a broad, slowly varying background from nonresonant single-photon excitation of other electronic states. Figure 2 is a plot of the energy-peak position of the $3d^{+2}$ transition thus obtained as a function of the magnetic-field strength. The figure clearly shows the avoided crossing around the LO-phonon energy, for which the coupling between the $3d^{+2}$ state and the LO phonon is strongest. Note that the data between 267 and 296 cm^{-1} are extremely difficult to obtain with single-photon excitation be-

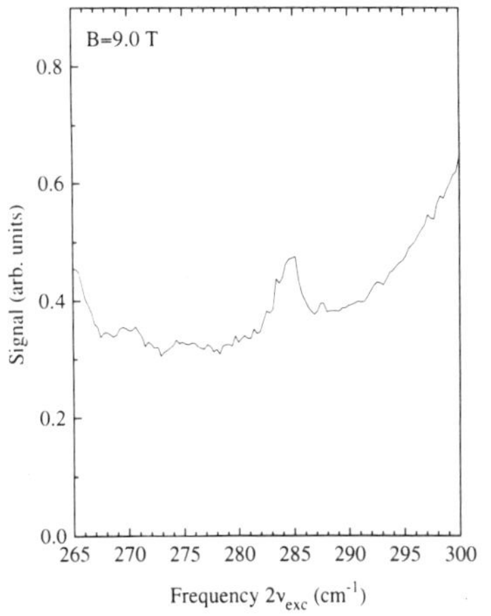

QWC3 Fig. 1. Measured photoconductivity signal versus twice the excitation frequency for a magnetic field of 9.0 T. The small peak around 284 cm^{-1} can be identified as the $3d^{+2}$ transition excited with two photons.

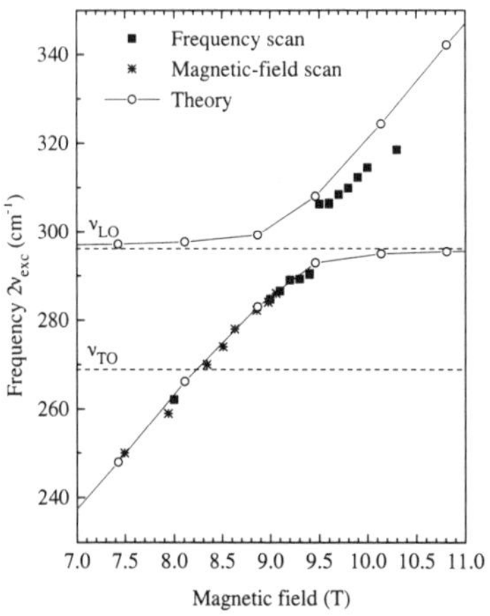

QWC3 Fig. 2. Measured and calculated $3d^{+2}$ peak position versus magnetic field. The plot contains data points obtained from both magnetic-field scans at a fixed laser frequency (stars) and frequency scans for a fixed magnetic-field strength (squares). The area in between the LO and TO phonon frequencies (dashed lines) is the Reststrahlen band. The lines connecting the calculated points are guides to the eye.

cause of their position in the middle of the Reststrahlen band.

In Fig. 2 we also plot a theoretical calculation that includes band-nonparabolicity effects. The calculation is in excellent agreement with the measurements for the lower branch of the avoided crossing, but it is less so for the upper branch. The latter is probably due to the occurrence of a second avoided crossing around 330 cm^{-1}, where the $3d^{+2}$ should cross the ($2p^{-1}$ + 1LO-phonon), not included in the present calculation.

**Department of Physics, University of Nottingham, University Park, Nottingham NG7 2RD, U.K.*

1. J.-P. Cheng., B. D. McCombe, J. M. Shi, F. M. Peeters, and J. T. Devreese, Phys. Rev B **48**, 7910 (1993).
2. A. J. van der Sluijs, K. K. Geerinck, T. O. Klaassen, W. Th. Wenckebach, and C.T. Foxon, J. Appl. Phys. **75**, 3698 (1994).

QWC4 **9:00 am**

Far-infrared time-resolved measurement of the TO-phonon free-induction decay in polycrystalline GaAs

P. C. M. Planken, H. P. M. Pellemans, P. C. van Son, T. O. Klaassen, W. Th. Wenckebach, *Department of Applied Physics, Delft University of Technology, P.O. Box 5046, 2600 GA Delft, The Netherlands*

Transverse-optical (TO) phonons and lattice vibrations in polar materials such as GaAs show strong infrared absorption and reflection around the resonance frequency. They form an example of an (an)harmonic ladder system in the solid state, in which the possibility of climbing this ladder in a sequential manner by using optical excitation accounts at least partially for the large absorption coefficients. In the past, time-resolved studies of TO phonons (and phonon-polaritons) have been mostly limited to Raman studies with visible-light pulses[1] because of the difficulty of obtaining ultrashort pulses in the far-infrared wavelength region.

Here, we present the first results of time-resolved measurements on TO phonons in GaAs by using direct far-infrared excitation with pulses of ≤1-ps duration from the Dutch Free-Electron Laser for Infrared EXperiments (FELIX) at the FOM-Institute "Rijnhuizen." In particular, we report on the time-resolved measurement of the free-induction-decay emission from GaAs TO phonons around 270 cm^{-1} by using a cross-correlation setup.

The sample is a 200-nm-thick layer of polycrystalline GaAs on a high-resistivity Si substrate. It has a linear transmission of 30% at the TO-phonon frequency. The experimental setup is shown in Fig. 1. Laser pulses from FELIX tuned around 270 cm^{-1} are split in two by using a Mylar beam splitter. One pulse undergoes a variable delay, while the other one is sent through the GaAs sample. The two pulses are then recombined on a second Mylar beam splitter. The whole setup is enclosed in a vacuum chamber to reduce strong water-vapor absorption. In Fig. 2(b) we plot the transmission of the second beam splitter as a function of the delay between the two pulses, with the sample removed from the setup. The figure shows a signal that oscillates with a period given by the laser frequency. It represents the field autocorrelation of the pulses, from which we obtain a pulse duration of approximately 1 ps, for bandwidth-limited pulses.[2] In Fig. 2(a) we plot the transmission of the second beam splitter versus delay of the two pulses when one laser pulse propagates through the sample and is resonant with the TO phonon at 270 cm^{-1}. The signal looks markedly different from the result in Fig. 2(b) because oscillations in the signal are still present for a time separation for which no oscillations are observed in the autocorrelate. It is clear that the TO phonon continues to emit light for a few picoseconds after the main (unperturbed) part of the pulse has propagated through the sample. This is the so-called free-induction emission. The free-induction component has a π phase shift with respect to the electric field of the laser pulse. As expected, this shows up as a node around 1 ps in the cross correlation measurement in Fig. 2.

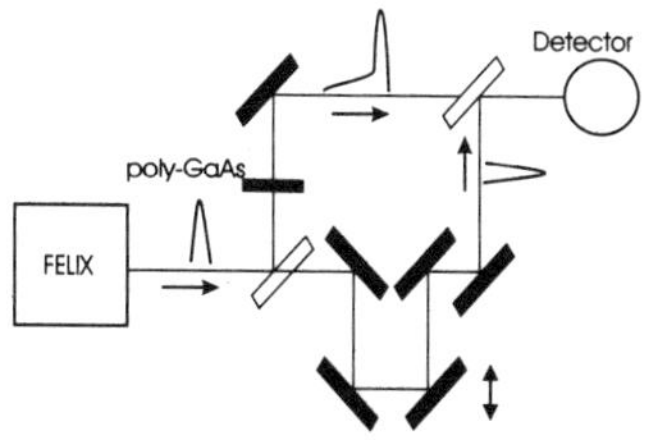

QWC4 Fig. 1. Experimental setup for measuring the free-induction-decay emission from TO phonons in polycrystalline GaAs by using electric-field cross-correlation.

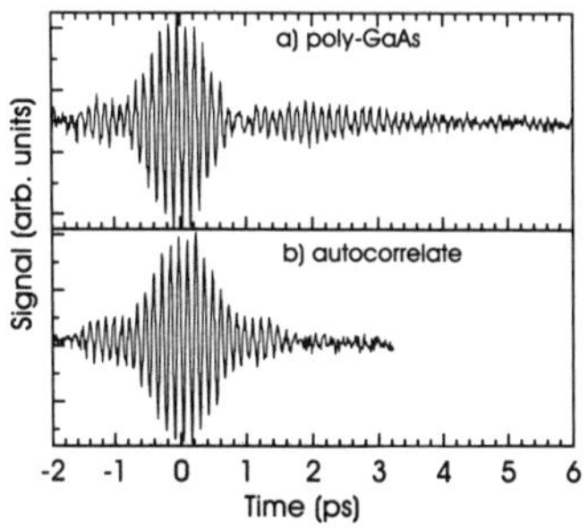

QWC4 Fig. 2. (a) Measured TO-phonon free-induction-decay emission obtained by using electric-field cross correlation. (b) Measured electric-field autocorrelation of the FELIX pulses. We deduce a pulse duration of approximately 1 ps from the autocorrelate. Both (a) and (b) were measured with the laser frequency at 267 cm^{-1}, the TO-phonon frequency in GaAs.

We plan to increase the far-infrared field strength in the sample and hope to reach a regime in which we can study nonlinear effects, such as multilevel coherences and Rabi oscillations, in this ladder system.

1. D. H. Auston and M. C. Nuss, IEEE J. Quantum Electron. **24**, 184 (1988); T. P. Dougherty, G. P. Wiederrecht, and K. A. Nelson, Ferroelectrics **120**, 79 (1991).
2. R. J. Bakker, C. A. J. van der Geer, D. A. Jaroszynsky, A. F. G. van der Meer, D. Oepts, P. W. van Amersfoort, V. Anderegg, and P. C. van Son, Nucl. Instrum. Methods **A331**, 79 (1993).

QWC5 **9:15 am**

Terahertz emission from coherent phonons in tellurium

H. Auer, T. Dekorsy, C. Waschke, H. G. Roskos, H. J. Bakker, H. Kurz, V. Wagner, P. Grosse,* *Semiconductor Technology II, Rhine–Westphalia Technical University, D-52056 Aachen, Germany*

Coherent optical phonons have been observed in several semiconductor materials in femtosecond time-resolved reflectivity experiments.[1,2] Here, we report on the observation of terahertz radiation emitted from coherently exited infrared-active phonons in single-crystal tellurium.[3]

The IR activity of phonons in the single-element semiconductor Te arises from the strong electronic polarizability and the lack of inversion symmetry. Te crystallizes in a hexagonal lattice (space group D_3^4 or D_3^6) with three atoms per unit cell arranged in a helix along the c axis. The lattice vibrations consist of a fully symmetrical, only-Raman-active A_1 mode (3.6 THz), two degenerate Raman- and IR-active E modes ($E'_{TO/LO}$: 2.76/3.09 THz, $E''_{TO/LO}$: 4.22/4.26 THz), and one only-IR-active A_2 mode ($A_{2,TO/LO}$: 2.6/2.82 THz). The internal polarization is either perpendicular (E modes) or parallel (A_2 mode) to the c axis. The laser source is a Kerr-lens mode-locked Ti:sapphire laser. The terahertz emission is detected in a standard setup.[4] The terahertz experiments are complemented by optical pump–probe experiments, which verify the excitation of the coherent phonon modes with frequencies different from those of the previously observed A_1 mode.[2]

Figure 1 depicts the electric field emitted from the $c_\perp$ surface. The signal consists of a strong initial emission followed by a periodic modulation that persists over several picoseconds. The strong peak results from the ultrafast buildup of a strong photo-Dember field.[5] The Dember field is driven by the strong carrier gradient at the surface and the difference between the electron and the hole diffusion coefficients. Because of the high excess energy of electrons and holes (Te band gap, 0.33 eV; excitation energy, 1.7 eV), large differences in the transient electron and hole temperatures are established. The diffusion coefficients D_i and the mobilities μ_i are related to the temperature T_i of the carrier distributions through the Einstein relation $D_i(\tau)/\mu_i(\tau) \sim kT_i(\tau)$, where the index i is for electrons and holes. The screening of depletion-layer fields as the source of the radiation[4] can be neglected because the charged surface states have a low density and the band gap is small. It should be mentioned that the amplitude of the initial terahertz signal due to the Dember field is only weaker by a factor of 5 than the emission from a polar semiconductor with strong surface fields (e.g., InP) when it is excited with equal laser intensity.

Figure 2 depicts the amplitude of the Fourier transform of the time-domain signal. The frequency spectrum reveals a broad peak at approximately 500 GHz. For high frequencies, the signal decreases to zero at 2.62 THz before it reaches a maximum at about 2.9 THz. For clarity, this part of the spectrum is also shown enlarged by a factor

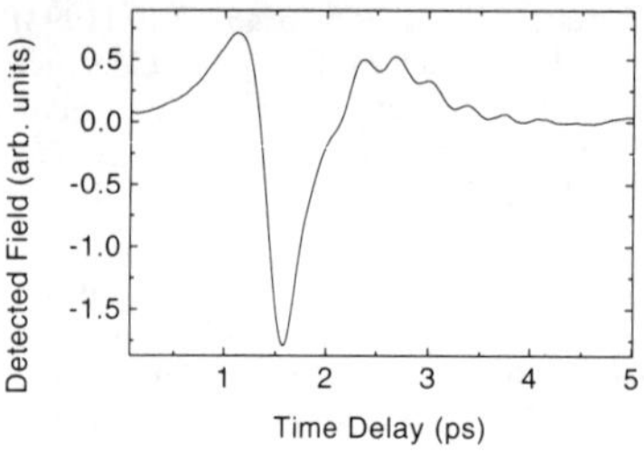

QWC5 Fig. 1. Detected electric field emitted from the $c_\perp$ surface of Te.

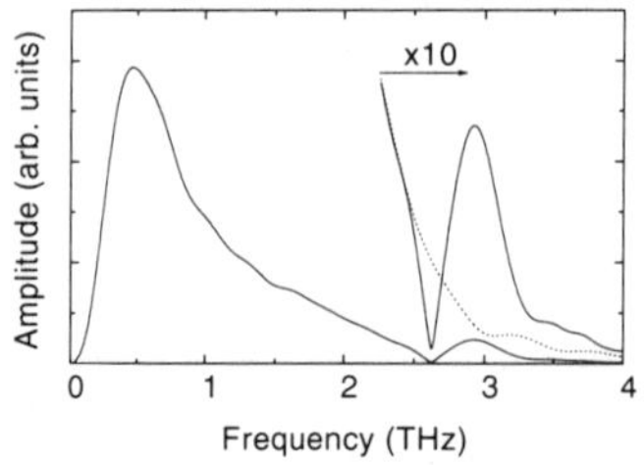

QWC5 Fig. 2. Amplitude of the Fourier transform of the time-domain signal. The high-frequency part depicted is enlarged by a factor of 10. Dashed curve gives the high-frequency part of the antenna response determined from the emission of a broadband emitter (InP, also ×10).

of 10. In addition, the high-frequency spectrum of a broadband emitting surface (InP, also ×10) is shown for comparison, representing the antenna response. The emission peak at 500 GHz results from the carrier-transport dynamics leading to the buildup of the Dember field. The high-frequency features correspond to the oscillatory part of the time-domain signal. A strong deviation of the terahertz emission from Te as compared to that of the broadband emitter is clearly observed in this frequency range. The origin of the oscillatory contribution in this frequency range is identified with the A_2 phonon, which has a polarization perpendicular to the exicted surface. Experiments performed with a wire-grid polarizer placed into the beam path of the terahertz radiation show that the oscilllatory contribution is only present in the p-polarized component of the radiation. Therefore, the source of the oscillatory contribution is a dipole oscillating perpendicular to the excited surface, i.e., the coherent A_2 mode. The coherent excitation of the not-Raman-active A_2 mode is explained by the ultrafast *buildup* of the photo-Dember field. This is a new excitation mechanism for coherent phonons, which has to be distinguished from the excitation mechanism based on the *screening* of built-in fields.[1]

We discuss selection rules for the excitation of coherent phonons in Te by comparing the emission from surfaces at various orientations. Furthermore, we discuss the relevance of propagation and outcoupling effects on the detected terahertz radiation.

This work is supported by the German Research Association.

*Institute of Physics, Rhine–Westphalia Technical University, D-52056 Aachen, Germany

1. G. C. Cho, W. Kütt, and H. Kurz, Phys. Rev. Lett. **65**, 764 (1990); W. A. Kütt, W. Albrecht, and H. Kurz, IEEE J. Quantum Electron. **QE-28**, 2434 (1992).
2. T. K. Cheng, *et al.*, Appl. Phys. Lett. **57**, 1004 (1990); T. K. Cheng, *et al.*, Appl. Phys. Lett. **59**, 1923 (1991); H. J. Zeiger *et al.*, Phys. Rev. B **45**, 768 (1992).
3. T. Dekorsy *et al.*, Phys. Rev. Lett., in press.
4. X.-C. Zhang and D. H. Auston, J. Appl. Phys. **71**, 326 (1992); H. G. Roskos, Lithuanian J. Phys. **34**, 175 (1994).
5. H. Dember, Phys. Z. **32**, 554 (1931).

QWC6 **9:30 am**

Nondiffusive excitonic transport in GaAs: evidence for polariton propagation

A. C. Schaefer, J. Erland,* D. G. Steel, *Harrison M. Randall Laboratory of Physics, University of Michigan, Ann Arbor, Michigan 48109*

Optical-grating experiments have been shown to be a powerful probe of carrier diffusion.[1] In these experiments two beams interfere in a material, forming a grating that is probed by a third beam. The decay of the optically created population grating occurs through many channels, such as recombination γ_{rec} and spatial diffusion. For diffusion independent of position and concentration, the decay due to spatial diffusion is governed by Fick's Law and the diffusion coefficient D is extracted by measuring the grating decay Γ versus the grating spacing Λ, with $\Gamma = \gamma_{rec} + (4\pi^2 D/\Lambda^2)$.

In semiconductors, the grating method of measuring diffusion has been used with much success. However, it was demonstrated earlier by our group[2] that in high-quality GaAs at low temperatures (< 30 K) the grating decay is not quadratic with the inverse grating spacing, indicating nondiffusive behavior. The explanation of this behavior was resolved. Here we present a systematic study of exciton motion in GaAs and show that polariton propagation accounts for the nondiffusive behavior. This model is supported by preliminary experimental data showing that purely diffusive behavior is characterized by a dephasing rate greater than a critical value.

This study was conducted by using the cw backward FWM geometry, as shown in the inset in Fig. 1. The grating spacing is determined by the angle between the two front beams, and the decay rate is measured as the FWHM of the Lorentzian line shape obtained as a function of $\delta = \omega_1 - \omega_2$. Figure 1 shows the decay of the grating for lh (1s) excitons as a function of $1/\Lambda$, which is clearly subquadratic. A model that takes into account propagation of polaritons over a characteristic distance (α, where $\alpha \approx \Lambda$) between scattering events can account for the subquadratic behavior. This model, which is a variation of one developed by Wong and Kenkre,[3] gives exponential grating decay and provides a fit of the data, as shown in Fig. 1. As expected, in the limit $\alpha \to 0$ the model describes diffusive motion.

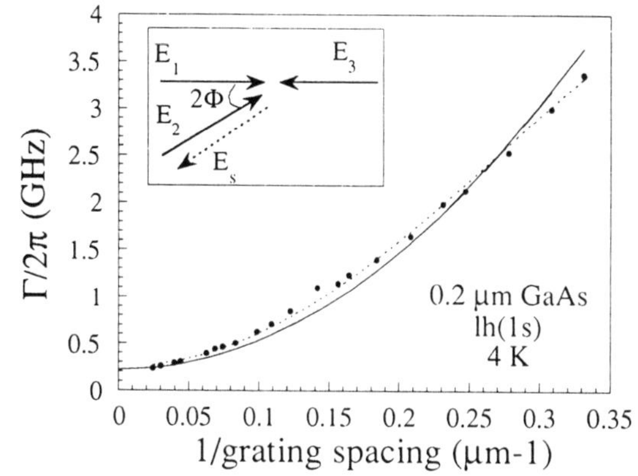

QWC6 Fig. 1. Lorentzian linewidth $\Gamma/2\pi$ as a function of the inverse grating spacing at lh exciton. Solid curve is a quadratic fit, and dashed curve is a fit to model mentioned in text. Data are taken at 4.5 K with an excitation density of 8×10^{14} cm^{-3}. Inset shows FWM geometry.

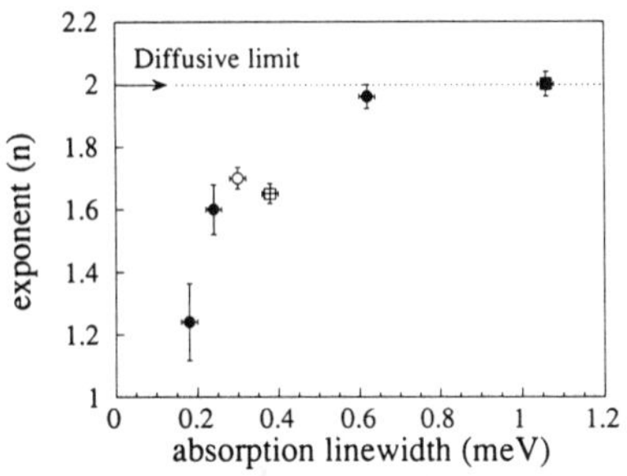

QWC6 Fig. 2. Exponent n in a power-law fit of the grating-decay data $[\Gamma_D = (1/\Lambda)^n]$ versus the absorption linewidth (FWHM): filled circles, sample A, at lh(1s), linewidth varied by varying temperature; solid square, sample A, at hh(1s); open circle, sample B, at lh(1s); open square, sample C, at hh(1s).

Support for this explanation is given by the dependence of the exciton motion on the dephasing rate (given by the homogeneous absorption linewidth), because the dephasing time determines the polariton propagation distance α. For dephasing rates less than a critical value, the motion is clearly nondiffusive. For dephasing greater than this value, the motion is diffusive. This is shown in Fig. 2. Here, a power-law fit of the grating decay $\Gamma = A + (B/\Lambda^n)$ (used only for illustration) indicates whether the motion is purely diffusive ($n = 2$) or nondiffusive ($n < 2$). Figure 2 includes data from several samples (where the same qualitative behavior was found) and for both hh and lh excitons.

Another possible explanation for motion that does not obey Fick's Law is a density-dependent diffusion coefficient. Density-dependent diffusion has been reported in GaAs[4]. However, this observation was found by excitation of hot carriers and with a subsequent change in recombination rate. Under resonant excitonic excitation, we find only a 10% variation of D over an order of magnitude of density. Also, any model for the motion must predict grating decays exponential in time, because the Lorentzian FWM line shapes are measured.

This work was supported by the U.S. Army Research Office and the U.S. Air Force Office of Scientific Research. J. E. acknowledges the Danish Research Academy. D. Citrin is acknowledged for helpful discussions.

*Permanent address: Department of Physics, University of Odense, DK-5230, Odense M, Denmark

1. J. R. Salcedo *et al.*, Phys. Rev. Lett. **41**, 131 (1978).
2. H. Wang *et al.*, in *International Quantum Electronics Conference*, 1990 Technical Digest Series, Vol. 8 (Optical Society of America, Washington, D.C., 1990).
3. Y. M. Wong and V. M. Kenkre, Phys. Rev. B **22**, 3072 (1980).
4. L. M. Smith *et al.*, Phys. Rev. B **39**, 1862 (1989).

QWC7 **9:45 am**

Raman effect in AlGaAs waveguides for all-optical applications

Y. H. Kao,* M. N Islam, J. M. Saylor,** R. E. Slusher,** W. S. Hobson,**
Department of Electrical Engineering and Computer Science, University of Michigan, Ann Arbor, Michigan 48109

By measuring the absolute Raman scattering cross section, we obtain the low-frequency depolarized Raman gain spectrum in $Al_{0.24}Ga_{0.76}As$ at a pump wavelength of 1.55 μm. We develop a model that shows that the Raman effect leads to cross talk between orthogonally polarized subpicosecond pulses and to spectral distortion. For example, for π-phase-shift conditions, the Raman effect causes a 30% energy exchange between orthogonally polarized 300-fs pulses. Therefore, the Raman effect limits the performance of semiconductor waveguides using subpicosecond pulses in all-optical switching or quantum-optics applications.

The absolute spontaneous Raman cross section is measured by using GaAs as a reference scatterer. The experiments are performed in the backscattering configuration at room temperature. The pump source is a cw argon-ion laser at a wavelength of 0.515 μm. An I_2 absorption-cell filter is used to reduce scattering light from the pump laser. The sample holder is placed inside a helium-gas cell during experiments to eliminate the Raman scattering from air at low frequencies.

By relating the absolute Raman cross section to the Raman gain coefficient,[1] we obtain the depolarized Raman gain spectrum at a pump wavelength of 1.55 μm (Fig. 1). The gain coefficient at 30 cm^{-1} is about 5.13×10^{-11} cm/W, which agrees

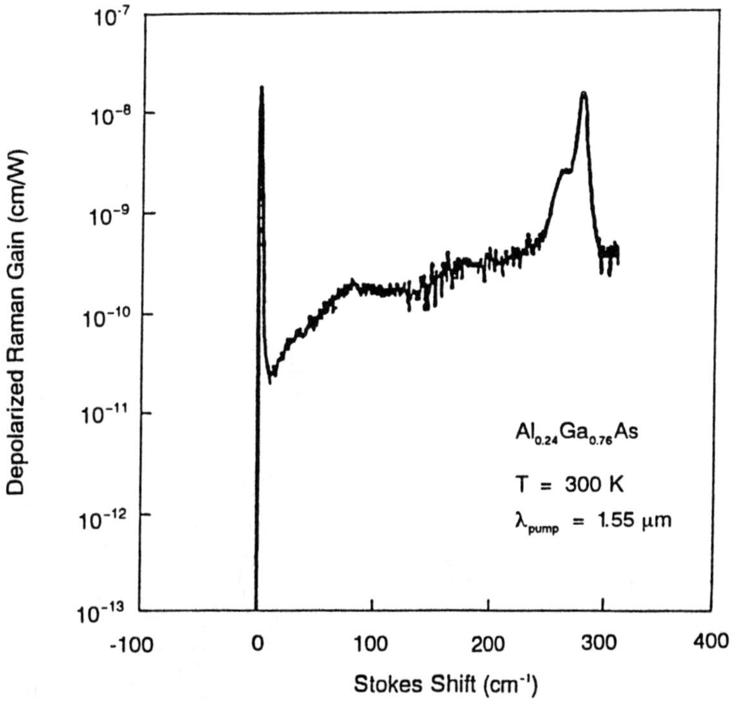

QWC7 Fig. 1. Depolarized Raman gain spectra for $Al_{0.24}Ga_{0.76}As$ at a pump wavelength of 1.55 μm at room temperature.

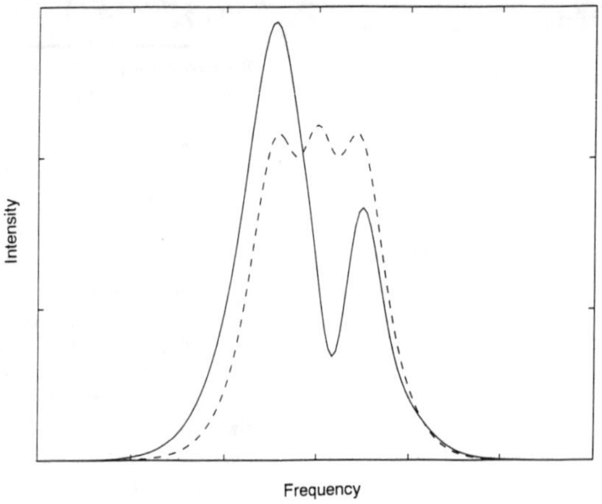

QWC7 Fig. 2. Frequency spectra for the pulse propagating in AlGaAs waveguides under π-phase-shift conditions. Dashed curves show the pulse spectra without the Raman effect, and the solid curves show the spectra with the Raman effect.

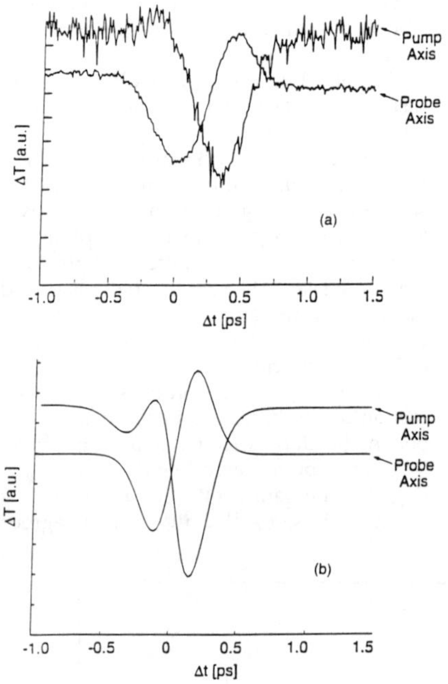

QWC7 Fig. 3. (a) Time-resolved pump–probe data for an $Al_{0.2}Ga_{0.8}As$ bulk waveguide using 285-fs pulses at 1.67 μm with the polarizer at the output along the pump axis and probe axis. The pump is orthogonally polarized to the probe. (b) Numerical result for two orthogonally polarized pulses under the same conditions as (a).

with the estimated value 5.5×10^{-11} cm/W from a previous pump–probe experiment.[2]

To understand the impact of the Raman effect on subpicosecond pulse interactions, we model the coupled progagation equations in AlGaAs waveguides. The equations include self-phase modulation (SPM), cross-phase modulation, anisotropic two-photon absorption,[3] and three Raman terms. The Raman terms are derived by using a derivation similar to that for fibers,[4] but by imposing a cubic-symmetry condition for AlGaAs.

Substituting the measured Raman gain coefficients into the propagation equations and using π-phase-shift conditions, we find that the Raman terms cause spectral distortion and energy exchange between orthogonal axes. Since low frequency extracts energy from high frequency because of the Raman effect, it induces spectral distortion, which is shown in Fig. 2. On the other hand, the amount of energy transfer is enhanced by the SPM and is inversely proportional to the pulse widths, because pulses of wider spectral width experience higher Raman gain. For example, the energy transfer is 30% and 9% with and without SPM, respectively, and it is 30% and 9% for 300-fs and 1-ps pulses, respectively. We further use this model to simulate the pump–probe experiment, and in Fig. 3 we compare the results to previous experimental data.[2] The consistent results show that this model successfully describes two orthogonally polarized pulses propagating in semiconductor waveguides.

In summary, we have reported the depolarized Raman gain spectrum in $Al_{0.24}Ga_{0.76}As$ at a pump wavelength of 1.55 μm with a value of 5.13×10^{-11} cm/W at 30 cm^{-1}. Our pulse-propagation model, which is consistent with previous experimental data,[2] shows that the Raman effect depletes the higher-frequency component of the pulse and that the induced energy exchange is enhanced by SPM. Because the Raman effect is inversely proportional to pulse width, it limits the performance of semiconductor waveguides in short-pulse (subpicosecond) applications.

**Department of Physics, University of Michigan, Ann Arbor, Michigan 48109*

***AT&T Bell Laboratories, Murray Hill, New Jersey 07974*

1. R. H. Stolen and E. P. Ippen, Appl. Phys. Lett. **22**, 276 (1973).
2. M. N. Islam, C. E. Soccolich, R. E. Slusher, A. F. J. Levi, W. S. Hobson, and M. G. Young, J. Appl. Phys. **71**, 1927 (1992).
3. M. D. Dvorak, W. A. Schroeder, D. R. Andersen, A. L. Smirl, and B. S. Wherrett, IEEE J. Quantum Electron. **QE-30**, 256 (1994).
4. C. R. Menyuk, Opt. Lett. **16**, 566 (1991).

JWA Tutorial — 8:00 am, Hyatt E/F

Confined Solid-State Structures (Fibers and Waveguides) Compared to Bulk Gain Lasers

Stephen G. Grubb, *AT&T Bell Laboratories, Presider*

JWA1 (Tutorial) — **8:00 am**

Confined solid-state structures (fibers and waveguides) compared to bulk gain lasers

David C. Hanna, *University of Southampton, U.K.*

The surge of interest in solid-state lasers over the past decade has been led by bulk lasers, with fiber lasers next to join in, and planar waveguide lasers the most recent addition. Each of these three types of laser geometry has its own attractive features which point to various areas of application where each type might be expected to excel. However, there are also areas of competition between the different types of laser with, as yet, no clear favorite. This tutorial will examine some of these areas of competition, as a way of reviewing the relative benefits of a confined, waveguide geometry versus a bulk geometry. Thus, while bulk lasers are seen as the natural candidates for high-power lasers, it has to be noted that the cladding-pumped fiber geometry has allowed fiber lasers to reach average powers of 10 W, with potentially much more to be achieved. Similarly, single-frequency lasers, often seen to be the domain of bulk lasers with short resonators, either ring lasers or microchips, are now seeing competition from short fiber lasers equipped with in fiber gratings, and even DFB fiber lasers. These and other areas of potential competition between guided and bulk devices will be examined. The balance of competition undergoes sudden shifts as new techniques, appear, e.g. the arrival of fiber gratings, the availability of quasi-phase-matched material in bulk form, the arrival of an improved technique for focusing the aspect from high-power diode sources, etc. So, the conclusion from this examination may not produce a clear-cut answer. In fact, the conclusion will likely raise yet more questions, more avenues to research. . . .

JWB Tutorial — 9:00 am

Hyatt E/F

Linear and Nonlinear Optical Techniques for Combustion Measurements

Dennis K. Killinger, *University of South Florida, Presider*

JWB1 (Tutorial) — **9:00 am**

Linear and nonlinear optical techniques for combustion measurements

Larry A. Rahn, *Sandia National Laboratories*

Laser-based diagnostic methods have revolutionized combustion research over the past two decades. These methods enable nonintrusive quantitative measurements of species concentration, temperature, velocity flow fields, and particulates with high spatial and temporal resolution in combustion environments. While much progress has been made in the complex field of combustion science, the advances in many important basic and applied problem areas is still limited by our ability to make experimental measurements. These challenges continue to motivate the further development of laser-based diagnostics and the laser and detector technologies on which they depend.

This tutorial will review the motivation for laser-based combustion diagnostics and point out some of the challenging problems that we face today. The underlying principles and limitations of linear and nonlinear optical spectroscopic diagnostics are reviewed. Methods included are absorption, Rayleigh, Raman, fluorescence, multiphoton ionization, coherent anti-Stokes Raman, resonant four-wave mixing, and some of their variants. The applicability of these methods and the associated requirements for laser technology will be explored. Finally, some of the current research topics in combustion diagnostics will be discussed, including scalar imaging techniques, thermal grating effects, and degenerate four-wave mixing.

QWD 1:00 pm

2nd and 3rd Level of the Convention Ctr.

Poster Session

Spatio-Temporal Dynamics in Nonlinear Optics

QWD1

Interaction between solitons with second-order nonlinearities

Daniel-Marian Baboiu, George I. Stegeman, Lluis Torner,* *Center for Research and Education in Optics and Lasers, University of Central Florida, Orlando, Florida 32826*

It was recently shown, both experimentally[1] and theoretically,[2] that coherent interactions through quadratic nonlinearities lead to very interesting phenomena with potential applications to all-optical devices and parametric processes. Unlike cubic nonlinear processes, the mutual trapping of the fundamental and the second-harmonic waves when they propagate simultaneously in a material with $\chi^{(2)}$ nonlinearity, i.e., a spatial soliton, imposes supplementary conditions; thus the stationary solutions are either zero- or one-parameter families.

As a coherent process, the evolution of solitary waves in materials with second-order nonlinearities is very sensitive to the phase relation between the components. This happens not only for the relation between the fundamental frequency and the second harmonic, but also for the relation between different pulses propagating in the same structure. Here we study interactions between stationary spatial solitons that consist of an in-phase fundamental and second-harmonic beams at the input. In Fig. 1 one can see the evolution of two in-phase identical solitons, separated by five pulse widths, in a structure without walk-off between the fundamental and the harmonic waves. As one might expect, the two pulses attract each other.

The presence of a moderate walk-off drastically modifies the evolution of the pulses: now not only do they not attract each other, but there is a slight repulsion between them. After the pulses propagate 20 normalized units, the separation between the solitons increases from five pulse widths to six pulse widths (Fig. 2).

An even more dramatic change is observed when the pulses are not in phase. Figure 3 illustrates the case of two soliton pulses with opposite phases at the input. After propagating less than 2 normalized units, the pulse with the phase shifted by π decays into radiation, part of its energy being captured by the other pulse. The numerical simulation was carried out here without walk-off, illustrating a nonsymmetrical evolution in a symmetrical system.

More interesting phenomena can be observed if we look at phase differences between the two pulses other than 0 and π and at nonstationary pulse shapes (e.g., Gaussian). They will have similar evolution patterns. If the pulses are not the stationary solutions, special precautions have to be taken to allow the nonsolitonic radiation to freely escape from the region of interest. The most interesting results are presented.

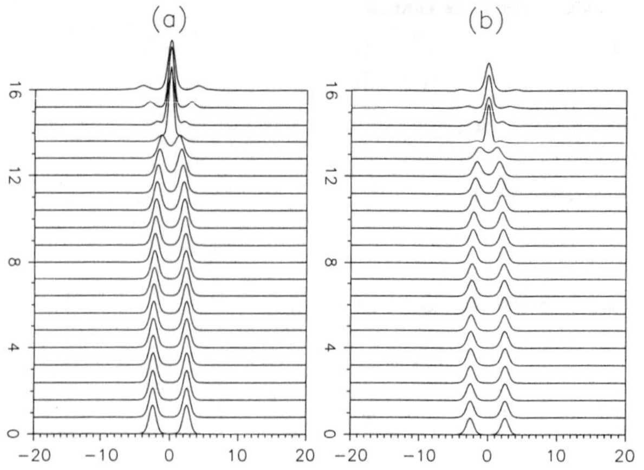

QWD1 Fig. 1. The interaction of two in-phase cascaded solitons in the absence of walk-off: (a) fundamental frequency; (b) second harmonic.

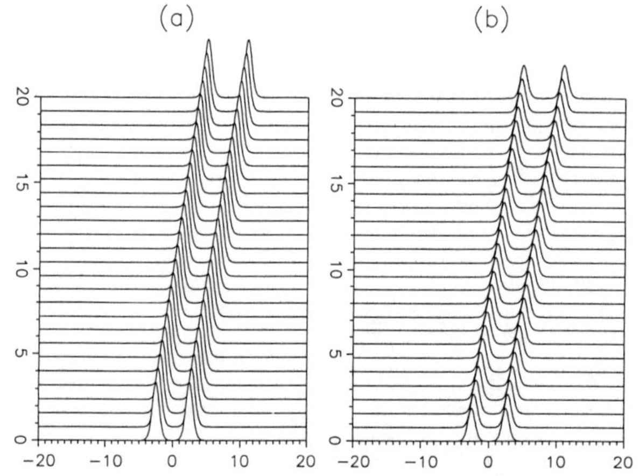

QWD1 Fig. 2. Same as Fig. 1 but in the presence of walk-off.

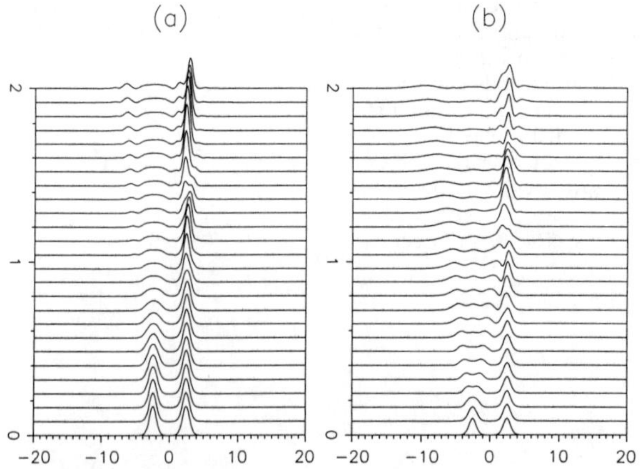

QWD1 Fig. 3. Evolution of the (a) fundamental frequency and (b) second harmonic for a pair of soliton pulses with opposite phases, in the absence of walk-off.

**Department of Signal Theory and Communications, Polytechnic University of Catalonia, P. O. Box 30002, 08080 Barcelona, Spain*

1. G. I. Stegeman, M. Sheik-Bahae, E. Van Stryland, and G. Assanto, Opt. Lett. **18**, 13 (1993).
2. C. R. Menyuk, R. Schiek, and L. Torner, "Solitary waves due to cascading," J. Opt. Soc. Am. B, in press.

QWD2

Nonlinear optical propagation in fractal structure

M. Bertolotti, F. Moretti, C. Sibilia, P. Masciulli,* *Department of Energetics, University of Rome I, Via Scarpa 16, I-00161 Rome, Italy*

We present the input–output characteristic of a fractal nonlinear-optical (intensity-dependent refractive index) Fabry–Perot resonator. The one-D dielectric integrated resonator is obtained by alternating two nondispersive planar dielectric layers of unperturbed refractive index n_2 and n_1 ($n_2 > n_1$) of thicknesses such that their optical paths are the same. The initiator is the layer of refractive index n_2. If L is the optical path of the initiator, the generator is obtained by substituting the central part of the initiator having an optical path of $L/3$ with the layer of refractive index n_1 and thickness such that the optical path is $L/3$ again. The resonator is obtained by iterating the operation on the left- and right-third optical paths and by stopping the iteration at the Nth step.

We consider the resonator embedded between two equal semi-infinite nondispersive layers of refractive index n less than both n_2 and n_1. The incident light is assumed to be a plane TE wave with normal incidence. The transmission properties of this linear resonator have been studied[1,2] by using the transfer-matrix formalism. The isolated transmission peaks in a forbidden frequency gap[1,2] suggest the possibility of obtaining optical switching at low threshold intensity. In the case of a nonlinear device, we have therefore studied the same resonator with layers of refractive index n_2 made by a Kerr-type medium. As a consequence, the transmitted intensity is a nonlinear function of the incident intensity at a given electromagnetic frequency. The calculations of the transmitted intensity versus the incident one are based on a generalization of the dummy method.[3] The material's losses are taken into account by considering complex refractive indices. Multistable behavior has been found, with switch input power depending on which of the two basic layers (of linear refractive index n_1 or n_2) is nonlinear. The considered resonator requires less input-switch power than does the analogous periodic structure. The simulations show that it is possible to have a multistable behavior with a relatively low first-switch threshold.

The spatial field profile inside the structure is also calculated. The calculation has also been performed by taking into account the finite spatial dimension of the input beam.

**Department of Chemical Engineering, Processes and Materials, University of Palermo, Via le Delle Scienze, I-90128 Palermo, Italy*

1. M. Bertolotti, P. Masciulli, and C. Sibilia, Opt. Lett. **19**, 777 (1994).
2. M. Bertolotti, P. Masciulli, and C. Sibilia, in *Proc. Integrated Optics Conference*, Lindau, Germany, April 1994.
3. H. M. Gibbs, *Optical Bistability: Controlling Light with Light* (Academic, New York, 1985), Chap. 2.

QWD3

Coincidentally performed measurements of all-optical bistabilities in transmittance, reflectance, and luminescence

Bruno Ullrich, Takayoshi Kobayashi, *Department of Physics, University of Tokyo, 7-3-1 Hongo, Bunkyo-Ku, Tokyo 113, Japan*

It is well known that the excitation of ZnSe and CdS with the blue and green lines, respectively, of an argon laser causes all-optical bistability (all-OB) in the features of the excitation, i.e., in the transmittance (all-OBT) and in the reflectance (all-OBR). By means of coincident measurements of all-OBRs and all-OBTs of thin ($\leq$10-μm) ZnSe and CdS films, we observed the *a priori* highly unexpected feature that even if an all-OBT is observed, no all-OBR takes place, as shown in Fig. 1. The observed phenomenon can be explained by the peculiar dependence of the reflectance on temperature, which differs considerably—in contrast to dependence of the transmittance on temperature—from the theoretically expected shape, and by considering the bistability in luminescence (all-OBL), which was discovered recently.[1,2] Particularly, the results can be interpreted by the expression

$$T + R + A' + E = 1, \quad (1)$$

where T is the transmittance, R is the reflectance, and A' is the absorbance converted into nonradiative transitions and/or heat and not into a luminescent emisson E.

In addition, by means of simultaneously performed investigations of all-OBL and all-OBR of thin CdS and ZnSe films, we demonstrate that all-OBLs in the near-infrared region are independent of the geometrical configuration of the sample; i.e., the bistable emissions in the reflected and the transmitted beams are the same.

In conclusion, for the first time to our knowledge, we present coincidentally performed investigations of all-OBT, all-OBR, and all-OBL of semiconducting thin films. The results give insights into the spatial origins of the bistabilities; i.e., all-OBL remains bulk dependent even when the emissions are studied in a reflection geometry.

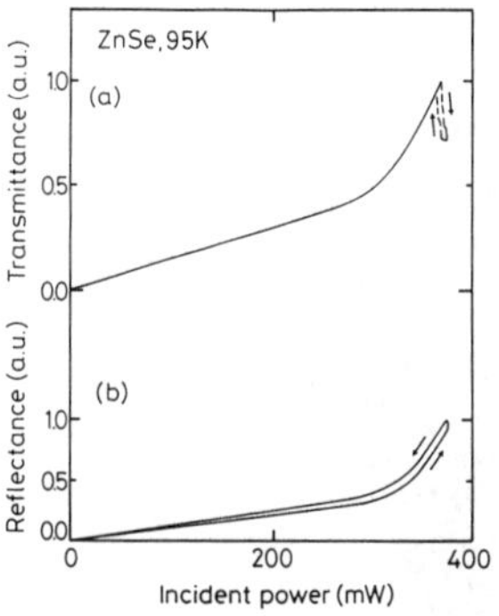

QWD3 Fig. 1. Coincident measurements of (a) transmittance and (b) reflectance of the thin ZnSe film at 95 K. Excitation was with the 488-nm line of an argon laser. The surprising fact is that even if an absorptive bistable switch in the transmittance is observed, no feature in the reflectance takes place.

1. B. Ullrich, A. Kazlauskas, S. Zerlauth, and T. Kobayashi, J. Cryst. Growth **138**, 234 (1994).
2. B. Ullrich and T. Kobayashi, in *Proc. International Conference on Nonlinear Optics*, Waikoloa, Hawaii, 1994, p. 258.

QWD4

Spatiotemporal instabilities of femtosecond pulses in dispersive nonlinear media

E. J. S. Fonseca, Solange B. Cavalcanti, J. M. Hickmann, *Department of Physics, Federal University of Alagoas, Maceió, AL 57072/970, Brazil*

Spatial and temporal instabilities are known to occur when light propagating in a Kerr medium is modulated spatially and temporally, respectively. The simultaneous presence of diffraction and dispersion in nonlinear media gives rise to spatiotemporal instabilities.[1] The analysis in previous research accounted for group-velocity dispersion (GVD) in the slowly-varying-envelope approximation (SVEA). However, the self-focusing of femtosecond pulses in dispersive media is not adequately described by the SVEA.[2] This paper deals with the theoretical analysis of spatiotemporal instabilities of femtosecond pulses propagating in a nonlinear medium. More specifically, in addition to including up to fourth-order dispersion, the present investigation takes into account the variation on the z-projected group velocity of off-axis rays by considering the wave-vector $k(\Omega)$-series expansion, keeping terms of $O(\Omega^2)$, instead of $O(\Omega)$, as in Ref. 1, to obtain,

$$k_z - k_0 - \frac{\Omega}{v_g} = \frac{1}{2}\Omega^2\beta_2 - \frac{k_T^2}{2k_0} \times \left[1 - \frac{\Omega^2}{k_0 v_g} - \frac{\Omega^2}{k_0}\left(\frac{\beta^2}{2} - \frac{1}{k_0^2 v_g^2}\right)\right], \quad (1)$$

where v_g is the group velocity, β_2 is the group-velocity dispersion, k_0 is the propagation constant, and $k_T = (k_x^2 + k_y^2)^{1/2}$ with k_x, k_y, and k_z being the Cartesian components of k. The new terms represent a coupling between GVD and diffraction. Following along the lines of Ref. 3, we base the investigation of the instability on a linear stability analysis around a quasi-homogeneous stationary solution of the wave equation by inserting

$$E(x, y, z, t) = E_0 \operatorname{sech}\left(\frac{t}{t_0}\right) \times \exp\left(-\frac{x^2 + y^2}{\omega_0^2}\right) \exp(i n_2 k z E_0^2) \times [1 + \alpha(x, y, z, t)] \quad (2)$$

in the propagation equation. Here n_2 is the Kerr coefficient, t_0 is the pulse width, and ω_0 is the beam waist. It is found that small perturbations grow exponentially with the

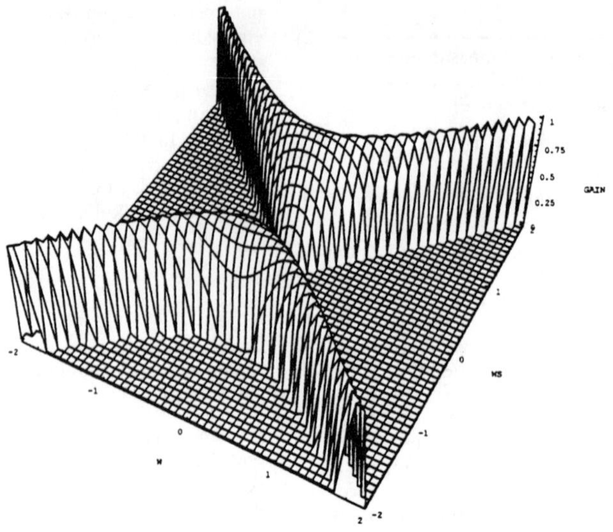

QWD4 Fig. 1. Normalized gain spectrum as a function of normalized spatial and temporal modulation frequencies, with the coupling parameter $\alpha = 0$.

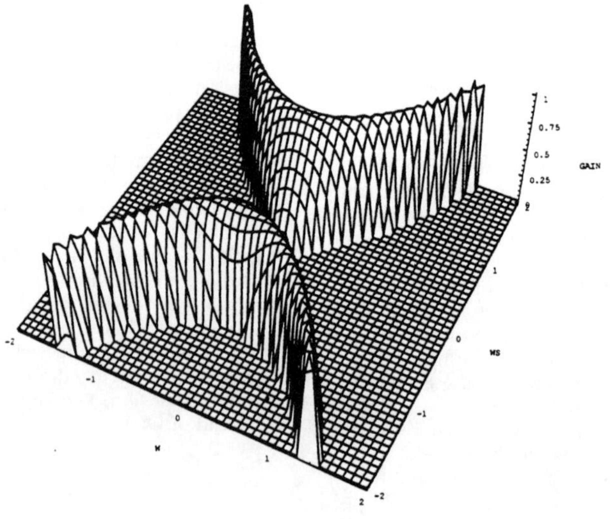

QWD4 Fig. 2. Normalized gain spectrum as a function of normalized spatial and temporal modulation frequencies, with the coupling parameter $\alpha = 0.2$.

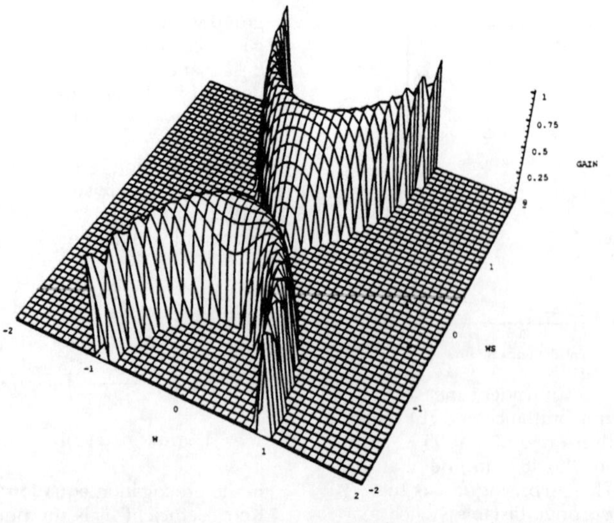

QWD4 Fig. 3. Normalized gain spectrum as a function of normalized spatial and temporal modulation frequencies, with the coupling parameter $\alpha = 0.75$.

gain coefficient

$$g(\Omega, \Omega_s) = |\beta_2|\Omega_c^2 \left[-\mathrm{sgn}(\beta_2)\frac{\Omega^2}{\Omega_c^2} + \frac{\Omega_s^2}{\Omega_c^2} + \frac{\alpha\Omega_s^2\Omega^2}{\Omega_c^4} \right]^{1/2} \times \left[\frac{\mathrm{sgn}(\beta_2)\Omega^2}{\Omega_c^2} - \frac{\Omega_s^2}{\Omega_c^2} - \frac{\alpha\Omega_s^2\Omega^2}{\omega_c^2} + \mathrm{sgn}(n_2) \right]^{1/2}, \quad (3)$$

where Ω_c is the critical modulation frequency and $\alpha = \Omega_c^2/(k_0^2 v_g^2)$. Figures 1–3, which plot the gain spectrum as a function of Ω and Ω_s for various values of the coupling parameter, illustrate the extent to which the coupling parameter influences the initial growth of the instability.

The most interesting feature is that, as far as the instability is concerned, higher-order dispersion terms are negligible when compared to the coupling effect, so that the dominant effect here arises from the GVD–diffraction coupling contribution.

1. L. W. Liou, X. D. Cao, C. J. McKinstrie, and G. P. Agrawal, Phys. Rev. A **46**, 4202 (1992).
2. J. E. Rothenberg, Opt. Lett. **17**, 1340 (1992).
3. A Höök and M. Karlsson, Opt. Lett. **18**, 1388 (1992).

QWD5

Polarization instability and multistability in gas lasers with weakly anisotropic cavities

L. P. Svirina, *Institute of Physics, Belarus Academy of Sciences, F. Skaryna Prospect 70, Minsk 220072, Belarus*

Nonstationary behavior of polarization parameters without any time-dependent external influence (polarization instability) was observed experimentally in single-mode He–Ne (Refs. 1 and 2) and He–Xe (Ref. 2) lasers at the $j_1 = 1 \rightarrow j_2 = 2$ transition and with linear phase anisotropy of the cavity as well as in a He–Ne (Ref. 3) laser at the $j_1 = 1 \rightarrow j_2 = 1$ transition and with circular phase anisotropy of the cavity.

It has been shown[4] that polarization instability in a single-mode gas laser occurs at comparable values of active-medium and empty-cavity anisotropies (weakly anisotropic cavities) and results from the competition of two standing waves. The nonlinear interaction of the emitted waves with the active medium (with or without external fields) induces the active-medium anisotropy and at definite types of the empty-cavity anisotropy leads to the instability of the laser eigenstates.

For $j \rightarrow j + 1$ transitions (weak coupling), for which the active medium eigenstates are two circular orthogonally polarized waves, and for linear phase anisotropy of the empty cavity, polarization instability manifests itself in spontaneous pulsations of emitted waves[1] intensities, ellipticities, and azimuths.[4] For $j \rightarrow j$ transitions (strong coupling), for which the active medium eigenstates are two linear orthogonally polarized waves, and for circular phase anisotropy of the cavity, only rotation of azimuths occurs at constant intensities and ellipticities whose values are determined by the value of the cavity anisotropy.[5] Two types of transitions between the steady orthogonal states of polarization (through polarization instability and through bistability) observed experimentally in the vicinity of the line center at the $j_1 = 1 \rightarrow j_2 = 2$ (Ref. 1) and the $j_1 = 1 \rightarrow j_2 = 1$ (Ref. 3) transitions have been explained.[4,5] The dependence of the laser operation at different tunings on the initial states of polarization of two standing waves (polarization multistability) has been predicted.[4,5]

For elliptical orthogonal eigenstates of the empty cavity, polarization instability occurs at both the $j \rightarrow j + 1$ and the $j \rightarrow j$ transitions (at small and at large values of ellipticity, respectively), and it manifests itself as oscillations of intensities and ellipticities and as rotation of azimuths. Polarization multistability has also been found.

Polarization instability in the presence of a longitudinal magnetic field on the active medium at different tunings gives a large variety of one- and two-frequency regimes, corresponding to limit cycles of the first and the second kinds.[4]

Polarization instability and multistability considered in the present work in the low-gain He–Ne laser are expected to be observed in other laser systems at comparable values of active-medium and empty-cavity anisotropies.

1. A. Le Floch, G, Ropars, J.-M. Lenormand, and R. Le Naour, Phys. Rev. Lett. **52**, 918 (1984).
2. G. P. Puccioni, G. L. Lippi, N. B. Abraham, and F. T. Arecci, Opt. Commun. **72**, 361 (1989).
3. J. C. Cotteverte, F. Bretenaker, and A. Le Floch, Opt. Lett. **16**, 572 (1991).
4. L. P. Svirina, Opt. Spectrosk. **77**, 124 (1994); Opt. Commum. **111**, 370 (1994).
5. L. P. Svirina, in *Technical Digest, European Quantum Electronics Conference*, Amsterdam, 1994, p. 213.

QWD6

Theoretical study of soliton amplification in erbium-doped optical fibers

Y. Zhao, L. M. Cheong, K. T. Chan, *Department of Electronic Engineering, Chinese University of Hong Kong, Shatin, N.T., Hong Kong*

In soliton-based communication systems, amplifiers are needed to compensate for fiber loss so as to maintain long-span nondistorted propagation. The erbium-doped fiber amplifier (EDFA) is the prime candidate for the amplification of solitons because of its well-known merits. Papers on soliton amplification[1,2] have been mainly devoted to femtosecond soliton amplification and therefore are not directly applicable to the optimum design of soliton amplifiers used in picosecond soliton communication. The emphasis of our work is on picosecond soliton amplification. A new model has been established. So that the model should be as practical as possible, excited-stated absorption is considered for what we believe to be the first time; this will enable the model to be used in cases with different pumping wavelengths.

Our model is based on the energy-level diagram of erbium shown in Fig. 1, whose effective state diagram is described by Fig. 2. In these figures R_{ij}, A_{ij}, S_{ij}, W_{24}, and W_{13} are the nonradiative transition, spontaneous emission, signal-stimulated transition, excited-state-absorption rate, and ground-state pump-absorption rate, respectively. The decay rate of $|\bar{2}\rangle$ is given by $\gamma_2 = A_{21} + R_{21} + W_{24} \approx A_{21} + \tilde{\alpha} W_{13}$. W_2 is the transition rate to the effective upper energy level $|\bar{2}\rangle$ due to pump action, which can be expressed as $W_2 = W_{13}[(\rho_{11} + q_2\tilde{\alpha}\rho_{22})q_1 + \tilde{\alpha}(1 - q_2)\rho_{22}]$, where $\tilde{\alpha} = W_{24}/W_{13}$ is the coefficient of excited-state absorption, $q_1 = R_{32}/(R_{32} + A_{31} + R_{31})$ is the fast nonradiative quantum-transfer efficiency from $|3\rangle$ to $|2\rangle$, $q_2 = R_{43}/(R_{43} + R_{42} + A_{42})$ is similar to q_1 but from $|4\rangle$ to $|3\rangle$, and ρ_{11} and ρ_{22} are the populations in levels $|\bar{1}\rangle$ and $|\bar{2}\rangle$, respectively, which satisfy $\rho_{11} + \rho_{22} = N_0 f(r, \theta)$. Here N_0 is the erbium concentration at the fiber axis and $f(r, \theta)$ is the doping shape function. To gain an insight into the evolution of the soliton shape and to extend this model to the study of femtosecond soliton amplification, we use the semiclassical theory based on the combined description of the population-density matrix and Maxwell's equations. The pump evolution along the fiber is also important, especially for the case of low-doping distributed amplification in which the long fiber length results in the inhomogeneous allocation of pump power. Here the pump equation is described by rate equations. This model is composed of three main equations: the soliton-propagation equation, the population-inversion evolution equation, and the pump equation. These equations are all normalized with respect to the soliton unit. In these equations, parameters involved in the optimization and design of soliton amplification, such as the doping profile, the light-wave mode, and the spectral broadening, are all considered. This model is a general form, from which we can deal with the cases of lumped amplification and distributed amplification.

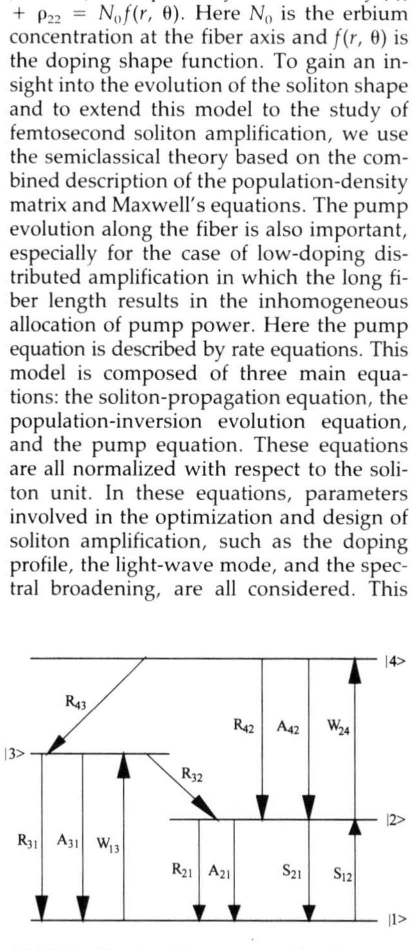

QWD6 Fig. 1. Energy-state level system of erbium.

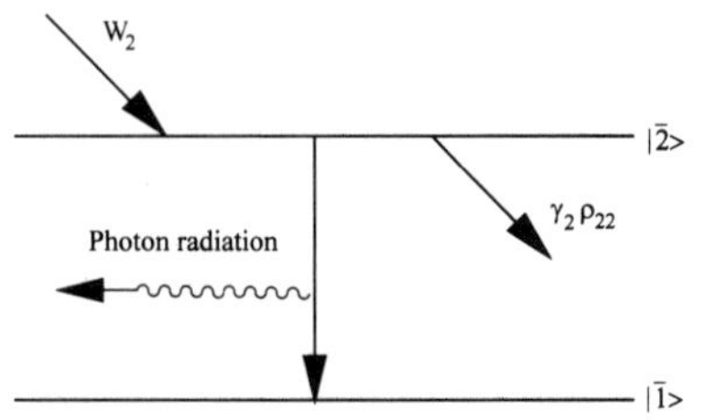

QWD6 Fig. 2. Effective two-energy-state level system corresponding to Fig. 1.

Combining this model with the initial elementary soliton injection, we have obtained the optimal pumping condition for transparency amplification; *viz.*, the fiber loss is just offset by the gain. When the parameters given by Nakazawa *et al.*[3] are used, the calculated pump power at the zero-net-gain point in soliton propagation in low-doping fiber is in agreement with the experimental results. Our analysis also shows that the pump power is not a function of soliton width and fiber dispersion under the condition of picosecond soliton-transparency amplification.

1. G. P. Agrawal, Phys. Rev. A **44**, 7493 (1991).
2. B. Gross and J. T. Manassah, Opt. Lett. **17**, 340 (1992).
3. M. Nakazawa, Y. Kimura, and K. Suzuki, IEEE J. Quantum Electron. **26**, 2103 (1990).

QWD7

Pulse propagation in an optically dense medium: refractive-index enhancement

A. S. Manka, M. Scalora, C. M. Bowden, *Weapons Sciences Directorate, AMSMI-RD-WS-ST, U.S. Army Missile Command, Redstone Arsenal, Alabama 35898-5248*

We numerically study the effect of two coherent copropagating pulses incident on an optically dense collection of four-level atoms. The injected pulses have different frequencies that are nearly resonant with the allowed dipole transitions. Because the transient solution is paramount, we cannot use the steady-state absorption spectrum as an appropriate indicator of probe absorption or amplification. We use instead the imaginary part of the polarization at the probe's frequency as a time-dependent measure of the probe field's gain. Because the gain is a function of time, this polarization by itself is not sufficient to describe whether the probe is amplified. Therefore, we consider the integrated energy density of the probe field to assess the parameters for which the energy at the probe frequency increases. Furthermore, we can drop the weak-coupling and weak-probe-field approximations and study regimes that have not been addressed before, including strong probe-field strengths for which standard linear treatments are not valid and optically dense mediums for which local-field effects cannot be ignored.

We have numerically demonstrated that probe-field amplification is possible in a four-level medium with Raman pumping without population inversion in which all fields are treated as propagating pulses.[1] The existence of gain is primarily attributed to the creation of coherence in the lower, nearly degenerate levels. For this reason we believe that two-photon processes are unlikely for small probe fields and do not contribute significantly to the single-photon gain. For a sparse medium we conclude that, in contrast to the steady-state case, the maximum gain decreases at the sidebands with increasing pump strengths. In the nonlinear regime, in which pump and probe are of comparable magnitudes, competition between the upper and lower coherences may lead to loss of gain.

Although we have predicted an increase in probe energy, it is small except when the probe coupling is large, i.e., at higher densities. It has been recently proposed by some of us that in the regime of high atomic densities ($N \geq 10^{13}$), in which near-dipole–dipole (NDD) or local-field effects are important, the complex susceptibility is significantly enhanced by cross coupling of the polarizations.[2] Here, we consider NDD effects to look for enhancements in the probe gain and index of refraction. We find that the main effect of NDD interactions is to increase the refractive index.

1. A. S. Manka, M. Scalora, J. P. Dowling, and C. M. Dowden, "Pulse propagation in a Raman pumped, four-level medium that exhibits inversionless gain," submitted to Opt. Commun.
2. A. S. Manka, J. P. Dowling, C. M. Bowden, and M. Fleischhauer, Phys. Rev. Lett. **73**, 1789 (1994).

QWD8

Nonlinear dynamics of a semiconductor ring laser: steady states, instabilities, and chaos

Ya. I. Khanin, I. V. Koryukin, P. A. Khandokhin, P. Mandel,* *Institute of Applied Physics, Russian Academy of Sciences, 46 Uljanov Street, 603600 Nizhny Novgorod, Russia*

The dynamic behavior of a class-B bidirectional ring laser is determined mainly by the interaction of the counterpropagating waves with the self-induced inversion grating. Therefore, all the phenomena that affect either the amplitude or the phase of the inversion grating could appreciably influence the dynamics of the laser. The carrier spatial diffusion and the gain line structure are among these factors. To consider their role we have generalized the model of a class-B ring laser[1,2] to account for both the carrier-diffusion coefficient d and the linewidth-enhancement factor (Henry's α factor[3]).

There are three types of steady-state solutions existing in the laser model we consider: (1) trivial solutions, (2) single-mode solutions, and (3) two-mode solutions, which correspond to absence of lasing, unidirectional laser action, and bidirectional laser action, respectively. A linear stability analysis leads to the conclusion that a stable two-mode regime with equal mode amplitudes is possible in a finite domain of d and α (region TM on the bifurcation diagram of Fig. 1). Such a result was obtained by analyzing numerically the characteristic roots and by numerical simulations of the model.

Other areas in Fig. 1 are (1) a region of stable single-mode lasing (SM) and (2) a domain of instability (U) where no steady-state solution exists. One can see that the stability boundaries of the TM and SM regions are close but do not coincide. So there is a bistability region on the diagram in which both single- and two-mode stable solutions coexist. The second boundary of

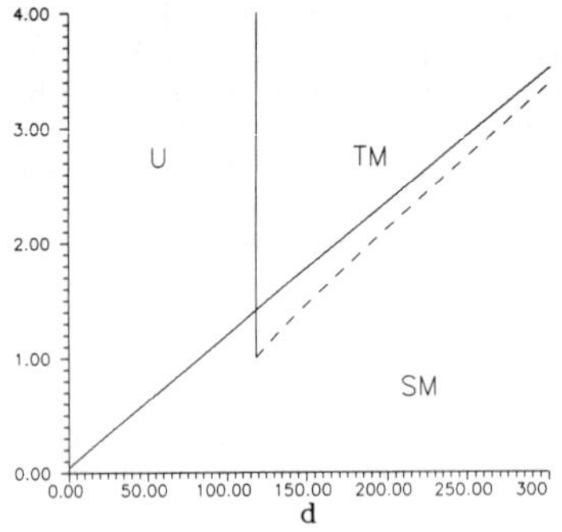

QWD8 Fig. 1. Bifurcation diagram in the plane of the control parameters (diffusion coefficient and linewidth-enhancement factor).

the TM region is independent of α and gives a critical value of d separating the two-mode and the instability domains. Below this value the two-mode regime becomes unstable through a supercritical Hopf bifurcation and leads to quasi-sinusoidal antiphased oscillations of amplitudes. These oscillations may be irregular or chaotic if diffusion is slow. Taking into account the linear mode–mode coupling by backscattering on the optical defects of the cavity leads to small changes in the stability boundaries of all regimes and to a transformation of the single-mode solution to the quasi-unidirectional solution (i.e., a large amplitude in one direction and a small amplitude in the other direction). Furthermore, a critical value of the linear coupling coefficient appears. Above it, the quasi-unidirectional solution becomes unstable through a subcritical Hopf bifurcation and the two-mode solution loses its stability by changing of a sign of the real characteristic root.

Summarizing, we found that (1) diffusion of atomic excited states could strongly affect the dynamics of a semiconductor ring laser by decreasing the effective mode–mode coupling, (2) there is a region of stable bidirectional lasing with equal mode amplitudes for nonzero α factor and diffusion, and (3) the linewidth-enhancement factor and linear backscattering are important in the dynamical behavior of this laser.

* *Theoretical Nonlinear Optics, Free University of Brussels, C.P. 231, B-1050 Brussels, Belgium*

1. P. A. Khandokhin and Ya. I. Khanin, J. Opt. Soc. Am. B **2**, 226 (1985).
2. H. Zeghlache, P. Mandel, N. B. Abraham, L. M. Hoffer, G. L. Lippi, and T. Mello, Phys. Rev. A. **37**, 479 (1988).
3. C. H. Henry, IEEE J. Quantum Electron. **18**, 36 (1982).

QWD9

Karhunen Löeve expansion analysis of the spatiotemporal dynamics of a phase-conjugate resonator

D. Korwan, G. Indebetouw, *Department 0435 of Physics, Virginia Polytechnic Institute and State University, Blacksburg, Virginia 24061*

The spatiotemporal dynamics of optical resonators have recently received much at-

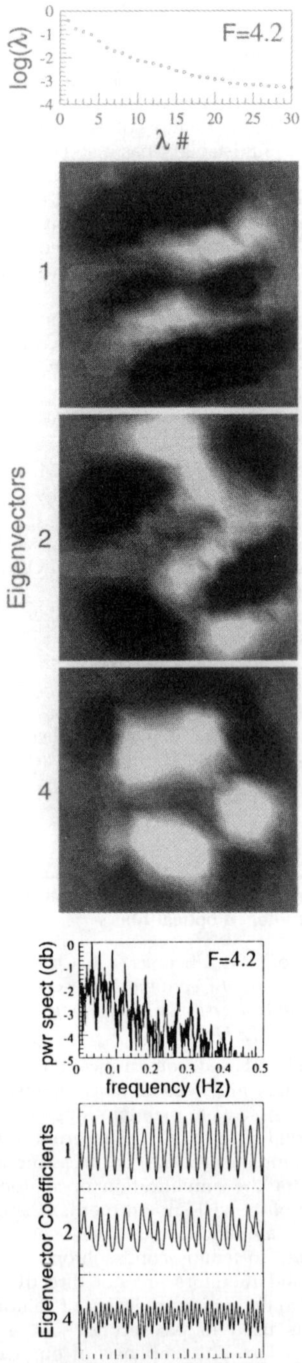

QWD9 Fig. 1. Chaotic dynamics at $F = 4.2$: (a) eigenvalues and three eigenvectors; (b) power spectrum of a local intensity time series and the temporal evolution of the eigenvectors' coefficients.

Wednesday Posters

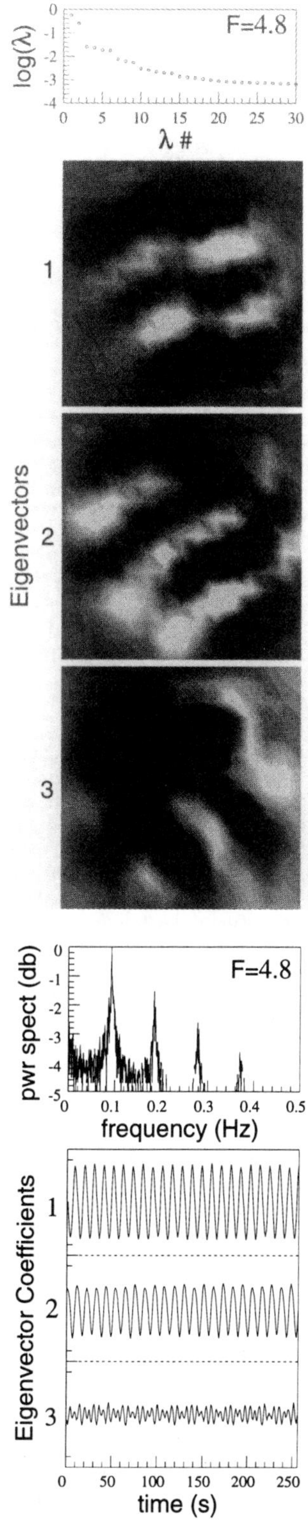

QWD9 Fig. 2. Same as Fig. 1 for the periodic dynamics at $F = 4.8$.

tention, partly because of their potential applications and because they offer a convenient means of studying the nonlinear dynamics of extended systems. Attention has been focused on two types of resonators, the laser and the photorefractive ring.[1] In both cases, an expansion of the resonator field in terms of empty cavity modes is a useful way of analyzing the spatiotemporal dynamics. However, for a phase-conjugate resonator a modal expansion is less justifiable. Because the phase-conjugate mirror is active only when providing gain and because it adapts itself dynamically to the cavity field, there are no empty cavity eigenmodes.

We analyzed the cavity intensity distribution by using a singular-value decomposition (Karhunen Löeve). The eigenvectors are interpreted as the most significant spatial patterns of the field, and their eigenvalues are a measure of their significance.

For increasing Fresnel numbers, the field amplitude changes from a uniform time-independent state ($F < 1.8$) to various configurations containing an increasing number of dynamic phase defects. For $2 < F < 3$ the motion is periodic with one pair of vortices repeatedly nucleating at the center and vanishing at the boundary. Generally, there are two significant eigenvectors with coefficients oscillating in phase quadrature. These observations confirm an earlier attempt at identifying characteristic modes by using holographic correlation.[2] The ratio λ_2/λ_1 of the two eigenvalues increases from ~ 0.01 near threshold ($F \sim 2$) to ~ 0.5 at $F \sim 3$, where a new bifurcation occurs.

For $3 < F < 4$ there are typically four eigenvectors with significant eigenvalues. Their spatial structures map the trajectories of two vortex pairs around a four-spot pattern. For $F > 4$ the motion can be periodic, quasi periodic, or chaotic. Figures 1 and 2 show an aperiodic state with five significant eigenvectors at $F = 4.2$, which locks into a periodic state with like eigenvectors at $F = 4.8$.

The dynamics were also studied by using the Bragg mismatch as a control parameter. Interesting trends, such as a considerable slowing and an increase of the spatial complexity as the Bragg matching condition is approached, are analyzed.

1. M. Brambilla *et al.*, Phys. Rev. A **49**, 1427 (1994); F. T. Arecchi, in *NATO 41st Scottish University Summer School*, R. G. Harrison *et al.*, eds. (1992) Vol. 65; D. Hennequin *et al.*, J. Opt. Soc. Am. B **11**, 676 (1993).
2. G. Indebetouw and D. Korwan, J. Mod. Opt. **41**, 941 (1994).

QWD10

Numerical analysis of intense-laser-pulse spatial and frequency instability in a resonant two-level medium

A. A. Panteleev, A. N. Starostin, D. I. Chekhov,* A. G. Leonov,* S. Rotin,*
Branch of I. V. Kurchatov Institute of Atomic Energy, Troitsk 142092, Moscow region, Russia

The spectrum–angular instability of laser radiation in a resonant medium is well known.[1,2] In particular, cone radiation red shifted with respect to the laser frequency was observed in a number of experiments at high-frequency laser detunings $\Delta = \omega_0 - \omega_L < 0$, where ω_L is the laser frequency and ω_0 is the atomic transition frequency. It has been theoretically shown[3] and confirmed experimentally[4] that this instability takes place for positive and negative values of Δ.

Several theoretical models have been offered for explaining the origin of the cone radiation (see references in Ref. 2). We suppose that the model of nondegenerated four-wave mixing (NFWM) is the most adequate description of mentioned instability. By using this approach and numerical calculations, good agreement between theory and experiment has been obtained[5] for the steady-state case. Some effects have been explained[1] by a non-steady-state simulation. However, calculations[1,5] have been carried out only for the case $\Delta < 0$ and for small propagation lengths.[1] In addition, the relaxation processes in Ref. 1 have not been taken into account.

This paper presents results of non-steady-state instability modeling and, in contrast to Ref. 1, explains all basic experimental data, both at $\Delta < 0$ and at $\Delta > 0$.[4] We carried out our calculation by solving the Maxwell–Bloch equation system

$$\left(\frac{1}{c}\frac{\partial}{\partial t} + \frac{\partial}{\partial z} - \frac{i}{2k}\Delta_\perp\right) E = i\frac{2\pi\omega_L}{c} N\mu_{12}\rho_{21},$$

$$i\frac{\partial}{\partial t}\rho_{22} = V_{21}\rho_{12} - V_{12}\rho_{21} - i\gamma_\perp\rho_{22},$$

$$\rho_{11} + \rho_{22} = 1,$$

$$i\frac{\partial}{\partial t}\rho_{12} = V_{12}(\rho_{22} - \rho_{11}) - i\gamma_\parallel\rho_{12}, \; \rho_{21} = \rho_{12}^*,$$

$$(t, x, y) \in R^3, \; z \in [0, L], \quad \Delta_\perp = \frac{\partial^2}{\partial x^2} + \frac{\partial^2}{\partial y^2},$$

where $V_{12} = \mu_{12}E/2\hbar$, μ_{12} is the dipole moment, k_L is the wave vector of the ω_L wave, $\gamma_\parallel$ and $\gamma_\perp$ are the longitudinal and transverse relaxation constants, respectively, and N is the density of the medium. The boundary condition is

$$E|_{z=0} = E_0 \exp\left(-\frac{x^2 + y^2}{2\alpha} - \frac{t^2}{2\tau} - it\Delta\right),$$

where α is the transverse laser-beam diameter and τ is the laser-pulse duration.

Numerical modeling was carried out on an IBM PC with a 40-MHz 486DLC microprocessor. We used special Crank–Nicolson schemes with variable steps and a program called DUMKA, which is designed for solving stiff differential equations by explicit stable methods.[6] We created adaptive grids in the space transversal layer (x, y), and we used an algorithm with a variable approximation step of depth z.

Two-dimensional calculations of the modification of the scattered-laser-pulse frequency–angular characteristics ($\tau = 0.1$ ns; laser intensity, $I = 0.64$ MW/cm^2, $N = 1.7 \times 10^{14}$ cm^{-3}) for propagation in a resonant medium along the z axis and at two symmetrical laser detunings $\Delta = \pm 48$ GHz are presented in Figs. 1 and 2. Parameters of an atomic transition correspond to the sodium D_2 line, $\gamma_\parallel = 5\gamma_\perp = 5\gamma$, where γ is the spontaneous relaxation rate. One can see that there are both red-shifted and blue-shifted components of scattering at small values of z. As the pulse propagates in the medium, both shifted components and laser components experience substantial frequency and spatial broadening. Note that at large values of z the main parts of scattered red spectra are shifted away from ω_0

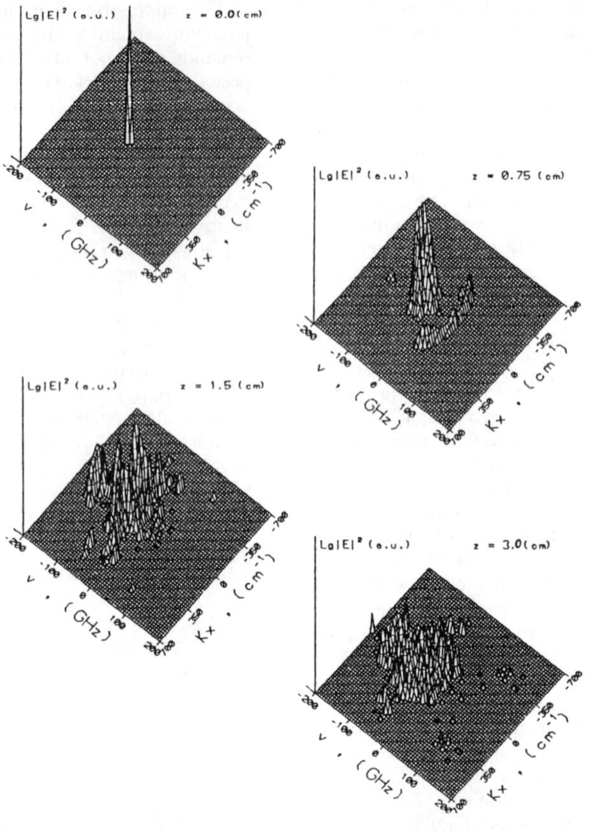

QWD10 Fig. 1. Calculated modification of scattered-laser-pulse frequency–angular characteristics for propagation in a resonant medium along the z axis for a laser detuning of $\Delta = -48$ GHz.

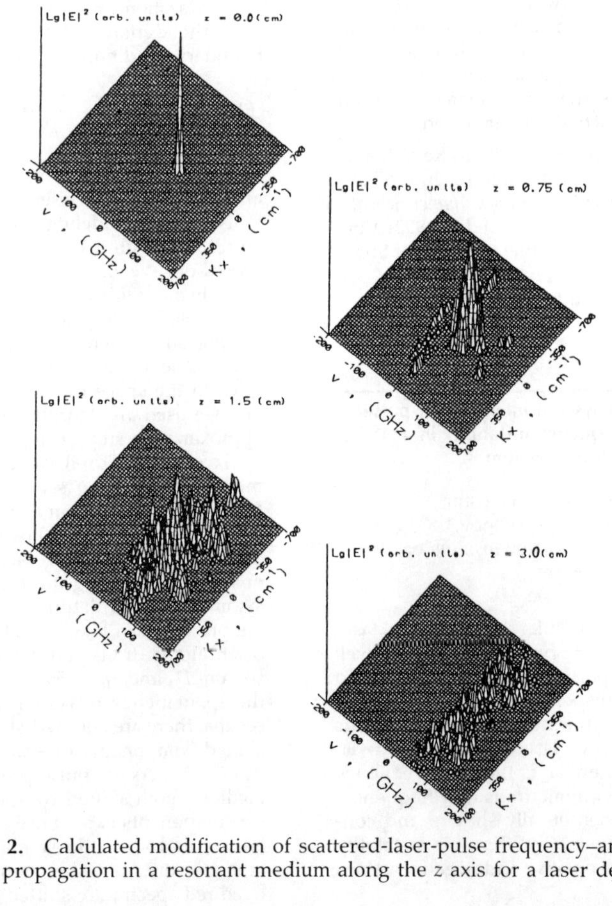

QWD10 Fig. 2. Calculated modification of scattered-laser-pulse frequency–angular characteristics for propagation in a resonant medium along the z axis for a laser detuning of $\Delta = +48$ GHz.

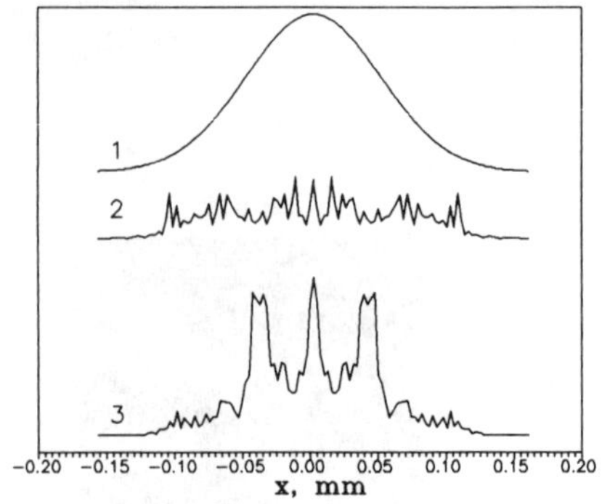

QWD10 Fig. 3. Curve 1, spatial profile of laser beam ($z = 0$). Curves 2 and 3, integrated within τ, the spatial profiles of scattered laser beam at $z = 30$ mm for $\Delta = +48$ GHz (curve 2) and $\Delta = -48$ GHz (curve 3). Other parameters are the same as those in Figs. 1 and 2.

relative to ω_L. The most interesting result is the existence of small-scale instability at large values of z, at both $\Delta < 0$ and $\Delta > 0$ (see Figs. 1 and 2), which confirms the theoretical prediction in Ref. 3 and the experimental observation in Ref. 4.

Figure 3 shows, integrated within τ, the spatial profiles of scattered laser beam at the large distance $z = 30$ mm. Figure 3 shows that the rings appear in the laser-beam section for both positive and negative Δ, which also correspond to the data of Ref. 4.

In conclusion, it should be noted that in spite of using two-dimensional calculations and short pulse duration (restricted by the computer), our numerical results show that the whole complex of the earlier experimental data can apparently be explained on the basis of the Maxwell–Bloch equation system (the NFWM model is a consequence of it) without use of additional or alternative theoretical approaches.

*Moscow Institute of Physics and Technology, Dolgoprudny 141700, Moscow region, Russia

1. M. E. Grenshow and C. D. Cantrel, Phys. Rev. A **39**, 126 (1989).
2. L. You, J. Mostowsky, and J. Cooper, Phys. Rev. A **46**, 2903 (1992).
3. A. A. Panteleev, V. A. Roslyakov, A. N. Starostin, and M. D. Taran, Zh. Eks. Teor. Fiz. **97**, 1777 (1990) [Sov. Phys. JETF **70**, 1003 (1990)].
4. D. I. Chekhov, D. V. Gaidarenko, A. G. Leonov, A. A. Panteleev, and A. N. Starostin, Opt. Commun. **105**, 165 (1994).
5. J. F. Valley, G. Khitrova, H. M. Gibbs, J. W. Grantham, and X. Jiajin, Phys. Rev. Lett. **64,** 2362 (1990).
6. V. I. Lebev, *Calculation Processes and Systems*, 8th ed. Nauka, Moscow, 1991.

QWD11

Evolution of amplified solitons

A. V. Mikhailov, *L. D. Landau Institute for Theoretical Physics, Kosygina 2, Moscow 117940, Russia*

In many practical cases optical solitons in fibers can be adequately described by the nonlinear Schrödinger equation[1] (NLS),

$$i\frac{\partial E}{\partial Z} + \frac{\partial^2 E}{\partial T^2} + 2|E|^2 E = 0. \qquad (1)$$

Here E represents the normalized complex envelope of the electric field, and Eq. (1) is written in the frame moving with the group velocity. The important feature of the soliton solution of the NLS is a definite relation between amplitude E_0 and width $T_0 \sim 1/E_0$ of a soliton. To compensate the fiber loss, amplifiers (localized in space) are used.[2,3] The effect of amplification is just a renormalization of the input amplitude: $E_{out}(Z_0, T) = \gamma E_{in}(Z_0, T)$, where $\gamma > 1$ is a coefficient of the amplification ($\gamma < 1.5$; otherwise the amplified pulse train would contain more solitons than the original one.[4]

We study both numerically and analytically the evolution of pulses (amplified solitons), taking $\hat{E}(0, T) = E_{out}(Z_0, T)$ (where $E_{in}(Z_0, T)$ is a single- or multiple-soliton solution) as an initial condition for the NLS of Eq. (1). The amplified pulses are not a simple set of solitons because the width–amplitude relation is violated. Each act of amplification generates a nonsoliton radiation that disperses out of the pulses and interacts with solitons. As a result of this interaction, the amplitudes of solitons oscillate in Z, and these oscillations slowly decay. But, what is more important, we have found that the above interaction changes the soliton separation times in finite shifts and therefore affects the arrival times.

Our analysis is based on an asymptotically exact solution of Eq. (1) that contains both radiation and one soliton and that we have obtained by the inverse-transform method. It follows from the solution obtained that the maximum amplitude oscillates with a frequency that approaches $\Omega = (2\gamma - 1)^2 E_0^2$. The amplitude of the oscillations decays as $1/\sqrt{Z}$ (see Fig. 1). The coincidence of the numerical results with our analytical solution is such that we can not even distinguish between the curves.

Two- or multiple-soliton solutions on an asymptotic nonsoliton-radiation background can be obtained as well. Suppose we have two amplified solitons separated by time interval $T_1 E_0 >> 1$. Each pulse emits a radiation that affects the parameters of another soliton. The most important result of that interaction is a shift of solitons toward each other. The shift is finite and depends on the coefficient of amplification γ. The asymptotic (total) shift is (Fig. 2)

$$\Delta T = \frac{2}{E_0(2\gamma - 1)} \log\left[\frac{\Gamma^2(\gamma)}{\Gamma(2\gamma - 1)}\right]. \qquad (2)$$

This result is in excellent agreement with our computer simulations Fig. 3).

The numerical analyses were made by the use of a modified split-step method.

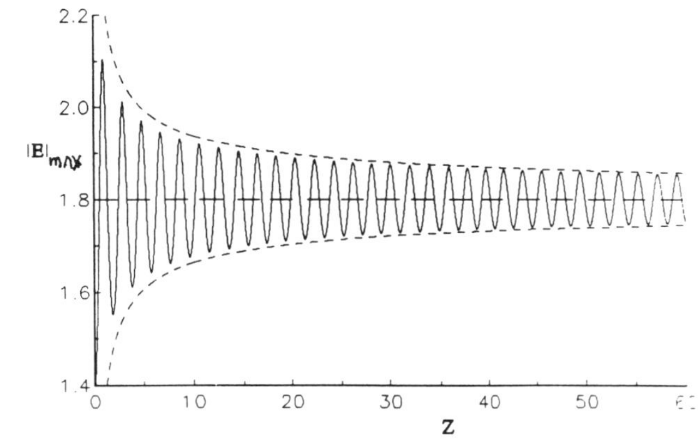

QWD11 Fig. 1. Maximum soliton amplitude $|E_{max}|$ as a function of Z for initial data $E_{out}(0, T) = 1.4\ \mathrm{sech}(T)$, $\gamma = 1.4$. Solid curve is a numerical result, and dashed curve is the boundary of oscillations, which follows from the solution obtained.

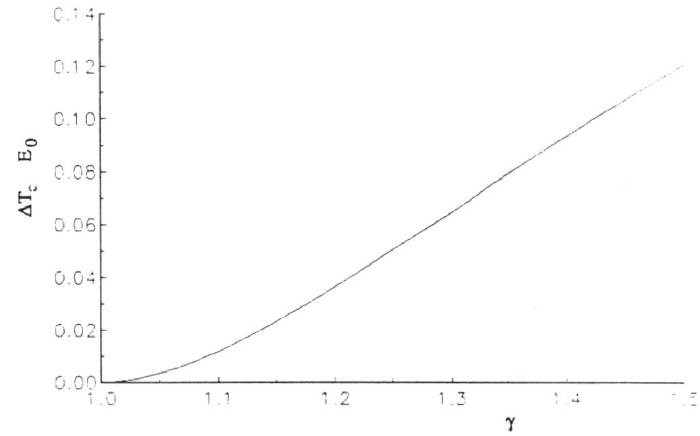

QWD11 Fig. 2. Time shift of solitons due to interaction with radiation as a function of the amplification coefficient γ.

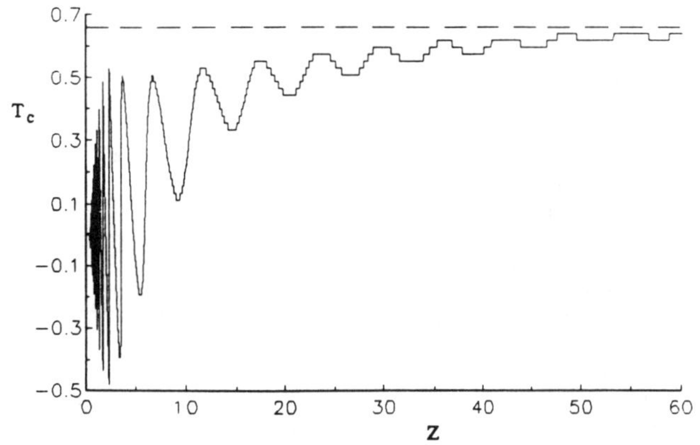

QWD11 Fig. 3. Dashed line is the theoretically predicted time shift of the maximum of the solitons, and the solid curve is the result of a computer simulation.

We used the periodic boundary conditions $E(T, Z) = E(T + L, Z)$, $-L/2 < T < L/2$. The value of L should be large enough so as not to influence the solutions of our problem. To avoid such an influence we also used some dumping at the ends. We have used $L = 360$, $N = 16384$, where N is the number of Fourier modes. The mesh size was taken as $\delta Z = (\delta T)^2$, $\delta T = L/N$. The runs were verified by calculation of the spectral transform. The maximum relative error of the isospectrality was less than 0.1%.

This work has been supported in part by the Russian Foundation of Basic Research and by International Scientific Foundation grant MLY000.

1. A. Hasegawa and F. Tappet, Appl. Phys. Lett. **23**, 142 (1973).
2. Y. Kodama, A. Maruta, and A. Hasegawa, submitted to J. European Opt. Soc. B.
3. L. F. Mollenauer, E. Lichtman, M. J. Neubelt, and G. T. Harvey, Electron. Lett. **29**, 910 (1993).
4. J. Satsuma and N. Yajima, Prog. Theor. Phys. Suppl. N **55**, 284 (1974).

QWD12

Possible mechanisms of second-harmonic generation in Si/GeSi periodic structures

D. V. Petrov,* V. T. Mashanov, O. P. Pcheljakov, V. V. Trifutin, *Institute of Semiconductor Physics, Russian Academy of Sciences, Siberian Division, Novosibirsk 630090, Russia*

Optical second-harmonic generation (SHG) in periodic structures on Si has been given particular attention recently (see, for example Ref. 1). We present results on SHG by reflection from sequences of Si layers and Ge_xSi_{1-x} solid-solution layers on the surface of Si.

The structures were grown by molecular-beam epitaxy on the surface of Si(111). The parameters of the structures are given in Table 1. Samples SL1–SL3 are superlattices, and sample F1 is the Ge_xSi_{1-x} solid-solution film. The superlattice periodicity was confirmed by TEM and RBS.

The experimental setup we used is similar to that of conventional surface-reflection SHG experiments. We used excitation at 1.06 nm from an Nd:YAG laser that produced 10-ns pulses at a 12.5-Hz repetition rate. A pump power of 5–50 kW/cm^2 per pulse in a spot of a 3-mm diameter was found to be sufficiently low to avoid laser-induced changes in the surface morphology.

The dependence of the second-harmonic intensity I_{SH} on the sample rotation angle was measured for four independent polarization combinations of the pump wave and its second harmonic. All results were normalized to the maximum intensity by *s* polarization of both pump and SH for GaAs(111) as a reference material.

The angular dependencies of I_{SH} for SL1, shown in Fig. 1, imply the existence of a threefold-symmetry axis. Samples of SL2, SL3, and F1 have basically the same angular dependencies but differ in peak amplitudes. The relative peak amplitudes of I_{SH} for *s* polarization of the pump and the SH are given in Table 1. The efficiencies of SHG for the superlattices and for the solid-solution film are of the order of 2 to 30 times that for a surface of pure Si(111).

One of the interesting features of SHG is the correlation between the SH intensity and the total number of Ge atoms within the penetration depth *L*. The latter can be estimated as follows. At a wavelength of 532 nm the absorption coefficients α_{Ge} and α_{Si} are 5.63×10^5 and 0.13×10^5 cm^{-1}, respectively. The effective absorption coefficient of the Si/Ge_xSi_{1-x} superlattice can be written as $\alpha = \alpha_{GeSi}(d_{GeSi}/D) + \alpha_{Si}(d_{Si}/D)$, where d_{GeSi} and d_{Si} are the thicknesses of the Ge_xSi_{1-x} and the Si layers, respectively, and $D = d_{GeSi} + d_{Si}$. For the estimations we anticipate that the absorption coefficient of the solid solution may be written as $\alpha = \alpha_{Si} + (\alpha_{Ge} - \alpha_{Si})x$, where *x* is the concentration of Ge. Then the total number of Ge atoms within an SH-field penetration depth $L \sim 1/\alpha$ is proportional to $G = xLd_{GeSi}/D$. The values of *L* and *G* are listed in Table 1. As shown in Fig. 2, the SHG efficiency increases with *G* for superlattices.

An order–disorder transition in the semiconductor-strained-superlattice Si/GeSi system that doubles the periodicity in the [111] direction has been observed.[2] The structure considered here can be described as a superposition of domains with a pseudodiamond configuration. In each domain the stacking in the [111] direction consists of (111) planes composed entirely of one atom type in the sequence SiSiGeGeSiSi. . . . This structure is stabilized by the presence of coherent, built-in strain. Obviously, with increasing numbers of Ge atoms, the number of noncentrosymmetric domains increases as well. Therefore, we assume that the observed increase of the SHG efficiency for these samples probably could be explained by the mechanism described in Ref. 2. The inhomogeneous mechanical strains, induced by mismatching of lattice parameters of Si and Ge_xSi_{1-x} and by threading defects and dislocations arising in the samples during growth, can induce a large number of these domains. It seems likely that because of its larger thickness, the solid-solution film has considerably fewer defects than do the superlattices. Therefore, we should expect it to have lower SHG efficiency. In fact, as shown in Fig. 1, the relevant value for the solid-solution film with the highest factor of *G* does not agree with the dependence we observe in superlattices.

As seen from Table 1, there is no simple correlation between the number of interfaces within the penetration depth *L* and the enhancement of the SH intensity. Thus we cannot explain the strong increase of SHG by the mechanism suggested in Ref. 3.

Table 1

sample	SL1	SL2	SL3	F1	Si(111)
thickness of SiGe (A^o)	300	180	150	10000	
thickness of Si (A^o)	900	1200	600	0	
x	0.3	0.3	0.3	0.5	
number of periods	9	20	15	1	
number of interfaces	19	41	31	2	
penetration depth L(μm)	0.18	0.29	0.22	0.035	
G	139	115	131	174	
total thickness of SL (μm)	1.08	2.76	1.125	1	
relative SH intensity (10^{-4})	4.1	1.0	2.2	0.27	0.14

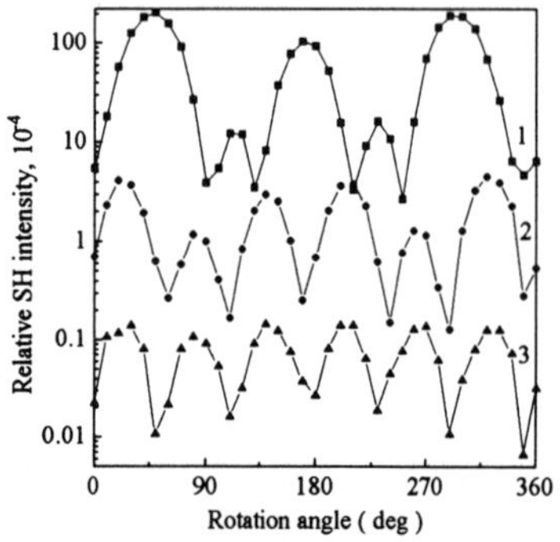

QWD12 Fig. 1. Angular dependence of SH-generation efficiency for SL1 (curves 1 and 2; *p–p* polarization) and for Si(111) (curve 3; *s–s* polarization).

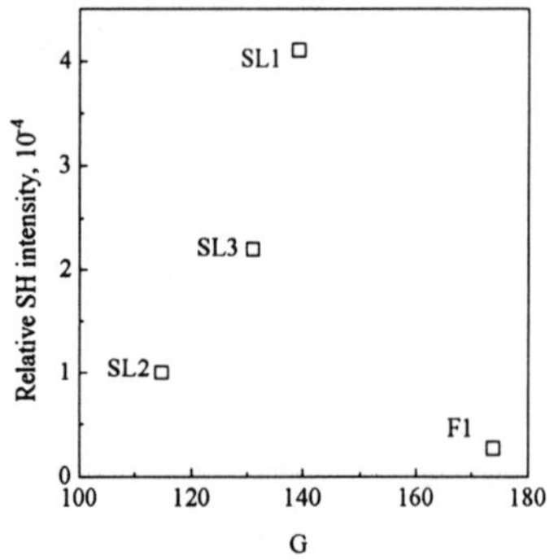

QWD12 Fig. 2. SH-generation efficiency versus *G*.

*Department of Physics, Federal University of Pernambuco, 50670-901, Recife, PE, Brazil

1. O. A. Akcipetrov *et al.*, in *Proc. CLEO/Europe '94*, p. 36.
2. A. Ourmazd, J. C. Bean, Phys. Rev. Lett., **55**, 765 (1985).
3. E. Ghahramani, D. J. Moss, J. E. Sipe, Phys. Rev. Lett., **64**, 2815 (1989).

QWD13

Femtosecond time-resolved Z-scan measurements of optical nonlinearities in solids and liquids

K. Y. Tseng, K. S. Wong, *Department of Physics, Hong Kong University of Science and Technology, Clear Water Bay, Kowloon, Hong Kong*

By using the single-beam Z-scan technique, both the sign and the magnitude of the nonlinear refractive index n_2 and the nonlinear absorption coefficient β can be readily determined.[1] However, this single-beam technique cannot distinguish the contributions to optical nonlinearities from different mechanisms, such as bound electronic motion, two-photon free-carrier generation, molecular Kerr reorientation and thermal effects. By using two laser beams (i.e., pump and probe beams) and introducing a temporal delay between them, a two-color picosecond time-resolved Z-scan was used to distinguish the various contributions.[2] Instead of using two-color light, we have extended the technique to a single-color time-resolved Z-scan by using cross-polarized pump-probe beams with femtosecond time resolution. Both techniques were successfully implemented by using a regenerative Ti:sapphire laser amplifier.

In order to obtain time-resolved Z-scan measurements, a single-beam Z-scan experiment must first be performed to locate the spatial positions of the peak and the valley.[2] Figure 1 shows such a spectrum for bulk ZnSe at $\lambda = 790$ nm, and the Z-scan curve indicates a positive optical nonlinear refraction n_2 at this wavelength. By measuring the time-dependent change in transmittance between the peak and the valley, i.e., $\Delta T_{p-v}(t)$, the dynamics of n_2 can be investigated. Figure 2 shows a single-color time-resolved Z-scan spectrum for ZnSe with pump and probe beams at $\lambda = 790$ nm; a change of sign in $\Delta T_{p-v}(t)$ near $t = 0$ is clearly shown. At picosecond and nanosecond time scales, n_2 is negative, and it decays exponentially with a time constant of $\tau = 4.9$ ns. The long decay time constant and the slight two-photon absorption observed at this excitation level (0.95 GW/cm^2) suggest that the negative n_2 is likely due to thermal contribution. The inset in Fig. 2 is an expanded view of the resolution-limited response (i.e., laser pulse width of ~200 fs) of $\Delta T_{p-v}(t)$ around the zero time delay. We believe that this ultrafast response is due to bound electronic motion, which gives rise to the positive n_2 seen in the single-beam Z-scan measurements (i.e., Fig. 1). A similar ultrafast bound electronic contribution was also observed in CaF_2 [Fig. 3(a)]. No slow component is seen in this case because the multiphoton generation of free carriers and the thermal contribution to the optical nonlinearity are negligible because of the large band-gap energy of CaF_2. Figure 3(b) shows the time-resolved Z-scan of liquid CS_2; a relaxation time of 1.17 ps due to molecular Kerr reorientation was observed, which is in good agreement with results obtained by using other time-resolved techniques.[3] A fast oscillating pattern near the zero delay was observed, and experiments are underway to investigate the possible origin of this feature.

In summary, we have implemented both single- and two-color time-resolved Z-scan techniques to measure the optical nonlinearities in solids and liquids. By using these techniques, dynamical processes with relaxation time constants ranging from a 200 fs to a few nanoseconds can be resolved.

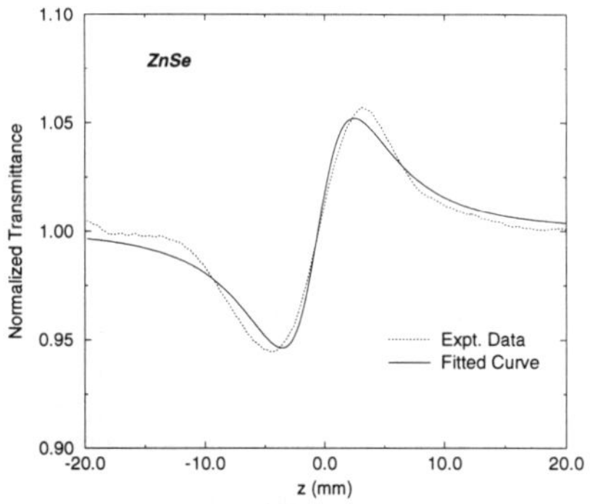

QWD13 Fig. 1. Single-beam Z-scan spectrum of ZnSe pumped at 0.95 GW/cm^2 at $\lambda = 790$ nm. A value of $n_2 = 2.33 \times 10^{-11}$ esu was obtained from the theoretical fitted curve.

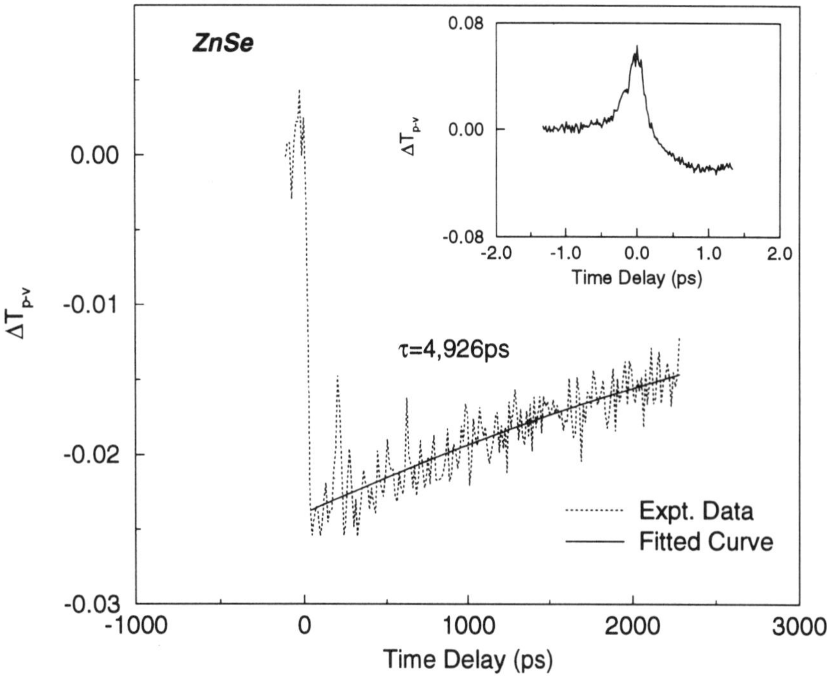

QWD13 Fig. 2. Time-resolved Z-scan spectrum of ZnSe at $\lambda = 790$ nm. A relaxation time constant of $\tau = 4.926$ ps due to thermal effects was obtained by using an exponential fit to the experimental curve. The inset shows the sign change of $\Delta T_{p-v}(t)$ and the instantaneous response of optical nonlinearity to the incident optical field around zero time delay.

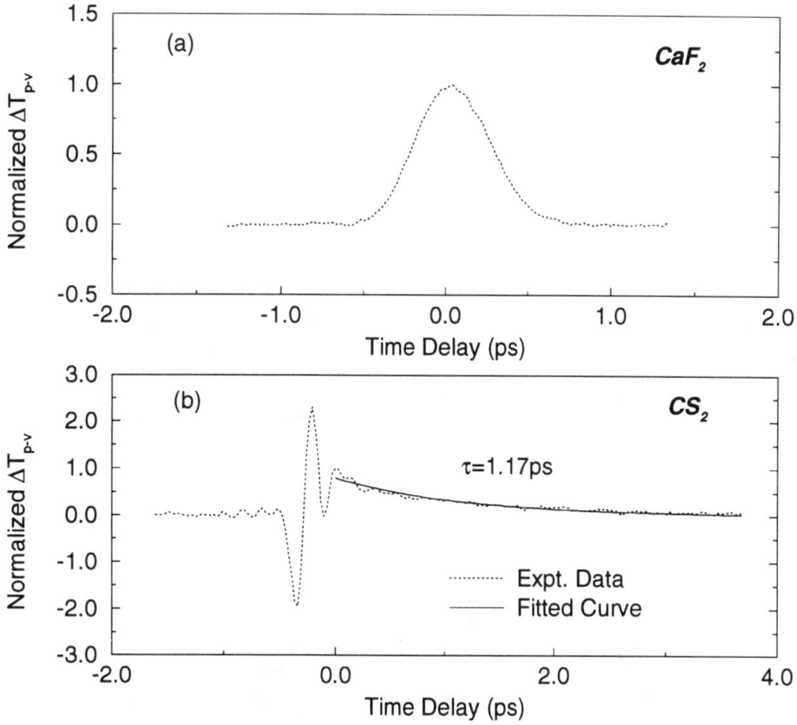

QWD13 Fig. 3. (a) Two-color ($\lambda_{pump} = 395$ nm and $\lambda_{probe} = 790$ nm) time-resolved Z-scan spectrum of CaF_2. (b) Single-color time-resolved Z-scan spectrum of CS_2 at $\lambda = 790$ nm.

1. M. Sheik-Bahae *et al.*, IEEE J. Quantum Electron. **QE-26**, 760 (1990).
2. J. Wang *et al.*, J. Opt. Soc. Am. B, **11**, 1009 (1994).
3. J. M. Halbout and C. L. Tang, Appl. Phys. Lett. **40**, 765 (1982).

QWD14

New degenerate-four-wave-mixing geometry in Kerr waveguides

C. Malouin, A. Villeneuve, R. A. Lessard, G. Vitrant,* *Center for Optics, Photonics, and Lasers, Pavillon Vachon, University of Laval, Ste.-Foy, Québec G1K 7P4 Canada*

Degenerate four-wave mixing (DFWM) is one of the most appropriate methods for characterizing third-order nonlinearity. We propose a new DFWM geometry in planar-waveguide geometry that is simpler than the previously reported technique.[1] Our geometry, shown in Fig. 1, consists of a probe beam incident from above the waveguide plane and two oppositely propagating guided pumps. Such DFWM signals are obtained in reflection and, what is particular to this geometry, also in transmission (see Fig. 1). Our theoretical model has shown that the reflectivity (DFWM energy divided by the probe energy) does not depend on the interaction length of the two pumps for waveguide modes far from cutoff. This result gives an evaluation of the nonlinearity by simply measuring the reflectivity and the guided pump energy decoupled out of the waveguide. We report here what we believe to be the first experimental observation of DFWM in waveguides with guided pumps and a free probe in a Kerr media.

Films of the planar waveguides were made of organic materials with side-chain-substituted polymer containing 4-dialkylamino-4'-nitro-stilbene (DANS). The thickness of the film was 2 μm. The light was coupled into the waveguide by two grating couplers spaced by 1 mm. The gratings were ion milled on fused-silica substrates.

The geometry was tested by using a Q-switched mode-locked Nd:YAG laser operating at 1.064 μm and producing 33-ps pulses. The linear coupling efficiency of the modes was between 4% and 6%. The DFWM reflected signal was measured versus one of the pump delays, and we found the response time of the nonlinearity to be smaller than the pulse duration. We were also able to capture the spatial profile of the signal with a CCD camera.

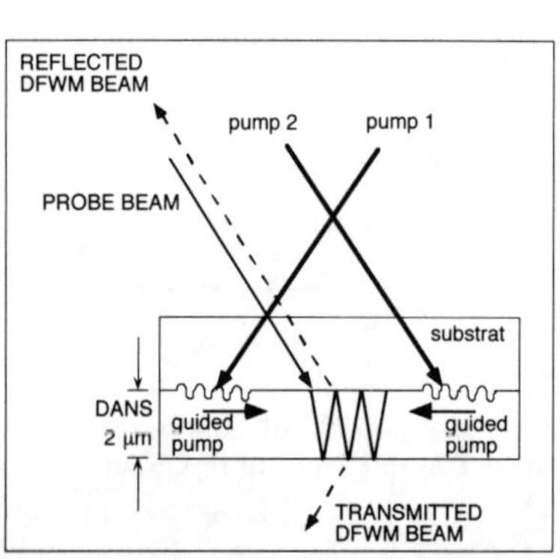

QWD14 **Fig. 1.** New four-wave mixing geometry consisting of a probe beam incident from above the film waveguide, which is pumped by two oppositely propagating guided modes. The probe is coming from the air side and is not guided.

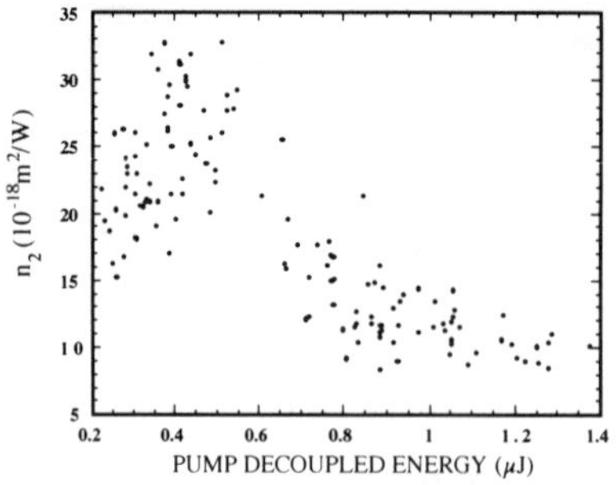

QWD14 **Fig. 2.** Nonlinear index of refraction n_2 as a function of decoupled pump energy.

The DFWM reflectivity was measured as a function of the incident energy. From this measurement, the nonlinear index n_2 was plotted (Fig. 2) as a function of the guided energy (measured as the decoupled energy). We can clearly see a dependence of n_2 on the incident energy. These values of n_2 compare well with those reported in the literature, i.e., $n_2 = 26 \times 10^{-18}$ m²/W (Ref. 2) to $n_2 = 9 \times 10^{-18} +$ m²/W (Ref. 3). The lower value of n_2 was measured at high energy, whereas the higher value was measured at low energy, which seems consistent with our results.

In conclusion, we have demonstrated a new DFWM geometry with guided pump beams and a free-space probe. It has advantages over the conventional DFWM in that the signal does not depend on the interaction length, but only on the probe reflectivity and the guided pump energy, simplifying the measurements. In addition to the reflected signal, a transmission signal can be observed. We have applied this technique to observe a saturation of the nonlinear index in DANS waveguides.

*LEMO (INPG-CNRS), ENSERG, BP 257, 38016 Grenoble, France

1. A. Gabel, K. W. DeLong, C. T. Seaton, and G. I. Stegeman, Appl. Phys. Lett. **51**, 1682 (1987).
2. M. J. B. Marques, "Nonlinear coupling to planar waveguides using diffraction gratings," Ph.D. dissertation (University of Porto, 1991).
3. G. Vitrant, L. Mayollet, B. Vögele, A. Rameix, R. Reineich, G. I. Stegeman, G. R. Möhlmann, W. H. G. Horsthuis, P. A. Chollet, and F. Kajzar, "Measurements of large nonresonant nonlinearities in doped polymers," to be published.

QWD15

Non-Gaussian noise fluctuations of the Stokes field for stimulated Brillouin scattering in optical fibers

Andrei A. Fotiadi, Eugenii A. Kuzin, *Laboratory for Quantum Electronics, A. F. Ioffe Physico-Technical Institute, Russian Academy of Sciences, Polytekhnicheskaya 26, St. Petersburg 194021, Russia*

The stochastic nature of initial stimulated Brillouin scattering (SBS) can lead to a deep modulation of Stokes intensity in fibers, even under conditions of strong pump depletion. It is known that at a moderate pump depletion, when parameter $T_1/T_2 >> 1$ (T_1 is the light propagation time through the effective amplification length and T_2 is the hypersound relaxation time), fluctuations correspond to the Gaussian statistics of temporal field distribution. As $T_1/T_2 \to 1$ the statistics of fluctuations can change. Gaeta and Boyd[1] were the first to show that at a sufficient pump depletion (by our estimate, parameter T_1/T_2 in Ref. 1 was ~4) the correlation function of the Stokes intensity $C(\tau) = (\langle I(t)I(t+\tau)\rangle - \langle I(t)\rangle^2)/\langle I(t)\rangle^2$ differs from the Gaussian by a more rapid initial fall and the presence of the region in which it is negative.

In our experiment[2] carried out at $T_1/T_2 \sim 1$ we observed an even more marked difference between $C(\tau)$ and the Gaussian (Fig. 1). A hormonic component with period $T \sim T_2$ dominates in it. Period T in the correlation functions calculated for different experimental realizations of the Stokes signal is a random value with mean $\langle T\rangle$ = 7.9 ns and dispersion $\sigma \sim 15\%$. Averaging the correlation functions over a series of realizations leads to a relaxing function $\langle C(\tau)\rangle$.

The analysis of the experimental results shows that the formation mechanism of fluctuations with non-Gaussian statistics can be connected with the property of selective amplification at SBS described by OS earlier.[3,4] According to our assumption, at $T_1/T_2 \sim 1$ a considerable variation of the initial Gaussian statistics of the Stokes signal occurs at the final stage of its amplification in the fiber, in the region in which the major pump-to-Stokes energy conversion occurs. At a strong pump depletion this region is, in essence, a selective SBS amplifier that is able to amplify efficiently only Stokes fluctuations with a characteristic time scale $\sim T_2$. The nonuniformity of amplification of fluctuations with different time scales by SBS leads to the change of their statistics.

To verify this hypothesis, we have performed computer simulation of the Stokes-wave dynamics in an SBS amplifier operating at $T_2/T_2 \sim 1$. As an input Stokes sig-

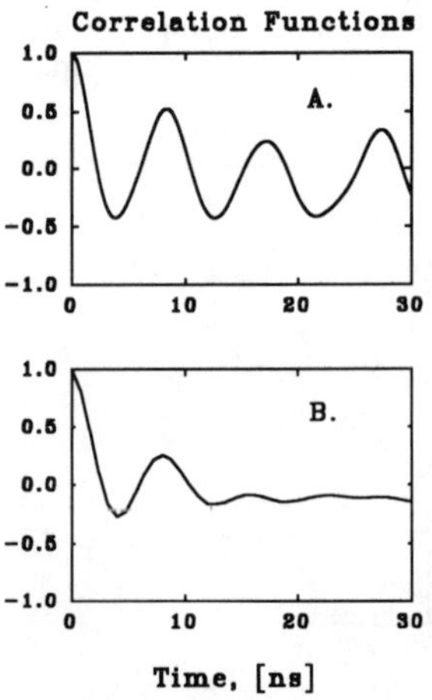

QWD15 **Fig. 1.** Normalized correlation function calculated from the experimental results at $T_1/T_2 \to 1$: example of the realization (curve A) and average of the set of eight realizations (curve B).

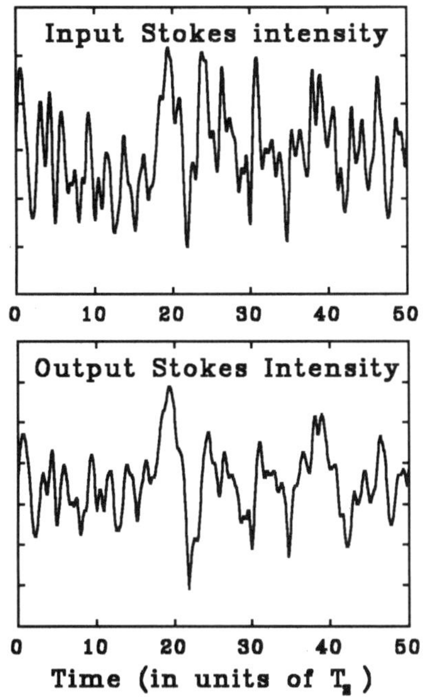

QWD15 Fig. 2. Example of the simulated realization of the input and ouput Stokes intensities in the SBS amplifier for $T_1/T_2 \to 1$.

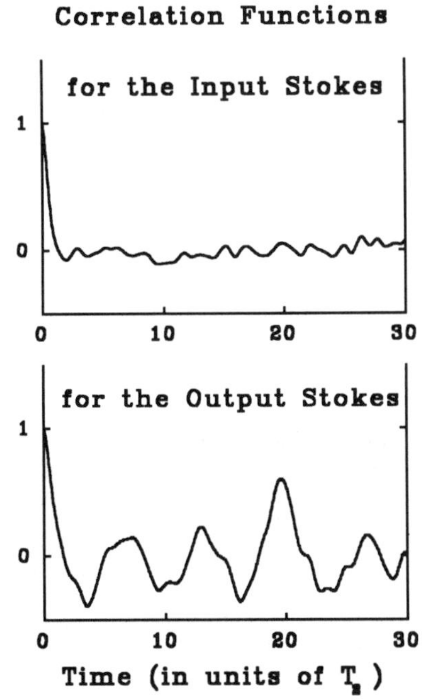

QWD15 Fig. 3. Correlation functions for input and ouput Stokes intensities averaged from the set of eight simulated realizations.

nal, a stochastic signal having Gaussian statistics with the spectrum width corresponding to SBS without depletion was used. The analysis of the set of realizations of the input and output Stokes signals (Fig. 2) has shown that the SBS amplifier enhanced most efficiently prolonged harmonic fluctuations with frequency $\sim 1/T_2$ rather than separate noncorrelated fluctuation peaks. As a consequence, a harmonic component is always present in the correlation function of the output Stokes signal and is preserved even in the correlation function averaged over a set of realizations (Fig. 3).

1. A. L. Gaeta and R. W. Boyd, Phys. Rev. A **44**, 3205 (1991).
2. A. A. Fotiadi and E. A. Kuzin, Sov. Phys. Tech. Phys., in press.
3. E. A. Kuzin, M. P. Petrov, and A. A. Fotiadi, in *Optical Phase Conjugation*, M. Gower and D. Proch, eds. (Springer-Verlag, New York, 1994), 74.
4. A. A. Fotiadi and E. A. Kuzin, in *Conference on Lasers and Electro-Optics*, 1994 Technical Digest Series, Vol. 8 (Optical Society of America, Washington, D.C., 1994), p. 84.

QWD16

Observation of Raman-assisted spatial cross-phase modulation in CS_2

Cid B. de Araújo, H. Ma, *Department of Physics, Federal University of Pernambuco, 50670-901 Recife, PE, Brazil*

We report changes in the spatial profile of an optical beam induced by another beam due to Raman-assisted cross-phase modulation (XPM). The effect is enhanced when the difference between the frequencies of the two incident beams is in resonance with the frequency of a vibrational Raman transition of CS_2 (ω_R = 655.7 cm^{-1}).

The two-color Z-scan technique[1,2] was used to characterize the phenomenon. The incident beams with parallel polarizations were focused onto a cell (10 mm long) containing CS_2. The probe-laser intensity was sufficiently weak not to induce a detectable nonlinear effect, whereas the pump-laser peak power ($\leq$40 kW) was strong enough to induce relevant XPM on the probe. The probe frequency ω_1 was fixed at 16 920 cm^{-1}, and the pump frequency ω_2 was varied in such way that $|\omega_1 - \omega_2| \approx \omega_R$. The laser's linewidths were $\leq$0.02 cm^{-1}.

The Z-scan measurements for an opened aperture placed in the far-field region were flat, indicating that nonlinear absorption is negligible for the laser frequencies and intensities used. The small-aperture two-color Z scan for two values of the Raman detuning, $\Delta = (\omega_2 - \omega_1) - \omega_R$, are shown in Fig. 1. The Z ordinate is measured along the beams' direction, and $Z < 0$ corresponds to locations of the sample between the focusing lens and its focal plane. The signal profile, with a peak followed by a valley, indicates a focusing of the probe beam induced by the pump beam. However, changes in the Z-scan profile were observed for different values of Δ, as shown in Figs. 1(a) and 1(b). The resonant enhancement due to the Raman nonlinearity is demonstrated in Fig. 2, in which the change in the transmitted intensity through the aperture is shown as a function of Δ. This is the first time to our knowledge that changes in the spatial characteristics of an optical beam induced by Raman-assisted XPM have been reported.

The curves in the figures illustrate the theoretical results obtained with the procedure introduced in Refs. 1–3. The influ-

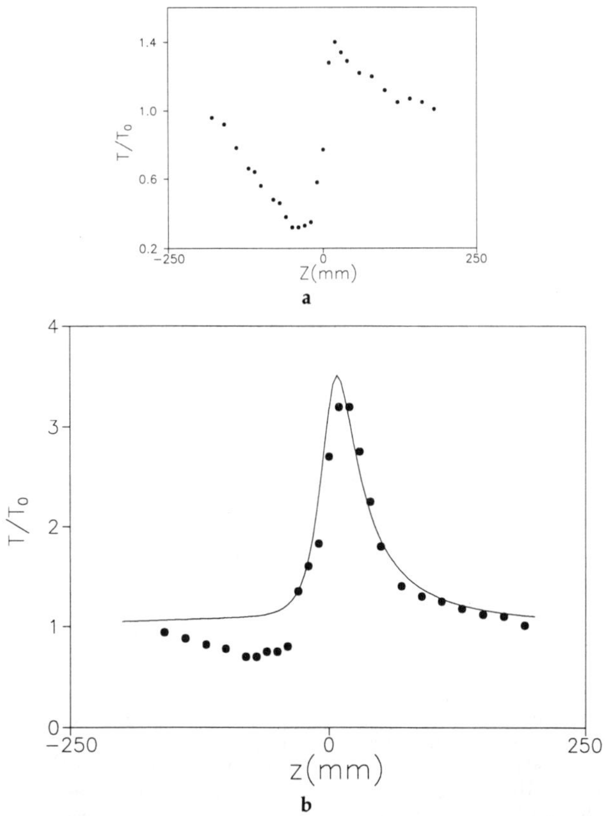

QWD16 Fig. 1. Normalized transmission as a function of Z, where T_0 is the transmission of the probe beam when the pump laser is blocked, for (a) Δ = 3000 cm^{-1} and (b) Δ = 0. The dots and the curve illustrate experimental and theoretical results, respectively.

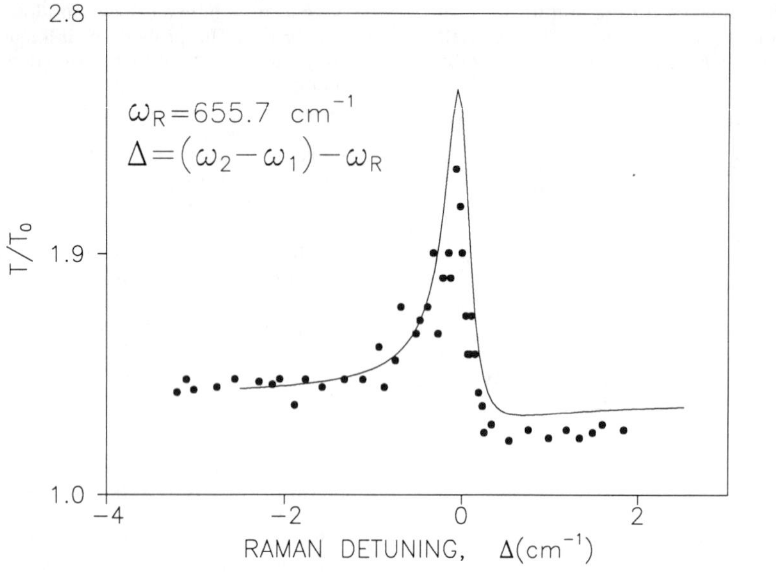

QWD16 Fig. 2. Normalized transmission as a function of the Raman detuning at Z = 32 mm.

ence of the Raman resonance is considered to be through the phase induced in the probe beam, which contains contributions from Re $\chi^{(3)}(\omega_1, \omega_2 - \omega_2, \omega_1)$ and Im $\chi^{(3)}(\omega_1, \omega_2, -\omega_2, \omega_1)$. The CS_2 parameters used to compare the data with the theoretical results were the Raman-dephasing relaxation rate (0.2 cm^{-1}), the nonlinear index of refraction n_2 = 1.2 × 10^{-11} esu, and the Raman strength (4.36 cm). Their values were obtained from Refs. 3 and 4.

The agreement between the experimental and the theoretical results shows that the Raman nonlinearity provides the main contribution for the XPM process when Δ ≈ 0. Moreover, this new variation of the Z-scan technique may allow new studies of Raman resonances in different materials.

1. H. Ma, A. S. L. Gomes, and C. B. De Araújo, Appl. Phys. Lett. **59**, 2666 (1991).
2. M. Sheik-Bahae, J. Wang, R. De Salvo, D. J. Hagan, and E. W. Van Stryland, Opt. Lett. **17**, 258 (1992).
3. M. Sheik-Bahae, A. A. Said, T. H. Wei, D. J. Hagan, and E. W. Van Stryland, IEEE J. Quantum Electron. **26**, 760 (1990).
4. H. Ma, A. S. L. Gomes, and C. B. de Araújo, Opt. Lett. **17**, 1052 (1992).

QWD17

Generation of dark pulses in additive-pulse mode-locked lasers

Gregg Sucha, *IMRA America, 1044 Woodridge Avenue, Ann Arbor, Michigan 48015*

Dark solitons have been produced by a number of means, including pulse shaping[1] and pulse evolution in dispersion-tapered fibers.[2,3] Here, I show numerical simulations that predict that dark pulses can be generated *directly* by an additive-pulse mode-locked (APM) laser. The APM laser, shown in Fig. 1, is a type of passively-mode-locked laser that comprises two interferometrically

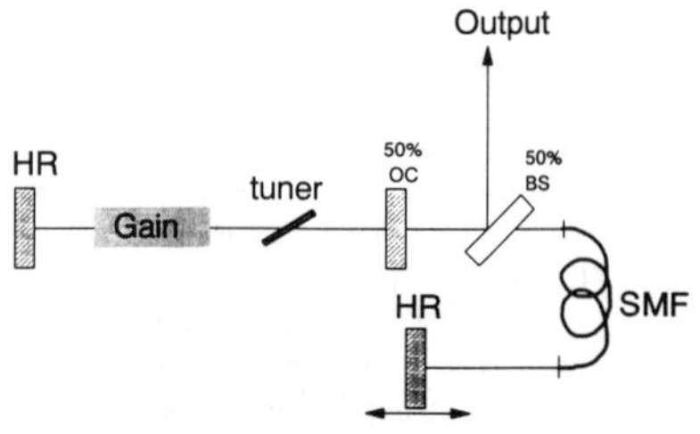

QWD17 Fig. 1. APM laser with single-mode fiber (SMF) in the control cavity. Phase bias Φ_0 is adjusted by changing the control-cavity length.

coupled cavities: a main cavity, containing gain and bandwidth limiting, and a control cavity, containing a Kerr nonlinearity implemented by a single-mode fiber. The phase bias $\Phi_0 = (2\pi/\lambda)\Delta L_{EXT}$ (where ΔL_{EXT} is the cavity-length mismatch) controls the laser behavior. The model used here assumes operation at the zero dispersion wavelength of the fiber in the control cavity. Thus, only self-phase modulation occurs in the fiber. It should be noted that this numerical model has also accurately predicted laser dynamics such as period-doubling and quasi-periodic behavior in the APM laser, as has been observed experimentally.[4]

Normally, bright pulses of picosecond or femtosecond duration are produced by setting the cavity-length mismatch to a phase bias in the range of $-2.5 < \Phi_0 < -1.5$ rad. In this range, the combination of nonlinearity and the interferometric recombination causes increasing cavity feedback with increasing intensity, thus giving the fast-saturable-absorber (FSA) action that is well known in the APM laser.[5] Dark pulses are produced by setting the cavity bias in the range $+1 < \Phi_0 < +2$, so that the same FSA mechanism actually *reduces* cavity feedback with increasing intensity. In addition to correct phase biasing, the cw background intensity and nonlinearity must be large enough to give nonlinear phase shift of $\phi_{nl} \approx \pi/2$.

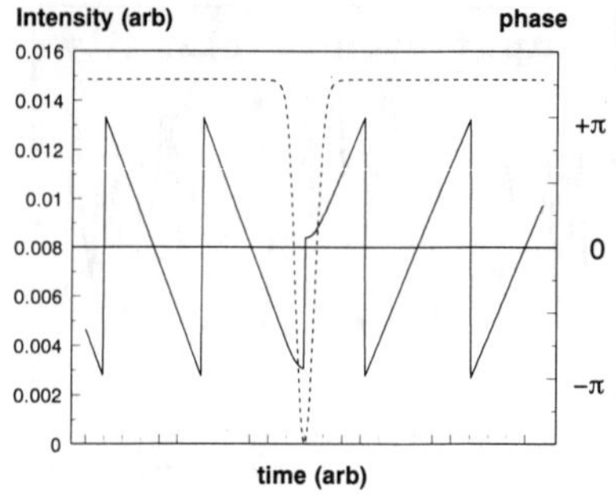

QWD17 Fig. 2. Dark-pulse solution for APM laser at bias of Φ_0 = +1.5 rad. Intensity (dashed curve) goes all the way to zero at the peak. Phase (solid curve) has a discontinuity of π at the peak.

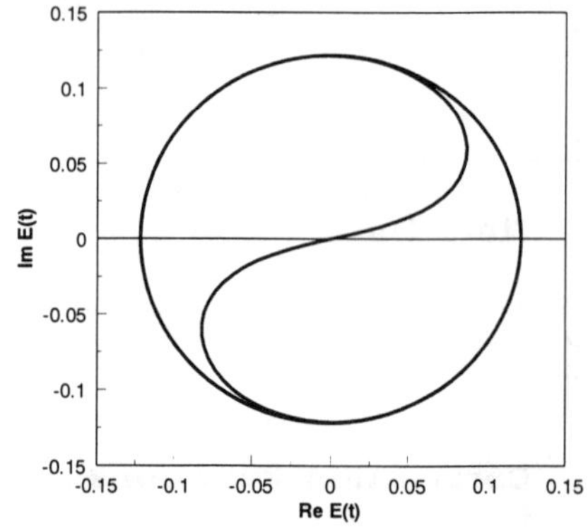

QWD17 Fig. 3. Phase structure of dark-pulse solution on a polar plot: Re $E(t)$ versus Im $E(t)$.

The dark-pulse solutions, shown in Fig. 2, are stable. Under certain conditions, they are accompanied by a smaller bright pulse that lies exactly half a cavity length away from the dark pulse and whose intensity can fluctuate from round trip to round trip. Figure 2 shows both the intensity and the phase structure of the pulse. The frequency of the cw background is higher on one side (increasing phase slope) and lower on the other side (decreasing phase slope). Interesting features of the phase structure can also be seen from a polar plot of the amplitude $E(t)$ in Fig. 3. Although the dark pulse from the APM laser has a phase discontinuity of π, similar to that of a dark soliton, its phase is otherwise quite different from that of a dark soliton. Thus far, the simulations have not generated any stable gray pulses.

Two things are remarkable about these results: first, dark pulses can be generated in the absence of any dispersion; second, they are generated by the inverse of the instantaneous FSA mechanism that produces bright pulses when the cavity is biased differently.

In a real laser containing 20 m of fiber, a cw intracavity power of 10 W would be appropriate for dark-pulse generation. Thus, APM lasers have the potential to generate high-power dark pulses with an average cw power of several watts.

1. A. M. Weiner, J. P. Heritage, R. J. Hawkins, R. N. Thurston, E. M. Kirschner, D. E. Leaird, and W. J. Tom-

linson, Phys. Rev. Lett. **61**, 2445 (1988).
2. P. V. Mamyshev, P. G. Wigley, I. Wilson, and G. Stegeman, in *Quantum Electronics and Laser Science*, 1993 Technical Digest Series, Vol. 12 (Optical Society of America, Washington, D.C., 1993), p. 289.
3. D. J. Richardson, R. P. Chamberlin, L. Dong, and D. N. Payne, Electron. Lett. **30**, 1326 (1994).
4. G. Sucha, S. R. Bolton, S. Weiss, and D. S. Chemla, in *Conference on Lasers and Electro-Optics*, 1993 Technical Digest Series, Vol. 11 (Optical Society of America, Washington, D.C., 1993).
5. E. P. Ippen, H. A. Haus, and L. Y. Liu, J. Opt. Soc. Am. B **6**, 1736 (1989).

QWD18

Measurements of excited-state diffusion coefficients by degenerate four-wave mixing

D. S. Glassner, B. Ai, R. J. Knize,
Department of Physics, University of Southern California,
Los Angeles, California 90089-0484

Diffusion of atoms in buffer gases is a fundamental transport phenomenon. Excited-state diffusion is important in testing intermolecular-scattering potentials, explaining light-induced drift and light-diffusive pulling, as well as in applications such as isotope separation. Although there have been many measurements of ground-state diffusion coefficients, excited-state coefficients have been measured only indirectly in pump–probe experiments. We have developed a new method to directly measure the diffusion of excited atoms by using degenerate four-wave mixing (DFWM).

A standard backward-DFWM configuration is used in our experiments. The grating spacing is dependent on the angle between the probe beam and the forward pump beams. In an atomic vapor, motion of the atoms will reduce or wash out the population gratings. The addition of a buffer gas will restrict the atomic motion so that this grating washout is reduced. For sufficient buffer-gas pressure and low input intensities, our two-level model that includes both ground-state and excited-state diffusion shows that the population difference between the ground and excited states, labeled with subscripts g and e, respectively, can be expressed as

$$N_e - N_g \propto \frac{1}{A + D_e k^2}\left(1 + \frac{D_e}{D_g}\right), \quad (1)$$

where A is the excited-state decay rate, D_e is the diffusion coefficient of the excited state, D_g is the diffusion coefficient of the ground state, and k is the wave vector of the grating formed by the forward pump and probe beams in DFWM. The contribution from the grating formed by the backward pump and probe beams can be described with a similar equation. Because the DFWM signal is proportional to the population difference squared and the angular dependence of the signal arises only from the grating wave vector k, it is seen from Eq. (1) that the angular dependence of the DFWM signal is dependent only on the diffusion coefficient and the decay rate of the excited state. A simple physical model explaining Eq. (1) is presented.

We have measured the diffusion coefficients of the potassium 4P states in xenon buffer gas and of the cesium 6P states in neon buffer gas. For potassium atoms, the two-level model described above must be modified because of fine-structure-mixing collisions. The observed angular dependence should be the same for using either the D_1 or the D_2 transition, and so only a

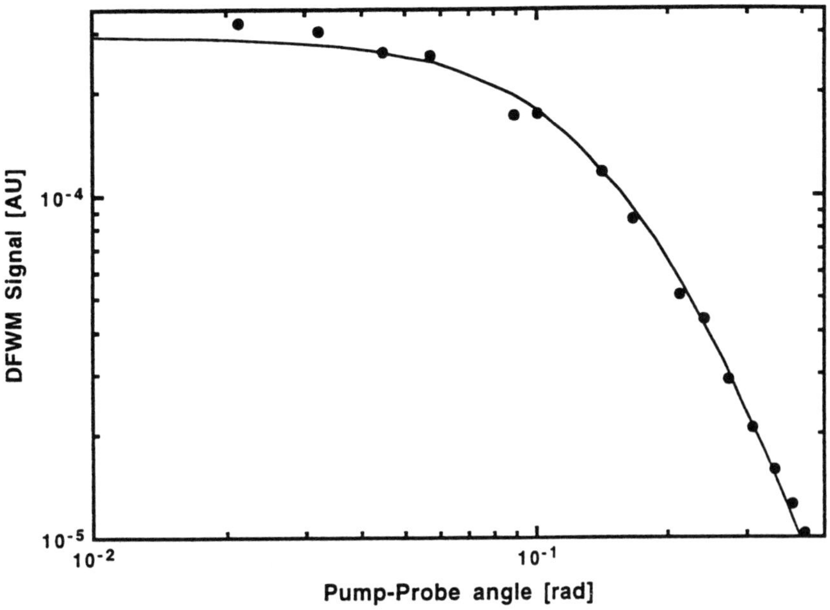

QWD18 Fig. 1. Measured DFWM signal versus the angle between the probe and the forward pump beams from a 1-mm-thick K-vapor cell with 252-Torr Xe at 155°C. The laser was tuned to the $4S_{1/2}$–$4P_{1/2}$ transition. Dots represent the experimental data, and curve is the best fit based on the use of Eq. (1) with 27 ns as the excited-state lifetime.

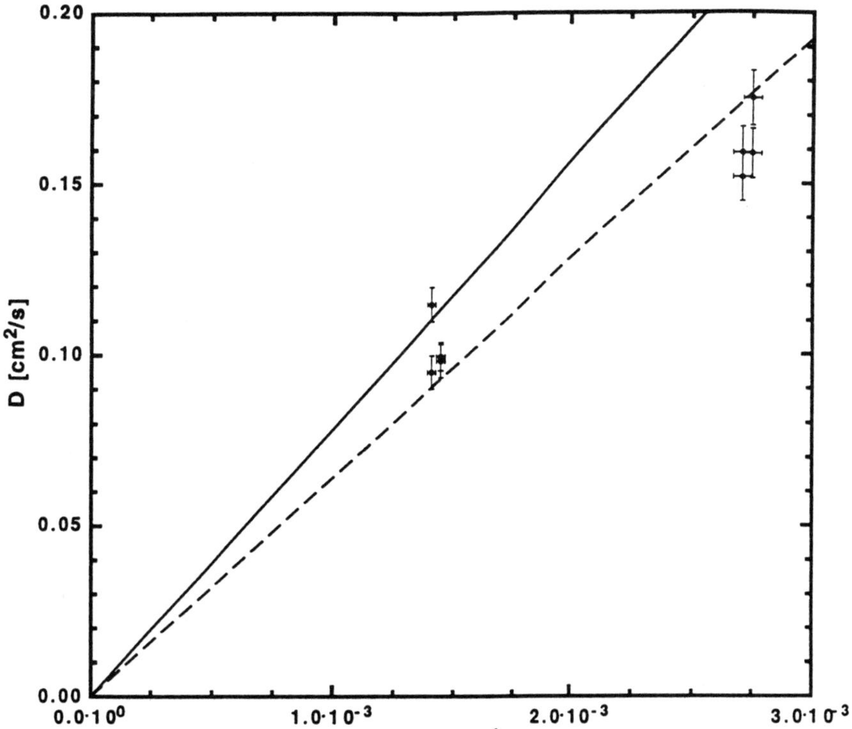

QWD18 Fig. 2. Measured diffusion coefficient versus the reciprocal of the xenon pressure in the K-vapor cell at 155°C. Dots with error bars are experimental results, and dashed line shows the best fit to the data. Solid line is calculated from a theory by Hamel *et al.*[1]

Wednesday Posters

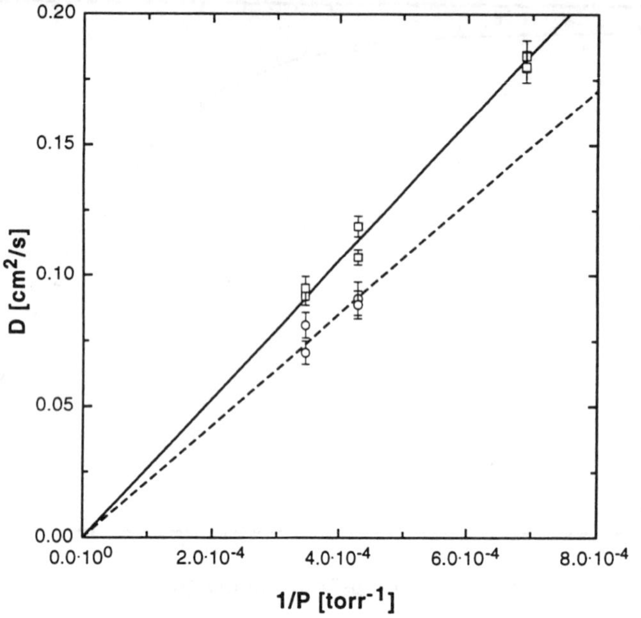

QWD18 Fig. 3. Measured diffusion coefficient versus the reciprocal of the neon pressure in a 1-mm-thick Cs-vapor cell at 96°C. Squares are experimental results at the D_1 transition ($6S_{1/2}$–$6P_{1/2}$), and circles are experimental results at the D_2 transition ($6S_{1/2}$–$6P_{3/2}$). Solid and dashed lines show best fits to the data.

weighted average of the diffusion coefficients of the $P_{3/2}$ and $P_{1/2}$ states can be determined. For cesium atoms in neon, fine-structure-mixing collisions are not as important, so that it is possible to determine the diffusion coefficients of the $P_{3/2}$ and $P_{1/2}$ states separately. Figures 1 and 2 show data for K in Xe at 155°C. Figure 3 shows the results for Cs in Ne at 96°C. The diffusion coefficient of K(4P) in Xe at 760 Torr and 155°C is 0.084(4) cm^{2}s^{-1}. The diffusion coefficients of Cs($6P_{1/2}$) and Cs($6P_{3/2}$) in Ne at 760 Torr and 96°C are 0.348(4) cm^{2}s^{-1} and 0.281(8) cm^{2}s^{-1}, respectively.

1. Hamel *et al.*, J. Phys. B **19**, 4127 (1986).

QWD19

Vibronic structure of the strongest two-photon band of poly-diacetylene toluene sulfonate

Brian L. Lawrence, William E. Torruellas, George I. Stegeman, Gregory Baker,* *Center for Research and Education in Optics and Lasers, University of Central Florida, 12424 Research Parkway, Orlando, Florida 32826*

The strong phonon coupling to the one-photon excitonic band is well known to produce the so-called phonon-mediated optical nonlinearity (PMON) in poly-diacetylene toluene sulfonate (PTS).[1–3] Indeed, the linear absorption spectrum of this one-dimensional electronic structure shows strong vibronic subbands that are, in principle, forbidden unless an anharmonicity of the electronic potential induces a Frank–Condon (F–C) displacement of the one-photon excited state (Fig. 1). On the other hand, until recently little was known about the location and strength of the theoretically predicted two-photon bands of such an inherently symmetric molecular system. The latter transitions are at the origin of the very large optical Kerr response observed within the optical gap of PTS. Also, nothing is known about the phonon (vibrational) coupling to the two-photon transitions.[4,5] It is therefore important to understand and properly assign such two-photon bands. Here, we intend to assign the strongest two-photon allowed transition located above the one-photon excitonic band.

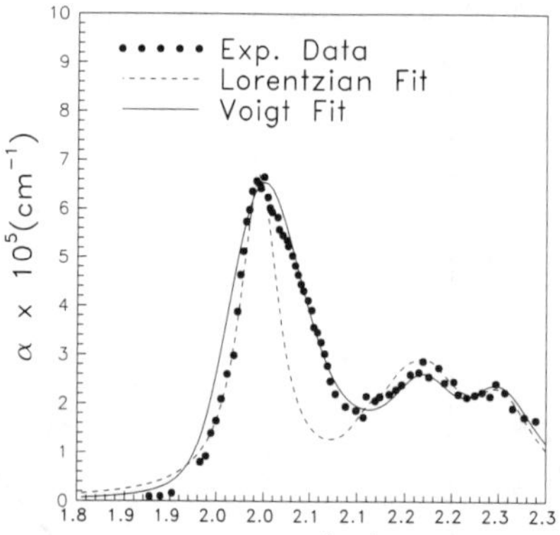

QWD19 Fig. 1. Linear absorption spectrum fitted to Lorentzian profiles, showing the first overtone of all four Raman modes.[7]

We reported last year the first two-photon absorption spectrum of PTS.[6] Such a spectrum shows a series of two-photon peaks. However, the strongest one by far is located approximately 0.5 eV above the one-photon exciton band. Associated with a sharp resonance at 2.65 eV, we observed what appeared to be a continuum of two-photon states (Fig. 2). We have performed tunable Z-scan measurements with laser pulses ranging from 65 ps to 150 fs. In all cases we have observed the same two-photon response, independent of pulse energy, thus eliminating the possibility of a two-step process. We have verified that indeed all of the observed resonances in the 2.4–3.5-eV range are strictly two-photon in nature. In other words, they involve an intermediate virtual (instantaneous) state.

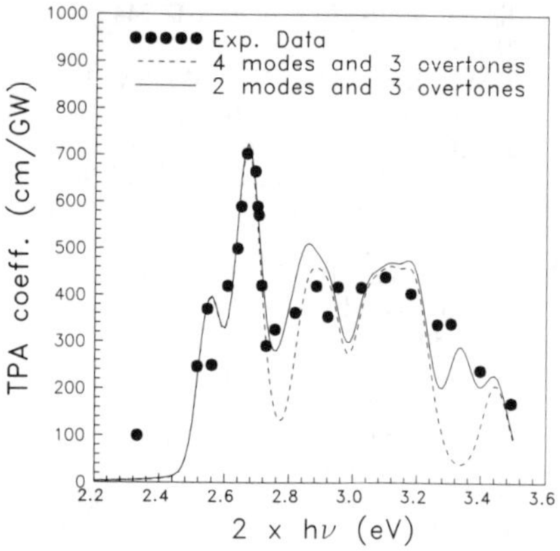

QWD19 Fig. 2. Two-photon absorption spectrum around 2.65 eV. The fits are to a three-level model that includes the linear superposition of two and four Raman modes.

The Raman ground-state and one-photon excited-state responses have been extensively studied.[7] As a result, it is well known that only four orthogonal Raman modes appear in the optical spectrum. Two of them are strongly dominant, namely, those associated with the carbon–carbon double- and triple-bond stretches. A fit to the linear absorption, taking into account the reported F–C displacements, results in excellent fits (Fig. 1). We have implemented a similar approach with the two-photon absorption spectrum, including only the two strongest modes observed in the ground state and the first one-photon excited states.[4,5] A reasonable fit is obtained when we include three overtones of the two strongest modes; only one transition dipole moment and two F–C displacements are fitted. We obtain an improved result, also shown in Fig. 2, when we linearly superpose the contributions of the other two weakest modes. The latter results show, as in the case of the one-photon exciton band, that the two-photon well is also displaced. A model describing the dynamics of the phonon–electron coupling is also discussed.

**Department of Chemistry, Michigan State University, East Lansing, Michigan 48824*

1. B. I. Greene, J. F. Mueller, J. Orenstein, D. H. Rapkine, S. Schmitt-Rink, and M. Thakur, Phys. Rev. Lett. **61**, 325 (1988).
2. G. J. Blanchard, J. P. Heritage, A. C. VonLehmen, M. K. Kelly, G. L. Baker, and S. Etemad, Phys. Rev. Lett. **63**, 887 (1989).
3. G. J. Blanchard and J. P. Heritage, J. Chem. Phys. **93**, 4377 (1990).
4. Y. Fujimura and S. H. Lin, J. Chem. Phys. **74**, 3726 (1981).
5. S. Aramaki, W. Torruellas, R. Zanoni and G. I. Stegeman, Opt. Commun. **85**, 527 (1991).

6. B. Lawrence, W. E. Torruellas, M. Cha, M. L. Sundheimer, G. I. Stegeman, J. Meth, S. Etemad and G. Baker, Phys. Rev. Lett. **73**, 597 (1994).
7. D. N. Batchelder, in *Polydiacetylenes: Synthesis, Structure, and Electronic Properties* D. Bloor and R. R. Chance, eds. (Kluwer, Dordrecht, The Netherlands, 1985).

QWD20

Measurements of the phase of surface second-harmonic generation in a total-internal-reflection geometry

Ekaterina Schwarzberg, Garry Berkovic, *Department of Materials and Interfaces, Weizmann Institute of Science, Rehovot 76100, Israel*

Nonlinear-optical techniques, particularly optical second-harmonic generation (SHG), have become valuable experimental tools in the study of surfaces. Useful analytical information about the surface depends on correct interpretation of the SHG signal. Knowledge of the phase of SHG is often very useful in this respect.

The basis of previous SHG phase measurements was to shine laser light on a surface so that some SH radiation was formed in either reflection or transmission, and then to combine this radiation interferometrically with a second SH signal created from another reference source. This procedure requires that in the reflection (transmission) from the surface of interest the harmonic and fundamental waves propagate collinearly. This is satisfied in the usual case of reflection at a gas/liquid or gas/solid interface. However, this is not so when SHG is created from a surface during total internal reflection (TIR) because the harmonic wave emerges at an angle different from that of the residual fundamental [see Fig. 1(a)]. SHG in the TIR geometry is very important because of large increase in sensitivity relative to simple reflection.

To achieve SHG interference in total internal reflection we have designed experimental setups to compensate for the inherent dispersion of SHG in TIR and thus to permit SHG interference and phase measurements. Using TIR from the hypotenuse of a regular 45° glass prism, we have successfully demonstrated two schemes. The first scheme uses a specially cut prism made from SF-11 glass to provide the right compensation, whereas the second uses two identical prisms [see Figs. 1(b) and 1(c)].

The two schemes have been analyzed and compared; the first scheme has the advantage of easier alignment and better interference contrast than the second scheme, but it suffers from the disadvantage of requiring a different compensating prism for each fundamental laser wavelength. The second scheme is, in contrast, polychromatic.

As a first application, we have checked the phase of the nonlinear Fresnel factors in TIR. Using the first SHG compensation scheme in TIR [Fig. 1(b)], we have verified the theoretical prediction[1] of the phase difference between two of the most important nonlinear Fresnel factors in TIR—those associated with the χ_{yzy} and χ_{zyy} components of the nonlinear susceptibility tensor.

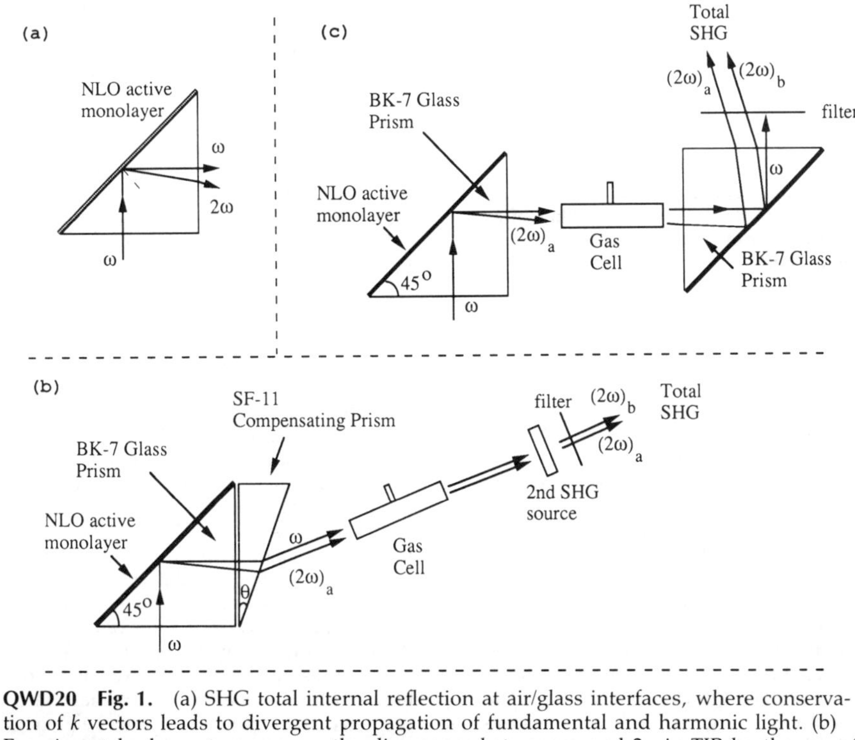

QWD20 Fig. 1. (a) SHG total internal reflection at air/glass interfaces, where conservation of *k* vectors leads to divergent propagation of fundamental and harmonic light. (b) Experimental scheme to overcome the divergence between ω and 2ω in TIR by the use of a high-index prism. For the 1064-nm light of the Nd:YAG laser, parallel emergence of ω and 2ω is obtained when the angle θ of the SF-11 prism is 17.5°. (c) Scheme to overcome the divergence by using two identical prisms. This setup is polychromatic, i.e., suitable for input lasers of variable wavelength.

1. B. U. Felderhof, A. Bratz, G. Marowsky, O. Roders, and F. Sieverdes, J. Opt. Soc. Am. B **10**, 1824 (1993).

QWD21

Second-harmonic generation in centrosymmetric magnetic films with domain walls

I. L. Lyubchanskii, *Donetsk Physico-Technical Institute, National Academy of Sciences of Ukraine, 340114 Donetsk, Ukraine*

Recent experimental observations of optical second-harmonic generation (SHG) in thin magnetic films of iron–yttrium garnet[1] have shown the strong dependence of SHG intensity on the presence of magnetic domain walls in the sample.

This paper is devoted to theoretical description of the possible mechanism of dipole-active SHG in thin centrosymmetric magnetic films with domain walls. The main idea is the following. Magnetic inhomogeneities such as domain walls lead to the appearance of a static electric polarization (and a static electric field) caused by the nonuniform magnetoelectric effect.[2] As a result, in magnetic films with domain walls there will appear a nonlinear polarization at double the frequency of the incident radiation. This nonlinear polarization can be described by third-order nonlinear-susceptibility-tensor components, some of which are different from zero, in a centrosymmetric medium.

Theoretical estimations of SHG intensity and its dependence on the domain-wall parameters were made.

1. R. V. Pisarev *et al.*, J. Phys. Cond. Mat. **5**, 8621 (1993).
2. V. Bar'yakhtar *et al.*, Sov. Phys. JETP Lett. **37**, 237 (1983).

QWD22

Interfacial atomic structure of a self-assembled alkyl thiol monolayer/Au(111)

M. S. Yeganeh, S. M. Dougal, R. S. Polizzotti, P. Rabinowitz, *Exxon Research and Engineering Co., Route 22 East, Annandale, New Jersey 08801*

Self-assembled monolayer (SAM) are potentially useful for corrosion inhibition, wear protection, and organic sensors. Perhaps the best characterized SAM system is $CH_3(CH_2)_{n-1}SH$, (C_n), adsorbed on an Au surface. The C_n/Au system, specifically its surface and chain orientation, have been widely studied.[1-3] Despite all the available experimental results, the location of sulfur chemisorption sites is known only by inference. The conventional belief has been that the S atoms reside at hollow sites of the gold substrate, generating a simple $(\sqrt{3} \times \sqrt{3})R30°$. The location of these sites is essential for a detailed evaluation of Au—S, Au—Au and S—S bonds at the interface.

In this work infrared-visible sum-frequency generation (SFG) was used to probe the structural symmetry of the C_{18}/Au(111) interface. The frequency dependence of the SFG signal at $\phi = 37°$ is shown in Fig. 1;

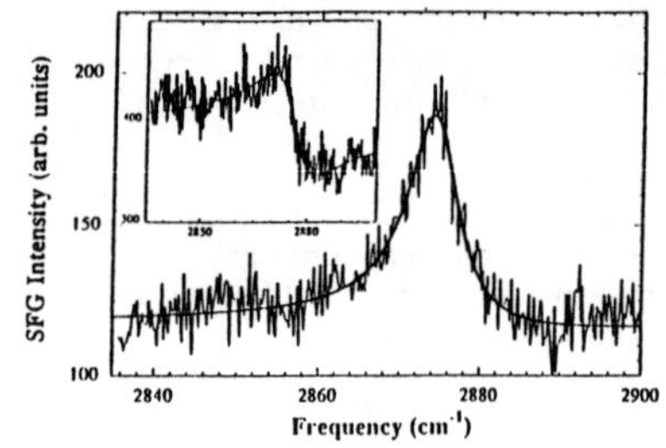

QWD22 Fig. 1. SFG spectrum of C_{18}/Au(111) in ppp-polarization at $\phi = 37°$. Insert: The same spectrum at $\phi = 7°$. The marked difference in the spectra results from the dependence of the relative phase between the background and resonance signal on the azimuthal angle, ϕ.

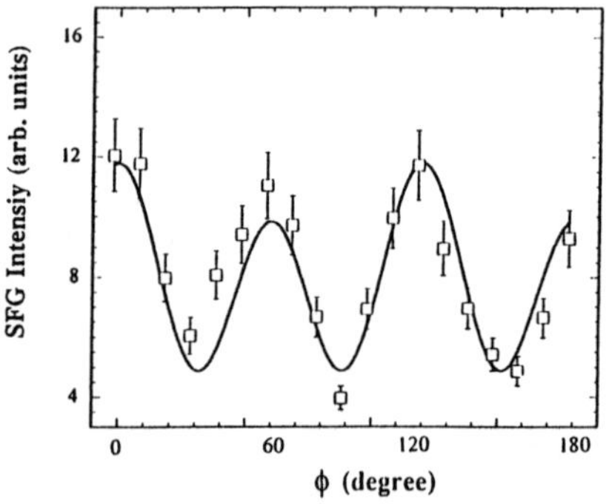

QWD22 Fig. 2. Resonance SGF intensity of the C_{18}/Au(111) as a function of the rotation angle, ϕ. Here ϕ is the angle between a mirror plane of the Au(111)(1 × 1) and the plane of the incidence. The solid line is the theoretical fit to the data.

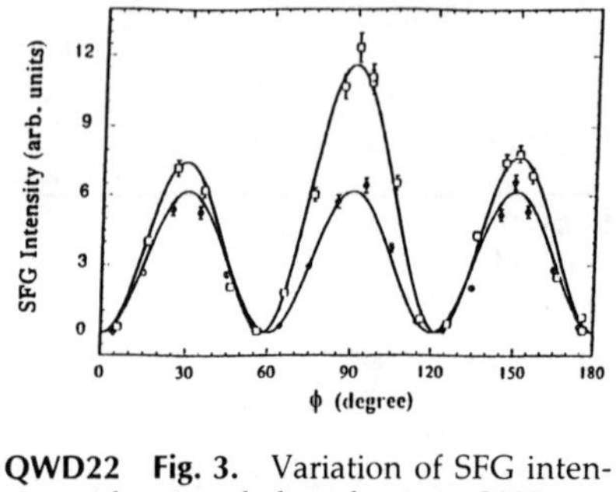

QWD22 Fig. 3. Variation of SFG intensity with azimuthal angle, ϕ, at 2800 cm^{-1} in sss-polarization for C_{18}/Au(111) (◇), and air/Au(111) (□). The solid lines are the theoretical fit based on (1 × 1)C_{3v} structure for C_{18}/Au(111) and ($\sqrt{3}$ × 23) structure, for air/Au(111).

where ϕ is the angle between a mirror plane of the Au(111)(1 × 1) substrate and the plane of incidence. The SFG spectrum of the sample was measured at different azimuthal angles, ϕ. Each spectrum was fit to a standard model consisting of a constant background and a Lorentzian resonance. In this way, the angular dependence of the background and the resonance signals was determined. Because the relative phase between the resonance and the background varies as a function of ϕ, this procedure is essential for separating the background and the resonance contribution from the total radiation intensity.

The resonance SFG signal is generated at the air/monolayer interface by CH_3 groups, and its dependence on the sample orientation is directly related to the position of the sulfur atoms at the Au interface. This results from the fact that the position of the S atoms, taken with a specific chain geometry and domain distribution, uniquely determines the position of the CH_3. The variation of the resonance SFG intensity as a function of the azimuthal angle is shown in Fig. 2. Our experiments reveal an unexpected three-fold rotational symmetry at the sulfur adsorbate sites of the C_{18}/Au(111) system. This result precludes the sulfur atoms being located solely at the hollow sites of the gold substrate. Such a configuration would give rise to an isotropic SFG signal. To generate such three-fold symmetry, the sulfur atoms must be located at either bridge and hollow sites, at two different bridge sites, or at two different hollow sites.

In addition to adsorbate site geometry, the effect of the thiol chemisorption on the surface stress of the gold substrate was also examined. Measured SFG rotational anisotropy of the gold substrate, shown in Fig. 3, is consistent with ($\sqrt{3}$ × 23) and (1 × 1) reconstruction patterns for air/Au(111) and C_{18}/Au(111). These measurements show that the surface lattice of clean Au(111) is stable in air and relaxes upon the chemisorption of S atoms.

1. R. G. Nuzzo and D. L. Allara, J. Am. Chem. Soc. **105**, 4481 (1983).
2. Camillon, C. E. D. Chidsey, G. Liu, and G. Scoles, J. Chem. Phys. **98**, 3503 (1993).
3. P. Fenter, P. Eisenberger, and K. S. Liang, Phys. Rev. Lett. **70**, 2447 (1993).
4. K. G. Huang, D. Gibbs, D. M. Zehner, A. R. Sandy, and S. G. J. Mochrie, Phys. Rev. Lett. **65**, 3313 (1990).

QWD23

Preparation of thin single-crystal films of a new organic nonlinear optical material (APDA)

Dan Zhi, M. Thakur, *Auburn University, Auburn, Alabama 36849*

We have used the shear method to prepare thin single-crystal films of a new organic second-order optical material, APDA (8-(4'-acetylphenyl)-1,4-dioxa-8-azspiro[4,5]decane). The films were prepared on quartz and oxide-coated silicon substrates. The thickness of the films is about 5 μm. Optical-microscopic studies have shown that the films have excellent optical uniformity. The film is colorless, with an absorption edge below 400 nm. In the first report[1] on this material, preparation of bulk crystals from the melt was discussed. The melt growth led to some defects that we do not observe in shear growth from solution. The material has a large second-order susceptibility that can be used in blue-light generation by the frequency doubling of diode lasers. The availability of the material in the form of thin films will allow fabrication of waveguides for phase-matched second-harmonic generation. More detailed studies of this new material are in progress.

1. K. Kagawa *et al.*, J. Cryst. Growth **139**, 309 (1994).

QWD24

Generation of low-frequency current and teraherz radiation by using χ^3 in semiconductors

J. B. Khurgin, *Department of Electrical and Computer Engineering, Johns Hopkins University, Baltimore, Maryland 21218*

Optical methods of generating ultrafast electrical signals have been the focus of attention of a number of researchers in the last few years. Although the second-order nonlinear optical processes have long been known to generate low-frequency signals, the third-order processes were not seriously considered until recently. The obvious advantage of the third-order processes is that they exist in all types of materials, including centrosymmetric and amorphous, and do not require an external bias. The nature of third-order optical rectification requires the simultaneous presence of two mutually coherent fields at frequencies ω and 2ω, respectively:

$$\mathbf{P}_{dc} = \chi^{(3)}(0;\ \omega,\ \omega,\ -2\omega)\mathbf{E}_\omega\mathbf{E}_\omega\mathbf{E}_{2\omega}. \quad (1)$$

Thus the sign of the nonlinear polarization does not depend on intrinsic asymmetry or the sign of the applied voltage, as in the case of second-order rectification, but does depend on the relative phases of the fundamental and second-harmonic fields; i.e., there is a possibility of optically controlling the sign of the nonlinear response.

More recently, it has been observed experimentally[1] that after the propagation of fundamental and second-harmonic signals through a Ge-doped glass fiber, the fiber becomes an efficient frequency-doubling medium. This is a dc-induced SHG. The exact nature of the process that created the permanent dc field remains controversial. It is unknown whether it is the third-order rectification,[2] or a coherent photogalvanic effect,[3] also known as directional photocurrent.[4]

In our research we have united two theories and have shown that the directional photocurrent can be described as nothing more than a third-order nonlinear process in the momentum gauge. We have evaluated the third-order conductivity $\sigma^{(3)}(0, \omega, \omega, -2\omega)$ and have demonstrated that it includes two components—real and virtual. We have also found the magnitude of each component for various materials. As an example, Fig. 1 shows the dispersion of the third-order conductivity for GaAs. Curve (a) corresponds to the real part of nonlinear conductivity associated with the preferential excitation of carriers with one sign of the momentum over the other. Curve (b) corresponds to the virtual excitation, or mixing, of three frequencies. We have also estimated the efficiency of low-frequency current generation for various methods of energy extraction, such as direct radiation into space, coupling into a transmission line, and a waveguiding geometry with phase matching. Our results show that up to few

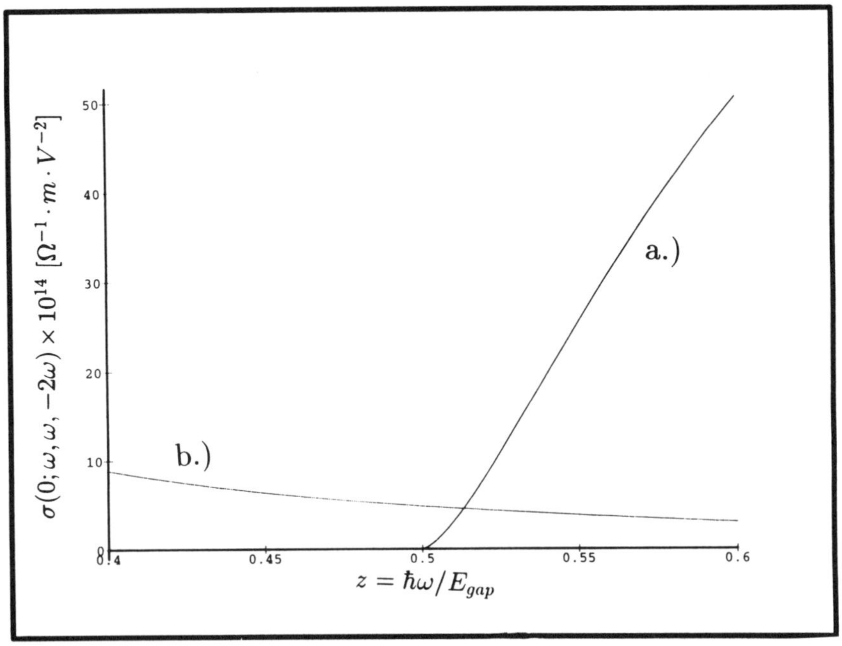

QWD24 Fig. 1. Dispersion of the nonlinear conductivity σ(0; ω, ω, −2ω) in GaAs near the half-band-gap: (a) real component; (b) virtual component.

microwatts of low-frequency power can be obtained from 1 W of near-IR light.

1. Y. Sasaki and Y. Ohmori, Appl. Phys. Lett. B **39**, 466 (1981).
2. N. M. Lawandy, Opt. Commun. B **85**, 369 (1991).
3. N. Baranova, A. N. Chudinov, and B. Ya. Zel'dovich, Opt. Commun. B **79**, 116 (1990).
4. E. Dupont, P. B. Corcum, H. C. Liu, Z. R. Wasiliewski, and M. Buchanan, in *Proc. 7th International Conference on Superlattices, Microstructures and Microdevices*, Banff, Alberta, Canada, 1994, p. 496.

QWD25

Pump depletion through two-wave mixing in semiconductors

C. W. Lowry, T. M. Brennan,*
B. G. Hammons,* *Laboratory for Physical Sciences, University of Maryland, College Park, Maryland 20740*

Two-wave mixing in a nonlinear medium can induce coherent energy transfer (CET) between a probe field and a pump[1–4] that can be strong enough to significantly affect both the probe and the pump (e.g., giving gain in absorptive media or loss in an amplifier). We use mode competition in a semiconductor laser as a sensitive test for *coherent* depletion of the lasing field by a weak injected probe. With injection locking, the injected field becomes the dominant competitor for optical gain, gain is suppressed for all modes, and the laser switches to the injected frequency. However, we find that below the injection-locking threshold, depletion of the strongly lasing field by a probe causes the laser to switch to a third mode, a next-best competitor that is different in polarization and frequency from both the original lasing pump mode and the injected probe field. The probe only acts to deplete the pump field and does not significantly change the material gain.

Our semiconductor vertical-cavity surface-emitting laser (VCSEL) naturally lases in two orthogonally polarized modes, a strong pump mode and a weak competitor. Figure 1 shows the VCSEL emission spectrum with the pump field lasing at ν_2, the weak competitor lasing at ν_{comp}, and the probe injected at ν_1. As the probe power is increased, the pump power is depleted, allowing the competitor mode to become dominant (top curve).

Noting that mode switching occurs at the same level of pump depletion regardless of ν_1, we measured the frequency dependence of pump depletion by adjusting the probe frequency and power. Figure 2 shows that the detuning $\Delta = \nu_2 - \nu_1$ goes to 42 GHz (such a rapid coherent process is supported by acceleration of the material response by the intense light fields in the medium[4]). Calculating the *intracavity* probe powers P_{cav} from emission spectra, we find that P_{cav} is linear with Δ [Fig. 2(b)]. This linear dependence of pump depletion on

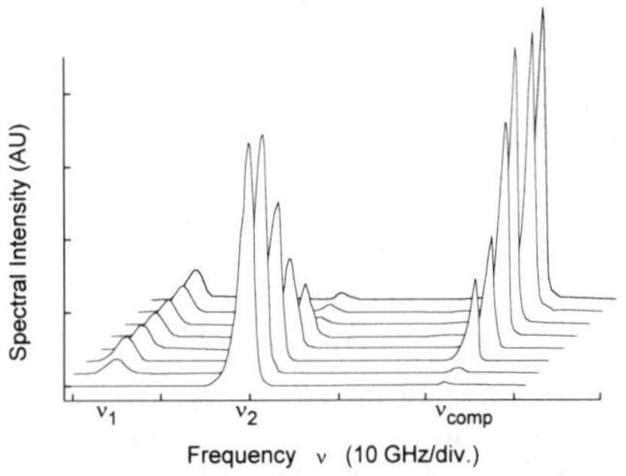

QWD25 Fig. 1. VCSEL output spectrum showing the pump mode lasing at ν_2, the weak competitor lasing at ν_{comp}, and the probe injected at ν_1. As the probe power is increased, the pump is depleted and the competitor dominates.

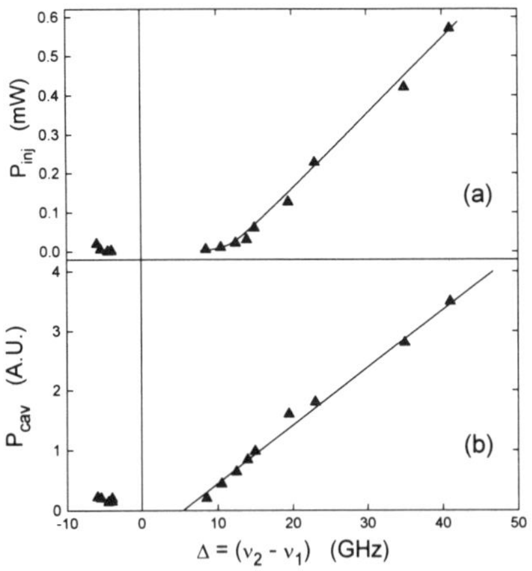

QWD25 Fig. 2. (a) Injected probe power P_{inj} required for CET-induced mode switching (triangles), showing switching with both positive and negative pump/probe detuning Δ. (b) *Intracavity* probe power P_{cav} required for CET mode switching.

detuning is in contrast to the quadratic dependences of other gain-depletion processes, such as injection locking[5] and spectral hole burning.[2]

In gain media, CET between pump and probe transfers energy in a direction away from the frequency of the nearest material resonance. In Fig. 1, $\nu_1 < \nu_2$ and pump energy is transferred to the probe, depleting the pump. In atomic media, a negatively detuned probe ($\nu_2 < \nu_1$) would enhance the pump (asymmetric gain modification). However, Fig. 2 shows that in semiconductors the pump is depleted by a positively or negatively detuned probe. Both symmetric depletion and linear dependence on Δ are explained by using quasi-equilibrium semiconductor theory.

In conclusion, we note that pump depletion in the presence of strong mode competition provides high-signal-gain, high-contrast polarization switching with frequency conversion, $\delta\nu = \nu_{comp} - \nu_1$, of up to 54.5 GHz. At lower |Δ| a very-low-power probe induces mode switching, giving $\delta\nu$ = 20.5 GHz with a pump/probe power ratio of 35.5 dB. A simple method for increasing $\delta\nu$ is also discussed.

**Department 1311, Sandia National Laboratories, Albuquerque, New Mexico 87185*

1. H. J. Eichler, P. Gunter, and D. W. Pohl, *Laser-Induced Dynamic Gratings* (Springer-Verlag, Berlin, 1986).
2. P. Meystre and M. Sargent III, *Elements of Quantum Optics*, 2nd ed. (Springer-Verlag, Berlin, 1992).
3. M. Sargent III, A. E. Paul, and S. W. Koch, Opt. Commun. **103**, 417 (1993).
4. C. W. Lowry *et al.*, Phys. Rev. Lett. **71**, 1534 (1993).
5. I. Petitbon, P. Gallion, G. Debarge, and C. Chabran, IEEE J. Quantum Electron. **24**, 148 (1988).

QWD26

Spectral shifts in dense collections of two-level atoms and in semiconductors due to the rotating-wave approximation

M. E. Crenshaw, C. M. Bowden, *Weapons Sciences Directorate, AMSMI-RD-WS-ST, Research, Development, and Engineering Center, U.S. Army Missile Command, Redstone Arsenal, Alabama 35898-5248*

For noninteracting two-level systems, the spectral shift resulting from making the rotating-wave approximation (RWA), known as the Bloch–Siegert shift, is negliglble at optical frequencies. In dense collections of two-level atoms and in a two-band semiconductor model, the Bloch-like equations of motion incorporate a parameter ε that represents the strength of the dipole–dipole interaction or the coherent Coulomb-exchange effects. Here, we discuss the resonance fluorescence spectra of dense collections of two-level atoms and the two-band semiconductor model and demonstrate that spectral shifts resulting from making the rotating-wave approximation can be substantial.

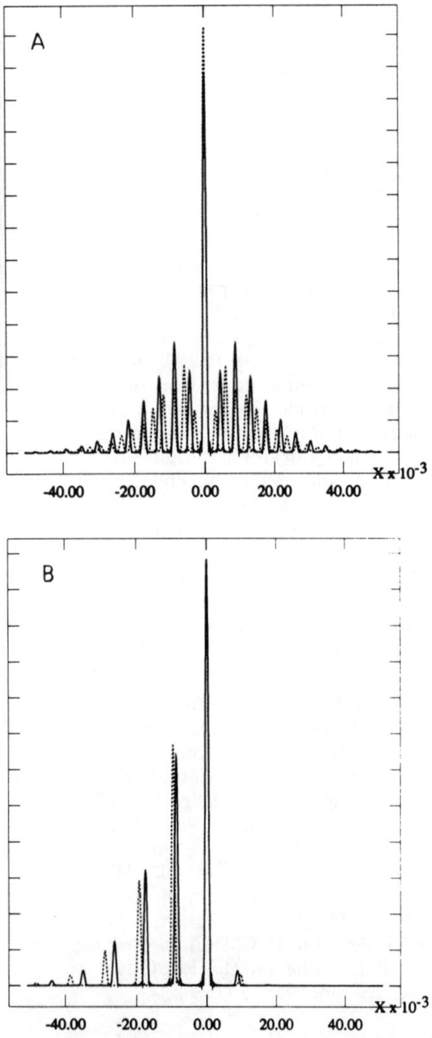

QWD26 Fig. 1. Spectra for a dense medium with (dashed curve) and without (solid curve) the RWA for (a) $\varepsilon = 1.988\Omega$ and (b) $\varepsilon = 2.012\Omega$.

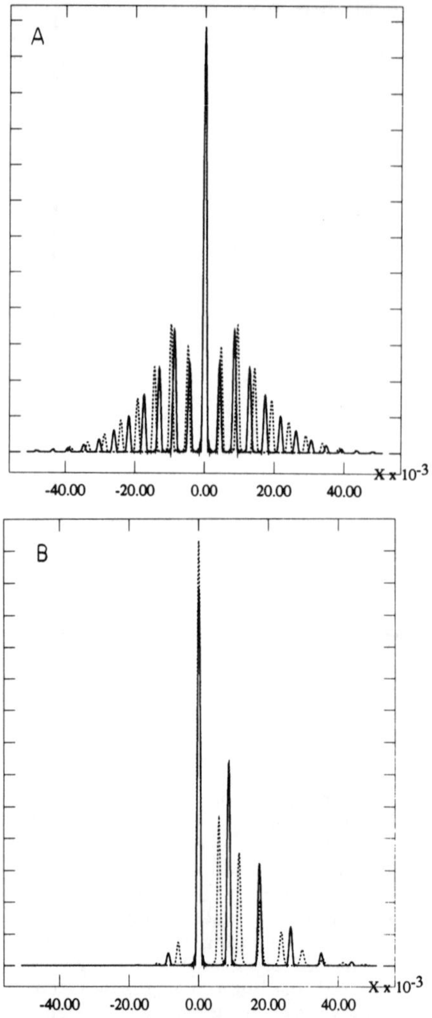

QWD26 Fig. 2. Spectra for a two-band semiconductor model with (dashed curve) and without (solid curve) the RWA for (a) $\varepsilon = -1.988\Omega$ and (b) $\varepsilon = -2.012\Omega$.

Using the modified optical Bloch equations in the RWA, Knorr, Süsse, and Welsch analyzed the resonance fluorescence spectrum of a dense collection of two-level atoms.[1] Their analysis reveals a transition from a symmetric spectrum to an asymmetric spectrum when the parameter ε, which represents the strength of the near-dipole–dipole coupling, is equal to twice the Rabi rate Ω of the field. We have found that in the region $\varepsilon \approx 2\Omega$ the features of a spectrum calculated with the RWA are shifted relative to features calculated without the RWA. This is demonstrated by Fig. 1, which shows the spectra with and without the RWA for $\Omega = 0.01\omega$ with ε less than and greater than 2Ω. It should be noted that for $\Omega = 0.01\omega$, the Bloch–Siegert shift is negligible and could not be discerned at the scale of Fig. 1.

We found that similar spectral shifts occur in two-band semiconductors. Figure 2 shows RWA and non-RWA spectra calculated by using the semiconductor Bloch equations.

In summary, both a dense medium and a semiconductor exhibit large Bloch–Siegert-like shifts based on spectra calculated from Bloch-like equations of motion. However, in a dense medium the dipole–dipole interaction parameter couples the polarization to Maxwell's wave equation, so that propagation effects will likely influence the spectrum. It is likely that such propagation effects can be substantially reduced in a semiconductor.

1. A. Knorr, K.-E. Süsse, and D.-G. Welsh, J. Opt. Soc. Am. B **9**, 1174 (1992).

QWD27

Numerical analysis of two-photon absorption and induced-free-carrier absorption in GaAs

Xueru Zhang, Miao Yang, Yinglin Song, Ruibo Wang, Chunfei Li, *Department of Physics, Harbin Institute of Technology, Harbin 150001, China*

The two-photon-absorption (TPA) coefficient is an important parameter of semiconductors. However, there is a wide disparity in the reported values of the TPA coefficient, especially in the results obtained with nanosecond pulsed lasers for which photogenerated carrier absorption is expected to dominate. To study the influence of free-carrier absorption (FCA), we analyzed theoretically TPA and FCA in GaAs with a nanosecond laser source.

If the slowly-varying-envelope approximation is used, the propagation equation of the irradiance for a thin sample is

$$\frac{dI}{dz} = -(\alpha + \beta I + \sigma N)I, \tag{1}$$

where α is the residual linear absorption, β is the TPA coefficient, σ is the FCA cross section, and N is the carrier density. The photogenerated carrier concentration is governed by

$$\frac{dN}{dt} = \frac{\beta I^2}{2\hbar\omega} - \frac{N}{\tau}, \tag{2}$$

where τ is the recombination time of free charge carriers. The diffusion process is neglected in this equation because the diffusion time of free carriers in bulk GaAs is longer than 1 ns. It is almost impossible to solve the nonlinear differential equations (1) and (2) to obtain an analytical solution for I and N. Therefore, we used iteration. Assuming that the last term including N in Eq. (1) is much smaller than the first two terms and can be neglected, we can obtain $I^{(1)}$ and $N^{(1)}$ from Eqs. (1) and (2), respectively. For the second iteration step, we obtain $I^{(2)}$ by substituting $N^{(1)}$ and $I^{(1)}$ into the right-hand side of Eq. (1), and we obtain $N^{(2)}$ by substituting $I^{(2)}$ into Eq. (2). We perform the second step until we obtain a constant I and N. In our simulation we selected the following parameters: $\sigma = 3 \times 10^{-18}$ cm^2, $\beta = 25$ cm/GW, $\tau = 20$ ns, $\alpha = 0.55$ cm^{-1}, the path length of sample $L = 0.1$ cm, and the pulse width of laser $T = 10$ ns. The output irradiance intensity versus the input irradiance is shown in Fig. 1. From Fig. 1 we find that when $I_{in} > 45$ MW/cm^2, FCA cannot be neglected and that when $I_{in} > 300$ MW/cm^2, FCA is dominant. It is interesting to use a pure TPA process with a larger β' to fit the result above (the solid curve in Fig. 1). For the best fit, $\beta' = 85$ cm/GW (circles in Fig. 1).

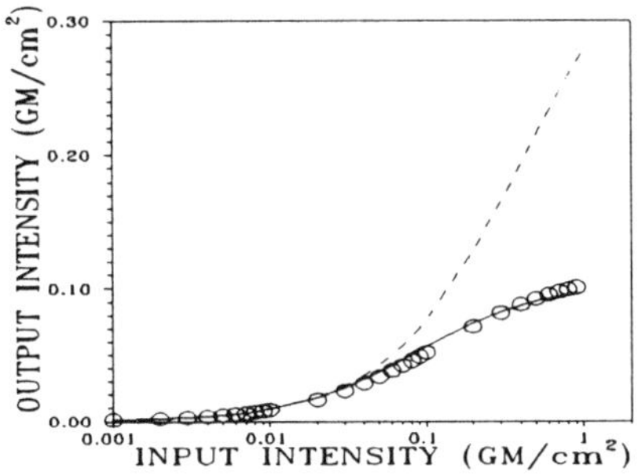

QWD27 Fig. 1. Theoretical variation of output irradiance as a function of input irradiance. The solid curve is obtained with our theory, the dashed curve is the result of pure two-photon absorption, and the circles are the best fit with pure two-photon absorption but a larger $\beta' = 85$ cm/GW.

In conclusion, when the laser pulse width is similar to or longer than the carrier lifetime, the enhanced nonlinearity based on FCA is important and cannot be neglected. This is one of the main origins of the wide disparity in the reported values of the two-photon-absorption coefficient of GaAs. This method can be used to analyze the other direct-band-gap semiconductors, such as InSb, ZnSe, and CdSe.

QWD28

Isomer effects of high-temperature electro-optic chromophores

R. F. Shi, S. Yamada, Y. M. Cai, O. Zamani-Khamiri, A. Panackal, A. F. Garito, *Department of Physics and Laboratory of Research on the Structure of Matter, University of Pennsylvania, Philadelphia, Pennsylvania 19104*

We have recently discovered a new class of high-temperature electro-optic (EO) chromophores, 1,8-naphthylene-benzimidazoles, that possess large second-order optical coefficients along with high thermal, chemical, and photo stabilities.[1,2] In an effort to optimize the structure, we have found that isomers that have very similar structures exhibit remarkable differences in both their linear and nonlinear optical properties. We report here theoretical and experimental results of the effects of isomers on second-order optical susceptibilities, as evidenced in the case of our high-temperature EO chromophores.

The structures of our model systems, SY177 and its isomer SY177P, are shown in Fig. 1 (inset). They differ only slightly in structure in that the position of the electron donor group is different. In spite of the strong structural similarity, the optical spectra are very different, as shown in Fig. 1. In SY177 the first excited state S_1 is strongly coupled to the ground state S_0, whereas in SY177P the coupling between S_1 and S_0 states is very weak. The second excited state S_2 in SY177P, on the other hand, was found to have a strong coupling with the ground S_0 state. This confirmed our theoretical predictions that whereas the S_1 state in SY177 is the main excited state in the second-order optical processes, in SY177P the corresponding state should be S_2.

Both structures were found to have the same ground-state dipole moment $\mu = 3.6$ D. However, they have markedly different second-order optical susceptibilities $\beta_x(-2\omega; \omega, \omega)$. Table 1 lists the dipole moments, calculated and measured $\mu\beta_x(-2\omega; \omega, \omega) + \langle\gamma(-2\omega; \omega, \omega, 0)\rangle\, 5kT$, and calculated and measured $\beta_x(-2\omega; \omega, \omega)$ for the two isomers. The measured values were obtained through standard dc-induced second-harmonic-generation (DCSHG) measurements at $\lambda = 1907$ nm in 1,4-dioxane solutions. The agreement between theory and experiment is quite satisfactory for both isomers, demonstrating the effectiveness of our theoretical methods in simulating the isomer effects in this class of chromophores. Most importantly, the measured value of $\beta_x(-2\omega; \omega, \omega)$ for SY177 is more than 4 times that for SY177P, which is indeed remarkable given the structural similarity of the two isomers.

QWD28 Table 1 Measured dipole moments, calculated and measured values of $\mu\beta_x(-2\omega; \omega, \omega) + \langle\gamma(-2\omega; \omega, \omega, 0)\rangle\, 5kT$, and calculated and measured values of $\beta_x(-2\omega; \omega, \omega)$ (at $\lambda = 1907$ nm ($\hbar\omega = 0.65$ eV) for SY177 and SY177P.

Structure	SY177	SY177P
μ (D)	3.6	3.6
$(\mu\beta + <\gamma> 5kT)^{calc}$ $(10^{-48}\, esu)$	413	98
$(\mu\beta + <\gamma> 5kT)^{exp}$ $(10^{-48}\, esu)$	413	85
β_x^{calc} $(10^{-30}\, esu)$	93	25
β_x^{exp} $(10^{-30}\, esu)$	93	21

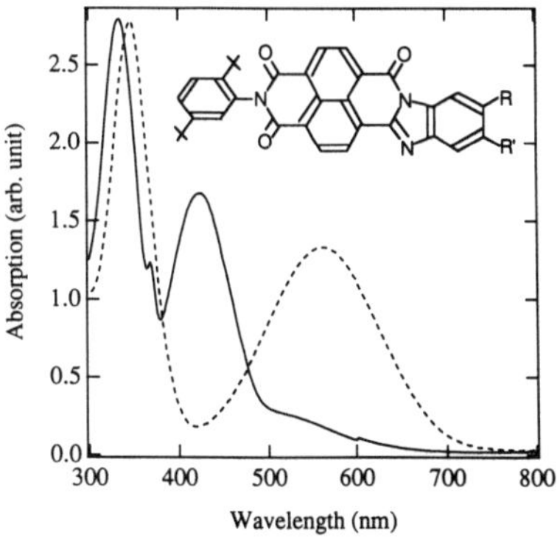

QWD28 Fig. 1. Ground-state linear absorption spectra for SY177 (dashed curve) and SY177P (solid curve) in solution with 1,4-dioxane. Inset: schematic molecular structure diagram of SY177 (R = NH_2, R' = H) and SY177P (R' = NH_2, R = H).

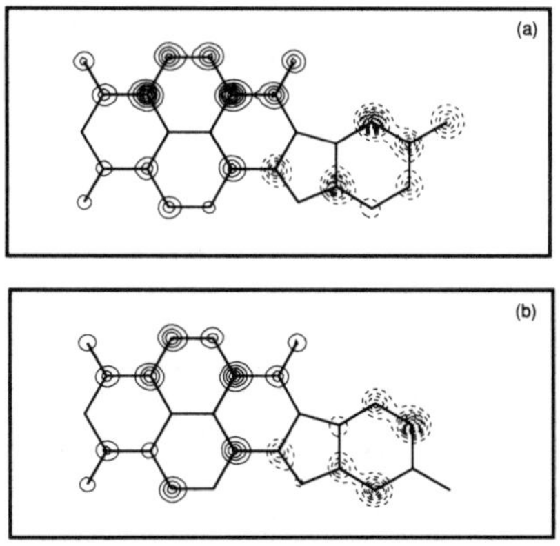

QWD28 Fig. 2. Contour diagrams of the electron-charge-density-distribution difference between key electronic states for (a) SY177 and (b) SY177P. Solid curves and dashed curves correspond to increased and decreased charge density, respectively, and the contour cut is taken at 0.4 Å above the molecular plane. The dipole-moment differences are 12.38 D for SY177 and 5.08 D for SY177P.

These dramatic differences can be directly understood by comparing the contour diagrams of the electron-density-distribution difference between the ground state and the key excited state for each model structure. As shown in Fig. 2(a), for SY177, upon virtual excitation charge is clearly transferred from the NH_2 donor group across the length of the chromophore, leading to a large dipole-moment difference. In contrast, as shown in Fig. 2(b) for SY177P, the NH_2 region is not even involved and the transfer of charge occurs over a shorter length and to a lesser degree. The key transition moments for SY177 were also found to be larger than those for SY177P. These results demonstrate that computer-aided designs and simulations are able to provide us with microscopic design rules for the EO chromophores before the actual synthesis.

1. R. F. Shi, M. H. Wu, S. Yamada, Y. M. Cai, and A. F. Garito, Appl. Phys. Lett. **63**, 1173 (1993).
2. S. Yamada, Y. M. Cai, R. F. Shi, M. H. Wu, and A. F. Garito, Mat. Res. Soc. Proc. **328**, 523 (1994).

QWD29

Cross-phase modulation in bulk and MQW AlGaAs waveguides at half the band gap

A. Villeneuve,* J. U. Kang, G. I. Stegeman, J. S. Aitchison,** D. C. Hutchings,** *Center for Research and Education in Optics and Lasers, University of Central Florida, Orlando, Florida 32828*

In many all-optical switching applications, such as polarization switching,[1] demultiplexing,[2] and soliton steering,[3] the value of the cross-phase modulation (XPM) relative to the self-phase modulation (SPM) determines the power requirements and the behavior of devices. We have found the XPM between TE and TM modes, or between TM and TE Modes, to be approximately equal to the SPM (TE) for bulk AlGaAs. However, in the MQW case we found the XPM to be equal to the SPM when a TM-polarized pump was used and to be equal to half the magnitude of the SPM when a TE-polarized pump was used.

The experiments were performed by using a color-center laser, producing 3–6-ps pulses, tunable in the 1.55-μm spectral region. The bulk AlGaAs waveguides consisted of a [001]-grown 1.5-μm-thick $Al_{0.18}Ga_{0.82}As$ high-index core surrounded by $Al_{0.24}Ga_{0.76}As$ upper and lower cladding regions. The MQW samples consisted of 7-nm-thick $Al_{0.1}Ga_{0.9}As$ wells between 10-nm-thick $Al_{0.24}Ga_{0.76}As$ barriers. Both waveguide structures were single moded at 1550 nm and had a band gap around 750 nm.

To characterize the XPM, we measured the spectral broadening imposed on a weak cross-polarized signal by a strong pump. A polarizer was used after the sample to remove the pump beam. These values were compared to the SPM of the strong pump only. The measured ratio of the relative magnitudes of XPM and SPM, defined as R = XPM/SPM, was measured as $R = 0.95 \pm 0.1$. The same result was found by using both TE and TM pumps.

For cubic symmetry, the ratio of perpendicular XPM to SPM is given by

$$R = \frac{\Delta n^{\perp}_{XPM}}{\Delta n_{SPM}} = \frac{1 - \delta' + \sigma'\left(2\sum_i |e_i|^2|p_i|^2 - \frac{1}{2}\right)}{1 + \sigma'\left(\sum_i |e_i|^4 - 1\right)},$$

where σ' is the nonlinear-refraction anisotropy parameter that relates n_2 for different crystalline orientations and δ' is the nonlinear-refraction dichroism parameter that relates the various values of n_2 for linear and circular polarizations.[4] **e** and **p** are unit polarization vectors for the excite and probe beams, respectively, and the summations take place over the crystal axes x, y, and z. In the case of an isotropic material and Kleinmann symmetry, $\sigma' = 0$ and $\delta' = 1/3$, which provide a ratio of 2/3, as is the case for optical fibers at communication wavelengths. By using an anisotropic-band-structure model for low-temperature GaAs at $\hbar\omega = E_g/2$, the values $\sigma' = -0.82$ and $\delta' = 0.075$ are obtained,[4] resulting in a predicted ratio of 0.95, in agreement with current experimental observations.

In the case of MQW sample we measured a value of $R = 1 \pm 0.05$ when the pump was polarized perpendicular to the plane of the wells (TM). However, when the pump was polarized parallel to the wells, the measured value was $R = 0.5 \pm 0.05$. Typical results are shown in Fig. 1(a) for a TM-polarized pump and in Fig. 1(b) for a TE-polarized pump. In each case the solid curve represents the experimental results and the dashed curve represents simulations carried out with a beam-propagation algorithm. This additional anisotropy is not surprising, given that even the linear susceptibility is anisotropic (i.e., uniaxial birefringence) for MQWs.

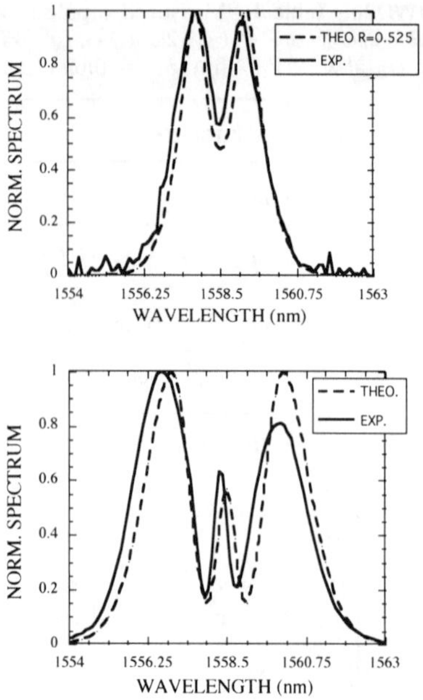

QWD29 Fig. 1. Experimentally measured (solid curve) and calculated (dashed curve) spectra for (a) probe and (b) pump beam in the MQW sample, obtained by using a TE-polarized pump and a TM-polarized probe.

In conclusion, we have measured the ratio of XPM to SPM for both bulk and MQW samples in the half-band-gap spectral region. The proximity of the two-photon-absorption resonance to the operating wavelength results in a predicted ratio significantly different from the isotropic Kleinmann result appropriate to optical fibers.

*Also with *Department of Physics, University of Laval, Québec G1K 7P4, Canada*

** *Department of Electronic and Electrical Engineering, University of Glasgow, Glasgow G12 8QQ, U.K.*

1. C. C. Yang, A. Villeneuve, G. I. Stegeman, C. H. Lin, and H. H. Lin, Opt. Lett. **18**, 1487 (1993).
2. A. Villeneuve, K. Al-Hemyari, J. U. Kang, C. N. Ironside, J. S. Aitchison, and G. I. Stegeman, Electron. Lett. **29**, 721 (1993).
3. P. V. Mamyshev, A. Villeneuve, G. I. Stegeman, and J. S. Aitchison, Electron. Lett. **30**, 726 (1994).
4. D. C. Hutchings and B. S. Wherrett, "Theory of anisotropy of ultrafast nonlinear refraction in zinc-blende semiconductors," in preparation; D. C. Hutchings, J. S. Aitchison, B. S. Wherrett, G. T. Kennedy, and W. Sibbett, in *Nonlinear Guided Waves*, 1995 OSA Technical Digest Series (Optical Society of America, Washington, D.C.), in press.

QWD30

Pump–probe spectroscopy in J-aggregates: interacting one-dimensional excitons

Frank C. Spano, *Department of Chemistry, Temple University, Philadelphia, Pennsylvania 19122*

Recent measurements of the differential pump–probe absorption spectrum, $\bar{D}(\omega)$, in low-temperature J-aggregates display exciton saturation and a higher-energy induced absorption[1–4] (see Fig. 1). The entire spectrum undergoes significant blue shifting with increasing pump intensity.[2,4] In this presentation a theory based on one-dimensional interacting Frenkel excitons is used to explain these observations.

The Hamiltonian for a Frenkel chain with N coupled two-level molecules includes near-neighbor energy transfer J and an exciton–exciton interaction term A that arises from the change in the molecular permanent dipole moment upon excitation. This was recently measured to be 10–20 D per aggregate for PIC Br (Ref. 5) and is expected to produce biexcitons, although none have been detected so far. We first consider the Hamiltonian without such interactions. By using the Jordan–Wigner transformation, all 2^N multiexciton states are evaluated exactly by diagonalizing only a single N-by-N matrix.[6] Without any adjustable parameters, a subsequent calculation of $\bar{D}(\omega)$ reproduces all of the important experimental observations, including the intensity dependent blue shift.[7] Figure 1 shows a comparison of theory and experiment.

How can one reconcile the success of a noninteracting exciton theory with the probable presence of large interactions? When exciton–exciton interactions were included, it was found that the differential pump–probe spectrum neatly divides into two parts: a noninteracting exciton part $\bar{D}_1(\omega)$ (identical to that plotted in Fig. 1) and a contribution from the exciton–exciton interactions $\bar{D}_2(\omega)$ that contains possible biexciton resonances. For the J-aggregate configuration ($J < 0$) and relatively weak inhomogeneous broadening $W << |J|$, $\bar{D}_1(\omega)$ and $\bar{D}_2(\omega)$ spectrally interfere only when A lies in the range $4J < A < 2J + \delta$, where $\delta = |J|\sqrt{W/|J|}$ is a small energy window. For A outside this range, the most intense

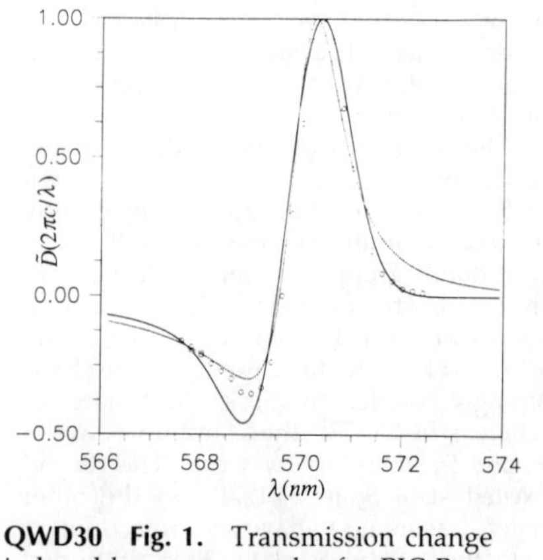

QWD30 Fig. 1. Transmission change induced by a pump pulse for PIC Br. Open circles are data from Ref. 3, and solid curve is the theoretical line shape without exciton–exciton interactions.

portion of the spectrum cannot be distinguished from $\bar{D}_1(\omega)$. This suggests three possibilities for the apparent success of the noninteracting exciton theory: (1) the exciton–exciton interactions are large and negative, $A << J$, so that the biexciton state is far red shifted from the two-exciton band; (2) the exciton–exciton interactions are weak, $-2J > A > 2J - \delta$, so that biexcitons are not formed; (3) $A > 2J$, so that the oscillator strength to the resulting (blue-shifted) biexciton is too small for detection. At this point further experiments are needed to establish the correct interpretation.

1. H. Fidder, J. Knoester, and D. A. Wiersma, J. Chem. Phys. **98**, 6564 (1993).
2. K. Minoshima, M. Taiji, K. Misawa, and T. Kobayashi, Chem. Phys. Lett. **218**, 67 (1994).
3. S. Kobayashi and F. Sasaki, J. Lumin. **58**, 113 (1994).
4. A. Johnson, S. Kumazaki, and K. Yoshihara, Chem. Phys. Lett. **211**, 511 (1993).
5. K. Misawa, K. Minoshima, H. Ono, and T. Kobayashi, Chem. Phys. Lett. **220**, 251 (1994).
6. F. C. Spano, Phys. Rev. Lett. **67**, 3424 (1991).
7. F. C. Spano, Chem. Phys. Lett. **220**, 365 (1994).

QWD31

Biexciton effects in the two-photon absorption spectrum of a one-dimensional polymer

Frank B. Gallagher, Frank C. Spano, *Department of Chemistry, Temple University, Philadelphia, Pennsylvania 19122*

The promise of optical computing has spawned a search by experimentalists for materials with large nonlinear optical responses and an effort by theorists to understand the origin of these responses. Conjugated polymers with large electron delocalization, such as polyacetylene and polysilane, are prime candidates because of their relative abundance and chemical tunability. Excitonic models of varying complexity have gained acceptance in explaining many experimental results in these systems.[1–5] For example, recent nonlinear-optical measurements of polysilanes show strong evidence for biexcitons.[1,2]

The simplest model of interacting excitons is based on the standard semiconductor Hamiltonian in one dimension:

$$h = \sum_{p=1}^{N} (E_e \hat{c}_p^\dagger \hat{c}_p + E_h \hat{d}_p^\dagger \hat{d}_p) - \sum_{p=1}^{N} [t_e(\hat{c}_p^\dagger \hat{c}_{p+1} + \hat{c}_{p+1}^\dagger \hat{c}_p) + t_h(\hat{d}_p^\dagger \hat{d}_{p+1} + \hat{d}_{p+1}^\dagger \hat{d}_p)] + \frac{1}{2} \sum_{pp'} V(p - p')[\hat{c}_p^\dagger \hat{c}_{p'}^\dagger \hat{c}_{p'} \hat{c}_p + \hat{d}_p^\dagger \hat{d}_{p'}^\dagger \hat{d}_{p'} \hat{d}_p - 2\hat{c}_p^\dagger \hat{d}_{p'}^\dagger \hat{d}_{p'} \hat{c}_p], \quad (1)$$

where $\hat{c}_p^\dagger(\hat{d}_p^\dagger)$ creates an electron (hole) at site p and periodic boundary conditions are imposed. The first two sums include the effective one-particle energies of the electrons and holes. The electron (hole) transfer integral is t_e (t_h), and $V(p - p')$ is the pairwise Coulombic potential. Using Eq. (1), we calculated the static third-order susceptibility γ in a previous work.[3] Here we present new results for the frequency-dependent two-photon absorption (TPA), $\gamma(-\omega; \omega, -\omega, \omega)$, with special emphasis on the role of biexcitons.

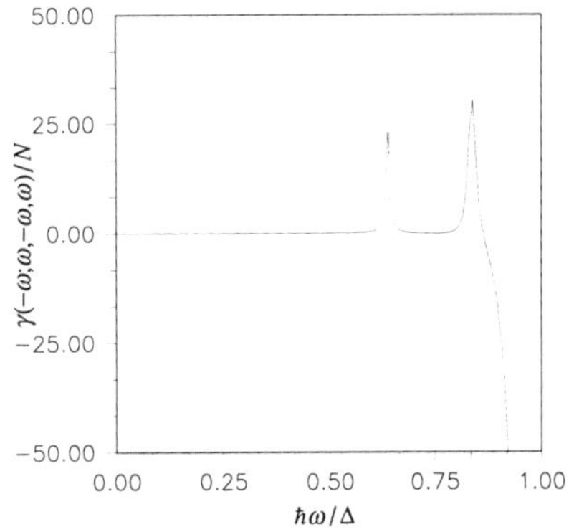

QWD31 Fig. 1. Polymer limit of $\gamma(-\omega; \omega, -\omega, \omega)/N$ in arbitrary units as a function of the one-photon energy in units of the optical band gap Δ. The low-energy and high-energy peaks correspond to a charge-transfer exciton and a biexciton, respectively. $V_0 = V_1 = 10\Delta$, and $t = (t_e + t_h)/2 = 0.1\Delta$.

The Hamiltonian in Eq. (1) contains a two-electron–hole-pair basis that scales as $O[N^3]$, making it difficult to investigate the polymer limit. However, restricting the radius of charge transfer excitons to be smaller than some predefined value results in a basis that scales as $O[N]$. This is justified, for example, when $V_1 >> t$, in which case charge-transfer states separated by more than one lattice unit can be ignored. Such restrictions preserve all states that are key to understanding the influence of excitons and biexcitons on the nonlinear optical response, well into the polymer limit.

The relative strength between on-site V_0 and nearest-neighbor V_1 Coulombic interactions is the major factor that determines biexciton formation. The roles of electron–hole hopping parameters t_e and t_h are to moderate the mixing between optically active bond exciton states and the optically dark charge-transfer states. For $V_0 \approx V_1 >> t$, biexcitons exist but are optically dark. As t increases to $t \approx V_0 - V_1$, the mixing between the bond and the charge-transfer states is enhanced, leading to prominent biexciton peaks in the TPA spectrum, as are shown in Fig. 1. A detailed numerical investigation of these effects is presented. We emphasize the various conditions for biexciton formation with applications to polysilanes and polygermanes.

Currently, the behavior of γ as the system evolves from an oligomer to a polymer is a central issue in the theoretical community. The strength of the present approach is that double exciton states are included exactly and that large sizes can be studied without sacrificing size consistency. Previous work has been restricted to oligomers[1,4] or has used CI methods, which are not size consistent.[5]

1. Z. G. Soos and R. G. Kepler, Phys. Rev. B **43**, 11908 (1991).
2. A. Tilgner, H. P. Trommsdorff, J. M. Zeigler, and R. M. Hochstrasser, J. Chem. Phys. **96**, 781 (1992).
3. F. B. Gallagher and F. C. Spano, Phys. Rev. B **50**, 5370 (1994).
4. F. Guo, D. Guo, and S. Mazumdar, Phys. Rev. B **49**, 10102 (1994).
5. V. A. Shakin and S. Abe, Phys. Rev. B **50**, 4306 (1994).

JWC Tutorial — 2:30 pm
Hyatt E/7

Practical OPOs from the UV to the Mid-Infrared

Hiroshi Komine, *Northrup Corporation, Presider*

JWC1 (Tutorial) — **2:30 pm**

Practical OPOs from the UV to the mid-infrared

Walter R. Bosenberg, *Lightwave Electronics Corporation*

Optical parametric oscillators (OPOs) are efficient frequency converters that generate tunable radiation in spectral regions and over ranges that are unavailable from conventional lasers. New nonlinear materials and pump lasers have created an explosion of interest in OPOs, which are now beginning to be used in many exciting applications. This tutorial provides a basic understanding of parametric frequency conversion devices. Important concepts, terminology, and design issues are discussed. The state of the art in devices is reviewed. The tutorial provides timely and insightful information to specialists as well as researchers in neighboring fields such as: material scientists and crystal growers; laser scientists; application and systems engineers; and dielectric mirror coating specialists.

JWD — 2:30 pm
Room 308

Joint Session on Far-Infrared Sources I

Eli Kapon, *Swiss Federal Institute of Technology, Presider*

JWD1 (Invited) — **2:30 pm**

Unipolar quantum cascade intersubband infrared lasers and LEDs

F. Capasso, J. Faist, C. Sirtori, A. L. Hutchinson, D. L. Sivco, A. Y. Cho, *AT&T Bell Laboratories, Murray Hill, New Jersey 07974*

The recently demonstrated quantum cascade laser (QCL) relies on only one type of carrier (unipolar laser) and on quantum jumps of electrons between discrete conduction band energy levels of quantum wells.[1,2] As such, the wavelength can be tailored over a very wide range by varying layer thicknesses. In the original structure the laser transition is between states centered in neighboring wells to facilitate population inversion, i.e., the transition is *diagonal*.[1,2] In this design, however, the luminescence linewidth is relatively broad due to the interface roughness because electrons traverse several heterointerfaces in the photon emission process, thus reducing the peak gain. To circumvent this problem we designed the structure of Fig. 1, where electrons make a *vertical* radiative transition essentially in the same well. This should reduce considerably the threshold current density. To prevent electron escape in the continuum, the digitally graded superlattice injector is designed to act as a Bragg reflector for electrons in the higher excited state and to siphon out electrons from the lower states via a miniband facing the latter (Fig. 1). The lower state of the laser transition is separated by an optical phonon ($\approx$30 meV) from the n = 1 state to ensure population inversion. Each of the 25 active regions (see Fig. 1) consists of a 4.5 nm GaInAs well coupled to a 3.6-nm well by a 2.8-nm AlInAs barrier. Injection into the n = 3 is through a 6.5-nm AlInAs barrier and electrons escape out of the n = 1 state through a 3.0-nm AlInAs barrier. The waveguiding cladding regions are similar to those of the diagonal transition structure.[1,2] Current pulses of 30 ns with a 20-kHz repetition rate were injected into the ridge mesa devices. Figure 2 displays the peak optical power versus drive current. The peak optical power is ~80 mW at 80 K. The measured slope efficiency is 300 mW/A per facet

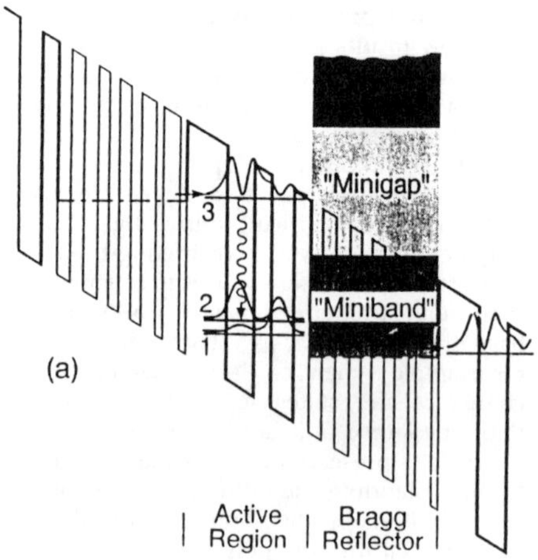

JWD1 Fig. 1. Energy diagram of one period of the QC laser with electron Bragg reflector showing the *vertical* laser transition (wavy arrow); $E_3 - E_2 = 271$ meV.

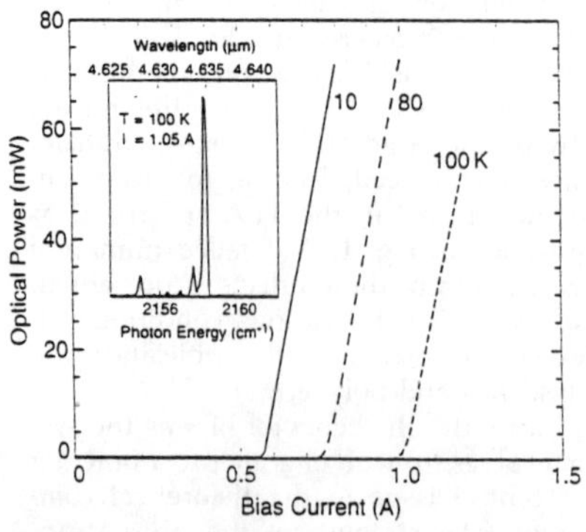

JWD1 Fig. 2. Peak optical output power from a single facet vs. drive current at various heat sink temperatures for the QC laser of Fig. 1. The device is 2.4-mm long. Inset: high-resolution spectra of the laser above threshold. The radiation is linearly polarized normal to the layers.

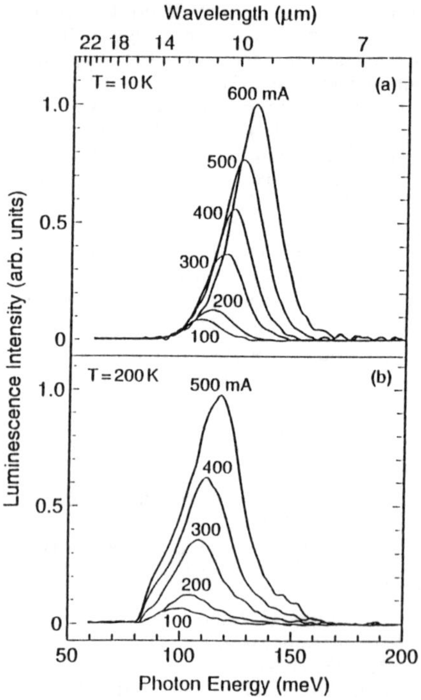

JWD1 Fig. 3. Intersubband electroluminescence spectrum of a $\lambda \sim 10$ μm QC LED with *diagonal* transition for various drive currents at 10 K and 200 K. The radiation is polarized normal to the layers. The integrated optical power at 500 mA is $\approx$5 mW.

and is essentially temperature independent and corresponds to a differential quantum efficiency per period of 4.5×10^{-2} when corrected for the collection efficiency of the apparatus. The threshold density has a value $J_{th} = 1.7$ kA/cm^2 at 10 K and 3 kA/cm^2 at 100 K.

We have also recently reported an intersubband unipolar LED at wavelengths in the 8–13-μm range based on the QC scheme with diagonal transition,[1,2] but without waveguiding layers and with only 5 stages.[3] The structure is grown by MBE in the AlInAs/GaInAs system and the separation of the two excited states defining the laser transition is 128 meV. The structure was processed into mesas with a 45° wedge to couple light out. Figure 3 shows the luminescence spectrum of the device. The peak of the electroluminescent spectrum blue-shifts linearly with drive current due to the Stark effect. From these data and simple electrostatic considerations we obtained the intersubband nonradiative electron lifetime of the upper state of the laser transition ($\approx$2.5 ps), in good agreement with the calculation.[3]

1. J. Faist, F. Capasso, D. L. Sivco, C. Sirtori, A. L. Hutchinson, and A. Y. Cho, Science **264**, 553 (1994); Electron. Lett. **30**, 865 (1994).
2. J. Faist, F. Capasso, D. L. Sivco, A. L. Hutchinson, C. Sirtori, Sung N. G. Chu, and A. Y. Cho, Appl. Phys. Lett., Dec. 5, 1994.
3. C. Sirtori, F. Capasso, J. Faist, D. L. Sivco, A. L. Hutchinson, and A. Y. Cho, Appl. Phys. Lett. (in press).

JWD2 **3:00 pm**

Intersubband semiconductor lasers without population inversion

Rajeev J. Ram, Ken L. Campman, Holger Schmidt, Atac Imamoglu, Arthur C. Gossard, John E. Bowers, *Department of Electrical and Computer Engineering, University of California, Santa Barbara, California 93106*

We propose a new class of intersubband lasers and amplifiers that achieve net gain without population inversion as well as parametric amplifiers that realize resonantly enhanced $\chi^{(2)}$ without absorption. These schemes are based on unipolar double-quantum-well (DQW) structures in which the gain occurs at a transition between the conduction-band subbands (Fig. 1). For net gain to be achieved without inversion, the emission and absorption processes must be made nonreciprocal. Establishing a steady-state coherence between two upper states of a three-state system can result in nonreciprocal emission and absorption profiles if there is one-way pumping into the narrow quantum well.[1] One-way pumping is achieved when the emitter Fermi level is several kT above the narrow well resonance: tunneling back into the emitter is forbidden by the Pauli exclusion principle. In this case, absorption is reduced by destructive interference of the frequencies that lie between the upper two states, whereas emission is enhanced by constructive interference in the same frequency range. There has been considerable interest in laser action in atomic systems without population inversion.[1] Our work presents what we believe to be the first discussion of lasing without inversion in an electrically pumped semiconductor. We have discussed the possibility of achieving gain without inversion in a DQW structure.[2] This paper presents what we believe to be the first discussion of laser oscillation without inversion and of enhanced frequency conversion due to coherence in an intersubband structure.

The specific layer profile under consideration is shown in Fig. 1 The $|3\rangle$–$|1\rangle$ subband energy spacing is 119 meV, which corresponds to an optical-transition wavelength of 10.4 μm. The separation between the two upper subbands is 3 meV. The lifetime of both upper states is determined by tunneling into the collector and is designed to be approximately 300 fs, which

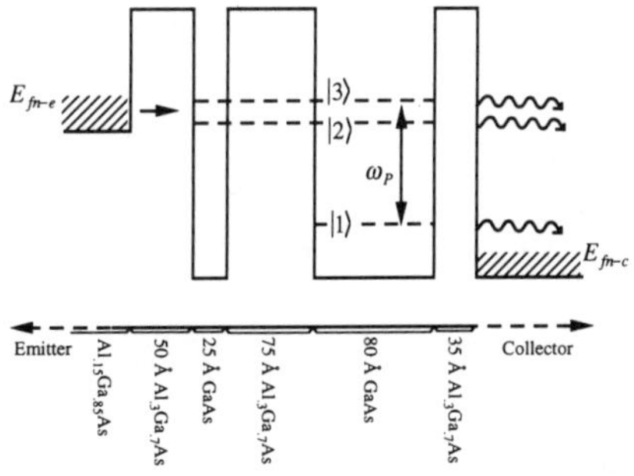

JWD2 Fig. 1. Energy-level diagram and layer profile for the DQW structure analyzed. There is an applied bias of 0.16 V.

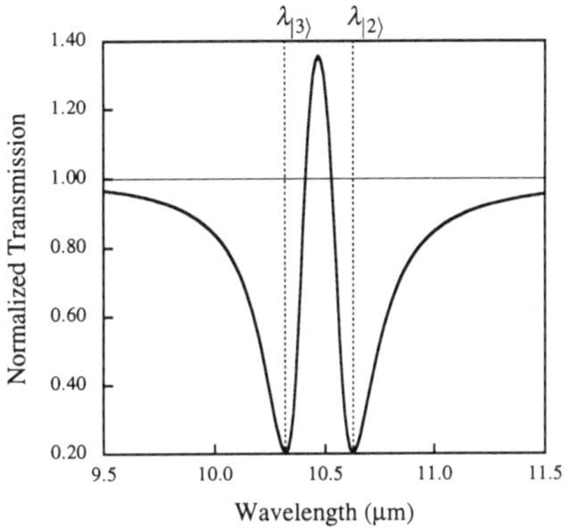

JWD2 Fig. 2. Normalized transmission profile of a probe field as a function of the probe wavelength for a 1-mm-long DQW structure, giving a peak gain of $e^{0.3}$.

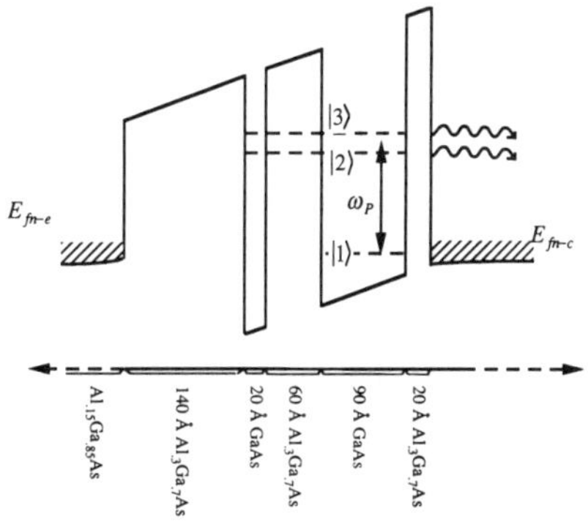

JWD2 Fig. 3. Energy-level diagram and layer profile for the DQW structure used for enhanced frequency conversion. Transparency is achieved at zero applied bias.

is significantly smaller than the dominant dephasing time (1 ps at these low densities) resulting from carrier–carrier scattering. Using the parameters presented above, we find a material gain per DQW of 375 cm^{-1} at carrier densities of 2.7 × 10^{10} cm^{-2} and current densities of J = 15 kA/cm^2. This operating point corresponds to a lower-state carrier density that is 1.1 times the *combined* upper-state densities. The calculated transmission profile for an intersubband amplifier is shown in Fig. 2. There are transmission minima at the subband resonances and a peak (with gain) in between these minima. Fourier–transform infrared (FTIR) spectroscopy is being used to study the gain profiles for multiple DQW samples. These data are presented, along with measurements of low-temperature luminescence and electrical characterization of resonant tunneling through these triple-barrier structures.

This approach can also be used to make lasers without population inversion. Achieving sufficient gain for oscillation requires a cascade of DQW structures: In their recent work Faist *et al.*[3] have used a cascade of 25 triple-quantum-well structures (with population inversion) to increase their gain to approximately 60 cm^{-1}. In our structures, achieving sufficient modal gain to compensate for mirror losses (13 cm^{-1}) and internal losses (9 cm^{-1}) requires at least 22 DQW pairs. An important limitation is the small lateral waveguide confinement factor (<66%): because of small refractive-index discontinuities and a large wavelength, waveguide dimensions for lateral mode confinement are prohibitively large. The threshold conditions for pulsed electrical injection are calculated to be J_{th} = 15 kA/cm^2 and I_{th} = 1.5 A for a 10-μm ridge-waveguide structure.

The transparency established by the destructive interference of upper-subband absorption can also be used to enhance frequency conversion in semiconductors. The constructive interference of emission in the upper subbands allows resonant enhancement of $\chi^{(2)}$. This enhancement is achieved without increased absorption, which would cause pump losses for downconversion or signal loss for harmonic generation. The proposed structure, shown in Fig. 3, is a direct analogue of the atomic scheme proposed in Ref. 4.

1. A. Imamoglu and S. E. Harris, Opt. Lett. **14**, 1344 (1989).
2. A. Imamoglu and R. J. Ram, Opt. Lett. **19**, 1744 (1994).
3. J. Faist, F. Capasso, D. L. Sivco, C. Sirtori, A. L. Hutchinson, and A. Y. Cho, Science **264**, 553 (1994).
4. S. E. Harris, J. E. Field, and A. Imamoglu, Phys. Rev. Lett. **64**, 1107 (1990).

JWD3 (Invited) **3:15 pm**

Coherent terahertz generation using low-temperature-grown GaAs photomixers

E. R. Brown, K. A. McIntosh, K. B. Nichols, O. B. McMahon, W. F. DiNatalo, T. M. Lysczarz, Alin Pine,* Rick Suenram,* *MIT Lincoln Laboratory*

This talk summarizes our recent work on photomixing coherent sources made by depositing interdigitated-electrode structures on epitaxial layers of low-temperature-grown GaAs and pumping the structures with two cw single-frequency lasers.

* *NIST-Gaithersburg*

JWD4 **3:45 pm**

Far-infrared p-Ge laser using a regular permanent magnet

Kijun Park, Robert E. Peale,* Henry Weidner,** Jin J. Kim,† E. E. Haller,‡, *Center for Research and Education in Optics and Lasers (CREOL), Department of Electrical and Computer Engineering, University of Central Florida, Orlando, Florida 32816*

We report emission near 100 μm from a semiconducter laser based on far-infrared (FIR) transitions from light to heavy hole bands in p-Ge. This lasing mechanism has been studied for a number of years outside the USA.[1] Our new idea, reported here, is to substitute a *regular permanent magnet* for the traditional super-conducting magnet. The B-field can be tuned by changing the pole spacing of the magnet. Our optimized magnet design provides a uniform magnetic field. This new laser system is useful for many practical applications.

Figure 1 shows a schematic of the Ge laser system with a regular magnet system.

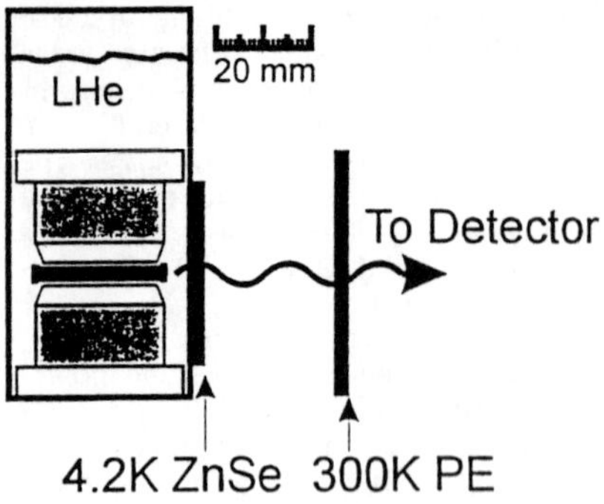

JWD4 Fig. 1. Schematic of the laser setup. The p-Ge sample is in between the poles of the permanent magnet.

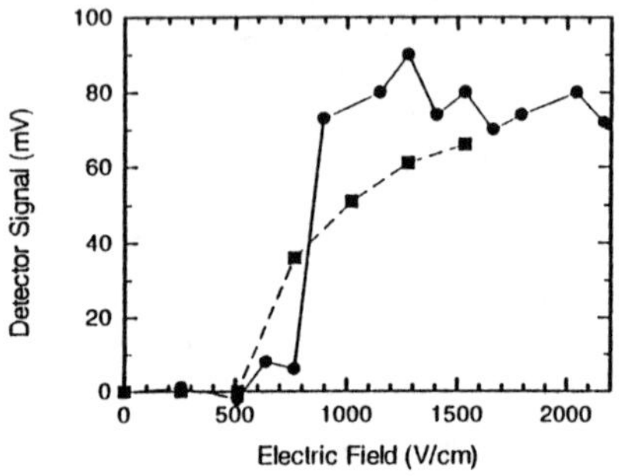

JWD4 Fig. 2. Measured signal as a function of electric field pulse magnitude for a p-Ge rod in the Voigt configuration with a Cu back mirror alone (solid square) and with an added #500 metal mesh output coupler (solid circle); B = 0.7 T.

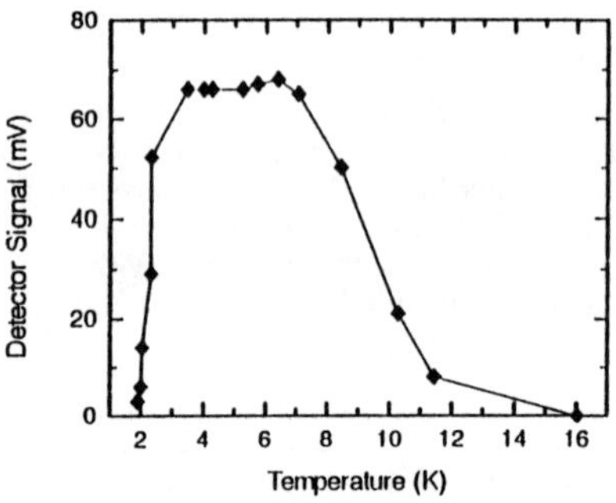

JWD4 Fig. 3. Measured signal as a function of temperature for p-Ge emission with Cu back mirror, E = 1530 V/cm and I ~ 400 A.

A pulsed electric field accelerates holes perpendicular to a magnetic field. Population inversion is achieved when E/B is tuned to prevent light holes from reaching the inelastic optical phonon scattering threshold while allowing heavy holes to do so. Thus light holes remain "hot" while heavy holes scatter back to **k** = 0. Because the gain region is broadband, the laser wavelength can be tuned by use of selective external resonators.

The polished $3.5 \times 3.9 \times 25$ mm^3 rectangular bar of 3×10^{14} cm^{-3} Ga-doped is placed between the regular permanent magnet (35 MG-Oe NdFeB alloy and soft iron) poles. The crystal is oriented so that **B**‖[001]. A Cu back mirror is compressed with a thin Teflon film to one end of the bar and the other end is coated with 20-μm Teflon film. The Teflon serves as an antireflection coating. Electric pulses (1-μs duration at 1 Hz) are applied to [1–10] faces through Ohmic contacts such that **E** is perpendicular to **B**. Emission passes through a ZnSe window at 3 K and a polyethylene window at 300 K. The signal is collected by a gold-coated off-axis ellipsoidal mirror and focused onto a 4.2-K Si bolometer through another 300-K polyethylene window and a 100-μm long pass filter.

The emission from p-Ge is shown in Fig. 2 as a function of applied electric field. The square symbols show emission without an output coupler. This inefficient cavity relies on reflection from the front surface of the Ge rod for feedback. The round symbols show emission with a #500 metal mesh output coupler. The threshold for emission is much sharper, and the output higher even though the output coupler orientation and mesh size are not optimized. With the mesh the emission is also more highly collimated. This strongly suggests that the emission is stimulated.

The temperature dependence of the emission is shown in Fig. 3. The peak is found around 5 K. The decrease in the signal below 3 K has not been previously reported, since to our knowledge all previous studies were performed at 4.2 K or above. Our tentative model involves strain splitting of the valence band at **k** = 0 so that a small amount of thermal energy is required for holes to enter the light hole band. This suggests that externally applied stress could serve to increase the temperature range for observing laser emission.

*Center for Research and Education in Optics and Lasers (CREOL) and Department of Physics, University of Central Florida, Orlando, Florida 32816
**Department of Physics, University of Central Florida, Orlando, Florida 32816
†Center for Research and Education in Optics and Lasers (CREOL), Department of Electrical and Computer Engineering, and Department of Physics, University of Central Florida, Orlando, Florida 32816
‡Lawrence Berkeley Laboratory, University of California, Berkeley, California 94720

QWE — 2:30 pm, Hyatt A

Nonlinear Optical Materials II

Frank W. Wise, *Cornell University, Presider*

QWE1 — 2:30 pm

Evidence of structural transformations of amorphous DMOP-PPV polymer waveguides in nonlinear-coupling experiments

F. Michelotti,* T. Gabler,** H. Hörhold,† R. Waldhäusl,** A. Bräuer,‡ *Department of Energetics, University of Rome "La Sapienza," National Research Council INFM and GNEQP, Via A. Scarpa 16, I-00161 Rome, Italy*

Among conjugated polymers, the family of the polyphenylene vinylenes (PPV) has recently attracted much attention because of its high fluorescence quantum yield[1–4] and nonlinear optical response.[5] The first excited state is an excitonic $1B_u$ level,[6] classifying the transition from the ground state ($1A_g$) as dipole allowed and giving rise to a high luminescence quantum yield.[4]

We report on the nonlinear coupling in slab poly(1,4-phenylene-1,2-bis) (4-methoxyphenyl) vinylene (DMOP-PPV) waveguides at $h\nu$ = 1.49 eV (τ_{FWHM} = 5 ps, f_{rep} = 9 KHz). In these measurements upcon-

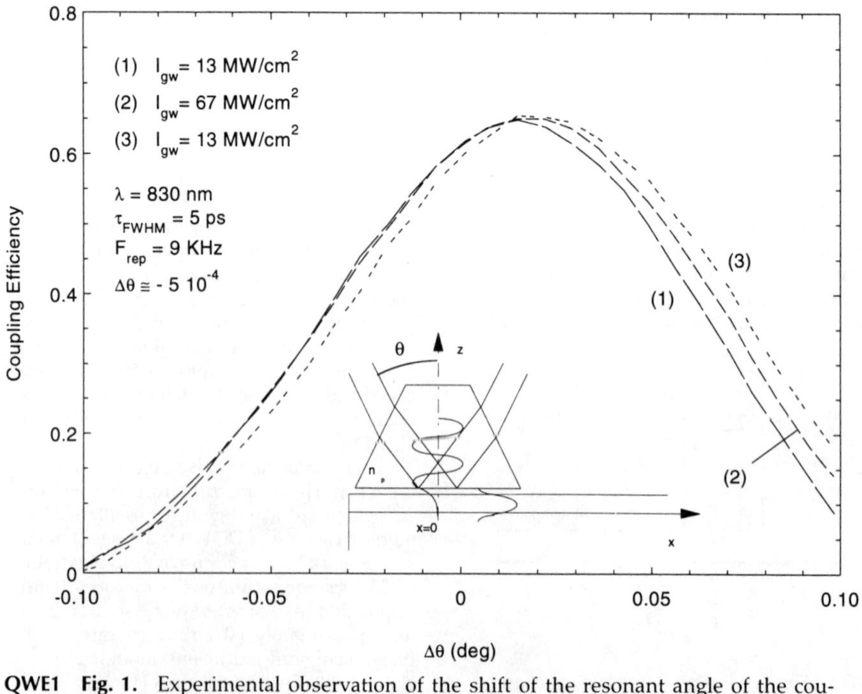

QWE1 Fig. 1. Experimental observation of the shift of the resonant angle of the coupling curve. The angular shift is $\Delta\theta = -7.6 \times 10^{-3}$, corresponding to a quasi-permanent change of the refractive index of -5.7×10^{-4}. Inset shows the measurement geometry.

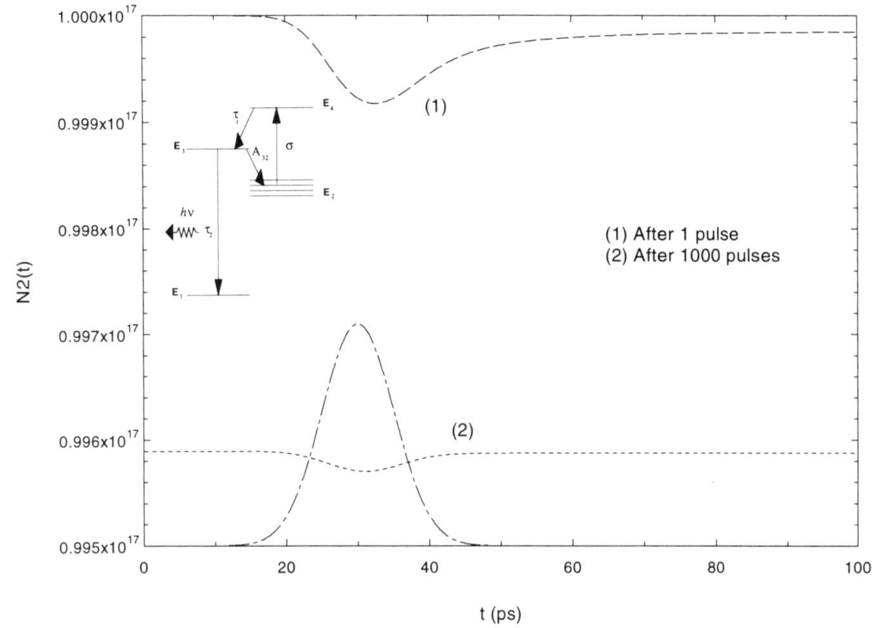

QWE1 Fig. 2. Behavior of the population density N_2 of the level E_2 during pulse excitation. Inset shows the four-level system that takes into account trap levels between the ground state and the first excited state.

verted fluorescence at photon energies $h\nu = 2.4$ eV has been observed by exciting with both pulsed and cw lasers.

In Fig. 1 we show the angular behavior of the coupling efficiency $\eta(\theta)$ versus the input angle θ for a one-prism-coupling experiment (setup in the inset of Fig. 1). Curve (1) was obtained for a peak intensity in the waveguide I_{gw} = 13 MW/cm^2; increasing the guided-wave peak intensity to I_{gw} = 67 MW/cm^2, we observed that, after that some radiation was coupled in the waveguide, curve (2) shifts toward positive angles; the shift is quasi permanent, as is proved by curve (3), obtained again at the low input intensity. The angular shift corresponds to a negative change of refractive index $\Delta n = -5.7 \times 10^{-4}$. For high-intensity excitation a bright green strike, due to luminescence, is observed along the waveguide, superimposed on the propagating infrared beam. Illuminating the sample below the prism with a UV incoherent lamp, we observed a shift back of the resonance curves. Two-photon absorption was excluded as a cause for the upconverted fluorescence by monitoring the power of the infrared light coming out at the edge of the waveguide as a function of the input intensity; no deviation from the linear behavior was detected up to waveguide input intensities of 0.5 GW/cm^2.

We propose that for stationary conditions under daylight irradiation, the presence of chain defects can lead to trapping of charged polarons in separated segments of polymer chains. According to the model, excitation with photon energy below the $1A_g$–$1B_u$ transition can lead to untrapping of charged polarons with polaron-couple formation; such polaron pairs can then recombine to give a molecular exciton, which finally decays in the ground state giving rise to fluorescence. These complex mechanisms can be represented by a four-level system (inset of Fig. 2, where E_1 is the ground state, E_2 is the trap level for polarons, E_4 is an energy level high in the DOS of the $1B_u$ band, and E_3 is the localization energy in the DOS of the $1B_u$ band).[3] We have solved numerically the four-level rate equations, starting with completely filled trap states $N_2(0) = N_T$ and empty E_3 and E_4 levels and illuminating with radiation that is resonant with the transition $E_2 \rightarrow E_4$. For N_T we used the value $N_T = 10^{17}$ cm^{-3}, which is the concentration of intrinsic defects in a semiconductor.

The time behavior of the level E_2 population is shown in Fig. 2. As can be seen at the end of the first pulse [curve (1)], the population of level E_2 is depleted because some of the untrapped carriers have left the system through the fluorescence channel. Curve (2) shows the situation after 1000 pulses. Because of the population decrease of level E_2, the model predicts a decrease of the absorption coefficient, defined as $\alpha = \sigma[N_2(t) - N_4(t)]$. Contemporarily, in the frame of a trapped-carrier model, we expect the refractive index to be proportional to N_2 and so to decrease quasi permanently after each pulse arrival.

*Current address: France Telecom, National Center for Telecommunications Studies, 196 Avenue Henri-Ravera, F-92225 Bagneux Cedex, France
**Institute of Applied Physics, Friedrich Schiller University, Jena, Germany
†Chemistry Section, Friedrich Schiller University, Jena, Germany
‡Fraunhofer Institute for Applied Optics and Precision Engineering, Jena, Germany

1. N. F. Colaneri, D. D. C. Bradley, R. H. Friend, P. L. Born, A. B. Holmes, and C. W. Spanquer, Phys. Rev. B **42**, 670 (1990).
2. E. L. Frankevich, A. A. Lymasev, I. Sokdik, F. E. Karasz, S. Blumstengel, R. H. Boughman, and H. H. Hörhold, Phys. Rev. B **46**, 9320 (1992).
3. R. Kursting, U. Lemmer, R. F. Mahrt, K. Leo, H. Kurz, H. Bassler, and E. O. Göbel, Phys. Rev. Lett. **70**, 3820 (1993).
4. Z. G. Soos, D. S. Galvão, and S. Etemad, Adv. Mater. **6**, 280 (1994).
5. U. Bartuch, A. Bräuer, P. Dannberg, H. H. Hörhold, and D. Rahbe, Int. J. Optoelectron. **7**, 275 (1992).
6. D. D. C. Bradley, A. R. Brown, P. L. Burns, J. H. Burroughes, R. H. Friend, A. B. Holmes, K. D. Mackay, and R. N. Marks, Synth. Met. **41–43**, 3135 (1993).

QWE2 2:45 pm

Femtosecond hyper-Rayleigh scattering of spatial-orientational correlations between chromophores

Koen Clays, Marvin Wu, André Persoons, *Laboratory of Chemical and Biological Dynamics, Center for Research in Molecular Electronics and Photonics, Department of Chemistry, University of Leuven, Celestijnenlaan 200D, B-3001 Leuven, Belgium*

Hyper-Rayleigh scattering (HRS) has been used to probe the spatial-orientational distribution function of nonlinear optical chromophores in the solid state. The fluctuations in this function cause a hyper-Rayleigh scattering signal that fluctuates as a function of position. The decay of the autocorrelation function of this fluctuating signal is characterized by a spatial correlation length. This length indicates the degree of spatial correlation between the chromophores.

Since the introduction of hyper-Rayleigh scattering as a technique for measuring the first hyperpolarizability (second-order polarizability) of molecules in solution,[1,2] the importance of fluctuations in the orientational distribution has been known.[3] For isotropic solutions, a combination of the spatial and temporal fluctuations causes the instantaneous and local deviation from macroscopic centrosymmetry, which allows the observation of a second-order HRS signal. The intensity of the scattered light can then be related to a value for the first hyperpolarizability of the molecules in solution.

After the molecular hyperpolarizability is measured, the individual molecules have to be assembled in a macroscopic, noncentrosymmetric structure. The efficiency of a phase-matched second-order NLO process is highly dependent on the length over which the coherent nonlinear interaction can be maintained. In artificially ordered systems, such as poled polymers and LB films, this length is determined largely by the length over which the order is constant. A scheme to measure the degree of spatial correlation between the chromophores would therefore be instrumental in the development of reliable and reproducible devices.

The relations between incoherent second-order nonlinear light scattering and coherent second-harmonic generation and the implications for the study of microscopic ordering are discussed. Experimental aspects that determine the spatial resolution (Gaussian beam theory with expansion, collimation, and focusing) with femtosecond pulses (femtosecond optics with zero group-velocity dispersion and minimal cubic distortion) are addressed. Various polymeric sample types (chromophores dispersed in various polymer matrices or covalently linked to different polymer backbones) and sample preparation methods (*in situ* polymerization or solvent evaporation) are described. The dynamics of inducing long-range order by poling are explained by *in situ* hyper-Rayleigh scattering.

In a preliminary study two typical ensembles of chromophores were investigated. As expected in a single-crystalline arrangement with a high degree of orientational order, a constant intensity of the scattered light was observed. The short correlation length observed in an on-average isotropic, centrosymmetric, polymeric sample results from spatial fluctuations in the orientational distribution function of the dispersed chromophores.

The measurements presented here were performed on bulk samples. As such, they indicate the degree of spatial orientational correlation in the bulk and are useful for bulk macroscopic devices. It is believed that the same nonlinear scattering technique is applicable in thin-film and waveguide formats, as long as only incoherently scattered light is collected.

1. K. Clays and A. Persoons, Phys. Rev. Lett. **66**, 2980 (1991).
2. K. Clays and A. Persoons, Rev. Sci. Instrum. **63**, 3285 (1992).
3. K. Clays, A. Persoons, and L. De Maeyer, in Modern Nonlinear Optics, Part 3, M. Evans and S. Kielich, eds. (Wiley, New York, 1994), p. 455.

QWE3 **3:00 pm**

Photorefractive composites containing bifunctional chromophores

Yue Zhang, Saswati Ghosal, Ryszard Burzynski, *Laser Photonics Technology, Inc., 1576 Sweet Home Road, Amherst, New York 14228*

Photorefractivity arises from the combined effects of photoconductivity and electro-optic response simultaneously present in a material.[1] Photoconductivity in organic and polymeric materials involves photocharge generation and charge-transporting processes.[2] We report the preparation of a new series of polymeric photorefractive composite materials based on bifunctional chromophores, i.e., molecules that serve as charge-transporting agents as well as second-order nonlinear-optical (NLO) chromophores. The structures of two chromophores are shown in Fig. 1. The charge-transporting properties stem from the triphenylamine moiety, and the second-order NLO response stems from the push–pull structure linked by different conjugation paths. These chromophores have been incorporated into poly-(*n*-butylmethacrylate) (PBMA), which serves only as a binder. When sensitized by C_{60}, photorefractivity has been observed with high diffraction efficiency, large net two-beam coupling gains, and moderate response times.

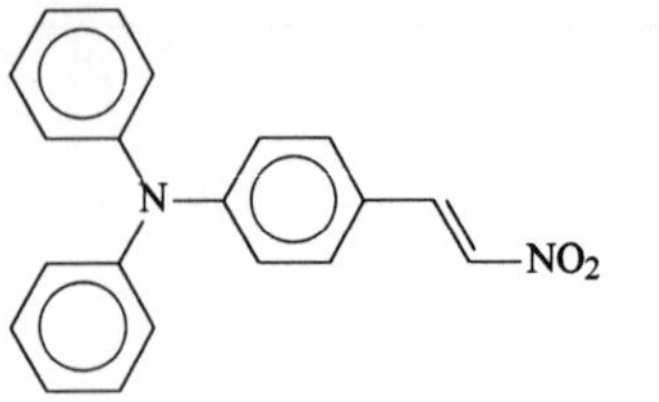

DPANST

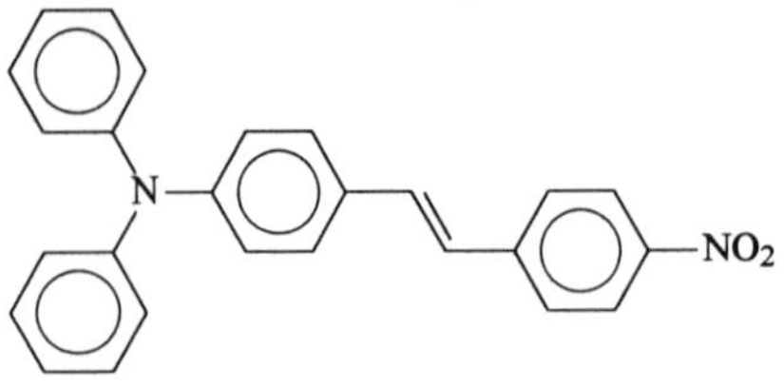

DPANS

QWE3 Fig. 1. Chemical structures of the bifunctional chromophores DPANST and DPANS.

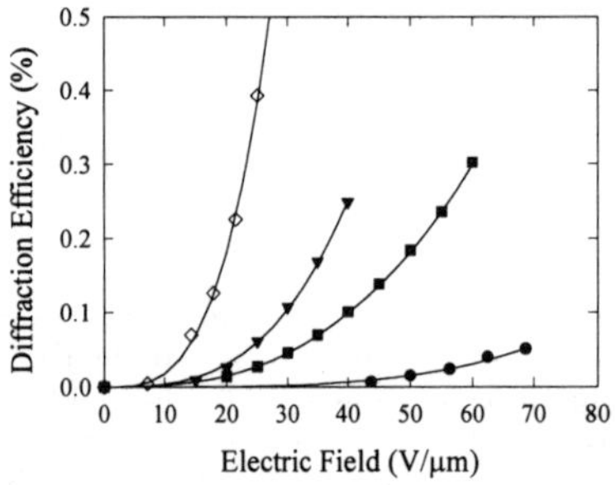

QWE3 Fig. 2. Electric-field dependence of the four-wave-mixing diffraction efficiencies from composites containing various chromophore concentrations: filled circles, 10-wt. % DPANST; filled squares, 22-wt. % DPANST; filled triangles, 36-wt. % DPANST; open diamonds, 33-wt. % DPANS.

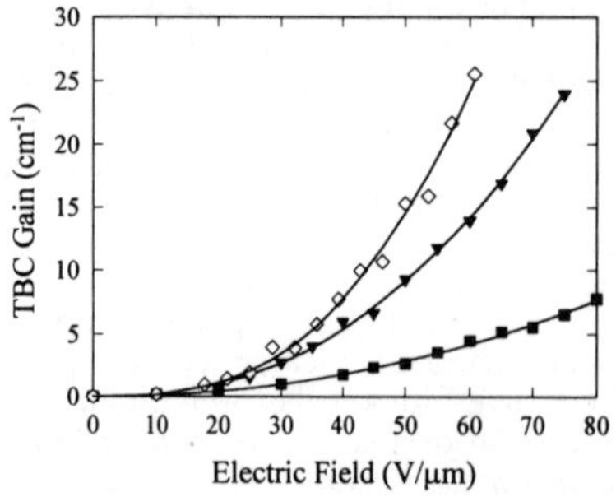

QWE3 Fig. 3. Electric-field dependence of the two-beam-coupling gains from composites containing various chromophore concentrations: filled circles, 10-wt. % DPANST; filled squares, 22-wt. % DPANST; filled triangles, 36-wt. % DPANST; open diamonds, 33-wt. % DPANS.

DPANST was synthesized in two steps from triphenylamine, which was formulated and was then reacted with nitromethane in the presence of equivalent amount of ammonium acetate,[3] whereas DPANS was synthesized by reacting monoformylated triphenylamine with 4-nitrophenylacetic acid in the presence of piperidine.[4] Samples were prepared by sandwiching composites containing selected amounts of DPANST (or DPANS), PBMA, and C_{60} between ITO-covered glass substrates at elevated temperatures. Various concentrations of DPANST (or DPANS) were used to study the charge-transporting abilities and electro-optic responses of the composites. Electro-optic coefficients and photoconductivity sensitivity measurements show that the charge mobilities and electro-optic coefficients in these composites increase with the chromophore concentrations.[5]

The photorefractive properties of these composites were studied by degenerate four-wave mixing and by two-beam coupling with laser beams at 633 nm. The two-beam-coupling gains were measured by monitoring the writing-beam intensities after the sample before and after the electric field was applied. Figures 2 and 3 display the electric-field dependence of the four-wave-mixing diffraction efficiencies and the two-beam-coupling gains for the composites containing various concentrations of the two chromophores. Results show that diffraction efficiency and two-beam-coupling gain increase with the chromophore concentration. It is seen that composites containing DPANS show higher diffraction efficiencies and two-beam-coupling gains than do those containing DPANST at the same concentration; this is consistent with the higher first hyperpolarizability of DPANS. A maximum diffraction efficiency of 12% was obtained in a composite containing 33 wt. % of DPANS at an applied field of 57 V/μm. Net two-beam-coupling gains have been obtained in composites with 33-wt. % chromophores.

In summary, bifunctional chromophores that serve as charge-transporting agents and as second-order NLO compounds have been developed. High-performance photorefractive composites have been developed from these bifunctional chromophores.

This research was supported in part by U.S. Air Force Office of Scientific Research contract F49620-92-C0061 and an award from the New York State Science and Technology Foundation.

1. P. Gunter and J. Huignard, eds., *Photorefractive Materials and Their Applications*, Vols. 61, 62 of Topics in Applied Optics (Springer-Verlag, Berlin, 1990), Vols. I, II.
2. J. Mort and D. M. Pai, eds., *Photoconductivity and Related Phenomena* (Elsevier, Amsterdam, 1976).
3. T. Kurihara, H. Kanbara, H. Kobayashi, K. Kubodera, S. Matsumoto, and T. Kaino, Opt. Commun. **84**, 149 (1991).
4. A. Ulman, C. S. Willand, W. Kohler, D. R. Robello, D. J. Williams, and L. Handley, J. Am. Chem. Soc. **112**, 7083 (1990).
5. Y. Zhang, S. Ghosal, M. K. Casstevens, and R. Burzynski, Appl. Phys. Lett., in press.

QWE4 **3:15 pm**

Efficient upconversion fluorescence in new microcrystals of charge transfer

H. Z. Wang, X. G. Zheng, W. D. Mao, Z. X. Yu, Z. L. Gao, G. Q. Yang,* P. F. Wang,* S. K. Wu,* *Institute for Lasers and Spectroscopy, Zhongshan University, Guangzhou 510275, China*

Charge separation is an important physical process. Since the early 1980's, a series of charge-transfer compounds have been synthesized as molecular models for the investigation of charge separation in photosynthesis. Recently, we synthesized some new intramolecular charge-transfer compounds with an absorption band at a shorter wavelength. We have found that some microcrystals of these new compounds possess an extremely high efficiency of upconversion fluorescence, or frequency doubling. In this paper, upconversion fluorescence in two new intramolecular charge-transfer compounds (NDPB and OTMP) is reported.

The relative yields of upconversion fluorescence for NDPB and OTMP are linearly proportional to the squared pump-laser intensities (as shown in Fig. 1). For comparison, the upconversion fluorescence of an infrared detector (Spectra-Physics 0433-3720) and Rhodamine-6G (Rh6G), which has been used as a two-photon pump lasing dye, is measured under the same conditions (0.1-mJ mode-locked Nd:YAG laser, 1.06-μm wavelength). Our experimental results show that the upconversion fluorescence efficiency of the new intramolecular charge-transfer microcrystals is much higher than those of the infrared detector and of Rh6G. Typically, the upconversion fluorescence efficiency of NDPB microcrystal is 24 times that of the infrared detector and 38 times that of Rh6G.

As shown in Fig. 2, the peak wavelengths of two-photon upconversion fluorescence for NDPB and OTMP are longer than those of fluorescence by single-photon absorption.

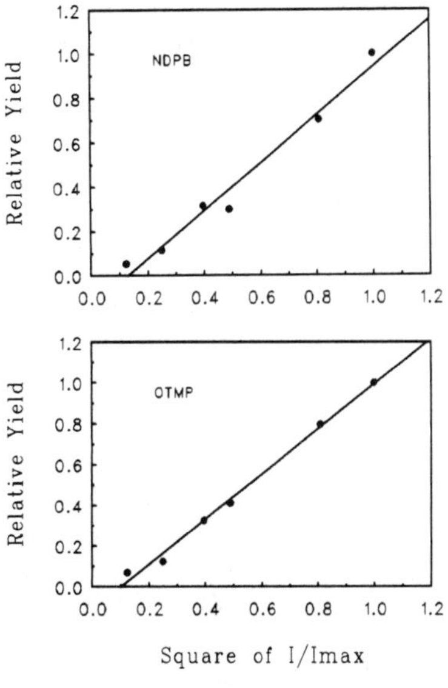

QWE4 Fig. 1. Relative quantum yields of two-photon upconversion fluorescence of NDPB and OTMP microcrystals versus squared pump intensities.

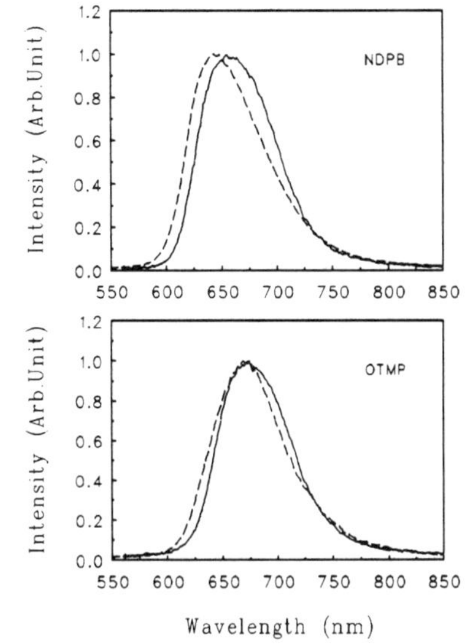

QWE4 Fig. 2. Fluorescence spectra of NDPB and OTMP microcrystals corresponding to double-photon (solid curve) and single-photon (dashed curve) pumping.

It has been demonstrated that exciton strings exist in organic charge-transfer crystals. At high pump densities, multiexciton-state strings appear. The nonlinear optical properties of charge-transfer microcrystals are enhanced by one-dimensional exciton strings.

This work demonstrates conclusively that charge-transfer exciton strings enhance the nonlinear optical effects and that the new charge-transfer crystals are good candidates for nonlinear optical materials.

**Institute of Photographic Chemistry, Chinese Academy of Sciences, Beijing 100101, China*

QWF 2:30 pm
Hyatt B

Applications of Laser Cooling

Phillip L. Gould, *University of Connecticut, Presider*

QWF1 (Invited) **2:30 pm**

Improved frequency-domain lifetime measurement of the Na 3*p* level in a magneto-optic-cell trap

C. W. Oates, K. R. Vogel, J. L. Hall, *Joint Institute for Laboratory Astrophysics, University of Colorado and National Institute of Standards and Technology, Boulder, Colorado 80309-0440*

Renewed interest in the lifetimes of the excited states of alkali atoms arises from their importance in the interpretation of experiments such as those dealing with parity violation in heavy atoms. Agreement between measured and calculated values varies for the alkalis—for sodium, in particular, there is an ~1% (4σ) discrepancy over the 3*p* lifetime.[1–3]

We report here major improvements in our development of a general method for making lifetime measurements. Such experimental investigations are usually conducted in the time domain, but our approach uses measurements of the transition *linewidth* to infer the excited-state lifetimes. Clearly, all other line-broadening mechanisms must be reduced to or characterized at the desired uncertainty level. To this end we use a magneto-optic-cell trap[4] to provide an ultracold atomic sample over which we have a high degree of control.

After the natural linewidth (10 MHz FWHM), the largest contributor to the line-shape width is the velocity distribution of the molasses-cooled atoms (~500 kHz through the first-order Doppler shift). Our previous measurements with a UV probe[5] indicated a potentially troublesome Lorentzian component (120 kHz) of this velocity distribution. Therefore, we have used velocity-sensitive stimulated Raman transitions[6] to synthesize our own distributions. About 10^7 atoms are trapped and are then postcooled (to <40 μK with optical molasses) before being optically pumped to the $F = 1$, $m = 1$ ground state. Counter-propagating Raman beams then pump a narrow (<60-kHz), carefully characterized velocity class back to the $F = 2$ ground state. Finally, we perform spectroscopy on the $(F = 2, m = 2)$–$(F' = 3, m' = 3)$ cycling transition by using a highly-frequency-stabilized circularly polarized 589-nm laser probe in a standing-wave configuration. Our previous scheme probed all magnetic sublevels, limiting our measurements at the 1% level. Probing a two-level system eliminates optical-pumping effects and suppresses the neighboring $F' = 2$ peak, thereby allowing us to achieve uncertainties of <0.5%. The observed low-power line shape is a pure Lorentzian, with corrections totaling less than 0.5%. We also observe novel line-shape asymmetries attributable to heating and cooling effects in a standing-wave configuration, in good agreement with precise Monte Carlo modeling. Figure 1 shows a preliminary scan of this transition, and Fig. 2 shows the measured linewidth versus probe power.

Other broadening mechanisms (laser

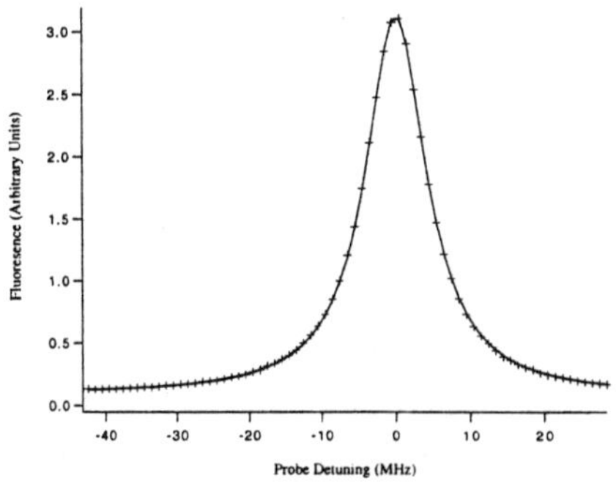

QWF1 Fig. 1. Fluorescence profile of velocity-selected Na $3s^2S_{1/2}$ $(F = 2, m = 2) \Rightarrow 3p^2P_{3/2}$ $(F' = 3, m' = 3)$ transition measured at 50-μW/cm² probe intensity. Crosses indicate experimental data, and the curve represents a nonlinear least-squares fit of the data to a Lorentzian.

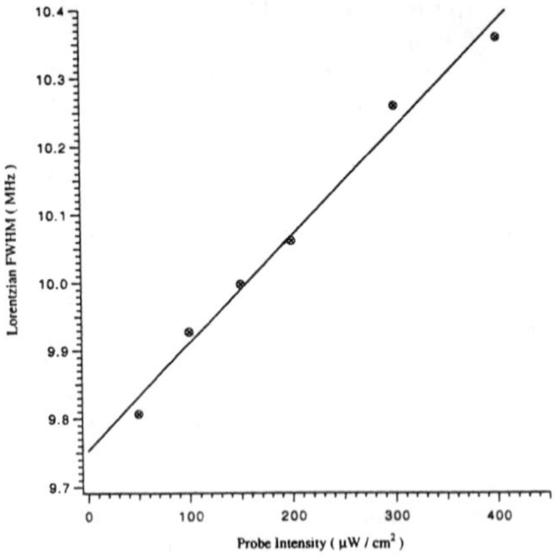

QWF1 Fig. 2. Preliminary plot of measured linewidth (FWHM) versus probe power for 50-μs probe interaction time. After the natural linewidth, power broadening is the dominant broadening mechanism in this low-power regime.

linewidth, power broadening, measurement-beam-induced velocity shifts, etc.) have been sufficiently reduced or characterized, and so the prospects are good for measurements below the 0.5% level by using this technique. Latest results are presented.

1. W. R. Johnson, M. Idrees, and J. Sapirstein, Phys. Rev. A **35**, 3218 (1987).
2. A. Gaupp, P. Kuske, and H. J. Andra, Phys. Rev. A. **26**, 3351 (1982).
3. C. Guet, S. A. Blundell, and W. R. Johnson, Phys. Lett. A **143** 384 (1990).
4. C. Monroe, W. Swann, H. Robinson, and C. Wieman, Phys. Rev. Lett. **65**, 1571 (1990).
5. M. Zhu, C. W. Oates, and J. L. Hall, Opt. Lett. **18**, 1186 (1993).
6. M. Kasevich, D. S. Weiss, E. Riis, K. Moler, S. Kasapi, and S. Chu, Phys. Rev. Lett. **66**, 2297 (1991).

QWF2 3:00 pm

Laser cooling of vibration, rotation, and translation in molecules by using a multiple-single-frequency laser

W. C. Stwalley, J. T. Bahns, P. L. Gould, *Department of Physics, University of Connecticut, Storrs, Connecticut 06269-3046*

One of three priorities of the 1994 National Research Council Report *Atomic, Molecular and Optical Science* is "to promote research . . . to better control and manipulate atoms, molecules, charged particles, and light." Although many techniques are available for atoms, charged particles, and light, we propose here new techniques for rapidly cooling (more generally, manipulating) the vibrational, rotational, and translational energy of simple molecules. A concrete example is manipulating Cs_2 by using the $B^1\Pi_u - X^1\Sigma_g^+$ bands. The key component is an optically pumped amplifier (using the same molecule, here Cs_2) to provide the synchronized multiple-single-frequency laser required to optically couple a significant number of rovibronic levels while rapidly (and simultaneously) tuning over all coupled Doppler profiles (as in atomic translational cooling). Proposed molecular-beam experiments providing total cooling in ~12 ms are described.

QWF3 3:15 pm

Thermodynamic Dicke phase transitions in atom traps

Charles M. Bowden, Jonathan P. Dowling, *U.S. Army Missile Command, AMSMI-RD-WS-ST, Redstone Arsenal, Alabama 35898-5428*

We consider a system composed of a collection of identical two-level atoms interacting with the electromagnetic field in the dipole approximation and with an externally applied steady-state coherent driving field. The atoms interact only through the field and are all contained within a cubic resonance wavelength. The system is treated as a quantum-statistical ensemble in the rotating frame.[1–3] The Hamiltonian in this frame is explicitly time dependent, and the exact analog of a spin temperature can be defined under the well-known Dicke isomorphism between a collection of spinors and a collection of two-level atoms.

The thermodynamic-equilibrium properties of this system are examined, and a nonlinear relationship between the applied field and the actual field in the volume containing the atoms is derived. The expression is a nonlinear function of the effective, or spin, temperature and has the form of a first-order phase transition, with the macroscopic atomic polarization as an order parameter. In other words, below a certain critical temperature the system makes a rather spectacular transition to a single macroscopic electric dipole.

This work is a complement to recent theoretical and experimental investigations into observing Bose condensation in atom traps.[4–8] We discuss possible implementation of this phenomenon in such traps.

1. R. Gilmore and C. M. Bowden, Phys. Rev. A **13**, 1898 (1975).
2. R. Gilmore and C. M. Bowden, J. Math. Phys. **17**, 1617 (1976).
3. C. M. Bowden and C. C. Sung, Phys. Rev. A **19**, 2392 (1979).
4. J. Javanainen, Phys. Rev. Lett. **72**, 2375 (1994).
5. R. J. C. Spreeuw *et al.*, Phys. Rev. Lett. **72**, 3162 (1994).
6. G. Lenz, P. Meystre, and E. M. Wright, Phys. Rev. A **50**, 1681 (1994).
7. A. J. Moerdijk and B. J. Verhaar, Phys. Rev. Lett. **73**, 518 (1994).
8. M. Lewenstein *et al.*, Phys. Rev. A **50**, 2207 (1994).

QWF4 (Invited) 3:30 pm

Radioactive ^{79}rubidium in a magneto-optical trap

L. A. Orozco, *Department of Physics, State University of New York, Stony Brook, New York 11794-3800*

The strong confinement of atoms in 6D phase space makes the magneto-optic trap (MOT) a promising tool for work with small samples of radioactive isotopes, and recently there has been considerable interest in using them for weak interaction precision measurements in radioactive atoms.[1,2] However, the capture velocity for trapping is typically at most 1/10 of the thermal velocity. Special attention has to be paid to loading the atoms into the MOT efficiently. One way of doing this is to laser-decelerate the incoming atomic beam.[2] We have chosen the vapor-cell approach.[3] During wall collisions the atoms rethermalize and have many chances to cross the trap region. Use of non-stick Dryfilm coatings on the inside of the cell prevents the sticking of alkali atoms to the glass walls.[4].

We have trapped radioactive ^{79}Rb (radioactive half-life 23 min.) produced by the reaction ^{51}V(^{31}P, 2np)^{79}Rb with a 90 MeV ^{31}P beam from the Stony Brook Tandem Van de Graaff accelerator. The ^{79}Rb reaction products recoil out of the V target and are stopped in a Au catcher heated close to its melting point. The Rb diffuses to the surface, is surface ionized, and extracted by an electrostatic lens system that delivers the ions to the high-vacuum trap region, about 1 m away. An Yttrium-lined tube heated to 800 °C neutralizes the Rb and ejects it as an atomic beam into the glass cell containing the trap. We use six large (4 cm diameter) laser beams of 20 mW each to achieve a high capture velocity. The laser is locked to a stable Rb reference cell and is frequency-offset with acousto-optic modulators. With help of the Dryfilm coating, sticking on the walls was reduced and we observed some 80 ^{79}Rb atoms in the trap. We estimate the overall efficiency as 10^{-5} atoms in the trap per atom/sec produced in the target. To suppress the strong background of scattered laser light we modulate the repumper frequency and a lock-in amplifier detects the resulting modulation in trap brightness.

The capture efficiency is being improved and next step will be to trap isotopes of francium produced at the Stony Brook Heavy Ion LINAC. In this case the Au-catcher will also function as the target.

Work done in collaboration with J. A. Behr, S. B. Cahn, A. Ghosh, G. Gwinner, J. E. Simsarian, G. D. Sprouse, P. Voytas, F. Xu. Supported in part by NSF and NIST.

1. G. Gwinner, J. A. Behr, S. B. Cahn, A. Ghosh, L. A. Orozco, G. D. Sprouse, and F. Xu, Phys. Rev. Lett. **72**, 3795 (1994).
2. Z-T. Lu, C. Bowers, S. J. Freedman, B. K. Fujikawa, J. L. Mortara, S. Q. Shang, K. P. Coulter, and L. Young, Phys. Rev. Lett. **72**, 3791 (1994).
3. C. Monroe, W. Swann, H. Robinson, and C. Wieman, Phys. Rev. Lett. **65**, 1571 (1990).
4. M. Stephens, K. Lindquist, and C. Wieman, Hyperfine Interactions **81**, 203 (1993). M. Stephens, C. Wieman, Phys. Rev. Lett. **72**, 3787 (1994).

QWG **2:30 pm**
Hyatt C/D

Transient Phenomena in Condensed Matter

Erich P. Ippen, *MIT, Presider*

QWG1 **2:30 pm**

Two-dimensional nonlinear optics using femtosecond spectral interferometry

L. Lepetit, M. Joffre, G. Chériaux, A. Migus, D. Hulin, *Laboratory of Applied Optics, ENSTA, Ecole Polytechnique, National Center for Scientific Research, URA 1406, F-91120 Palaiseau, France*

The nonlinear second-order optical response of a material is not fully characterized until the susceptibility tensor $\chi^{(2)}(\omega_1, \omega_2)$ is known for all values of frequencies ω_1 and ω_2. More than merely characterizing the material optical properties, the two-dimensional response provides essential information on the excited states of the system. This is illustrated in Fig. 1, in which $|\chi^{(2)}(\omega_1, \omega_2)|$ is plotted for a three-level system. The most intense features result from double resonances, which are directly sensitive to the coupling between the excited states. A more conventional measurement, such as frequency-resolved second-harmonic-generation efficiency, corresponds only to a section $\omega_1 = \omega_2$ of the response function.

To measure the nonlinear response function $\chi^{(2)}(\omega_1, \omega_2)$, we use a method analogous to two-dimensional NMR. We focus onto the sample a sequence of two phase-locked pulses[1] separated by a time delay τ. If a thin sample is assumed, the electric field generated by second-order processes can be written

$$\varepsilon_2^{\tau}(\omega) \propto \int \chi^{(2)}(\omega_1, \omega - \omega_1)\varepsilon_1(\omega_1) \times \exp(i\omega_1\tau)\varepsilon_1(\omega - \omega_1)\, d\omega_1, \quad (1)$$

where $\varepsilon_1(\omega)$ is the incident-pulse spectrum. As in Fourier NMR, the response function $\chi^{(2)}(\omega_1, \omega - \omega_1)$ can be obtained through a Fourier transform of $\varepsilon_2^{\tau}(\omega)$ with respect to τ if it is assumed that we know the incident-pulse electric field $\varepsilon_1(\omega)$. However, this method implies that $\varepsilon_2(\omega)$ must be monitored in both amplitude and phase. This last measurement, based on femtosecond spectral interferometry, is achieved in a Mach–Zender interferometer.[2] The whole experimental setup is shown in Fig. 2. Note that phase-sensitive photon-echo experiments can be performed with almost the same experimental setup.

To demonstrate our method we use a moderately thick transparent nonlinear crystal, such as KDP. In this case the nonlinear response $\chi^{(2)}$ can be assumed to be frequency independent, but the phase-matching factor

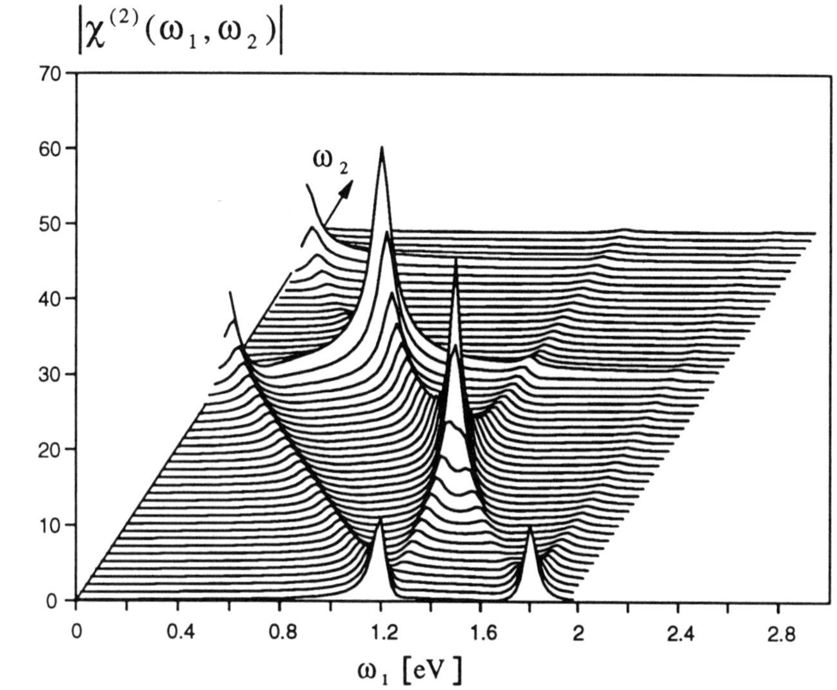

QWG1 Fig. 1. Two-dimensional plot of $|\chi^{(2)}(\omega_1, \omega_2)|$ as computed from the sum-over-state expression on a three-level model. The two transition energies are 1.2 and 1.8 eV.

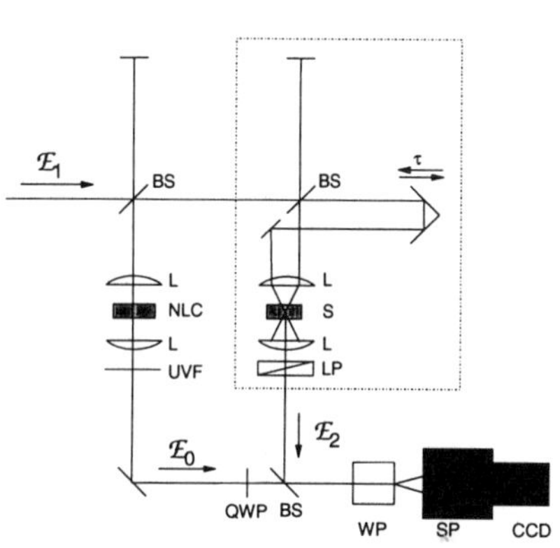

QWG1 Fig. 2. Phase-sensitive design for the measurement of $\chi^{(2)}$: ε_1, incident laser pulse; ε_0, frequency-doubled reference pulse; ε_2, pulse emitted by the sample placed in the autocorrelator (outlined with a dotted line) and frequency resolved in amplitude and phase in the spectral interferometer; BS, beam splitter; L, lens; NLC, nonlinear crystal; UVF, UV filter; S, sample; LP, linear polarizer; QWP, quarter-wave plate; WP, Wollaston prism; SP, spectrometer; CCD, charge-coupled-device array detector.

$$\Xi^{(2)}(-\omega; \omega_1, \omega_2) = \frac{\exp\{i[k_{o,e}(\omega_1) + k_{o,e}(\omega_2) - k_{o,e}(\omega)]L\} - 1}{k_{o,e}(\omega_1) + k_{o,e}(\omega_2) - k_{o,e}(\omega)}, \quad (2)$$

where L is the crystal thickness, can be considered as a model two-dimensional nonlinear response. Furthermore, its theoretical value is accurately known and can be compared with the experiment. Experimental results, obtained by using a Ti:sapphire oscillator delivering a 100-nm-wide spectrum, are presented.

1. N. F. Scherer, R. J. Carlson, A. Matro, M. Du, A. J. Ruggiero, V. Romero-Rochin, J. A. Cina, G. R. Fleming, and S. A. Rice, J. Chem. Phys. **95**, 1487 (1991).
2. F. Reynaud, F. Salin, and A. Barthelemy, Opt. Lett. **14**, 275 (1989).

QWG2 **2:45 pm**

Spectral signatures of excitonic nonlinearities measured by transient four-wave mixing

J. Erland,* D. Birkedal, J. M. Hvam, *Microelectronics Center, Technical University of Denmark, DK-2800 Lyngby, Denmark*

Transient four-wave mixing (TFWM) has in recent years been a powerful tool for the study of the optical excitation and nonlinear response of excitons in bulk semiconductors[1] and semiconductor structures.[2] Optical nonlinearities due to band filling and homogeneous broadening due to exciton–exciton interactions[3] have been studied with ultrafast lasers. Of special interest has been the latter nonlinearity as the origin of a TFWM signal for negative time delay due to scattering of polarization (i.e., local-field effects), excitation-induced dephasing,[4] or biexciton formation.[5] Interference between excitations and fields at different optical frequencies can be significant in the wave-mixing process for excitation by femtosecond (broadband) pulses. As a consequence, a clear interpretation of the experimental data can often be difficult.

We present spectrally resolved TFWM data from a GaAs multiple-quantum-well sample with four different well widths of nominal thicknesses of 80, 100, 130, and 160 Å excited with 100-fs pulses from a self-mode-locked Ti:sapphire laser. The TFWM spectra with the laser centered at the heavy-

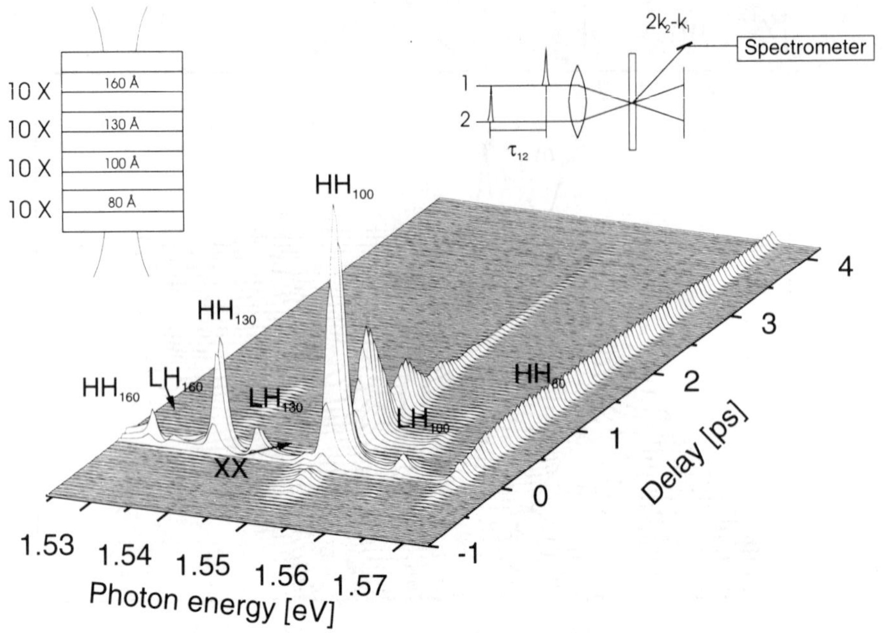

QWG2 Fig. 1. TFWM spectra as a function of time delay for a GaAs multiple-quantum-well sample with various well widths (80, 100, 130, and 160 Å; 10 periods of each), kept at 5 K. Light-hole (LH) and heavy-hole (HH) resonances are indicated. Only for the HH_{100} resonances is the biexciton (XX) clearly observed. The 100-fs pulsed excitation is centered around the HH_{100} resonance. The left inset shows the GaAs sample with barriers of 150-Å $Al_{0.3}Ga_{0.7}As$, and the right inset shows the experimental configuration.

hole exciton resonance of the 100-Å quantum well (HH_{100}) measured as a function of time delay between the incident pulses are shown in Fig. 1. The HH_{80} decay is relatively long and represents the dephasing at low density. The maximum for finite delay demonstrates that the response appears as a photon echo, and details from the spectral behavior reveal the (in)homogeneous linewidths.[6] A pronounced signal is observed for negative delay.

From the last three types of quantum wells (100, 130, and 160 Å), we observe both the HH and the LH resonances, but in each case with distinctly different spectral and time-delay behaviors. The 100-Å resonances exhibit pronounced beats from the coherent superposition of states, and the spectral behavior gives evidence for a quantum beat as opposed to a polarization beat.[7] Also observed is a signal attributed to biexciton formation (XX). Even the signal for negative delay is modulated by the beat, but it is spectrally separated from the HH_{100} biexciton signal.

In the last two quantum wells (130 and 160 Å), we observe decays of the order of the exciting laser pulse, suggesting a surprisingly fast dephasing. However, from the obtained spectra it can be concluded that this interpretation is wrong because we do not observe a corresponding broadening of the resonances. Instead, because of the spectrally broad laser profile, a wave packet is formed from a coherent superposition of many excited (exciton and continuum) states, and the corresponding polarization decays (because of destructive interference) in a time τ_{wp} proportional to the inverse spectral width of the wave packet.[8] Notice that for this excitation condition the signal at negative delay vanishes.

In conclusion, complicated excitation dynamics can thus be studied with this technique, taking advantage of time resolution on the input side of the TFWM signal (through time delay between the short pulses) and spectral resolution on output side of the TFWM signal. In addition to this, we present a detailed analysis of the behavior of the TFWM spectra at negative delay for various excitation conditions.

**Permanent address: Department of Physics, Odense University, DK-5230 Odense M, Denmark*

1. Wang *et al.*, Phys. Rev. Lett. **71**, 1261 (1993).
2. Gobel *et al.*, Phys. Rev. Lett. **64**, 1801 (1990).
3. Wegener *et al.*, Phys. Rev. A **42**, 5675 (1990).
4. Wang *et al.*, Phys. Rev. A **49**, 1551 (1994).
5. Pantke *et al.*, Phys. Rev. B **47**, 2413 (1993).
6. Erland *et al.*, Phys. Rev. B, in press.
7. Lyssenko *et al.*, Phys. Rev. B **48**, 5720 (1993).
8. Feldmann *et al.*, Phys. Rev. Lett. **70**, 3027 (1993).

QWG3 **3:00 pm**

Observation of optical coherent transients by current switching a semiconductor diode laser

R. M. Macfarlane, M. Zhu,* *IBM Almaden Research Center, San Jose, California 95120*

We report a new, simple technique for measuring coherent transients, such as the photon echo and free-induction decay, by the application of small current pulses to switch the frequency of a diode laser. We have used this technique to make what we believe to be the first measurements of optical free-induction decay in the Tm^{3+} ion. An analysis of the heterodyne beat signal can also give, in real time, the frequency history of the laser during the preparation, switching, and detection periods. There have been a number of reports on frequency switching and heterodyne detection of coherent transients by using intracavity phase modulators, extracavity acousto-optic modulators, or electric or magnetic field switching of the energy levels of the material system.[1]

The materials chosen for study were $LaF_3:Tm^{3+}$ and $YAG:Tm^{3+}$, specifically the $^3H_6(1) \rightarrow {}^3H_4(1)$ transition at 796.30 nm (LaF_3) and at 793.17 nm (YAG). Because of the long coherence times of these materials at 1.6 K, an external-cavity diode laser was used. For the observation of a free-induction decay, the waveform shown in Fig.

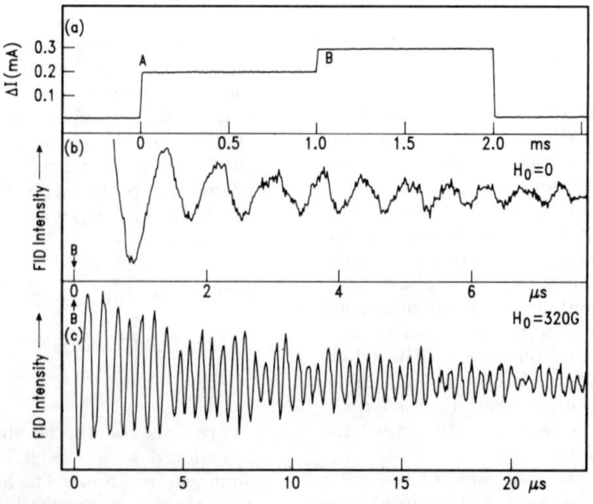

QWG3 Fig. 1. Optical free-induction decay on the $^3H_6(1) \rightarrow {}^3H_4(1)$ transition of the $LaF_3:Tm^{3+}$ transition at 796.30 nm and 1.6 K. (a) Current step applied to the external-cavity diode laser. (b) Optical free-induction decay measured at zero applied field. (c) Optical free-induction decay measured in an external magnetic field of 320 G.

1(a) was applied to the laser injection current. The step at A shifts the frequency of the laser out of the steady-state hole and into resonance with a new subset of Tm^{3+} ions. These ions are then prepared for 1 ms, after which a frequency switch of 1.5 MHz is applied at B. The width of the frequency packet excited is a combination of the homogeneous width of the transition Γ_h and the frequency jitter of the laser during the preparation. After the frequency switch, the coherent emission from the sample beats with the laser at 1.5 MHz, as shown for LaF_3:Tm^{3+} in Fig. 1(b). This is a single shot selected in a period in which the laser-frequency jitter is very small. The decay of the envelope of this beat gave a dephasing time of T_2 = 6.7 μs (Γ_h = 47 kHz), in good agreement with the value of 7 μs from a photon-echo measurement[2] showing laser jitter of <5 kHz. In general, the laser jitter was much higher, preventing the averaging of multiple shots. This was demonstrated by a hole-burning experiment, which gave a frequency stability of ~400 kHz when averaged over ten 20-ms periods. In an external magnetic field the coherence-decay time increases to 26 μs (12 kHz), as shown in Fig. 1(c). About half of this width is laser jitter, as was also the case for an FID measured in YAG:Tm in zero field, which decayed in 22 μs.

Using a different current waveform to generate a pair of 0.6-μs pulses, we observed a heterodyne-detected photon echo in YAG:Tm.

We have shown that optical coherent transients can be measured by the very simple technique of frequency switching an external-cavity diode laser with a small current step. The method could be applied to time-domain information storage and signal processing using stimulated photon echoes.[3] It can also provide a real-time measurement of laser-frequency stability up to the time scale of the sample coherence, and it is extendable to much higher frequencies.

*Hewlett Packard Research Laboratories, Palo Alto, California 94304

1. R. M. Macfarlane, in *Optical Properties of Excited States in Solids*, B. Di Bartolo, ed. (Plenum, New York, 1992), p. 399.
2. R. M. Macfarlane, Opt. Lett. **18**, 1958 (1993).
3. T. W. Mossberg, Opt. Lett. **7**, 77 (1982); W. R. Babbitt and T. W. Mossberg, Opt. Commun. **65**, 185 (1988).

QWG4 **3:15 pm**

Transient pulse response of $In_{0.2}Ga_{0.8}As$/GaAs microcavity lasers

P. Michler, A. Lohner, W. W. Rühle, G. Reiner,* *Max Planck Institute for Solid State Research, Heisenbergstrasse 1, D-70506 Stuttgart, Germany*

We report on the transient emission of microcavity lasers at 20 K after femtosecond optical excitation. The two investigated structures contain one and three 8-nm $In_{0.2}Ga_{0.8}As$ quantum wells (QWs) located in the center of λ cavities formed by AlAs/GaAs Bragg-reflector mirrors. The bottom mirror is n-doped and has 30.5 layer pairs, and the top mirror is p-doped and consists of 20 layer pairs. The cavity features a 105-nm-wide stop band centered at the resonance wavelength, which varies from 1020 nm at the wafer center to 860 nm at the edge because of growth inhomogeneities. The QW band-gap energy is nearly independent of the position on the wafer (Δλ < 2 nm) and is centered at 922 nm at 20 K. The microcavity-laser-emission wavelength is tuned with respect to the gain spectrum of the QWs by probing various points on the wafer. The QWs are directly excited by 80-fs laser pulses at 821 nm with photon energy below the GaAs barriers. The time evolution of the emission is traced by an upconversion technique.

Figure 1(a) shows the temporal evolution of the emission at 925 nm. The intensity rises to a maximum in 36 ps and decays initially fast with a time constant of 9 ps followed by a slower decay. This response reflects the buildup of the carrier density above threshold, a subsequent fast stimulated recombination, then a transition into an amplified-spontaneous-emission regime or simply a spontaneous-emission regime. The full width at half maximum (FWHM) of the emission pulse is 17.5 ps. Figure 1(b) shows the time evolution at a different spot on the wafer where the cavity resonance is at 884 nm. Lasing emission starts with a rise time of 2.3 ps, reaches the maximum in 31 ps, and decays monoexponentially with a decay time of 6 ps. The FWHM of the emission pulse is 12 ps. The longer time constants at higher wavelengths demonstrate the dependence of the microcavity laser speed on the energy dependence of the differential gain: On the high-energy side of the gain spectrum, the shift of the Fermi level with density results in a larger change of gain with density, and on the low-energy side, band-gap renormalization changes gain less with density. These results are in agreement with recently published data on InGaAs/InGaAsP distributed-feedback lasers.[1]

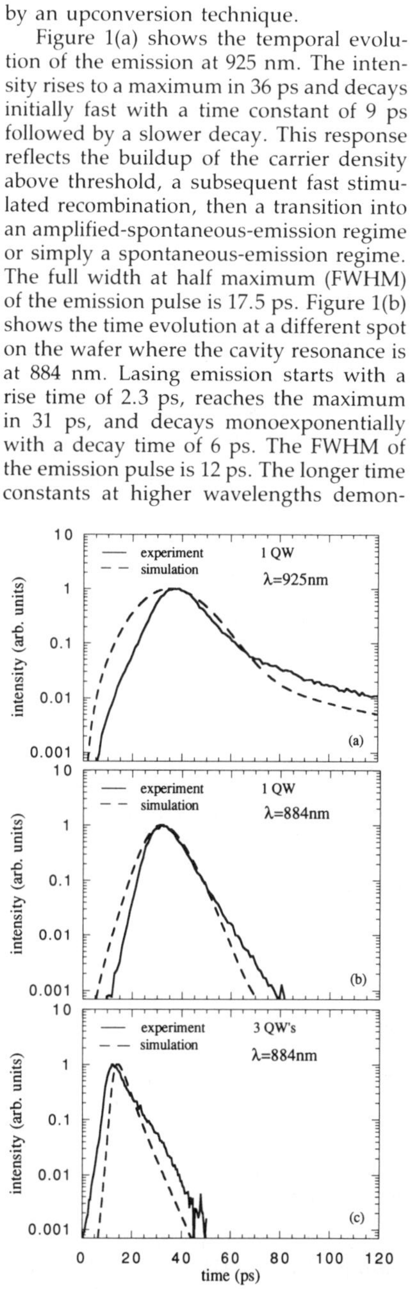

QWG4 Fig. 1. Measured (solid curves) and calculated (dashed curves) time evolution of the emission of $In_{0.2}Ga_{0.8}As$/GaAs microcavity lasers after pumping with femtosecond optical pulses.

Next, we compare the time behavior of a microcavity containing one QW [Fig. 1(b)] with a microcavity containing three QWs [Fig. 1(c)]: The rise time, fall time, and FWHM are 1, 5.8, and 6.8 ps, respectively. An increase in the number of QWs appreciably fastens the pulse response because of the increase of the confinement factor, which increases with the size of the active gain region.

Laser rate equations, including spontaneous and stimulated emission as well as confinement factors of the optical fields (which are different for the two structures) and the variation of the differential gain with photon energy, are numerically solved. The parameters are chosen appropriate to our microcavity structures and our experimental conditions. The results are shown as dashed curves in Fig. 1. Our phenomenological model satisfactorily explains the faster response with increasing photon energy of the emission and with increasing number of QWs in the cavity.

Department of Optoelectronics, University of Ulm, Albert-Einstein-Allee 45, D-89069 Ulm, Germany

1. J. Kovac, H. Schweizer, H. Schmidt, C. Kaden, M. Klenk, R. Weinman, and E. Zielinski, Proc. Soc. Photo-Opt. Instrum. Eng. **2142**, 348 (1994).

QWG5 **3:30 pm**

Switching of superradiance in semiconductor superlattices

S. J. Lee, J. B. Khurgin, S. Li, *Department of Electrical and Computer Engineering, Johns Hopkins University, Baltimore, Maryland 21218*

Recently, excitonic superradiant decay in semiconductor quantum wells (QWs) has been studied by a number of researchers[1] and has been verified experimentally by fast radiative lifetimes obtained in a single GaAs quantum well (~10 ps.).[2]

In this study we investigate spontaneous-emission properties of the excitonic states in the superlattice (SL) undergoing the Wannier–Stark (WS) transition in the presence of an external electric field. In an SL [Fig. 1(a)], as the period (barrier) increases, the Coulomb-attraction energy between the electrons and holes becomes greater than the interwell coupling energy V. As a result, the Wannier exciton in the z direction becomes Frenkel excitons, indicating that an electron and a hole are always in the same well, but it is not known in which one [see Fig. 1(b)]. The oscillator strength (OS) of the exciton then becomes $f_x = Nf_0\phi_r^2(0)$, where N is the number of periods and $\phi_r(0)$ is the in-plane part of the excitonic wave function. We can see that there are two OS enhancement factors—

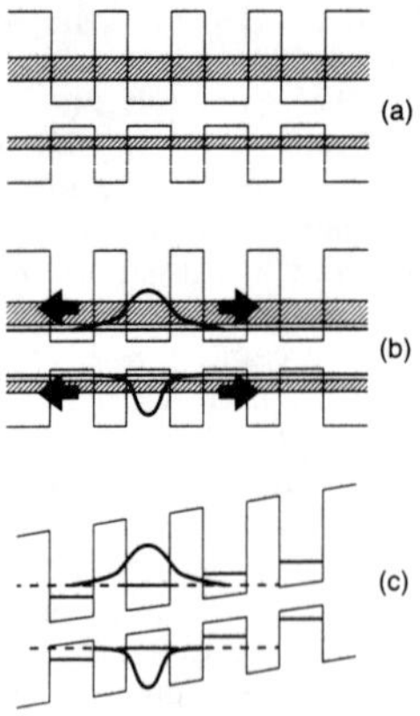

QWG5 Fig. 1. Superlattice (a) with minibands of width $4V$ and (b) with $Q > V$; excitons are confined in well regions but are free to move around in the structure. (c) superlattice with $F \neq 0$; excitons are confined in a single well.

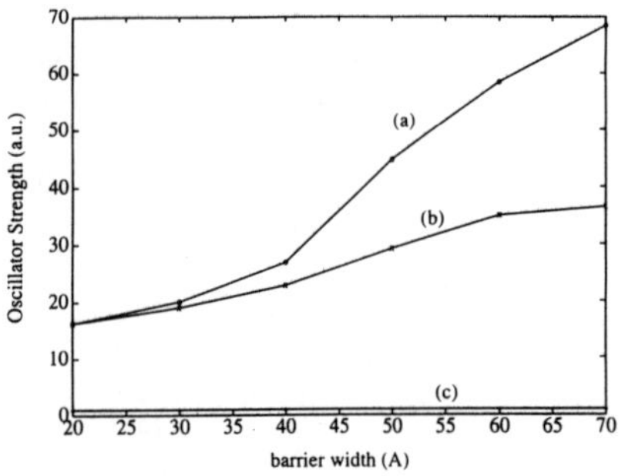

QWG5 Fig. 2. Oscillator strength ratio f/f_0 versus number of periods for a barrier width of 50 Å. f denotes the case in which the Coulomb attraction is considered, and f_0 denotes the case in which the Coulomb attraction is neglected. (a) No external bias; (b) external bias applied.

ϕ_r^2 due to the formation of Wannier excitons in the parallel plane and N due to Frenkel excitons in the z direction.

When a sufficiently strong electric field F is applied to the SL, the WS transition takes place and the excitonic wave function becomes localized in one particular well, with no coherent coupling between the adjacent states. Thus, the system can be described by N separate WS ladder wave functions. The OS of each state is $f_{WS} = f_0\phi_r^2(0)$. The total OS of N states in the WS ladder is equal to the OS of the superradiant Frenkel state, indicating that the absorption strength is preserved, but for each particular state the superradiance enhancement is completely gone, thereby slowing down the radiative decay of the WS ladder states by a factor of N. Therefore, sudden imposition of the relatively weak external field can cause a large increase in the radiative lifetime of the SL, opening the possibility of modulating the spontaneous emission on a picosecond scale, which can be far more efficient than switching of the OS in a single QW.[3]

Our calculations are based on the linear combination of 1s exciton states with a single-well basis[4] and on the tight-binding method in the nonzero-field case for the WS ladder.[5] Figure 2 shows the OS enhancement ratio after the excitonic effect is considered for both the $F = 0$ and $F \neq 0$ cases. It is easily seen that both longitudinal and transverse excitonic components increase the OS dramatically as a linear function of N. However, when the field is applied, the enhancement due to the number of periods is canceled out. It is also found that the excitonic effect is more prominent as the barrier is increased.

In practice, the enhancement is limited by the SL coherence length, which depends on the presence of scattering mechanisms, and factors such as temperature and sample quality may be significant. Because the calculated and measured single-well decay time is ~20 ps,[6] the corresponding value for the SL should be of the order of a picosecond for a coherence length of 10 periods. Thus, aside from modulation of spontaneous emission by the electric field, the enhancement factor thus obtained could provide a useful measure of the coherence length and various scattering events in SLs.

1. E. Hanamura, Phys. Rev. B **38**, 1228 (1988).
2. B. Deveaud, F. Clérot, N. Roy, B. Sermage, and D. S. Katzer, Surf. Sci. **263**, 491 (1992).
3. I. Suemune, T. Takoka, M. Yamanishi, and Y. Lee, IEEE J. Quantum Electron. **QE-22**, 1900 (1988).
4. S. Li and J. B. Khurgin, Phys. Rev. B **46**, 12535 (1992).
5. J. Bleuse, G. Bastard, and P. Voisin, Phys. Rev. Lett. **60**, 220 (1988).
6. L. C. Andreani, Solid State Commun. **77**, 641 (1991).

QWG6 **3:45 pm**

Two-photon photoemission study of two-dimensional scattering processes on Na-covered Cu surfaces

X. Y. Wang, R. Paiella, R. M. Osgood, Jr., *Columbia Radiation Laboratory, Columbia University, New York, New York 10027*

Two-dimensional electronic systems have been the subject of considerable interest for the last few decades. An interesting excited-state two-dimensional electronic system is that of the image-potential states on highly polarizable single-crystal surfaces. Bound states result from an electron confined between the projected band gap of the crystal and the potential of the image charge at the surface—to give a Rydberg series of states with binding energies E_n typically $\leq$0.8 eV below the vacuum level. Although image states have been extensively studied on various single-crystal metal surfaces in recent years, the scattering mechanism for image-state electrons has not been extensively investigated. Here, we present an angle-resolved two-photon photoemission (2PPE) investigation of the scattering of the $n = 1$ image-state electrons by ionized species, introduced through Na adsorption at a Cu(111) surface. The use of angle-resolved photoemission allows us to selectively study the electron dynamics as a function of a given, unique lateral velocity.

The experiments were performed with 20-ns, p-polarized laser pulses that were generated by a three-stage excimer-laser-pumped tunable dye laser and were frequency-doubled to give an incident photon energy of 4.47 eV. This photon energy allows resonant excitation from an intrinsic surface state to the $n = 1$ image state. Image-state quenching and the broadening of the photoemitted image-state electron energy distribution were studied as a function of coverage and $k_\parallel$. In particular, the image-potential peak, at normal emission under constant excitation intensity, was studied on Cu(111) with Na coverages ranging from $\Theta = 0$ to $\Theta = 9/1000$ ML. The results showed a significant linewidth broadening and a sharp decrease in the photoemission intensity with Na coverage; for coverages as low as 0.012 ML, the image state appeared to be completely quenched. To characterize the quenching and linewidth-broadening mechanisms, an angle-resolved 2PPE study was performed. The image-state linewidth broadening as a function of the electron momentum parallel to the Cu(111) surface is shown in Fig. 1 for a clean surface and for a Na-dosed surface. The Lorentzian natural linewidths shown were derived by numerical deconvolution of measured peak widths with the known detector Gaussian line shape. The linewidth is clearly seen to increase with $\hbar k_\parallel$ (the electron wave vector parallel to the surface) and Na coverage Θ in all cases. The linewidth-broadening mechanism is basically attributed to lifetime and nonlifetime effects. The lifetime effect directly relates to the loss of the electrons from the image state. The nonlifetime broadening comes from weakened conservation of momentum $\hbar k_\parallel$ caused by elastic scattering processes through defects and impurity, i.e., localization of the image electron. The measured linewidth as shown in Fig. 1 can be fitted with the expression

$$\Gamma = \Gamma_{\text{lifetime}} + \Gamma(\Delta k_\parallel)$$
$$= \Gamma_{\text{lifetime}} + \hbar^2 k_\parallel(\Delta k_\parallel^s + \Delta k_\parallel^{\text{res}})/m^*. \quad (1)$$

The fitted results give scattering cross sections of about 100 Å. Results for $n = 2$ have also been obtained.

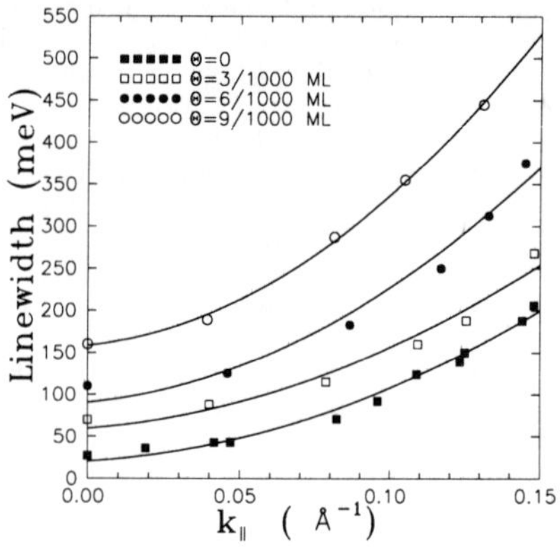

QWG6 Fig. 1. Image-state ($n = 1$) linewidth broadening as a function of the electron momentum parallel to the surface on clean and Na-adsorbed Cu(111) at various coverages Θ.

JWE 4:30 pm Room 308

Session on Far-Infrared Sources II

Ramon U. Martinelli, *David Sarnoff Research Center, Presider*

JWE1 (Invited) 4:30 pm

InAsSb strained quantum-well lasers emitting in the 4-μm band

H. K. Choi, G. W. Turner, H.Q. Le, *Lincoln Laboratory, Massachusetts Institute of Technology, Lexington, Massachusetts 02173-9108*

Recently, the development of high-performance III–V diode lasers emitting in the mid-infrared (2–5 μm) band has generated increased interest.[1-10] Such lasers are very desirable for many applications such as remote sensing, pollution monitoring, and molecular spectroscopy.

Strained quantum-well (QW) diode lasers consisting of GaInAsSb active and AlGaAsSb barrier/confining layers grown on GaSb substrates have exhibited excellent room-temperature performance at ~2 μm, with pulsed threshold current density J_{th} as low as 143 A/cm^2 and cw output power of 1.3 W.[1] However, realizing GaInAsSb/AlGaAsSb QW lasers with emission wavelengths beyond 3 μm has been difficult because of the decreasing valence-band offset and refractive index step between the active and barrier layers. In addition, there is a large miscibility gap in the GaInAsSb alloy, which makes it very difficult to grow alloy compositions that have somewhat favorable valence-band offsets.

In this paper, we will describe recent progress in strained QW lasers emitting at ~4 μm. Because InAsSb is stable at the growth temperature, we selected it as the well material, and tried to find an adequate barrier material that provides good carrier confinement. We first selected InAlAs barriers, which provide a valence-band offset of ~90 meV, but a conduction-band offset of only ~30 meV. These devices emitting at 4.5 μm operated pulsed up to 85 K.[5] The performance might have been limited by the lack of adequate electron confinement and nonuniform hole distribution in the multiple quantum wells.

The InAlAsSb barriers provide a much larger conduction-band offset than InAlAs, while maintaining about the same valence-band offset. The InAsSb/InAlAsSb QW lasers emitting at 3.9 μm have operated pulsed up to 165 K, with J_{th} of 78 A/cm^2 at 80 K. The characteristic temperature T_0 is 30 K up to 120 K. These devices operated cw up to 123 K, with output power of 30 mW/facet at 80 K. These results are better than those for InAsSb/AlAsSb double-heterostructure lasers, which had the maximum cw operating temperature of 105 K and T_0 of 20 K.[5]

This work was sponsored by Phillips Laboratory, the Department of the Air Force.

1. H. K. Choi *et al.*, IEEE Photon. Technol. Lett. **6**, 7 (1994).
2. H. Q. *et al.*, Appl. Phys. Lett. **64**, 152 (1994).
3. S. R. Kurtz, *et al.*, Appl. Phys. Lett. **64**, 812 (1994).
4. A. N. Baranov *et al.*, Appl. Phys. Lett. **64**, 2480 (1994).
5. H. K. Choi et al., presented at Int. Semicond. Laser Conf. Paper M3.6, Maui, Hawaii, September 19–23, 1994.
6. H. Lee *et al.*, *ibid* Paper PD6.
7. Y. H. Zhang *et al.*, *ibid* Paper PD7.
8. S. Forouhar *et al.*, Electron. Lett. **28**, 14 (1992).
9. J. S. Major *et al.*, IEEE Photon. Technol. Lett. **5**, 594 (1993).
10. J. Faist *et al.*, Electron. Lett. **30**, 1360 (1994).

JWE2 5:00 pm

Midwave-IR semiconductor lasers with InAs/GaInSb Broken-gap superlattice active regions

T. C. Hasenberg, D. H. Chow, R. H. Miles, A. R. Kost, L. West, *Hughes Research Laboratories, 3011 Malibu Canyon Rd., Malibu, California 90265*

InAs/GaInSb broken-gap superlattices (BGSLs) are attractive candidates for mid-wave-infrared (MWIR) laser active regions because they possess sufficient degrees of freedom to tailor the valence band structure while keeping the strain low enough to prevent the formation of misfit dislocations. The valence band structure can be optimized to suppress Auger recombination and to reduce the density of states at k = 0, thereby reducing laser threshold. Here we report the first MWIR MQW laser diodes with InAs/$Ga_{0.75}In_{0.25}Sb$ BGSL quantum wells. The laser structures employ $Ga_{0.75}In_{0.25}As_{0.22}Sb_{0.78}$ barrier layers, and InAs/AlSb superlattice cladding layers with an AlSb blocking layer deposited directly on top of the p-type superlattice clad for enhanced electron confinement. The $Ga_{0.75}In_{0.25}As_{0.22}Sb_{0.78}$ barriers yield a type-I MQW with the BGSL active region, confining both electrons and holes, as well as enhancing the optical confinement factor of the structure due to their relatively large refractive index. The thicknesses of the InAs/$Ga_{0.75}In_{0.25}Sb$ superlattice constituent layers have been varied to tune the emission

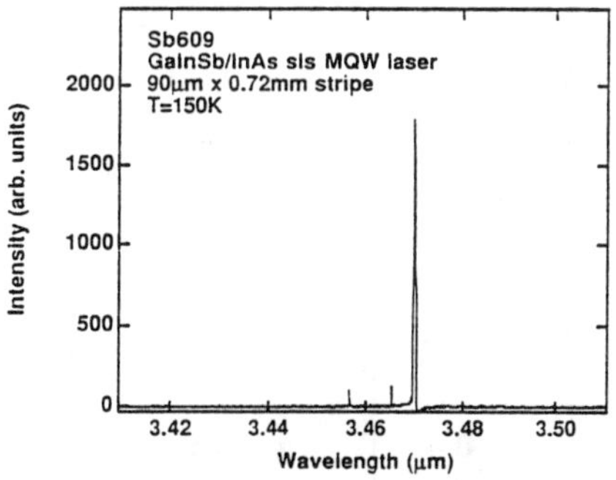

JWE2 Fig. 1. High resolution FTIR spectrum from a MQW laser diode with $Ga_{0.75}In_{0.25}As_{0.22}Sb_{0.78}$ barriers and $Ga_{0.75}In_{0.25}Sb$/InAs BGSL quantum wells. The maximum temperature of operation for the diode is 160 K.

wavelength of the devices, and a range of laser wavelengths have been demonstrated.

Sample Sb615 employed a 5 period MQW active region consisting of InAs (15 Å)/$Ga_{0.75}In_{0.25}Sb$ (33 Å) (4.5 period—starts and ends with $Ga_{0.75}In_{0.25}Sb$) BGSL wells and 400 Å $Ga_{0.75}In_{0.25}As_{0.22}Sb_{0.78}$ barriers. This device lases at 3.28 μm up to 150 K with a 77 K pulsed threshold current density of 134 Amps/cm^2.

Sample Sb609 possesses a five-well MQW structure where each $Ga_{0.75}In_{0.25}Sb$/InAs (35 Å/17 Å) superlattice quantum well consists of 4 periods plus 1 $Ga_{0.75}In_{0.25}Sb$ layer. Figure 1 depicts a high-resolution FTIR spectrum taken from the sample at 150 K, showing laser modes at 3.47 μm under pulsed operation. The pulsed threshold current density at 77 K is 406 A/cm^2.

Sample Sb614 is a six-well MQW structure similar to Sb609, except that the $Ga_{0.75}In_{0.25}Sb$ and InAs constituent layer thicknesses are 35 Å and 21 Å, respectively, in each superlattice period. This sample lased at 3.90 μm up to 84 K.

These results demonstrate that InAs/$Ga_{0.75}In_{0.25}Sb$ BGSL can be grown with sufficient optical quality to support lasing. Work directed towards the attainment of room temperature cw operation is in progress.

This work was supported by the Air Force Phillips Laboratory, contract number F29601-93-C-0037.

JWE3 5:15 pm

High power (1.3 W) optically pumped midwave-IR (3.4 μm) InAs/InAs$_x$Sb$_{1-x}$ superlattice lasers

Y.-H. Zhang, H. Q. Le,* *Hughes Research Laboratories, 3011 Malibu Canyon Road, Malibu, California 90265*

Midwave infrared lasers have great potentials in many applications such as pollution monitoring, medical diagnostics, explosive detection, and military counter-measures. This paper reports experimental results on optically pumped InAs/$InAs_xSb_{1-x}$ type-II superlattice (SL) laser structures grown by modulated-molecular-beam epitaxy (MMBE) InAs substrates.

The laser structure used for the present study was designed for optical pumping, and consists of a 0.2-μm thick InAs buffer, followed by a 1.0–1.5-μm thick $AlAs_{0.16}Sb_{0.84}$ cladding layer, an active region, a second $AlAs_{0.16}Sb_{0.84}$ cladding layer, and finally capped by 0.1-μm InAs. The active region consists of an InAs/$InAs_xSb_{1-x}$ SLs sandwiched by two 0.4-μm thick InAs layers, which function as optical absorbers. The $AlAs_{0.16}Sb_{0.84}$ cladding layers provide both optical and carrier confinements. These group-V alloys grown by MMBE have shown superior structural and optical properties compared to those grown by conventional MBE.[1]

These laser structures exhibited cw operation when optically pumped by a Ti: sapphire laser at 800 nm. The lasing wavelengths are around 3.4 μm at 95 K, as shown in Fig. 1. Distinct longitudinal modes are observed in high resolution optical spectra. The maximum cw operation temperature with 0.3-W pump power is 95 K. The equiv-

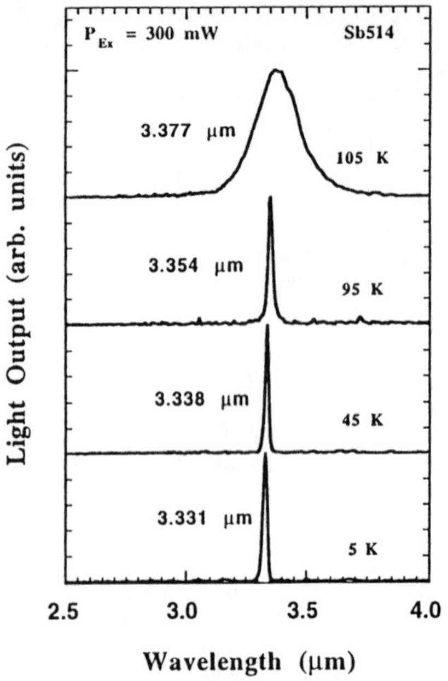

JWE3 **Fig. 1.** Optical spectra of the $InAs/InAs_xSb_{1-x}$ type-II SL laser at various temperatures under a constant pumping power of 0.3 W. Lasing is observed up to 95 K under this excitation power. The estimated equivalent threshold current density is 95 A/cm^2 at 95 K.

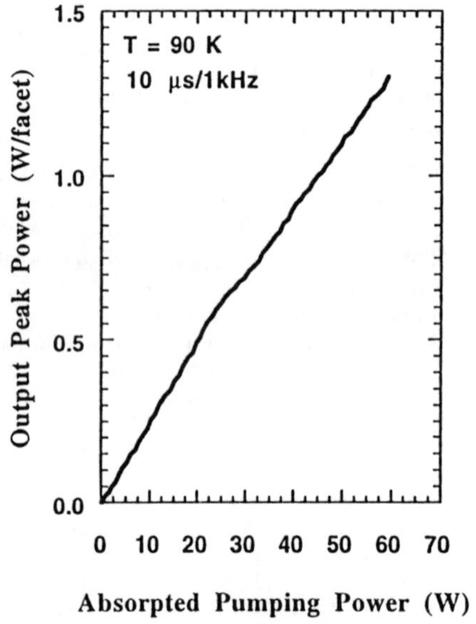

JWE3 **Fig. 2.** Light output peak power per facet vs. absorbed pump power for the studied laser structure at 90 K. The maximum peak power achieved as 1.3 W/facet.

alent threshold current density for a device with a cavity length of 1.4 mm is estimated to be 7.5 and 95 A/cm^2 at 5 and 95 K, respectively.[2]

High-power operation was studied using a 940-nm pump laser diode array.[3] For a broad area device with ~0.3-mm aperture and 3.1-mm long cavity, a peak power of 1.3 W/facet was obtained with 1% duty cycle at 90 K, as shown in Fig. 2. The pump pulse length is 10 µs. With 5% duty cycle, up to 50 mW/facet average power was also achieved at 95 K. The external differential quantum efficiency is 8%/facet. With 60-W absorpted pump power, the maximum pulse operation temperature was 195 K.

The Lincoln Laboratory portion of this work was sponsored by the U.S. Department of the Air Force.

Lincoln Laboratory, Massachusetts Institute of Technology, Lexington, Massachusetts 02173

1. Y.-H. Zhang and D. H. Chow, to appear in Appl. Phys. Lett.
2. Y.-H. Zhang, to appear in Appl. Phys. Lett.
3. H. Q. Le, G. W. Turner, S. J. Eglash, H. K. Choi, and D. A. Coppeta, Appl. Phys. Lett. **64**, 152 (1994).

JWE4 **5:30 pm**

Carrier recombination in GaInSb/InAs mid-IR lasers

Alan Kost, Linda West, T. C. Hasenberg, D. H. Chow, *Hughes Research Laboratories, 3011 Malibu Canyon Road, Malibu, California 90265*

GaInSb/InAs superlattices hold a number of significant potential advantages relative to GaInAsSb (or InAsSb) alloys as quantum-well materials for 2–5-µm laser diodes. The heavy- and light-hole bands in the superlattices can be separated from each other by more than an energy gap, greatly reducing Auger recombination cross sections. Energy gaps out to 5 µm and well beyond are easily obtained.[1] Tuning of the

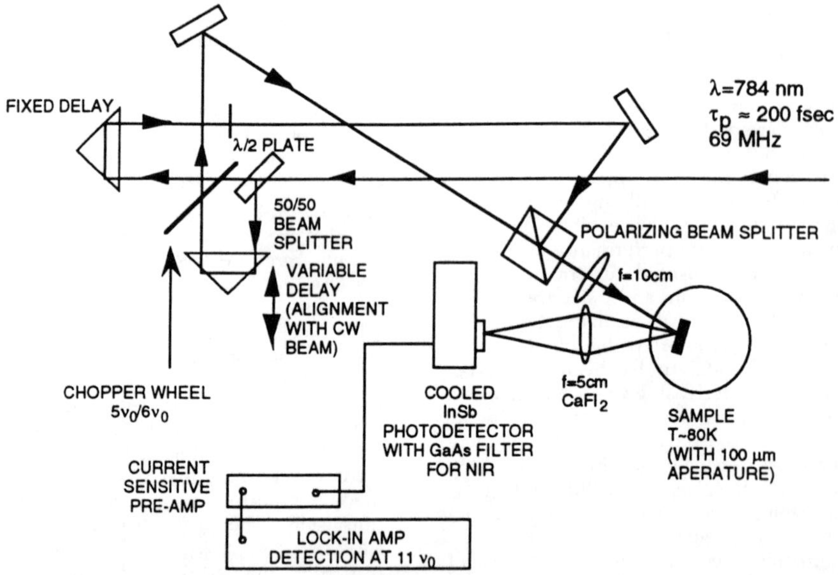

JWE4 **Fig. 1** Experimental setup to measure photocarrier decay by excitation correlation.

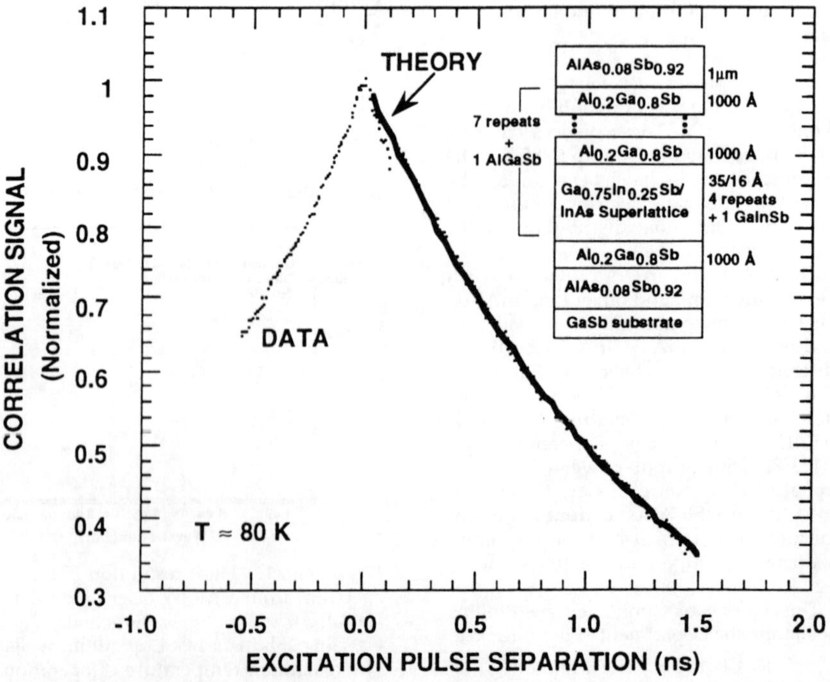

JWE4 **Fig. 2** Excitation correlation and best fit to the data.

superlattice energy gap from one wavelength to the next can be performed by changing the layer thicknesses.

A knowledge of electron–hole recombination channels is essential for predicting laser performance. Here we report studies of carrier recombination in GaInSb/InAs superlattices by using excitation correlation.[2–4] We describe numerical calculations to derive Shockley–Read, radiative, and Auger coefficients.

With excitation correlation, samples are excited with two short optical pulses separated in time by a variable amount τ. In the setup used for these measurements (Fig. 1), a mode-locked Ti:sapphire laser produces optical pulses ≈200 fs long at a repetition rate of 69 MHz. Photoluminescence is detected with a cooled InSb photodiode. Optical beams are chopped at frequencies of $5\nu_0$ and and $6\nu_0$, respectively, and a lock-in amplifier is used to detect the signal at the sum frequency of $11\nu_0$. Carrier decay can be measured with time resolution better than 1 ps.

Figure 2 shows the correlation signal for a multiple-quantum-well (MQW) structure with $Ga_{0.75}In_{0.25}Sb/InAs$ type-II superlattices for the wells. Luminescence is at 3.4 μm. The average power in the excitation beams was 45 mW focused to a 40-μm-diameter spot ($1/e$ diameter in intensity) so that the incident photon flux was 2×10^{14} cm^{-2} per pulse absorbed almost entirely in the MQW region. To obtain Shockley–Read, radiative, and Auger recombination rates, the data were fit by using a numerical model in which the photocarrier decay was taken to be $dn/dt = -c_1 n - c_2 n^2 - c_3 n^3$, with c_1, c_2, and c_3 as adjustable parameters. The best fit was found for $c_2 = c_3 = 0$ and $c_1 = 0.7\ ns^{-1}$, which implies that the carrier decay is exponential with a time constant of 1.4 ns. The results suggest that for these carrier densities the Shockely–Read recombination through traps is the dominant decay mechanism.

This work was supported in part by U.S. Air Force Phillips Laboratory contract F29601-93-C-0037.

1. D. H. Chow, R. H. Miles, J. N. Schulman, D. A. Collins, and T. C. McGill, Semicond. Sci. Technol. **6**, C47 (1991).
2. D. Rosen, A. G. Doukas, Y. Budansky, A. Katz, and R. R. Alfano, Appl. Phys. Lett. **39**, 935 (1981).
3. M. Jorgensen and J. M. Hvan, Appl. Phys. Lett. **43**, 460 (1983).
4. M. B. Johnson, T. C. McGill, and A. T. Hunter, J. Appl. Phys. **63**, (1988).

QWH 4:30 pm Hyatt A

Photonic Bandgap Structures

Eli Yablonovitch, *University of California–Los Angeles, Presider*

QWH1 (Invited) 4:30 pm

Photonic bandgap structures and devices

Pierre R. Villeneuve, Shanhui Fan, Ilya Kurland, Jerry C. Chen, J. D. Joannopoulos, *Massachusetts Institute of Technology, Cambridge, Massachusetts 02139*

Photonic bandgap (PBG) structures have the ability to control the density of electromagnetic states. They can affect the radiative dynamics within the structures and lead to significant change in the properties of optical devices. This has opened a new and fascinating area for potential applications in optoelectronic devices.

Several structures have been found to yield a complete 3D photonic bandgap.[1–5] We have attempted to design a structure that easily lends itself to fabrication at micron and submicron lengthscales. We found a simple layered structure that requires the etching of only one series of holes at normal incidence to generate a complete photonic bandgap. Our calculations show that the bandgap is not very sensitive to disorder which could arise during the fabrication process, such as element misalignment or surface roughness.

The introduction of a local defect in the PBG structure (either a point defect or line defect) can generate a state within the gap, localized around the defect. The introduction of point defect can give rise to a resonant cavity mode while a line defect can give rise to a guided mode. We have designed a new class of resonant microcavities which allows for efficient coupling into other components of optoelectronic circuits. These microcavities are integrated directly into the plane of the substrate and give rise to strong field confinement. By concentrating the field fluctuations into a small volume, we expect that the recombination rate of carriers will be increased. This will lead to the enhancement of spontaneous emission and will allow the microcavities to be modulated at very high speeds. These new microcavities may offer exciting possibilities for the fabrication of high-density and high-speed optical interconnects and ultra-low threshold single-mode microlasers.

We have also designed high-efficiency waveguide bends with a radius of curvature comparable to the free-space wavelength. Since waveguide bending is traditionally accomplished by total internal reflection based on the index contrast between the guide and its surroundings, large radii of curvature are usually required to guide light efficiently through a bend. By using PBG structures, we have managed to propagate light through sharp turns, with very low losses. The ability to perform tight bending will be required for the integration of a large number of optical components onto a small area of an optoelectronic integrated circuit.

1. E. Yablonovitch, T. J. Gmitter, and K. M. Leung, Phys. Rev. Lett. **67**, 2295 (1991).
2. C. T. Cha, K. M. Ho, and C. M. Soukoulis, Europhys. Lett. **16**, 563 (1991).
3. H. S. Sözüer and J. W. Haus, J. Opt. Soc. Am. B **10**, 296 (1993).
4. K. M. Ho, C. T. Chan, C. M. Soukoulis, R. Biswas, and M. Sigalas, Solid State Commun. **89**, 413 (1994).
5. H. S. Sözüer and J. P. Dowling, J. Mod. Opt. **41**, 231 (1994).

QWH2 5:00 pm

Two-dimensional photonic band structure in the near-infrared regime

Kuon Inoue, Mitsuo Wada, Kazuoka Sakoda, Akio Yamanaka, Masaki Hayashi, Joseph W. Haus,* *Research Institute for Electronic Science, Hokkaido University, Kita-ku, Sapporo 060, Japan*

A triangular lattice of air rods was fabricated for spectroscopic studies of photonic band structures in the infrared regime. The samples are composed of hollow rods embedded in a dielectric made from a PbO glass. Their physical size is about 1 mm in diameter and 1 mm in height. Two samples have been examined; one[1] has a lattice constant of 1.3 μm and a filling fraction is 0.54, and the other has a lattice constant of 1.02 μm and a filling factor of 0.43. To our knowledge, these are the smallest dielectric lattices whose band structure has been characterized to date. The dielectric contrast is 2.62, which has sufficient dispersion to create gaps in the photonic band spectra.

Transmittance spectra were measured over the range from 3700 to 2600 cm^{-1} by using an infrared spectrometer.[1] When light is propagated in the Γ–X direction, a distinct structure is obtained for different polarizations of the field. The wave vector of the X-point is $\pi(1, 1/\sqrt{3})$ in units of the lattice constant.

An opaque region was observed for both polarizations of the field, in agreement with the theoretical calculations. The photonic band gap in the spectrum depends on the polarization and the direction of propagation of the light. When the electric field is polarized parallel to the axis of the rod and propagation is in the Γ–X direction, the gap occurs over the range from 3850 to 4560 cm^{-1}, which is about 20% for this direction. For the orthogonal polarization the measured gap extends from 3660 to 4160 cm^{-1}, which is about 10%.

Calculations include a determination of Bragg scattering for the zeroth through the second order for the reflected and transmitted light in a finite lattice. The reflected and transmitted intensities calculated by this method are in good agreement with the experiments.

Future samples can be designed to give practical structures in the visible regime. Experiments are being designed to further characterize the samples and examine the role of surface structure in the experimental results. This technology can be applied to explore new aspects of interactions between radiation and matter. The present samples allow the use of picosecond pulses from infrared lasers to study this problem in detail.

Department of Physics, Rensselaer Polytechnic Institute, Troy, New York 12180-3590

1. K. Inoue, M. Wada, K. Sakoda, A. Yamanaka, M. Hayashi, and J. W. Haus, Jpn. J. Appl. Phys. **33**, L1436 (1994).

QWH3 — 5:15 pm

Pulse propagation near highly reflective surfaces: applications to photonic-band-gap structures and the question of superliminal tunneling times

Michael Scalora, Jonathan P. Dowling, Aaron S. Manka, Charles M. Bowden, Joseph W. Haus,* *Weapons Sciences Directorate, AMSMI-RD-WS-ST, U.S. Army Missile Command, Redstone Arsenal, Alabama 35898-5248*

The issues of propagation, energy flow, and field dynamics in electromagnetism have been of primary importance ever since the equations that describe these phenomena were first formulated in their entirety by Maxwell.[1] The validity of the theoretical predictions of Maxwell's equations rests on an overwhelming body of experimental evidence that has never been known to be in contradiction to the basic postulates of the theory, at least in the classical domain. These equations therefore form the basic underpinnings of all of modern optics, including nonlinear and quantum optics.

Although Maxwell's equations lay claim to all phenomena in classical optics, in practice many restrictive assumptions must be made in order to render the equations tractable. Chief among such approximations is the plane-wave limit that is typically applied in propagation problems. There is a danger, however, that the simplicity and omnipresence of the plane-wave approach can lead to a false sense of security and hence to a tendency to overlook some of the more interesting physics involved in, say, pulse propagation or transverse effects. For instance, the study of propagating fields in nonlinear media in the plane-wave approximation overlooks interesting diffraction effects that become apparent only when the plane-wave limit is relaxed.[2]

We show that for pulse scattering from a photonic-band-gap structure, interference terms affect the Poynting vector in a nontrivial fashion and also alter the usual notions of the momentum and the energy density of a traveling electromagnetic pulse.[3] We show that the magnetic field depends on the spatial curvature of the electric-field envelope and, near reflective surfaces, the two fields are spatially anticorrelated. In other words, where the electric field is a maximum, a minimum occurs in the magnetic field, and vice versa. We also show that the minimum time necessary for the field–matter interaction to occur cannot be smaller than the free-space-propagation interaction time. From our simulations, it appears that the tunneling of a pulse through a linear photonic-band-gap material takes place at a speed at, or nearly equal to, the vacuum velocity of light.

Department of Physics, Rensselaer Polytechnic Institute, Troy, New York 12180-3590

1. J. C. Maxwell, *Treatise on Electricity and Magnetism*, 3rd ed., 1904 (Dover, New York, 1954).
2. J. V. Moloney, M. R. Belic, and H. M. Gibbs, Opt. Commun. **41**, 379 (1982); F. P. Mattar and C. M. Bowden, Phys. Rev. A **27**, 345 (1983); J. W. Haus and M. Scalora, Phys. Rev. A. **42**, 3149 (1990); M. Scalora and J. W. Haus. J. Opt. Soc. Am. B **5**, 1003 (1991).
3. M. Scalora, J. P. Dowling, A. S. Manka, C. M. Bowden, and J. W. Haus, "Pulse propagation near highly reflective surfaces: applications to photonic bandgap structures and the question of superluminal tunneling times," submitted to Phys. Rev. A.

QWH4 — 5:30 pm

Photonic band gaps and molecular fluorescence

S. C. Kitson, W. L. Barnes, *Department of Physics, Exeter University, Exeter, Devon EX4 4QL, U.K.*

We present what we believe to be the first experimental evidence that shows the effect of a photonic band gap in the propagation of surface plasmon polaritons (SPPs) on molecular fluorescence. The potential for photonic materials has been widely discussed,[1] with particular interest in their use for controlling spontaneous emission.[2] In this work we show how a photonic band gap in surface, rather than bulk, modes may be used to control the emission spectrum of dye molecules.

In our experiment dye molecules (DCM) were spun onto a metal-coated photoresist grating and were excited by 488-nm light from an Ar^+ laser. The proximity of the dye layer to the metal means that the dominant decay route for the excited molecules is to couple to SPPs on the grating surface. The grating serves to couple the surface

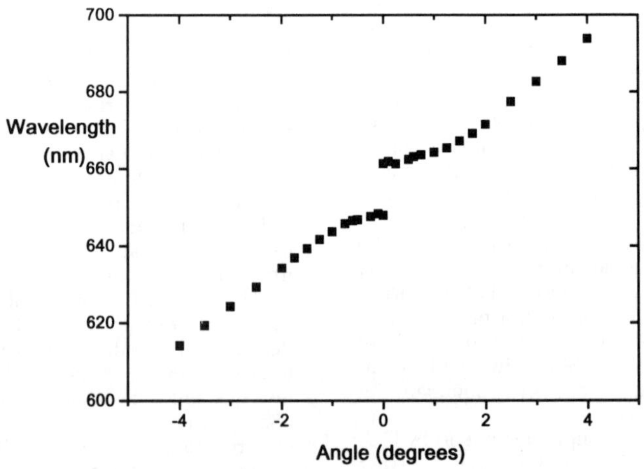

QWH4 Fig. 1. Dispersion curve measured by collecting the light emitted by the dye molecules via surface plasmons; the collection angle corresponds to the mode momentum. The nonsinusoidal profile of the grating creates a band gap.

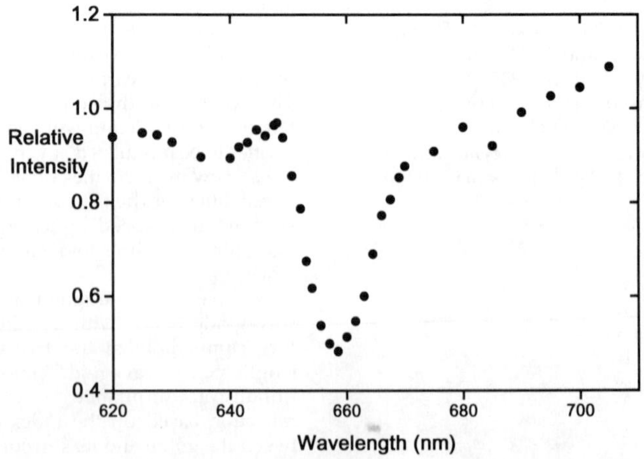

QWH4 Fig. 2. Emission spectrum of the surface plasmons excited by the dye molecules. The spectrum has been normalized to emission from a grating that does not exhibit a band gap. The dip clearly shows how decay of the excited molecule by SPP modes is inhibited in the vicinity of the gap.

modes to photons, resulting in visible radiation from the sample.

If the grating is nonsinusoidal, then the interaction of surface plasmons with the grating causes an energy gap to open up in the SPP dispersion curve[3] (Fig. 1). This gap results from the component of the surface profile that is at twice the spatial frequency of the grating. Bragg scattering by this high-spatial-frequency component sets up two SPP standing waves having different energies. By choosing the pitch and geometry of the grating carefully, we induced an energy gap within the emission band (~670 nm) of the dye, thus effectively blocking coupling of the dye to SPP modes.

By collecting the light coupled out from the molecules by SPP excitation, we have studied the effect of the band gap on the emission spectrum. Figure 2 shows the emission spectrum for the molecules above the grating normalized with respect to emission above a metallic grating with no energy gap. These results show that a photonic band gap has a significant effect on the fluorescent emission of molecules. Further work is now underway to look at the effect of other surface profiles.

The financial assistance of the DRA, Malvern, U.K., and the EPSRC is gratefully acknowledged.

1. E. Yablonovitch and T. J. Gmitter, Phys. Rev. Lett. **67**, 2295 (1991).
2. D. G. Deppe *et al.*, Electron Lett. **26**, 1665 (1990).
3. M. G. Webber and D. L. Mills, Phys. Rev. B **31**, 2510 (1985).

QWI — 4:30 pm, Hyatt B

Atom Optics

William Daniel Phillips, *National Institute of Standards and Technology, Presider*

QWI1 — **4:30 pm**

Atomic waveguide using evanescent waves in a hollow fiber

Haruhiko Ito, Keiji Sakaki,*
Takeshi Nakata,* Wonho Jhe,**
Motoichi Ohtsu, † *Kanagawa Academy of Science and Technology, KSP East, 3-2-1 Sakado, Takatsu-ku, Kawasaki 213, Japan*

The realization of atomic waveguides, which have been proposed by a few research groups,[1-3] is an important subject in atom optics. Atomic waveguides are expected to apply even to nanoscale crystal growth, by combination with the laser manipulation of atoms.[4] For this application, atomic waveguides are used for one-dimensional manipulation to carry atoms to a specified point on a substrate.

Our proposed atomic waveguide is composed of a hollow silica fiber with a thin cylindrical core. This type of atomic waveguide is advantageous because evanescent waves are used to produce the optical potential for guiding atoms. In particular, one can avoid laser-light perturbations of guided atoms due to heating effects. Evanescent waves have an effective thickness less than one wavelength and therefore interact only with atoms approaching the inner surface of the hollow fiber.

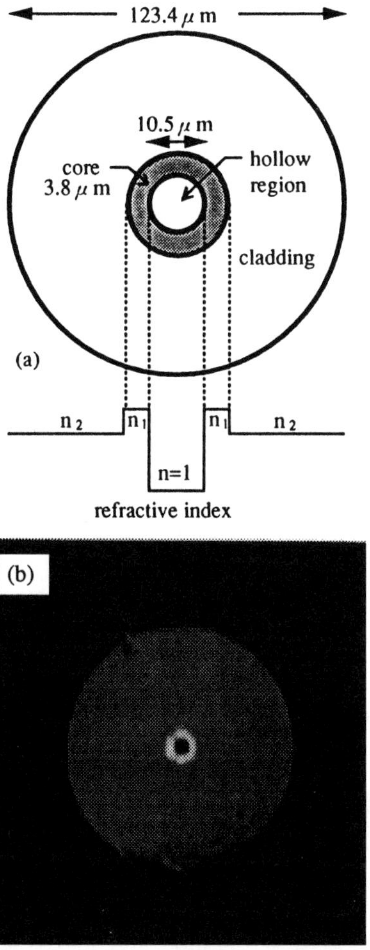

QWI1 Fig. 1. (a) Configuration of the hollow silica fiber with a thin cylindrical core. The refractive indices of the core and the cladding are n_1 and n_2, respectively. The relative refractive-index difference Δn is 0.18%. (b) Cross-sectional view for demonstrating the hollow-fiber configuration and cylindrical beam propagation. A He–Ne laser was used.

Figure 1 shows the configuration of the cylindrical-core hollow fiber fabricated on the basis of our design and a view of the cross section when a He–Ne laser beam was directed into the fiber. Figure 2 shows the dispersion curves for a wavelength of 780 nm, for which eight propagation modes are folded. Note that the fundamental mode is HE_{11}, as in the case of the conventional homogeneous-core fibers for optical transmission.

The atoms in the hollow region are kicked back by the strong gradient force due to the evanescent wave near the inner surface under blue-detuning conditions. The optical potential is evaluated by calculating the time-averaged Poynting vector. Figure 3(a) shows the optical potential for ^{85}Rb atoms in terms of the temperature representing thermal kinetic energy as a function of the distance from the inner surface. For simplicity, we supposed that the HE_{11} mode gives the principal contribution to the optical potential. Provided that the optimal blue-detuning Δ is chosen, an optical potential deeper than 1 mK is produced with a laser power of a few milliwatts, which is sufficient to guide atoms injected into the hollow fiber from an atomic beam or a magneto-optical trap. The optimal blue detuning changes according to the coupled laser power. Moreover, as shown in Fig. 3(b), the axial dependence of the He_{11} mode is periodic, so that the produced optical potential depends on the azimuth angle θ. The optimal detuning ought, therefore, to be determined such that the optical potential is sufficiently deep for all values of θ.

We will report our experiments using an atomic beam, including laser-beam tailoring for efficient coupling to the hollow fiber. We also touch on an atomic funnel for pouring atoms into the hollow fiber.

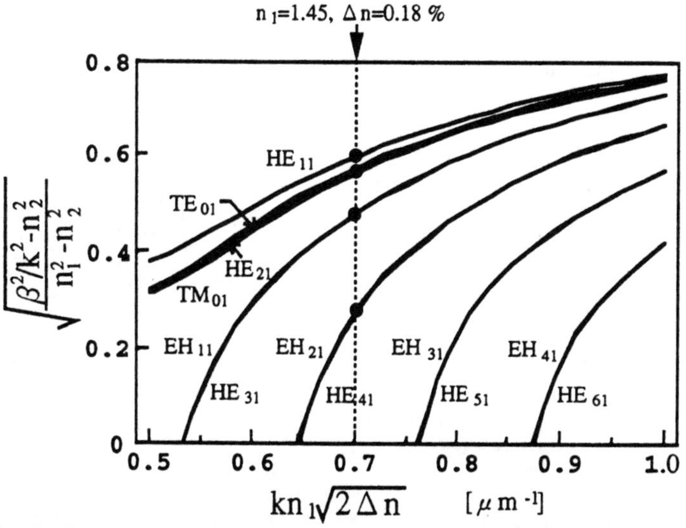

QWI1 Fig. 2. Dispersion curves of low-order modes in the cylindrical-core hollow fiber shown in Fig. 1(a), where β and k are the propagation constant and the wave number, respectively. Filled circles indicate the propagation modes of the hollow fiber for a wavelength of 780 nm. The vertical axis is the normalized propagation constant, and the horizontal axis is proportional to the normalized frequency.

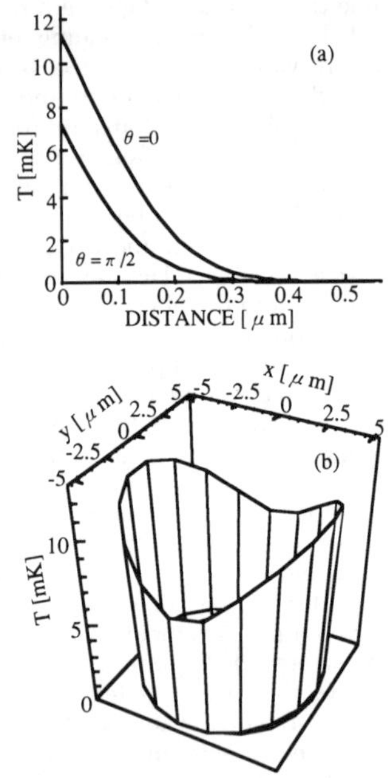

QWI1 Fig. 3. Optical potential produced by the evanescent wave (HE_{11}) for 10 mW of laser power. The blue-detuning Δ is taken to be 12 times as large as the natural linewidth Γ, which is almost optimal. The potentials are expressed by a temperature representing one-dimensional thermal kinetic energy: (a) cross-sectional profile as a function of the distance from the inner surface and (b) three-dimensional distribution. The azimuth angle θ is chosen such that the potential becomes maximum at θ = 0 and minimum at θ = π/2.

* *Interdisciplinary Graduate School of Science and Engineering, Tokyo Institute of Technology, 4259, Nagatsuta, Midori-ku, Yokohama 226, Japan*
** *Department of Physics, Seoul National University, Seoul 151-742, Korea*
† *Also with Interdisciplinary Graduate School of Science and Engineering, Tokyo Institute of Technology, 4259, Nagatsuta, Midori-ku, Yokohama 226, Japan*

1. M. A. Ol'Shanii, Yu. B. Ovchinnikov, and V. S. Letokhov, Opt. Commun. **98**, 77 (1993).
2. S. Marksteiner, C. M. Savage, P. Zoller, and S. L. Rolston, Phys. Rev. A **50**, 2680 (1994).
3. W. Jhe, M. Ohtsu, H. Hori, and S. R. Friberg, Jpn. J. Appl. Phys., in press.
4. H. Hori, S. Jiang, M. Ohtsu, and H. Ohsawa, in *Proc. International Quantum Electronics Conference*, 1992, Vienna, p. 48.

QWI2 **4:45 pm**

Adaptive imaging of atomic interference patterns

M. L. Marable, T. A. Savard, J. E. Thomas, *Department of Physics, Duke University, Durham, North Carolina 27708-0305*

We demonstrate adaptive imaging of atomic interference patterns by using Raman-induced resonance imaging in a high magnetic-field gradient.[1] Two different imaging methods are investigated. For the first method, a movable virtual optical slit is created in the imaging region. The slit is moved by changing the frequency of the transition-inducing fields. For the second method, a virtual optical grating with a variable period and position is created.

In the experiments (Fig. 1) ^{152}Sm atoms cross two optical polarization gratings separated by a distance L to create an atomic interference pattern a distance L downstream from the second polarization grating. This scheme is analagous to the standing-wave grating scheme of Chebotayev *et al.*[2] The gratings are created by using laser fields that propagate along the z axis with a polarization in the x–y plane. The fields emerge from two birefringent wedges with a small angle between the x- and y-polarized components, causing the $\hat{\sigma}_-$ intensity to vary along the x axis with a period of 20 μ. The two-level atomic system comprising the $|i\rangle = |^7F_1, M = 1\rangle$ ground state and the $|^7F_0\rangle$ excited state interacts resonantly with the fields emerging from the wedges. After the excited state spontaneously decays, the amplitude for the $|i\rangle$ state remains spatially modulated.

The interference pattern rephases in the center of a resonance imaging region located a distance L from the second polarization grating. In this region, two optical fields induce a Raman transition from the initial state $|i\rangle$ to a previously emptied final state $|f\rangle = |^7F_1, M = 0\rangle$. The state $|i\rangle$ is shifted by a magnetic-field gradient to make the Raman transition frequency dependent on position x along the measurement axis. Atoms emerging in state $|f\rangle$ are detected downstream by resonance fluorescence. The fluorescence intensity is measured as a function of the difference frequency of the transition-inducing fields to determine the atomic spatial profile.

When the spatial profile of the Raman inducing fields is Gaussian and the acceleration of the atoms due to the magnetic-field gradient is negligible, the spatial-resolution function, i.e., the shape of the region sampled by the imaging method, has a Gaussian spatial distribution, $\exp[-(x - x_0)^2/\Delta x^2]$. For a Zeeman-shift gradient of 1 GHz/cm and atomic-beam collimation of 1 mrad, Δx is in the micrometer region.[1] The atomic interference pattern for the conditions used in the experiments has a modulation depth of only a few percent. To improve the signal-to-background ratio, the center of the Gaussian distribution x_0 is shifted at 300 Hz between a fixed position x_B and a variable position x_A, where count rates B and A are measured, respectively. The ratio $(A - B)/B$ is measured as a function of the variable position x_A to obtain the data shown in Fig. 2(a). The period of the measured fringes is in good agreement with that expected on the basis of the wedge-angle design parameters.

For experiments in which a periodic distribution is measured, greater signal-to-background ratios can be obtained by using a spatial-resolution function that is periodic in space because this samples a greater number of atoms than does a single slit. This is accomplished in the experiments by using two tightly focused, spatially separated Raman transition-inducing regions. In this case, when the motion of the atoms along the x axis during the transit time between the two regions is negligible, the spatial-resolution function takes the form of a spatial Ramsey fringe, $\propto \cos[2\pi FT(x - x_0)/h]$, where F/h is the spatial tuning rate in units of hertz per centimeter (≃0.5–1 GHz/cm) and T is the atom transit time between the two Raman regions. Figure 2(b) demonstrates the increased signal-to-background ratio obtained for the measurement of the atomic spatial pattern when the wave vector $FT/\hbar$ is matched to that of the atomic interference pattern.

Adaptive-imaging methods can be extended by using more general Raman spatial (and/or frequency) distributions and

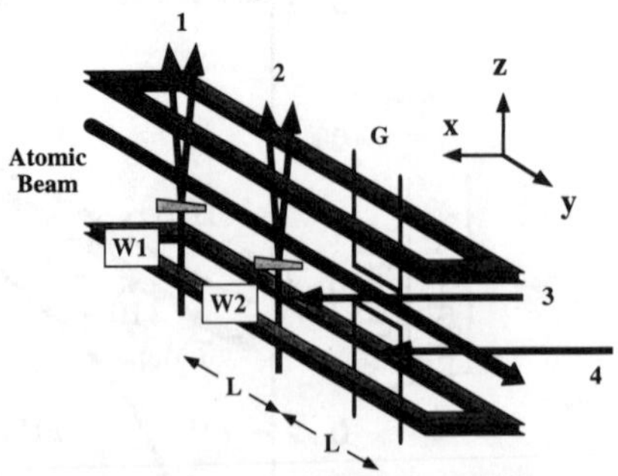

QWI2 Fig. 1. Atom interferometer. Atoms traverse two optical polarization gratings created by using birefringent wedges W1 and W2 to introduce a small angle between the x and y components of laser fields 1, 2 that propagate in the z direction. This results in an interference pattern that rephases in the center of a resonance imaging region comprising two gradient wires G, Raman transition-inducing fields 3, and a detection beam 4.

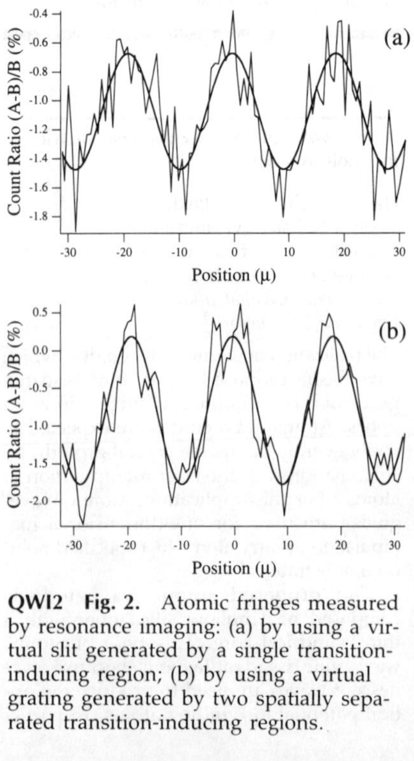

QWI2 Fig. 2. Atomic fringes measured by resonance imaging: (a) by using a virtual slit generated by a single transition-inducing region; (b) by using a virtual grating generated by two spatially separated transition-inducing regions.

higher field gradients to achieve suboptical-wavelength spatial resolution. Because the atomic spatial distribution can be arbitrarily shaped when this method is used, applications to adaptive atom optics, to create a virtual Fresnel lens, for example, can be envisioned.

1. K. D. Stokes, C. Schnurr, J. R. Gardner, M. Marable, G. R. Welch, and J. E. Thomas, Phys. Rev. Lett. **67**, 1997 (1991).
2. V. P. Chebotayev, B. Ya. Dubetski, A. P. Kasantsev, and V. P. Yakovlev, J. Opt. Soc. B **2**, 1791 (1985).

QWI3 (Invited) **5:00 pm**

Large angle atomic beam splitters

K. S. Johnson, J. R. Lawall, A. Chu, T. W. Lynn, K. K. Berggren, M. S. Shahriar, Mara Prentiss, *Department of Physics, Harvard University, Cambridge, Massachusetts 02138*

Many atom interferometry applications would benefit from efficient, large-angle atomic beam splitters. A variety of experiments and proposals have addressed the issue of creating a large-angle, efficient atomic beam splitter.[1–5] Recently Pfau *et al.* proposed and demonstrated the magneto-optical beam splitter, where the interaction of a three level atom with a light field of spatially varying polarization and a large magnetic field leads to beam splitting.[3,6,7] Grimm *et al.* have suggested that a similar beam splitter can be realized by driving a two-level atom with bichromatic standing wave light fields.[8] We demonstrate a beam splitter based on a three-level V system that does not require an external magnetic field. Furthermore, we will discuss a general formulation of the blazed grating interaction for three-level V systems and propose extensions applicable in a variety of atomic systems.

A blazed grating provides a phase accumulation that is triangular as a function of position. This phase grating can result in a momentum distribution dominated by two narrow, widely separated peaks. The blazed interaction used in the work of Pfau *et al.* involves and excited-state energy splitting due to an external magnetic field; we show that a blazed grating potential for three-level atoms can be obtained for degenerate excited states, as long as the two driving fields have different frequencies. The blazed grating interaction can be extended to a variety of combinations of excited-state energy splittings and differential detunings.

The two transitions of a V system are driven by standing wave laser fields, Ω_+ and Ω_-, detuned symmetrically above and below the atomic transition frequency. We assume that Ω_+ interacts only with the $|g\rangle \rightarrow |e_+\rangle$ transition and Ω_- only with the $|g\rangle \rightarrow |e_-\rangle$ transition. In the rotating frame, we may write $\Omega_\pm = \Omega\cos(kz + \chi/2)$, where χ is the relative spatial phase shift between the standing waves, k is the wavenumber of the light, z is the position in the standing wave, and Ω is the peak Rabi frequency. The interaction Hamiltonian is, in the $|e_+\rangle$, $|e_-\rangle$, $|g\rangle$ basis:

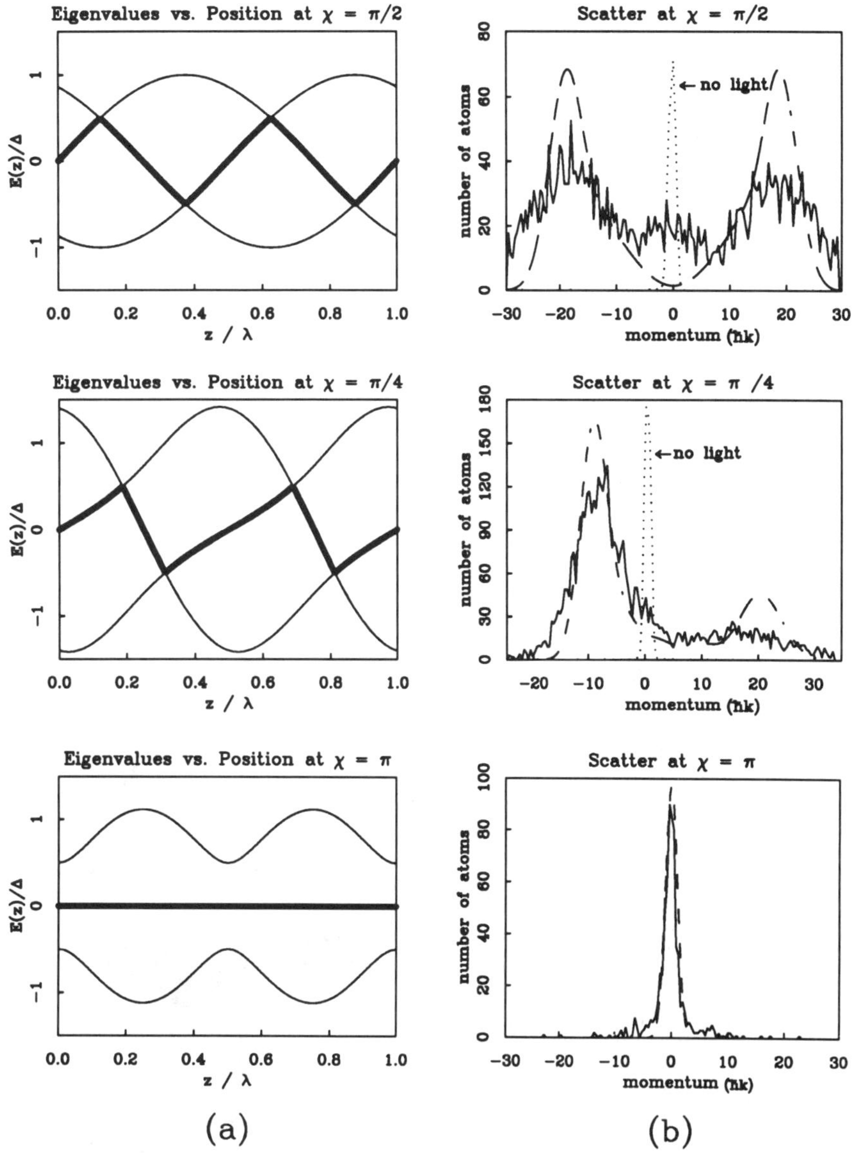

QWI3 Fig. 1. (a) Eigenvalues as a function of position in the laser field for phases of $\pi/2$, $\pi/4$, and π. In each case, the central, darker energy level is adiabatically connected to $|g\rangle$ as $\Omega \rightarrow 0$. (b) The dashed lines correspond to the theoretical far field distributions. Dotted lines represent the atomic beam in the absence of laser interactions. Solid lines correspond to the experimental data for phases of $\chi = \pi/2$, $\pi/4$ and π

$$H = \hbar/2\begin{pmatrix} \Delta & 0 & -\Omega_+ \\ 0 & -\Delta & -\Omega_- \\ -\Omega_+ & -\Omega_- & 0 \end{pmatrix} \quad (1)$$

where Δ is the difference between the frequencies of the two driving fields. For a resonant condition, $\Omega = \sqrt{2\Delta}/\sin(\chi)$, and for a particular spatial phase shift, $\chi = \pi/2$, the dressed state adiabatically connected to $|g\rangle$ as $\Omega \rightarrow 0$ has an energy that is nearly triangular as a function of position, as shown in Fig. 1(a). Thus, if an atom in $|g\rangle$ enters a Gaussian laser profile slowly, the atom will adiabatically follow into the dressed state associated with the triangular potential. We remark that in the cases of experimental interest here, the atoms do not move significantly in the transverse direction during the interaction (Raman-Nath or impulse approximation). If the interaction time is short enough then the effect of spontaneous emission will be small and the beam splitting will be coherent.

Figure 1(a) shows the dressed state energies as a function of position, and Fig. 1(b) shows the predicted far-field scatter patterns and the experimental data. The ratio of Δ and Ω is shown to satisfy the resonant condition for $\chi = \pi/2$ and $\chi = \pi/4$. The energies are plotted as a function of position in the standing wave. The dressed state adiabatically connected to $|g\rangle$ is the darker, central energy level in each case. The results of our numerical simulations for the final momentum distributions are given by the dashed lines. The data shown corresponds to a splitting of 38 times the photon momentum; we have observed splittings of 46 times the photon momentum.

We gratefully acknowledge the support of NSF grant PHY-9312572 and ONR grant ONR-N0014-91-J-1808.

1. M. Weitz, B. C. Young, and S. Chu, Phys. Rev. Lett. **72**, 2563 (1994).
2. D. S. Weiss, B. C. Young, and S. Chu, Phys. Rev. Lett. **70**, 2706 (1993).
3. T. Pfau, Ch. Kurtsiefer, C. S. Adams, M. Siegel, and J. Mlynek, Phys. Rev. Lett. **71**, 3427 (1993).
4. L. S. Goldner, C. Gerz, R. J. C. Spreew, S. L. Rolston, C. I. Westbrook, W. D. Phillips, P. Marte, and P. Zoller, Phys. Rev. Lett. **72**, 997 (1994).
5. J. Lawall and M. G. Prentiss, Phys. Rev. Lett. **72**, 993 (1994).
6. T. Pfau, C. S. Adams, and J. Mlynek, Europhys. Lett. **21**, 439 (1993).
7. C. S. Adams, T. Pfau, Ch. Kurtseifer, and J. Mlynek, Phys. Rev. A **48**, 2108 (1993).
8. R. Grimm, J. Soding, and Yu. B. Ovchinnikov, Opt. Lett. **19**, 658 (1994).

QWI4 5:30 pm

Magnetic imaging with slow atoms

W. Kaenders, F. Lison, A. Richter, R. Wynands, D. Meschede, *Institute for Quantum Optics, University of Hannover, Welfengarten 1, D-30167 Hannover, Germany*

The focusing of atoms carrying a magnetic dipole moment by magnetic hexapole lenses was demonstrated in the early 1950's.[1] The application of such lenses with focal length $f \sim v^{-2}$ (atomic velocity v) for imaging was never attempted, however, because of the large velocity spread of thermal atomic beams. With laser-cooling techniques it is now possible to prepare atomic beams at low velocities with a narrow velocity distribution. With NdFeB permanent magnets very strong hexapole fields ($d^2B/dr^2 = 2.5$ T/cm²) can be created for reasonable apertures of the order of 1 cm.[2] Hence, in combination with a slow atomic beam, focal lengths in the centimeter domain are readily obtained and can be used for imaging purposes.

We have prepared a cesium atomic beam with a velocity $v = 70$ m/s and $\Delta v/v = 0.1$ by chirp cooling.[3] A shutter for the atomic beam allows us to separate cooled and thermal atoms in the time domain. A PMMA mask with an asymmetric pattern of drilled holes of the order of 0.1 mm was illuminated (Fig. 1). The image was formed on a light screen a few centimeters away from the mask. This light screen was generated with transverse laser light resonant with the cesium D_2 line. The fluorescence pattern was registered by a CCD camera for further analysis.

In Fig. 2 we give an example of a magnified and a demagnified image of our mask (imaging ratios, 1.4 and 0.8). The optical axis is indicated in Fig. 2 by an enhanced intensity due to fast atoms that have left the chirp-cooling cycle too early. The image is subject to known aberrations from traditional light optics. In addition, there exist aberrations caused by the angular momentum of atoms about the optical axis that contribute to the image observed in our experiment.

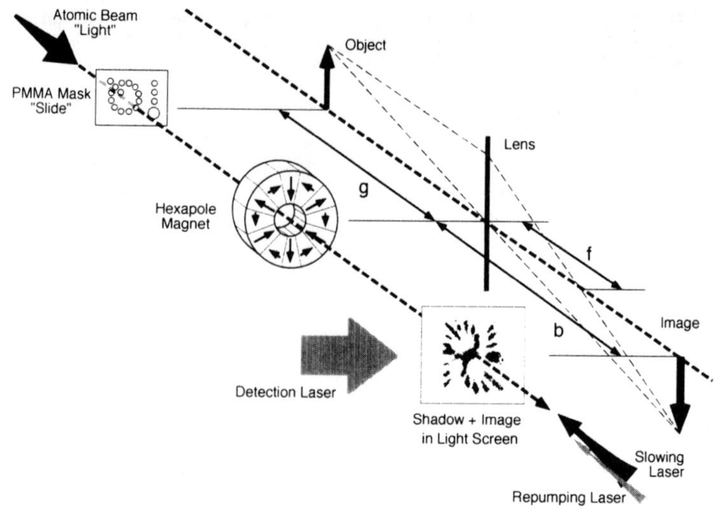

QWI4 Fig. 1. Experimental setup.

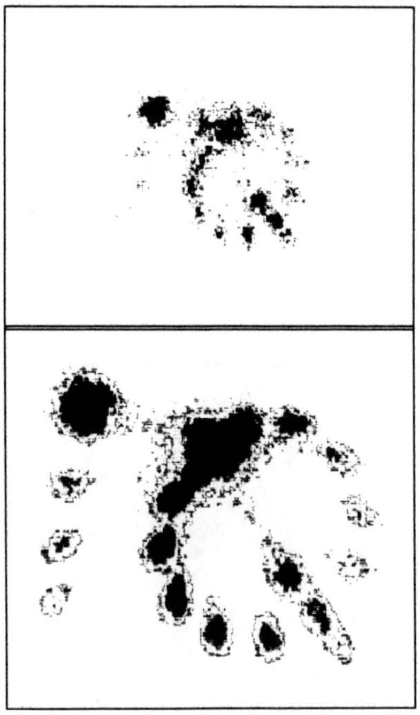

QWI4 Fig. 2. Magnified and demagnified images of mask.

1. H. Friedburg and W. Paul, Naturwissenschaften **38**, 159 (1951); H. Friedburg, Z. Phys. **130**, 493 (1951).
2. K. Halbach, Nucl. Instrum. Methods **169**, 1 (1980).
3. W. Ertmer, R. Blatt, J. L. Hall, and M. Zhu, Phys. Rev. Lett. **54**, 996 (1985).

QWI5 5:45 pm

Beam imbalance and crosswind elevating force in an optical supermolasses

G. I. Surdutovich, V. S. Bagnato, *São Carlos Institute of Physics/USP, C.P. 369, CEP 13560-970, São Carlos, SP, Brazil*

The insensitivity of the molasses decay rate to the imbalance ΔI between intensities of counterpropagating molasses beams is one of most serious discrepancies between experimental observations and classical theory predictions.[1,2] According to this theory, an imbalance $\varepsilon = \Delta I/I$ leads to a drift of atoms along the direction of the stronger beam with the velocity $v_{\text{drift}} = \varepsilon F/\beta$, where F is the force from one beam and Γ is the friction coefficient. For optimal damping conditions, $v_{\text{drift}} = \varepsilon\Gamma/2k$, where Γ is the linewidth of the transition and k is the light wave vector. As a result, a 1% imbalance gives a theoretical drift time of 0.2 s, whereas the experimentally observed drift-induced decay rate is a factor of 10 or 20 times smaller. We propose an explanation of this effect in the framework of the classical two-level atom theory, accounting for the rotational motion of atoms in a misaligned supermolasses device.

In magneto-optical traps, vortex forces give rise to various spatial patterns of the trapped atoms.[3,4] The rotational motion in a supermolasses setup, accompanied by rapid radial oscillations, leads to suppression of the radial diffusion and the dynamically induced stabilization of the atomic motion,[5] somewhat in analogy with the Kapitza pendulum-stabilization effect. But it is not a sole consequence of the rotational motion. For the considered effect it is sufficient to study an atomic motion in the x, y plane at small distances $r < w$ from the center of molasses, where w is half the width of the Gaussian beams. Then there exists a purely azimuthal component $F_\varphi = br$ of the vortex force. After placing the equation of the atomic motion in the convenient complex plane $z = x + iy$, we have

$$m\ddot{z} = -\beta\dot{z} + ibz. \tag{1}$$

Seeking a solution of the form $z = z_0 e^{\Omega t}$, we obtain the untwisting solution with

$$\Omega_+ = -\frac{\beta}{2m} + \left[\left(\frac{\beta}{2m}\right)^2 - \frac{ib}{m}\right]^{1/2} \cong \frac{mb^2}{\beta^3} \pm i\frac{b}{\beta} \tag{2}$$

Omitting the numerical factor and limiting ourselves to the case of moderate misalignments, $s \ll w$, we can express the ratio Ω'_+/Ω''_+ through the Doppler-cooling-limit energy and the recoil energy $\varepsilon_R = (\hbar k)^2/2m$.

For sodium molasses under $s/w = 0.1$ we obtain an estimation of this ratio as 5×10^{-3}, which corresponds to a slow untwisting spiral motion. At the same time, the diffusion departure time $t_D = (kw)^2/60\Gamma$ happens to be much less than the mechanical departure time

$$(\Omega'_+)^{-1} \cong \left(\frac{W}{s}\right)^2 \frac{\varepsilon_R}{\hbar\Gamma}\frac{(kW)^2}{\Gamma} \cong \frac{(kW)^2}{\Gamma},$$

i.e., rotational motion under a moderate misalignment only slightly influences the decay rate of the molasses. However, it changes qualitatively the mode of an atom's motion under action of the imbalanced pair of forces. An imbalance of the x pair leads to the appearance of an additional constant force along the x axis. As is evident from Eq. (1), it leads to a change of the solution $z \to z + i\delta y$ ($\delta y = \varepsilon F/b \cong \varepsilon W^2/4s$), i.e., to a constant shift of the center of the spiral motion in the direction perpendicular to the acting forces. Such an effect is a direct analogy of the aerodynamic effect due to a crosswind elevating force acting on a rotating cylinder in an air flow.[6] This estimation holds true only for not too large an imbalance ε, to afford the inequality $\varepsilon W/4s < 1$. For a larger imbalance we must take into account the nonlinear dependence of the vortex force from r in the form $F_\varphi = br - b_1 r^3$. Then the critical imbalance $\varepsilon_{\text{crit}}$ is determined from the relation $\delta y < \sqrt{3}W/4$. For $\varepsilon > \varepsilon_{\text{crit}} \cong 1.7\ s/w$ the abrupt increase in the departure rate of atoms from the supermolasses must be taken into account. A thresholdlike dependence of the decay rate from the imbalanced supermolasses has been observed (Ref. 2, Fig. 10).

1. S. Chu, M. Prentiss, A. Cable, and J. Bjorkholm, in *Laser Spectroscopy VIII*, W. Persson and S. Svanberg, eds. (Springer-Verlag, Berlin, 1987), p. 58.
2. P. D. Lett, W. D. Phillips, S. L. Rolston, C. E. Tanner, R. N. Natts, and C. I. Westbrook, J. Opt. Soc Am. B **6**, 2084 (1989).
3. D. Sesko, T. Walker, and C. Wieman, J. Opt. Soc. Am. B **8**, 946 (1991).
4. V. S. Bagnato, L. G. Marcassa, M. Oria, G. I. Surdutovich, and S. Zilio, Laser Phys. **2**, 172 (1992).
5. V. S. Bagnato, N. P. Bigelow, G. I. Surdutovich, and S. Zilio, Opt. Lett. (October 1994).
6. L. Landau and E. Lifshitz, *Mechanics of Continuous Media* (Nauka, Moscow, 1954).

QWJ — 4:30 pm — Hyatt C/D

Nonlinear Optical Processes in Quantum Wells

Jean-Pierre Huignard, *Thomson CSF-LCR, Presider*

QWJ1 (Invited) — **4:30 pm**

Optical frequency conversion in asymmetric quantum wells: interband vs intersubband

E. Rosencher, A. Fiore, V. Berger, B. Vinter, *Thomson/CSF Central Research Laboratory, 91404 Orsay, France*

Envelope wavefunctions of carriers in semiconductor quantum wells (QWs) typically extend over distances in the few nanometer range. The dipole elements of optical transitions between subband states have thus the same order of magnitude, i.e., the nanometer range instead of the Ångström range usually observed in bulk crystals. Huge second order optical susceptibilities are then expected and measured in intersubband transitions, provided the inversion symmetry is broken.

We will review our main results on intersubband optical rectification (OR), second-harmonic generation (SHG) and sum frequency in compositionally asymmetric multiquantum wells. The expected huge nonlinearities are indeed easily observed in the mid-infrared range ($\approx$10 μm) with SHG coefficients which are more than 3 orders of magnitude higher than in bulk GaAs, an already excellent nonlinear material. We will also specifically address the problems of optical saturation and power conversion due to the resonant nature of these transitions.[1,2,3]

As suggested by Khurgin,[4] intersubband transitions are also involved in interband optical transitions, since they participate to the overall valence to conduction band transitions. Large optical nonlinearities are thus expected in interband transitions in asymmetric QWs, i.e, in the near infrared to visible wavelength region. We will present a theoretical model from which, thanks to different sum rules, a very compact and manageable expression is found, allowing easy QW optimization procedures. Particularly, we will show that destructive interference effects due to high energy states drastically decrease the theoretical values obtained by simple models. We will illustrate this approach with two different sets of experimental results: (a) infrared photodetection by near (1 μm) and mid-infrared (10 μm) two photon absorption in compositionally asymmetric QWs.[5]; (b) second harmonic generation near GaAs midgap ($\approx$1.7 μm) in electrically biased asymmetric QWs. This will open the way to electrically tunable phase-matching techniques which will allow compact, room temperature, tunable and integrable mid-infrared coherent sources to be designed in semiconductor waveguides.

1. E. Rosencher and P. Bois, in *Intersubband Transitions in Quantum Wells* E. Rosencher, B. Vinter, and B. Levine, eds. (Plenum, London, 1992) p. 183.
2. E. Rosencher, J. Appl. Phys. **73**, 1909 (1992).
3. P. Boucaud, F. H. Julien, D. D. Yang, J. M. Lourtioz, E. Rosencher, and P. Bois, Optics Letters **16**, 199 (1991).
4. J. Khurgin. *Second-order intersubband nonlinear optical susceptibilities of asymmetric quantum well structures* 1-69 (Optical Society of America, Washington D.C.,1989).
5. E. Rosencher, B. Vinter, V. Berger, and N. Laurent, NonLinear Opt. (to be published).

QWJ2 — **5:00 pm**

Fundamental limits on quantum-well intersubband-transition-induced frequency conversion

Gilad Almogy, Amnon Yariv, *Department of Applied Physics, California Institute of Technology, Pasadena, California 91125*

The resonant enhancement and the controllable optical-dipole-matrix elements of intersubband transitions in quantum wells have led to the observation of extremely large optical nonlinear coefficients[1-3]. Second-harmonic nonlinear optical coefficients as high as $\chi^{(2)}(2\omega, \omega, \omega) \cong 7.6 \times 10^{-7}$ (m/v) have been observed,[2] and the third-harmonic nonlinear optical coefficient observed,[3] $\chi^{(3)}(3\omega, \omega, \omega, \omega) \cong 4 \times 10^{-14}(m/v)^2$, is to our knowledge the largest reported in any material system. Despite these extremely large nonlinearities, the expectations for an efficient device have failed to materialize. In fact, the highest second-harmonic conversion efficiency obtained to date, $\eta \cong 3 \times 10^{-4}$, is limited in part by the intersubband selection rules, the narrow epitaxial layers, and the surface limitations on the usable intensities.[4] In this paper we demonstrate that beyond these experimental issues there are fundamental limits arising from the basic nature of the nonlinear interaction. We believe these limitations must be addressed if intersubband nonlinear phenomena are to advance to the device stage.

For near-resonant frequencies the standard nonlinear coefficients, derived by means of a perturbative expansion, saturate and become inadequate at moderate intensities and hence for small conversion efficiencies. We use algebraic solutions of the Bloch equations within the rotating-wave approximation[5] to derive effective nonlinear coefficients that replace the perturbative terms. The saturation of parametric processes, second-harmonic generation (SHG) in particular, is studied. Figures 1(a) and 1(b) show the saturation of the diagonal (population) and the off-diagonal (induced-polarization) density-matrix terms in the presence of a strong pump (first-harmonic) intensity in a resonant three-level system. The saturation value of the second-harmonic susceptibility ρ_{02} is used to derive an upper limit on the second-harmonic intensity $I_{2\omega}$ that may be generated in a finite propagation length L. For the case of an all-resonant system with equal dipole-matrix elements and relaxation times, we obtain the simple expression

$$I_{2\omega}(L) \leq \frac{I_{sat}}{(\sqrt{T_2/T_1} + 3\sqrt{T_1/T_2})^2}(\alpha_\omega L)^2$$

$$\overset{(T_1 >> T_2)}{\cong} I_{sat}\left(\frac{\alpha_\omega L}{4}\right)^2, \quad (1)$$

where I_{sat} is the saturation intensity of the second harmonic, T_1 is the intersubband lifetime, T_2 is the intersubband relaxation time, and α_ω is the first harmonic's linear nonsaturated absorption coefficient. A numerical solution of the propagation equations using the nonperturbative susceptibilities is used to obtain Fig. 2, which shows the second-harmonic intensity and the conversion efficiency as functions of the pump amplitude and the detuning for a propagation of several extinction lengths ($\alpha_\omega L$ = 1, 10, 25). As the system approaches fully saturated behavior, the conversion intensity approaches the upper limit [Eq. (1)]. The conversion intensity, on the other hand, goes through a peak and then decreases.

Similar limits to those shown in Eq. (1) and Fig. 2 may be obtained for other parametric processes. The importance of these limitations is that the obtainable frequency-conversion efficiencies may be directly inferred from the on-resonance nonsaturated extinction $\alpha_\omega L$. Currently, it is difficult to obtain $\alpha_\omega L >> 1$. Thus, without increased interaction lengths, intersubband transitions will not lead to high conversion efficiencies, despite their large nonlinear coefficients. The advantages of intersubband transitions are thus mostly limited to harmonic-generation applications requiring short propagation lengths and weak pump intensities.

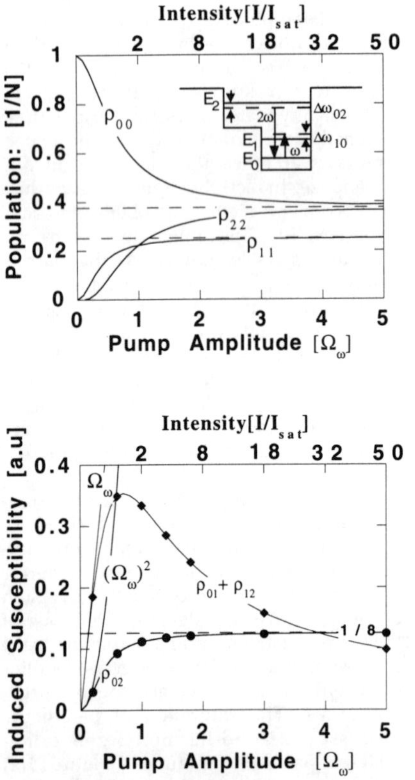

QWJ2 **Fig. 1.** Diagonal and off-diagonal density-matrix elements in a resonant three-level system subject to a strong pump field (with a normalized Rabi frequency of $\Omega_\omega = \Omega_R T_2$). The dashed lines show saturation values, the second-harmonic response ρ_{02} saturates to 1/8, and the first-harmonic response, $\rho_{01} + \rho_{12}$, saturates to 0.

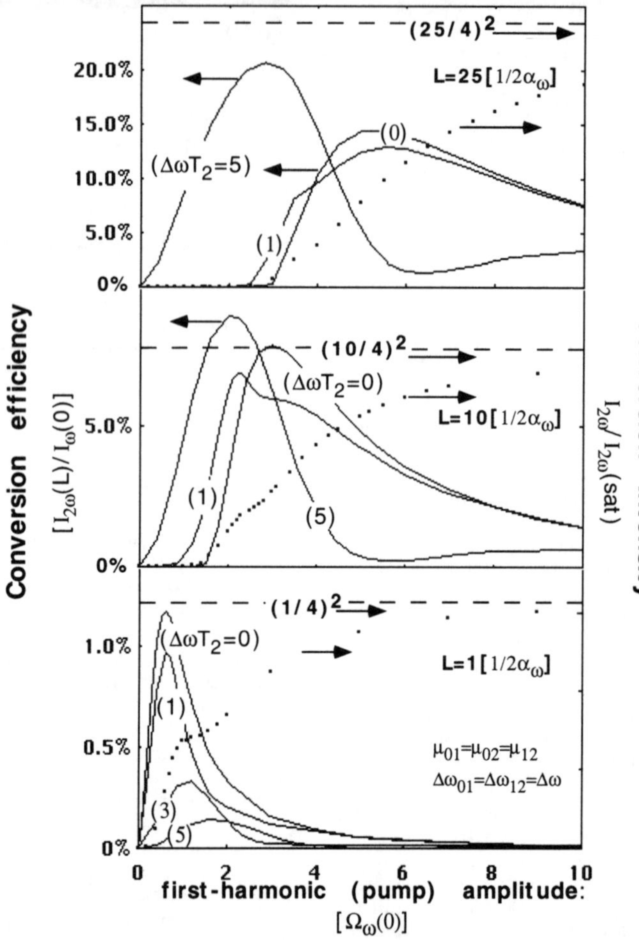

QWJ2 **Fig. 2.** Conversion efficiency (left axis) and converted intensity (right axis) versus the normalized Rabi frequency of the initial pump amplitude for several detunings and propagation lengths.

1. M. M. Fejer, S. J. B. Yoo, R. L. Byer, A. Harwit, and J. S. Harris, Phys. Rev. Lett. **62**, 1041 (1989).
2. E. Rosencher, P. Bois, J. Nagle, and S. Delaitre, Electron. Lett. **25**, 1063 (1989).
3. F. Capasso, C. Sirtori, and A. Y. Cho, IEEE J. Quantum Electron. **QE-30**, 1313 (1994).
4. P. Boucaud, F. H. Julien, D. D. Yang, J. M. Lourtoid, E. Rosencher, and P. Bois, Opt. Lett. **16**, 199 (1991).
5. T. A. DeTemple, L. A. Bahler, and J. Osmundsen, Phys. Rev. A **24**, 1950 (1981).

QWJ3 **5:15 pm**

Strain-compensated type-II asymmetric quantum wells for wavelength converter applications

S. J. B. Yoo, R. Bhat, M. A. Koza, C. Caneau, *Bellcore, 331 Newman Springs Road, Red Bank, New Jersey 07701*

Asymmetric quantum wells possess large optical nonlinearities[1] useful for a number of applications, including modulators, optical switches, and wavelength converters. Second-order nonlinearities due to intersubband transitions have been demonstrated in both GaAs-based and InP-based quantum wells. The measured nonlinear optical susceptibilities for second harmonic generations of mid-infrared radiation were 2 to 3 orders of magnitude larger than that of bulk GaAs.[2] For near-infrared applications, interband processes should also be considered. For high wavelength conversion efficiencies in near-infrared, the ratio (*F.M.*) between the nonlinear optical susceptibility and the interband absorption coefficient has to be large. In type-I quantum wells, *F.M.* is limited because the spatial distributions of the ground state electron and the heavy-hole wave functions are similar. Type-II asymmetric quantum wells, on the other hand, can achieve large *F.M.* resulting from relatively small overlap be-

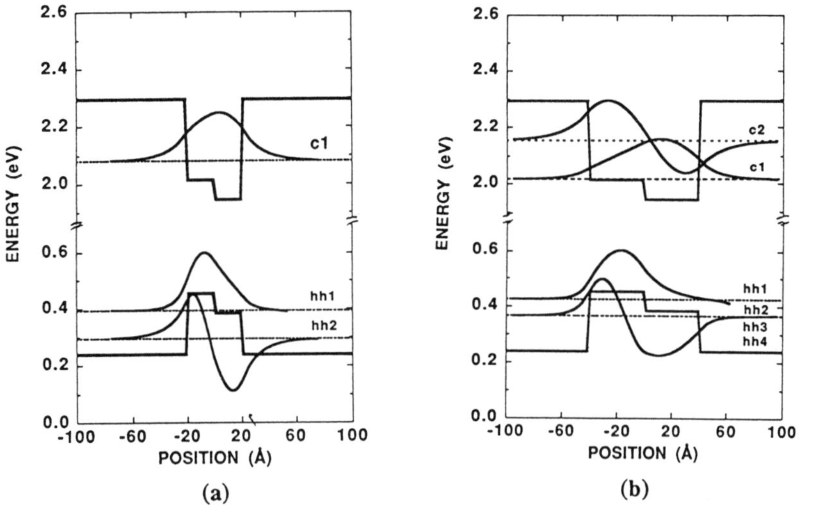

QWJ3 **Fig. 1.** Calculated energy bands and wave functions for (a) sample #783 and (b) sample #784. Light-hole states are omitted for clarity.

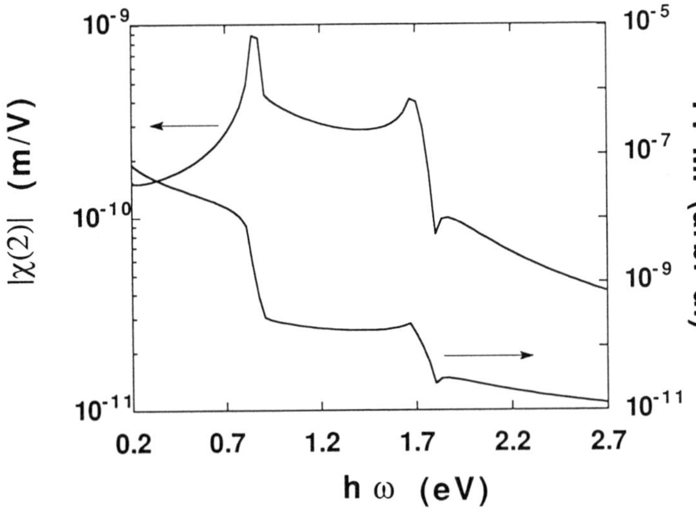

QWJ3 **Fig. 2.** Calculated spectra of nonlinear susceptibilities ($\chi^{(2)}$), and ratios (*FM*) between nonlinear optical susceptibilities and the interband absorption coefficients of sample #783. The peak value of $\chi^{(2)}$ is 9×10^{-10} m/V which is 5 times that of bulk GaAs and 2.5 times that of optimized type-I AlGaAs quantum wells in Ref. 4.

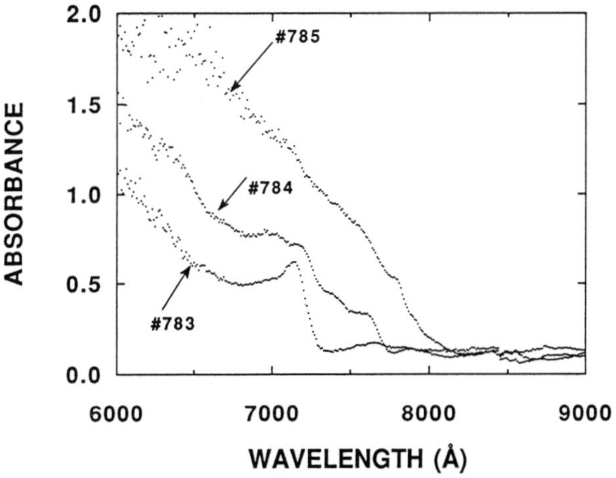

QWJ3 **Fig. 3.** Measured absorption spectra of samples #783, #784, and #785. The samples are prepared by removing GaAs substrates and depositing AR coatings on the exposed samples. Sample #783 shows an exciton feature at the band-gap, whereas samples #784 and #785 show very weak absorption at the vicinity of the band-gap. Absorbance is defined as $-\log_{10}$(Transmission).

tween the lowest-level electron wave function and the uppermost heavy hole wave function.[3] Type-II quantum wells found in lattice-matched material systems have limited choices of band-gap energies and band offsets. In this paper we discuss the realization of type-II quantum wells in a strain-compensated system.

We considered strain-compensated InGaAsP quantum wells with lattice-matched AlGaAs barriers grown on a GaAs substrate. This choice allows a reasonably large band offset (>60 meV) between two oppositely strained (±0.5%) InGaAsP layers of equal band-gap wavelengths in the 0.75 micron range. We have designed three quantum wells with different well thicknesses: 20 Å of 0.5% compressively strained and 20 Å of 0.5% tensile strained quantum wells (sample #783), 40 Å of 0.5% compressively strained and 40 Å of 0.5% tensile strained quantum wells (sample #784), and 80 Å of 0.5% compressively strained and 80 Å of 0.5% tensile strained quantum wells (sample #785). Figure 1 shows calculated energy-band diagrams and wave functions for sample #783 and #784. Samples with thicker quantum wells show smaller overlaps between the lowest electron wave function and the uppermost heavy-hole wave function. Figure 2 illustrates calculated nonlinear susceptibilities ($\chi^{(2)}$) and *F.M.* for sample #783. The peak value of $\chi^{(2)}$ is 9×10^{-10} m/V which is 5 times that of bulk GaAs and 2.5 times that of optimized type-I AlGaAs quantum wells in Ref. 4. The *F.M.* calculated at 20 meV below the lowest energy interband transition resonance is two order of magnitude larger than that of type-I quantum wells in Ref. 4. Three samples #783, #784, #785, were OMCVD grown according to the designs. The number of quantum well period is 80 for all samples. Figure 3 shows absorption spectra measured on the three samples with the GaAs substrates removed. Compared to sample #783 showing an exciton feature at the band-gap, samples #784 and #785 showed relatively weak absorption at the vicinity of the band-gap as predicted from overlap integral calculations. Photoluminescence measurements showed integrated intensities significantly weaker than that from Type-I AlGaAs quantum wells of a similar band-gap energy.

In summary, we report type-II quantum wells realized in a InGaAsP strain-compensated material system. The large ratio between the nonlinear optical susceptibility and the interband absorption coefficient imply its applicability for high-efficiency nonlinear optical wavelength-conversion devices.

1. M. K. Gurnick and T. A. De Temple, IEEE J. Quantum Electron. **19**, 791 (1983).
2. S. J. B. Yoo, M. M. Fejer, R. L. Byer, and J. S. Harris, Jr., Appl. Phys. Lett. **58**, 1724 (1991).
3. S. Scandolo, A. Baldereschi, and F. Capasso, Appl. Phys. Lett. **62**, 3138 (1993).
4. P. J. Harshman and S. Wang, Appl. Phys. Lett. **60**, 1277 (1992).

QWJ4 5:30 pm

Anizotropy and dispersion of optical rectification in symmetric quantum wells

J. B. Khurgin, *Department of Electrical and Computer Engineering, Johns Hopkins University, Baltimore, Maryland 21218*

Generation of far-infrared light by optically excited semiconductors has been reported by numerous groups of researchers.[1,2] Although various mechanisms, such as current surge and depletion field-induced optical rectification (OR), have been proposed, recent results[3] have shown that bulk OR accounts for a major portion of the observed teraherz signal. Recently we evaluated the OR coefficients of bulk zinc-blende semiconductors excited *above* the bandgap,[4] and the results are in agreement with experimental data.

In the present work the bond-orbital treatment is expanded to symmetric, unbiased quantum wells (QWs) and it is shown that the dispersion, anisotropy, and dynamics of the teraherz radiation are all altered drastically as the result of the bonding orbitals' mixing induced by the confinement.

GaAs QWs grown on (100), (011), and (111) surfaces were considered, and the OR coefficients in the system of coordinates associated with the direction of growth were evaluated. The results are shown in Fig. 1. One can see a drastic difference between the differently oriented QWs. In the (100)-oriented material, four bonding orbitals are equally mixed in the ground heavy-hole state. The OR coefficient is zero at the absorption edge and then grows rather slowly. Only two components, $\chi^{(2)}_{zxy}$ and $\chi^{(2)}_{xzy}$, are substantial. In the (011)-oriented material, two of the bonding orbitals contribute to the heavy hole, and the other two contribute to the light hole. As a result, large $\chi^{(2)}_{xxx}$ and $\chi^{(2)}_{xyy}$ components of the OR coefficient are present, and because of heavy-hole–light-hole splitting, a reversal of the sign of the OR coefficient is possible! The presence of a large $\chi^{(2)}_{xxx}$ component is beneficial for teraherz radiation generation, when the optical radiation impinging onto the surface is polarized in the plane of QWs.

In the (111)-oriented material, the light-hole state consists almost exclusively of only one bonding orbital and the heavy-hole state consists mostly of the other three orbitals. The OR coefficients were found to be the highest for QWs of this orientation, although they are not the most convenient for teraherz radiation in the surface-emitting configuration. On the other hand, the largest coefficients, $-\chi^{(2)}_{xzz}$ and $\chi^{(2)}_{zzz}$, are perfect for difference-frequency generation in the waveguide configuration.

It is also shown that strain can produce an effect similar to confinement; this can explain the reversal of sign in the terahertz radiation in Ref. 2.

1. S. L. Chuang, S. Schmitt-Rink, B. I. Greene, P. N. Saeta, and A. F. J. Levi, Phys. Rev. Lett. **68**, 102 (1992).
2. X.-C. Zhang, Y. Jin, K. Yang, and L.-J. Scholwalter, Phys. Rev. Lett. **69**, 2303 (1992), and references therein.
3. P. N. Saeta, B. I. Greene, and S. L. Chuang, Appl. Phys. Lett. **63**, 3482 (1993).
4. J. B. Khurgin, "Teraherz radiation via above the gap optical excitation of the semiconductor structures," J. Opt. Soc. Am. B, in press.

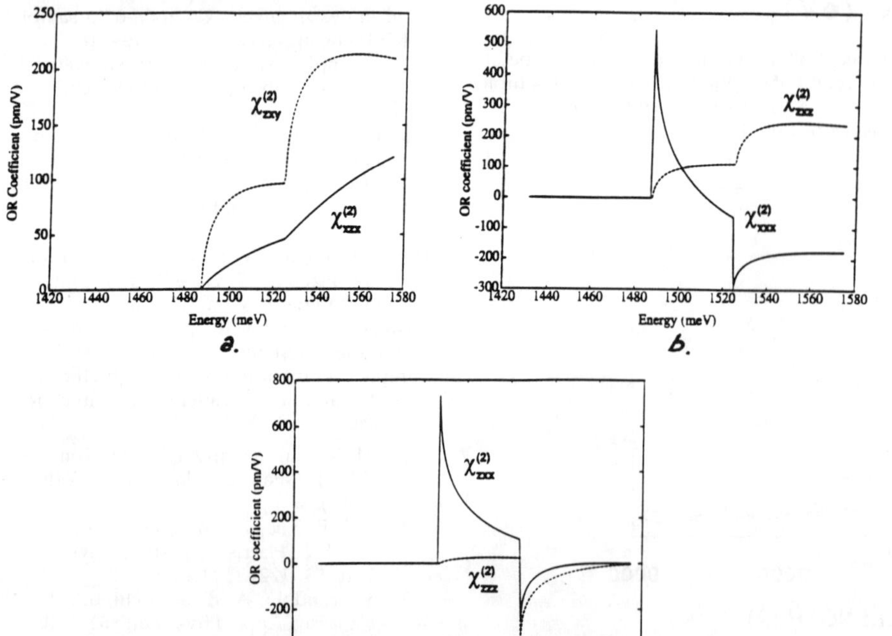

QWJ4 Fig. 1. Frequency dependence of the optical rectification coefficient in 100-Å GaAs QWs: (a) (100) orientation; (b) (011) orientation; (c) (111) orientation.

QWJ5 5:45 pm

Strain-enhanced second-order susceptibility of asymmetric InGaAs/GaAs/AlGaAs quantum wells

V. W. Lee, T. K. Gustafson, *Department of Electrical Engineering and Computer Sciences, University of California at Berkeley, 211-121 Cory Hall, Berkeley, California 94720-1772*

The interband difference-frequency susceptibility of strained asymmetric quantum wells (QW) is theoretically studied. For the particular case of an undoped InGaAs/GaAs/AlGaAs QW, an order of magnitude enhancement in the peak susceptibility is found when the effects of biaxial compressive strain are included in the calculations.

The QW material studied consists of 45.2 Å $In_{0.2}Ga_{0.8}As$ plus 33.9 Å GaAs surrounded by $Al_{0.2}Ga_{0.8}As$ barriers, all on top of a (111) GaAs substrate. The table in Fig. 1 lists the effective masses (in the direction perpendicular to the well), energy gap and Kane energy for each of those materials at 77 K. The GaAs and AlGaAs material are assumed to be unstrained. The valence band effective masses for the (111) biaxial compressively strained $In_{0.2}Ga_{0.8}$ are found following the method of Bir and Pikus.[1]

Conduction/valence band offset ratios of 60:40 for the (111) strained well (taken to be similar to that for (100) strain[2]) and 85:15 for the unstrained well[3] are assumed. The envelope function approximation together with the boundary conditions of wavefunction and probability current density conservation are used to calculate the energy levels, their corresponding wavefunctions and effective masses, and the dipole matrix elements.[4] The large disparity in the heavy-hole effective masses between strained $In_{0.2}Ga_{0.8}As$, GaAs and $Al_{0.2}Ga_{0.8}As$ results in heavy-hole wavefunctions which are more asymmetric for the strained than for the unstrained QW. Inclusion of the piezoelectric effect in (111) strained material[6] further asymmetrizes the strained well with respect to the unstrained well.

The QW difference-frequency susceptibility is calculated by using the density matrix approach and the parabolic band approximation. The three-level atomic susceptibility is summed over the transition frequencies distributed according to the joint QW density of states.[5] Only the near-resonant susceptibility terms are retained. An intersubband damping rate of 8×10^{12} s^{-1} and an interband damping rate of 2×10^{13} s^{-1} are also assumed.

Figure 2 shows the calculated difference-frequency susceptibility of the QW. An order of magnitude improvement in the peak susceptibility is predicted when the (111) biaxial compressive strain effects of the $In_{0.2}Ga_{0.8}As$ layer are included. This improvement is attributed to the enhanced product of dipole matrix elements, resulting from the enhanced asymmetry of the strained QW compared to the unstrained QW.

Future work will compare our estimates obtained from the parabolic band model to those obtained from the valence-band mixing model.

Material System	m_c (m_0)	m_{hh} (m_0)	m_{lh} (m_0)	E_g (eV)	E_p (eV)
$Al_{0.2}Ga_{0.8}As$	0.0836	0.56	0.0996	1.767	30.4
GaAs	0.067	0.51	0.082	1.51	28.8
InAs	0.0242	0.35	0.027	0.418	22.2
unstrained $In_{0.2}Ga_{0.8}As$	0.059	0.467	0.071	1.212	27.48
(111) strained $In_{0.2}Ga_{0.8}As$	0.059	0.917	0.053	1.333	26.69

QWJ5 Fig 1. Material constants for the various well layers.

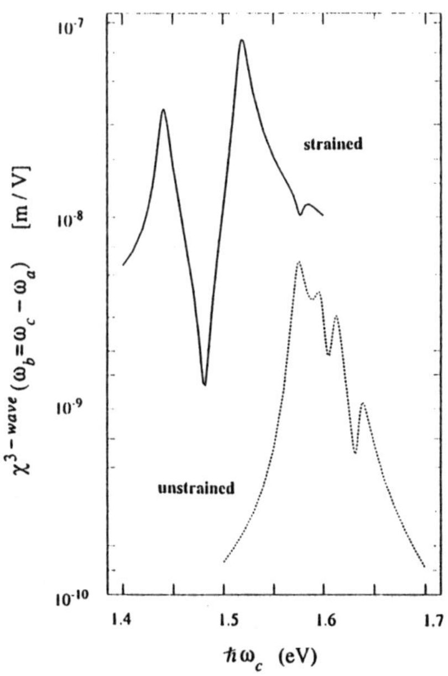

QWJ5 Fig. 2. Difference-frequency susceptibility $\chi^{(2)}$ ($\omega_b = \omega_c - \omega_a$) as a function of $\hbar\omega_c$, where $\hbar\omega_a$ is assumed to be resonant with the c1–hh1 transition.

1. G. L. Bir and G. E. Pikus, *Symmetry and Strain-induced Effects in Semiconductors*, (John Wiley & Sons, New York, 1974), Chap. 5.
2. J. P. Reithmaier *et al.*, Appl. Phys. Lett. **56**, 536 (1990).
3. D. J. Arent, K. Deneffe, C. Van Hoof, J. De Boeck, and G. Borghs, J. Appl. Phys. **66**, 1739 (1989).
4. G. Bastard, *Wave Mechanics Applied to Semiconductor Heterostructures* (Halsted, New York, 1988), Chap. 3.
5. V. W. Lee and T. K. Gustafson, Appl. Phys. Lett. **66**, 271 (1995).
6. E. A. Caridi, T. Y. Chang, K. W. Goossen, and L. F. Eastman, Appl. Phys. Lett. **56**, 659 (1989).

JThA 8:00 am
Room #317

Joint Session on Emerging Applications of Fluorescence to Biophysics and Cellular Imaging I

David R. Sandison, *Sandia National Laboratory, Presider*

JThA1 (Invited) 8:00 am

Digital microscope imaging of cell dynamics

D. L. Farkas, E. S. Wachman, L. D. Harris, A. H. Gough, D. L. Taylor, *Center for Light Microscope Imaging and Biotechnology, Carnegie Mellon University, Pittsburgh, Pennsylvania 15213*

Modern biological research has benefited from a renaissance of light microscopy, brought about by the convergence of new developments in fields as diverse as electronics, physical optics, molecular biology, computer science, and synthetic chemistry. Integration of these has transformed microscopy from an ancillary, static, 2-D tool into a highly useful, dynamic, 3-D research capability for biology and medicine. We believe that the understanding of certain fundamental biological functions by highly resolved, dynamic mapping of chemical and molecular events in living systems is within reach, based on novel, interdisciplinary imaging methods applied at the cellular level.

The use of multimode microscopy[1,3] allows the investigation of chemical and molecular events in living cells and tissues, through a synergetic combination of microscopic imaging methods in a computer-controlled instrument with automated acquisition. Improvements of this approach are based on addressing some of the main remaining challenges: (A) axial resolution: we significantly improved this by an interference-based patterning of the excitation field (standing wave microscopy)[2]; (B) mode switching speed: we approached this by the use of acousto-optical filters of our design for wavelength, intensity, and timing control at both the excitation and emission end of a microscope; and (C) image processing speed. By using high performance computing and communications, we can change digital light microscopy as follows: (a) demanding tasks of image processing that formerly were performed off-line, can now be done in near-real time and the results displayed while the experiment is still going on and while functional manipulations (e.g. by uncaging, photoactivation, perfusion) can be performed; (b) automating the detection of "events of interest," as specified by the microscope user, can trigger spatially and temporally intensive data-gathering only when something novel happens; and (c) we can explore new kinds of image analysis and visualization that were computationally unfeasible before. These are achieved in a workstation we call AIM (Automated Interactive Microscope).

The usefulness of all these advances will be measured by the types of problems that can be studied, and the quality of the results. We believe that with the right combination of technologies, new types of biological specimens can be studied quantitatively, detectivity and resolution can be greatly improved, new time domains (~milliseconds, important in fields such as neurosciences) can become accessible, and important applications (e.g. alternatives to animal testing[4] can benefit from digital light microscopy of live specimens.

Some of these promising new approaches, main achievements, and future directions are reviewed.

1. D. L. Farkas *et al.*, "Multimode light microscopy and the dynamics of molecules, cells and tissues," Annual Rev. Physiology, **55**, 785 (1993).
2. B. Bailey *et al.*, "Enhancement of axial resolution in fluorescence microscopy by standing wave excitation," Nature, **366**, 44 (1993).
3. D. L. Farkas *et al.*, "New waves in light microscopy," Proc. SPIE, **2137**, 2 (1994).
4. D. L. Taylor *et al.*, "Potential of machine-vision light microscopy in toxicologic pathology," Toxicologic Pathology, **22**, 145 (1994).

JThA2 (Invited) 8:30 am

High-speed fluorescence microscopy: lifetime imaging in the biomedical sciences

Brain Herman, Xue F. Wang, Pawel Wodnicki, Gerald W. Gordon, Seongwook Kwon, Pamela A. Diliberto, Ammasi Periasamy, *University of North Carolina*

The marriage of the optical microscope and environmentally sensitive fluorescent probes has allowed the biologist to study the chemical and molecular dynamics of intact cells, providing fundamental information regarding the homeostatic and pathologic functions of individual living cells and tissues. Time-resolved fluorescence microscopy permits the quantitation of dynamic events at very high temporal resolution and provides the ability to monitor interactions between cellular components with very high spatial resolution. We review the recent adaptation of fluorescence lifetimes to the fluorescence microscope and describe some biomedical applications of this new type of high-speed microscopy.

JThA3 (Invited) 8:45 am

Applications of electro-optic and laser technology to fluorescence microscopy

R. Nitschke, K. R. Spring, *Laboratory of Kidney and Electrolyte Metabolism, NHLBI, National Institutes of Health, 10/6N307, Bethesda, Maryland 20892*

A confocal attachment (Odyssey) to an inverted microscope was modified to better study living cultured epithelial cells stained with fluorescent dyes. Improvements to the instrument included elimination of light leaks, improved electronic shielding, reduction of thermal effects, and use of low dark current detectors. In addition, rapid changes in illumination wavelength and power were accomplished by replacing the original mechanical filter changer by an acousto-optic tunable filter attached to the argon laser light source. The addition of a liquid crystal tunable filter to one of the two photomultiplier detectors also permitted rapid spectral scanning of the fluroescence emission. High resolution, differential interference contrast transmitted light images were formed simultaneously by replacement of the photodiode-based transmitted light detector with a photomultiplier tube and dichroic mirror assembly. An illumination intensity of only 40 μW/cm^2 at the back focal plane of the microscope objective allowed high quality fluorescence and transmitted light images of living cells at video rates with minimal bleaching and photodynamic damage. Both excitation ratio imaging and emission spectral scanning of living epithelial cells were accomplished. The system performance was evaluated by optical sections of fluorescent beads and thin films.

JThA4 (Invited) 9:15 am

Applications of nonlinear microscopy in cellular biophysics

David W. Piston, *Department of Molecular Physiology and Biophysics, Vanderbilt University, Nashville, Tennessee 37232*

Two photon excitation provides attractive advantages over confocal microscopy for 3-D fluorescence imaging and photochemistry, such as release of caged compounds. Two-photon excitation arises from the simultaneous absorption of two photons in a single quantitized event whose probability is proportional to the square of the instantaneous intensity. For example, two red photons can cause the transition to an excited electronic state normally reached by absorption in the ultraviolet. In our fluorescence experiments, the final excited state is the same singlet state that is populated during a conventional fluorescence experiment. Thus, the fluorophore exhibits the same emission properties (e.g. wavelength shifts, environmental sensitivity) used in typical biological microscopy studies. The most important property of two-photon excitation is that the excitation is limited to the focal volume because of the intensity-squared dependence of the two-photon absorption. This inherent localization provides 3-D resolution and eliminates background equivalent to an ideal confocal microscope without requiring a confocal spatial filter. Confinement of excitation to the focal volume also minimizes photobleaching and photodamage, and allows for three-dimensionally resolved point photochemistry. Cellular biophysics applications include three-dimensionally resolved, dynamic quantification of intracellular calcium activity by Indo-1, redox imaging of cell autofluorescence arising from NADH, and imaging of nuclear division by Hoechst 33342. The pulsed nature of mode-locked excitation has also been utilized to efficiently realize fluorescence decay time imaging. Finally, point photorelease has been used to marked cells in developing embryos, map receptor distributions in neural cells, and trigger local calcium release in muscle cells.

JThA5 (Invited) 9:45 am

Fiber-optic metal ion biosensors based on fluorescence resonance energy transfer

Richard B. Thompson, Zhengfang Ge, Marcia W. Patchan, *Department of Biological Chemistry, University of Maryland School of Medicine, 108 N. Greene St., Baltimore, Maryland 21201*

Recently, we have described a fluorescence-based fiber-optic biosensor for zinc, which uses a metalloenzyme as the sensor transducer. Zinc is bound tightly and specifically to the active site of the enzyme carbonic anhydrase. If we add a fluorescent inhibitor, dansylamide, it binds to the zinc in the active site with concomitant large changes in its fluorescence quantum yield, emission wavelength, and lifetime. In the absence of zinc apocarbonic anhydrase does not bind the inhibitor and one observes the weak, red-shifted, short-lifetime emission of the free dansylamide. These fluorescence changes can be measured through optical fiber and correlated with the zinc concentration, with sup-ppb sensitivity, 50 dB dynamic range, and high selectivity.

However, dansylamide requires excitation in the ultraviolet, where fiber attenuation is very high, and we would prefer to excite in the visible or near-IR. Thus, we have changed our transduction scheme by attaching a covalent fluorescent label to the protein, and using a colored inhibitor, which also binds in a metal-dependent fashion. In the presence of metal the inhibitor binds and is thus in close proximity; it quenches the fluorescent label by resonance energy transfer and suffers a decrease in lifetime and intensity. In the absence of metal the inhibitor is free to diffuse away and the lable is unquenched. Recent results from this work will be shown.

The authors wish to thank the Office of Naval Research for support.

QThA 8:00 am
Hyatt A

Emerging Applications

John E. Sipe, *University of Toronto, Presider*

QThA1 (Invited) 8:00 am

Photo-assisted ferroelectric domain fixing for volume hologram storage

M. Zgonik, R. S. Cudney,* J. Fousek,** P. Günter, *Nonlinear Optics Laboratory, Institute of Quantum Electronics, Swiss Federal Institute of Technology, ETH-Hönggerberg, CH-8093 Zürich, Switzerland*

Photoreactive gratings can be fixed if a complementary "grating" of some kind is created, which does not decay with time or under illumination. One promising mechanism is the fixing in ferroelectric photorefractive crystals through the formation of ferroelectric domains which electrically compensate the space-charge field. In other words, the direction of the spontaneous polarization $\mathbf{P}_s$ of the crystal can be selectively reversed such that the bound charge located on the domain walls electrically compensates the light-induced space-charge distribution of the original grating. This effect was observed first in the 70's in $BaTiO_3$[1] and SBN ($Sr_xBa_{1-x}Nb_2O_6$).[2] Recently interest in this field has been revived and the research has been concentrated again on SBN[3-5] and $BaTiO_3$.[6,7]

Usually the selective reversal of the direction is accomplished by first creating a space-charge distribution by the usual manner (interfering two light beams in the crystal) and then applying an electric field in the direction opposite to the original poling field. Several models are feasible to describe the local switching of ferroelectric polarization: antiparallel domain generation wherever the total field (applied plus space-charge field) exceeds the coercive field; needle-shaped domains nucleated at one electrode propagate towards the other electrode until they are stopped by the space-charge distribution; deformation of the domain walls by the space-charge field such that this field is screened. Electrical, optical, and combined revealing of the fixed grating is possible.

Application in holographical storage will depend on the magnitudes of the usual photorefractive parameters; on the precision with which one can control the fine shape of the domains; and on the minimum domain size.

**Present address: División de Física Aplicada, CICESE, Apdo. Postal 2732, C.P. 22880, Ensenada, Baja California, Mexico*

***Present address: Institute of Physics, Academy of Sciences of the Czech Republic, Na Slovance 2, 180 40 Prague 8*

1. F. Micheron and G. Bismuth, Appl. Phys. Lett. **20**, 79 (1972).
2. F. Micheron and G. Bismuth, Appl. Phys. Lett. **23**, 71 (1973).
3. Y. Qiao, S. Orlov, D. Psaltis, and R. R. Neurgaonkar, Opt. Lett. **18**, 1004 (1993).
4. A. S. Kewitsch, M. Segev, A. Yariv, and R. Neurgaonkar, Opt. Lett. **18**, 1262 (1993).
5. M. Horowitz, A. Bekker, and B. Fischer, Opt. Lett. **18**, 1964 (1993).
6. R. S. Cudney, J. Fousek, M. Zgonik, P. Günter, M. Garrett, and D. Rytz, Appl. Phys. Lett. **63**, 3399 (1993).
7. R. S. Cudney, J. Fousek, M. Zgonik, P. Günter, M. Garrett, and D. Rytz, Phys. Rev. Lett. **72**, 3883 (1994).

QThA2 8:30 am

Use of a vertical-cavity surface-emitting laser as a logic gate

F. Prati, G. Tissoni, M. Travagnin, L. A. Lugiato, *Department of Physics, University of Milan, via Celoria 16, I-20133 Milan, Italy*

In the growing field of transverse nonlinear optics[1] the phenomenon of *spatial multistability* is one of the most relevant because of possible applications to optical information processing. It consists of the capability, displayed by several kinds of lasers, of emitting beams with different spatial configurations under the same parametric conditions. Here we demonstrate spatial bistability in a vertical-cavity surface-emitting laser[2] (VCSEL), and we suggest the exploitation of this property for the realization of an optical logic gate.

We consider an index-guided VCSEL with a square transverse section of the active region. We assume that the only lasing modes are the Gauss–Hermite modes TEM_{10} and TEM_{01}. The laser dynamics are described by a rate-equation model consisting of two equations for the mode amplitudes and one equation for the carrier density.[3] The band structure of the energy levels is simply described by the linewidth enhancement factor α.

The diagonals of the square (Fig. 1) are both stable for values of the injected current such that the laser is less than 5% above threshold. We assume the following values for the relevant parameters: recombination time of carriers, $\tau_r = 0.2$ ns; optical length of the cavity, $L = 6$ μm; reflectivity of the Bragg reflectors, $R = 0.998$. Our calculations show that the inclusion of carrier diffusion in the model has the effect of increasing the size of the bistability domain for modes TEM_{10} and TEM_{01}.

We studied numerically under which conditions a bistable laser that operates initially, say, with mode TEM_{01}, switches to mode TEM_{10} on injection of an appropriate signal. We found switching energies of about 20 fJ and switching times of 10–20 ns.[4] Figure 2 shows a typical switch process.

A NOR gate can be realized by using a bistable VCSEL. The device works in this way: Two input beams (I_1, I_2) are injected in correspondence with the two peaks of one mode, say, mode TEM_{01}; the detector that measures the output (O) and a reset beam (Re) are placed on the other diagonal of the square (Fig. 3). Both the individual input beams and the reset beam have sufficient intensity to cause switching from one to the other mode. We assign the value 1 to each of the inputs I_1 and I_2 when they are on, and we assign the value 0 when they are off. The detector measures 0 (no intensity) if the laser emits mode TEM_{01} and 1 (finite intensity) if the laser emits mode TEM_{10}. The reset beam, which is injected at the beginning of each operation, is necessary to ensure that initially the laser emits mode TEM_{10}. Then, if no beam is sent to either of the two inputs (i.e., $I_1 = I_2 =$

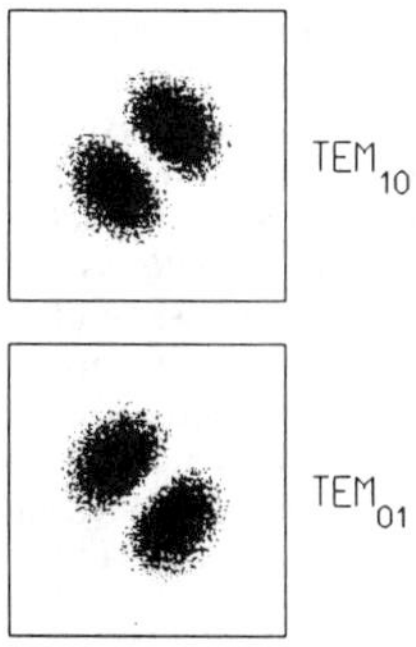

QThA2 Fig. 1. Intensity distribution of the TEM_{10} and TEM_{01} laser modes in the transverse plane. The square represents the transverse section of the active region.

Thursday, May 25

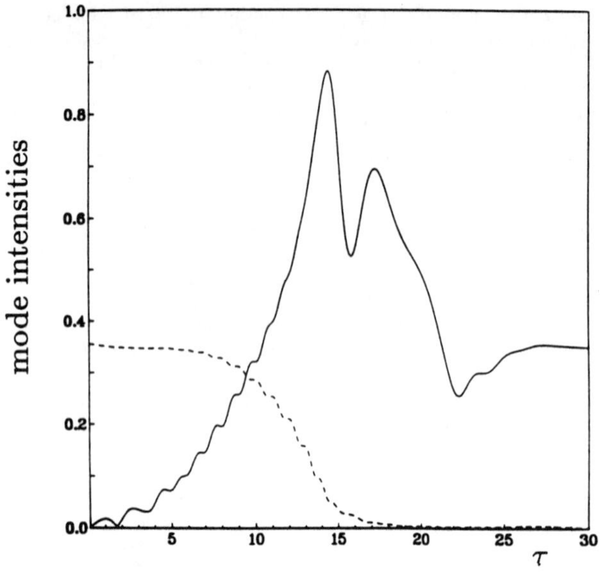

QThA2 Fig. 2. Dynamics of switching from one Gauss–Hermite mode to the other. The dashed curve represents the intensity of the mode that was emitted by the laser before injection of the other mode, whose intensity is represented by the solid curve. Time τ is measured in units of the recombination time τ_r.

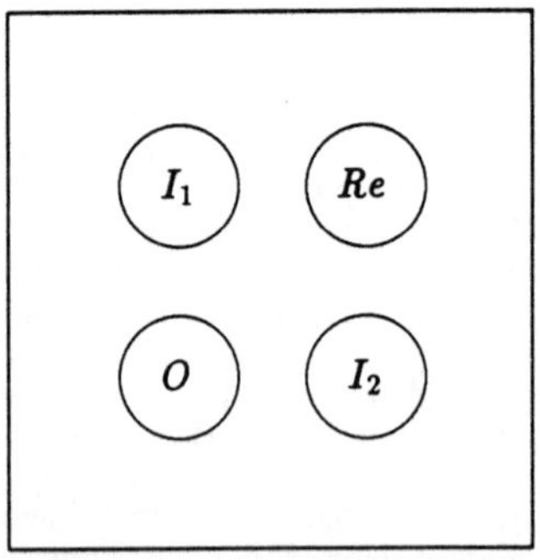

QThA2 Fig. 3. Scheme of the NOR logic gate. I_1 and I_2 are the input beams, O is the output, and Re is the reset beam.

0), the output is 1; in all other cases the laser switches to mode 0 and the output is 0; hence the system works as a NOR gate.

This research has been carried out in the framework of ESPRIT Basic Research Project 7118 TONICS.

1. L. A. Lugiato, Phys. Rep. **219**, 293 (1992).
2. J. L. Jewell, J. P. Harbison, A. Scherer, Y. H. Lee, and L. T. Florez, IEEE J. Quantum Electron. **27**, 1332 (1991).
3. F. Prati, A. Tesei, L. A. Lugiato, and R. J. Horowicz, "Stable states in surface-emitting semiconductor laser," in *Nonlinear Optical Patterns, Structures and Chaos*, special issue of Chaos, Solitons and Fractals, in press.
4. F. Prati, M. Travagnin, and L. A. Lugiato, "Optical switching and logic gates with vertical cavity surface emitting lasers," Opt. Lett., in press.

QThA3 **8:45 am**

Pyroelectric depth profiling of the optical nonlinearity caused by charge injection

A. Donval, G. Berkovic,
S. Bauer-Gogonea,* W. Wirges,*
S. Bauer,* R. Gerhard-Multhaupt,*
Department of Materials and Interfaces, Weizmann Institute of Science, Rehovot 76100, Israel

Over the last decade, the electric-field poling of polymer films containing hyperpolarizable organic dye molecules has received considerable attention as an efficient means of creating second-order nonlinear optical materials. A simple electrostatic model can explain symmetry breaking and the creation of bulk nonlinearity through dipolar alignment of the dye molecules in the electric field. Effects due to current flow and/or charge injection have usually not been considered.

Some studies,[1,2] however, have shown that significant charge injection can occur when poling is performed on polymer films by using electrodes in an in-plane geometry. In extreme cases, when poling is performed[1] by using high voltages but only moderate fields (2 kV across 2 mm), charge-injection effects can be stronger than dipolar alignment. Consequently, the strongest nonlinearity is perpendicular to the applied field direction and not parallel to it.

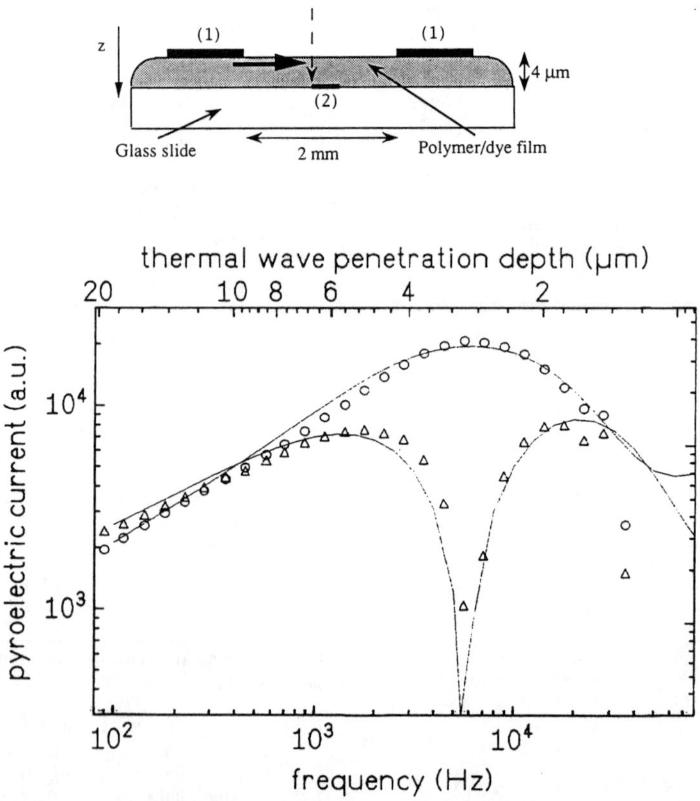

QThA3 Fig. 1. Top: In-plane geometry for poling. When high voltage is applied across the poling electrodes (1), charge injection (thick arrow) followed by diffusion (broken arrow) leads to polarization along the z axis. After poling, electrodes (1) are enlarged by evaporation of additional aluminum and electrode (2) is used to measure the pyroelectric response. Bottom: Frequency-dependent pyroelectric response of a 4-μm-thick film of polymethyl methacrylate containing 2% of the dye Disperse Red 1, after poling in the in-plane geometry. The data is fitted to a nonuniform polarization distribution with the real and imaginary parts of the polarization shown by the circles and triangles, respectively.

It has been suggested that a charge gradient is formed perpendicular to the film plane and that charged dimers and quadrupolar nonlinearities are responsible for the nonlinearity along the direction of the charge gradient.[3]

To further investigate the charge-injection mechanism, we have analyzed the spatial distribution of polarization along the thickness of such in-plane poled films [see Fig. 1 (top)]. This is performed by pyroelectric depth profiling.[4,5] Light from a low-power diode laser is partially absorbed by a metal electrode at the top of the polymer film, leading to heating of the polymer. Modulation of the laser leads to a periodic temperature oscillation that is measured as a periodic pyroelectric current. Variation of the modulation frequency (from dc to 5 MHz) changes the penetration depth of the heating pulse, with lower frequencies leading to deeper penetration. The frequency dependence of the pyroelectric response can be related to the depth profile of the so-called polarization function $r(z)$ through a fitting procedure that has been described previously.[5]

We have studied a 4-μm-thick sample of polymethyl methacrylate (PMMA) containing 2% of the dye Disperse Red 1. By following in-plane poling, the pyroelectric response shown in Fig. 1 was obtained. The data can be fitted by using a profile $r(z) = r_d \exp[-(1 - \zeta z/d)]$, showing the nonuniform polarization created perpendicular to the applied field as a result of charge injection. Polarization is maximum at the polymer–glass interface ($z = d$). When the same sample was poled in a sandwich geometry, the pyroelectric response was consistent with a uniform polarization $r(z)$.

In conclusion, pyroelectric analysis has confirmed that strong, nonuniform polarization is created by charge injection during in-plane poling of polymer dye films.

*Heinrich Hertz Institute, Einsteinufer 37, D-10587 Berlin, Germany

1. S. Yitzchaik, G. Berkovic, and V. Krongauz, J. Appl. Phys. **70**, 3949 (1991).
2. M. Stähelin, C. A. Walsh, D. M. Burland, R. D. Miller, R. J. Twieg, and W. Volksen, J. Appl. Phys. **73**, 8471 (1994).
3. S. Yitzchaik, G. Berkovic, and V. Krongauz, Nonlinear Opt. **4**, 265 (1993).
4. S. Bauer-Gogonea, S. Bauer, W. Wirges, and R. Gerhard-Multhaupt, J. Appl. Phys. **76**, 2627 (1994).
5. S. Bauer, J. Appl. Phys. **75**, 5306 (1994).

QThA4 9:00 am

Bichromatic harmonic-optical-field effects in the photoinduced etching of borosilicate glasses

Jae H. Kyung, E. Sauvain,* N. M. Lawandy, *Department of Physics and Division of Engineering, Brown University, Providence, Rhode Island 02912*

Recently, we have observed that silicate-based glasses containing approximately 10% or more boron by weight exhibit selective etching when exposed to intense 532-nm laser pulses.[1] We believe that a two-photon process is the dominant channel leading to charge production, diffusion, and recombination in these glasses and that it results in an inhomogeneous charge distribution. The preferential etching is believed to occur at special reactive sites where trapped charges produce enhanced etching rates.[2]

In this work we present results obtained with simultaneous exposure of a sample of borosilicate glass to intense IR (1.064-μm) and green (532-nm) laser pulses. A mode-locked (76-MHz) and *Q*-switched (1-kHz) Nd:YAG laser was used to produce intensities of a few gigawatts per square centimeter in the green and several tens of gigawatts per square centimeter in the IR at the beam focus of approximately 70-μm diameter at the $1/e^2$ intensity. The sample was directly exposed for several minutes to solid-Gaussian IR and green beams that were perfectly overlapped and were colinearly polarized, and it was subsequently etched in a 12% HF solution. The resulting etch pattern was no longer a ring, as was previously observed when only 532-nm light was used to encode the glass, but was instead found to be either a double-lobe pattern or an elongated spot. Figure 1 shows an optical-microscope image of the ring pattern obtained with only 532-nm exposure, and Figs. 2(a) and 2(b) show images of the double-lobe and elongated etch patterns obtained after simultaneous exposure at 1.064 μm and 532 nm. We have also observed that the size of the etch pattern induced by green and IR encoding is measurably larger than that due to green-only encoding. Further studies with contact atomic force microscopy revealed that the depth of the etch spot is shallower when bichromatic exposure is used.

In addition, exposure to only 1.064-μm laser pulses at even higher intensities does not induce selective etching in these glasses.[1] Furthermore, when third-harmonic radiation at 355 nm was used in conjunction with the IR pulses, the etched spots no longer exhibited elongation. This suggests that the mechanism leading to the double-lobe or elongation pattern may be related to an interaction of the fundamental and its second-harmonic radiation. This may provide a further insight into the seeded second-harmonic generation observed in some glasses and into a possible way of mapping the charge distribution related to this effect.

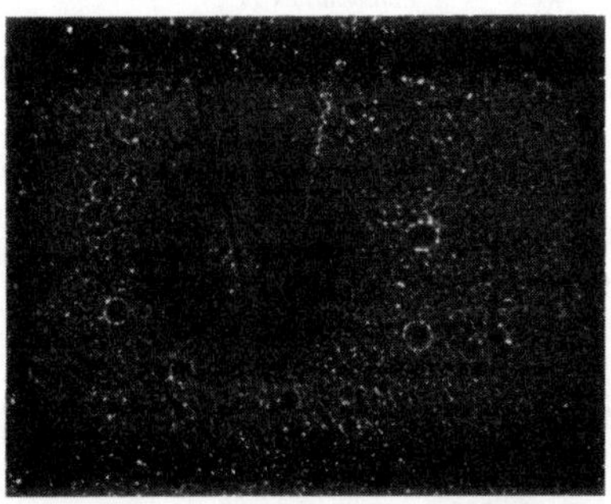

QThA4 Fig. 1. Optical-microscope image of a ring etch pattern in SK5 glass induced by exposure to a solid-Gaussian beam of 532-nm radiation for 10 min at an average power of 300 mW.

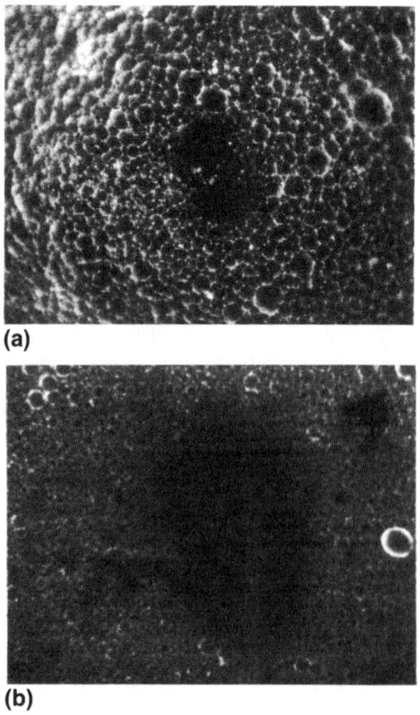

QThA4 Fig. 2. (a) Optical-microscope image of a double-lobe etch pattern after simultaneous exposure to 1.064-μm ($P_{ave} = 1.0$ W) and 532-nm ($P_{ave} = 100$ mW) radiation for 5 min. (b) Typical elongated etch pattern also observed after bichromatic exposure to 1.064-μm and 532-nm radiation. Note that (a) and (b) were taken at different magnifications.

*LEPES, National Center for Scientific Research, B.P. 166, F-38042 Grenoble, Cedex 9, France

1. E. Sauvain, J. H. Kyung, and N. M. Lawandy, "Multiphoton-induced micron scale etching in silicate-based glasses," Opt. Lett., in press.
2. E. Sauvain *et al.* in *Proc. 7th Annual Meeting of the Lasers and Electro-Optics Society* (Institute of Electrical and Electronics Engineers, New York, 1994), Vol. 2, postdeadline paper PD2.4.

QThA5 9:15 am

Novel optical comb generator

G. Macfarlane, A. Bell, E. Riis, A. I. Fergurson, *Photonics Group, Department of Physics and Applied Physics, University of Strathclyde, 107 Rottenrow, Glasgow G4 0NG, U.K.*

A new optical standard may come from advances in laser cooling and trapping of atoms. To evaluate whether these systems would provide a suitable standard, they must be linked with the microwave standards, most conveniently by means of the 3.39-μm methane-stabilized He–Ne laser. Various interesting transitions are only terahertz away from harmonics of 3.39 μm, and therefore a method for measuring these small differences to high accuracy is required. Such a device for this purpose is an optical comb generator. Such a system is devised of an electro-optic modulator in an optical cavity. The cavity is designed to be resonant not only with the carrier frequency but also with the modulated sidebands produced.[1,2]

A single-cavity system has the problem of low efficiency. This is due to the fact that the transfer of power from the carrier frequency to the sidebands, by the electro-optic modulator, looks like a loss to the cavity. Because a high-finesse cavity is required to produce many round trips and hence a wide frequency comb, little light can be coupled into the cavity. However, by reducing the finesse, coupling light into a single cavity can occur, but this leads to a narrower comb being produced.

The system reported here uses a coupled-cavity design (Fig. 1) to overcome this problem. The front cavity is used to increase the input coupling to the second cavity, which can, in principle, be 100% and was 60% for our system when the modulator was switched on. The electro-optic modulator used has a microwave resonant cavity with a lithium tantalate crystal that is driven at 2.7 GHz, and it has a modulation index of 2.4 for 1.4 W of rf power. This setup produces a comb of light that has a power of 12 mW for 350 mW of incident power. It can be seen from the output of an optical spectrum analyzer (Fig. 2) that there is macroscopic power in the 190th mode of 110 nW. The overall span, which is greater than 1 THz, is limited by the absorption in lithium tantalate at the laser wavelength of 780 nm. This absorption also limits the throughput of the cavities, but both of these problems could be reduced by changing to a higher wavelength.

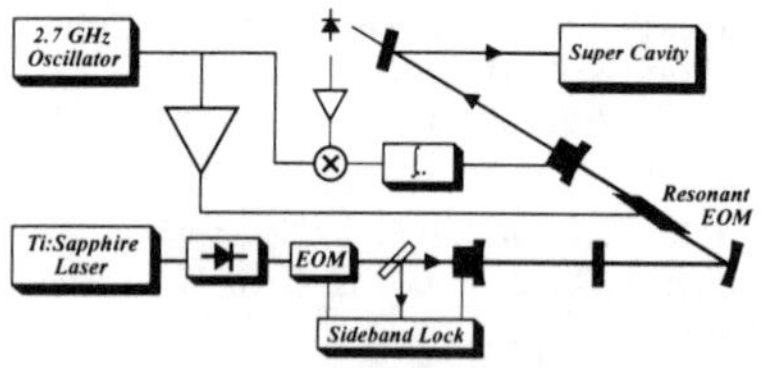

QThA5 Fig. 1. Schematic diagram. Light from the Ti:sapphire laser is isolated from the cavities by a Faraday rotator. The light is locked to the first cavity by the use of a sideband lock, with the modulation provided by the electro-optic modulator (EOM). The second cavity is locked by using a first-derivative technique with an avalanche photodiode to detect this signal.

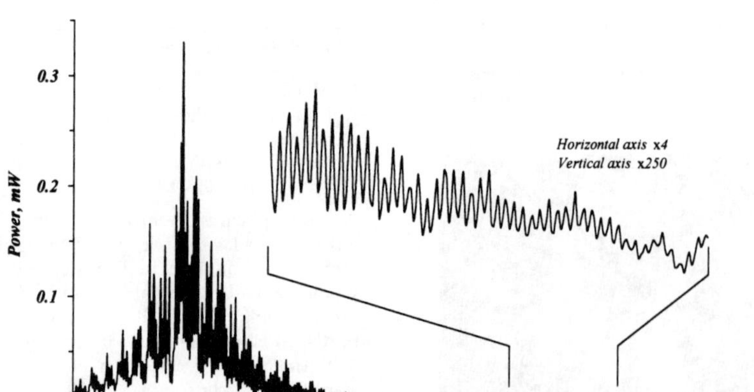

QThA5 Fig. 2. Output from the optical spectrum analyzer with a maximum mode number on one side of 190. This gives a frequency comb of 380 modes. The modulation in the spectrum is due to interference between the fiber end of the supercavity and the focusing lens.

1. M. Kourogi, K. Nakagawa, and M. Ohtsu, IEEE J. Quantum Electron. **29**, 205 (1993).
2. L. R. Brothers, D. Lee, and N. C. Wong, Opt. Lett. **19**, 245 (1994).

QThA6 **9:30 am**

Generation of 2.5-THz radiation by difference-frequency mixing in GaAs/GaAlAs multiple-quantum-well waveguides

Babak N. Saif, Bernard D. Seery,* Jacob Khurgin,** Colin Wood,† *Swales & Associates, 5050 Powder Mill Road, Beltsville, Maryland 20705*

NASA's Earth and space-science missions have motivated the investigation of methods to produce coherent radiation in the 1–3-THz spectral region, corresponding to 100–150-μm wavelengths, for use as local-oscillator sources in high-resolution-laser heterodyne spectrometers. Scientific applications of this technology include measurements of the hydroxyl radical (OH) at 2.5 THz in Earth's upper atmosphere, which is important because of its role in the destruction of ozone. In addition, astrophysicists are interested in the spectroscopy of the interstellar molecular cloud in this previously unexplored region of the electromagnetic spectrum. However, these measurements necessitate a space-based instrument because the lower atmosphere is largely opaque in this spectral region.

Conventional high-frequency-microwave local-oscillator methods and superconductor–insulator–superconductor (SIS) mixer technology are presently limited to frequencies less than 1 THz. Carbon-dioxide-dumped molecular lasers can produce adequate amounts of narrow-linewidth far-infrared radiation; however, these large, complex laser systems are ill suited to the harsh launch and space environments. The research effort described here is to design and develop a compact, robust semiconductor mixer device.

Recently, there have been two research groups in addition to ours investigating difference-frequency mixing in GaAs photovoltaic and photocurrent devices.[1,2] We have measured in excess of 12 μW of far-infrared radiation, an order of magnitude higher than reported previously, due largely to the higher intrinsic nonlinear conversion efficiency, better phase matching, and better energy-coupling efficiency in our waveguide device.

Our GaAs/GaAlAs multiple-quantum-well (MQW) mixer device was grown by molecular-beam epitaxy (MBE). The large dipole-moment characteristic of intersubband transitions was engineered by growing epitaxial layers that comprise the compositionally asymmetric multiple quantum wells. The quantum wells, shown schematically in Fig. 1, were optimized to have only two intrasubband transitions, separated by 150 meV, within the conduction band. The asymmetry resulted in an estimated second-order nonlinear susceptibility $\chi^{(2)}(\omega_1 - \omega_2, \omega_1, \omega_2)$ of 10 nV/m. To provide efficient interaction between the pumps and the nonlinear medium, field confinement and a long interaction length were deemed necessary. These requirements suggested a waveguide implementation, providing a means for efficient phase matching. Because of the long wavelength (118 μm) of the output radiation, phase matching required 25–50-μm thick epitaxial layers.

To our knowledge, our 35-μm device is the thickest free-standing MQW fabricated to date. A Boman Fourier-transform infrared (FTIR) spectrometer was used to verify the presence of the prescribed intrasubband energy levels.

Once the MBE growth process was complete, the sample was thinned by polishing to 35 μm and was cleaved to 500 μm (length) × 5000 μm (width). Two CO_2 lasers, separated in frequency by 2.5 THz, were used as input pumps to the MQW mixer. The respective wavelengths were 9.67 and 10.55 μm. An infrared two-element beam expander/collimator assembly expanded and collimated the beams before their incidence on the concave mirror used to focus the light into the waveguide (see Fig. 2). A quartz crystal with an attenuation of 10^6, placed in front of the pyroelectric detector, was used to discriminate between the desired far-infrared radiation and unwanted infrared pump radiation. The experimental results are shown in Fig. 3, which is a plot of the far-infrared output radiation versus the input infrared laser power. A quarter-wave plate was used to verify that the effect observed was in fact the far-infrared radiation signature and not spurious thermal radiation resulting from the lack of a thermal heat sink. Both lasers must be TM polarized to excite intersubband transitions, according to theory (see Fig. 3). When the lasers are orthogonally polarized, the output power drops off dramatically, as can be seen in the lower data set in the figure. This, and the fact the output power is proportional to the product of the input laser powers, unmistakably demon-

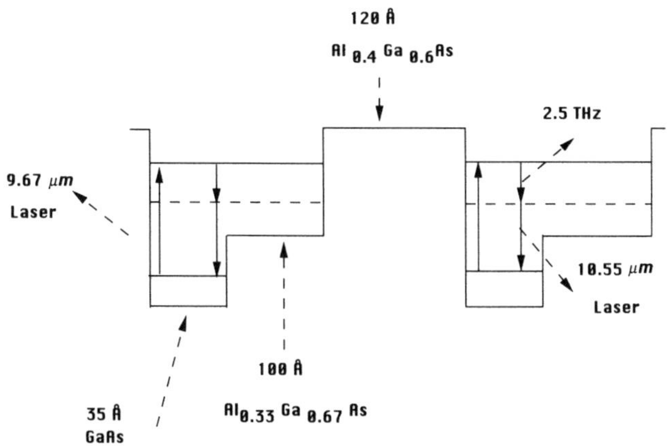

QThA6 Fig. 1. Design of the GaAs/GaAlAs asymmetric multiple quantum wells. The GaAs layer is 35 Å wide, the $Al_{0.4}Ga_{0.6}As$ barrier layer is 120 Å wide, and the $Al_{0.33}Ga_{0.67}As$ well is 100 Å wide. The arrows inside the wells depict the absorption and emission of radiation.

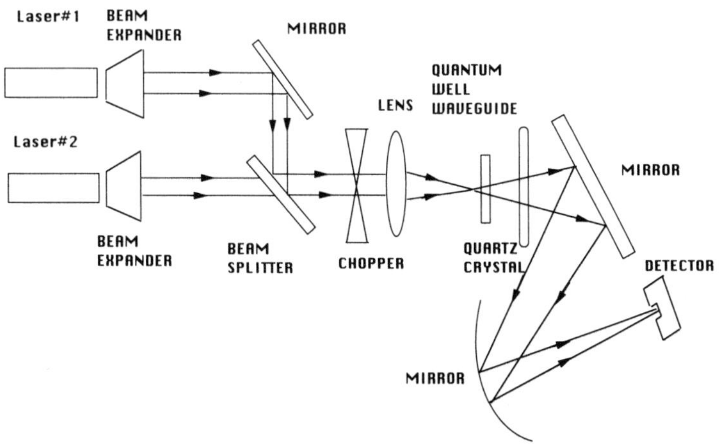

QThA6 Fig. 2. Difference-frequency-experiment hardware configuration. The beams from two cw CO_2 lasers and infrared beam expander/collimator assemblies are made colinear and are chopped for later synchronous detection with the pyroelectric detector. A quartz plate and polyethylene window, with a combined attenuation of 10^6, are used to separate out the far-infrared radiation from the infrared pump light. The *f*-number of the detector fore-optics is chosen to tailor the spot size to match the detector. Spherical optics were used for expediency; however, cylindrical optics are preferred.

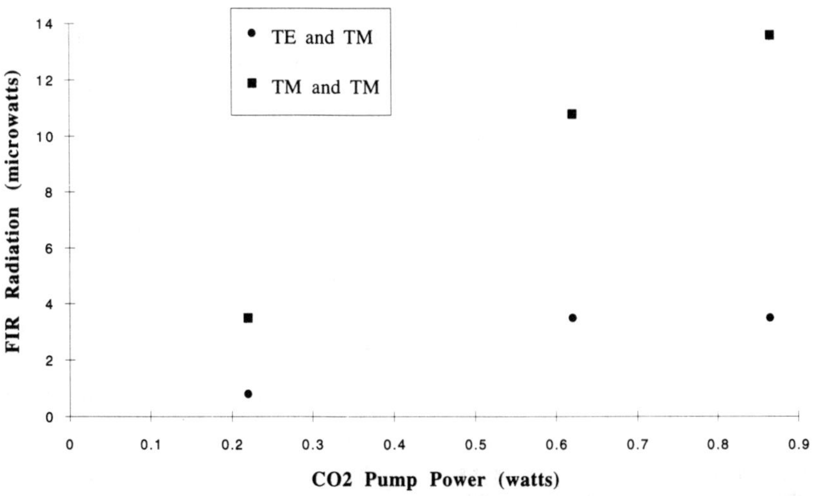

QThA6 Fig. 3. Far-infrared output radiation versus input power of the 9.67-μm laser pump. Three data points are displayed for the cases for which the pump laser polarizations are both TM polarized (squares) and orthogonal (circles). Theory suggests that nonlinear mixing using intrasubband states should disappear for orthogonally polarized sources, and indeed it does, leaving only the thermally generated component.

strates the nonlinear-mixing origin of the far-infrared output radiation.

In conclusion, we have demonstrated a novel two-level, compositionally asymmetric, nonresonant MQW waveguide structure for the generation of far-infrared radiation. In excess of 12 μW of power has been measured at the output wavelength of 118 μm. At these power levels, practical device applications, such as local-oscillator sources, can be considered.

This work was supported by B. D. Seery of the Goddard Space Flight Center under a grant from the Director's Discretionary Fund, 1993–1994.

*Space Technology Division, NASA Gooddard Space Flight Center, Greenbelt, Maryland 20771

**Department of Electrical and Computer Engineering, Johns Hopkins University, 105 Barton Hall, 3400 North Charles Street, Baltimore, Maryland 21218

†Laboratory for Physical Sciences, 850 Greenneed Drive, College Park, Maryland 20740

1. C. Sirtori, F. Capasso, J. Faist, L. N. Pfeiffer, and K. W. West, in *Conference on Lasers and Electro-Optics*, 1994 Technical Digest Series, Vol. 8 (Optical Society of America, Washington, D.C., 1994), paper QMB6.
2. E. R. Brown, K. A. McIntosh, F. W. Smith, M. J. Manfra, and C. L. Dennis, Appl. Phys. Lett. **62**, 1206 (1993).
3. J. J. Khurgin, Opt. Soc. Am. B. **6**, 1673 (1989).

QThA7 **9:45 am**

Polarization flipping in a cw He–Ne Raman laser

X.-W. Xia, W. J. Sandle, R. J. Ballagh, D. M. Warrington, *Department of Physics, University of Otago, P.O. Box 56, Dunedin, New Zealand*

We report a novel Raman polarization-flipping phenomenon that occurs within the 1*s*–2*p* manifold of Ne I when Raman and pump are both resonant in a single cavity, a situation that produces a wide range of stimulated electronic Raman Stokes and anti-Stokes lines exhibiting interesting competition effects.[1] The pump laser, mode matched to the cavity of a commercial He–Ne laser (the standard 633-nm output is not scientifically important here and is used only to stabilize the cavity length) interacts passively with the 588.2-nm ($1s_5 \rightarrow 2p_2$) Ne transition. The corresponding Raman line, 603 nm ($2p_2 \rightarrow 1s_4$), actively lases in the He–Ne cavity. The Raman polarization is always linear and is aligned with one of the cavity principal axes, which are determined by residual mirror birefringence. However, the direction of the polarization can suddenly flip between these orthogonal states as the dye laser is scanned. The effect is reminiscent of the polarization behavior of longitudinal modes of the He–Ne laser itself as the length of the laser cavity is changed and the laser output frequency scans through the Doppler gain profile. Here, we stress, the cavity length remains fixed and the dye laser is scanned through a relatively small frequency range–that comprising a *single* cavity resonance.

We believe that the mechanism responsible is similar to the one operating in previous Raman studies in alkali vapors.[2,3] The small-signal Raman gain for general elliptical Stokes polarization, generated by a linearly polarized pump, has been shown to be maximized when the Stokes wave is also linearly polarized.[4] The stimulated Stokes radiation, generated in a medium without a cavity, has the polarization with greatest Raman gain, so that the angle between the Raman- and pump-polarization directions is fixed by the relative magnitudes of the (symmetric and antisymmetric) gains. When the antisymmetric term is greater, the maximum gain occurs for Raman polarization perpendicular to the pump. This is the normal situation in our experiment.

However, to explain the polarization flipping that occurs in our doubly-resonant, single-cavity experiment, we must include the effect of the Raman frequency shifts and the Raman gains. Because the effective cavity detunings for the symmetric and antisymmetric emissions will differ, the balance between polarizations may shift as the pump is tuned through a cavity profile. The small-signal gains for the symmetric and antisymmetric contributions must, of course, be very nearly equal for this to occur. This restricts polarization flipping to a small range of laser–atom detunings, as is experimentally confirmed.

1. X.-W. Xia, W. J. Sandle, R. J. Ballagh, and D. M. Warrington, Opt. Commun. **96**, 99 (1993).
2. S. Barak and S. Yatsiv, Phys. Rev. A **3**, 382 (1971).
3. V. M. Arutunian, T. A. Papazian, A. V. Karmenian, and S. M. Sarkissian, Phys. Lett. **47A**, 324 (1974).
4. D. Cotter and D. C. Hanna, J. Phys. B **9**, 2165 (1976).

QThB — 8:00 am — Hyatt B

Novel Spectroscopy

Hiroshi Takuma, *University of Electro-Communication, Presider*

QThB1 (Invited) — 8:00 am

Ca^- formation in collisions between Rydberg and ground state Ca atoms

D. W. Duquette, *Behlen Laboratory of Physics, University of Nebraska, Lincoln, Nebraska 68588-0111*

Selective laser excitation of atomic Rydberg levels can provide a unique tool for collisional spectroscopy of weakly bound systems. We have observed the relative rate as a function of n for formation of positive and negative Ca ions in the collision:

$$Ca^*(4snd\,^1D) + Ca(4s^2\,^1S) \rightarrow Ca^-(4s^24p\,^2P) + Ca^+(4s^2S). \quad (1)$$

Preliminary calculations by Fabrikant[1] show a cross section "on resonance" of 10^{-10} cm^2 at thermal kinetic energies. The cross section decreases rapidly with energy "mismatch" between the Rydberg and negative ion, being down by an order of magnitude only a few meV off resonance.

We have observed this resonant charge transfer from laser-prepared Ca Rydberg atoms. The striking feature of our results for both positive and negative time-of-flight selected ions is that the normalized counts are enhanced by more than an order of magnitude above nearby levels for only n = 25. Negative ion production in Rydberg atom collisions has been observed previously in a Na cell,[2] but without a dramatic dependence on Rydberg energy. The Na^- binding energy of approximately 0.5 eV is too large for there to be any significant probability of resonance with a Na Rydberg level. Ca^- has a very small binding energy, likely $<$0.030 meV,[3,4,5] where the spectral density of Ca Rydberg levels is high.

A difference in the magnitudes of the positive and negative ion signals is puzzling, though we believe it is explained by differences in detection efficiency as well as photodetachment of Ca^- prior to detection. Ancillary measurements have also ruled out all likely processes other than Eq. (1).

Further work, both experimental and theoretical, needs to be performed. Other systems with small energy (other negative ions, low-lying atomic metastable levels, vibrational levels of a molecular ground electronic state) can be used with different atoms in Rydberg levels. A critical test of the current observations using a different Rydberg atom in collision with Ca is in development.

This work is supported by the National Science Foundation, Grant PHY-9109164.

1. I. I. Fabrikant, Phys. Rev **A48**, 5, R3411 (1993).
2. M. Ciocca, M. Allegrini, E. Arimondo, C. E. Burkhardt, W. P. Garver, and J. J. Leventhal, Phys. Rev. Lett. **56**, 7, 704 (1986).
3. C. W. Walter and J. R. Peterson, Phys. Rev. Lett. **68**, 15 (1992).
4. M-J. Nadeau, X-L. Zhao, M. A. Garwan, and A. E. Litherland, Phys. Rev. **A46**, 7 (1992).
5. H. van der Hart, C. Laughlin, and J. E. Hansen, Phys. Rev. Lett. **71**, 1506 (1993).

QThB2 (Invited) — 8:30 am

Observation of laser-induced resonant de-excitation in $\bar{p}$-He^{2+}-e^- atomcules

Norio Morita, *Institute for Molecular Science, Myodaiji, Okazaki 444, Japan*

A recent discovery of anomalous survival of antiprotons in dense helium media suggested the creation of $\bar{p}$-He^{2+}-e^- compounds. If this unusual compound is really created in the helium media, it is the first long-lived exotic atom containing hadrons other than normal protons and neutrons. Moreover, from the viewpoint of atomic and molecular spectroscopy, this new compound is a quite attractive object of study, because it has both characters of atoms and molecules; in other words, it is neither an atom nor a molecule, but a completely new entity (for this reason, we propose to call it an "atomcule"). Therefore, we started the laser spectroscopic investigation of the compound in order to obtain definite evidence of its creation.

The helium medium used was low temperature gas (5–10 K) at an atmospheric pressure, and was irradiated by a low-energy antiproton beam from LEAR (Low Energy Antiproton Ring) of CERN. As a result, we successfully observed laser-induced resonant transitions in the $\bar{p}$-He^{2+}-e^- compound, using pulsed dye lasers pumped by excimer lasers. The observed transitions occur from metastable states having large principal and angular-momentum quantum numbers (denoted hereafter by n and l, respectively) of the antiproton, to lower short-lived states, which proceed to ionization through fast Auger transition. After the ionization, the $\bar{p}$-He^{2+} ion immediately annihilates to generate high-energy particles, such as pions, through reactions between $\bar{p}$ and He^{2+}, because the $\bar{p}$-He^{2+} ion is hydrogenic and small angular-momentum orbitals are easily mixed into $\bar{p}$'s orbital through perturbations from surrounding helium atoms. Detecting those high-energy particles, we observed resonance profiles of the laser de-excitation. We found out two transitions with vacuum wavelengths 597.259 and 470.725 nm, which are assigned to the (n = 39, l = 35) → (38,34) and (37,34) → (36,33) transitions, respectively, in the ground electronic state. These wavelengths are in good agreement with theoretical values predicted by several theoreticians with different calculation methods. This fact is direct proof of the creation of this new exotic atom.

We also attempted to obtain the lifetime of the relevant levels. Varying the irradiation timing of a laser pulse, which is tuned on resonance, we obtained the time dependence of the population in the initial metastable state. On the other hand, using two laser pulses with a variable time separation, we also observed the population feeding to the initial state from higher atomic levels. Making appropriate rate equations describing the population dynamics of the relevant levels and fitting these equations to the experimental results, we determined the decay rates of those levels. The decay rates thus obtained are in good agreement with the theoretical values that were calculated, taking into consideration the correlated motion of the electron and antiproton in the atomcule. Those values are as much as three times larger than those calculated ignoring the correlation. This fact not only further confirms the creation of the atomcule, but also proves the significance of the correlation effect.

This study was carried out in collaboration with nuclear physics groups of University of Tokyo, Technische Universität München, KFKI Research Institute for Particles and Nuclear Physics (Budapest) and CERN.

QThB3 **9:00 am**

FM spectroscopy with transform-limited nanosecond laser pulses

E. E. Eyler, S. Gangopadhyay,
N. Melikechi, J. C. Bloch,* R. W. Field,*
Department of Physics and Astronomy, University of Delaware, Newark, Delaware 19716

Frequency-modulation spectroscopy is among the best techniques for sensitive absorption measurements with cw lasers, but its extension to pulsed lasers has been problematic. Some years ago, Gallagher and coworkers used very large FM sideband offsets, typically 10–20 GHz, to obtain FM spectra by using a multimode dye laser.[1] We report an improved approach, using pulsed amplification to produce transform-limited, temporally smooth pulses that are especially well suited for FM spectroscopy and other techniques requiring a well-defined optical phase. This approach offers much higher spectral resolution, a better-defined optical phase, and a much smaller modulation frequency that allows more sensitive and flexible detection. We are also exploring new methods for digital signal processing that will extend the capabilities of conventional arrangements using balanced mixers.

In the new pulsed-FM method, an electro-optic modulator operating near 600 MHz is used to impose balanced blue and red sidebands on the output of a single-mode cw dye laser. Pulsed amplification in a chain of capillary-type dye cells preserves the sidebands almost intact, producing smooth pulses with energies in the millijoule range, with a Fourier power spectrum that overlaps the sidebands only at about the 1% level. These pulses can be used directly for FM spectroscopy or can be converted to the ultraviolet region by using frequency doubling or nonlinear mixing.

We initially investigated the absorption of I_2 in the visible red region. A sensitivity approaching 1 part in 10^4 was obtained readily by using a double-balanced mixer to detect differential absorption of the blue and red FM sidebands. Considerable further improvement is anticipated as we incorporate better signal-processing methods, including pulse-by-pulse normalization and digital signal processing to provide optimal phase-sensitive detection. The principal noise source is the fluctuating high-frequency tail of the power spectrum of the laser pulse envelope. Although weak, this tail extends out to 1 GHz and beyond. We have discovered that it arises largely from a fast transient structure introduced by the pulsed amplifier, arising from interstage feedback and other stray processes. Work to understand and reduce these effects is ongoing, and we are optimistic that sensitivities at the parts-per-million level will be achieved.

By generating the third harmonic of a pulse-amplified laser, we have also obtained FM spectra in the deep UV at about 215 nm, where conventional FM spectroscopy with cw sources is impossible. Figure 1 shows the results of our first experiment, in which a fast UV phototube was used to measure the FM absorption of a cell containing a few torr of NO gas. Extension further into the UV is possible by using four-wave mixing in gases, and there are also good prospects for obtaining sensitivity sufficient to measure two-photon absorption. The pulsed-FM technique offers a combination of high resolution, high absorption sensitivity, and broad UV tunability that should find widespread application.

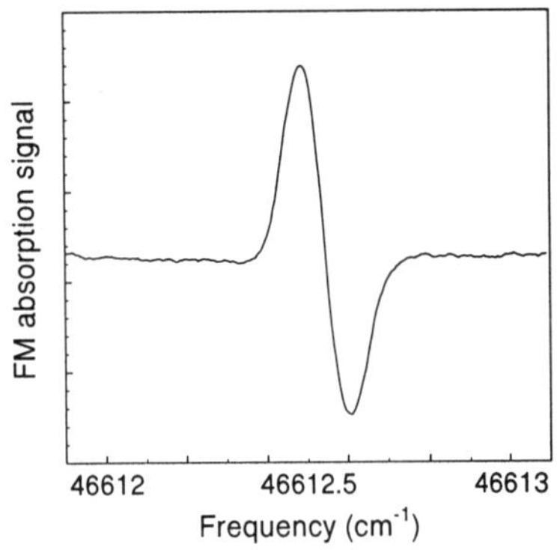

QThB3 Fig. 1. Pulsed-FM absorption spectrum of the $S_{21}(7)$ transition in the (1–0) γ band of NO at 214.5 nm, obtained by using a pulse-amplified, frequency-tripled dye laser. Resolution is determined by the Doppler width of the room-temperature NO-gas cell; laser resolution is about 100 MHz.

**Department of Chemistry, Massachusetts Institute of Technology, Cambridge, Massachusetts 02139*

1. N. H. Tran, R. Kachru, P. Pillet, H. B. van Linden van den Heuvell, T. F. Gallagher, and J. P. Watjen, Appl. Opt. **23**, 1353 (1984), and references therein.

QThB4 **9:15 am**

Intracavity phase spectroscopy of samarium

Scott Diddams, Briggs Atherton,
Jean-Claude Diels, *Department of Physics and Astronomy, University of New Mexico, Albuquerque, New Mexico 87131*

The fact that the two counterpropagating pulses in a mode-locked dye ring laser can be completely decoupled in frequency makes such a laser a sensitive probe of phase differences between the counterpropagating pulses. The difference in center frequency between the two counterpropagating pulses can be resolved down to the noise limit of 100 Hz. This limit corresponds to a phase difference between the counterpropagating pulses of 10^{-6}, or an effective cavity length difference of 10^{-12} m.[1,2]

A passively-mode-locked ring dye laser is modified as shown in Fig. 1 to create a linear tail into which a heated cell (5 cm long) containing atomic samarium is inserted. The presence of the wave plates and the calcite polarizing beam splitter ensure that the counterpropagating pulses from the ring are orthogonally polarized (circularly right and left) when they travel through the cell. With the entire interaction region surrounded by three orthogonal sets of Helmholtz coils, stray magnetic fields are eliminated and a uniform field can be applied along the direction of laser propagation. The laser is tuned to resonance with the $J = 1 \rightarrow J' = 0$ transition of samarium at 654 nm, and the longitudinal magnetic field is scanned linearly about zero to Zeeman-split the degenerate ground state, creating a three-level Λ structure. The left- and right-circularly-polarized pulses each interact with only one branch of the Λ structure. These interactions transfer population between the $m_j = 1$ and $m_j = -1$ ground states via the excited state and create a coherence between the two ground states. The result of this coherence, an enhancement of the normal Faraday rotation, is observed in the measurement of the beat frequency between the counterpropagating pulses.

An example of the experimental results

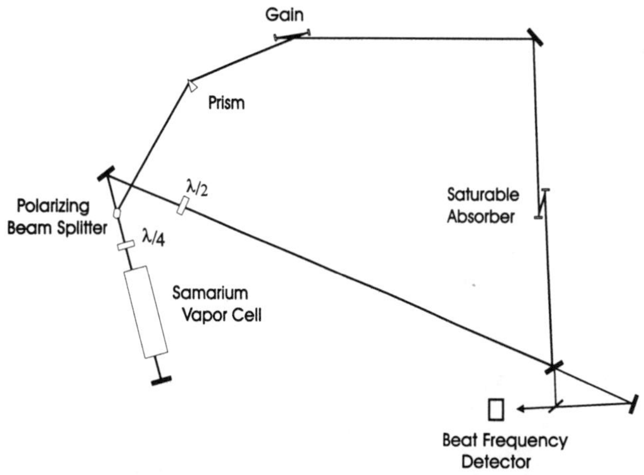

QThB4 Fig. 1. Modified passively-mode-locked ring dye laser. The two counterpropagating pulses are made to travel through a linear tail, where they interact with samarium vapor. Phase differences introduced between the counterpropagating pulses are manifest in a beat frequency between the pulses.

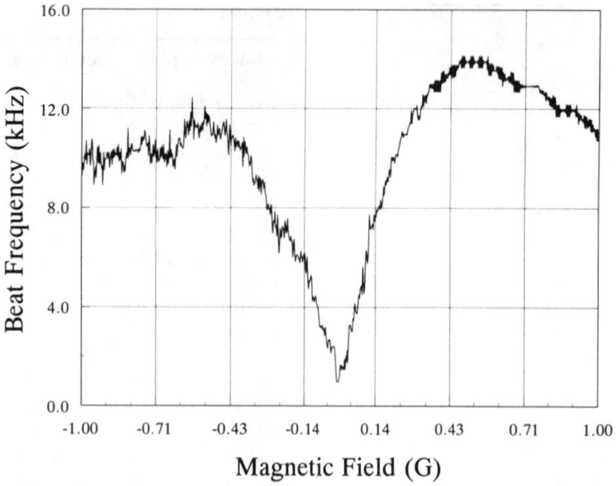

QThB4 Fig. 2. Change in beat frequency as a function of the magnetic field applied to the samarium vapor. The beat frequency values were obtained from the Fourier transform of the digitally sampled waveform.

is shown in Fig. 2. The beat frequency is observed to drop to its bias value at zero field when both pulses experience no rotation. The half width of the V-shaped curve is 0.45 G, which corresponds to a lower-level splitting of approximately 1.9 MHz. We note that this width is much less than the bandwidth of the pulse (200 GHz) as well as the Doppler-broadened linewidth of the transition (800 MHz). Such results are characteristic of nonlinear Faraday rotation and are in qualitative agreement with those of Drake *et al.*[3] In addition, accurate control of the difference frequency between the counterpropagating pulses (with an intracavity Pockels cell[2]) offers the possibility for observing new types of resonances. Any hyperfine atomic splitting between 100 Hz and 1 MHz that can be made to correspond to the difference frequency yields a resonance signal in the beat frequency.

1. M. L. Dennis, J.-C. Diels, and M. Lai, Opt. Lett. **16**, 529 (1991).
2. M. Lai, J.-C. Diels, and M. L. Dennis, Opt. Lett. **17**, 1535 (1992).
3. K. H. Drake, W. Lange, and J. Mlynek, Opt. Commun. **66**, 315 (1988).

Thursday, May 25

QThB5 **9:30 am**

Laser-frequency-noise measurement by absorption on an atomic resonance

A. P. Willis, A. I. Ferguson,* D. M. Kane, *School of Mathematics, Physics, Computing, and Electronics, Macquarie University, Sydney, NSW 2109, Australia*

The frequency noise of an external-cavity diode laser has been measured by converting the frequency fluctuations to amplitude fluctuations through absorption on an atomic transition. The frequency noise has also been measured by the standard technique of conversion to amplitude fluctuation by means of transmission by a Fabry–Perot interferometer, for comparison. The two techniques are largely equivalent. This work makes an important link between noise spectroscopy[1,2] and the characterization of the frequency noise of laser diodes. In particular, it has been noted that using an external-cavity laser diode (ECLD) reduces the frequency noise and so gives very small signals in noise spectroscopy.[2] However, calculations of the frequency noise of an ECLD show an enhancement of the frequency noise at multiples of the longitudinal-mode-spacing frequency ν_{LMS} and a suppression at intermediate frequencies.[3] We have confirmed this experimentally by using an ECLD tuned to the Cs D_2 atomic resonance. Thus, an ECLD can be used for noise spectroscopy, provided that the detection frequency is tuned to a multiple of ν_{LMS}.

The ECLD consisted of a multiple-quantum-well laser diode (STC LT50-03U), a beam splitter to couple light out of the cavity, and a diffraction grating (1200 lines/mm, blazed for 750 nm) operated in the Littrow configuration. The length of the cavity was varied from about 10 to 50 cm, corresponding to a longitudinal-mode spacing of 1500 to 300 MHz. One output beam from the laser passed through a Cs cell before being focused onto a fast photodiode (BPW 28) connected to an rf spectrum analyzer (Tektronix 2754P). A reference beam propagated in free space and was focused onto a photodiode. The laser was operated without active frequency stabilization but gave stable single-frequency operation, as verified by the low level of

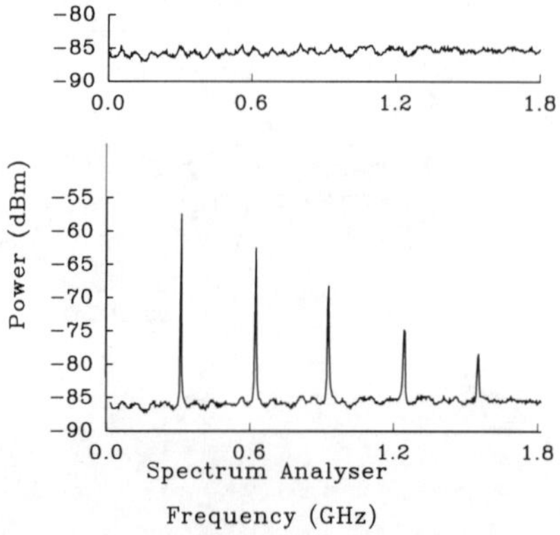

QThB5 Fig. 1. Rf beat spectrum for ECLD (longitudinal-mode spacing, 308 MHz): (a) laser off Cs resonance; (b) laser on Cs resonance.

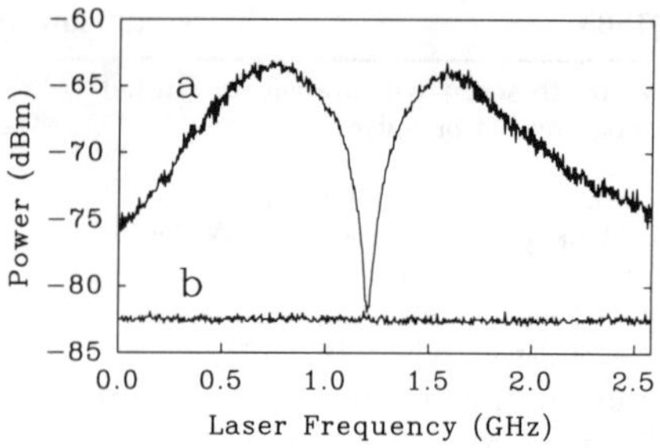

QThB5 Fig. 2. Rf beat power as a function of laser detuning (longitudinal-mode spacing 1.1138 GHz): (a) beam passing through Cs cell; (b) beam bypassing Cs cell.

intensity noise shown in Fig. 1(a). When the laser frequency was tuned near to, but not centered on, the Cs resonance [Fig. 1(b)], very large rf beats were observed at multiples of ν_{LMS}. Up to 30-dBm enhancement in the intensity noise was observed because of frequency-noise conversion.

When the rf detection frequency was fixed at ν_{LMS} and the laser frequency was scanned across the atomic resonance [Fig. 2(a)], the beat intensity showed the double-peaked structure expected for diffusion of the laser frequency.[2] The reference beam [Fig. 2(b)] showed no enhancement of the noise. Further experiments using intracavity electro-optic modulation showed that controlled frequency modulation yields very similar signals. We present further results on how the frequency-noise characteristics of an ECLD are a more sensitive indicator of good laser alignment than are the amplitude-noise characteristics.

Permanent address: Department of Physics and Applied Physics, University of Strathclyde, Glasgow G4 ONG, U.K.

1. T. Yabuzaki, T. Mitsui, and U. Tanaka, Phys. Rev. Lett. **67**, 2453 (1991).
2. D. H. McIntyre, C. E. Fairchild, J. Cooper, and R. Walser, Opt. Lett. **18**, 1816 (1993).
3. T. Okoshi and K. Kikuchi, *Coherent Optical Fiber Communication* (Reidel, Dordrecht, The Netherlands, (1988), p. 114.

QThB6 **9:45 am**

Ultrastable, compact diode-pumped Nd:YAG laser for precision measurements

S. N. Bagayev, A. M. Belkin, A. S. Dychkov, V. A. Zhmud, A. A. Kaminskii, Yu. A. Matyugin, A. Yu. Nevskii, M. N. Skvortsov, S. A. Farnosov, *Institute of Laser Physics, Russian Academy of Sciences, Siberian Division, Prospect Lavrentyev, 13/3, Novosibirsk 630090, Russia*

We report the creation and investigation of a mobile, highly stable laser system based on diode pumping an Nd:YAG-laser. The active element of the laser was made of a crystal with an increased concentration of neodymium (1.3%). The intracavity frequency was doubled by using KTP. Short-term stability was provided by locking the laser frequency to the external reference interferometer through the use of an FM sideband technique. Phase modulation was performed by an electro-optical modulator. The reference interferometer was placed in an evacuated housing and was thermostabilized. Long-term stability was provided by locking the second harmonics of the laser frequency (λ = 532 nm) to the resonance of saturated absorption in I_2. The system of stabilization by I_2 resonances is analogous to the system we used for stabilizing an Ar_2 laser. It includes a luminescent I_2 cell that is 10 cm in length (the pressure of the iodine vapor is $\sim 10^{-2}$ Torr) and stabilization of radiation power incident on the cell. Power is stabilized by the use of an acousto-optical modulator, which also provides optical isolation. In this work we minimized the dimensions of the I_2 cell and the reference interferometer. The size of the optical system, including the laser, the cell, and the interferometer, is not more than 0.5 m^2. There is a real possibility of reducing the dimensions of the setup. The short- and long-term frequency stability of the Nd:YAG laser is 10^{-13}. The possibility of increasing stability by an order of magnitude has been considered. The results of investigations on the stabilization of tunable solid-state lasers at any frequency in the tuning region are presented.

Application of ultrastable crystal lasers to precision spectroscopic measurements in the UV and visible regions and to telecommunication systems is considered.

QThC — 8:00 am, Hyatt C/D

Emission Spectroscopy

Duncan G. Steel, *University of Michigan, Presider*

QThC1 (Invited) — 8:00 am

Coherent generation of hot electrons in band-to-acceptor luminescence

Tilmann Kuhn, Stefan Haas,* Fausto Rossi,* Alfred Leitenstorfer,** Andrea Lohner,** Thomas Elsaesser,** W. Klein,† G. Boehm,† G. Traenkle,† G. Weimann,† *Institut für Theoretische Physik, Universität Stuttgart, Pfaffenwaldring 57, 70550 Stuttgart, Germany*

On a semiclassical level, the dynamics of carriers generated by a short laser pulse in a semiconductor is described by the Boltzmann equation (BE). The carrier generation rate derived from Fermi's golden rule is proportional to the laser intensity with a spectral shape given by the laser spectrum. This approach, however, neglects all effects related to the coherence of the laser, as have been observed, e.g., in four-wave-mixing experiments. To describe coherence, the dynamics of the interband polarization has to be taken into account. This is accomplished by the semiconductor Bloch equations (SBE).[1] In this contribution we present a theoretical and experimental analysis of the role of coherence for the generation of nonequilibrium electrons in GaAs.[2] Time-integrated band-to-acceptor (BA) luminescence spectra calculated from an approach including the coherent interband polarization account for experiments which demonstrate initial carrier distributions much broader than the spectrum of the femtosecond excitation pulse.

In the experiment a 3 μm thick GaAs layer doped with Be acceptors of a concentration of 3×10^{16} cm^{-3} is excited at a photon energy of 1.73 eV by transform limited 150 fs pulses from a mode-locked Ti:sapphire laser. Time-integrated spectra are recorded with a double monochromator and a single-photon-counting multiplier.

In Fig. 1, calculated BA luminescence spectra are plotted for three different densities, Fig. 2 shows the experimental results for the same densities. The left column in Fig. 1 refers to the BE including a semiclassical generation rate with a spectral width identical to the laser pulse resulting in a first luminescence peak (marked in the figure) practically independent of the generated carrier density. In the model based on the SBE, the decay of the interband polarization due to phase-breaking scattering processes gives rise to an additional broadening of the generation rate. This effect is evident from the marked regions in the spectra in the right column and the linewidths plotted in Fig. 3. For densities higher than 10^{15} cm^{-3}, this mechanism makes the main contribution to the overall linewidth of the unrelaxed peak and of the phonon replicas. Furthermore, the simulation shows that up to densities of 2×10^{16} cm^{-3} the contribution of band-to-band luminescence to the overall emission is less than 10 percent.

In Fig. 3, the spectral widths of the different luminescence peaks is plotted versus density as derived from the data (squares). The density dependence of the spectral width is well explained by the coherent model (triangles). For a fixed density, the unrelaxed peak and the first and second phonon replica show identical spectral widths. This behavior, which is fully reproduced by the calculation, demonstrates that the main broadening is introduced in the generation process whereas subsequent carrier-carrier scattering plays only a minor role.

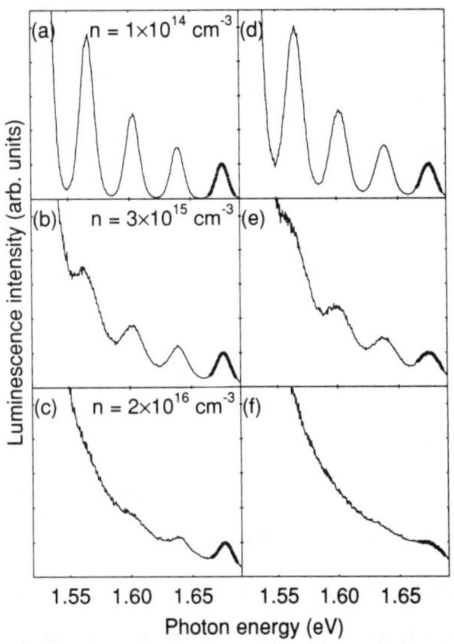

QThC1 Fig. 1. Theoretically calculated time-integrated BA luminescence spectra for three different densities, (a), (b), (c) from the BE with a semiclassical generation rate and (d), (e), (f) from the SBE.

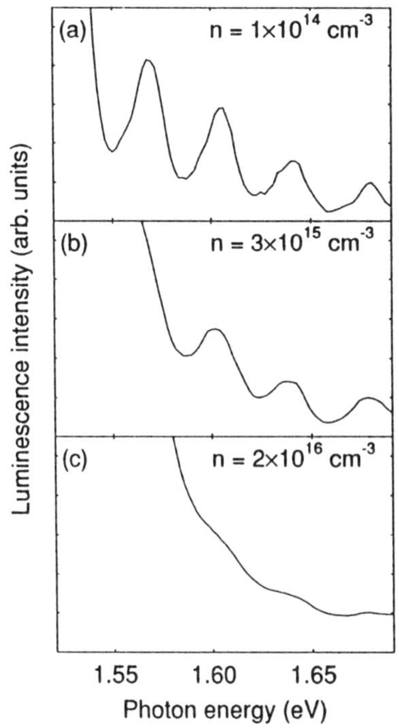

QThC1 Fig. 2. Measured time-integrated BA luminescence spectra for the same densities as in Fig. 1.

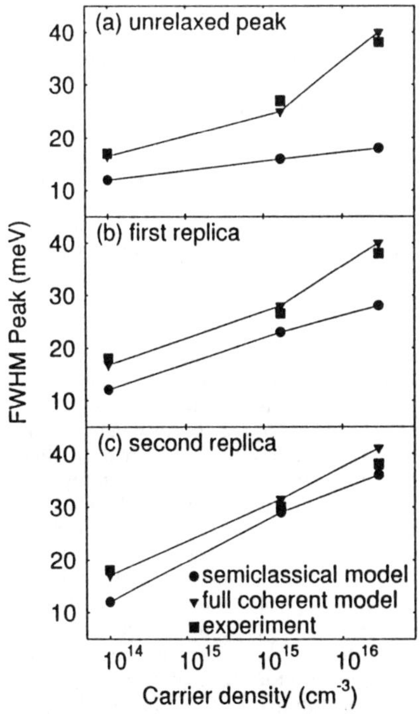

QThC1 Fig. 3. Linewidth of the unrelaxed peak and of the first and second phonon replica as a function of carrier density obtained from the experimental spectra and the theoretical calculations within the two models. The lines connecting the theoretical points are meant as a guide for the eye.

*Fachbereich Physik, Philipps-Universität Marburg, 35032 Marburg, Germany
**Physik Department E 11, Technische Universität München, 85748 Garching, Germany
†Walter-Schottky-Institut, Technische Universität München, 85748 Garching, Germany

1. F. Rossi, S. Haas, and T. Kuhn, Phys. Rev. Lett. **72**, 152 (1994).
2. A. Leitenstorfer, A. Lohner, T. Elsaesser, S. Haas, F. Rossi, T. Kuhn, W. Klein, G. Boehm, G. Traenkle, and G. Weimann, Phys. Rev. Lett. **73**, 1687 (1994).

QThC2 **8:30 am**

Biexcitons in GaAs/AlGaAs quantum wells: long radiative lifetimes

D. S. Citrin, *Center for Ultrafast Optical Science, University of Michigan, 2200 Bonisteel Boulevard, Ann Arbor, Michigan 48109-2099*

In bulk crystals the correlation of the biexciton internal motion leads to rapid luminescence-decay times.[1] Time-resolved photoluminescence (TRPL) on GaAs/AlGaAs quantum wells (QWs) reveals a decay transient associated with biexcitons.[2] Is this transient due to the intrinsic radiative decay of a ground-state biexciton (GSB) or to competing processes, such as dissociation?

Our theoretical results indicate that the former is unlikely when the interexciton separation Λ_M within the biexciton is small compared to the wavelength λ of the emitted light—the situation for the GSB in GaAs/AlGaAs QWs. One finds transition rates for biexciton decay into an exciton and a photon less than the free one-exciton radiative decay rate, in contrast to what is expected from the giant-oscillator-strength concept.[1] In bulk materials considerable additional phase space for the decay process is accessed because of the ability of the exciton remaining after one photon is emitted to absorb the recoil momentum.[3] The naive application of this notion, however, relies on a mistaken comparison of what is measured in TRPL on bulk and on high-quality low-dimensional systems at low temperature. The difference is due to the stationarity of exciton polaritons in bulk and to their intrinsic decay in QWs.

The biexciton wave function $\Psi_{\mathbf{K}_\parallel}$ for the state with an in-plane center-of-mass wave vector $\mathbf{K}_\parallel$ is a superposition of free-exciton pair states. If $\psi_{\mathbf{k}_\parallel}$ is a free one-exciton with wave vector $\mathbf{k}_\parallel$, the wave function of a biexciton with $\mathbf{K}_\parallel = 2\mathbf{k}_\parallel$ is $\Psi_{\mathbf{K}_\parallel} = \sum_{\mathbf{k}'_\parallel} c_{\mathbf{k}'_\parallel}\psi_{\mathbf{k}_\parallel+\mathbf{k}'_\parallel}\psi_{\mathbf{k}_\parallel-\mathbf{k}'_\parallel}$, where $c_{\mathbf{k}'_\parallel}$ are coefficients. Consider the decay process

$$\Psi_{\mathbf{K}_\parallel} \rightarrow \psi_{\mathbf{k}_\parallel+\mathbf{k}''_\parallel} + (\text{photon})_{\mathbf{k}_\parallel-\mathbf{k}''_\parallel,k_z}, \quad (1)$$

where k_z of the photon satisfies energy conservation. Using Fermi's Golden Rule and summing over final stages gives for process (1) the rate

$$\Gamma_M(\mathbf{K}_\parallel) = 2\sum_{\mathbf{k}'_\parallel} |c_{k'_\parallel}|^2\Gamma(\mathbf{k}_\parallel - \mathbf{k}'_\parallel), \quad (2)$$

where $\Gamma(\mathbf{k}_\parallel)$ is the rate for one-exciton decay,[4] and

$$\psi_{\mathbf{k}_\parallel} \rightarrow (\text{photon})_{\mathbf{k}_\parallel,k_z}. \quad (3)$$

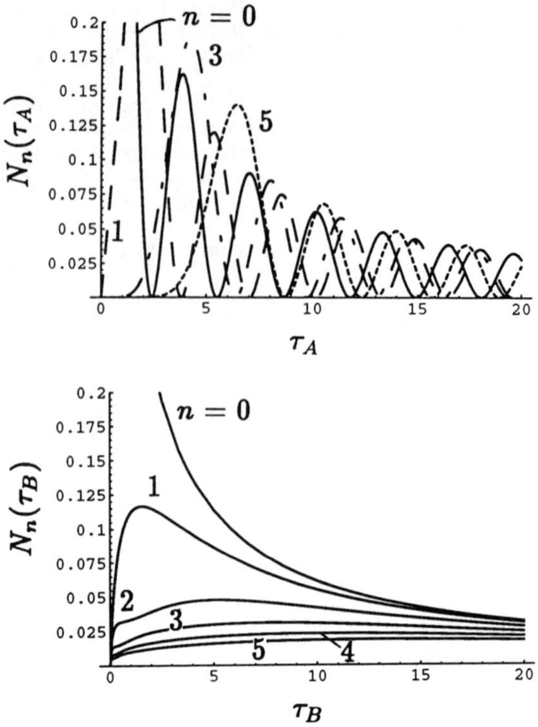

QThC2 Fig. 1.

Energy–momentum conservation dictates that $\Gamma(\mathbf{k}_\parallel)$ must vanish for $k_\parallel > \kappa$.[5] $\Gamma_M(\mathbf{K}_\parallel)$ reduces to the predictions given by polariton theory and the giant-oscillator strength in the respective limits. Equation (2), the $\Lambda_M \ll \lambda$ limit, is very similar to the case of excitons localized by static disorder[4]—the giant-oscillator strength.[1]

We assume $c_{k_\parallel} = \eta_M^{-1}\pi^{-1/2}\exp\{-k'^2_\parallel/2\eta_M^2\}$, where η_M^{-1} is the interexciton separation Λ_M. Equation (2) reads

$$\Gamma_M(\mathbf{K}_\parallel) = 2\frac{1}{\eta_M^2\pi}e^{-k_\parallel^2/\eta_M^2}\int_{k'_\parallel<\kappa} d^2k'_\parallel e^{-k'^2_\parallel/\eta_M^2} \times \Gamma(\mathbf{k}'_\parallel)I_0(2k'_\parallel k_\parallel/\eta_M^2), \quad (4)$$

where I_0 is a Bessel function. For GaAs/AlGaAs QWs we estimate $\Lambda_M \approx 400$–500 Å.[2] Since $\lambda \approx 2500$ Å, the appropriate limit is $\Lambda_m \ll \lambda$ ($\eta_M \gg \kappa$). We can set narrow bounds on $\Gamma_M(\mathbf{K}_\parallel)$, namely, $\exp(-\mu_M^2)\tilde{\Gamma}_M(\mathbf{K}_\parallel) \le \Gamma_M(\mathbf{K}_\parallel) \le \tilde{\Gamma}_M(\mathbf{K}_\parallel)$, where

$$\tilde{\Gamma}_M(\mathbf{K}_\parallel) = 2\frac{1}{\eta_M^2\pi}e^{-k_\parallel^2/\eta_M^2}\int_{k'_\parallel<\kappa} d^2k'_\parallel\,\Gamma(\mathbf{k}'_\parallel) \times I_0(2k'_\parallel k_\parallel/\eta_M^2) \quad (5)$$

and $\mu_M = \kappa/\eta_M$. Equation (5) gives

$$\tilde{\Gamma}_M^{(T)}(\mathbf{K}_\parallel) = 4\mu_M^2 e^{-k_\parallel^2/\eta_M^2}\beta^{-1}\sinh\beta\,\Gamma_{0,\parallel}, \quad (6)$$

$$\tilde{\Gamma}_M^{(L)}(\mathbf{K}_\parallel) = 4\mu_M^2 e^{-k_\parallel^2/\eta_M^2}\beta^{-2} \cdot (\cosh\beta - \beta^{-1}\sinh\beta)\Gamma_{0,\parallel}, \quad (7)$$

$$\tilde{\Gamma}_M^{(Z)}(\mathbf{K}_\parallel) = 4\mu_M^2 e^{-k_\parallel^2/\eta_M^2}\beta^{-1} \times [(1+\beta^{-2})\sinh\beta - \beta^{-1}\cosh\beta]\Gamma_{0,\perp}, \quad (8)$$

where $\beta = 2\mu_M k_\parallel/\eta_M$. The limit $\eta_M \gg \kappa$ gives the giant-oscillator-strength result for all $k_\parallel$. We wish to see how fast the fundamental biexciton radiative-decay process can be with respect to one-exciton decay. The maxima of $\tilde{\Gamma}_M^{(T,Z)}(\mathbf{K}_\parallel)$ occur near $k_\parallel = \kappa$. They are approximately $\tilde{\Gamma}^{(T)}(2\kappa) \sim 2e^{-\mu_M^2}2\mu_M^2\Gamma_{0,\parallel}$ and $\tilde{\Gamma}^{(Z)}(2\kappa) \sim 2(2/3)e^{-\mu_M^2}2\mu_M^2\Gamma_{0,\perp}$. Since $\eta_M \gg \kappa$, we have $\tilde{\Gamma}_M^{(T,Z)}(\mathbf{K}_\parallel) \ll \Gamma_{0,(\parallel,\perp)}$. For the L polariton[4] the maximum in $\tilde{\Gamma}_M^{(L)}(\mathbf{K}_\parallel)$ occurs near $\mathbf{k}_\parallel = \mathbf{0}$ to give ($\eta_M \gg \kappa$) $\tilde{\Gamma}_M^{(L)}(\mathbf{0}) \sim 2(1/3)2\mu_M^2\Gamma_{0,\parallel}$. Thus, $\tilde{\Gamma}^{(L)}(\mathbf{K}_\parallel) \ll \Gamma_{0,\parallel}$. This shows that for typical parameters describing GSBs in GaAs/AlGaAs QWs, the radiative-decay rates are all less than the characteristic one-exciton decay rates and that the extremely rapid transients observed in cross-polarized TRPL with resonant excitation by Wang *et al.*[2] are not likely to be due simply to the fundamental biexcitonic radiative processes.

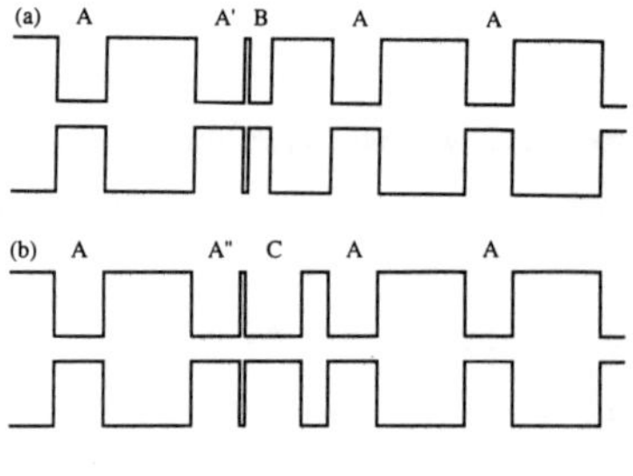

QThC2 Fig. 2.

This work was supported by National Science Foundation grant STC PHY 8920108.

1. A. A. Golovin and E. I. Rashba, Pis'ma Zh. Eksp. Teor. Fiz. **17**, 690 (1973) [JETP Lett. **12**, 478 (1973)].
2. H. Wang, J. Shah, T. C. Damen, and L. N. Pfeiffer, in *International Quantum Electronics Conference*, 1994 Technical Digest Series, Vol. 9 (Optical Society of America, Washington, D.C., 1994) postdeadline paper QPD22; *Ultrafast Phenomena*, 1994 Technical Digest Series, Vol. 7 (Optical Society of America, Washington, D.C., 1994), postdeadline paper

PD13; Solid State Commun. **91**, 869 (1994).
3. E. Hanamura, J. Phys. Soc. Jpn. **39**, 1506 (1975); **39**, 1516 (1975).
4. D. S. Citrin, Phys. Rev. B **47**, 3832 (1993).
5. V. M. Agranovich and A. O. Dubovskii, Pis'ma Zh. Eksp. Teor. Fiz. **3**, 345 (1966) [JETP Lett. **3**, 223 (1966)].

QThC3 **8:45 am**

Spontaneous-emission lifetime of Wannier excitons in mesoscopic semiconductor quantum disks

Mitsuru Sugawara, *Fujitsu Laboratories, Ltd., 10-1 Morinosato-Wakamiya, Atsugi 243-01, Japan*

Wannier excitons in semiconductor quantum-confinement structures are expected to show rapid spontaneous emission because of their coherent nature and the breakdown of translational symmetry. Theories of the radiative lifetime have been presented for two structures: quasi-two-dimensional semiconductor quantum wells[1] and spherical microcrystals.[1,2] A key to deriving the thoeretical formulas is the selection rule between the radiation field with wave vector **q** and exciton center-of-mass motion with wave vector **K**. The selection rule **q** = **K** holds in quantum wells for the in-plane component since exciton motion is free in the macroscopic two-dimensional plane. In the microcrystals, which are much smaller than the exciton resonant wavelength, the electric-dipole approximation $\exp(i\mathbf{q}\cdot\mathbf{R}) \cong 1$ holds for the center-of-mass motion. Mostly because of this difference in applied selection rules, the formulas for the two structures are different.[1] We may wonder how theories for the two limiting microscopic and macroscopic cases connect with each other in the intermediate mesoscopic region.

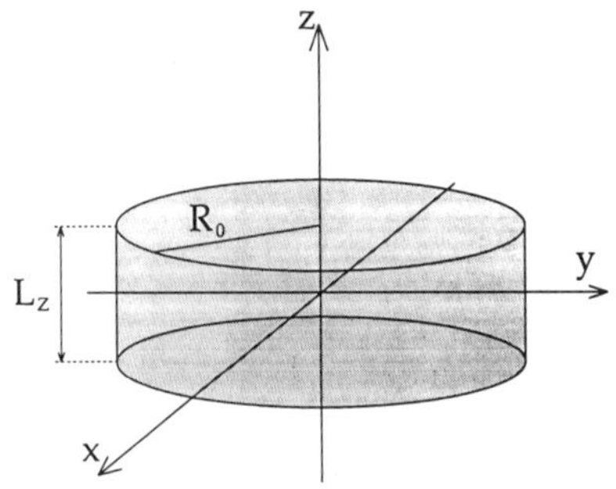

QThC3 Fig. 1. Quantum disk whose height is L_z and whose radius is R_0. The origin of the coordinate axis is the center of the disk, and the *z* axis is perpendicular to the layer.

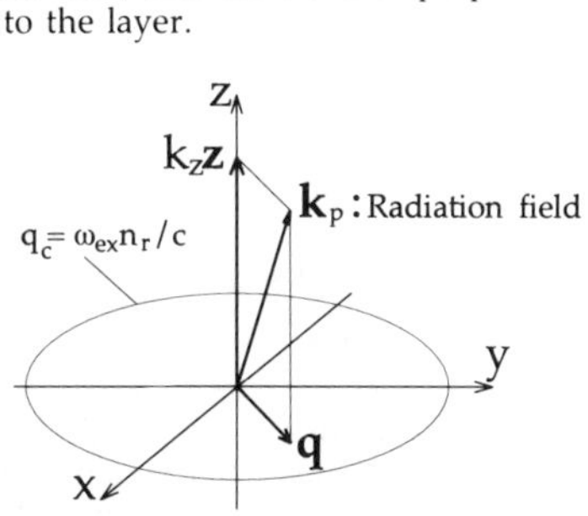

QThC3 Fig. 2. Spontaneous radiation field with the wave vector of $\mathbf{k}_p$ = ($\mathbf{q}$, $k_z\mathbf{z}$). Spontaneous emission is prohibited when the wave vector of exciton center-of-mass motion is above $q_c = \omega_{ex} n_r/c$, where ω_{ex} is the resonance frequency and n_r is the refractive index.

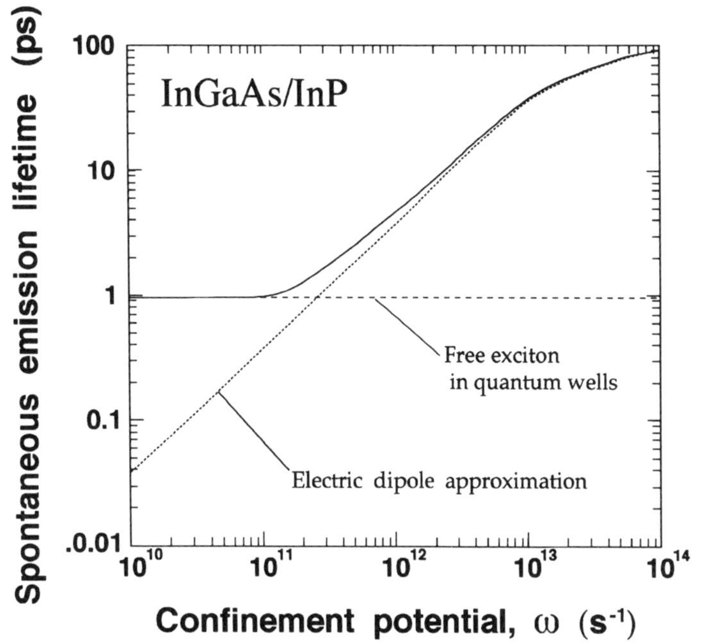

QThC3 Fig. 3 Solid curve is the calculated spontaneous-emission lifetime of the ground-state electron–heavy-hole excitons in $In_{0.53}Ga_{0.47}As$/InP quantum disks. The horizontal dashed line is the lifetime of free excitons in quantum wells. The dotted curve corresponds to the lifetime calculated by using the electric-dipole approximation.

This paper reports a theoretical formula of spontaneous-emission lifetime of excitons in mesoscopic semiconductor quantum disks. The disk, with width L_z less than the exciton Bohr radius with radius R_0, confines excitons by the potential barriers of surrounding regions (Fig. 1). We have macroscopic quantum wells as R_0 increases above the exciton resonant wavelength, and we have microscopic quantum dots as R_0 decreases below the Bohr radius. The radiation field is quantized in the crystal that contains the disk and that is much larger than the exciton resonant wavelength.

Basic ideas for deriving the exciton spontaneous-emission lifetime are as follows: (1) We should take into account the selection rule **q** = **K** for the in-plane component (Fig. 2). (2) Only excitons with a wave vector inside a critical value q_c can radiate spontaneously because of the selection rule (Fig. 2). (3) By expanding the localized center-of-mass motion wave function into a Fourier series, we can apply the selection rule with the radiation field to confined excitons. We have derived a theoretical formula by counting the optical mode number in the *z* direction and by summing Fourier components for the wave vector less than q_c.

Figure 3 shows the spontaneous-emission lifetime of the electron–heavy-hole exciton, calculated for an $In_{0.53}Ga_{0.47}As$ (L_z = 10 nm) quantum disk surrounded by InP. We assumed harmonic-oscillator confinement potential ($\propto\omega^2$) for the in-plane confinement. As ω exceeds 1×10^{11} s^{-1}, the lifetime gradually increases from a constant value. This occurs because localized center-of-mass motion wave function in **K** space spreads outside q_c and the amplitude of Fourier components that satisfy the selection rule decreases. The horizontal dashed line in Fig. 3 is the lifetime of free excitons in quantum wells, and the dotted curve is the lifetime calculated from the electric-dipole approximation. Our formula shows that exciton spontaneous-emisson lifetime increases continuously from macroscopic quantum wells to microscopic quantum dots through mesoscopic regions.

1. E. Hanamura, Phys. Rev. B **38**, 1228 (1988).
2. T. Itoh, Nonlinear Opt. **1**, 61 (1992).

QThC4 (Invited) **9:00 am**

Spontaneous emission of excitons in GaAs quantum wells: the role of momentum scattering

Hailin Wang, Jagdeep Shah, T. C. Damen, L. N. Pfeiffer, *AT&T Bell Laboratories, Holmdel, New Jersey 07733*

The radiative dynamics of excitons have been studied in detail in quasi-two-dimensional systems, such as quantum wells (QW), by using time-resolved photoluminescence (PL). The important regime immediately following resonant excitation, however, has yet to be investigated because of the limited time resolution in earlier studies. In this paper we report studies of resonantly excited PL with femtosecond resolution in GaAs QWs. The much-improved time resolution reveals the crucial role of momentum scattering in exciton spontaneous emission in quasi-two-dimensional systems. The gradual rise of the PL provides a direct measurement of exciton momentum-scattering rates. The persistence of quantum beats in the PL further allows us to obtain, for what we believe to be the first time, information on the relationship between exciton dephasing and momentum scattering. In addition, a surprisingly fast decay rate for biexcitons is also observed.

We achieved femtosecond resolution for resonantly excited PL by upconverting the PL in a nonlinear crystal with output from a synchronously pumped optical parametric oscillator (see Fig. 1). Excitons were resonantly excited with 120-fs pulses centered at 6 meV below the heavy-hole resonance. The samples were kept at 10 K. We summarize the main experimental results in the following

Observation of the rise time in resonantly excited PL is shown in Fig. 2. Conceptually, excitons created by the excitation pulse have a momentum $k_{\|}$, the photon momentum along the plane of the QW. The excitons cannot emit a photon in directions other than the transmitted or reflected direction of the excitation pulse unless the exciton momentum is randomized through momentum scattering. The rise time provides a quantitative measure of the exciton momentum-scattering rate. As shown in Fig. 2(a), the rise time increases with increasing exciton densities, indicating that exciton–exciton scattering is the leading contribution to momentum scattering under the experimental conditions. Within the density range we have used, the momentum-scattering rate is proportional to the exciton density.

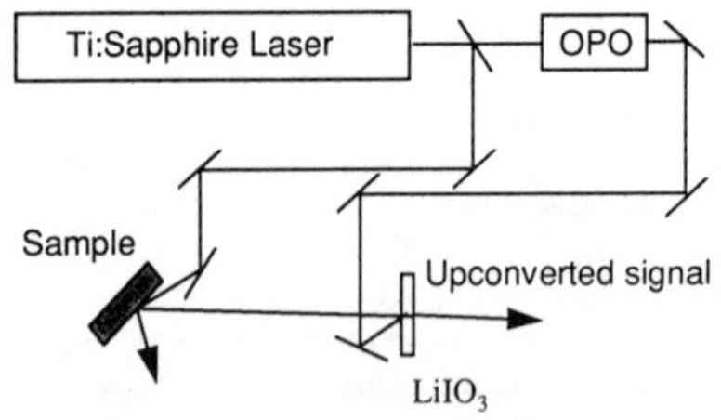

QThC4 Fig. 1. Experimental setup for the upconversion measurement.

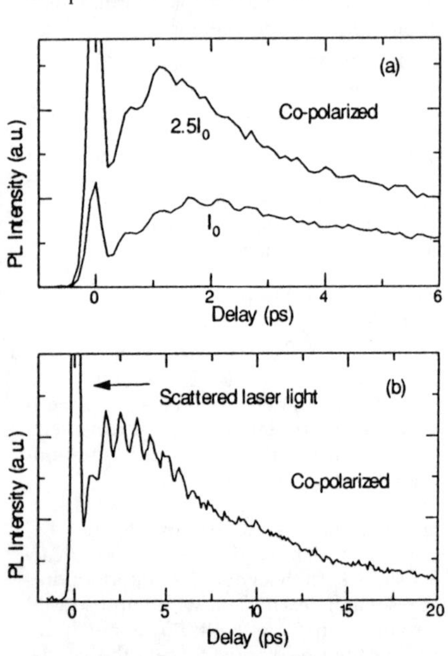

QThC4 Fig. 2. PL excited resonantly with a circularly polarized field: (a) 130-Å QW, where I_0 corresponds to an exciton density of 5×10^9 cm^{-2}; (b) 175-Å QW.

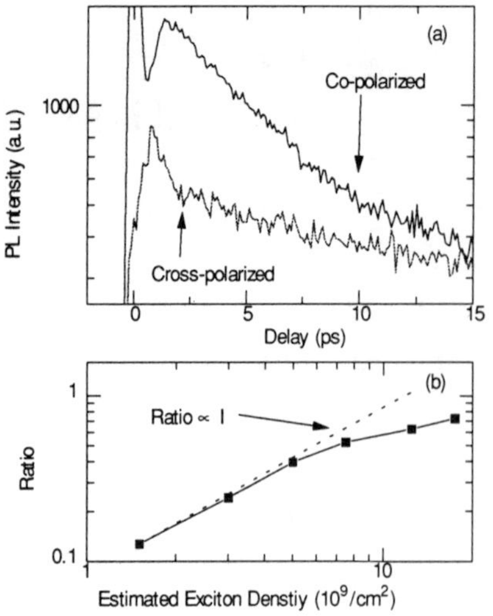

QThC4 Fig. 3. (a) PL from a 130-Å QW excited resonantly with a linearly polarized field. (b) Intensity dependence of the ratio of the peak intensities of the cross- and co-polarized PL.

The presence of quantum beats in PL, along with the finite rise time, allows a simultaneous measurement of the exciton-dephasing rate and the momentum-scattering rate. The PL in Fig. 2 features an initial temporal oscillation. The period of the oscillation corresponds to the heavy- and light-exciton splitting. The persistence of the quantum beats after the rise of the PL is very surprising because the excitons must go through momentum scattering before they can emit photons in the direction of the detection. The result in Fig. 2 shows that the momentum-scattering rate can be considerably larger than the exciton-dephasing rate, indicating a breakdown of the impact approximation for the exciton and suggesting the presence of correlation between scattering events.

The polarization dependence of the PL reveals a surprisingly fast decay rate for biexcitons in QWs. Figure 3(a) shows both co- and cross-polarized PL excited with linearly polarized pulses. Within the excitonic picture no cross-polarized PL is expected in the absence of exciton-spin relaxation (time scale, 50 ps). The strong signal for the cross-polarized PL in Fig. 3(a) is the result of the unique polarization-selection rule for the biexcitons and is due to the biexcitonic radiative recombination, as confirmed by the intensity dependence in Fig. 3(b). The extremely short lifetime (2.5 ps) for the biexciton PL disagrees with a recent theoretical calculation and raises the question of the validity of the current theoretical model for biexcitons.

We thank Spectra-Physics for the loan of an optical parametric oscillator.

JThB 10:30 am
Room 317

Joint Symposium on Emerging Applications of Fluorescence to Biophysics and Cellular Imaging II

David R. Sandison, *Sandia National Laboratory, Presider*

JThB1 (Invited) 10:30 am

Biomedical applications of time-resolved fluorescence

Joseph R. Lakowicz, Henryk Szmacinski, *Center for Fluorescence Spectroscopy, Department of Biochemistry, University of Maryland School of Medicine, 108 N. Greene St., Baltimore, Maryland 21201*

Time-resolved fluorescence spectroscopy is presently regarded as a research tool in biochemistry, biophysics, and chemical physics. However, advances in laser technology and the development of long-wavelength probes are resulting in the rapid migration of lifetime-based methods to the clinical chemistry lab and the patient's bedside. The advantages of lifetime-based fluorescence sensing are apparent from the various possible schemes for fluorescence sensing (Fig. 1). At present, most fluorescence assays are based on the intensity measurements, in which the intensity of the probe molecule changes in response to the analyte (far left). Intensity measurements are simple and accurate in the laboratory, but are often inadequate in real-world situations. The sample may be turbid, the optical surfaces may be misaligned or imperfect, and the probe concentration may vary from sample to sample. The problems of intensity-based sensing can be avoided using wavelength-ratiometric probes, that is, fluorophores which display spectral changes in the absorption or emission spectrum upon binding with the analytes. In this case, the analyte concentration can be determined independently of the probe concentration by the ratio of intensities at two excitation or two emission wavelengths. However, few such wavelength-ratiometric probes are available.

The difficulties of intensity-based measurements may be circumvented by the use of time-resolved or lifetime-based sensing. A wide variety of quenchers and/or molecular interactions result in lifetime changes of fluorophores. In contrast, changes in spectral shape are the exception rather than the rule. We now know of lifetime probes for Ca^{2+}, Mg^{2+}, pH, Cl^-, K^+, and O_2. Lifetime can be measured using time-domain (TD) or frequency-domain (FD) methods (Fig. 1). In time-domain fluorometry, the sample is excited by a pulse of light followed by measurement of the time-dependent intensity. In frequency-domain fluorometry the sample is excited with amplitude-modulated light (Fig. 1, right). The lifetime can be found from the phase angle delay and demodulation of the emission relative to the modulated incident light.

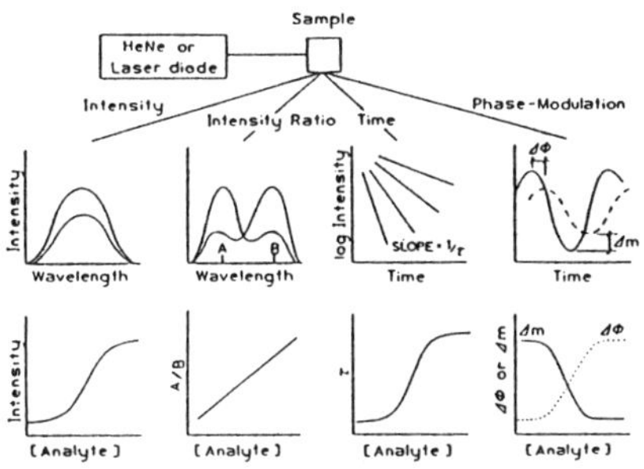

JThB1 Fig. 1. Schemes for fluorescence sensing, from left to right; intensity, intensity ratio, time-domain and phase-modulation.

Phase and modulation measurements can be performed with simple and low cost instrumentation.

Optical detection of blood gases (pH, PCO_2, and pO_2) is highly desirable for use in hospitals. Optical sensing of blood gases now seems possible using lifetime-based methods and a simple laser source. Noninvasive monitoring of glucose now appears possible based on our current understanding of lifetime-based sensing and the optical properties of tissues. The red wavelength of laser diodes are only weakly absorbed by skin, but of course the light is highly scattered.

JThB2 (Invited) **10:45 am**

Two-photon induced fluorescence of biomolecules

Ignacy Gryczynski, Joseph R. Lakowicz, *Center for Fluorescence Spectroscopy, Department of Biological Chemistry, University of Maryland, School of Medicine, 108 North Green St., Baltimore, Maryland 21201*

We studied two-photon induced fluorescence (TPIF) of several bioprobes, such as fluorescent amino acids (tyrosine and tryptophan), membrane probe DPH or DNA stains DAPI and Hoechst.

The fluorescence spectra and intensity decays of studied fluorescence standards and biomolecules are usually similar for one and two-photon excitation. A significant difference between one- and two-photon excitation has been observed for fluorescence anisotropy. Presently, we do not know any simple dependence between fluorescence anisotropies achieved with one- and two-photon excitation. Several studied molecules like DPH or PPO have higher two-photon anisotropy, resulting from higher two-photon photoselection. Some molecules with mixed transitions (indole derivatives) display lower two-photon anisotropies. Other molecules with single transition (tyrosines) do not follow the photoselection rules and displayed a high one-photon anisotropy and a low two-photon anisotropy (see figures below). Time-resolved measurements of fluorescence anisotropy revealed similar correlation times for one- and two-photon excitation.

Recently we found that frequency doubled output of R6G dye laser (280–310 nm) is very suitable for TPIF study of alkanes. This excitation allowed us to overcome the

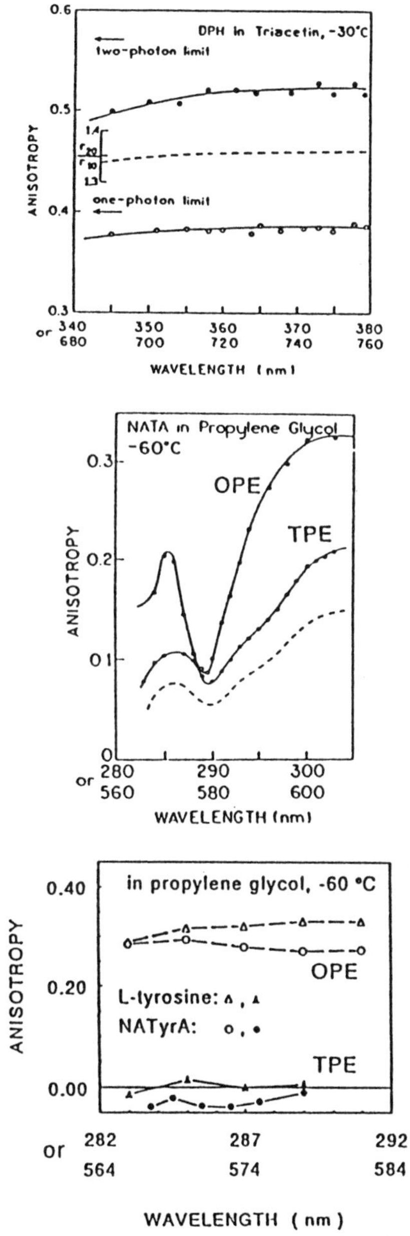

JThB2 Fig. 1.

difficulties associated with vacuum UV direct excitation. The efficient dynamic quenching of alkanes by alcohols was detected and measured over a wide range of fluorophores and quencher concentrations. We believe that direct two-photon excitation of detergents and membranes should be possible.

We have also developed several convenient reference compounds with short, well determined lifetimes, high quantum yields and high two photon cross sections. Scintilators, bis-MSB, PPO or TPB are convenient lifetime standards for two-photon excitation within the wavelength range provided by rhodamine 6G dye laser (560–600 nm). For longer wavelengths (Pyridine dye, 690–760 nm) we prefer styrene and stilbene derivatives or coumarins.

JThB3 (Invited) **11:15 am**

Bioprocess monitoring use fluorescence lifetime-based sensing

Govind Rao, *Department of Chemical and Biochemical Engineering and Medical Biotechnology Center of the Maryland Biotechnology Institute, University of Maryland, Baltimore County, Baltimore, Maryland 21228*

Advances in genetic engineering have in the past two decades generated a wealth of novel molecules that have redefined the role of microbes in solving environmental, pharmaceutical, industrial and agricultural problems. While some products have painstakingly entered the marketplace the difficulties of doing so and complying with federal mandates of "safety, purity, potency, effectiveness and consistency" has shifted focus to the word "engineering" from words like "genetic" and "enzymatic." This transition from the laboratory to production, the basis of bioprocess engineering, involves a careful understanding of the conditions most favored for optimal microbial production and the duplication of these conditions during scaled-up production.

The smallest and most complex unit of a bioprocess is the cell. The ability of the cell to live and carry out its task of bioconversion is extremely sensitive to its environment. Our ability to control the bioconversion is therefore contingent upon how closely one can monitor and control this complex environment. The availability of sensors which can be integrated on-line into the bioreactor, require minimal maintenance and calibration, have high sensitivity and be readily sterilized would dramatically improve bioprocesses. Our work combines recent advances in probe chemistry, optics and electronics to give robust and low cost sensing devices for these analytes.

Although the requirements for clinical monitoring are more stringent and regulated, similar concerns prevail. The technology we describe is particularly applicable to making measurements in biological media since it is mostly independent of its interfering effects. Moreover, the availability of red excitable probes encourage the possibility of through tissue sensing. Consequently, our discussions based on bioprocess monitoring also apply to clinical situations.

QThD 10:30 am
Hyatt A

Nonlinear Optical Propagation Effects

Michael D. Duncan, *U.S. Naval Research Laboratory, Presider*

QThD1 10:30 am

Quantum fluctuations as the origin of laser-beam filamentation

E. M. Nagasako, R. W. Boyd,
G. S. Agarwal,* *Institute of Optics, University of Rochester, Rochester, New York 14627*

According to traditional theories, the filamentation[1] of a laser beam as it passes through a nonlinear material is a consequence of the spatial growth[2,3] of weak wave-front perturbations initially present on the beam. To a certain extent these perturbations can be removed by passing the beam through a spatial filter before it enters the medium. However, quantum fluctuations in the field amplitudes of the transverse side modes impose perturbations that cannot be removed by spatial filtering. These quantum fluctuations can lead to the filamentation of a beam with an otherwise perfect wave front. To model this behavior, we begin by expanding a monochromatic field in its transverse modes as

$$\hat{\mathbf{E}}^{(+)}(\mathbf{r}, t) = \left[E_0 + \int d^2q(\hbar\omega_0 k_0/4\pi^2)^{1/2}\hat{a}(\mathbf{q}, z)\exp(i\mathbf{q}\cdot\mathbf{r})\right] \times \exp(i\gamma z)\exp(ik_0 z - i\omega_0 t), \quad (1)$$

where E_0 represents the incident plane-wave laser field and $\hat{a}(\mathbf{q},z)$ is the amplitude of the side mode with transverse wave vector $\mathbf{q}$. We assume that all of the side modes are initially in the vacuum state. Using the paraxial wave equation and linearizing the resulting equations in the amplitude of the perturbation, we have obtained an expression for the total intensity I_{sm} of the side modes as a function of the nonlinear index of refraction of the medium, the length of the medium, and the applied field intensity. In the regime in which $\gamma z > 1$, this expression can be reduced to

$$I_{sm} = \frac{nc}{8\pi^2}\hbar\omega_0 k_0^2 \gamma \exp(2\gamma z), \quad (2)$$

where $\gamma = (2\pi/\lambda_0)n_2 I_{laser}$. According to the present model, filamentation does not occur if I_{sm} is much smaller than the intensity of the incident laser beam. This condition places a limit on the size of the quantity $(2\pi/\lambda_0)n_2 I_{laser}z$. For any given nonlinear optical medium, there is thus a fundamental limit on the intensity of a laser beam that can propagate without the occurrence of filamentation.

We have also performed numerical simulations of the transverse intensity pattern that is formed as the filaments begin to develop. To do so, we model the vacuum fluctuations by introducing a half photon, with random phase, per mode. The simulations show the growth of large transverse intensity fluctuations from small initial perturbations. An example of such a simulation is shown in Fig. 1.

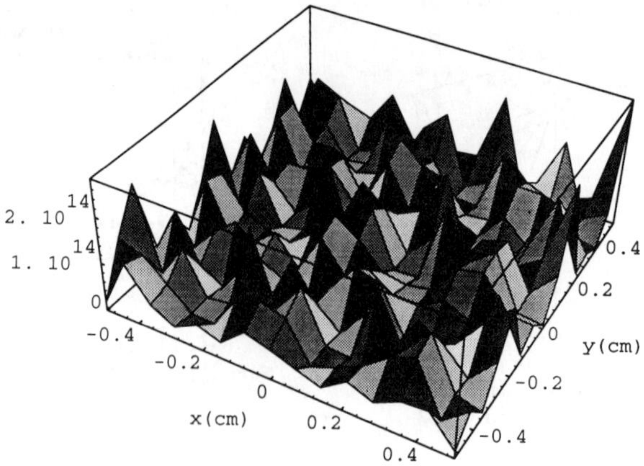

QThD1 Fig. 1. Transverse intensity distribution of a laser beam after it passes through a nonlinear optical medium for which $(2\pi/\lambda_0)n_2 I_{laser}z = 15$.

In summary, we have shown that, even in the absence of classical wave-front perturbations, a laser beam propagating through a nonlinear optical medium is subject to breakup through the process of filamentation initiated by quantum-mechanical vacuum fluctuations.

**School of Physics, University of Hyderabad, Hyderabad 500 134, India*

1. V. I. Bespalov and V. I. Talanov, JETP Lett. **3**, 471 (1966).
2. R. Y. Chiao, P. L. Kelley, and E. Garmire, Phys. Rev. Lett. **17**, 1158 (1966).
3. E. S. Bliss, D. R. Speck, J. F. Holzrichter, J. H. Erkkila, and A. J. Glass, Appl. Phys. Lett. **25**, 448 (1974).

QThD2 10:45 am

Filamentation of femtosecond UV pulses in air

Xin Miao Zhao, Patrick Rambo,
Jean-Claude Diels, *Department of Physics and Astronomy, University of New Mexico, Albuquerque, New Mexico 87131*

Filamentation of high-peak-power infrared (800-nm) femtosecond laser pulses in air has been reported [1] and has been used for triggering high-voltage lightning discharges.[2] We report here a low-threshold filamentation of UV (248-nm) femtosecond pulses in air.

At 800 nm the dispersion of air is negligible for pulses of the order of hundreds of femtoseconds at distances shorter than the characteristic distance (of the order of 10 km). Therefore, only self-trapping in space is required to achieve confinement of the light in a filament. The situation is quite different at 248 nm, where the dispersion in air is sufficient to cause pulse broadening. A balance between dispersion and negative self-phase modulation is essential to ensure a distortionless propagation in time, i.e., a soliton.[3] This property provides a unique opportunity to observe the coexistence of a soliton in space and time.

The experiment is performed with our Ti:sapphire–excimer laser system. Ti:sapphire laser pulses at 744 nm are amplified to millijoule level with a four-stage Nd:YAG-pumped dye amplifier. After frequency tripling, the pulses are further amplified by an excimer laser amplifier that delivers pulses at 248 nm with up to 10 mJ of energy at 8 Hz. The pulses are approximately 500 fs in duration. Self-focusing and filamentation are observed at pulse energies in excess of 0.5 mJ. Because the beam has a nonuniform transverse profile, it breaks into several filaments. The filament diameter remains practically constant over several meters, as shown in Fig. 1. The

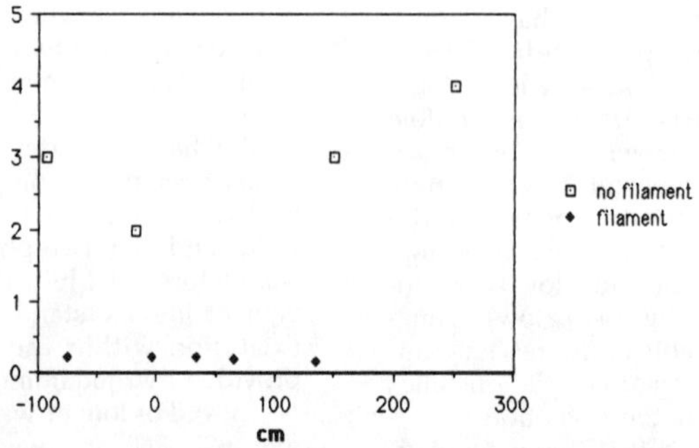

QThD2 Fig. 1 Laser-beam diameter versus distance. The open squares represent the propagation of the beam below the self-focusing threshold, and the filled diamonds represent the filamented beam.

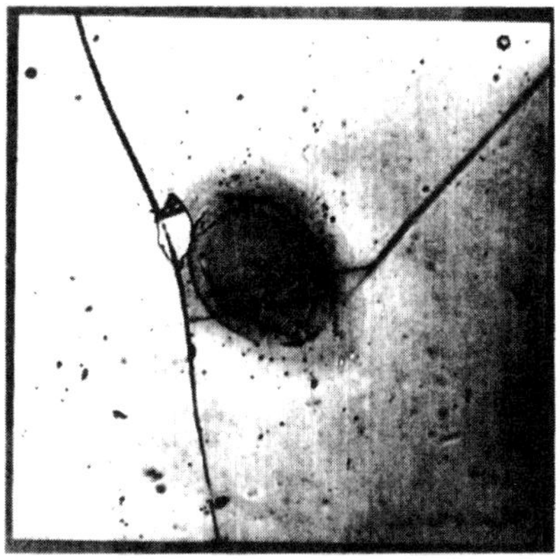

QThD2 Fig. 2. Holographic film damaged by filamented laser beam. The long straight lines are scratches on the old film.

average diameter of the filament is about 220 μm and has less than 0.5 mJ of energy trapped in it. The size is about three times that of the 80-μm filaments observed at 800 nm.[1] The peak power corresponding to a 0.1-mJ, 500-fs pulse at 500 fs is approximately ten times the critical power for self-trapping ($P_{crit} = \lambda^2/4\pi n_0 n_2$, where n_2 of air is 8×10^{-19} cm²/W). The trace of a filament on a plastic film is shown in Fig. 2.

At 248 nm the stability of the filaments results from a balance of self-focusing and defocusing in space and, simultaneously, dispersion and self-phase modulation in time. On the axis of the filament, the peak intensity of the pulse is sufficient to partially ionize the air. The index of refraction first increases, but after the negative contribution due to the free electrons dominates, the index of refraction decreases rapidly. On the edge of the filament a reduced intensity pulse induces only a positive nonlinear index. On the leading edge of the pulse the Kerr effect produces a weak positive lens. The larger size of filaments at 248 nm is thus consistent with the greater ionization at that wavelength. As time progresses, the contribution from the free electrons produces a toruslike index distribution with a minimum on the axis and a maximum on at radius r, as in a leaky hollow waveguide.

1. A. Braun, G. Korn, X. Liu, D. Du, J. Squier, and G. Mourou, in *Ultrafast Phenomena*, 1994 Technical Digest Series, Vol. 7 (Optical Society of America, Washington, D.C., 1994), paper ThA4.
2. X. M. Zhao, J.-C. Diels, J. Elizando, A Braun, D. Du, X. Liu, G. Korn, and G. Mourou, in *Ultrafast Phenomena*, 1994 Technical Digest Series, Vol. **7** (Optical Society of America, Washington, D.C., 1994), paper ThD24.
3. L. F. Mollenauer and R. H. Stolen, Opt. Lett. **9**, 13 (1984).

QThD3 **11:00 am**

Vector characteristics of femtosecond laser pulses propagating in single-mode optical fibers

Q. D. Liu, L. Shi, W. L. Sha, P. P. Ho, R. R. Alfano, *Institute for Ultrafast Spectroscopy and Lasers, New York State Center for Advanced Technology for Ultrafast Photonic Materials and Applications, Departments of Electrical Engineering and Physics, City College and Graduate School of The City University of New York, New York, New York 10031*

Since the 1980's the polarization instability that is due to nonlinear effects has been observed in high- and low-birefringence optical fibers by several groups.[1-3] We report on the measurement of intensity-induced polarization rotation and depolarization of linearly polarized femtosecond laser pulses in short nonbirefringement single-mode optical fiber. The experimental results are consistent with the proposed theoretical model, in which the nonlinear polarization rotation is attributed to self- and cross-phase-modulation processes in the optical fiber and the depolarization is due to rotation.

A femtosecond pulse train was derived from a coherent Mira 900-B Ti:sapphire laser operating at 780 nm. The laser pulses had a typical duration of 120 fs and a spectral bandwidth of 6 nm. The linearly polarized laser pulses were coupled into and out of a 1-m-long nonbirefringent single-mode optical fiber with two 10× objective lenses. The nonbirefringent single-mode optical fiber (Newport Corp. model FSV-10) has a core diameter of 4 μm and a cladding thickness of 125 μm. The output laser pulses were spatially separated by a polarizing prism into two orthogonal linearly polarized components. For a quasi-linearly polarized laser pulse, one of the components was at its maximum when the optical axis of the polarizing prism was parallel to the direction of the polarization of the laser pulse, while the other component was at its minimum. By rotating the polarizing prism, we measured the directions of the polarization of the output laser pulses at various input-pulse intensities. The rotation angle was read out from the rotational stage that held the polarizing prism. When the polarization direction was determined for a given input-pulse intensity, the polarization ratio was also measured by a meter that monitored the intensity of each component.

The experimental results for the rotation angle and the polarization ratio are displayed as points in Figs. 1 and 2. The polarization ratio and the rotation angle depend on the intensity of input laser pulses. A theoretical model has been developed to explain the intensity-induced rotation and depolarization. Right-handed and left-handed unit vectors were used as two orthogonal polarization eigenmodes in this model. A linearly polarized laser pulse is the superposition of the two circularly polarized components with zero phase difference. The solutions of the nonlinear coupling equations show a phase shift between two circularly polarized components in a quasi-linearly polarized laser pulse. The phase shift that causes the rotation of the linear polarization is proportional to the instantaneous intensity of the pulse profile. To be specific, the polarization state is everywhere linear within the pulse, but the plane of polarization rotates as a function of time. This leads to different degrees of polarization rotation within a pulse. Furthermore, this effect leads to the intensity-induced depolarization. The theoretical fits are displayed as curves in Figs. 1 and 2. This model has a good agreement with the experimental results.

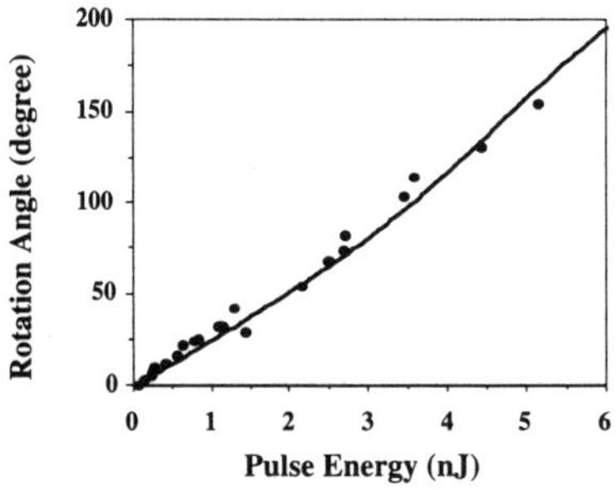

QThD3 Fig. 1. Polarization rotation angle of a femtosecond laser pulse propagating in an optical fiber as a function of incident-laser-pulse energy: points, experimental data; curve, theoretical fit.

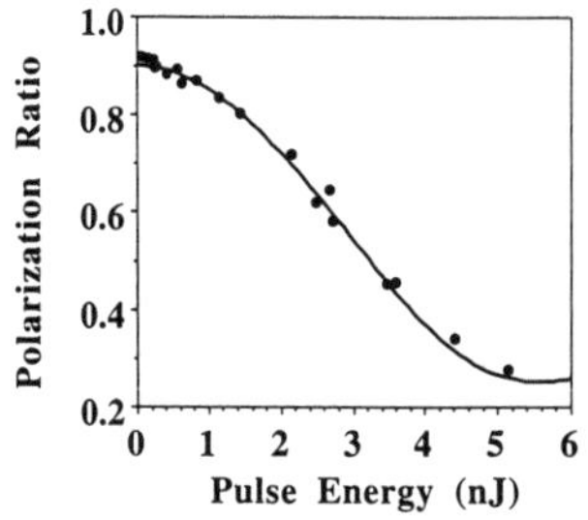

QThD3 Fig. 2. Polarization ratio of a femtosecond laser pulse propagating in an optical fiber as a function of the incident-laser-pulse energy: points, experimental data; curve, theoretical fit.

1. S. Trillo, S. Wabnitz, R. H. Stolen, G. Assanto, C. T. Seaton, and G. I. Stegeman, Appl. Phys. Lett. **49**, 1224 (1986).
2. S. F. Feldman, D. A. Weinberger, and H. G. Winful, Opt. Lett. **15**, 311 (1990).
3. R. Kashyap and N. Finlayson, Opt. Lett. **17**, 405 (1992).

QThD4 **11:15 am**

Tunable polarization modulation instability in weakly birefringent fibers

S. G. Murdoch, R. Leonhardt, J. D. Harvey, *Department of Physics, University of Auckland, Private Bag 92019, Auckland, New Zealand*

In the scalar description of wave propagation in optical fibers governed by the nonlinear Schrödinger equation (NLSE), modulation instability (MI) can occur only in the anomalous-dispersion regime, and this was the first form of MI to be ob-

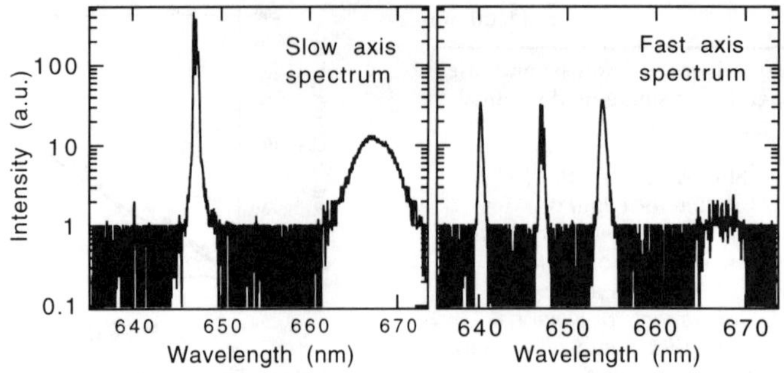

QThD4 Fig. 1. Spectrum of light emerging from each axis of 3 m of low-birefringence fiber wound on a 3-cm-diameter spool, when the pump was polarized along the slow axis. The peak power at the start of the fiber was 700 W.

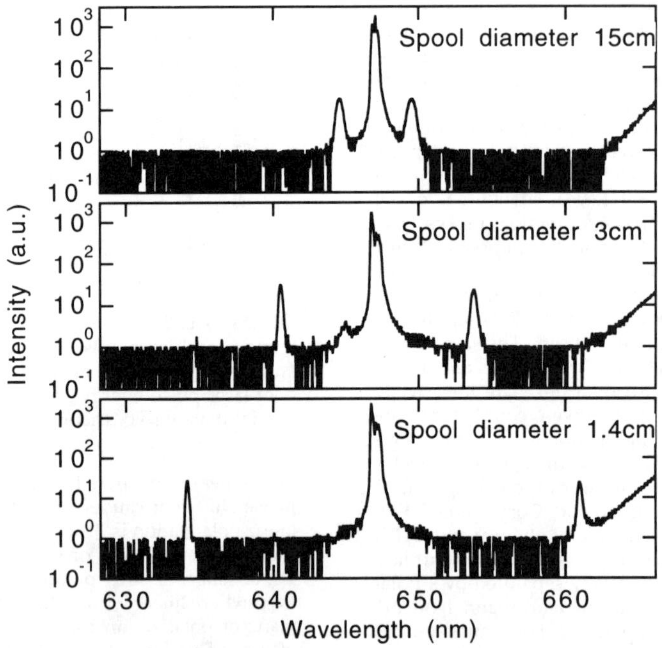

QThD4 Fig. 2. Spectrum of light emerging from 5 m of fiber wound on spools of various diameters when the pump light was polarized along the slow axis. The peak power at the start of the fiber was 390 W.

served.[1] The coherent interaction of two different circularly polarized waves of the same frequency propagating in a birefringent fiber has also been predicted to give rise to modulation instability in both the normal- and the anomalous-dispersion regimes.[2] MI has, however, been observed in the normal-dispersion regime only by pumping highly birefringent fibers with intense red or green pulses.[3,4] This form of MI arises from the incoherent interaction of two linearly polarized pulses. We have recently observed polarization modulation instability (PMI) in the normal-dispersion regime.[5] When pumping is on the slow axis of the fiber, there is no low-power threshold for the instability and the peak gain is always at a finite detuning from the pump. We report here the observation of PMI and the fabrication of a ring laser that generates amplitude-modulated light pulses whose frequency is continuously tunable over 10 THz.

In these experiments the pump laser used was a mode-locked cavity-dumped krypton-ion laser, operating at 647 nm, which can produce 60-ps pulses with a peak power of in excess of 1 kW at a repetition rate of 1.2 MHz. The fiber used was standard low-birefringence He–Ne single-mode fiber, whose birefringence could be altered by changing the radius of the spool on which it was wound. Figure 1 shows the spectrum of the light on the fast and the slow axes emerging from the end of the fiber, confirming the predicted polarization of the PMI sidebands. Figure 2 shows the spectra obtained when the fiber was wound onto spools of diameter 15, 3, and 1.4 cm, respectively. The modulation was also observed directly by forming the optical autocorrelation function of the superposition of the two sidebands. Using this technique, we have directly measured modulation frequencies of up to 20 THz.

The dependence of PMI gain on the induced birefringence has been used to construct a ring laser by using a fiber wound on a conical spool. This laser has gain for a range of modulation frequencies and can be used to generate modulated pulses whose frequency can be tuned over several terahertz by using the different round-trip times of the induced sidebands in the cavity.

1. K. Tai, A. Hasegawa, and A. Tomita, Phys. Rev. Lett. **56**, 135 (1986).
2. S. Wabnitz, Phys. Rev. A **38**, 2018 (1988).
3. P. Drummond, T. A. B. Kennedy, J. M. Dudley, R. Leonhardt, and J. D. Harvey, Opt. Commun. **78**, 137 (1990).
4. J. E. Rothenberg, Phys. Rev. A **42**, 682 (1990).
5. S. G. Murdoch, R. Leonhardt, and J. D. Harvey, submitted to Opt. Lett.

QThD5 **11:30 am**

Very large group velocities due to coherent population trapping

O. Schmidt, Z. Hussein, R. Wynands, D. Meschede, *Institute for Quantum Optics, University of Hannover, Welfengarten 1, D-30167 Hannover, Germany*

It has been suggested[1] that in the vicinity of nonabsorbing (dark) resonances interesting properties of the corresponding index of refraction could be expected and potentially also be applied, for instance, for high-sensitivity magnetometers[2] or for the generation of low group velocities.[3] A three-level atom is usually at the heart of such considerations.

For the sake of experimental simplicity we have used the D_2 line of cesium atoms with a large number of participating sublevels for an investigation of the index of refraction in the vicinity of such a dark resonance. We have found that this complication does not fundamentally alter the expected behavior. The index of refraction was measured with the help of a Mach–Zehnder interferometer (Fig. 1).

The probe-beam intensity in both interferometer arms was carefully equilibrated. A cesium-vapor cell was inserted simultaneously into both arms, and a pump beam inducing nonlinear dark resonances (and hence nonlinear contribution to the index of refraction) was superposed with one arm. A servo loop was used to make the intensity identical in both output ports of the interferometer by means of a piezo-driven mirror. A more detailed analysis shows that the corresponding control voltage is proportional to nonlinear contributions of the index of refraction if piezo nonlinearities of the order of 10% are neglected.

The frequency difference between the pump and the probe lasers is near the hyperfine-level spacing, $\Delta = 9.19$ GHz. It is perfectly controlled by phase-locking techniques. In Fig. 2 we show as a function of relative detuning the resulting measurement for the index of refraction and its imaginary counterpart, the absorption coefficient, which can be registered simultaneously.

Both curves are well described by a Lorentzian with a width of 864 kHz. At the line center of the dispersive resonance a gradient $dn/df = 3.6 \times 10^{-2}$ Hz^{-1} is measured, corresponding to a group velocity $\vartheta_g = c/1280$ in our case. At a cesium base pressure of less than 10^{-5} mbar this dis-

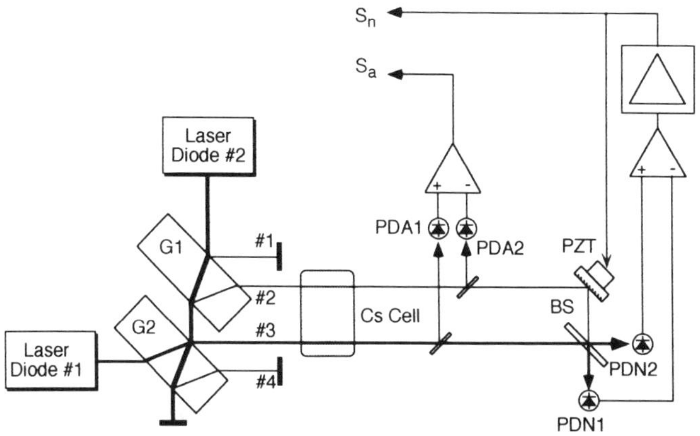

QThD5 Fig. 1.

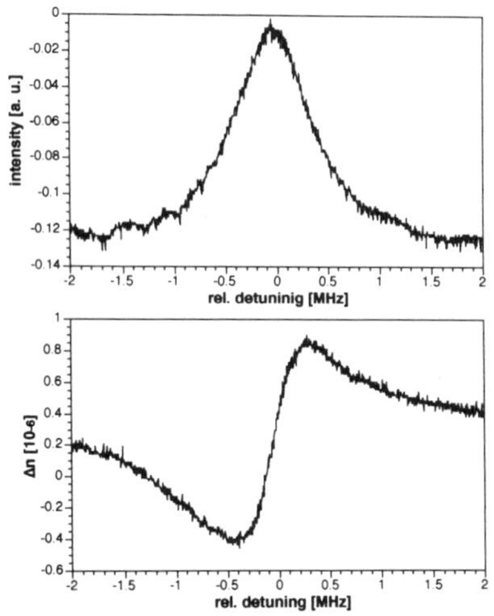

QThD5 Fig. 2.

persion is more than 5 orders of magnitude higher than the dispersion of glass.

1. M. O. Scully, Phys. Rev. Lett. **67**, 1855 (1992); M. Fleischhauer *et al.*, Phys. Rev. A **46**, 1468 (1992).
2. M. O. Scully and M. Fleischhauer, Phys. Rev. Lett. **69**, 1360 (1992).
3. S. E. Harris *et al.*, Phys. Rev. A **46**, R29 (1992).

QThD6 **11:45 am**

Dispersion-induced beam narrowing in a self-defocusing medium

Andrew T. Ryan, Govind P. Agrawal, *Institute of Optics, University of Rochester, Rochester, New York 14627*

The multidimensional nonlinear Schrödinger equation (NSE) is derived from Maxwell's equations for the case of an intensity-dependent (Kerr-type) index of refraction (resulting in a cubic nonlinearity in the field dependence of the polarization) and a frequency-dependent or dispersive susceptibility. Much work has been done by using the NSE to model the behavior of an optical field propagating in nonlinear dispersive Kerr media. Such interesting phenomena as multidimensional wave collapse,[1] pulse splitting,[2–4] and pulse compression in the normal-dispersion regime[5] are among the effects that have recently been predicted by the NSE. Analytic solutions exist for the one-dimensional NSE, but their applicability is limited to pulses in fibers and to cw beams in waveguides. In numerically solving the multidimensional NSE we are limited only by the paraxial and slowly-varying-envelope approximations used in its derivation.

The multidimensional NSE governs both the spatial (diffractive) and the temporal (dispersive) variations of the optical field; moreover, it implies a spatiotemporal coupling between diffraction and dispersion that does not exist in a linear medium. As a result of this coupling, the spatial behavior of long pulses is different from that of short pulses; i.e., the spatial (beam) width exhibits a dependence on pulse width and material dispersion. Figure 1 shows the effects of this coupling explicitly by plotting the evolution of the spatial profile of the beam at the peak of the pulse for two cases: (a) a long pulse (negligible dispersion) and (b) an ultrashort pulse (strong normal dispersion). Apart from the pulse width, all simulation parameters are identical. In both cases an unchirped field that is initially

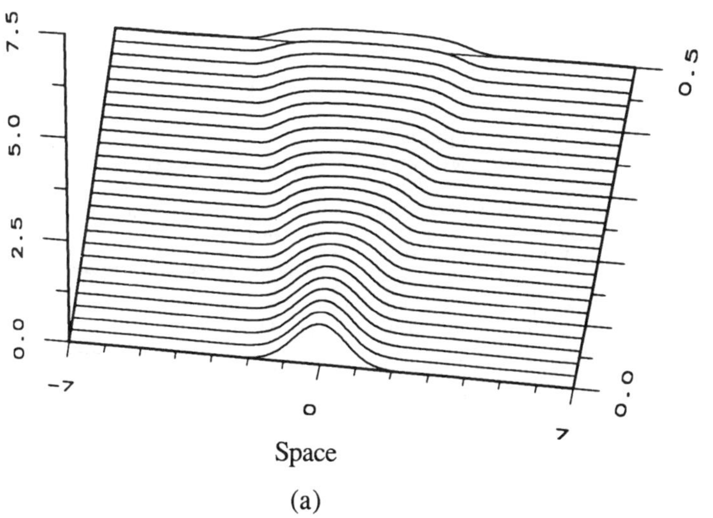

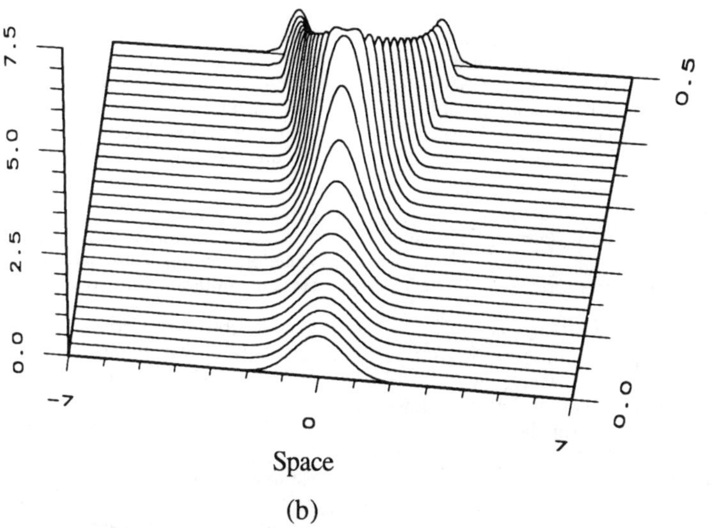

QThD6 **Fig. 1** Evolution of an initially Gaussian field in a self-defocusing medium: (a) beam broadening in a long pulse (negligible dispersion); (b) beam narrowing in an ultrashort pulse (strong normal dispersion). Spatial profiles are through the peak of the pulse.

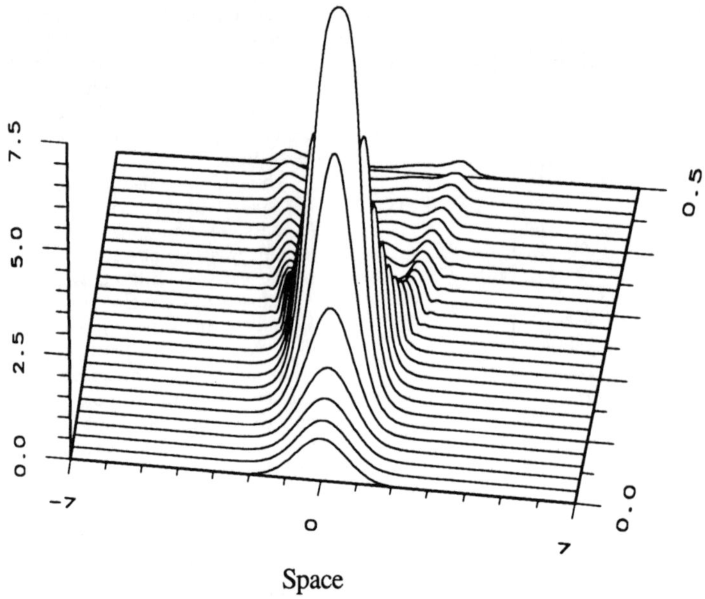

QThD6 Fig. 2 Effect of frequency chirp on the spatial profile of an initially Gaussian field. The beam is narrowed by a factor of 2 from its input width and by more than twice that for an unchirped field [see Fig. 1(b)].

Gaussian in both space and time is launched into a defocusing nonlinear medium in which the peak power is chosen such that a fourth-order soliton would be supported in a self-focusing medium. Figure 1(a) shows the expected beam broadening of a defocusing medium, and Fig. 1(b) shows that the inclusion of dispersion leads to an initial reduction of the FWHM of the spatial profile of the field, followed by a beam splitting. Beam narrowing is thus a direct consequence of this spatiotemporal coupling.

Obviously, beam narrowing can be achieved simply by controlling the phase curvature of the input beam. Figure 2 shows how temporal phase modulation (frequency chirp) can be used to enhance the dispersion-induced narrowing of the beam. Conversely, we can also show that spatial phase modulation can be used to enhance the pulse compression found in this regime. Both of these results are direct consequences of the spatiotemporal coupling implied by the multidimensional NSE. The choice of spatial or temporal modulation parameters is crucial in either case but a complete discussion is not feasible here.

1. Y. Silberberg, Opt. Lett. **15**, 1282 (1990).
2. P. Chernev and V. Petrov, Opt. Lett. **17**, 72 (1992).
3. J. E. Rothenberg, Opt. Lett. **17**, 583 (1992).
4. G. G. Luther, A. C. Newell, and J. V. Moloney, Physica D **74**, 59 (1994).
5. A. T. Ryan and G. P. Agrawal, "Pulse compression and spatial phase modulation in normally dispersive Kerr media." to be published in Opt. Lett.

QThE 10:30 am Hyatt B

Microcavity Effects I

Yoshi Yamamoto, *Stanford University*, *Presider*

QThE1 (Invited) 10:30 am

Cavity QED in the evanscent zone of quartz microspheres

H. J. Kimble, H. Mabuchi, *Norman Bridge Laboratory of Physics, 12-33, California Institute of Technology, Pasadena, California 91125*

The extremely small volumes and high quality factors for the whispering gallery modes of quartz microspheres offer the potential for achieving strong coupling in cavity QED. Schemes for QND measurement and for binding atoms in orbit around the resonator will be discussed, as will our experimental progress.

QThE2 (Invited) 11:00 am

Quantum nonlinear dressed excitons in a single-quantum-well microcavity

J. M. Jacobson, H. Cao, S. Pau, G. Björk, I. Chuang, Y. Yamamoto,* E. Hanamura,** *ERATO Quantum Fluctuation Project, Edward L. Ginzton Laboratory, Stanford University, Stanford, California 94305*

Recently there has been tremendous interest in analogs to atom–cavity interactions carried out in monolithic-semiconductor-quantum-well microcavities.[1–3] These papers have realized exciton–microcavity field interactions that may be treated as linear coupled harmonic oscillators, including the so-called vacuum Rabi splitting and oscillatory (or reversible) spontaneous emission. In this work we realize a system of quantum nonlinear harmonic oscillators by exciting a single-quantum-well microcavity to high exciton density. In this regime excitons undergo Coulombic exciton–exciton interactions (biexciton formation). In the presence of this interaction the Heisenberg equations of motion become nonlinear. Alternatively, in the dressed-state picture, transitions that were forbidden now become allowed. The result is the onset of additional spectral sidebands.

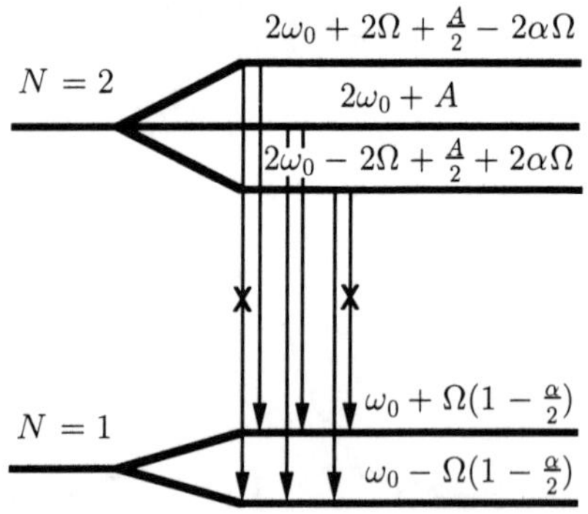

QThE2 Fig. 1. Dressed-state level diagram of the $N = 2$ and $N = 1$ manifolds. The level energies are written to lowest order in the Coulomb and phase-space-filling nonlinear coefficients, A and α, respectively. The X marks transitions that are forbidden in the absence of a nonlinearity.

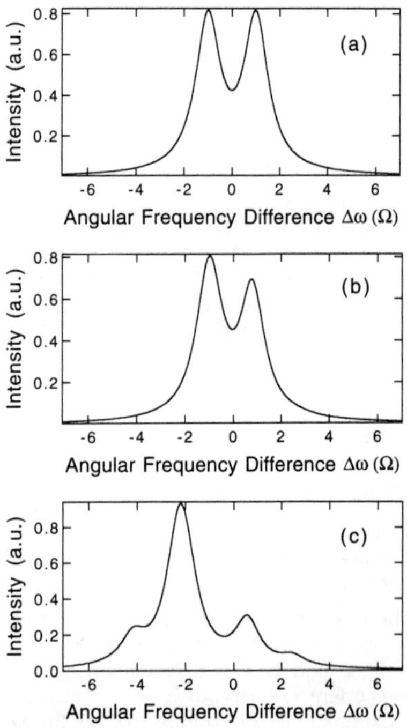

QThE2 Fig. 2. Theoretical spectra from the ($N = 2$)-to-($N = 1$)-dressed-state transitions. The horizontal axis $\Delta\omega = \omega - \omega_0$ is in Ω. The spectra are computed numerically, retaining all terms of the nonlinearity. Broadening is added phenomenologically. (a) No nonlinearity, $\alpha = 0$, $A/\Omega = 0$; (b) phase-space filling and small biexciton effect, $\alpha = 0.1$, $A/\Omega = -0.13$; (c) phase-space filling and large biexciton effect, $\alpha = 0.1$, $A/\Omega = -1.33$.

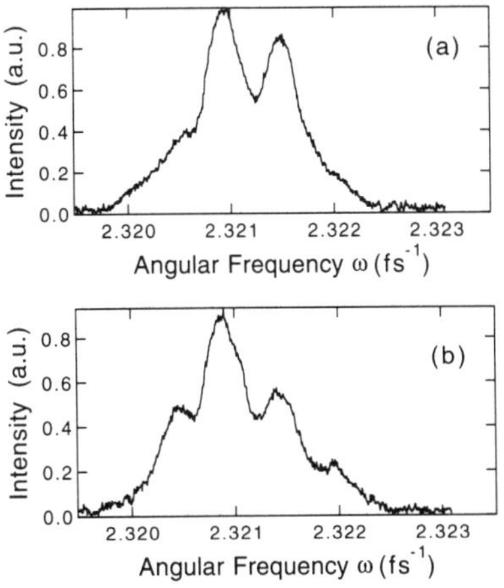

QThE2 Fig. 3. Experimental spectra for (a) circularly polarized and (b) linearly polarized excitation.

The Hamiltonian for the exciton–cavity system is given by

$$\hat{H} = \hbar\omega_p \hat{a}^{\dagger}\hat{a} + \hbar\omega_e \hat{B}^{\dagger}\hat{B} + \hbar A \hat{B}^{\dagger}\hat{B}^{\dagger}\hat{B}\hat{B} + \hbar\Omega(\hat{a}^{\dagger}\hat{B} + \hat{a}\hat{B}^{\dagger}), \quad (1)$$

where $\hat{a}$ and $\hat{B}$ are the photon and exciton operators, respectively, and $\hbar A$ gives the binding energy of the biexciton.

Figure 1 shows the resulting dressed-state manifold for the lowest-lying levels. Forbidden transitions (marked by X) become allowed under the influence of the nonlinearity. Figures 2(a) –2(c) show the predicted spectral output for no, small, and large attractive biexciton nonlinearity, respectively. We note that for a repulsive nonlinearity the spectra would be flipped about the $\Delta\omega = 0$ axis. Figure 3 shows the emission from the quantum-well microcavity for circularly and linearly polarized excitation of the same intensity. The onset of sidebands is clearly visible.

We note that whereas a linear quantum system (either atom–cavity or exciton–cavity) may be treated either classically or quantum mechanically, the present system is fundamentally quantum mechanical in that the nonlinearity requires that proper commutator ordering be maintained in order to properly calculate the details of the emitted spectra.

**Also with NTT Basic Research Laboratories, Atsugishi, Kanagawa, Japan*

***Department of Applied Physics, University of Tokyo, Hongo, Bunkyo-ku, Tokyo 113, Japan*

1. C. Weisbuch, M. Nishioka, A. Ishikawa, and Y. Arakawa, Phys. Rev. Lett. **69**, 3314 (1992).
2. T. B. Norris, J.-K. Rhee, C. Y. Sung, Y. Arakawa, M. Nishioka, and C. Weisbuch, in *Quantum Electronics and Laser Science*, 1993 Technical Digest Series, Vol. 12 (Optical Society of America, Washington, D.C., 1993), postdeadline paper QPD19-1/42.
3. J. Jacobson, S. Pau, H. Cao, G. Björk, and Y. Yamamoto, in *International Quantum Electronics Conference*, 1994 Technical Digest Series, Vol. **9** (Optical Society of America, Washington, D.C., 1994), postdeadline paper QPD21-50.

QThE3 **11:30 am**

Alteration of the spontaneous-emission rate in GaAs/AlGaAs/AlAs multilayer structures

M. Tocci, M. J. Bloemer,* M. Scalora,* J. P. Dowling,* C. M. Bowden,* *University of Alabama in Huntsville, Huntsville, Alabama 35899*

The emission rate of a semiconductor p-i-n light-emitting diode can be altered by surrounding the emitting region with a one-dimensional virtual cavity.[1] The cavity used in this experiment is actually a pair of high-reflectance Bragg mirrors, tuned so that the reflectivity band edge coincides with the wavelength of peak emission from GaAs.

We begin by using a pulse-propagation model that treats both the spatial and temporal dynamics of the electromagnetic field and atomic polarization. The results of this numerical experiment[2] show that the total power emitted can be enhanced for frequencies near the photonic band edge by more than an order of magnitude and can be reduced by nearly 3 orders of magnitude for frequencies near the center of the forbidden gap (Fig. 1).

A matrix-transfer method[3] was next used to find the complex transmission coefficients for a realistic semiconductor structure, and a novel method for finding spontaneous-emission rates was devised by using the results of the matrix-transfer model. The spontaneous-emission enhancement (over the free-space rate) was multiplied by the bulk-emission rate of GaAs (as used in a typical p-i-n diode) to predict the actual emission rate of such a cavity-confined emitter. This calculation was performed for

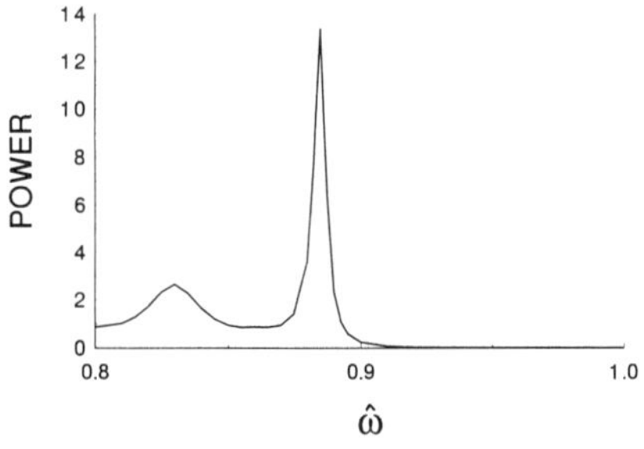

QThE3 Fig. 1. Power emitted from a radiating dipole located between two Bragg reflectors, normalized to the free-space emission rate. This plot was made by assuming an emitter in the center layer of a 1-D high-reflectance stack with layer thicknesses $\lambda_0/4n_l$, where n_l is the index of refraction of the layer and $\lambda_0 = c/2\pi\omega_0$ is called the "monitor wavelength" of the stack, so that at an incident wavelength of λ_0 the stack acts as a very good reflector. In this plot $\hat{\omega} = \omega/\omega_0$ is the normalized frequency of emitted light. The high-reflectivity band edge of the reflector is at $\hat{\omega} \simeq 0.88$. It is at this reflectivity band edge that the power emitted from the dipole is seen to increase dramatically. Likewise, the power emission is seen to drop well below the free-space rate inside the high-reflectivity band, that is, when $\hat{\omega} > 0.9$.

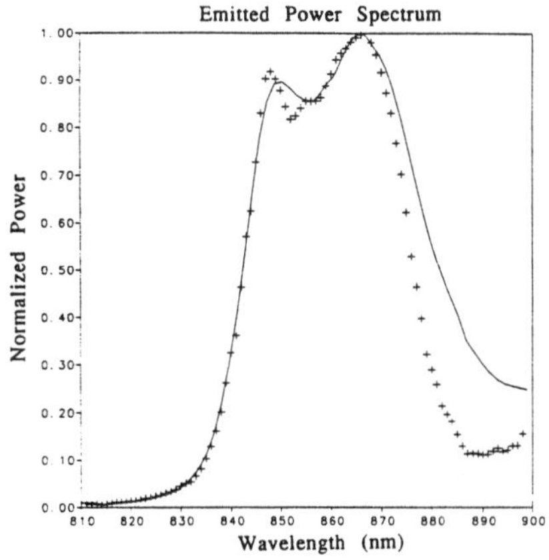

QThE3 Fig. 3. Normalized power emission from the structure shown in Fig. 2. The curve represents the calculated results from the matrix-transfer model, and the crosses represent experimental data.

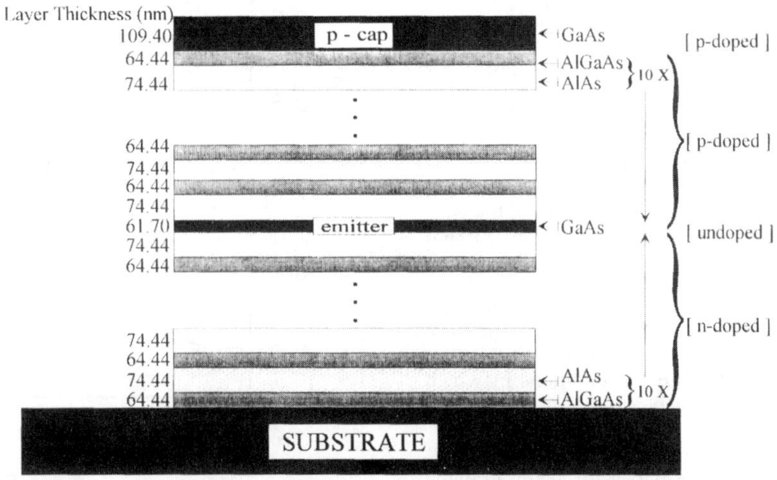

QThE3 Fig. 2. Schematic representation of the multilayer structure used in the experiment and in the matrix-transfer modeling. The structure consists of an n-doped (Si = 2 × 10^{18} cm^{-3}) GaAs substrate followed by 10 periods of alternating n-doped (Si = 1 × 10^{17} cm^{-3}) $Al_{x=0.2}GaAs$ and AlAs, then a single layer of intrinsic (undoped) GaAs followed by 10 periods of alternating p-doped (Be = 1 × 10^{17} cm^{-3}) AlAs and $Al_{x=0.2}GaAs$, and then a final layer of p-doped (Be = 2 × 10^{18} cm^{-3}) GaAs.

two cases: (1) the reflectivity band edge occurring at roughly the same wavelength as that of the peak emission from bulk GaAs and (2) the reflectivity band gap occurring at the GaAs peak.

We then had epitaxially grown an actual sample of the device we had modeled. The materials, doping scheme, and thicknesses are presented in Fig. 2. The power emitted was measured as a function of wavelength by using a diffraction-grating spectrometer, and the experimental results agree well with those calculated previously. The results of the experiment are shown, along with the calculated emission rate, in Fig. 3.

Weapons Sciences Directorate, AMSMI-RD-WS-ST, U.S. Army Missile Command, Redstone Arsenal, Alabama 35898-5248

1. E. M. Purcell, Phys. Rev. A **69**, 681 (1946).
2. M. Scalora, J. P. Dowling, M. Tocci, M. J. Bloemer, and C. M. Bowden, in feature issue of Appl. Phys. B: Fundamental Systems in Quantum Optics, in press.
3. J. He and M. Cada, Appl. Phys. Lett. **61**, 2150 (1992).

QThE4 **11:45 am**

Spontaneous emission in planar semiconductor microcavities displaying the vacuum Rabi splitting

I. Abram, J. L. Oudar, *Bagneux Laboratory, France Telecom, CNET/PAB, 196 Avenue Henri Ravera, 92220 Bagneux, France*

The recent observation of exciton–photon mode splitting in the spectra of quantum wells embedded in monolithic epitaxially grown semiconductor microcavities[1,2] has been interpreted as the vacuum Rabi splitting of the exciton resonance, attesting to a strong coupling between the cavity modes and the excitons. By analogy to the quantum-electrodynamic effects observed in atomic physics inside single-mode cavities, it is thought that this strong coupling can be exploited to modify the spontaneous-emission dynamics of the excitons, thus opening the way, for example, for controlling the initiation of laser action in vertical-cavity surface-emitting microlasers and significantly lowering their threshold.

We calculate the spontaneous-emission characteristics of a quantum-well exciton embedded in the center of a planar microcavity, and we compare them with the emission characteristics of a single-mode cavity with lateral confinement of the electromagnetic field. The results indicate that because the modal structure of a planar microcavity consists of a spectrally dense continuum, the exciton decays into the cavity continuum with a time constant close to the free-space spontaneous emission lifetime, even under conditions in which its spectrum displays vacuum Rabi splitting.

This behavior contrasts with the case of a single-mode cavity with lateral in-plane confinement of the field, in which the exciton level is coupled to the single discrete mode of the cavity. In that situation, vacuum Rabi splitting is the signature of a strong coupling between the exciton and the cavity mode. The strong coupling gives rise to a coherent oscillatory exchange of the energy between the exciton and the cavity, so that if the decay constants are suitably adjusted, the strongly coupled exciton–cavity system can decay preferentially into the de-excitation channel of the cavity, giving rise to a directional beam emerging from the cavity, rather than to the omnidirectional spontaneous emission associated with the radiative decay of the exciton.

From the point of view of technological applications, this analysis indicates that simple planar microcavities cannot be used to bring about a significant modification of the spontaneous-emission characteristics of the excited material they contain, because of the existence of a spectral continuum in their modal structure. In order for one to observe cavity quantum-electrodynamic effects associated with strong coupling and vacuum Rabi splitting, the spectral continuum of the planar cavity must be broken up into a small number of sparsely spaced discrete modes, thus approximating the single-mode situation. This can be achieved by confining the field also in the two dimensions of the cavity plane, by laterally patterning the semiconductor wafer.

1. C. Weisbuch, M. Nishioka, A. Ishikawa, and Y. Arakawa, Phys. Rev. Lett. **69**, 3314 (1992).
2. I. Abram, S. Iung, R. Kuszelewicz, G. LeRoux, C. Licoppe, J. L. Oudar, E. V. K. Rao, J. Bloch, R. Planel, and V. Thierry-Mieg, Appl. Phys. Lett., in press.

QThF **10:30 am**
Hyatt C/D

Localization, Disorder, and Scattering

Jurgen Kuhl, *Max Planck Institute, Presider*

QThF1 **10:30 am**

Fast exciton localization in gallium selenide

V. Mizeikis,* V. G. Lyssenko,** J. Erland,† D. Birkedal, J. M. Hvam, *Mikroelectronics Center, Building 345, Technical University of Denmark, DK-2800 Lyngby, Denmark*

Stacking faults in the layered compound semiconductor gallium selenide introduce one-dimensional disorder that localizes excitons along the crystallographic c axis, whereas in other two dimensions the material retains the crystalline structure and the excitons are extended.[1,2]

We study the effects of the disorder and localization in GaSe by transient degenerate four-wave mixing (TDFWM) in the excitonic absorption region, inhomogeneously broadened by the stacking-fault disorder. Samples of ε-GaSe with a thickness of ≈15 μm were excited resonantly by picosecond laser pulses (≈3-ps autocorrelation FWHM). The TDFWM signal was spectrally resolved and was recorded time integrated as a function of the delay t_{12} between the pump and probe pulses.

We find that exciton localization, faster than the pulse duration, saturates lower states in the one-dimensional disordered band. As a consequence, the TDFWM spec-

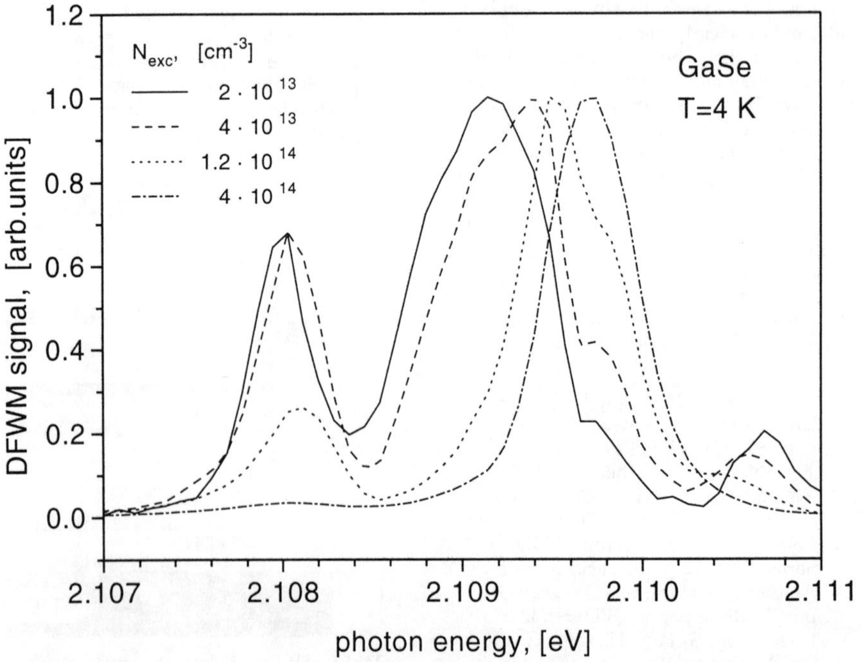

QThF1 **Fig. 1.** Normalized TDFWM spectra at various exciton densities. The fine structure in the spectra is due to statistic predominance of some stacking-fault sequences. We concentrate on the behavior around the main peak at ≈2.1095 eV. The relative contribution to the TDFWM signal from the states on the low-energy side of the main peak decreases with increasing exciton density.

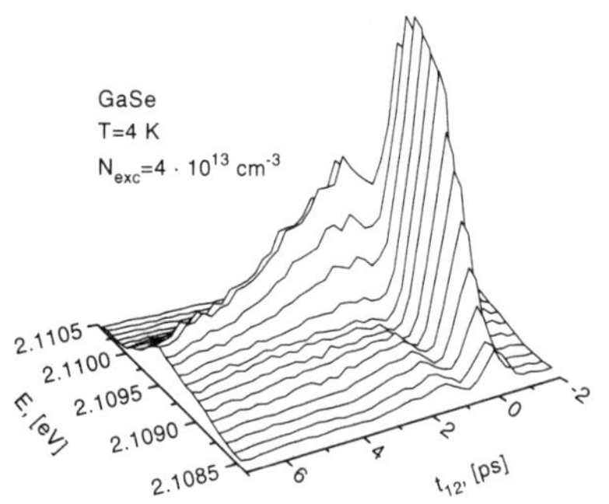

QThF1 Fig. 2. Spectrally resolved TDFWM signal (in linear scale) around its main peak, as a function of t_{12}. The time-integrated photon-echo signal, with a characteristic maximum at positive delay times, is weak in the lower part of the spectrum.

tra shift to the higher energies with increasing exciton density (Fig. 1).The excitons, created in the high-energy part of the disordered one-dimensional band, rapidly relax to the lower (localized) states before they recombine, whereas the localized excitons predominantly retain the same energy.[2] Because of the comparatively long exciton lifetime in GaSe (≈80 ps), the intraband relaxation saturates states in the lower part of the band, reducing their contribution to the TDFWM signal and shifting the maximum to higher energies.

The illustration of such behavior is given in Fig. 2. The TDFWM signal, having characteristic photon-echo features,[3] is strong only at higher energies. In the low-energy part, only the signal due to the pump–probe overlap at $t_{12} = 0$ is intense.

At higher exciton densities the TDFWM spectra are shifted to the higher energies, where the excitonic states are substantially delocalized and inhomogeneous broadening is small. In this case the signal dependence on t_{12}, observed on the high-energy side of the TDFWM spectrum, is close to that expected for the free polarization decay.[3]

**Permanent address: Department of Solid State Electronics, Vilnius University, Sauletekio 9, Building III, 2054 Vilnius, Lithuania*

***Permanent address: Institute of Microelectronics Technology and Superpure Materials, Chernogolovka, Moscow District 142432, Russia*

† Permanent address: Department of Physics, Odense University, Campusvej 55, DK-5230 Odense M, Denmark

1. J. Forney, K. Maschke, E. Mooser, J. Phys. C **10**, 1887 (1977).
2. V. Capozzi, K. Maschke, Phys. Rev. B **34**, 3924 (1986).
3. J. Erland *et al.*, Phys. Rev. B, in press.

QThF2 **10:45 am**

Microroughness localization of exitons in GaAs/AlGaAs multiple quantum wells studied by ultrafast transient four-wave mixing

D. Birkedal, V. G. Lyssenko, J. Erland, J. M. Hvam, *Microelectronics Center, Building 345, Technical University of Denmark, DK2800 Lyngby, Denmark*

The dephasing of excitons in quantum wells has been intensively studied by transient four-wave mixing (TFWM) in the past. The TFWM signal has been shown to be in the form of photon echoes[1] (PE) or free induction decays[2] (FID) or to be in a form in which no clear distinction can be made between PE and FID.[3,4] In this paper we show that the dephasing mechanisms of the excitons are linked to the microscopic nature of the interfaces. Using spectrally resolved TFWM, we are able to distinguish between dephasing mechanisms characteristic of the different types of interface morphology.

The investigated sample is an MBE-grown GaAs/$Al_{0.3}Ga_{0.7}As$ multiple-quantum-well structure, with 20 wells of 110-Å GaAs excited by 150-fs pulses from a Ti:sapphire laser focused to approximately 40 μm at the sample surface. The TFWM signal was dispersed in a spectrometer and was detected in an optical multichannel analyzer with a resolution better than 0.1 meV.

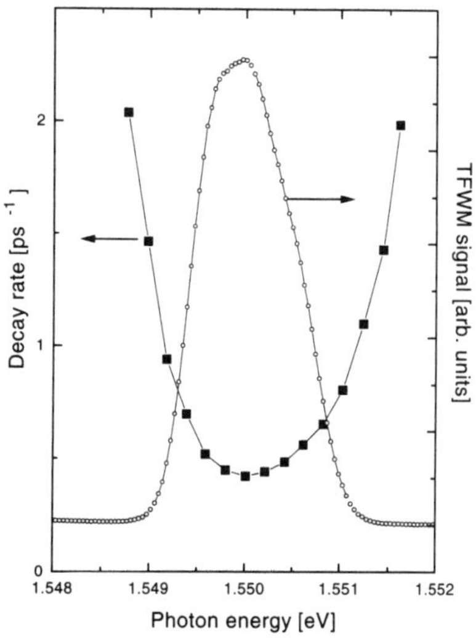

QThF2 Fig. 1. Spectrally resolved transient four-wave-mixing signal from the multiple-quantum-well sample in the region of the heavy-hole exciton. The spectrum shows heavy–light-hole quantum beats as a function of delay. Also shown is the spectral splitting of the signal for near-zero delay due to monolayer step formation.

Figure 1 shows the spectrally resolved TFWM signal as a function of delay τ. The spectrum shows an initial doublet with a splitting of 1.35 meV, decaying into a single line centered at the average value of the two initial peaks. In addition, modulation due to quantum beats between heavy- and light-hole excitons is seen.

The two initial peaks decay as FID originating from two atomically smooth areas larger than the exciton diameter, with a difference in thickness of one monolayer. The single peak observed for longer delays is a time-integrated PE from regions with roughness on a length scale smaller than the exciton diameter. The exciton energies are distributed between the exciton energies of the smooth areas, giving rise to an inhomogeneously broadened signal.

The excitons created in the large smooth areas are mobile extended states, resulting in a predominantly homogeneously broadened signal. However, the excitons created in the areas of microroughness (single atomic steps) are localized. We show in Fig. 2 the decay rate of the singlet for different photon energies, together with the TFWM signal at a 5-ps delay. A clear minimum in the decay rate is seen at the center of the line, showing the effect of microroughness localization. The states with energies in the wings of the line approach the characteristics of the delocalized states, which have faster dephasing times.

In summary, we have shown that the presence of microroughness leads to a minimum in the decay rate at the center of the TFWM line, as opposed to localization due to potential fluctuations on a length scale larger than the exciton diameter, for which the low-energy wing dephases the most slowly.[4]

1. L. Schultheis, M. Sturge, and J. Hegarty, Appl. Phys. Lett. **47**, 995 (1985).
2. L. Schultheis, A. Honold, J. Kuhl, K. Köler, and C. W. Tu, Phys. Rev. B **34**, 9027 (1986).

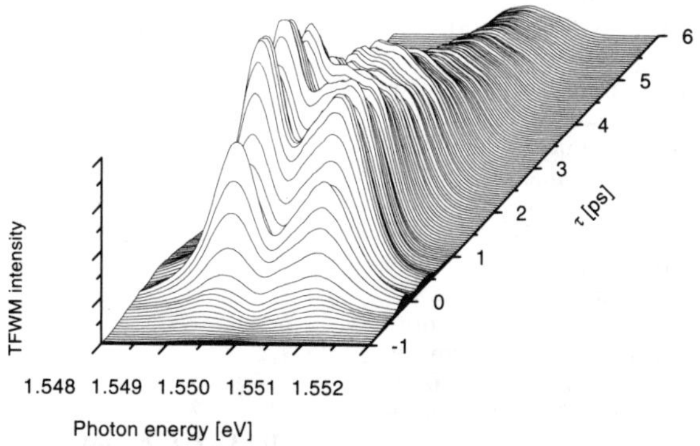

QThF2 Fig. 2. Decay rate of the TFWM signal for various spectral positions in the single line for long delays. Also shown is the spectral line shape of the TFWM signal at a delay of 5 ps.

3. M. Webb, S. Cundiff, and D. Steel, Phys. Rev. B **43**, 12658 (1991).
4. S. Cundiff, H. Wang, and D. Steel, Phys. Rev. B **46**, 7248 (1992).

QThF3 **11:00 am**

Direct observation of population-induced broadening in quantum-well intersubband transitions

Gilad Almogy, Yuanjian Xu, Andrew Tong, Ali Shakouri, Amnon Yariv, *Department of Applied Physics, California Institute of Technology, Pasadena, California 91125*

We report on the use of charge transfer between asymmetric double quantum wells (ADQWs) to directly study the broadening of intersubband transitions (ISBTs) as a function of the electron population. Although other mechanisms, such as field-induced scattering over interface roughness and impurities, optical and acoustic phonons, and width nonuniformity, play a role, it is shown that the broadening is directly related to subband occupation.

The high resolution necessary for studying the behavior of intersubband transitions under applied bias was obtained by the monolithic integration of a quantum-well infrared photodetector (QWIP) and the coupled wells under study. This characterization technique will be discussed in detail elsewhere. The asymmetric double quantum well (ADQW) consists of 50 periods of an undoped narrow well (9-nm GaAs), a 3-nm $Ga_{0.6}Al_{0.4}As$ barrier, and a 10.8-nm storage well nominally silicon doped to 2×10^{18} cm^{-3} from 0.5 to 4.5 nm away from the barrier. The periods are separated by a 42.2-nm $Ga_{0.6}Al_{0.4}As$ layer. The wells (inset of Fig. 1) are designed to maximize the charge transfer with applied external bias.

The room-temperature absorption spectrum and its curve fit are shown in Fig. 1. Close to 25% of the total electron population is in the narrow well, which has a full-width-at-half-maximum (FWHM) broadening of $\Gamma \cong 9$ meV. The experiment is conducted at 10 K, reducing the narrow-well population to roughly 4% and the broadening to $\Gamma \cong 3.75$ meV. Application of external bias shifts the population from the wide well to the narrow well, leading to the linear increase of the integrated absorbance in the narrow well shown in Fig. 2(a). The maximum population transferred to the narrow well is estimated from the integated absorbance at 6×10^{11} cm^{-2} filling roughly 20 meV of the ground subband. When the opposite (depleting) bias is applied, from the bias-induced broadening of constant-population wells and from theoretical models[1] we deduce that the bias-induced-broadening contribution is roughly 1 meV. The rest of the broadening [Fig. 2(b)] is consistent with the filling of the ground subband of the narrow well. The contribution of single-particle nonparobilicity is negligible, but the effective nonparobilicity created by the momentum dependence of the exchange interaction[2] leads to the measured broadening. The contribution of the electron–electron scattering is expected to be minor[3] at these densities. It should be noticed that the large difference between the broadening at low tem-

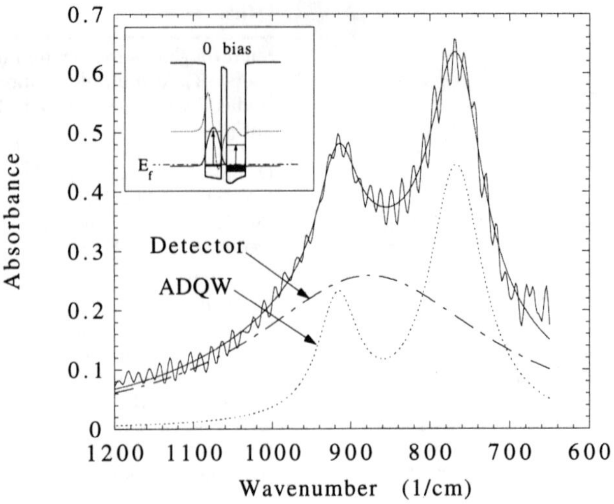

QThF3 Fig. 1. Room-temperature absorption of the asymmetric coupled quantum wells (ACQWs) together with the monolithic detector and their respective curve fits (obtained by etching away the detector). A self-consistent model of the coupled wells under no bias is shown in the inset.

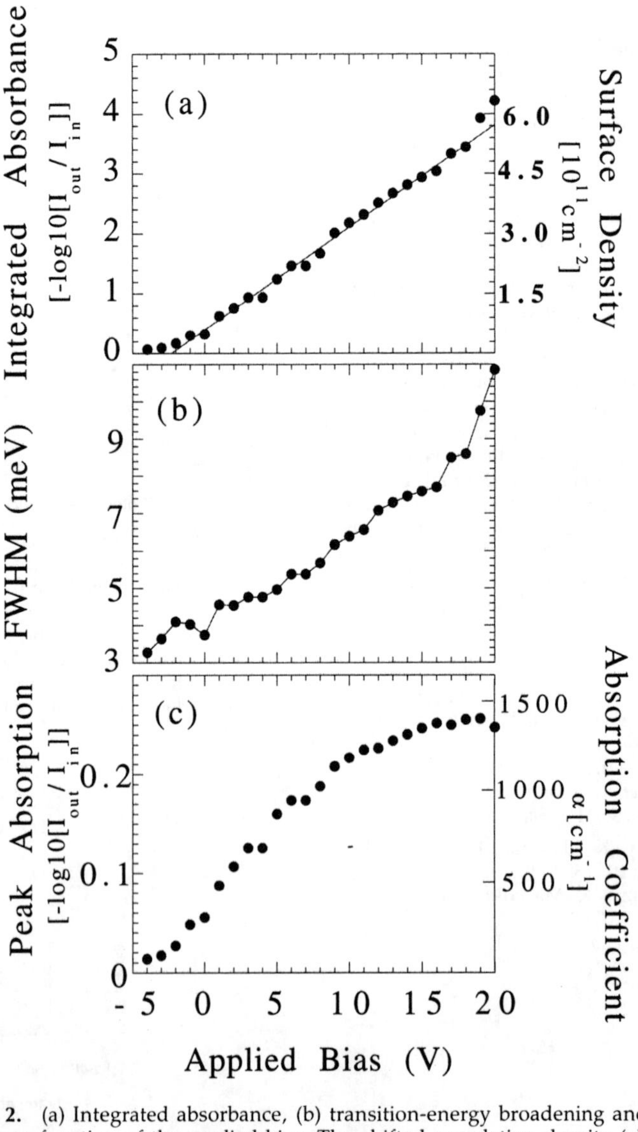

QThF3 Fig. 2. (a) Integrated absorbance, (b) transition-energy broadening and (c) peak absorbance as a function of the applied bias. The shifted population density (a) is estimated by using the calculated well parameters and by comparison with the absorption of the wide well.

perature (10 K) and at room temperature, much larger than typically measured for constant well populations, is consistent with the increased population in the narrow well and with the thermal spread to higher momentum states.

The competing contributions of the increased population and the increased broadening lead to the eventual saturation of the peak absorption height at a population density of under 10^{12} cm^{-2}, as shown in Fig. 2(c). Although the exact number may be material dependent, our results show that population-induced broadening is a fundamental limit on the magnitude of ISBT optical interactions and that the broadening may be significantly decreased by separating the donors from the wells.[4]

1. R. Ferreira and G. Bastard, Phys. Rev. B **40**, 1074 (1989).
2. J. W. Choe, O. Byungsung, K. M. S. V. Bandara, and D. D.Coon, Superlattices Microstruct. **10**, 1 (1991).
3. P. Sotirelis, P. Von Allmen, and K. Hess, Phys. Rev. B **47**, 12744 (1993).
4. E. B. Dupont, D. Delacourt, D. Papillon, J. P. Schnell, and M. Papuchon, Appl. Phys. Lett. **60**, 2121 (1992).

QThF4 11:15 am

Exciton scattering with bare electrons

M. Koch, R. Hellmann, G. Bastian, J. Feldmann, E. O. Göbel, P. Dawson,* *Department of Physics and Scientific Center for Materials Sciences, Renthof 5, D-35032 Marburg, Germany*

Transient four-wave-mixing (FWM) experiments using a prepulse have shown that the exciton dephasing rate due to exciton↔electron–hole-pair scattering is 10 times as efficient as exciton↔exciton scattering.[1] Here, we focus on the scattering of excitons with bare electrons. We have performed transient FWM and photoluminescence studies on mixed type-I–type-II GaAs/AlAs asymmetric double quantum wells.[2] The band scheme of such a structure is depicted in the inset of Fig. 1. These structures provide the unique possibility of investigating the scattering of excitons with bare charged electrons. They consist of a 2.5-nm narrow GaAs well (NW) and a 6.8-nm wide GaAs well (WW) separated by an AlAs barrier of varying thickness. With regard to the lowest quantized conduction-band and valence-band states, the NW and WW have type-II and type-I characters, respectively.

As schematically illustrated in the inset of Fig. 1, the FWM experiments are performed on the energetically lowest excitonic $n = 1$ transition of the WW by using 100-fs pulses from a mode-locked Ti:sapphire laser. We have performed two kind of experiments. First, a prepulse from the same Ti:sapphire laser is used to create an incoherent density of excitons in the WW in order to measure the dephasing rate due to exciton–exciton (X/X) scattering, as was done in Ref.[1] Second, a cw He–Ne laser is used to photocreate electron–hole pairs in the NW, which then spatially separate because of an efficient $\Gamma \rightarrow X \rightarrow \Gamma$ electron transfer from the NW to the WW. Because of the extremely long lifetime of these spatially separated electron–hole pairs, the

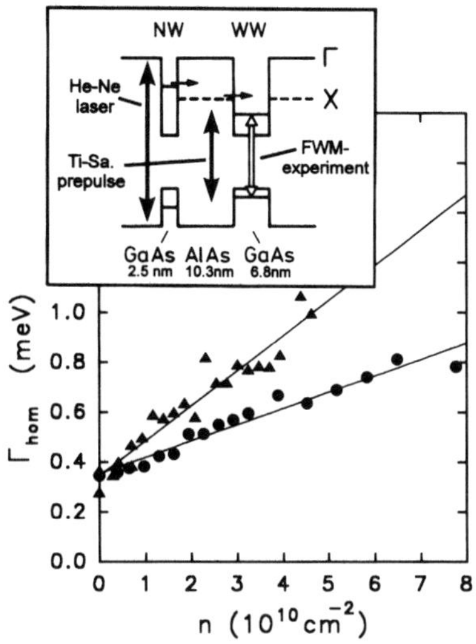

QThF4 Fig. 1. Experimentally determined homogeneous linewidth as a function of the exciton density and as a function of the density of bare electrons. The inset schematically illustrates the excitation conditions for the mixed type-I–type-II sample.

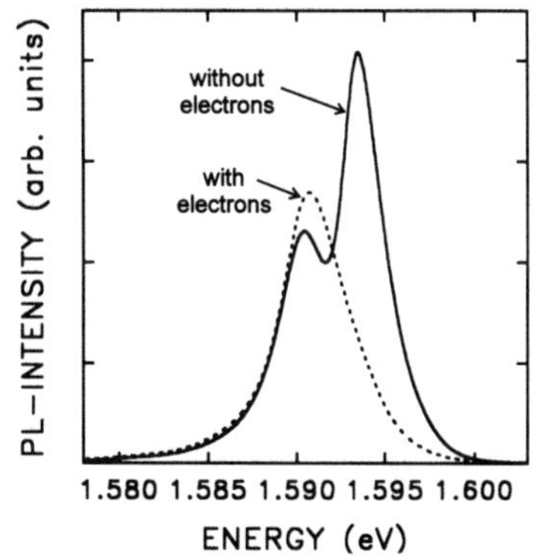

QThF4 Fig. 2. Photoluminescence spectra of the WW in the absence (solid curve) and in the presence (dashed curve) of bare electrons.

density of cold electrons in the WW can be easily controlled over a wide range by changing the intensity of the He–Ne laser. This experiment allows the determination of the dephasing rate due to exciton–electron (X/e$^-$) scattering in the absence of any holes in the WW. In Fig. 1 the experimentally determined homogeneous linewidth is plotted versus the density of excitons (circles) and versus the density of bare electrons (triangles). We find that X/e$^-$ scattering is twice as efficient as X/X scattering, in reasonable agreement with theoretical calculations.[3]

In addition, we observe that the creation of cold electrons in the WW has a drastic effect on the photoluminescence of WW excitons. The solid curve in Fig. 2 is the WW photoluminescence spectrum in the absence of bare electrons. The two peaks correspond to exciton transitions associated with distinct quantum-well islands. The presence of 6×10^{10}-cm^{-2} bare electrons (again photoexcited by a He–Ne laser) leads to the WW photoluminescence spectrum shown by the dashed curve in Fig. 2. The high-energy luminescence peak corresponding to quantum well islands having a narrower well width is quenched by the presence of bare electrons. We argue that excitons are localized on these islands because of disorder on a small spatial scale but become mobile in the presence of bare electrons because of X/e$^-$ scattering. Then excitons diffuse and relax to the low-energy islands and, in addition, migrate to nonradiative recombination centers leading, respectively, to only one luminescence peak and a decreased overall quantum efficiency.

*Department of Pure and Applied Physics, UMIST, Manchester M60 1QD, U.K.

1. L. Schultheis *et al.*, Phys. Rev. Lett. **57**, 1635 (1986).
2. P. Dawson *et al.*, Appl. Phys. Lett. **58**, 2889 (1991).
3. H. Haug, Z. Phys. B **24**, 351 (1976).

QThF5 11:30 am

Dephasing of localized and delocalized excitons in disordered $Cd_{1-x}Mn_xTe/Cd_{0.60}Mg_{0.40}Te$ quantum wells

R. Hellmann, S. T. Cundiff, M. Koch, J. Feldmann, E. O. Göbel, G. Mackh,* A. Waag,* G. Landwehr,* *Department of Physics and Scientific Center for Materials Sciences, Philipps University, Renthof 5, W-35032 Marburg, Germany*

Transient four-wave mixing (TFWM) has recently been used to study the dynamics of coherently excited states in semiconductor quantum wells.[1] It allows the direct measurement of the excitonic dephasing time T_2, which is determined by the interaction of the excitons with their environment. Excitons typically dephase by scattering with other quasi particles, such as ponons and other carriers. Additionally, disorder can be significant in determining the dephasing rate. It can lead either to localization of excitons, and thereby to a modification of their interaction with the environment, or to exciton scattering by the disorder potential. In semimagnetic semiconductors the strong exchange interaction between the excitons and the magnetic ions can also result in additional dephasing.

We present here low-temperature TFWM experiments on $Cd_{1-x}Mn_xTe/Cd_{0.60}Mg_{0.40}Te$ multiple quantum wells ($0 \leq x \leq 0.1$). Disorder arises from well-width fluctuations and, for $x \neq 0$, from compositional disorder inside the well. On the basis of the distinct spectral, temperature, and density dependences of the dephasing rate γ_{ph} in nonmagnetic structures ($x = 0$), we can spectrally classify excitons into localized and delocalized regimes below and above the absorption-line center.[2,3] Previous attempts to investigate γ_{ph} above the absorption-line center have been hindered by the significant shortening of the dephasing time in this regime due to excitation of higher-lying exciton and continuum states. We overcome this by using wide-gap II–VI materials, which inherently exhibit large exciton binding energies. In particular, we find that localized excitons dephase by phonon-assisted tunneling and activation to extended states, leading to an activation-type behavior of the temperature dependence of γ_{ph}. In this regime, i.e., below 1.660 eV, γ_{ph}

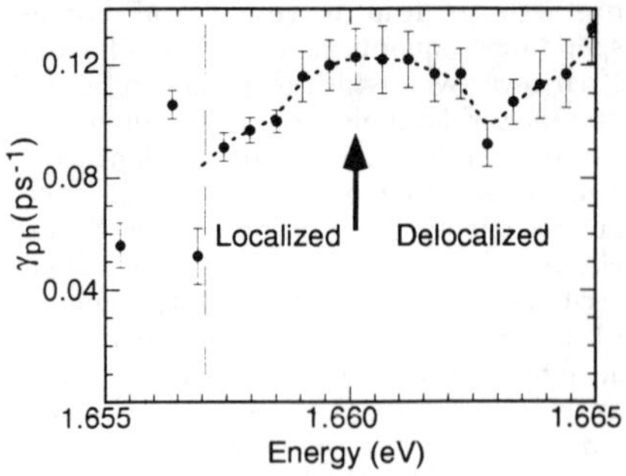

QThF5 Fig. 1. Dephasing rate γ_{ph} as a function of photon energy within the spectral region of the inhomogeneously broadened 1*s* heavy–hole exciton resonance (dashed curve is a guide to the eye). This resonance is split by large-scale interface islands. We focus here on the high-energy peak above 1.657 eV. The error bars account for the density dependence of γ_{ph}. The vertical arrow indicates the absorption-line center.

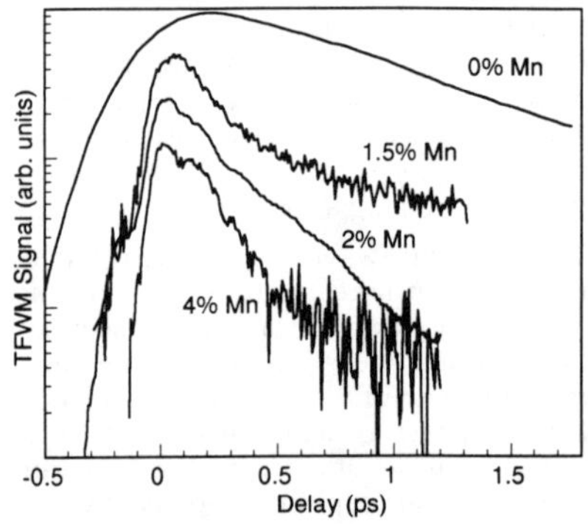

QThF5 Fig. 2. TFWM signal as a function of time delay for various Mn contents. The curves are normalized and are offset for clarity.

increases with increasing energy because of the increasing number of phonons with sufficient energy to activate the excitons (Fig. 1). Delocalized excitons predominantly dephase by acoustic-phonon-mediated intraband scattering. This leads to a linear temperature dependence and to a decrease of γ_{ph} with increasing energy above 1.660 eV because of a decreasing population of phonons as the energy increases.[2] With further increase in energy, γ_{ph} increases because of multiphonon processes.

In semimagnetic quantum wells ($x \neq 0$) the excitonic dephasing is strongly enhanced over that of nonmagnetic material (Fig. 2) and is relatively temperature and density independent. The later dependences indicate that exciton–phonon and exciton–exciton interactions are not the pertinent dephasing mechanisms. The nonexponential decay of the TFWM signal is due to the presence of slowly dephasing impurity bound excitons. The incorporation of manganese into the well results in (1) the introduction of spins associated with the half-filled 3*d* shell of the manganese, which strongly interact with the excitons through the exchange interaction, and (2) the well becoming an alloy. Hence, the changes in the dephasing can be caused by localization effects, alloy scattering, or the spin–spin interaction. We conclude that localization is not essential, because it often results in a reduced dephasing, contrary to our observations. Disentangling the relative contributions of alloy scattering and spin–spin interaction, which both are definitely present, will require further experiments comparing semimagnetic (CdMnTe) with nonmagnetic (CdMgTe) wells.

**Institute of Physics, University of Würzburg, W-97074 Würzburg, Germany*

1. See, e.g., Optics in Semiconductor Nanostructures, F. Henneberger, S. Schmitt-Rink, and E. O. Göbel eds. (Akademie Verlag, Berlin, 1993).
2. Takagahara, Phys. Rev B **32**, 7013 (1985).
3. R. Hellmann, S. T. Cundiff, M. Koch, J. Feldmann, E. O. Göbel, B. Kuhn-Heinrich, D. R. Yakovlev, A. Waag, and G. Landwehr, Phys. Rev. B, in press.

QThF6 11:45 am

Dephasing mechanisms and homogeneous linewidths of excitons in CdTe quantum wells

E. J. Mayer, N. T., Pelekanos, J. Kuhl, N. Magnea,* H. Mariette,* *Max Planck Institute for Solid State Research, Heisenbergstrasse 1, D-70569 Stuttgart, Germany*

We undertook a systematic degenerate-four-wave-mixing (DFWM) study of the coherent exciton dynamics in a series of CdTe/CdZnTe single quantum wells (QWs), with the QW thickness varying between 18 and 130 Å. The enhanced excitonic character and the shallow potential barriers (80 meV for the electrons and 25 meV for the heavy holes) are the two main features that distinguish this QW system from the intensely studied GaAs/AlGaAs one.[1] The low-temperature inhomogeneous exciton linewidths range between 1.3 and 2 meV, and the corresponding Stokes shift ranges between 0.5 and 1 meV, depending on the QW thickness. It is important to note that the exciton linewidth decreases for the narrower QWs. This behavior is contrary to expectations and is opposite to that observed for GaAs/AlGaAs systems. The DFWM measurements were two-pulse experiments, using 0.8-ps Ti:sapphire laser pulses of 4-meV spectral width tuned in resonance with the e_1h_1 heavy-hole QW exciton resonance.

In Fig. 1 we show a typical DFWM signal as a function of time delay for the 18-Å-QW sample for 10 and 25 K. In the 10-K curve we distinguish a fast and a slow decay time. We attribute the fast decay to free excitons and the slow decay to a small population of weakly localized excitons. This is supported by the fact that at higher temperatures (25 K) the slow-decay component disappears. In the rest of the paper we consider only the fast time decay.

We investigated the effect of exciton–acoustic-phonon interaction in dephasing the exciton by measuring the linear increase of the homogeneous linewidth as a function of temperature (in the range 10–25 K), which can be described by $\Gamma_{hom}(T) = \Gamma_{hom}(T = 0) + \gamma_{XP}T$. The measured

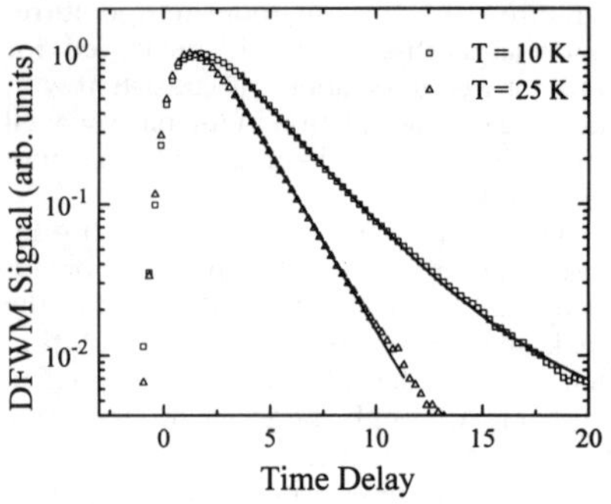

QThF6 Fig. 1. Diffracted intensity as a function of time delay from an 18-Å CdTe/CdZnTe single QW, for T = 10 K (open squares) and 25 K (open triangles); curves correspond to fits for a two- and a single-exponential decay, respectively.

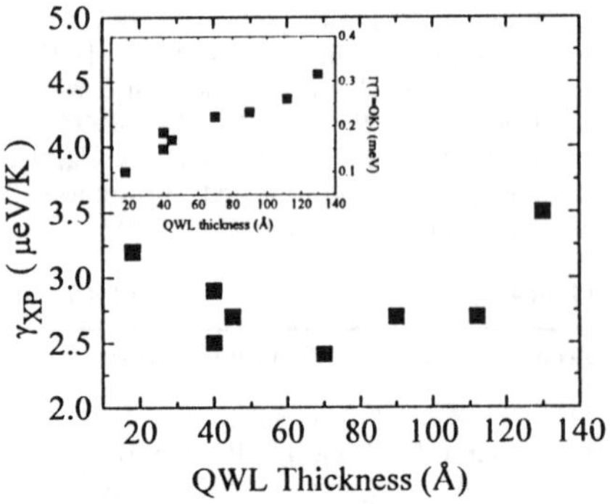

QThF6 Fig. 2. Exciton–acoustic-phonon coupling strength γ_{XP} for various QW thicknesses. The inset shows the low-temperature limit of the homogeneous linewidth for an incident exciton density of $n_{XX} = 0.5 \times 10^{10}$ cm^{-2}.

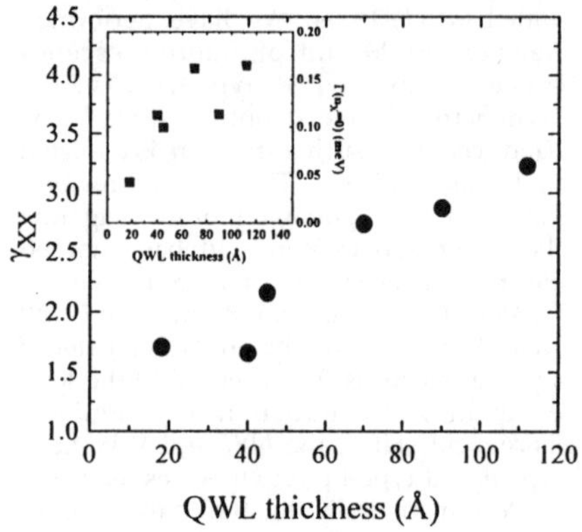

QThF6 Fig. 3. Exciton–exciton interaction strength γ_{XX} as a function of QW thickness. The low-density-limit values of the homogeneous linewidth for T = 10 K are plotted in the inset.

exciton–acoustic-phonon coupling parameter γ_{XP}, shown in Fig. 2, has no strong dependence on the QW thickness in the range between 40 and 112 Å with an average value of γ_{XP} = 2.5 μeV/K, which is a factor of 5 smaller than the value measured in a CdTe/CdMnTe QW.[2] By comparison with the scattered γ_{XP} values available for GaAs QWs, we conclude that the strength of the exciton–acoustic-phonon coupling in the CdTe QWs is consistently smaller than that in GaAs. This is in agreement with the deformation potential of CdTe being smaller than that of GaAs. In the inset of Fig. 2 we plot the low-temperature limit of the $\gamma_{hom}(T = 0\ K)$. These values decrease with decreasing QW width, indicating a reduction of other scattering mechanisms, such as exciton–exciton scattering, impurities, interface roughness, and alloy fluctuations.

To probe the exciton–exciton interaction, we monitored the DFWM-signal decay for various incident laser powers, keeping $T = 10$ K. For densities up to $n_{XX} = 10^{10}\ cm^{-2}$, we observe a linear increase of the homogeneous linewidth described by $\gamma_{hom}(n_{XX}) = \Gamma_{hom}(n_{XX} = 0) + \gamma_{XX} a_B^2 E_B n_{XX}$. Assuming an increase of the Bohr radius α_B from 45 to 70 Å and a decrease of the exciton binding energy E_B from 19 to 13 meV for QW widths between 18 and 112 Å, we show in Fig. 3 the calculated parameter γ_{XX}, representing the strength of the exciton–exciton interaction. The resulting γ_{XX} values increase with QW thickness and generally agree with values observed in GaAs QWs.[1] In the inset of Fig. 3 we show the low-density limit of $\gamma_{hom}(n_{XX} = 0)$. The observed trend is a strong decrease of $\Gamma_{hom}(n_{XX} = 0)$ with decreasing QW width. This indicates, given the relatively constant value of the exciton–phonon coupling γ_{XP}, an unexpected decrease in the influence of the remaining dephasing mechanisms, such as scattering by interface roughness and alloy fluctuations, with decreasing QW thickness. This behavior could tentatively be attributed to increasing exciton localization by the QW disorder for the narrower QWs. However, such increased QW disorder should result in larger inhomogeneous exciton linewidths and Stokes shift for the narrower QWs, which, as we noted above, is not the case.

*Microstructures of Semiconducters II-VI Joint Group, CEA-CNRS, BP 85, F-38042 Grenoble, France

1. J. Kuhl *et al.*, Advances in Solid State Physics **29**, 157 (1989).
2. R. Hellmann *et al.*, Phys. Rev. B **48**, 2847 (1993).

QThG 1:00 pm

2nd and 3rd Level of the Convention Ctr.

Poster Session

Laser Spectroscopy

QThG1

Ultrastable lasers for verifying the big-bang hypothesis from Earth

M. A. Ivanov, *Department of Physics, Belorussian State University of Informatics and Radioelectronics, Pyatrus Brovka Street 6, 220027 Minsk, Belarus*

The author formulates an alternative conjecture to the big-bang hypothesis,[1,2] namely, that there exists a local physical process that is the cause of red shifts.

Let us assume that a graviton background, similar to the microwave one, exists. Photon scattering is possible on gravitons $\gamma + h \rightarrow \gamma + h$, where γ is a photon and h is a graviton. Energy losses of a photon in such a process can be written as $dE = -aEdr$, where E is energy, a is a constant, and r is a distance to the source. Actually, the gravitational charge of a photon must be proportional to E, which gives the factor E^2 in the interaction cross section. Normalization of a photon wave function gives the factor E^{-1}. As a first approximation we have for the frequency shift $\Delta\nu/\nu = -ar_0$, where r_0 is the distance from the source to an observer.

The conjecture will be checked in the near future. For small distances the red shift is $\Delta\nu/\nu \approx -1.6 \times 10^{-18} r_0$, where ν is the frequency of radiation (if one takes r_0 in light seconds). With a local cause for the shift, light trajectory is not important. If Γ is the width of the laser-radiation line and ν is the center frequency, then a frequency shift $\Delta\nu/\nu \sim \Gamma/\nu$ will be observable. Let us assume that the suggestion by Dehmelt and Nagorney[3] is realized and that an ultrastable laser will be created with $\Gamma \sim 0.01c$ and $\gamma/\nu \sim 10^{-17}$. Then for verifying the conjecture from Earth it will be necessary to create an optical delay line with a delay ~10 s to observe an effect.

Probably the vacuum tubes and auxiliary equipment that are planned for the LIGO[4] or similar project for detection of gravitation waves from compact sources could be used to solve the problem.

1. G. Gamow, Phys. Rev. **70**, 572 (1946).
2. S. Weinberg, *The First Three Minutes: A Modern View of the Origin of the Universe* (Basic, New York, 1977).
3. H. Dehmelt, W. Nagorney, Proc. Natl. Acad. Sci. USA **85**, 7426 (1988).
4. R. E. Vogt, "The U.S. Laser Interferometer Gravitational-Wave Observatory (LIGO) project," in *Proc. Sixth Marsel Grossmann Meeting on General Relativity*, Kyoto, Japan, June 1991, LIGO 91-7 (September 1991).

Thursday Posters

QThG2

Photon echo of a single light pulse

G. N. Nikolaev, *Institute of Automation and Electrometry, Russian Academy of Sciences, Siberian Branch, 630090 Novosibirsk, Russia*

The photon echo of a single light pulse in a magnetic or an electric field is predicted. This photon echo can take place even though inhomogeneous broadening is absent. The reversible relaxation of macroscopic optical polarization is also absent in this case, and, consequently, the ordinary photon echo cannot take place.

The origin of such an unusual photon echo is as follows: The magnetic or electric field splits the resonance frequency of an atom. For the anomalous Zeeman effect or the quadratic Stark effect the number of the splitting resonance frequency is proportional to the full angular momenta of the combining atomic states. The interaction of light of a given polarization with each atomic oscillator takes place independently. The set of such oscillators of every atom plays the role of inhomogeneous broadening. So, because of this set of different oscillators, photon echo can take place. Moreover, it can take place even though the distribution of atoms with regard to the external and internal degrees of freedom is uniform.

In the case of ordinary photon echo, the rephasing of oscillators is a result of the second π-pulse of light. In our case this aim can be reached without applying the π-pulse by means of change of magnetic or electric field. The maximum intensity of the photon echo occurs when the integral over the field with respect to the time that has elapsed after the light pulse is equal to zero.

The shape of the photon echo strongly distinguishes different types of optical transitions; this can be used for experimental identification of combining atomic states.

The mechanism of photon echo discussed above extends the range of conditions for echo observation. In particular, it can take place in atomic or molecular gases at high pressure or in the infrared and far-infrared regions of the spectrum for atoms or for ions cooled in a trap.

QThG3

Strong-field photoionization of helium

Barry Walker, K. Schafer,* B. Sheehy, P. Agostini,** K. Kulander,* L. F. DiMauro, *Brookhaven National Laboratory, Building 555, Upton, New York 11973*

This study examines the photoelectron spectra (PES), the electron angular distributions, and the ion yields of helium in an intense laser field of 10^{14}–10^{16} W/cm^2. The strong-field double ionization of helium is measured over 12 orders of magnitude and is compared with a classical two-electron ionization model.

The experiment exposes room-temperature helium in an ultrahigh-vacuum environment to 160-fs, 780-nm pulsed laser radiation. The PES for the ionization of helium (Fig. 1) extend over the transition from predominately multiphoton ionization into the tunneling ionization regime. The gross feature of the ionization is a break in *all* the electron distributions at twice the ponderomotive energy U_p. The electron angular distributions show that electrons above and below the break contain two different angular dependencies (Fig. 2). The electrons

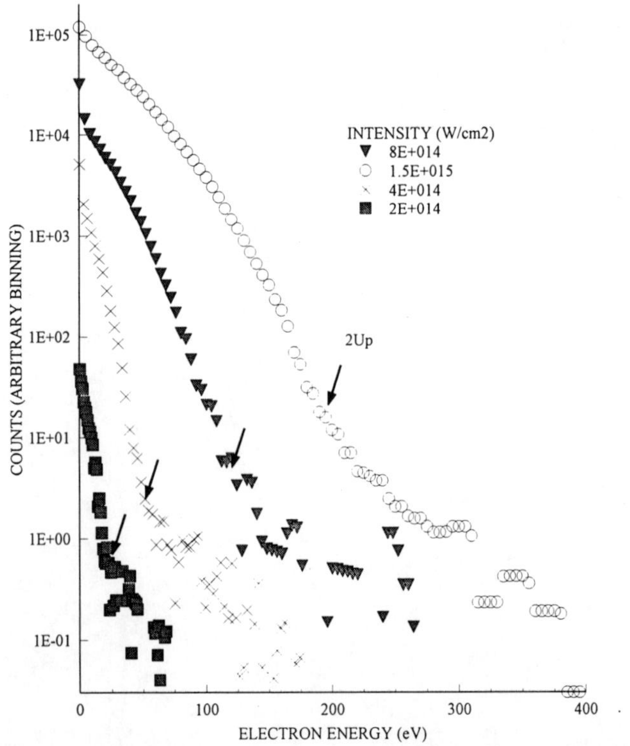

QThG3 Fig. 1. PES of helium at several intensities. The break point of $2U_p$ is noted with an arrow for each intensity.

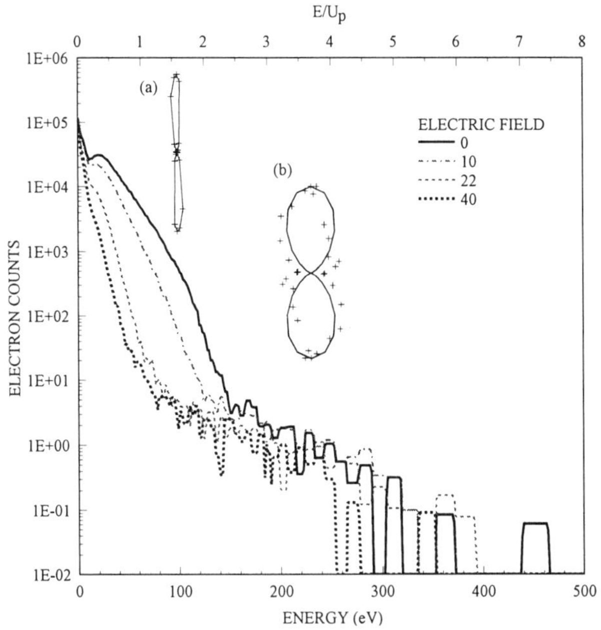

QThG3 Fig. 2. Angular distributions for the photoelectrons at 1×10^{15} W/cm^2 shown (a) above and (b) below $2U_p$. The raw PES emphasize the presence of two distinct distributions.

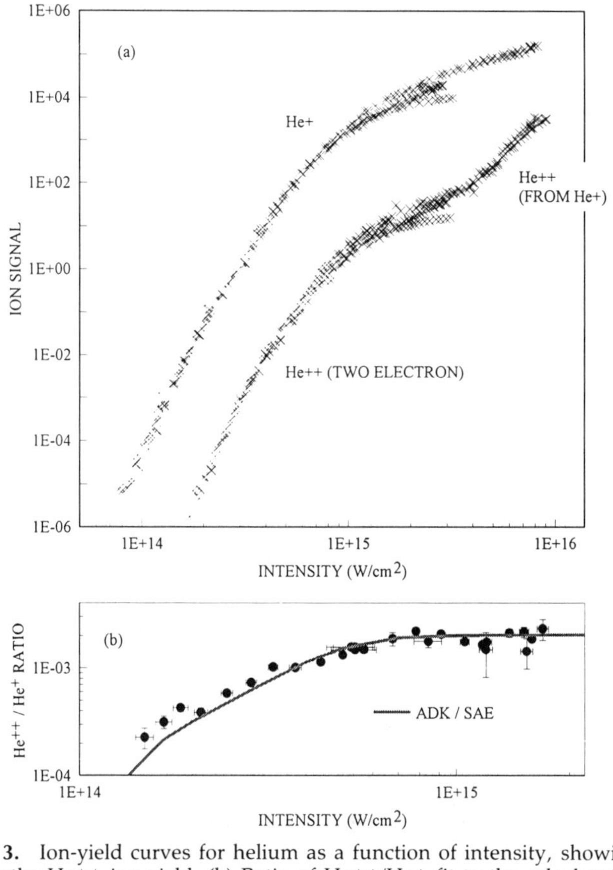

QThG3 Fig. 3. Ion-yield curves for helium as a function of intensity, showing two distinct rates for the He^{++} ion yield. (b) Ratio of He^{++}/He^{+} fit to the calculated tunneling yield divided by the single active-electron-Schrödinger-equation total-ion yield.

below $2U_p$ have dynamics dominated by the field and an energy ultimately limited by the field-imparted drift energy. Higher-energy electrons have a nearly isotropic distribution consistent with free-electron elastic scattering off the ion core. Overall, the photoelectron spectra neatly divided into prompt and rescattering ionization-energy regions. Also evident from the spectra (Fig. 1) is an increase in the amount of scattered electrons relative to the total electron spectrum at lower multiphoton intensities as opposed to the scarce yield with tunneling ionization. Ultimately, the highest-energy electrons are observed for the ionization of He^+. Ionization of He^+ near the saturation intensity produces electron energies exceeding 1000 V, corresponding to more than 600 photons absorbed during the ionization.

The ion yields as a function of intensity (Fig. 3) show that ionization occurs by three pathways: single-electron ionization of neutral helium, two-electron ionization of neutral helium, and ionization of the He^+ ion. The dynamic range of the experiment has shown no double-ionization threshold, as predicted by a classical two-step rescattering model using field-free e–2e inelastic cross sections. Furthermore, by dividing the two-electron He^{++} yield by the total H^+ ion yield, we find that the double ionization is proportional to the ac-tunneling rate of the neutral. The two-electron ionization mechanism then is distinguished by its scaling with the tunneling component of the ionization, even when multiphoton ionization dominates.

The ionization of helium shows very clearly the nature of ionization as one approaches the tunneling regime. Ion spectra have shown that a simple two-electron atom in the strong tunneling field produces double ionization and that this ionization cannot be fully explained by current classical scattering models.

This work was supported by the U.S. Department of Energy.

*Lawrence Livermore National Laboratory, Livermore, California 94550
**Saclay Research Center, France

QThG4

4π electron–ion image detector

J. Zhu, W. T. Hill III, *Institute for Physical Science and Technology, University of Maryland, College Park, Maryland 20742*

Contemporary intense-field dissociative ionization process in diatomic systems, Coulomb explosions, always involve the ejection of multiple electrons, resulting in atomic nuclei in several ionization stages with a variety of kinetic energies. This has led to some ambiguity in the interpretation of experimental results and hence the development of an appropriate model of intense-field dissociative ionization. Part of the problem lies in the importance of asymmetric dissociation events or the proper sequencing of ionization and dissociation events. By asymmetric we mean

$$AB^{(p+q)+} \rightarrow A^{p+} + B^{q+},$$

where $|p - q| \geq 2$. Some investigators have observed only symmetric Coulomb explosions, whereas others have seen asymmetric explosion events at appreciable levels.[1] The bulk of the experiments revealing only symmetric events have used covariance-mapping techniques on the atomic ions. This technique can unequivocally pair atomic ions that originate from the same molecular parent.[2] This technique, however, when applied only to the ions, is incapable of determining when ionization occurs. That is, a strong covariance will exist whether the electrons are removed before or after dissociation.[1] If, however, covariance mapping were applied to the ions and electrons

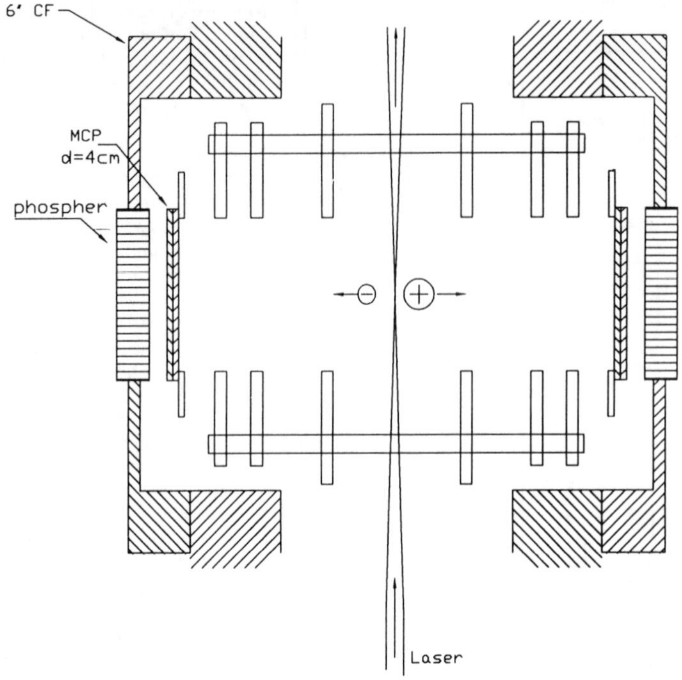

QThG4 Fig. 1. Sketch of the electron–ion image detector.

simultaneously, it would be possible to eliminate this ambiguity. This paper describes what is to the best of our knowledge the first detector capable of establishing a covariance between ions and electrons.

A sketch of the detector is shown in Fig. 1. It consists of an interaction region placed between two microchannel plates (MCPs) spaced by about 11.5 cm. A spatially uniform static electric field fills the region between the MCPs and forces the electrons and the positive ions in opposite directions. When the polarization of the radiation field is perpendicular to the axis of the detector (i.e., into and out of the page), the charged particles will initially be directed into and out of the page. Charged particles will then hit their respective detectors at a radius, measured from the center of the MCP, and an angle, measured from the polarization direction, that depend only on their charge, energy, and angular distribution. A phosphor screen is used to convert the electrons emerging from the back of the MCP into a two-dimensional image that is finally detected with a CCD camera.[3] The covariance between the ions and the electrons can be extracted by digitizing the electron and the ion images simultaneously. Although it was designed for our intense-field studies, this detector can also be used for low-field studies in which multiple fragments are created.

1. W. T. Hill III, J. Zhu, D. L. Hatten, Y. Cui, J. Goldhar, and S. Yang, Phys. Rev. Lett. **69**, 2646 (1992).
2. L. J. Frasinski, K. Codling, and P. A. Hatherly, Science **246**, 1029 (1989).
3. The philosophy of this detector is similar to the electron detector described by H. Helm, N. Bjerre, M. J. Dyer, D. L. Huestis, and M. Saeed, Phys. Rev. Lett. **70**, 3221 (1993).

QThG5

Nonlinear phenomena in intracavity-laser spectroscopy

E. A. Sviridenkov, S. A. Kovalenko, N. A. Raspopov, D. D. Toptygin, S. E. Vinogradov, *Lebedev Physics Institute, Leninsky Prospect 53, Moscow 117924, Russia*

Intracavity-laser spectroscopy (ICLS), which has been developing for 25 years,[1,2] is the most sensitive technique for absorption analysis. The first papers on the subject indicated that the general limit of ICLS sensitivity is determined by spontaneous emission and that the corresponding minimum absorption coefficient that may be detected is $k_{min} = 1/(cT_{ph}M)$, where c is the velocity of light, T_{ph} is the photon lifetime in the cavity, and M is the number of photons per mode. In the presence of absorption within the cavity in mode q with absorption coefficient k_q, the emission J_q in this mode (if radiation intensities in other modes is kept constant) should reduce according to

$$J_q = J_0 \exp(-k_q ct), \qquad (1)$$

where t is the time that has passed since the emission started. Equation (1) holds true until $t < MT_{ph}$. For common lasers this time is of the order of a second, which permits one to measure absorption coefficients down to 10^{-11} cm^{-1}. The first experiments[3] demonstrated a sensitivity of 10^{-7} cm^{-1}. Then, with amplitude stabilization of cw dye lasers, a sensitivity of 10^{-8} cm^{-1} was achieved.[4] The search for the limiting factors of ICLS sensitivity has led to understanding of the importance of nonlinear phenomena in the dynamics of multimode lasers. We have investigated both theoretically and experimentally the effect of stimulated Mandel'shtam–Brillouin scattering and four-wave parametric mixing in an active medium. Nonlinear interaction in an active medium results in a regime of dynamic chaos.

The effect of Mandel'shtam–Brillouin scattering was weakened by using a unidirectional ring laser. The dynamics of emission in such a laser change from a chaotic regime to a regime of periodic oscillations. Theoretical calculations of laser dynamics with uniform modes show that four-wave mixing (FWM) should result in phase relationships that minimize the effect of FWM itself. A laser with compensated cavity dispersion made it possible for us to achieve a regime of stable generation for a duration of 10^{-1} s, to reach an ICLS measurement sensitivity of 10^{-1} cm^{-1}, and to detect absorption with coefficients down to 10^{-11} cm^{-1} (see Ref. 5).

ICLS is employed in spectroscopy for investigations of highly excited molecular spectra, weak forbidden transitions, and two-photon and Raman absorption. The high sensitivity of ICLS has been used in investigations of the kinetics of chemical reactions with radicals NH_2, PH_2, HCO, HNO, etc. Certain reactions have been substantially revised, e.g., the reaction of the photo-oxidation of ammonia. Additional uses of ICLS include plasma diagnostics and the measurement of level populations, vibrational and rotational temperatures, and electrical fields in plasma.

1. L. A. Pakhomycheva, E. A. Sviridenkov, A. F. Suchkov, L. V. Titova, and S. S. Churilov, JETP Lett. **12**, 43 (1970).
2. N. C. Peterson, M. G. Kurylo, W. Braun, A. M. Bass, and R. A. Keller, J. Opt. Soc. Am. **61**, 746 (1971).
3. T. P. Belikova, B. R. Dorofeev, E. A. Sviridenkov, and A. F. Suchkov, Sov. J. Quantum Electron. **5**, 722 (1975).
4. V. M. Baev, T. P. Belikova, E. A. Sviridenkov, and A. F. Suchkov, Sov. Phys. JETP **47**, 21 (1978).
5. S. E. Vinogradov, A. A. Kachanov, S. A. Kovalenko, and E. A. Sviridenkov, JETP Lett. **55**, 581 (1992).

QThG6

Very-high-resolution molecular-beam cw-laser spectrometer for molecular overtones

M. Becucci, A. Callegari, J. W. Dolce, J. E. Gambogi, K. K. Lehmann, B. Meyer, C. J. O'Brien, G. Scoles, *Department of Chemistry, Princeton University, Princeton, New Jersey 08544*

In order to achieve both higher resolution and higher sensitivity, a high-resolution molecular-beam spectrometer based on a 1.5-μm cw color-center laser[1] has been improved by the addition of an intracavity electro-optic frequency-stabilization system and a 1.5-μm resonant power-buildup cavity. The cavity is a high-resolution Fabry–Perot interferometer made with a modified Newport SuperCavity™ SR100, in which the two cavity mirrors are bonded to a single piezoelectric tube (PZT). Typical finesse exceeds 1000 over the 1.45–1.65-μm range, and the efficiency is about 50%. The molecular beam crosses the cavity, through a hole in the PZT, perpendicular to its axis at the laser-beam waist in order to minimize Doppler broadening. About 15 W of circulating power in the cavity are obtained upon injection of 25 mW through an optical fiber. This gives a power-buildup factor of about 600. The fiber feed and the monolithic construction make the device insensitive to the vibrations of the molecular-beam apparatus.

In principle, by using the SuperCavity a very narrow instrumental linewidth can be obtained, because the only limiting factors are transit-time broadening and residual Doppler broadening due to the finite beam divergence (about 2.5 MHz each). To fully exploit this potential the laser jitter has been reduced from 6 MHz to less than 500 kHz by locking the laser to the transmission peak of the SuperCavity through a $LiNbO_3$ electro-optic crystal. This crystal is located inside the laser cavity so that the effective length of the latter, and thus the laser frequency, can be easily controlled by the voltage applied to the crystal. The laser frequency is modulated at 75 kHz. The modulation in the SuperCavity transmission signal is lock-in detected with a 3-ms time constant. The corresponding error signal, which is proportional to the amount of detuning from the SuperCavity peak, is amplified and is applied as a correction to the voltage across the crystal, and this achieves the desired jitter reduction. In turn, the closed-loop error signal is filtered through a 1-Hz low-pass filter, is integrated, and is sent to the SuperCavity PZT so that the latter can follow the laser on a long time scale. In this way the correction due to the crystal is kept well within the 150 MHz allowed by the maximum available crystal voltage. Moreover, the laser frequency can be continuously scanned over many wave numbers at a speed as high as 20 MHz/s.

With this system a typical linewidth of 5 MHz has been achieved and the second overtone of the Si–H stretch of both silane and methyl silane has been resolved with a typical signal-to-noise ratio of 20.

1. Kerstel *et al.*, J. Chem. Phys. **95**, 8282 (1991).

QThG7

Methane spectrum obtained by using an all-diode-based difference-frequency generator

Michael P. McCann, Edesly Canto-Said, Peter G. J. Wigley, G. J. Dixon, *Center for Research and Education in Optics and Lasers, University of Central Florida, Suite 400, 12424 Research Parkway, Orlando, Florida 32826-3271*

Methane is a gas of interest for several reasons. Methane is the second most abundant gaseous source of carbon after carbon dioxide. Methane is an indicator of anaerobic activity and thus is prevalent in wetlands. Methane is a greenhouse gas and contributes 20% to global warming. Diode lasers are an attractive light source for the detection of methane. They are inexpensive and are easily modulated. Thus diode lasers easily lend themselves to derivative spectra and to frequency-modulation spectroscopy. The mid-IR region has been the domain of lead-salt diodes. These diodes are operated at liquid-nitrogen temperatures, a characteristic that has limited their use.

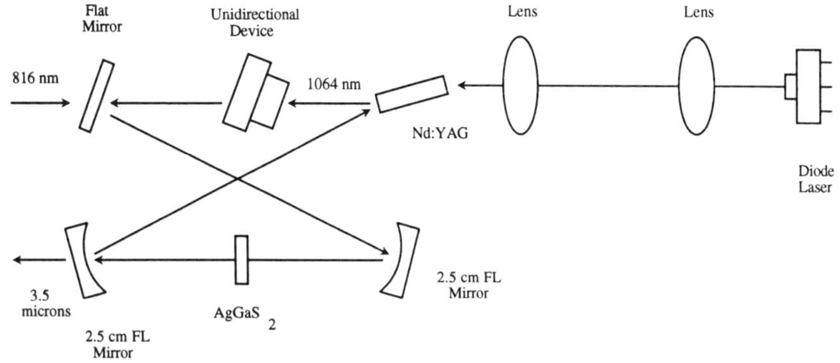

QThG7 Fig. 1. Difference-frequency generator.

By using two thermoelectrically cooled diodes, laser light has been generated from 2.8 to 4.1 μm. For the first time to our knowledge, difference-frequency generation pumped by two diode lasers has produced microwatts of coherent light. A difference-frequency generator (DFG) was constructed, as shown in Fig. 1. The nonlinear crystal, silver thiogallate, is pumped intracavity by a 3-W diode-pumped Nd:YAG ring laser. The coherent 1.06-μm light and the 816-nm coherent light from a single-striped diode are used for the difference-frequency generation. The single-striped diode was current tuned from 815 to 818 nm. Thus current-tunable DFG light was obtained from 3.49 to 3.55 μm.

An absorption spectrum in methane was obtained in the above wavelength region. This wavelength region corresponds to the ν_1 band in methane. The absorption spectrum was obtained by using a 10-cm cell with barium fluoride windows and a methane pressure of 1 atm. A derivative spectrum from 3.523 to 3.544 μm is shown in Fig. 2. Experiments are currently underway to determine the detection limits.

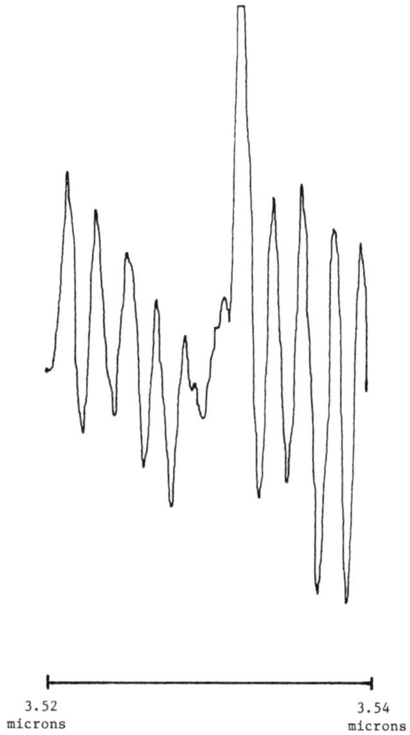

QThG7 Fig. 2. Derivative spectrum of methane, obtained by using the difference-frequency generator.

QThG8

Stark-shift measurement of the $(6s)^2\ {}^1S_0 \rightarrow (6s6p)\ {}^1P_1$ barium transition

J. Li, W. A. van Wijngaarden, *Department of Physics, York University, Toronto, Ontario M3J 1P3, Canada*

Precise measurements of Stark shifts have permitted new stringent tests of atomic theory.[1-3] These measurements have produced much more accurate data than have been obtained in experiments that determine radiative lifetimes, polarizabilities, or oscillator strengths. This work reports the measurement of the Stark shift of the $(6s)^2\ {}^1S_0 \rightarrow (6s6p)\ {}^1P_1$ transition in barium. The final result has an uncertainty that is less than one-tenth that of the most accurate value found in the literature.

The response of an atom that does not have any hyperfine structure, such as ^{138}Ba (71.7%), to an electric field E is given by the Hamiltonian

$$H = -\left[\alpha_0 + \alpha_2 \frac{3m^2 - J(J+1)}{J(2J-1)}\right]\frac{E^2}{2}, \tag{1}$$

where α_0 and α_2 are the scalar and tensor polarizabilities, respectively, J is the electronic angular momentum, and m is its azimuthal component. The electric field shifts the frequency for the transition between the $(6s)^2\ {}^1S_0$ and the $(6s6p)\ {}^1P_1$ states by an amount KE^2, where the Stark-shift rate is

$$K = -\frac{1}{2}\left[\alpha_0({}^1P_1) - 2\alpha_2({}^1P_1) - \alpha_0({}^1S_0)\right]. \tag{2}$$

Here, m has been set to zero because only the $m = 0$ sublevel of the ${}^1P_1$ state is populated in our experiment.

The apparatus shown in Fig. 1 consists of a ring dye laser that excites an atomic beam. The laser frequency was tuned across the transition while fluorescence produced by the radiative decay of the excited state was monitored by two photomultipliers (PM). Data was collected by using two lock-in amplifiers that were externally triggered by a signal generator. The frequency scan

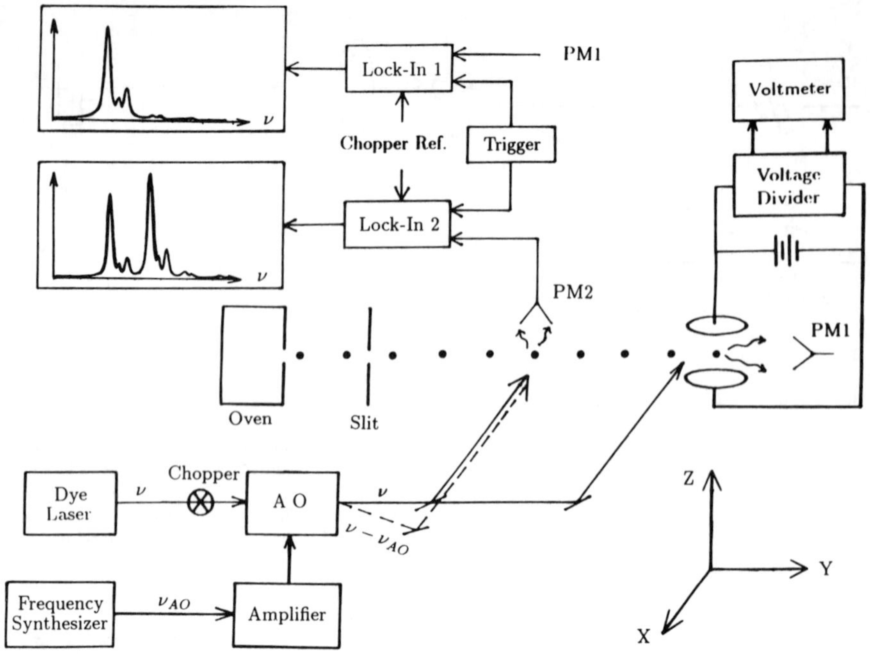

QThG8 Fig. 1. Schematic diagram of apparatus.

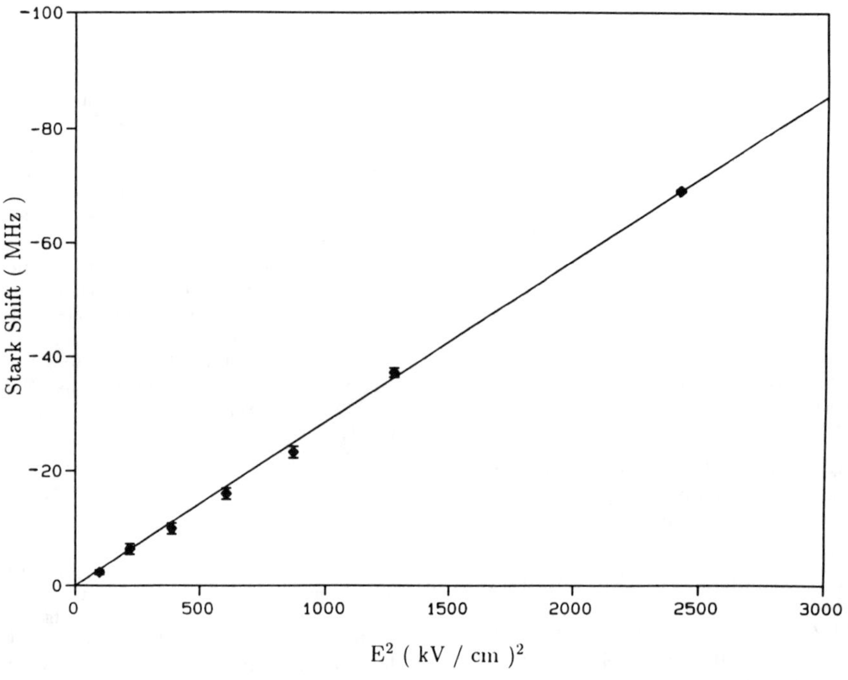

QThG8 Fig. 2. Stark shift versus electric field squared. The uncertainty of the Stark shift measured at the highest electric field is substantially smaller than the diameter of the data point.

was calibrated by using an acousto-optic modulator (AO) to frequency-shift part of the laser beam by 300.000 MHz. The frequency-shifted and unshifted laser beams were superimposed and excited the atoms in the field-free region. The laser beams intersected the atomic beam orthogonally to eliminate first-order Doppler shifts.

Each atomic transition therefore generated two peaks in the fluorescence spectrum that were separated by the modulation frequency. Each frequency scan was individually calibrated because the laser-scan speed varied by up to 0.1% from scan to scan. The average frequency interval represented by each data point was 0.185 MHz. The Stark shift was then found by measuring the frequency separating the ^{138}Ba peaks that were excited by the laser that was not frequency shifted, in the field and the field-free regions.

The electric field was generated by applying a voltage across two stainless-steel plates having a diameter of 3 in. (7.62 cm). The plate spacing was determined to be 0.4001 ± 0.0001 in. (1.01625 ± 0.00025 cm) by using machinists' blocks whose thickness had an uncertainty of less than 1×10^{-5} in. (2.54×10^{-5} cm). The voltage was continuously monitored by using a voltage divider that reduced the voltage by a factor of 5000 with an accuracy of 0.01%. The reduced voltage was measured by a voltmeter having an uncertainty of less than 0.002%. The electric field between the plates was numerically modeled, and its uniformity at the intersection of the laser and atomic beams was found to be better than one part in 10^5.

A total of 370 separate wavelength scans were taken at various voltages. Figure 2 shows a plot of the average Stark shift measured at each voltage versus the square of the electric field. Each error bar is one standard deviation of the data about the average value. Most of the data (260 scans) were taken at the maximum voltage of 50 kV. The Stark-shift rate K was found by fitting the Stark shift to a line KE^2, yielding $K = -28.532 \pm 0.108$ kHz/(kV/cm)^2. This is consistent with the value of -28.3 ± 1.2 kHz/(kV/cm)^2 obtained by Kopfermann.[4]

The authors thank the Natural Science and Engineering Council of Canada and York University for financial support.

1. C. E. Tanner and C. Wieman, Phys. Rev. A **38**, 162 (1988).
2. L. R. Hunter, D. Krause, K. E. Miller, D. J. Berkeland, and M. G. Boshier, Opt. Commun. **94**, 210 (1992).
3. W. A. van Wijngaarden, E. A. Hessels, J. Li, and N. E. Rothery, Phys. Rev. A **49**, R2220 (1994).
4. H. Kopfermann, Naturforschung und Medizin in Deutschland, **12**, 1 (1953).

QThG9

High-resolution microwave spectroscopy of the sodium-atom Rydberg states

S. F. Dyubko, M. N. Efimenko, V. A. Efremov, S. V. Podnos, *Laboratory of Molecular and Atom Beam Spectroscopy, Department of Quantum Radiophysics, Kharkov State University, Svobody Square 4, Kharkov 310168, Ukraine*

We present the results of a spectroscopic investigation of laser-excited sodium-atom Rydberg states using an atom-beam spectrometer with field ionization. A detailed description of the spectrometer and the measurement methods is presented in Refs. 1 and 2. Measurement results of transition frequencies with a relative error less than 10^{-8} between levels with quantum numbers in the range of $24 \leq n \leq 30$ are presented in Table 1.

It is necessary to note that the frequencies measured in Ref. 4 and the quantum-defect constants known from Ref. 3 are significantly different from the values measured in this investigation. Apparently, the interaction region between excited atoms and the microwave field is insufficient to protect against the electrostatic field in the spectrometer used in Ref. 4. This leads to a Stark frequency shift in the direction of higher frequencies. The shift is 1.5–2 MHz. For the frequency shift and for transition types nS–nP ($25 \leq n \leq 35$), the weak electrostatic field with $E = 0.5$ V/cm is sufficient. In actuality, we transformed the transition frequencies to the values in Ref. 4

QThG9 Table 1. High-resolution microwave spectroscopy of sodium-atom Rydberg states.

Transition type	Measured frequency, MHz	Calculated frequency, Mhz
$25S_{1/2}$ -- $25P_{1/2}$	237437,10	237437,11
$25S_{1/2}$ -- $25P_{3/2}$	237819,19	237819,21
$26S_{1/2}$ -- $26P_{1/2}$	209960,69	209960,66
$26S_{1/2}$ -- $26P_{3/2}$	210299,40	210299,16
$28S_{1/2}$ -- $28P_{1/2}$	166520,70	166520,73
$28S_{1/2}$ -- $28P_{3/2}$	166790,11	166790,04
$29S_{1/2}$ -- $29P_{1/2}$	149247,80	149247,78
$29S_{1/2}$ -- $29P_{1/2}$	149489,38	149489,48
$24D_{5/2}$ -- $25P_{3/2}$	75551,420	75551,422
$25D_{5/2}$ -- $26P_{3/2}$	66873,318	66873,323
$26D_{5/2}$ -- $27P_{3/2}$	59474.,849	59474.,850
$27D_{5/2}$ -- $28P_{3/2}$	53128,400	53128,405
$29S_{1/2}$ -- $30S_{1/2}$	2 * 147541,96	2 * 147541,94

only when we applied an electrostatic field of $0.3 \leq E \leq 0.5$ V/cm in the interaction region.

Use of a microwave frequency synthesizer and an automated experiment-driving system provided high-precision experimental results. The properties of the synthesizer are a narrow line (<7 KHz), an error of frequency determination less than 7 KHz, and a frequency range of 50–600 GHz. These properties provided a good signal-to-noise ratio in the experiment. At present we are accumulating the Rydberg-state atoms' spectra for inclusion in a computer data bank of microwave spectroscopic information.

1. S. F. Dyubko *et al.*, IEEE J. Quantum Electron. **18**, 410 (1991).
2. V. G. Gerasimov, S. F. Dyubko, *et al.*, in *Proc. Physics and Techniques of Millimeter and Submillimeter waves*, Kharkov, Ukraine, 1991, p. 2 (in Russian).
3. R. Goy, Infrared and Millimeter Waves, **8**, 342 (1983).
4. C. Fabre, Ann. Phys. Fr., 1982, p. 2547.

QThG10

Sub-Doppler cooling of trapped ions

G. Birkl, J. A. Yeazell,* R. Rueckerl,** H. Walther,** *National Institute of Standards and Technology, Phys A167, Gaithersburg, Maryland 20899*

A primary objective of laser-cooling research is the study of quantum collective effects for many-particle systems. For trapped ions, the goal is to produce an ensemble of particles in a single quantum state of the external trapping field. Improvements in the ability to cool the ions to the lowest possible temperatures are crucial to this research. Here, we explore techniques for cooling ions and, in particular, ionic crystals below the Doppler cooling limit.

In contrast to neutral particles, the Coulomb interaction between charged ions leads to additional heating effects that limit the achievable temperatures for more than a single particle in a three-dimensional quadrupole trap. In a quadrupole storage ring,[1] ions are strongly bound in only two dimensions (secular frequency, 1 MHz). Along the third dimension (z direction) the ions are only weakly bound, the vibrational frequency being only 12 kHz. String-type ionic crystals created along the z axis of this trap are not driven or heated by the external radio-frequency trapping field.[1]

The lin $\perp$ lin configuration of polarization-gradient cooling is used to cool the magnesium ion ($^{24}Mg^+$). The laser-beam axis and the z axis coincide (see Fig. 1), so that the cooling takes place in this dimension. Fluorescence light from the ions is spatially resolved, and the kinetic energies of the ions are derived from the observed spatial distribution of an ion during a 5-s exposure time. Since the z potential is harmonic in nature and is well characterized, the variance of this distribution is directly related to the thermal energy of the ions.

The ions were cooled below the Dopper limit[2] to a minimum temperature, in agreement with predictions for polarization-gradient cooling of neutral particles. For a two-ion crystal the lowest temperature along the z direction was found to be 280 ± 75 μK. The cooling of trapped ions does differ strongly from that of neutral atoms. Even such weak confinement produces large quantitative differences in the laser parameters leading to the minimum temperatures. This observation is supported by recent theories on the polarization-gradient cooling of trapped ions.[3] Ions with sufficiently low kinetic energies should be trapped in the potential wells created by the polarization-gradient field. Evidence for a strong localization of the ions was observed.

Progress on the implementation of a Raman-type cooling scheme for this system is discussed.

*Department of Physics, Pennsylvania State University, 104 Davey Laboratory, University Park, Pennsylvania 16802

**Max Planck Institute for Quantum Optics and Physics Section, Ludwig Maximilians University of Munich, D-85748 Garching bei Muenchen, Germany

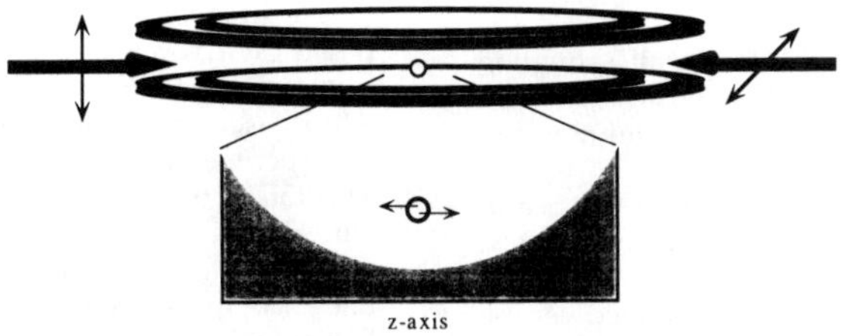

QThG10 Fig. 1. Experimental setup. The polarization-gradient field is established by counterpropagating linear polarized laser beams. These beams intersect the trap volume tangentially and coincide with the axis along which the ions are only weakly confined by a static electric field. The confinement in the radial direction is produced by the quadrupolar ring electrodes. The blowup depicts the harmonic potential that lies along the z axis. It is in this dimension that the cooling takes place. A photon-counting camera system observes the spatial position of the ion from a direction perpendicular to the z axis. The camera has a resolution of 0.4 μm.

1. I. Waki, S. Kassner, G. Birkl, and H. Walther, Phys. Rev. Lett. **68**, 2007 (1992); G. Birkl, S. Kassner, and H. Walther, Nature (London) **357**, 310 (1992).
2. G. Birkl, J. A. Yeazell, R. Rueckerl, and H. Walther, Europhys. Lett. **27**, 197 (1994).
3. S. M. Yoo and J. Javanainen, Phys. Rev. A **48**, R30 (1993); J. I. Cirac, R. Blatt, A. S. Parkins, and P. Zoller, Phys. Rev. A **48**, 1434 (1993); Y.-Q. Li and K. Moelmer, Laser Phys., in press.

QThG11

Transverse confinement of atoms in a gravitational cavity by a doughnut beam

S. Friebel, R. Deutschmann, M. Schiffer, G. Wokurka, W. Ertmer, *Institute for Quantum Optics, University of Hannover, Welfengarten 1, D-30167 Hannover, Germany*

Recent progress in preparing ensembles of cold atoms holds great promise for the realization of cavities for atomic de Broglie waves. Those atomic cavities should provide discrete levels, long storage times, and the possibility of achieving a high degeneracy of atoms in a certain mode. The first proposal for such a cavity was analogous to optical cavities with their physical mirrors replaced by light-induced mirrors.[1] For laser-cooled atoms, placing a single *curved* atomic mirror in the earth's gravitational field has also been proposed.[2] First experiments with atoms bouncing on such a device have been performed.[3,4] One key problem of these configurations is the limitation of the confinement time by the transverse motion of the atoms out of the reflecting spot.

Here we propose a modified configuration of a gravitational cavity consisting of a single horizontal *plane* mirror and a TEM_{01}^* doughnut beam propagating perpendicular to the mirror (see Fig. 1). This provides an additional transverse confining potential that allows the controlled manipulation of both the external and internal atomic states.

The dipole-force potential[5] of the doughnut mode at the beam waist w_0 is given by

$$V(r) = \frac{1}{2} V_0 \frac{r^2}{w_0^2} \exp\left(-2\frac{r^2}{w_0^2}\right),$$

where V_0 ($V_0 \sim I_0/\Delta$ for small saturation and large detunings Δ) is positive for blue detuning, thus rendering a confining potential.

By using the separability of the Schrödinger equation in cylindrical coordinates, the atomic motion of the bound states can be decomposed into an accelerated vertical motion and a transverse harmonic motion. In the paraxial limit, the eigenmodes of the cavity, corresponding to three-dimensional standing de Broglie waves, consist of Airy functions along the vertical axis[2] and of 2-D harmonic oscillator eigenfunctions in the transverse directions. A typical density profile for the ground state of transverse motion is shown in Fig. 2. The energy levels are determined, and the transverse mode spacing is found to be

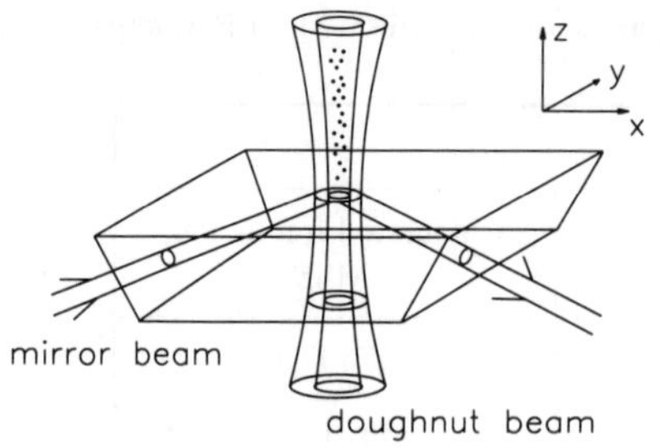

QThG11 Fig. 1. Sketch of the cavity: the mirror potential is provided by an evanescent laser wave. The TEM_{01}^* doughnut beam for the transverse confinement is fed through the glass block.

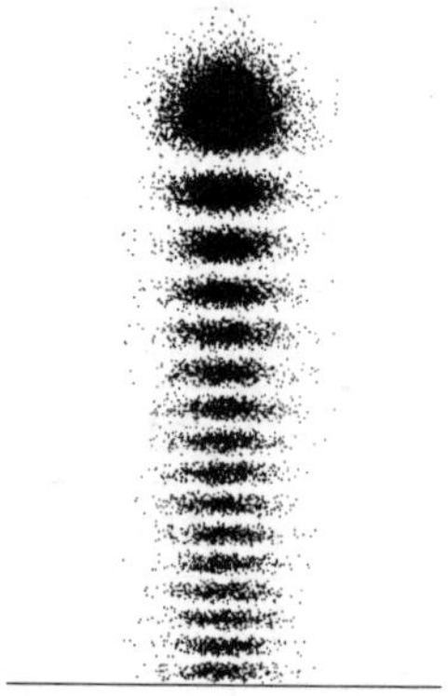

QThG11 Fig. 2. Calculated probability-density profile of the transverse fundamental mode and the 16th excited state of the vertical motion of the gravitational cavity.

$$\omega_t = \left(\frac{V_0}{mw_0^2}\right)^{1/2}.$$

It can be adjusted by varying the depth of the dipole-force potential. The treatment is also extended outside the paraxial domain. The number of bound states and their populations are calculated.

Losses due to spontaneous processes, tunneling, and nonadiabatic transitions to unbound states are estimated. To minimize these losses, which destroy the coherence of the atomic motion, it is necessary to use very large detunings. This requires very high power in the doughnut beam.

We discuss the feasibility of the proposed cavity on the basis of its characteristic parameters for typical elements for which laser cooling has been achieved.

Experimental investigations for the realization of this scheme are in progress in our group. For instance, a doughnut beam was generated with high efficiency by using a computer-generated blazed phase hologram.[6]

1. V. I. Balykin *et al.*, Appl. Phys. B **48**, 517 (1989).
2. H. Wallis *et al.*, Appl. Phys. B **54**, 407 (1992).
3. M. Kasevich *et al.*, Opt. Lett. **15**, 607 (1990).
4. C. G. Aminoff *et al.*, Phys. Rev. Lett. **71**, 3083 (1993).
5. J. P. Gordon *et al.*, Phys. Rev. A **21**, 1606 (1980).
6. H. He *et al.*, J. Mod Opt., in press.

QThG12

High-resolution spectroscopy on laser-trapped atoms: signal-to-noise analysis

J. H. Müller, V. Rieger, D. Bettermann, F. Ruschewitz, J. L. Peng, K. Sengstock, W. Ertmer, *Institute for Quantum Optics, University of Hannover, Welfengarten 1, D-30167 Hannover, Germany*

High-resolution spectroscopy on laser-trapped and cooled atoms has evolved into an effective tool for precision measurements and frequency standards research.[1] In these applications the achievable signal-to-noise ratio (S/N) is of crucial importance. The strong fluorescence on the cycling-trap transition is frequently used to monitor the small excitation rate on a much weaker clock transition (electron shelving and atom recycling).[2] In this case the change of the stationary-trap fluorescence rate is taken as a measure of the weak-line excitation probability. Because this fluorescence from the spatially-well-localized ensemble can be collected with high efficiency, the S/N is dominated mainly by the noise properties of the atom number, i.e., the atomic shot noise. The dynamics of the trapping process introduce long-term correlations into the fluctuations of the number of atoms. This leads to a nonwhite spectral density of the atomic shot noise. The correlations occur on time scales that are comparable to typical integration times during measurements. Therefore, the exact shape of the noise spectral density has to be taken into account when the S/N is determined.

In a theoretical model of the shot-noise contribution of atoms stored in a magneto-optical trap (MOT), the correlation time for the atom-number fluctuations is determined by the lifetime of the trap, leading to a characteristic plateau region in the S/N dependence on the integration time (Fig. 1). For short integration times this behavior is modified because of the shot noise of the detected photons. A comparison of the theoretical predictions with experimental results obtained with a magnesium MOT shows good agreement. A quantitative analysis yields important parameters from the observed noise of the fluorescence signal, e.g., the absolute number of trapped atoms and the overall light collection efficiency. Understanding the influence of the trap dynamics on the spectroscopically observed signal points the way to optimizing the trap parameters for high-resolution spectroscopy.

A completely different approach is to perform measurements of the weak-line excitation probability repetitively with integration times that are short compared to the correlation time of the atom-number fluctuations. If a large number of photons per atom can be detected in integration times short compared to the trap lifetime, the actual number of trapped atoms can be estimated from the fluorescence signal with an uncertainty smaller than the mean value of the long-term fluctuations. Because of the strong correlations, the number of trap-

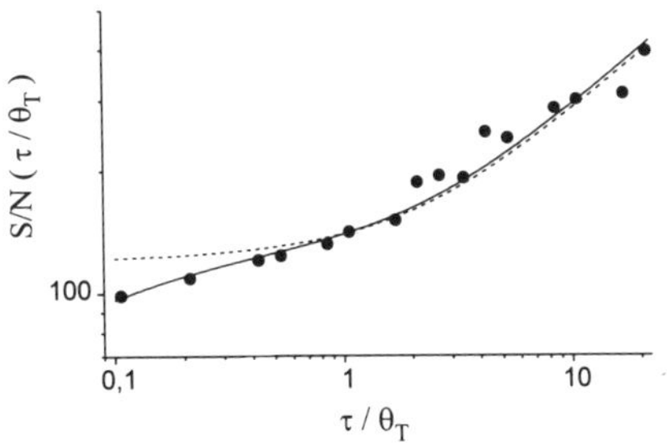

QThG12 Fig. 1. S/N of the trap fluorescence of Mg atoms as a function of integration time τ. The time is given relative to the independently determined trap lifetime $\theta_T = 0.53$ s. Circles are experimentally determined points, and the solid curve is a fit of the full theory to the data. The fitted parameters are the number of trapped atoms and the fluorescence collection efficiency. The dashed curve shows the theoretical S/N resulting only from atomic shot noise and indicates the plateau region reached for short integration times.

ped atoms participating in the subsequent weak-line excitation process will only slowly diffuse away from this value. Performing this as part of each measurement cycle yields data that can be used to normalize the signal count. The conditions under which this measurement strategy yields results superior to those of the first-discussed procedure of long-time averages have been investigated.

This research was supported by the German Research Association.

1. See, e.g., special issue of Appl. Phys. B **59** (1994), R. Blatt and W. Neuhauser, eds.
2. K. Sengstock *et al.*, Opt. Commun. **103**, 73 (1993).

QThG13

Calculation of optical trapping forces for micrometer-sized particles with a complex refractive index

Shunichi Sato, Yasunori Harada, Yoshio Waseda, *Institute for Advanced Materials Processing, Tohoku University, Katahira 2-1-1, Aoba-ku, Sendai 980-77, Japan*

Many experiments and force calculations[1,2] relating to optical trapping for transparent particles have been carried out. Recently, optical trappings of micrometer-sized iron particles with a scanning laser beam[3] and of nanometer-sized gold particles[4] have been reported. We have also demonstrated optical trapping of micrometer-sized metal particles with a fixed laser beam.[5] However, metal particles have a complex refractive index, and force calculations for these absorptive particles have, to our knowledge, not yet been reported.

In this paper we describe calculations of optical trapping forces for particles that have a complex refractive index. The model used for the calculations was developed by Ashkin[1] on the basis of ray optics for transparent particles. Here we improve this model by the addition of a complex component to the refractive index. It is found that the calculations agree with experimental results for micrometer-sized metal particles.[5] Furthermore, the possiblity of optical trapping is predicted for semiconductor materials with a large refractive index and a small extinction coefficient.

According to the model used, the trapping force F is given by

$$F = QnP/c, \qquad (1)$$

where Q is a nondimensional efficiency parameter, P is the power of the incident laser, n is the refractive index of the surrounding medium, and c is the velocity of light in free space. Thus, the force can be estimated by the calculated value of Q.

Figure 1 shows Q calculated for a gold particle as a function of position of the laser beam focus. Each arrow represents the magnitude and the direction of Q. The starting point of the arrow corresponds to the position of the laser beam focus. The hemisphere shows the left half of a trapped sphere. Thus, arrows in the left and right directions mean that the particle will be pulled into and pushed from the laser beam axis, respectively.

In the previous experiments[5] the metal particles were pulled into the beam axis and were then trapped two dimensionally when the beam focus was located near the bottom of the particle. Then the particles were pushed onto the bottom of the sample cell. When the focus is located near the top of the particle, the particle is pushed away horizontally, normal to the beam axis. In this paper it is found that the predictions by the calculation agree qualitatively with the previously observed results.[5]

Calculations were also performed for materials with a large refractive index, which are considered to be difficult to trap. For example, calculated results for a GaAs sphere at various laser wavelengths are shown in Fig. 2. It is clearly shown that a GaAs sphere is pushed downward with a 1.064-μm laser beam (two-dimensional trapping) and is pulled upward from the bottom of the sample cell with a 0.8-μm beam, indicating the possibility of three-dimensional trapping. It should be noted in the last case that the extinction coefficient at 0.8 μm is large. By choosing the laser wavelength properly, similar results are also obtained for some kinds of semiconductors with a large refractive index and a small extinction coefficient.

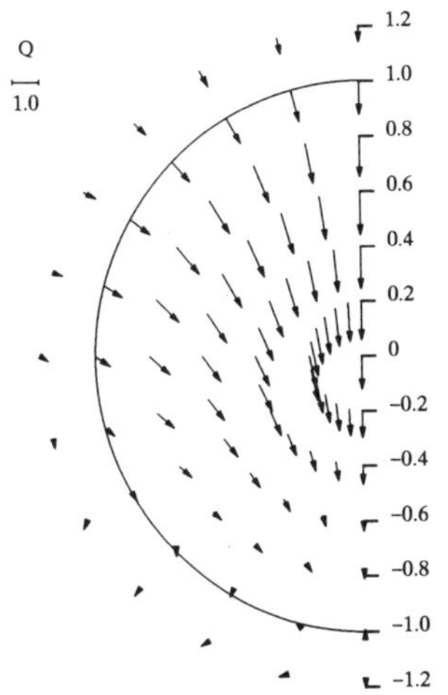

QThG13 Fig. 1. Calculated Q as a function of the position of the laser beam focus for a gold sphere at a laser wavelength of 1.064 μm ($n = 0.28 - 7.5i$). In the calculation it is assumed that the circularly polarized laser beam is focused at an angle of 65°. The radius of the sphere is assumed to be 3 μm. The numbers of the right-hand side of the figure show the positions normalized to the radius.

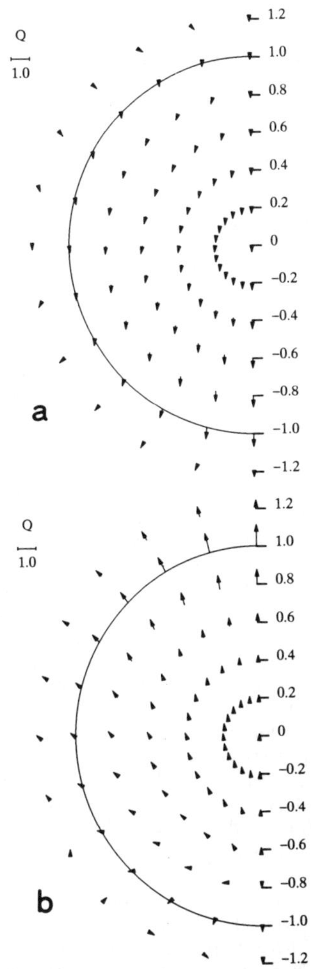

QThG13 Fig. 2. Calculated Q as a function of the position of the focus obtained for a GaAs sphere at a wavelength of (a) 1.064 μm ($n = 3.48 - 0.0001i$) and (b) 0.8 μm ($n = 3.68 - 0.086i$).

1. A. Ashkin, Biophys. J. **61**, 569 (1992).
2. W. H. Light, G. J. Sonek, and M. W. Berns, Appl. Opt. **33**, 1735 (1994).
3. K. Sasaki, M. Koshioka, H. Misawa, N. Kitamura, and H. Masuhara, Appl. Phys. Lett. **60**, 807 (1992).
4. K. Svoboda and S. M. Block, Opt. Lett. **19**, 930 (1994).
5. S. Sato, Y. Harada, and Y. Waseda, Opt. Lett. **19**, 1600 (1994).

QThG14

Observation of rectified forces in a three-level cascade system

T. T. Grove, B. C. Duncan,
V. Sanchez-Villicana, P. L. Gould,
Department of Physics, U-46,
University of Connecticut,
Storrs, Connecticut 06269-3046

Rectified dipole forces[1] offer the possibility of providing deep potential wells for neutral atoms. In the intense standing wave, a two-level atom can experience a large dipole force due to the large intensity gradient. However, this force changes sign on the optical wavelength scale, preventing the formation of deep potential wells. If a second, nearly commensurate, standing wave is added, its ac Stark shift causes the effective detuning of the first field to vary on the optical wavelength scale. The resulting change in sign of the detuning can counteract the change in sign of the intensity gradient, causing the force to maintain its sign over many wavelengths (i.e., the force becomes rectified), thereby forming a deep potential well. This rectified force is periodic with the beat wavelength between the two laser fields. In a three-level atom the same basic mechanism applies, but with the advantage that the two frequencies of the bichromatic standing wave can now be tuned close to resonance with the two transitions of the atom.

We have observed these rectified dipole forces by illuminating a sample of cold Rb atoms with a bichromatic standing wave tuned near resonance with the $5S_{1/2} \rightarrow 5P_{3/2}$ (780-nm) and $5P_{3/2} \rightarrow 5D_{5/2}$ (776-nm) transitions. This combination yields a beat wavelength of 71 μm. The cold atoms are produced by a magneto-optical trap (MOT).[2] In the experiment the MOT is alternated with the bichromatic standing wave in order to maintain cooling and three-dimensional confinement. We observe that atoms do accumulate in the potential wells formed by the rectified forces, forming a fringe pattern at the 71-μm beat wavelength. The contrast of these fringes is a measure of the ratio of potential well depth to temperature. If we maintain two-photon resonance but vary the detuning from the intermediate ($5P_{3/2}$) state, we see a variation of contrast in qualitative agreement with theory.[3] An important prediction is that when the sign of this detuning is changed, the fringe pattern should shift by one-half cycle. This is indeed the case, as shown in Fig. 1. We have also investigated the dependence of fringe contrast on laser intensity and on the duty cycle of the rectified fields relative to the MOT. The results agree reasonably well with theoretical expectations.

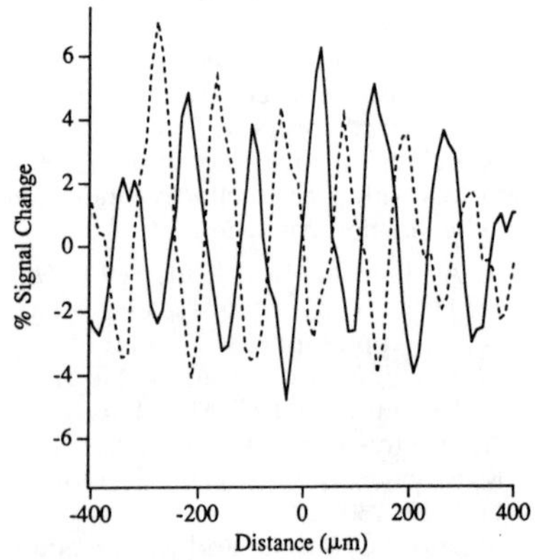

QThG14 Fig. 1. Fringe pattern produced by cold atoms accumulating in the periodic rectified potential. A higher signal level indicates more atoms. Note that the maximum contrast (peak-to-peak) is ~10%. The two curves are for opposite signs of detuning from the intermediate state: solid curve, $\Delta_1 = +14.7$ MHz; dashed curve, $\Delta_1 = -14.7$ MHz.

1. A. P. Kazantsev and I. V. Krasnov, JETP Lett. **6**, 332 (1987).
2. E. L. Raab, M. G. Prentiss, A. E. Cable, S. Chu, and D. E. Pritchard, Phys. Rev. Lett. **59**, 2631 (1987).
3. T. T. Grove and P. L. Gould, Laser Phys., in press.

QThG15

Optical techniques for measuring photonic band structure

İ. İnanç Tarhan, George H. Watson,
Department of Physics and Astronomy,
University of Delaware, Newark, Delaware 19716

Materials that exhibit a full photonic band gap within a certain frequency range open the possibility of studying new physics within the gap. The concept of a three-dimensional photonic band structure was introduced by Yablonovitch as an extension of a familiar one-dimensional optical device, a Fabry–Perot resonator constructed from dielectric mirrors.[1] Of great interest is the likelihood that disorder in a photonic semiconductor will lead to strong localization of photons in the band tails, analogous to localization of electrons in amorphous semiconductors.[2] Structures exhibiting full photonic band gaps in the microwave,[3] millimeter,[4] and submillimeter[5] regimes have been fabricated; scaling these structures down to the optical regime remains a technical challenge.

One way of realizing a three-dimensional photonic crystal (although it would not have a full photonic band gap) is with colloidal crystals that are prepared from charged monodisperse polystrene microspheres (diameter, 0.1–0.5 μm) dispersed in water. Under suitable conditions these microspheres form a face-centered cubic lattice depending on the particle density, the diameter of particles, and the free-ion concentration in the colloidal solution. The lattice parameter can be optimized within the range of the optical probe by adjusting the particle density. Under these conditions Bragg reflection of light is obtained, analogous to x-ray diffraction from atomic crystals.

The single crystallinity of these colloidal crystals is checked by using Kossel line patterns.[6,7] Kossel lines are also used as visual guides in orienting the crystal to locate the high-symmetry points on the first Brillouin zone; see Fig. 1. The dye-laser wavelength is adjusted so that the intersection of the 111, 111, and 002 Kossel lines is obtained. The crystal, initially oriented so that the

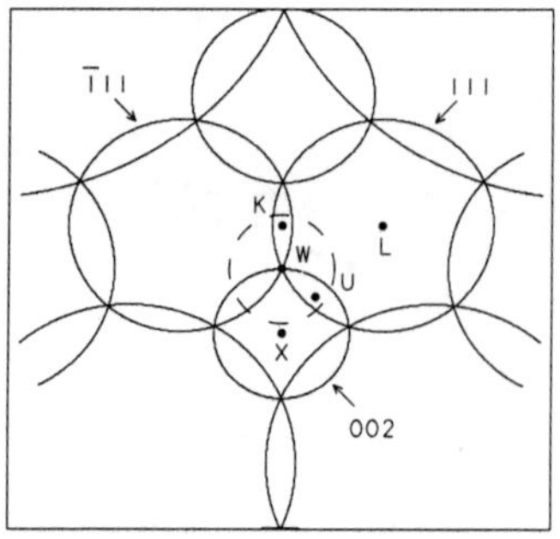

QThG15 Fig. 1. Simulated Kossel line pattern[7] obtained along the [012] direction (*W* point) for a face-centered cubic lattice with a lattice parameter of 0.486 μm and a laser wavelength of 0.582 μm.

incoming beam lies along the [111] direction (*L* point, normal to glass sample cell surface), is rotated to the [012] direction (*W* point), where the intersection point of the three Kossel lines blocks the transmitted beam. The experimentally accessible area for this orientation is indicated by the broken circle in Fig. 1.

Two complementary techniques for measuring photonic band structure in colloidal crystals are discussed. Interferometric measurement of the phase shift acquired for light transmitted through a colloidal crystal reveals its photonic band structure. A modified Mach–Zender interferometer has been developed for accurately measuring the phase shifts to determine the resulting photon dispersion along the band edges. Transmission measurements of the stop band provide a convenient technique to determine band gaps and their depths at the symmetry points on the Brillouin-zone edge.

This work was supported by the National Science Foundation (DMR 9113618).

1. E. Yablonovitch, Phys. Rev. Lett. **58**, 2059 (1987).
2. S. John, Phys. Rev. Lett. **58**, 2486 (1987).
3. E. Yablonovitch, T. J. Gmitter, and K. M. Leung, Phys. Rev. Lett. **67**, 2295 (1991).
4. E. Özbay, E. Michel, G. Tuttle, R. Biswas, M. Sigalas, and K.-M. Ho, Appl. Phys. Lett. **64**, 2059 (1994).
5. E. Özbay, E. Michel, G. Tuttle, R. Biswas, K.-M. Ho, J. Bostak, and D. M. Bloom, Opt. Lett. **19**, 1155 (1994).
6. P. Pieranski, E. Dubois-Violette, F. Rothen, and L. Strzelecki, J. Phys. **42**, 53 (1981).
7. T. Yoshiyama, I. Sogami, and N. Ise, Phys. Rev. Lett. **53**, 22 (1984).
8. S. Weber, Ch. Schetelelich, and V. Geist, Cryst. Res. Technol. **29**, 727 (1994).

QThG16

Coherent population trapping and the enhancement of three-wave-mixing efficiency in isotropic media

V. G. Arkhipkin, S. A. Myslivets, *Institute for Physics, Russian Academy of Sciences and Krasnoyarsk University, 660036 Krasnoyarsk, Russia*

Coherent population trapping (CPT) has been widely investigated.[1,2] In the resonant interaction of two laser fields with three-energy-level systems, atoms can be trapped into the coherent superposition of the ground and excited levels, whereas the intermediate level remains unoccupied. Under these conditions the absorbing medium becomes transparent for powerful radiation. The interest in this phenomenon is due to its many applications. We show that trapped states are significant in the sum-frequency mixing of two arbitrary intense fundamental fields in three-level systems under both single- and two-photon resonances.

Three-wave mixing in an isotropic medium is possible under certain conditions.[3,4] The second-order susceptibility experiences great enhancement in both single- and two-photon resonances. On the other hand, one-photon-resonance absorption imposes severe limitations on the density of nonlinear atoms. In the presence of strong optical radiation, the resonant nonlinear susceptibility decreases because of saturation and ac Stark splitting. In this paper we show how CPT influences the resonant three-wave mixing of strong optical fields in an isotropic medium, and we define the conditions for a great improvement in the conversion efficiency over that for weak fields.

We study the sum-frequency mixing of two arbitrary intense fundamental waves in a three-level system under both single- and two-photon resonances. Consider the cascade configuration of levels: $\varepsilon_2 > \varepsilon_1 > \varepsilon_0$, where ε_i are the energies of the levels. Strong fields at frequencies $\omega_1 \simeq \omega_{10}$ and $\omega_2 \simeq \omega_{21}(\omega_1 + \omega_2) = \omega_{20}$ (ω_{ij} are the transition frequencies) couple levels 0 and 2. The weak field generated at the frequency $\omega_s = \omega_1 + \omega_2$ does not change the population of the levels.

The radiation power generated at frequency ω_s is $P \propto |\chi^{(2)}(\omega_s)E_1E_2|^2N^2$, and the absorption of the pump fields is defined by $\mathrm{Im}[\chi^{(1)}(\omega_i)]$, where $\chi^{(1)}(\omega_i)$ and $\chi^{(2)}(\omega_s)$ are the linear and second-order atomic susceptibilities, respectively, E_1 and E_2 are the fundamental waves' amplitudes, and N is the density of the medium. The susceptibilities are dramatically modified in strong fields. They are calculated from the set of density-matrix equations (steady state).

$$\begin{aligned}
\Delta_{10}\rho_{10} &= -i[G_{10}(\rho_0 - \rho_1) + G_{12}\rho_{20}],\\
\Delta_{21}\rho_{21} &= -i[G_{21}(\rho_1 - \rho_2) - \rho_{20}G_{01}], \qquad (1)\\
\Delta_{20}\rho_{20} &= -i[G_{21}\rho_{10} + \rho_{21}G_{10}],
\end{aligned}$$

$$\begin{aligned}
\Gamma_1\rho_1 &= -i[G_{12}\rho_{21} - \rho_{12}G_{21}\\
&\quad + G_{10}\rho_{01} - \rho_{10}G_{01}] + \Gamma_2\rho_2,\\
\Gamma_2\rho_2 &= -i[G_{21}\rho_{12} - \rho_{21}G_{12}], \qquad (2)\\
&\quad \rho_0 + \rho_1 + \rho_2 = 1.
\end{aligned}$$

$G_{10} = -d_{10}E_1/2\hbar$, $G_{21} = -d_{21}E_2/2\hbar$ are the Rabi frequencies; $\Delta_{10} = -i\Omega_{10} + \Gamma_{10}$, $\Delta_{21} = -i\Omega_{21} + \Gamma_{21}$, $\Delta_{20} = -i(\Omega_{10} + \Omega_{21}) + \Gamma_{20}$, $\Omega_{10} = \omega_1 - \omega_{10}$, $\Omega_{21} = \omega_2 - \omega_{21}$, d_{ij} are pertinent matrix elements; Γ_{10}, Γ_{21}, Γ_{20} are the relaxation rates of the off-diagonal elements and $\Gamma_{1,2}$ are the decay rates for the atomic populations in states $|1\rangle$ and $|2\rangle$.

The solutions for the off-diagonal elements ρ_{ij} are

$$\begin{aligned}
\rho_{20} &= G_{21}G_{10}[\Delta_{10}(\rho_1 - \rho_2)\\
&\quad - \Delta_{21}(\rho_0 - \rho_1)]D^{-1},\\
\rho_{10} &= -iG_{10}[(\rho_0 - \rho_1)(\Delta_{20}\Delta_{21} + |G_{10}|^2)\\
&\quad + (\rho_1 - \rho_2)|G_{21}|^2]D^{-1}, \qquad (3)\\
\rho_{21} &= -iG_{21}[(\rho_1 - \rho_2)(\Delta_{21}\Delta_{10} + |G_{21}|^2)\\
&\quad + (\rho_0 - \rho_1)|G_{10}|^2]D^{-1},\\
D &= \Delta_{10}\Delta_{21}\Delta_{20} + \Delta_{10}|G_{10}|^2 + \Delta_{21}|G_{21}|^2.
\end{aligned}$$

Expressions for $\chi^{(1)}(\omega_i)$ and $\chi^{(2)}(\omega_s)$ are defined by using the standard rules.

$$\begin{aligned}
\chi^{(2)} &= -i\chi^{(2)}_{20}[(\rho_1 - \rho_2)\\
&\quad \cdot(\Delta_{10} - (\rho_0 - \rho_1)\Delta_{21}]D^{-1}\Gamma_{20}\Gamma_{10} \qquad (4)\\
\chi^{(1)}(\omega_1) &= -i\chi^{(1)}_{10}[(\rho_0 - \rho_1)(\Delta_{20}\Delta_{21}\\
&\quad + |G_{10}|^2) + (\rho_1 - \rho_2)|G_{21}|^2]D^{-1}\Gamma_{10}. \qquad (5)
\end{aligned}$$

Here $\chi^{(1)}_{10}$ and $\chi^{(2)}_{20}$ are the resonance susceptibilities in weak fields.

The on-diagonal elements ρ_i (atomic populations) can be found from Eqs. (2) by using Eqs. (3). We do not treat them because of their complexity.

We studied the atomic populations, $\rho_{0,1,2}$; the absorption, $\mathrm{Im}[\chi^{(1)}(\omega_{1,2})]$; the dispersion, $\mathrm{Re}[\chi^{(1)}(\omega_{1,2})]$; and the nonlinear atomic polarization, $\chi^{(2)}(\omega_s)g_1g_2$, $g_1 = |G_{10}|/(\Gamma_{20}\Gamma_{10})^{1/2}$, $g_2 = |G_{21}|/(\Gamma_{20}\Gamma_{21})^{1/2}$, as functions of the detuning Ω_{10} and the Rabi frequency g_1 under various conditions. The absorption and the dispersion at the frequency ω_s are very weak because the transition 2–0 is quadrupole or magnetic dipole. Some of the results are shown in Figs. 1 and 2.

Figure 1 shows $\rho_{0,1,2}$, $\mathrm{Im}\,\chi^{(1)}(\omega_1)$, $\mathrm{Re}\,\chi^{(1)}(\omega_1)$, and $\chi^{(2)}(\omega_s)g_1g_2$ as functions of detuning Ω_{10} for $\Omega_{21} = 0$. Figure 1(a) demonstrates coherent population trapping ($\rho_1 \ll \rho_0, \rho_2$) under two-photon resonance. Absorption of the fundamental wave with frequency ω_1 decreases considerably under CPT, but the nonlinear atomic polarization $\chi^{(2)}g_1g_2$ has a maximum [Fig. 1(b)]. It is much greater than that for weak fields.

Figure 2 shows the ratio $\eta = P_s/P_w$ (P_s and P_w are the powers of generated radiation for the strong and weak fundamental waves, respectively) as a function of normalized Rabi frequency g for the case $g_1 = g_2$ and both single- and two-photon resonances. It exhibits a great enhancement in the efficiency of two-photon resonant three-wave mixing using CPT. It is thus clear that the generated radiation is enhanced by using strong fundamental fields and a negligible absorption of the generated field.

In conclusion, our work shows the tre-

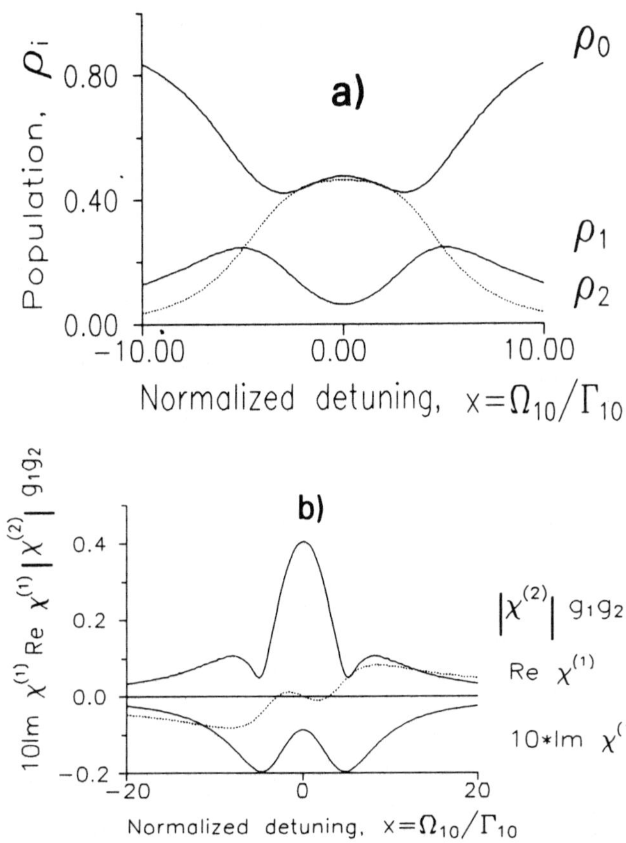

QThG16 Fig. 1. (a) Atomic population, $\rho_{0,1,2}$; (b) absorption, $\mathrm{Im}(\chi^{(1)}\omega_1)$, dispersion, $\mathrm{Re}(\chi^{(1)}(\omega_1)$ and nonlinear atomic polarization, $\chi^{(2)}g_1g_2$, versus normalized detuning, $x_{20} = \Omega_{20}/\Gamma_{20}$, for $\Omega_{21} = 0$. The parameters are $g_1 = |G_{10}|/(\Gamma_{10}\Gamma_{20})^{1/2} = 10$, $g_2 = |G_{21}|/(\Gamma_{20}\Gamma_{21})^{1/2} = 10$, $\Gamma_{10}/\Gamma_{21} = 1$, $\Gamma_{10}/\Gamma_{20} = 7$, $\Gamma_{20}/\Gamma_2 = 0.5$, and $\Gamma_1/\Gamma_2 = 7$.

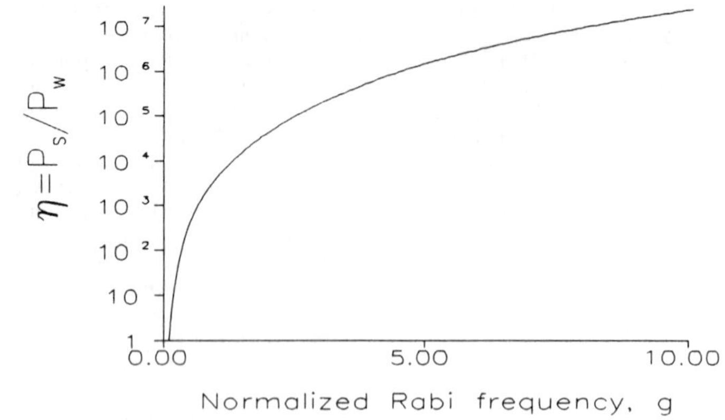

QThG16 Fig. 2. Dependence of η on normalized Rabi frequency g for $\Omega_{10} = \Omega_{20} = 0$ and $g_1 = g_2 = g$. All other parameters are the same as for Fig. 1. The Rabi frequency of the weak field is 0.1.

mendous possibilities for enhancing the efficiency of resonant three-wave mixing in an isotropic medium by a proper choice of the fundamental fields. The results can be easily generalized for the case of a lambda configuration of levels.

1. B. D. Agap'ev, M. B. Gorny, B. G. Matisov, and Yu. V. Rozhdesvenskii, Usp. Fiz. Nauk **163**, 1 (1993).
2. J. H. Eberly, M. L. Pons, and H. R. Haq, Phys. Rev. Lett. **72**, 56 (1994).
3. V. G. Arkhipkin and A. K. Popov, Usp. Fiz. Nauk **153**, 423 (1987).
4. A. J. Poustie and M. N. Dunn, Phys. Rev. A **47**, 1365 (1993).

QThG17

High-resolution optical hole burning and double resonance of holmium centers in CaF_2

Z. Hasan, M. Solonenko, G. A. Khodaparast, *Department of Physics, Temple University, Philadelphia, Pennsylvania 19122*

Optical-hole burning and associated double-resonance techniques give unsurpassed resolution to the frequency-domain spectroscopy of solids. The present work not only is a demonstration of the strength of the technique in simplifying a complicated spectrum, Ho^{3+} in CaF_2, but it also gives fundamental information about this system in great detail.

In very dilute crystals, ~0.0007-at. % Ho^{3+}, the dominant centers formed have either a tetragonal (C_{4v}) or a trigonal (C_{3v}) site symmetry, depending on the position of the charge-compensating fluorine ion. The trigonal and tetragonal crystal fields are of similar strength; therefore, spectra from these two centers overlap. The combination of the large electronic g values and the large nuclear spin, $I = 7/2$, of Ho gives exceptionally large hyperfine splittings to the electronic levels. This further complicates the optical spectra.

First, high-resolution site-selective spectroscopy was used to isolate the spectra as much as possible. Zeeman measurements were then performed to determine the symmetry of the centers and the states involved in the optical transitions.

In the trigonal center the ground state is an electronic doublet of the $^5I_8$ multiplet. It is split by large hyperfine interaction. We have studied the $^5I_8$–$^5F_5$ transition. Transient optical hole burning in the hyperfine levels of the ground state has been observed because of population redistribution in these levels. Holes are narrow and last for a fraction of a second. An optical–microwave double-resonance technique has been used to determine the hyperfine interaction parameter, the nuclear quadrupole interaction parameter, the electronic g factor, and the nuclear magnetogyric ratio of this state: $A/2 = 3.8500 \pm 0.0005$ GHz; $D = -6.0100 \pm 0.0001$ MHz; $g = 9.8$; $\gamma_\parallel = 7.3$ kHz/G, $\gamma_\perp = 5.62$ kHz/G.

In the tetragonal center the ground crystal-field state is a singlet (A_1); however, it is interacting strongly with another closely lying singlet (A_2). In combination, these two singlets behave as a pseudodoublet. This gives very interesting hyperfine and higher-order interactions observable in this state. The high-resolution spectrum of the $^5I_8$ (A_1)–$^5F_5$ (A_1) transition is shown in Fig. 1. There are only four hyperfine lines, as opposed to eight. The other singlet has the remaining four lines. The hole-burning spectrum of the first doublet in Fig. 1 is shown in Fig. 2. Holes last for a fraction of a second and are due to population redistribution in both hyperfine and superhyperfine levels. This spectrum identified the hyperfine levels in the ground state. A resolved spectrum of the optical hole, which gives superhyperfine resonances of the fluorine ions in the vicinity of the holmium ion, is shown in Fig. 3. Optical–rf double resonance has been used to study superhyperfine and other higher-order interactions. It confirms the assignments on the basis of the hole-burning spectrum.

In summary, high-resolution hole-burning and double-resonance techniques have completely clarified the spectra associated with two dominant centers of holmium. Furthermore, they have given us detailed information about electronic and nuclear interactions.

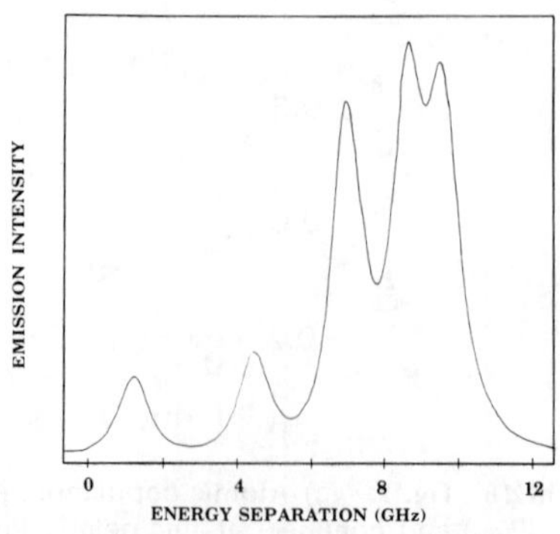

QThG17 Fig. 1. High-resolution spectrum of the $^5I_8$ (A_1)–$^5F_5$ (A_1) transition.

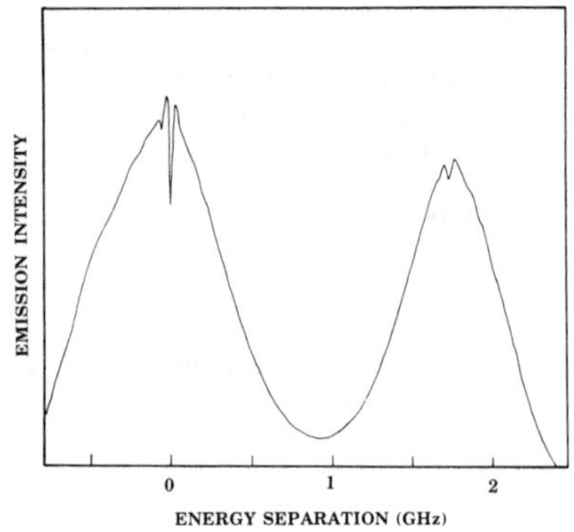

QThG17 Fig. 2. Hole-burning spectrum of first doublet in Fig. 1.

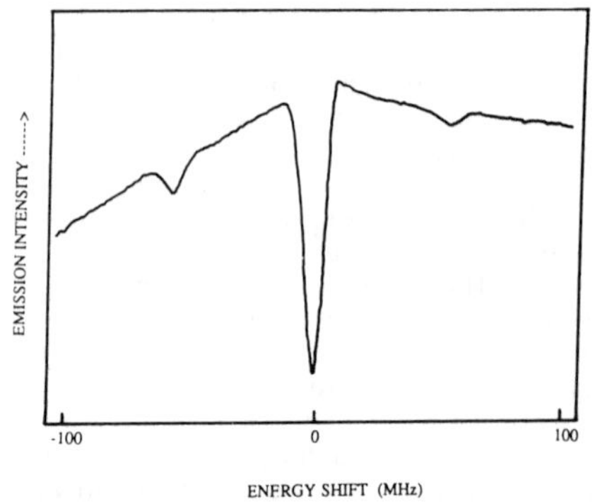

QThG17 Fig. 3. Resolved spectrum of optical hole.

QThG18

Upconversion and cooperative energy transfer in erbium in cubic crystalline hosts

Z. Hasan, L. Biyikli, M. Shishkov, G. A. Khodaparast, *Department of Physics, Temple University, Philadelphia, Pennsylvania 19122*

Cubic crystalline hosts have been the proving ground for spectroscopic investigations in solids. Both experiments and theoretical models to interpret the data are simple in these high-symmetry systems. The same argument makes cubic systems more suitable for studies of excitation dynamics and electron–lattice coupling, which are of great significance for laser media. However, energy-transfer and upconversion studies are confined to noncubic systems because the parity-selection rules generally give prohibitively small transfer rates in a cubic environment. In this paper we present our results on energy transfer and upconversion in a cubic host.

Upconversion by cooperative energy

transfer, stepwise absorption, and two-photon absorption have been observed for erbium in cubic Cs_2NaYCl_6 crystals under pulsed excitation with a laser. With this mode of excitation of the $^4F_{9/2}$ state in the red, blue emission from the $^2G_{9/2}$ state and green emission from the $^4S_{3/2}$ state have been observed. These upconversions are quite strong; at their onset the crystal totally changes color because of upconverted emission. The concentration dependence reveals that in crystals containing <10 at. % of erbium, such an upconverted signal is not detectable with pulsed excitation. On the other hand, under cw excitation with a laser, only the green upconversion is observed and it occurs for all concentrations of erbium. We also observed two-photon absorption in the $^2H_{9/2}$ state in the ultraviolet ($E \sim 36\ 000\ cm^{-1}$). However, a resonant one-photon absorption in the $^4S_{3/2}$ state at low temperatures complicates the TPA spectrum.

A rate-equation analysis of the time behavior of the upconverted signal has been used to determine energy pathways for the upconversion. On the basis of the time evolution and the concentration dependence of the upconverted signal, a two-ion cooperative-energy-transfer process has been proposed for the blue emission. In this process two erbium ions in the $^4I_{9/2}$ state cooperatively transfer energy such that one of the ions is further excited to $^2G_{9/2}$ state from which the blue emission originates. The green emission under pulsed-excitation conditions is a complex mixture of several processes that populate the $^4S_{3/2}$ state. In the cw excitation the green upconversion is a result of the stepwise-absorption process seen in many other systems. A systematic study of upconversion as a function of temperature and concentration has been performed. At high temperatures multiphonon relaxations deplete the relevant states and the upconversion stops.

Reasons for such a strong upconversion in these centrosymmetric systems are not fully understood.[1] If we assume a dipole–dipole interaction for which the transfer rate has an r^{-6} dependence on the donor–acceptor separation, it appears that the high concentration of Er allowed in these crystals more than compensates for the weak dipoles (typical of an ion at a center of inversion symmetry). Complete crystal-field calculations have been performed, and a good knowledge of crystal-field energy levels and eigenfunctions in this system has greatly helped in understanding various upconversion processes and pathways. A detailed calculation of transfer rates with a microscopic theoretical model is underway. To our knowledge this is the first report of a cooperative energy transfer in a cubic system.

1. F. S. Richardson, private communication.

QThG19

Measurement of free-carrier nonlinearities in ZnSe by using the Z-scan technique with a nanosecond laser

Kyu-Haeng Lee, Wook-Rae Cho, Seung-Han Park, Ung Kim, *Department of Physics, Yonsei University, Seoul 120-749, Korea*

Z-scan techniques have been used to measure the nonlinear refractive index and the absorption coefficient of a large number of direct-gap semiconductors, such as GaAs, InSb, CdSe, and ZnSe.[1-3] The origins of the nonlinearities are mainly attributed to two-photon transitions, Raman transitions, the ac Stark shift, and free charge carriers due to two-photon absorption.[1] Picosecond Z-scan and degenerate-four-wave-mixing measurements showed that two-photon transitions, Raman transitions, and the ac Stark shift are the dominant mechanisms for nonlinear refraction up to an irradiance of ~0.5 GW/cm². However, the nonlinearity caused by the two-photon-excited free carriers becomes significant at much higher irradiance levels.[1]

In this paper we report the observation of self-defocusing and nonlinear absorption due to the free charge carriers generated by two-photon absorption at a low power density (~30 mW/cm²) in a polycrystalline ZnSe by using the Z-scan technique with nanosecond laser pulses. A frequency-doubled, *Q*-switched Nd:YAG laser was used as a light source. The pulse duration and the repetition rate were 25 ns (full width at half maximum) and 5 Hz, respectively. The spatial profile of the pulse was nearly Gaussian, which was ascertained by scanning a 10-μm pinhole across the beam. The sample thickness was 2.7 mm.

Using the slowly-varying-envelope approximation, we can separate the wave equation into equations for the phase and the irradiance,[1,3]

$$\frac{d\Delta\phi}{dz'} = k(\gamma I + \sigma_r N), \tag{1}$$

$$\frac{dI}{dz'} = -(\alpha_0 + \beta I + \sigma_{ab} N)I', \tag{2}$$

where k is the magnitude of the wave vector in free space, α_0 is the linear absorption, σ_r is the change in index of refraction per unit photoexcited-charge-carrier density, σ_{ab} is the total carrier cross section (holes plus electrons), and N is the number density of these carriers.

The equation for the generated carrier

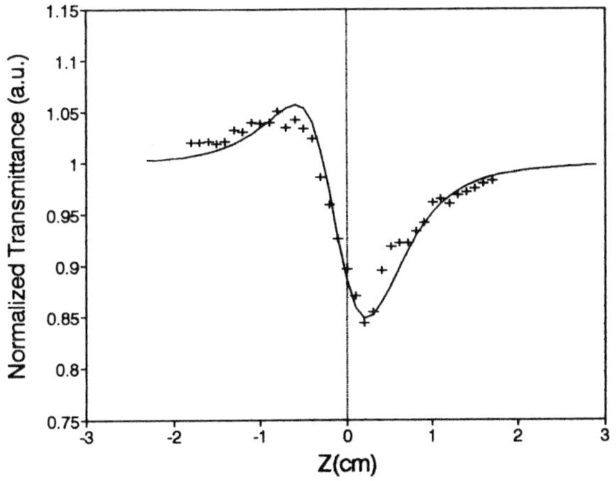

QThG19 **Fig. 1.** Normalized 50%-aperture Z-scan data on a 2.7-mm ZnSe sample measured with 25-ns laser pulses and γ = 532 nm at I_0 = 30 MW/cm². The solid curve is the theoretical fit with $\sigma_r = 0.6 \times 10^{-21}\ cm^3$.

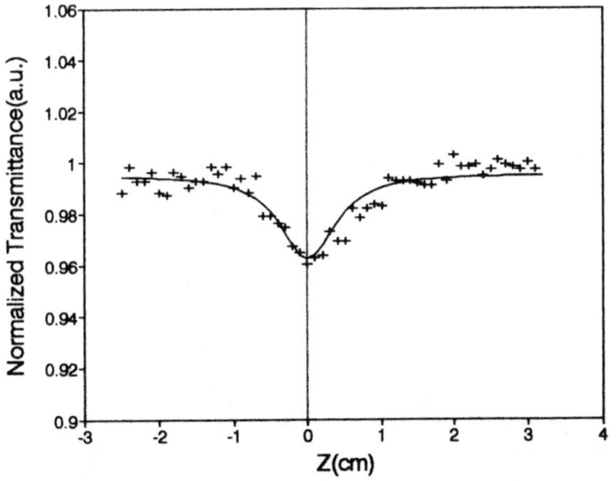

QThG19 **Fig. 2.** Normalized open-aperture Z-scan data at I_0 = 35 MW/cm². The solid curve is the theoretical fit with $\sigma_{ab} = 0.8 \times 10^{-18}\ cm^2$.

is given by

$$\frac{dN}{dt} = \frac{\beta I^2}{2\hbar\omega}, \qquad (3)$$

indicating that one electron–hole pair is generated for every two absorbed photons.

The diffusion process is neglected in this equation because the diffusion time of free carriers is longer than the pulse width of our laser.[4] An iterative method was used to solve these equations, and it was assumed that the last term in Eq. (2) is much smaller than the other terms.

Figures 1 and 2 show the normalized closed-aperture and open-aperture Z-scan data with 25-ns laser pulses, respectively. In contrast to the data obtained from the picosecond Z-scan experiment, we observed self-defocusing and nonlinear absorption due to the free charge carriers even at the low power density of ~30 MW/cm^2. The solid curves in Figs. 1 and 2 are theoretical fits with β = 5.8 cm/GW and γ = -6.8×10^{-14} cm^2/W, obtained by the picosecond Z-scan technique.[1] We determined that $\sigma_r = 0.6 \times 10^{-21}$ cm^3 and $\sigma_{ab} = 0.8 \times 10^{-18}$ cm^2 from the fits.

1. A. A. Said, M. Shiek-Bahae, D. J. Hagan, T. H. Wei, J. Wang, and E. W. Van Stryland, J. Opt. Soc. Am. B **9**, 405 (1992).
2. M. Shiek-Bahae, A. A. Said, T. H. Wei, D. J. Hagan, and E. W. Van Stryland, IEEE J. Quantum Electron. **QE-26**, 760 (1990).
3. E. W. Van Stryland, H. Vanherzeele, M. A. Woodal, M. J. Soileau, A. L. Smirl, S. Guha, and T. F. Boggess, Opt. Eng. **24**, 613 (1985).
4. J. R. Milward, W. Ji, A. K. Kar, C. R. Pidgeon, and B. S. Wherrett, J. Appl. Phys. **69**, 2708 (1991).

QThG20

Dynamics of infrared-to-visible upconversion in $Cs_3Lu_2Br_9$:1%Er^{3+}

Markus P. Hehlen, Gabriela Frei,* Hans U. Güdel,* *Los Alamos National Laboratory, Chemical Science and Technology Division, Mailstop E535, Los Alamos, New Mexico 87545*

Various processes can lead to near-infrared (NIR) to visible (VIS) upconversion. The knowledge of their relative efficiency is an important parameter in optimizing a desired upconversion laser pumping scheme. $Cs_3Lu_2Br_9$:1%Er^{3+} is an ideal system to study NIR-to-VIS upconversion dynamics[1]: It is one of a family of dimer compounds $Cs_3Lu_2X_9$:Er^{3+} (X = Cl, Br, I) which show intense VIS upconversion luminescence when excited in the NIR.[2] The low phonon energies in these halide lattices strongly suppress multiphonon relaxation and, in contrast to oxide and fluoride systems, radiative relaxation is dominant for most of the Er^{3+} excited states. In addition, cross-relaxation processes, which efficiently quench luminescence from excited states such as $^2H_{9/2}$ and $^4F_{7/2}$ in concentrated $Cs_3Er_2Br_9$, are suppressed in $Cs_3Lu_2Br_9$:1%Er^{3+}.[3] Using cw or pulsed NIR excitation of the $^4I_{9/2}$ and $^4I_{11/2}$ multiplets VIS upconversion luminescence mainly from $^2H_{9/2}$ and $^4F_{7/2}$ is observed, respectively (Fig. 1). After a first excitation, upconversion to these states can either occur radiatively by a second absorption or nonradiatively by an energy transfer from an excited neighboring ion. In $Cs_3Lu_2Br_9$:1%Er^{3+}, both excited-state absorption (ESA) and energy-transfer upconversion (ETU) were found to be active and could be identified on the basis of the upconversion transients. The short excitation laser pulse duration $\delta t \approx 10$ ns imposes a significant restriction on the time scale for an ESA process. Whereas nonradiative ETU can still proceed after the excitation pulse, the sequential two-photon absorption in the radiative ESA process can only take place within the time δt. As the time constants which are involved in ETU processes are usually significantly longer than δt = 10 ns, the observation of upconversion luminescence intensity within δt is a direct indication of an ESA process. This is illustrated in Fig. 2 for the upconversion of $^4I_{11/2}$ excitation to $^4F_{7/2}$. Whereas the maximum of $^4F_{7/2}$ population is reached instantaneously within δt for excitation B, clearly indicating an ESA process, there is a competition between ESA and ETU for excitation A, with $^4F_{7/2}$ being populated mainly after the laser pulse. On the basis of a complete crystal-field analysis for $Cs_3Lu_2Br_9$:1%Er^{3+},[4] the relative efficiency of ESA and ETU can be directly related to the energetic resonance of the respective processes. Whereas nonresonant ESA during δt = 10 ns is strongly suppressed for energy mismatches >12 cm^{-1}, ETU is still efficient. The energy mismatches for various $^4I_{11/2}$ excitation wavelengths suggest ETU and ESA to be the dominant upconversion processes for excitation A and B, respectively (Fig. 3), in agreement with the observed transients (Fig. 2). Time-resolved upconversion spectroscopy can thus provide essential information on both the nature of the upconverting centers and the active upconversion processes.

**Institut für anorganische Chemie, Universität Bern, Freiestrasse 3, 3000 Bern 9, Switzerland*

1. M. P. Hehlen, G. Frei, and H. U. Güdel, Phys. Rev. B **50**, (1994), in press.
2. M. P. Hehlen, K. Krämer, H. U. Güdel, R. A. McFarlane, and R. N. Schwartz, Phys. Rev. B **49**, 12475 (1994).
3. M. P. Hehlen, K. Krämer, and H. U. Güdel, J. Lumin. **60&61**, 142 (1994).
4. M. P. Hehlen, H. U. Güdel, J. R. Quagliano, J. Chem. Phys. **101**, (1994), in press.

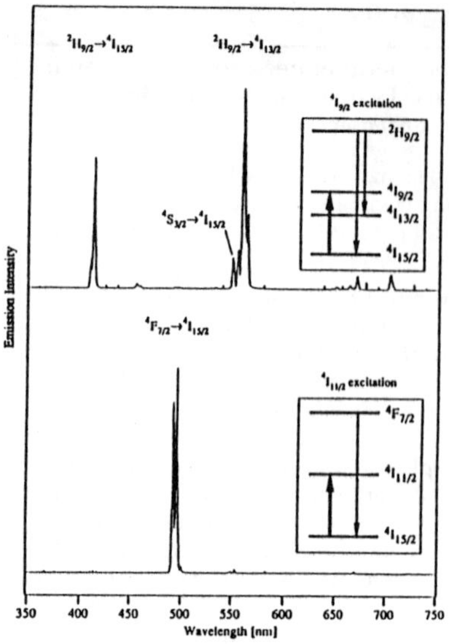

QThG20 Fig. 1. Unpolarized upconversion luminescence spectra of $Cs_3Lu_2Br_9$:1%Er^{3+} at 10 K using pulsed $^4I_{9/2}$ and $^4I_{11/2}$ excitation, respectively. The insets schematically show the relevant excitation and luminescence transitions.

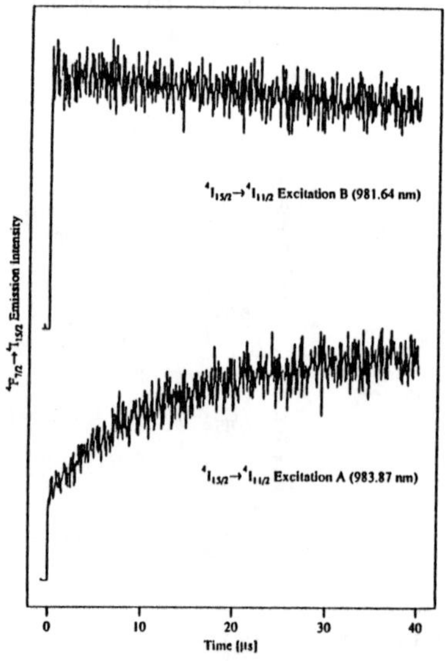

QThG20 Fig. 2. Rise characteristics of $^4F_{7/2}$ upconversion transients under $^4I_{11/2}$ excitation of $Cs_3Lu_2Br_9$:1%Er^{3+} at 10 K. The excitation wavelengths A and B correspond to different $^4I_{15/2} \rightarrow {}^4I_{11/2}$ crystal-field transitions (see Fig. 3).

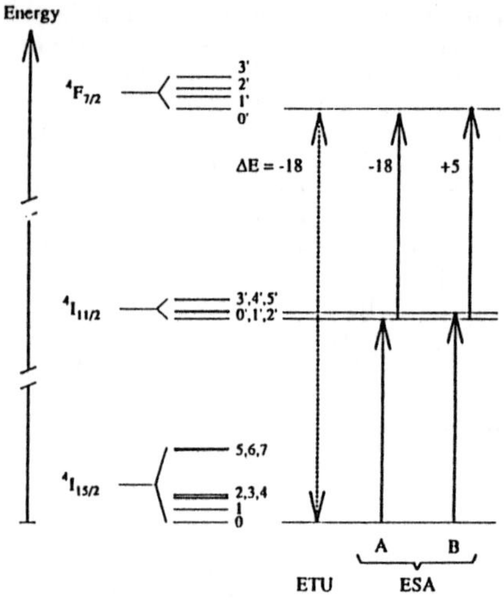

QThG20 Fig. 3. Energy mismatches ΔE (cm^{-1}) for $^4I_{11/2} \rightarrow {}^4F_{7/2}$ excited-state absorption (ESA) and $^4I_{11/2} \rightarrow {}^4F_{7/2}$, $^4I_{15/2}$ energy-transfer upconversion (ETU) in $Cs_3Lu_2Br_9$:1%Er^{3+}. The labels A and B refer to excitation wavelengths used to obtain the upconversion transients shown in Fig. 2.

QThG21

Subrecoil-cooling schemes for cesium

Ralph Dum, *ENS, Laboratoire Kastler, 24, Rue Lhomond, 75231 Paris 05 France*

Models for subrecoil cooling that use velocity-selective coherent population trapping[1] have been restricted to the hyperfine transitions $J = 1 \rightarrow J = 1$; typically, experiments have been performed on helium or sodium.[1] By using a bichromatic laser field on the $J = 3 \rightarrow J = 4$ and the $J = 4 \rightarrow J = 4$ hyperfine transitions, we obtain a velocity-selective trapping state for cesium. The specific form for (the positive frequency part of) the laser field is

$$E^{(+)}(y, t) = \exp(-i\omega_1 t)[\exp(-ik_1 y)\mathbf{e}_z + \exp(ik_1 y)\mathbf{e}_x] + \exp(-i\omega_2 t)\cos k_2 y \mathbf{e}_z, \quad (1)$$

where $\omega_{1,2}(k_{1,2}, k_1 \approx k_2)$ are the frequencies (wave numbers) of the lasers acting on the $J = 4 \rightarrow J = 4$ and the $J = 3 \rightarrow J = 4$ transitions, respectively, and $\mathbf{e}_{x,z}$ are the polarization vectors. This configuration allows for a velocity-selective trapping state as a linear superposition of the $m = 0$ state of the $J = 4$ hyperfine level and the $m = \pm 1$ states of the $J = 3$ hyperfine level. To clarify this point, assume that we construct a linear superposition Ψ_D of the ground-state hyperfine levels in $J = 3$ and $J = 4$ such that the transition amplitudes to any excited-state sublevel ($J = 4$) interfere destructively. Furthermore, the coefficients of these linear superpositions are linear in the Rabi frequency, so that Ψ_D not only is a noncoupled state but also is an eigenstate of the kinetic energy eigenoperator. The validity of the following underlying assumptions is shown: (a) The lasers operate on their respective transitions due only to the large energy separation of the $J = 3$ and the $J = 4$ ground states (9.2 GHz). (b) The $J = 5(3)$ hyperfine level is not excited; i.e., the energy separation between the $J = 4$ and the $J = 5(3)$ levels (251 and 201.5 MHz, respectively) is large compared to the laser detuning from the $J = 3(4) \rightarrow J = 4$ resonance; typical lifetimes of the dark state of the order of milliseconds are estimated.

We discuss the efficiency of cooling, i.e., the times to accumulate a substantial part of the atoms in the dark state, and we compare the theoretical results with a planned experiment.

Using cesium instead of helium could lead to much higher phase-space densities and thereby to interesting quantum collective effects. Furthermore, by using techniques of adiabatic passage[2] the atoms can be prepared in the $m = 0$ state, an ideal situation for atomic clocks.

Extensions to two- and three-dimensional cooling schemes and the limitations due to the mass of cesium atoms (gravity) are discussed.

1. A. Aspect *et al.*, Phys. Rev. Lett. **61**, 826 (1988).
2. P. Marte *et al.*, Phys. Rev. A **44**, R4118 (1991).

QThG22

InGaAs nanostructure formed by the Stranski–Krastanow growth mode on GaAs (311)A

P. O. Vaccaro, M. Hirai, K. Fujita, T. Watanabe, *ATR Optical and Radio Communications Research Laboratories, 2-2 Hikaridai, Seika-cho, Soraku-gun, Kyoto 619-02, Japan*

Recently there have been reports of InGaAs quantum dots formed by the Stranski–Krastanow growth mode by using molecular-beam epitaxy (MBE) on GaAs (100) substrates.[1] There is also a report of InGaAs quantum dots grown by MOCVD on GaAs (100) tilted between 0° and 41° toward (111)B.[2] These structures are formed because the highly strained InGaAs epilayer decreases its energy by the formation of islands after a layer-by-layer growth. The formation of nanostructures in the Stranski–Krastanow growth mode on substrates different from (100) is an interesting possibility to be investigated. In this paper we show the formation of InGaAs wirelike structures on GaAs (311)A substrates.

The samples were grown on GaAs (311)A and (100) substrates by MBE. The structure consisted of a 400-nm GaAs buffer layer, a 3-nm $In_{0.15}Ga_{0.85}As$ reference quantum well, a 120-nm GaAs spacer layer, a 1.7-nm $In_{0.5}Ga_{0.5}As$ layer, and 90-nm GaAs cap layer. A set of samples without the cap layer was also grown.

Figure 1 shows the surface of the samples without the cap layer, observed by atomic-force microscopy. The (100) sample shows the small InGaAs islands already reported on this surface.[1] On the other hand, the (311)A sample shows a completely different structure. The surface has a wirelike morphology with a height of about 1 nm and with a 25-nm periodicity. Figure 2 shows the photoluminescence (PL) spectrum at 12 K for samples with a GaAs cap layer. The (100) sample shows a narrow peak at 1.481 eV (FWHM = 2.4 meV) from the reference $In_{0.15}Ga_{0.85}As$ quantum well and a wide peak at 1.316 eV (FWHM = 39.4 meV) from the island layer. The large FWHM originates from the size distribution of the islands. These results compare well with the characteristics of the best samples published by Leonard *et al.*[1] The (311)A sample also shows a narrow peak at 1.482 eV (FWHM = 2.2 meV) from the reference quantum well and a wide peak at 1.352 eV (FWHM = 23.6 eV) from the wirelike layer. The stronger PL intensity of the (311)A sample is also observed in samples with conventional quantum wells, and it is not related to the different nanostructures. Figure 3 shows the dependence of PL-peak height on temperature for the nanostructures. The (100) sample's PL decreases as temperature increases, and it vanishes above 210 K. The

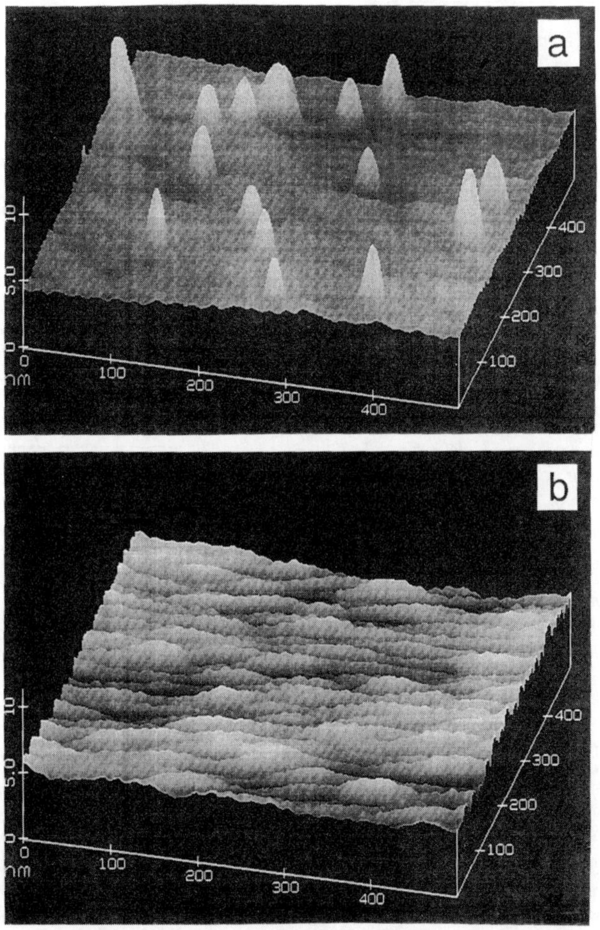

QThG22 Fig. 1. Surface of the $In_{0.5}Ga_{0.5}As$ layer observed by atomic-force microscopy: (a) GaAs (100) substrate; (b) GaAs (311)A substrate.

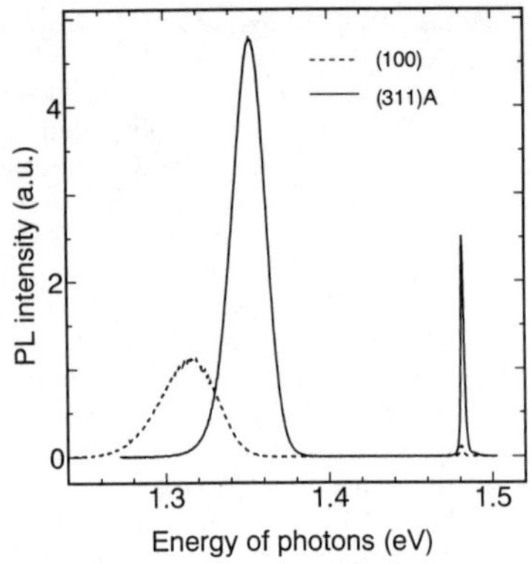

QThG22 Fig. 2. Low-temperature photoluminescence spectra. The samples were excited by using an Ar^+ laser (wavelength, 488 nm; power, 5 mW).

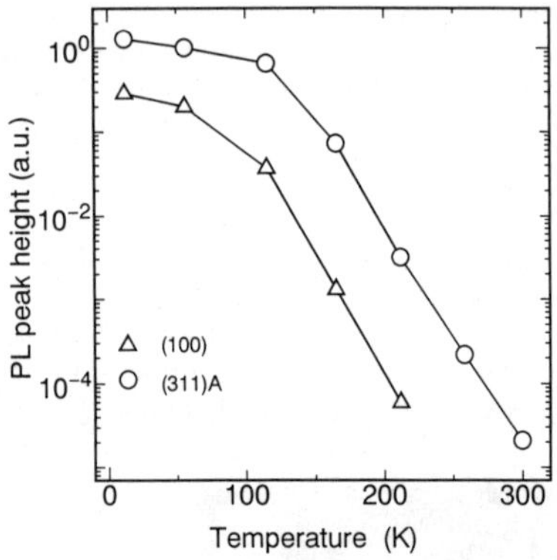

QThG22 Fig. 3. Dependence of photoluminescence-peak height on temperature. The (311)A sample shows little decrease of intensity up to 115 K.

(311)A sample shows little reduction of PL intensity up to 115 K. Further increases in temperature produce decreases in PL intensity, but it remains easily detectable up to room temperature. The PL measurements indicate that the material in the nanostructures formed on the (311)A substrate has a high optical quality. We are currently investigating the influence of indium composition and growth parameters on these nanostructures. We are also studying the time resolution and polarization dependence of PL.

1. D. Leonard, M. Krishnamurthy, S. Fafard, J. L. Merz, and P. M. Petroff, J. Vac. Sci. Technol. **B12**, 1063 (1994).
2. J. Oshinowo, M. Nishioka, S. Ishida, and Y. Arakawa, submitted to Jpn. J. Appl. Phys.

QThG23

Theory of optical gain in strongly confined semiconductor quantum dots

Y. Z. Hu, N. Peyghambarian, S. W. Koch,*
Optical Sciences Center, University of Arizona, Tucson, Arizona 85721

Optical gain has been demonstrated in strongly confined CdSe quantum dots. A gain bandwidth of approximately 0.5 eV has been experimentally demonstrated in these structures.[1] As a consequence of the Pauli blocking in the lowest conduction band, only the two-electron–hole-pair-excitation (biexciton) and the one-electron–hole-pair-excitation (exciton) states contribute to the gain. In this presentation, for the first time to our knowledge, we show theoretically that the biexciton-to-exciton transitions are important in forming gain in quantum dots (QDs).

The Coulomb interaction strongly modifies optical selection rules in QDs.[2] To illustrate this point, a schematic plot of the energy spectrum in QDs is given in Fig. 1. Without the Coulomb interaction and valence-band-mixing effects, only electrons and holes with identical quantum numbers can recombine, as denoted by the solid arrows. These optical selection rules do not hold in the presence of the Coulomb interaction. In Fig. 1 we use the dotted arrows to indicate the dipole transitions induced by the Coulomb interaction or valence band mixing.

To quantify these arguments, we performed numerical studies of gain dynamics in strongly confined QDs. Using the numerical matrix-diagonalization method,[3] we calculated the eigenenergies of exciton and biexciton states and optical dipole moments. The gain spectra were obtained by integrating the optical Bloch equations. An example of calculated optical gain spectra in small QDs is shown in Fig. 2. The solid curve is the calculated linear absorption. The dashed curves are the computed absorption/gain spectra corresponding to an increasing biexciton population. In QDs the gain below the absorption edge is predominately caused by the transitions between the ground-state biexciton and excited exciton states. It is also interesting that the high-frequency side of the gain can be close to or even beyond the pump photon energy. Hence, realization of a broad region in QDs is predicted, as observed experimentally.[1]

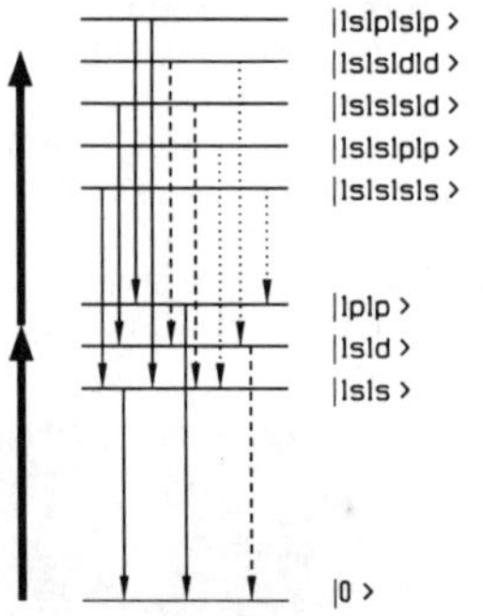

QThG23 Fig. 1. Schematic plot of relevant exciton and biexciton levels and dipole allowed transitions. The solid arrows indicate the dipole allowed transitions in the strong-confinement approximation. The dashed and dotted arrows indicate the transitions induced by the valence-band-mixing and Coulomb interaction, respectively. $|0\rangle$ denotes the ground state, and exciton and biexciton states are expressed by $|n_e, n_h\rangle$ and $|n_{e1}, n_{e2}, n_{h1}, n_{h2}\rangle$, respectively.

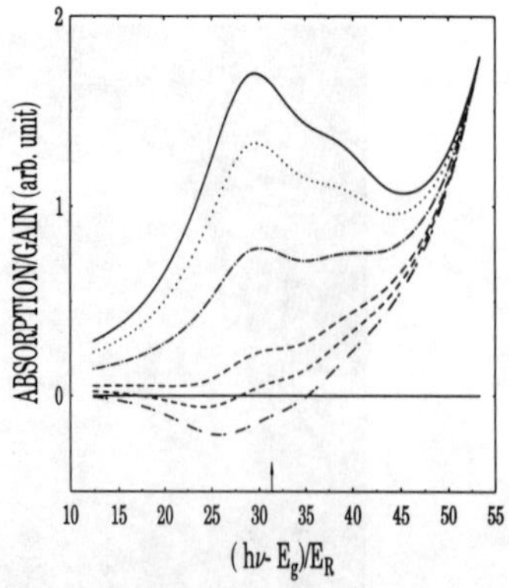

QThG23 Fig. 2. Computed absorption/gain spectra in a CdS quantum dot with $R = 0.5a_B$. The curves from top to bottom correspond to increasing pump intensity. E_R and E_g are the exciton Rydberg energy and band-gap energy, respectively.

We acknowledge financial support from the National Science Foundation, the U.S. Army Research Office, the U.S. Air Force Office of Scientific Research, the Ballistic Missile Defense Organization, and OCC at the University of Arizona.

Department of Physics, Philipps University, 35112 Marburg, Germany

1. H. Giessen, B. Fluegel, G. Mohs, N. Peyghambarian, and U. Woggon, submitted for this conference.
2. Y. Z. Hu, S. W. Koch, M. Lindberg, N. Peyghambarian, E. L. Pollock, and F. F. Abraham, Phys. Rev. Lett. **64**, 1805 (1990).
3. Y. Z. Hu, M. Lindberg, and S. W. Koch, Phys. Rev. B **42**, 1713 (1990).

QThG24

Time dynamics and optical response unique to triple quantum wells

M. S. Tobin, G. W. Bryant,*
J. L. Bradshaw, R. P. Leavitt, J. T. Pham,
Microphotonics Devices Branch, AMSRL-EP-EE, U.S. Army Research Laboratory, 2800 Powder Mill Road, Adelphi, Maryland 20783-1197

Long-lived spatially indirect excitons have been observed in asymmetric coupled GaAs/AlGaAs double quantum wells by others (see, for example Refs. 1 and 2). We are investigating asymmetric triple-quantum-well (TQW) systems; these systems permit greater flexibility in that electron and hole tunneling take place across different barriers and consequently can be independently controlled. Depending on the particular design, the TQW system should be adaptable to very long carrier lifetimes with an enhanced polarizability or might be useful for fast carrier sweepout in device applications. We have previously reported results[3] from the investigation of asymmetric TQW GaAs/AlGaAs structures designed so that photogenerated electrons resonantly tunnel between the central well and one of the side wells while photogenerated holes resonantly tunnel between the central well and the other side well at the same applied bias. We have found that the value of band offset is critical in the successful modeling of the structures with simultaneous resonances.

Sample	166	277	420
well 1	50 A	50 A	70 A
well 2	100 A	100 A	150 A
well 3	33 A	33 A	44 A
barrier 1	25 A	100 A	25 A
barrier 2	15 A	100 A	15 A

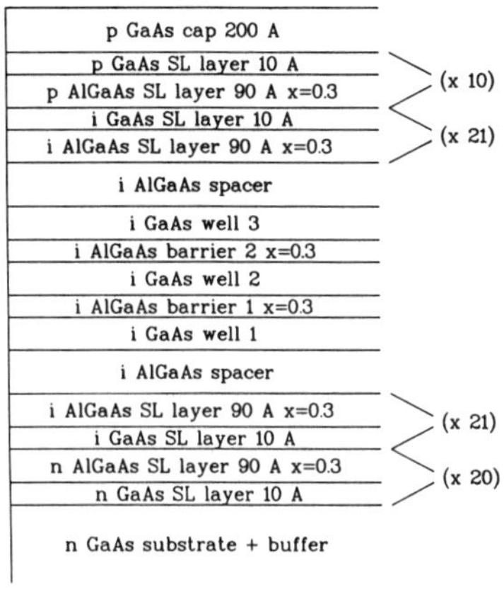

QThG24 Fig. 1. Growth parameters for the three samples.

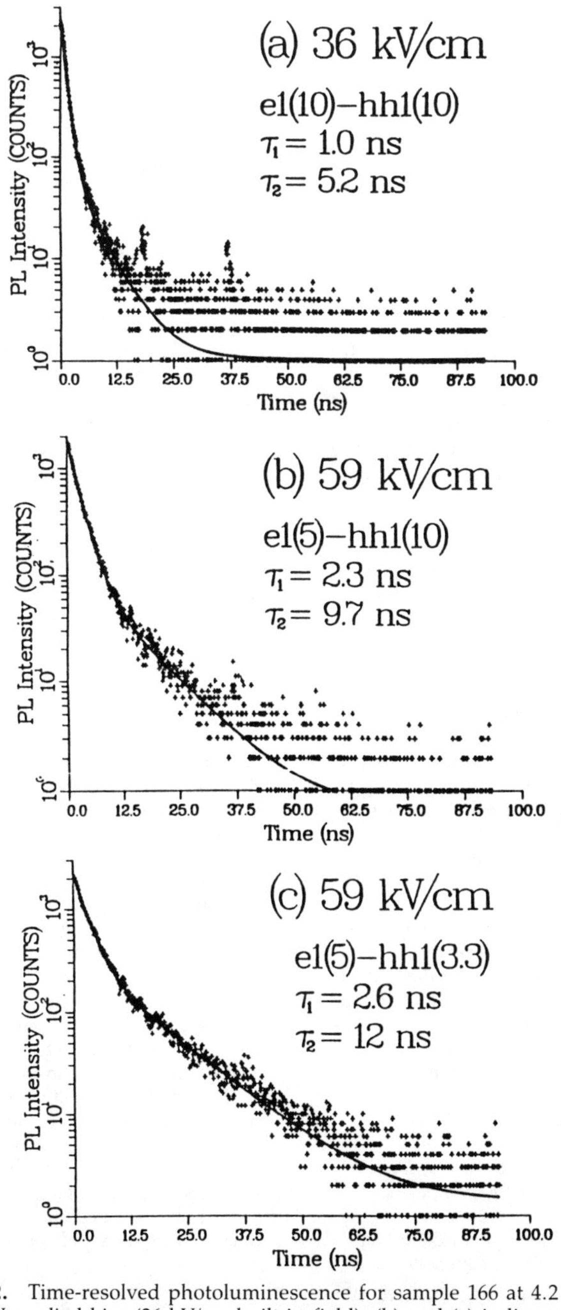

QThG24 Fig. 2. Time-resolved photoluminescence for sample 166 at 4.2 K: (a) direct transition at 0-V applied bias (36-kV/cm built-in field); (b) and (c) indirect transitions as labeled at −1.1-V applied bias (59-kV/cm total field). A styryl-9M pump laser at 1.575 eV was used.

In this paper we compare the results of photocurrent and time-resolved photoluminescence (PL) measurements for three TQW samples. The TQW parameters are indicated in Fig. 1. Sample 166 was the sample for which we previously reported simultaneous electron and hole resonances. We observed an enhancement of carrier lifetime with reverse bias past resonance such that the observed transitions are indirect (Fig. 2). However, eventually fields are reached such that the lifetime begins to decrease and PL quenching is observed. We have investigated several other samples to find the quenching mechanism with the aim of designing samples with longer carrier lifetimes. Sample 277 was grown with well widths identical to those of sample 166 but with uncoupled wells in order to examine the PL from each well independently. We observed significant PL quenching for the 3.3-nm well of sample 277 with modest applied electric fields, which suggests that tunneling from that well may be the cause of the quenching in sample 166. Sample 420 was growth with wider wells to provide better confinement of the carriers to the TQW system. Results for all three cases are presented and used to identify carrier dynamics unique to TQWs that can be tailored and exploited in optoelectronic devices.

*Currently at *National Institute for Standards and Technology, Gaithersburg, Maryland 20899*

1. T. B. Norris, N. Vodjdani, B. Vinter, C. Weisbuch, and G. Mourou, Phys. Rev. B **40**, 1392 (1989).
2. J. E. Golub, K. Kash, J. P. Harbison, and L. T. Florez, Phys. Rev. B **41**, 8564 (1990).
3. G. W. Bryant, J. L. Bradshaw, R. P. Leavitt, M. S. Tobin, and J. T. Pham, Appl. Phys. Lett. **63**, 1357 (1993).

QThG25

External field-dependent enhancement of internal Coulomb interactions in time-resolved four-wave mixing

P. Leisching, W. Beck, W. Schäfer,* K. Leo,** K. Köhler,† H. Kurz, *Institute for Semiconductor Technology II, Rhine–Westphalia Technical University, D-52056 Aachen, Germany*

Time-resolved four-wave mixing (TRFWM) has been extensively used to study the influence of Coulomb interactions on the free-induction decay of excitons in semiconductor quantum-well systems (see, e.g., Refs. 1–4). We show that the ratio between the Coulomb-mediated and the pure phase-space occupation signal can be changed by more than an order of magnitude by changing the external field. The two signal contributions to the diffracted signal can be separated by variation of the excitation density, and they show completely different behaviors as a function of the applied bias.

The experiments are performed on a high-quality $GaAs/Al_{0.3}Ga_{0.7}As$ superlattice of 35 periods of 97-Å wells and 17-Å barriers.[3,4] We use a standard TRFWM three-beam setup (see, e.g., Ref. 1: pulse 2 is delayed by τ relative to pulse 1, where the time delay between pulses 2 and 3 is denoted by t. The experiments are performed in a copolarized configuration, and the sample is kept at 10 K. The broadband 18-meV (130-fs) laser pulse is centered at 805 nm, slightly below the ($n' = 0$ heavy-hole)-to-($n = 0$ electron) WS transition, hh_0, in both experiment and theory. Under this condition only the hh_{-1} and hh_0 transitions are excited above an applied bias of about 5 kV/cm.

In Fig. 1 the TRFWM data at an excitation density of 10^{10} cm^{-2} and an electric field of 7 kV/cm are depicted. We observe two pronounced peaks in the signal: The first peak is attributed to the scattering of the pulse off the transient grating. The second peak is significantly delayed by about 500 fs; i.e., it can be associated with the evolution of a polarization wave.[1-3] The beating in t and τ in the decay of the signal

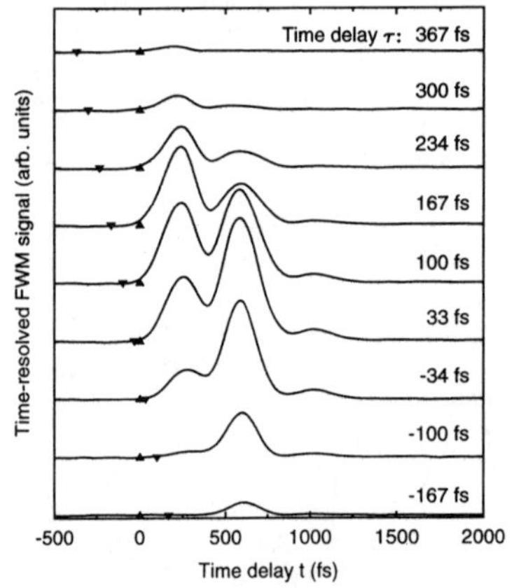

QThG25 Fig. 1. TRFWM signal for various delays τ at an excitation density of 10^{10}-cm^{-2} carriers per period. The down/up triangles symbolize the arrival on the sample of pulse 1 and pulse 2, respectively.

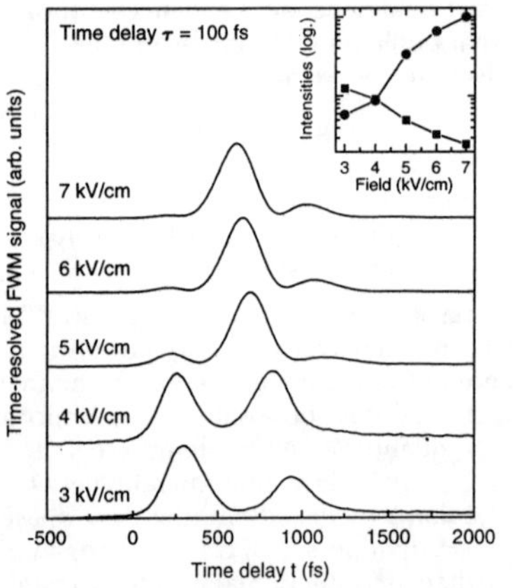

QThG25 Fig. 2. Experimentally observed field dependence in the WS regime of the TRFWM signal at an excitation density of 7×10^9 cm^{-2}. The data are normalized, and the experimental error in the fields due to screening effects is about 10%. The relative strength of the two contributions to the signal is depicted in the inset. The squares symbolize the peak amplitude of the prompt contribution, and the circles symbolize the delayed part.

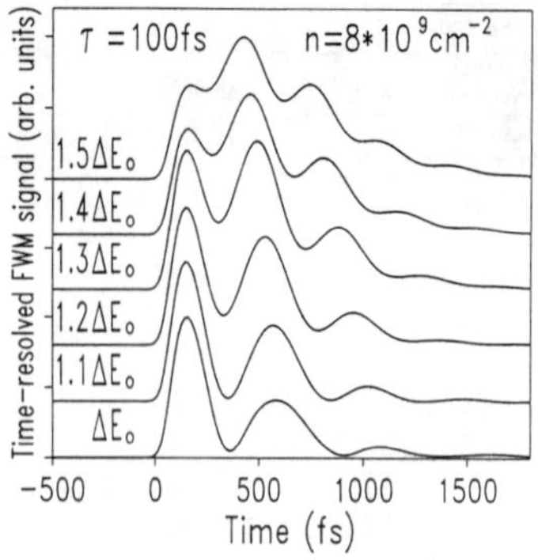

QThG25 Fig. 3. Calculated field dependence of the two signal contributions. The potential drop $\Delta E_0 = eFd = 6.5$ meV, where d is the lattice period and F is the electric field, was tuned by a factor of 1.5 at a constant excitation density.

at about 450 fs is due to Bloch oscillations. At a lower excitation density of 3×10^9 cm^{-2}, only the delayed peak is observed.[3,4] The localization of electron–hole pairs within the wells, and thus the localization of their binding energy, crucially depends on the external field. To demonstrate the field control of the Coulomb interaction and that ratio, we fixed the delay to $\tau = +100$ fs (to measure the maximum of the prompt signal) and varied the applied voltage at an intermediate excitation density of 7×10^9 cm^{-2}. As shown in Fig. 2, increasing the applied field from 3 to 7 kV/cm changes the period of the beating and shifts the delayed signal peak to lower delays because of the decrease of the oscillation period. Even much more drastically, the ratio of the two different signal contributions changes, as is shown in the inset of Fig. 2. Increasing the field leads to an increasing localization of the hh_0 exciton wave function, which dominates the signal. In the limit of an isolated quantum well, the hh_0 exciton binding energy (which is a measure of the Coulomb interaction) increases, thus increasing the exciton–exciton interaction. Correspondingly, the exciton radius shrinks, thus lowering the phase-space-occupation effect. Therefore, the ratio of the delayed signal to the prompt signal is drastically enhanced with increasing field.

A quantitative description of this effect requires a consistent treatment of many-body Coulomb interactions and the action of the static external field (for details see Refs. 2 and 3). In Fig. 3 the results of a theoretical model based on a numerical solution of the semiconductor Bloch equations in a Wannier–Stark-state basis are depicted. We obtain an excellent qualitative, and even quantitative, agreement with the experiments.

*Jülich Research Center, D-52425 Jülich, Germany

**Institute for Applied Photophysics, Dresden Technical University, D-01062 Dresden, Germany

†Fraunhofer Institute for Applied Solid State Physics, D-79108, Freiburg, Germany

1. D. S. Kim, J. Shah, T. C. Damen, W. Schäfer, F. Jahnke, S. Schmitt-Rink, and K. Köhler, Phys. Rev. Lett. **69**, 2725 (1992); S. Weiss, M.-A. Mycek, J.-Y. Bigot, S. Schmitt-Rink, and D. S. Chemla, Phys. Rev. Lett. **69**, 2685 (1992).
2. W. Schäfer, F. Jahnke, and S. Schmitt-Rink, Phys. Rev. B **47**, 1217 (1993).
3. P. Leisching, W. Beck, W. Schäfer, K. Leo, K. Köhler, and K. Kurz, submitted to Phys. Rev. Lett.
4. P. Leisching, T. Dekorsy, C. Waschke, W. Beck, H. G. Roskos, K. Leo, K. Köhler, and H. Kurz, in *Ultrafast Phenomena*, 1994 Technical Digest Series, Vol. 7 (Optical Society of America, Washington, D.C., 1994).

QThG26

Dynamic Wannier–Stark effect in semiconductor superlattices

J. B. Khurgin, S. J. Lee, N. M. Lawandy,*
Department of Electrical and Computer Engineering, Johns Hopkins University, Baltimore, Maryland 21218

We investigate the possibility of observing the transition between the localized states and the extended states in semiconductor superlattices (SLs) in the presence of a strong external ac field, i.e., the dynamic Wannier–Stark (DWS) effect.[1]

Let us consider the SL in Fig. 1(a). The adjacent wells in this structure consist of materials with different band gaps. This can be achieved by varying the Al mole fractions in $Al_xGa_{1-x}As$, for example. The SL parameters can be controlled so that the condition-band subbands are nearly in resonance while the valence-band energy levels are well separated ($\Delta E^{(0)}_{v,12} \sim 50$ meV). When a strong optical field is applied [Fig. 1(b)], the energy levels become dressed and the ac Stark effect moves the conduction-band energy level in well 1 by $(\mu_{cv}\varepsilon)^2/\delta E$ and moves the level in well 2 by $(\mu_{cv}\varepsilon)^2/(\delta E + \Delta E_{v,12} - \Delta E^{(0)}_{c,12})$, where $\mu_{cv} = e\langle r_{cv}\rangle$ is the dipole matrix element, ε is the optical field, and δE is the excitation detuning. The lower limit on detuning is put on by absorption that might be caused by the miniband width and the line broadening. Because $\Delta E_{v,12}$ can be large, it is clear that the DWS shift of the level in well 2 is substantially larger than that in well 1. Thus, the energy-level separation becomes wider and the miniband widths are substantially reduced [see Fig. 1(b)]. Considering the dispersion relation of the minibands, as shown in Fig. 2, we see that the DWS effect results in a wider gap at $ka = \pi/2$ and a smaller miniband effective mass. Since the conductivity across the SL is decreased because of the optically induced miniband effective-mass change, a practical scheme to use this structure as an optical detector/switch can be considered.

To detect the effective-mass change, we can apply a dc field to cause current flow in the SL growth direction. As long as the dc field is small enough not to cause energy-level shifts, the Drude conductivity formula can be used. Then the ac component of the current caused by the DWS effect can be written as

$$J_{WS}(P) = \frac{1}{6\sqrt{3}} \frac{VNa\langle r_{cv}\rangle^2 e^3 \tau \eta_0}{\hbar^2} \times \left(\frac{1}{\delta E} - \frac{1}{\delta E + \Delta E_{v,12}}\right) P, \quad (1)$$

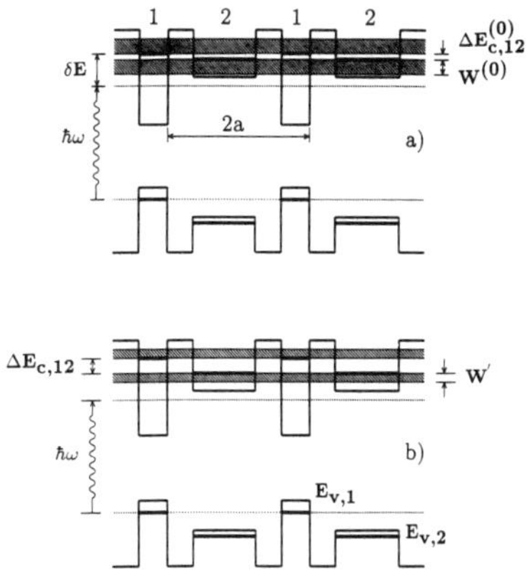

QThG26 Fig. 1. Energy levels and bands in an SL (a) without and (b) with the optical field. 2a is the unit period of the SL.

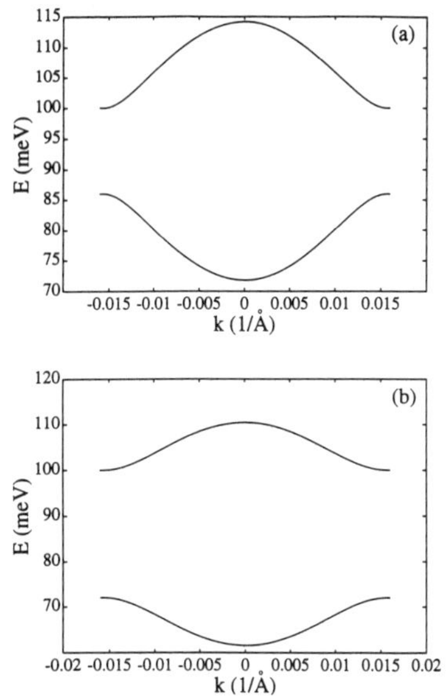

QThG26 Fig. 2. Miniband structure (a) with and (b) without the optical field. Note that the field increases the energy gap at $ka = \pi/2$.

where P is the power density of light, N is the density of conduction electrons, V is the interwell coupling energy, τ is the momentum relaxation time, and $\eta_0 = 377\ \Omega$. This type of current generation can be used in optical detection in a waveguide, whereby the responsivity can be expressed as

$$R = \frac{\rho\alpha_0}{2\sqrt{3}} Na\langle r_{cv}\rangle^2 \frac{e\tau}{\hbar} V \times \left(\frac{1}{\delta E} - \frac{1}{\delta E + \Delta E_{c,12}}\right)\left(\frac{L}{T}\right), \quad (2)$$

where α_0 is the fine-structure constant and L and T are the waveguide length and thickness, respectively. The dimensional factors enter only when the light propagates along the interface. When the L/T ratio is equal to 1000, R can be as high as 0.1 A/W, a figure similar to that of a photodiode. Since no light absorption is involved in the process, the DWS detection scheme is suitable for nondemolition measurement. Compared to the virtual-conductivity scheme,[2] far greater responsivity can be obtained in the DWS scheme, in which the energy is supplied by the dc source.[3] Compared to the quantum-wire-interferometer method,[4] the DWS device seems far easier to implement because of the confinement necessary in one direction and the less stringent requirement of electron coherence. Also, by using a differential scheme, the dc component can be canceled out, substantially reducing the noise in this device.

Division of Engineering and Department of Physics, Brown University, Providence, Rhode Island 02912

1. R. W. Boyd and M. Sargent III, J. Opt. Soc. Am. B **5**, 99 (1988).
2. E. Yablonovich, J. P. Heritage, D. E. Aspnes, and Y. Yafet, Phys. Rev. Lett. **62**, 28 (1989).
3. A. Shimizu and M. Yamanishi, Phys. Rev. Lett. **72**, 3343 (1994).
4. A. Shimizu, K. Fujii, and M. Okuda, Phys. Rev. B **40**, 1357 (1989).

QThG27

Relaxation dynamics of carriers interacting with confined LO phonons in quantum-dot structures

H. Nakayama, Y. Arakawa, *Institute of Industrial Science, University of Tokyo, 7-22-1 Roppongi, Minato-ku, Tokyo 106, Japan*

Much effort has been devoted to the realization of quantum-dot structures because of possible applications to lasers and other optical devices.[1] Recent progress in crystal-growth techniques has made it possible to realize quantum-dot structures as small as 15 nm.[2] Although the bottleneck that restricts a carrier's cascade down toward ground level through the emission of LO phonons has been pointed out,[3] luminescence is clearly observed even in such small structures.

The purpose of this paper to show theoretically that the bottleneck problem is not serious, even if only LO-phonon emission is taken into account in the quantum-dot structures. In the analysis, instead of Fermi's golden rule, coupled equations in Wigner–Weisskopf theory[4] are used, with attention to the uncertainty relation between time and energy. The results show that electrons can interact with LO phonons in a wide energy range (>50 meV), and, therefore, the bottleneck problem can almost be eliminated.

If the carrier-relaxation rate is calculated simply by using the Fermi's golden rule, the calculated transition rate is too short (~1 fs) from the viewpoint of the uncertainty principle, because the energy range of electrons that can interact with LO phonons is very narrow (~0.1 meV). This results mainly from the strong coupling between LO phonons and electrons. To solve this contradiction, we should start from coupled equations instead of from Fermi's golden rule.

The time-dependent Schrödinger equation is solved by using the wave function, which is described as linear combinations of states, in terms of time-dependent constants $C(t)$ as follows:

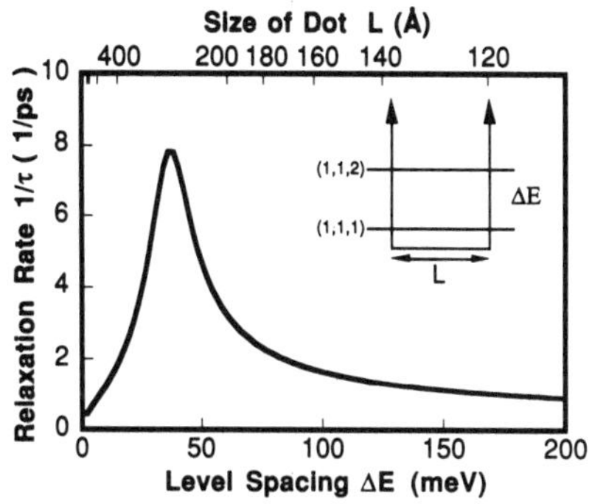

QThG27 Fig. 1. Relaxation rate of quantum-dot structures from the second subband to the first subband.

$$\psi = C_{u,\{0\}}(t)|u\rangle|\{0\}\rangle + \sum_n C_{l,\{1_n\}}(t)|l\rangle|\{1_n\}\rangle, \quad (1)$$

where the subscripts u and l represent upper and lower states in the electron system, respectively, and $\{1_n\}$ represents an excitement of the nth phonon mode. If Eq. (1) is substituted into the time-dependent Schrödinger equation and the result is multiplied by the appropriate bra vector, the expression becomes

$$\frac{\partial}{\partial t} C_{u,\{0\}}(t) = -i\omega_u C_{u,\{0\}}(t) - \sum_n i g_n \sqrt{n_{ph} + 1}\, C_{l,\{1_n\}}(t), \quad (2)$$

$$\frac{\partial}{\partial t} C_{l,\{1_n\}}(t) = -i\omega_n C_{l,\{1_n\}}(t) - i g_n^* \sqrt{n_{ph} + 1}\, C_{u,\{0\}}(t), \quad (3)$$

where g_n is the coupling constant describing an interaction with a confined phonon[5] and n_{ph} is the number of photons. In the calculation, GaAs/AlGaAs quantum dots with finite cubic potential are assumed.

Figure 1 shows the relaxation rate from the second subband to the first subband of the quantum-dot structures as a function of size of the quantum dot (upper scale) and the energy spacing (lower scale). The result shows a wide energy range of electrons that can interact with LO phonons. In addition, the calculated relaxation rate is much higher than the spontaneous-emission rate (~$10^9\ s^{-1}$), which leads to sufficient carrier injection from upper levels into the ground level in the quantum-dot lasers. It should be noted that our definition treats the transition probability as a classical transition in the calculation, because the transition probability has the property of Rabi oscillation.

In conclusion, we describe the interaction between LO phonons and electrons theoretically by using coupled equations instead of using Fermi's golden rule. The results indicate spectrum broadening originating from strong coupling between electrons and LO phonons, which almost eliminates the bottleneck problem. This should encourage those who are making an effort to realize quantum-dot lasers.

1. Y. Arakawa and H. Sakaki, Appl. Phys. Lett. **40**, 939 (1982).
2. J. Oshinowo, M. Nishioka, S. Ishida, and Y. Arakawa, Appl. Phys. Lett. **65**, 1421 (1994).
3. H. Benisty, C. M. Sotomayor-Torrès, and C. Weisbuch, Phys. Rev. B **44**, 10945 (1991).
4. W. H. Louisell, *Quantum Statistical Properties of Radiation* (Wiley, New York, 1973), p. 285.
5. R. M. de la Cruz, S. W. Teitsworth, and M. A. Stroscio, Superlattices Microstruct. **13**, 481 (1993).

QThG28

Picosecond dynamics of hot excitonic molecules in CdSe: energy relaxation and radiative decay

A. S. Vengurlekar, S. S. Prabhu, S. K. Roy, Jagdeep Shah,* *Tata Institute of Fundamental Research, Bombay 400005, India*

We study the picosecond time evolution of luminescence due to excitons and hot excitonic molecules (biexcitons) at 8 K in CdSe. We find that from an initial (<50 ps) temperature of 40 K, the biexcitons cool to 15 K in about 250 ps with an average energy loss rate (ELR) of a few 10^7 eV/s, mainly from acoustic-phonon emission. We also deduce the biexciton and exciton lifetimes to be 10 and 600 ps, respectively.

At low temperatures and high densities, photoexcited electron–hole pairs are known[1,2] to form excitons and excitonic molecules (biexcitons) in semiconductors such as CuCl, CdS, and CdSe and in GaAs quantum wells. Study of the picosecond dynamics of the formation, thermalization, cooling, and recombination of excitons and biexcitons in these semiconductors is of much current interest. Many of these processes are yet to be fully understood. In this presentation we report what we believe to be some of the first measurements of picosecond luminescence due to biexcitons[3] and the longitudinal-optical (LO) phonon-related Stokes sidebands of excitons in CdSe by using upconversion luminescence spectroscopy. Using these, we deduce the time evolution of the effective temperature of the hot biexcitons and obtain the energy relaxation rates of the hot biexcitons. We calculate the biexciton ELRs by using a theory of phonon emission by the biexcitons, and we find satisfactory agreement with experiments. The time evolution of the energy-integrated luminescence due to biexcitons shows that it rises with a time constant of 12 ps, reaches a peak in about 50 ps, and then decays with a time constant of 240 ps. To obtain the biexciton and the exciton lifetimes, we compare the time evolution of the exciton and biexciton luminescence with a simple theory describing the coupled exciton–biexciton dynamics.

In our experiments we excite CdSe crystals, maintained at 8 K in a cryostat, with 1.8-ps pulses from a rhodamine 6G dye laser pumped by an Nd:YAG laser. The upconversion luminescence is measured with a time resolution of 2.5 ps and an energy resolution of 2.5 meV. The carrier excitation density n_o is varied from 0.8 to 8×10^{17} cm^{-3}. The features due to biexcitons and the bound excitons in the luminescence spectra are identified by the excitation-density dependence of time-integrated and picosecond excitation-correlation luminescence spectroscopy. Figure 1 illustrates the time evolution of the luminescence spectra due to radiative annihilation of the biexcitons with $n_o = 8 \times 10^{17}$ cm^{-3}. Assuming that the biexcitons thermalize among themselves at a time-dependent temperature $T_M (> T_L$, the lattice temperature), we compare the spectra with a theoretical expression for the emission intensity spectra.[2] As shown in Fig. 1, the agreement is quite good. From the time evolution of the biexciton temperature T_M thus obtained, we deduce the hot biexciton energy loss rate $d\epsilon_M/dt = -3kdT_M/dt$, as shown in Fig. 2. Comparison with a calculation of the ELRs based on phonon

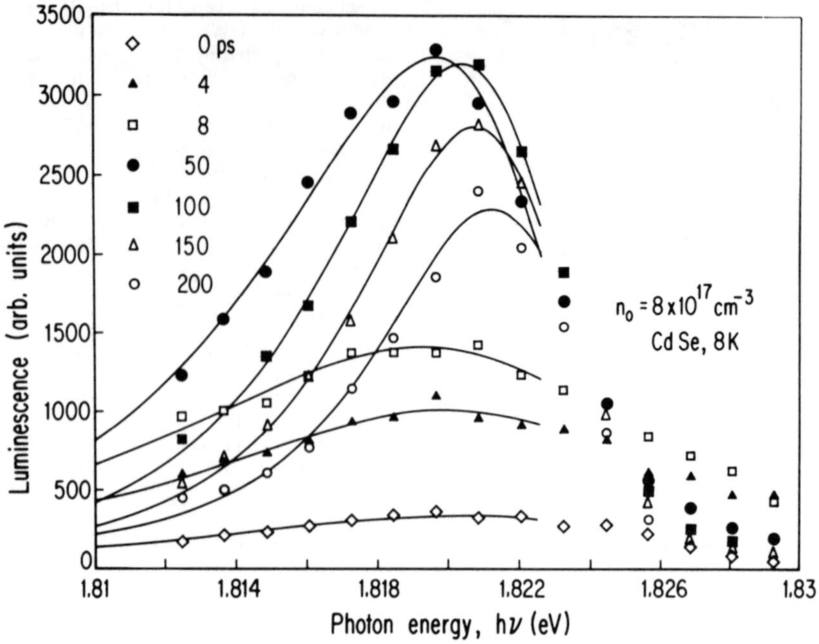

QThG28 Fig. 1. Time evolution of the biexciton-related luminescence spectra (points) compared with theory (curves) at $n_o = 8 \times 10^{17}$ cm^{-3}.

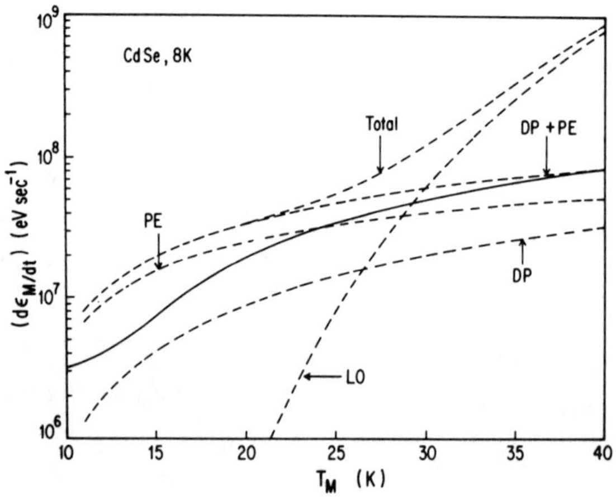

QThG28 Fig. 2. Experimental ELRs (solid curve) of hot biexcitons compared with theoretical calculations (dashed curves).

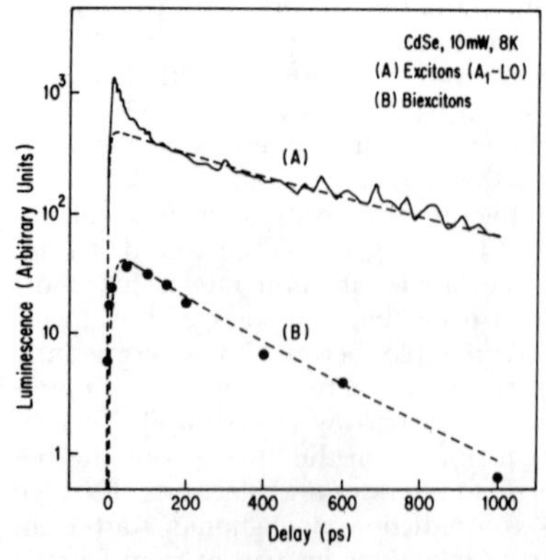

QThG28 Fig. 3. Time evolution of the biexcitons (points) and of excitons (solid curve) compared with solutions to equations describing coupled exciton–biexciton dynamics (dashed curves).

Thursday Posters

emission shows that the ELRs are mainly due to acoustic phonon emission through piezoelectric and deformation-potential interactions. Figure 3 shows the time evolution of the total biexciton luminescence and of the A − LO Stokes band of the free exciton A compared with a simple theory of the coupled exciton–biexciton dynamics.[1]

This work is partly supported by National Science Foundation grants INT-9022623 and INT-9201350.

*AT&T Bell Laboratories, Holmdel, New Jersey 07733

1. C. Klingshirn and H. Haug, Phys. Rep. **70**, 315 (1981).
2. M. Ueta, H. Kanzaki, K. Kobayashi, Y. Toyozawa, and E. Hanamura, *Excitonic Processes in Solids* (Springer-Verlag, Berlin, 1986).
3. A. S. Vengurlekar, S. S. Prabhu, S. K. Roy, and J. Shah, Phys. Rev. B **50**, 8348 (1994); Phys Rev. B, in press.

QThG29

Time-resolved investigation of the quasi-particle dynamics of high-temperature superconductors

Ch. Jaekel, M. Schröer, H. G. Roskos, H. Kurz, *Institute for Semiconductor Technology, Rhine-Westphalia Technical University–Aachen, Sommerfeldstrasse 24, 52056 Aachen, Germany*

The analysis of carrier relaxation in high-T_c superconductors leads to a deeper insight into the coupling mechanisms of the charge carriers. Time-resolved experiments are performed in a standard pump–probe setup with Ti:sapphire and colliding-pulse-mode-locked (CPM) dye lasers as sources for infrared and 620-nm-light pulses, respectively. The samples are 200-nm-thick HTSC films, dc sputtered or laser ablated on $SrTiO_3$ substrates and with their c axes perpendicular to the substrate surface. The transition temperatures range from 85 to 92 K.

The transient reflectivities of BSCCO and YBCO at various temperatures are depicted in Fig. 1. The transients of YBCO show an increase and a change of sign below T_c, which we attribute to the generation and recombination of quasi particles. BSCCO exhibits a more complex temperature behavior. We observe a change of sign already above T_c. This may result from an admixture of BSCCO-2223 as found by a TEM analysis of our samples. Below T_c, the transients are principally the same as in YBCO but with a different sign.

In our model for the reflectivity and transmission changes, carriers are excited by the incident laser pulses. Below T_c, Cooper pairs are broken in a cascading process. Subsequently, the quasi particles begin to recombine to Cooper pairs. These recombination times allow us to determine the electron–phonon coupling constant. In Fig. 2 the measured recombination times τ_r for YBCO and BSCCO are depicted. The solid curve is a fit with a dynamical BCS model[1]

$$\tau_r = 1.15 \times \tau_{e-ph} \frac{T}{T_c}\left(1 - \frac{T}{T_c}\right)^{-1/2}, \quad (1)$$

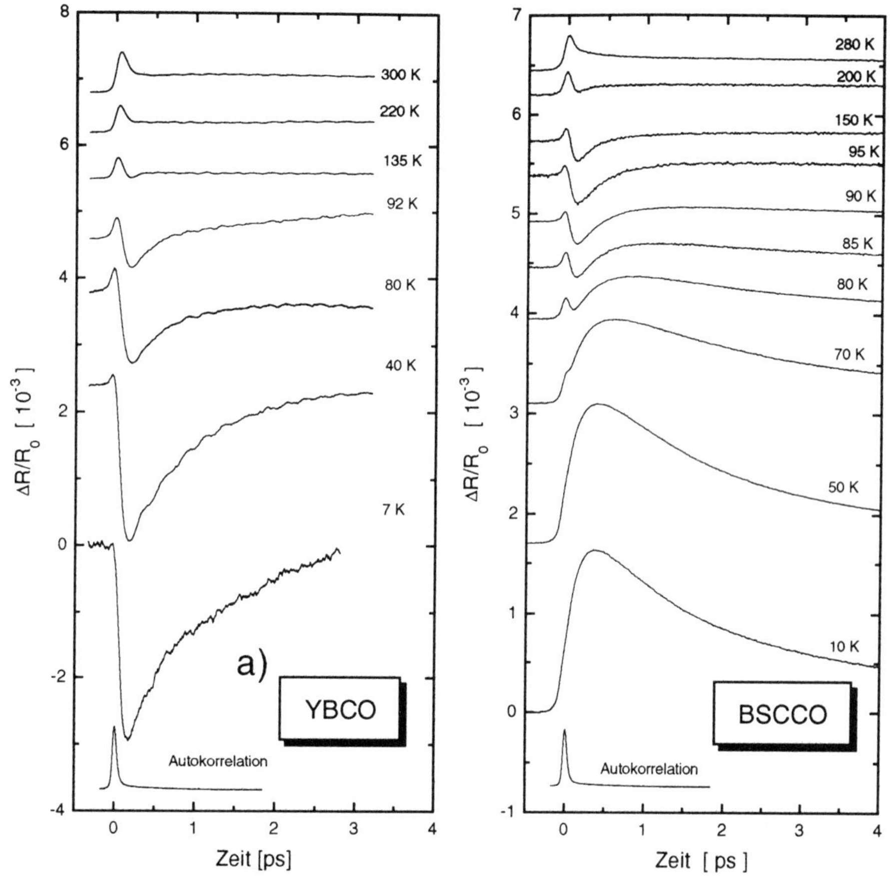

QThG29 Fig. 1. Transient reflectivity at various temperatures for YBCO and BSCCO.

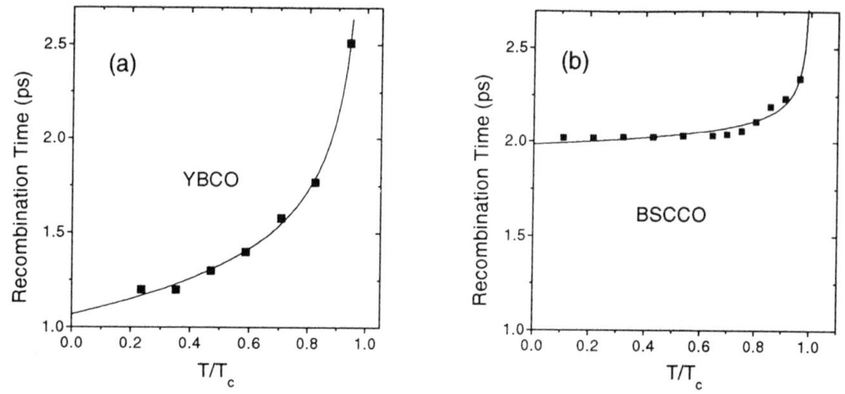

QThG29 Fig. 2. Recombination times τ_r for YBCO and BSCCO. The solid curve is a fit with Eq. (1).

where τ_{e-ph} is the characteristic electron–phonon scattering time. In both materials we see a tendency for τ_r to diverge near T_c, but an additional constant τ_0 has to be added to Eq. (1) to account for the nonvanishing relaxation time as $T \rightarrow 0$ K. The recombination times for BSCCO are generally somewhat larger than those of YBCO. We find $\tau_0 \sim 1$ ps for YBCO and $\tau_0 \sim 2$ ps for BSCCO. The temperature dependence for BSCCO is weaker, resulting in a τ_{e-ph} = 0.06 ps compared to τ_{e-ph} = 0.32 ps in YBCO.

By tuning the photon energy, we can determine the energetic distance between the Fermi level and the upper Hubbard band (charge-transfer gap), where a minimum or a change of the sign of the transients occurs. For YBCO we obtain a value of 2.1 eV (Ref. 2), for BSCCO of 1.55-eV gap energy. This is consistent with cw-reflectivity measurements.

This work was supported by German Federal Ministry for Research and Technology grant 13N6288.

1. See e.g., A. Frenkel, Phys. Rev. B **48**, 9717 (1993).
2. W. Albrecht, T. Kruse, K. Leo, and H. Kurz, Appl. Phys. A **56**, 463 (1993).

QThG30

Second-conduction-band relaxation of a photoexcited hot-electron distribution in GaP

Michael A. Cavicchia, R. R. Alfano, *Institute for Ultrafast Spectroscopy and Lasers, New York State Center for Advanced Technology for Photonic Materials and Applications, Department of Physics, City College and Graduate School, City University of New York, Convent Avenue at 138th Street, New York, New York 10031*

By using a 500-fs 293-nm pump pulse, the dynamics of photoexcited electrons in the $k \neq 0$, X_6 and X_7 satellite valleys of the first and second conduction bands, respectively, were studied. The kinetics of carriers in these valeys were probed by a 500-fs IR (2.5–5.5-μm) pulse that minitored the induced absorption due to the free-electron populations, see Fig. 1. The technique of using an ultrafast IR probe pulse is seen to be unique in obtaining information about the relaxation of hot electrons. Between 2 and 30 ps the hot-electron distribution relaxes and the electrons redistribute among the conduction-band valleys by intra- and intervalley scattering among the X_6, X_7, L, and Γ valleys (see Fig. 1).

Figure 2 displays the full temporal behavior of the absorption for γ_{pump} = 293 nm and γ_{probe} = 4.7 μm. It is seen that a quasi equilibrium is reached within 4.5 ps. Photons at λ_{probe} = 4.7 μm are not energetic enough to cause electron transitions from the X_6 valley to the X_7 valley. Only

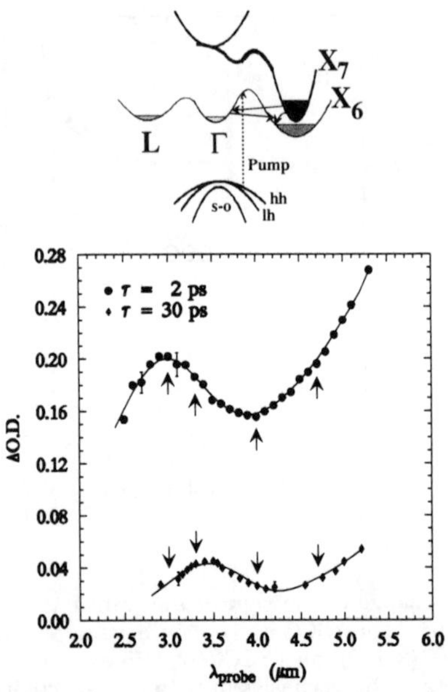

QThG30 **Fig. 1.** Induced IR absorption spectra in GaP, λ_{pump} = 293 nm, at delay times of τ = 2 ps and τ = 30 ps. Typical error bars are shown. The solid curves are theoretical fits of the data. The arrows indicate the probe wavelengths at which the complete temporal behavior was measured. The upper part is a schematic band-structure diagram of GaP showing the UV-pump and IR-probe transitions. The nonequilibrium X_7 valley population is depleted by intervalley scattering to the X_6, Γ, and L valleys.

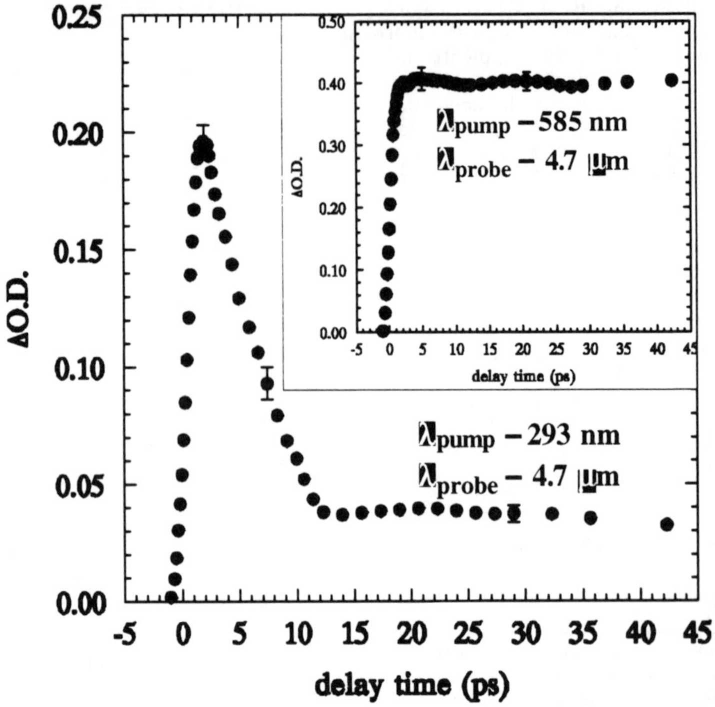

QThG30 **Fig. 2.** Transient induced IR absorption when λ_{pump} = 293 nm and λ_{probe} = 4.7 μm. The excess absorption is due to electrons in the X_7 valley. Inset is the transient IR absorption when λ_{pump} = 585 nm and λ_{probe} = 4.7 μm.

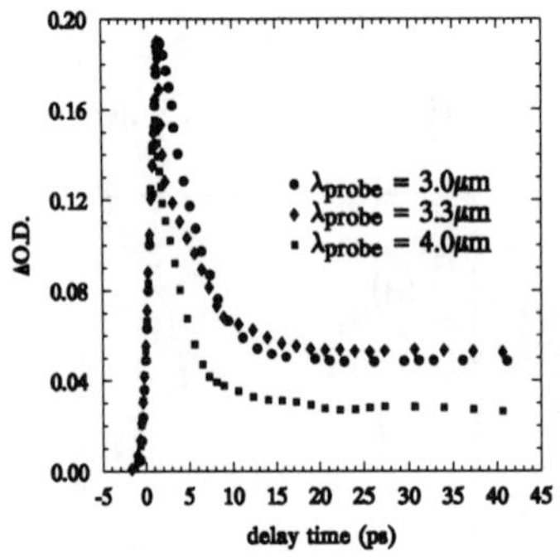

QThG30 **Fig. 3.** Transient induced IR absorption when λ_{pump} = 293 nm and λ_{probe} = 3.0, 3.3, and 4.0 μm.

free-carrier absorption exists. When transfer to the second conduction band is eliminated by using λ_{pump} = 585 nm, the peak at early delay times, detected with 293-nm excitation, is not observed (inset, Fig. 2).

The following explains the temporal profiles in Fig. 2. For λ_{pump} = 585 nm, electrons are photoexcited only to the X_6 valley by an indirect absorption and do not obtain sufficient kinetic energy to undergo intervalley scattering to the X_7 valley. No redistribution occurs between the X_6 and X_7 valleys. Therefore, for 293-nm excitation we attribute the 4.5-ps decay in Fig. 2 to depopulation of the satellite, X_7, *second*-conduction-band valley. It is shown that the excess FCA due to pumping by 293 nm is due to the electrons in the second conduction band, which are undergoing rapid intervalley scattering back to the first conduction band. This strong scattering significantly reduces the mobility of electrons in the second conduction band.

Knowledge of the cooling of the hot-electron distribution is obtained by measuring the transient absorption at the probe wavelengths indicated by arrows in Fig. 1. The temporal behaviors of the absorption for probe wavelengths of 3.0, 3.3, and 4.0 μm are shown in Fig. 3. The different decay behaviors result from the depopulation of the X_7 valley and the cooling of the electron temperature. As the electron distribution relaxes, the peak of the absorption band moves to longer wavelengths, passing from 3.0 to 3.5 μm. This causes the absorption at 3.3 μm to initially decay slowly for 5 ps and then to rapidly decay for another 8 ps, whereas the decay of the absorption at λ_{probe} = 3.0 μm is characterized by a single decay component of about 6 ps.

Time-resolved UV-pump–IR-probe absorption measurements provide, for the first time to our knowledge, a *direct* observation of the relaxation of a hot-electron distribution in the satellite, X_7, *second*-conduction-band valley in GaP.

This work is supported by the National Aeronautics and Space Administration and the New York State Technology Foundation.

QThG31

Ultrafast carrier dynamics in porous silicon

P. M. Fauchet, J. von Behren, K. B. Ucer, Y. Kostoulas, *Department of Electrical Engineering, University of Rochester, Rochester, New York 14627*

The discovery in 1990 that porous silicon is a strong visible-light emitter at room temperature[1] has challenged the conventional wisdom that silicon is not a useful semiconductor for light-emitting devices.[2]

Light-emitting porous silicon is made of a large density of nanometer-size silicon quantum dots whose surface is nearly perfectly passivated by silicon hydrides or silicon dioxide. Despite extensive research, there is still considerable debate concerning not only the mechanisms of the luminescence, but also the electronic band structure and the electron and hole dynamics in porous silicon. To answer these questions, we have performed femtosecond spectroscopic measurements by using a variety of femtosecond laser sources. In particular, we have monitored the induced absorption in red-light-emitting free-standing porous silicon thin films after above-band-gap excitation.

Whether the probe is tuned above or below the pump energy, or above or below the luminescence peak, induced absorption is observed (Fig. 1). The lack of a transient hole burning immediately after excitation and of any bleaching near the band edge indicates that porous silicon is still an indirect-gap semiconductor. This conclusion supports some very recent calculations of the band structure and matrix elements in nanocrystalline silicon.[3] The recovery dynamics are nonexponential and have a complicated dependence on the injected carrier density. A partial recovery takes place on a 1-ps time scale at high injected density and on a sub-100-fs time scale at low density. Experiments performed on degraded samples suggest that this initial transient is due to trapping at surface states that do not luminesce. The increase in effective trapping time with injected density, which occurs when on average more than one electron–hole pair is injected in each silicon quantum dot, is attributed to trap saturation. The induced absorption then displays a longer recovery, on a time scale of 10 ps or more. This slower transient becomes faster as the injected carrier density increases (Fig. 2), especially when on average more than one electron–hole pair is generated in each quantum dot, suggesting that it is produced by Auger recombination. Auger recombination is expected to become very efficient when electrons and holes are strongly confined, an assignment that is supported by recent estimates of the Auger recombination rate in silicon nanocrystals.[4]

This work was supported by Rochester Gas and Electric, the New York State Energy Research and Development Authority, the National Science Foundation Center on Photoinduced Charge Transfer, and Xerox.

1. L. T. Canham, Appl. Phys. Lett. **57**, 1046 (1990).
2. P. M. Fauchet *et al.*, Proc. Soc. Photo-Opt. Instrum. Eng. **2144**, 34 (1994).
3. M. Hybertsen, Phys. Rev. Lett. **72**, 1514 (1994).
4. M. Lannoo *et al.*, to be published.

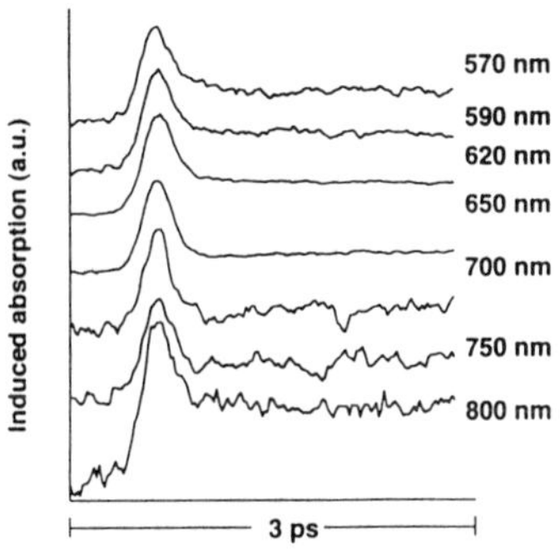

QThG31 Fig. 1. Induced absorption measured at various probe wavelengths after excitation at 2 eV. The measurements are performed at room temperature on a sample that luminesces near 1.7 eV and is ~50% transmitting at 2 eV. The measurements indicate that porous silicon is an indirect-gap semiconductor.

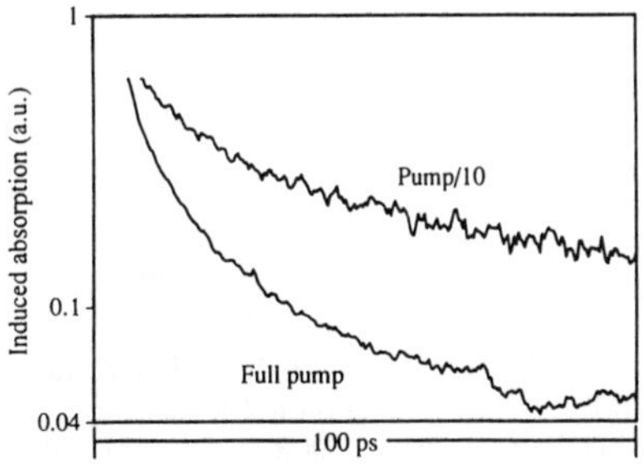

QThG31 Fig. 2. Recovery of the induced absorption (on a logarithmic scale) after injection at two different excitation levels. The faster recovery at high injection is attributed to Auger recombination.

QThG32

Photoluminescence characterization of crystal-silicon quantum-well layers

P. N. Saeta, A. C. Gallagher, *Joint Institute for Laboratory Astrophysics, Campus Box 440, University of Colorado, Boulder, Colorado 80309-0440*

The recent discovery of efficient luminescence from porous silicon[1] has stimulated great interest in developing efficient silicon-based light emitters for optoelectronic applications. Although both porous silicon and oxide-coated crystal-silicon nanoparticles[2] exhibit photoluminescence quantum efficiencies of several percent at room temperature, they have yet to be incorporated into efficient electroluminescent structures. Thin crystal-silicon quantum-well layers have the potential to provide high efficiency and electrical connectivity, as well as compatibility with conventional silicon processing.

To investigate the luminescence properties of thin two-dimensional crystal-silicon layers, we have prepared single quantum wells from SIMOX silicon-on-insulator substrates by thermally oxidizing the 200-nm-thick silicon layer in a quartz furnace. Initial oxidation is performed in a spatial temperature gradient to impress a gradual taper on the thickness of the silicon layer. This taper is then preserved during subsequent oxidation (1110°C, 1 atm, O_2) at uniform temperature. Hence, the final Si/SiO_2 interface has the low defect density of high-temperature thermal oxide. The structure is illustrated schematically in the left inset of Fig. 1. The quantum well was

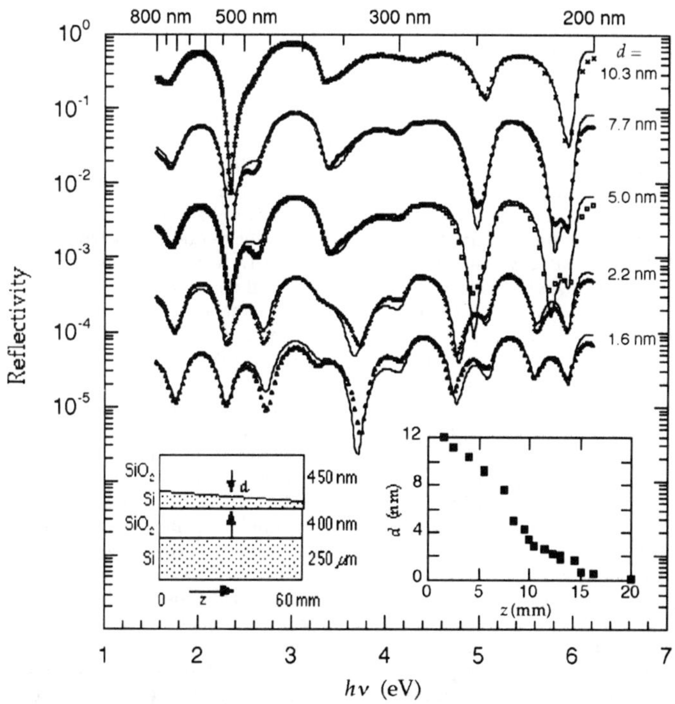

QThG32 Fig. 1. Reflectivity spectra of a silicon-quantum-well layer at five positions along the direction of the taper. Discrete symbols are data; smooth curves are fits to a three-layer model using bulk dielectric functions. Successive curves are offset for clarity. The layer thickness *d* obtained from the fits is indicated at the right and is summarized in the right inset.

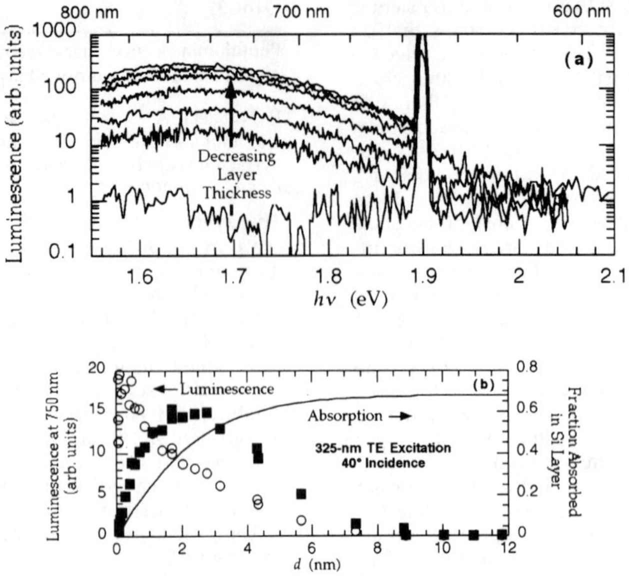

QThG32 Fig. 2. (a) Spectrally corrected photoluminescence spectra taken at several values of d for 325-nm (3.8-eV) excitation. (b) Peak luminescence intensity at 750 nm as a function of d (filled squares) and relative efficiency (open circles) obtained by dividing luminescence by calculated absorption in the layer (smooth curve).

characterized by near-normal reflection spectra, which are shown in Fig. 1. The deduced silicon layer thickness d as a function of position z along the taper, shown in the right inset, agrees with the taper estimated from the temperature-dependent thermal-oxidation rates of silicon.

Photoluminescence spectra were recorded as a function of d with cw excitation at 325 nm. Figure 2(a) shows a broad red luminescence peak centered at 750 nm that develops as the layer thickness decreases below 8 nm. The spectrum is seen to be essentially independent of layer thickness, suggesting the role of surface states in the emission. The filled squares in Fig. 2(b) show the detected luminescence at the 750-nm peak as a function of d, deduced from the reflection spectra. The intensity is seen to rise to a peak at 2.5-nm thickness and fall more gradually to our detection limit at a thickness of ~8 nm. However, when divided by the fraction of light absorbed in the layer [smooth curve in Fig. 2(b)], the relative efficiency (open circles) is seen to rise monotonically with *decreasing* layer thickness. Peak observed photoluminescence efficiencies are of the order of 10^{-4}, which is more than 2 orders of magnitude lower than that observed in oxide-passivated nanocrystals. Room-temperature time-resolved photoluminescence measurements show nonexponential decay with $1/e$ times of 30–50 μs.

The low efficiency could arise from two factors. First, large variations in the layer thickness could permit photoexcited carriers to diffuse to thick regions (>2.5 nm) in which quantum confinement energies are insufficient to produce visible luminescence. Second, in regions of nearly uniform thickness, the extended character of the layer permits carriers to diffuse over large distances. Because the luminescence decay times are long, if this distance exceeds the separation between fast nonradiative recombination centers, most carriers will recombine nonradiatively.

1. L. T. Canham, Appl. Phys. Lett. **57**, 1046 (1990).
2. W. L. Wilson, P. F. Szajowski, and L. E. Brus, Science **262**, 1242 (1993).

QThG33

Optical relaxation of nile blue

Yiping Zhang, S. R. Hartmann, Fred Moshary,* *Department of Physics, Columbia University, 538 West 120th Street, New York, New York 10027*

We used incoherent degenerate four-wave mixing (FWM) to study the optical relaxation of nile blue organic dye molecules in the polymer PVOH at 5 K. Laser noise sources with bandwidths of 30 and 600 cm^{-1} and center frequencies varying from 620 to 680 nm were used to resolve various features of the line shape. The FWM response varies dramatically over our excitation range. We, and others, find that the homogeneous relaxation rate increases with increasing excitation frequency. This increase has previously been explained as being due to either multilevel interference[1,2] or intramolecular dephasing.[3,4] We are able to explain both this increase and the other features of the FWM response by proposing a simple model with only two components. It supposes the existence of just one single-mode harmonic oscillator at 590 cm^{-1} with an additional optically active pseudo-local mode due to the interaction of the molecule with its local environment.[1] The 590-cm^{-1} mode has a Franck–Condon factor[5] of 0.2 and a damping time[6] of 1 ps. It gives rise to quantum beats. Our data shows that the S_1–S_0 electronic dephasing time T_2 is 60 ps. This value of T_2 agrees with that of other experiments carried out at the red edge of the absorption spectrum.[4] We fit our FWM data to obtain the line shape and coupling constant of the pseudo-local modes; see Fig. 1. We find that the local mode has an asymmetric line shape that peaks at 30 cm^{-1}, and it has an effective Debye–Waller factor of 0.3. These results agree with those of earlier hole-burning experiments,[7] which established the existence of the 30-cm^{-1} peaked phonon sideband with a non-Lorentzian line shape. We also determine the center frequency (662 nm) and the width (600 cm^{-1}) of the in-

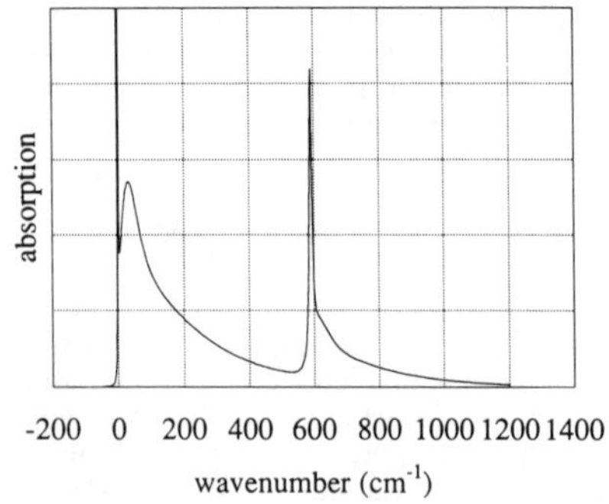

QThG33 Fig. 1. Single-molecule absorption line shape of nile blue in PVOH at 5 K.

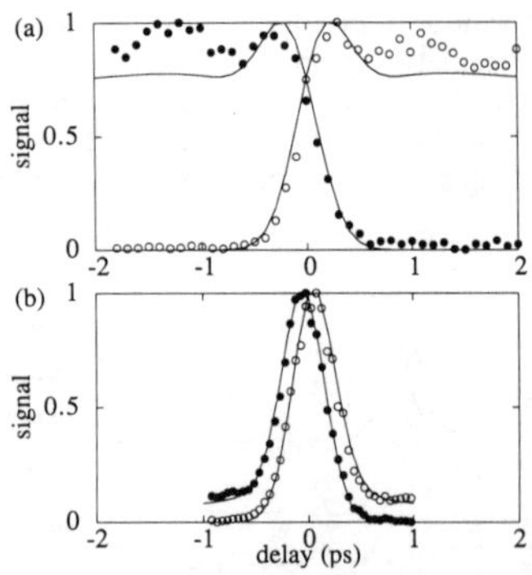

QThG33 Fig. 2. Normalized FWM data and theoretical fit obtained by using narrowband sources (40 cm^{-1} wide) centered at (a) 674 nm and (b) 634 nm. Data in (a) and (b) are vertically displaced. Open and filled circles represent signals in the $2k_2 - k_1$ and $2k_1 - k_2$ directions, respectively, and curves represent calculations.

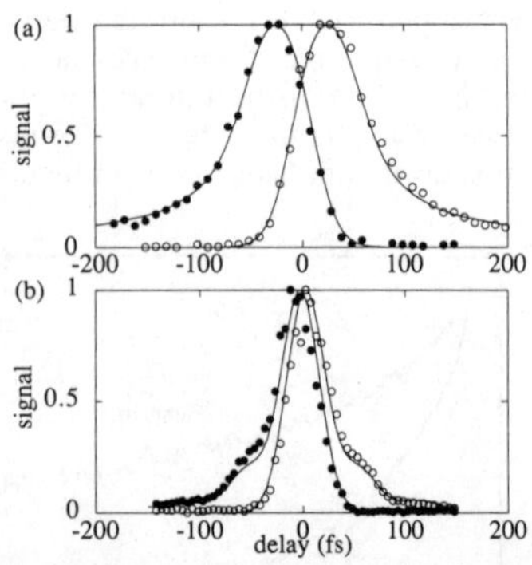

QThG33 Fig. 3. Normalized FWM data and theoretical fit obtained by using broadband sources (500 cm^{-1} wide) centered at (a) 677 nm and (b) 640 nm. Data in (a) and (b) are vertically displaced. Filled and solid circles represent signals in the $2k_2 - k_1$ and $2k_1 - k_2$ directions, respectively, and curves represent calculations.

homogeneous distribution associated with the zero-phonon line.[4]

Our data and theoretical calculations are shown in Figs. 2 and 3, where narrowband sources (40 cm^{-1} wide) at 674 and 634 nm and broadband sources (500 cm^{-1} wide) at 677 and 640 nm are used. All of the experimental features we observe are thus explained by the proposed simple two-mode model.

*Department of Electrical Engineering, City College of the City University of New York, 138th Street and Convent Avenue, New York, New York 10031

1. C. J. Bardeen and C. V. Shank, Chem. Phys. Lett. **203**, 535 (1993).
2. A. M. Weiner, S. D. Silvestri, and E. P. Ippen, J. Opt. Soc. Am. B **2**, 654 (1985).
3. M. Fujiwara, R. Kuroda, and H. Nakatsuka, J. Opt. Soc. Am B **2**, 1634 (1985).
4. S. Nakanishi, H. Ohta, N. Makimoto, H. Itoh, and M. Kawase, Phys. Rev. B **45**, 2825 (1992).
5. M. K. Lawless and R. A. Mathies, J. Chem. Phys. **96**, 8037 (1992).
6. H. L. Fragnito, J. Bigot, P. C. Becker, and C. V. Shank, Chem. Phys. Lett. **160**, 101 (1989).
7. B. L. Fearey, T. P. Carter, and G. J. Small, J. Phys. Chem. **87**, 3590 (1983).

QThG34

Quantum beating among the s, p, and d orbitals of atomic Rb

H. C. Tran, P. C. John, J. G. Eden, *Everitt Laboratory, Department of Electrical and Computer Engineering, University of Illinois, Urbana, Illinois 61801*

Radially localized wave packets in atoms were described theoretically in 1986[1,2] and were observed shortly thereafter in rubidium.[3] In this work we produce wave packets by two-photon excitation of the *ns* and $(n-2)d$ states of K and Rb ($n = 8$ for K, $n = 10$ and 11 for Rb). Although the wave-packet behavior does exhibit quantum beats between the *ns* and the $(n-2)d$ states, Fourier analysis also demonstrates the existence of frequency components associated with the $ns-(n-1)p$ and the $(n-2)d-(n-1)p$ energy differences.

As illustrated in Fig. 1, irradiation of Rb vapor with a femtosecond pulse produces intense, spatially coherent UV emission. Specifically, two-photon excitation of the *ns* and the $(n-2)d$ states of Rb and K with ~620-nm pulses from a CPM four-stage amplifier system generates wave packets composed of *ns*, $(n-2)d$, and $(n-1)p$ states. The temporal behavior of the wave packet is detected by monitoring the coherent UV radiation produced in a four-wave-mixing scheme in which an 80–100-fs red pulse is scattered by the radial wave packet established by an earlier pulse.

Varying the time delay between the pump and probe pulses while monitoring the intensity of the coherent Rb emission at 311.3 and 315.8 nm (10*p*, 11*p* $^2P \rightarrow 5s\ ^2S_{1/2}$) yields the data shown in Fig. 2. A periodic signal having a nominal frequency of 2.5 THz (period of 400 fs) is observed. As illustrated in the inset of Fig. 2, Fourier analysis shows several frequency components, all of which correlate with energy differences between excited states of Rb. The primary components are associated with 11*s*–9*d* and 12*s*–10*d* energy defects (95.27 and 66.96 cm^{-1}, respectively). This can be understood as quantum beats between pairs of *ns* ^2S and $(n-2)d\ ^2D$ states. Also present are frequency components at 161.84 and 261.40 cm^{-1}, which are assigned to the 9*d*–10*p* and the 11*s*–10*p* energy separations in Rb, respectively. This suggests the presence of $(n-1)p\ ^2P$ character in the wave packet. Long-time characteristics of the wave-packet evolution and similar phenomena observed in K vapor are discussed.

1. J. Parker and C. R. Stroud, Jr., Phys. Rev. Lett. **56**, 716 (1986).
2. G. Alber, H. Ritsch, and P. Zoller, Phys. Rev. A **34**, 1058 (1986).
3. A. ten Wolde, L. D. Noordam, A. Lagendijk, and H. B. van linden van den Heuvell, Phys. Rev. Lett. **61**, 2099 (1988).

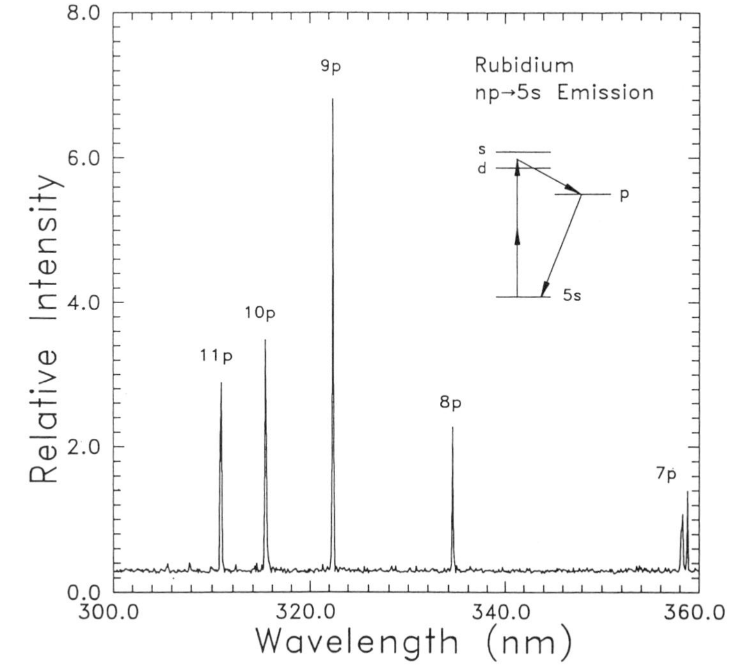

QThG34 **Fig. 1.** Forward emission spectrum of Rb. Inset shows the basic four-wave-mixing pumping scheme.

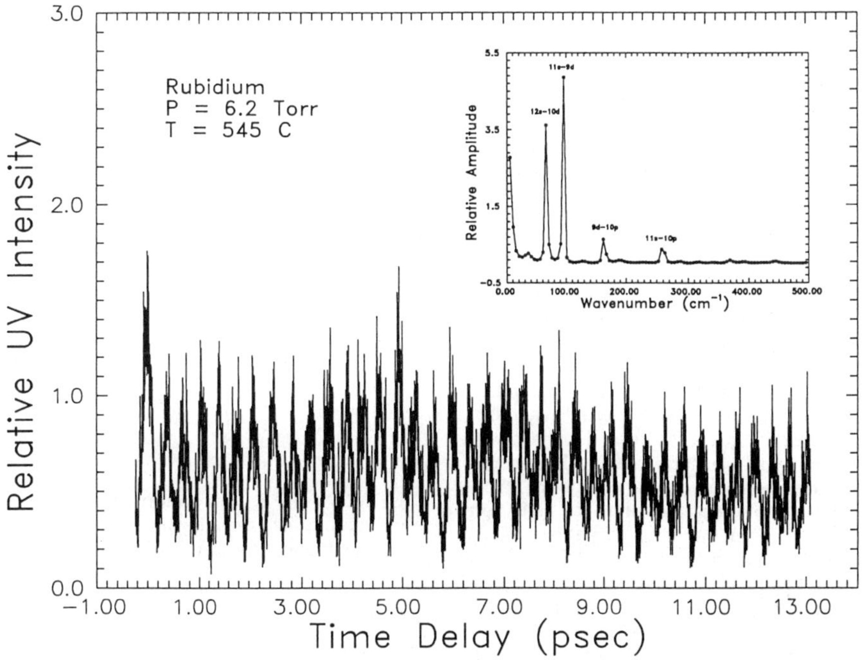

QThG34 **Fig. 2.** Relative UV intensity of the Rb (11*p*, 10*p*, $^2P \rightarrow 5s\ ^2S_{1/2}$) as a function of time delay between the pump and probe laser pulses. Inset shows the Fourier transform of these data with the four strongest peaks identified.

QThG35

Diffraction effects in the spatial and spectral properties of terahertz radiation from bulk GaAs excited by ultrashort pulses

N. Sarukura, Z. Liu, Y. Segawa, S. Koshihara,* K. Shimoyama,** Y. Kondo,** Y. Shibata,† T. Takahashi,† S. Hasebe,† M. Ikezawa,† *Photodynamics Research Center, Institute of Physical and Chemical Research, Nagamachi Koeji 19-1399, Aoba-ku, Sendai, Miyagi 980, Japan*

The mechanism and application to optoelectronics of terahertz radiation induced by lasers have been studied intensively.[1,2] The mechanism of terahertz radiation from semiconductors has been explained by the dipole induced by the surface surge current or by optical rectification,[3] and this phenomenon can be applied as an interface between ultrafast optics and electronics. It is also an attractive light source for studying time-resolved far-infrared spectroscopy. For this purpose, the radiation properties should be known in detail. In this paper we describe the spatial and spectral properties and diffraction effects of terahertz radiation from bulk GaAs excited by ultrashort laser pulses.

The excitation sources were mode-locked Ti:sapphire lasers with a 1-ps or a 100-fs pulse duration. Terahertz radiation was collected by a concave mirror and was detected by a silicon bolometer. The emitter was a nondoped bulk GaAs sample with a (100) surface. The radiation intensity depends on the three-halves power of the excitation intensity, unlike for an optical-rectification process. Spectra were obtained by Fourier transformation of the autocorrelation of the radiation from a nitrogen-purged Martin–Puplett interferometer. Broad spectra peaked at around 0.2 THz were obtained for both 1-ps and 100-fs pulse excitation. Frequency components over 1 THz were observed for the 100-fs excitation case. We found that the center frequency is lower than the value expected when optical rectification is assumed. Furthermore, this radiation was horizontally polarized and was independent of the excitation polarization.[4] This indicates that the radiation has lost the memory of excitation polarization, unlike in a coherent process.

The angular distributions, examined by rotating the detector, were recorded for various incidence angles (Fig. 1). The radiation was detected only in the reflection and transmission directions, suggesting that the radiation retains the memory of the excitation direction. The excitation-spot-size dependence of the beam divergence was measured by using the same configuration. The beam-divergence angle (FWHM) depends on the inverse of the excitation spot size. This dependence shows the diffraction effect (Fig. 2), which was also seen in the spectral shape (Fig. 3). As the excitation spot size decreased, the longer-wavelength component was diffracted and the shorter-wavelength component was emphasized. From this information, the excitation spot size in the probe emitter should be large enough in pump-and-probe experiments to avoid distortion in the white spectrum and low spatial resolution.

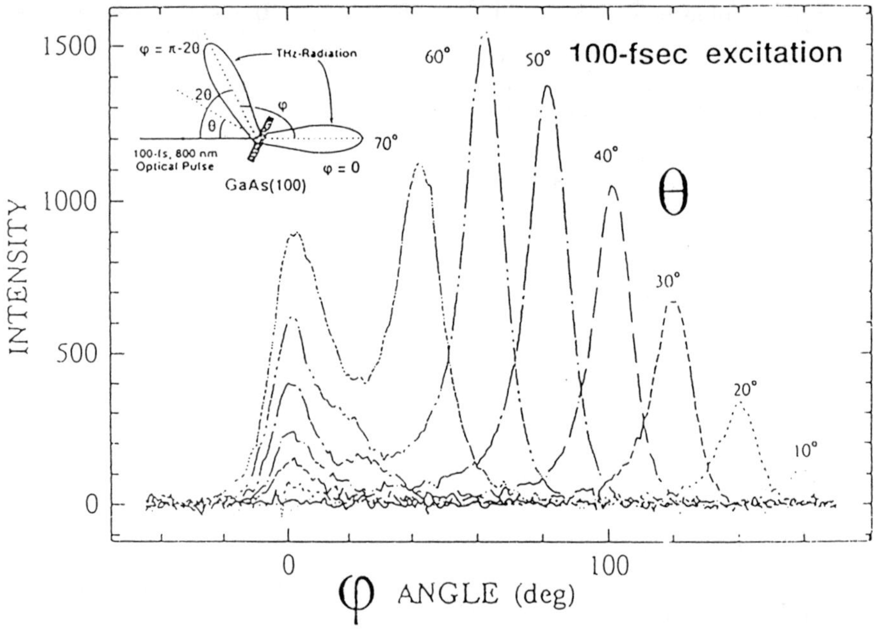

QThG35 **Fig. 1.** Angular distribution of terahertz radiation excited by 100-fs pulses for various incidence angles. The radiation was detected only in the reflection and transmission directions.

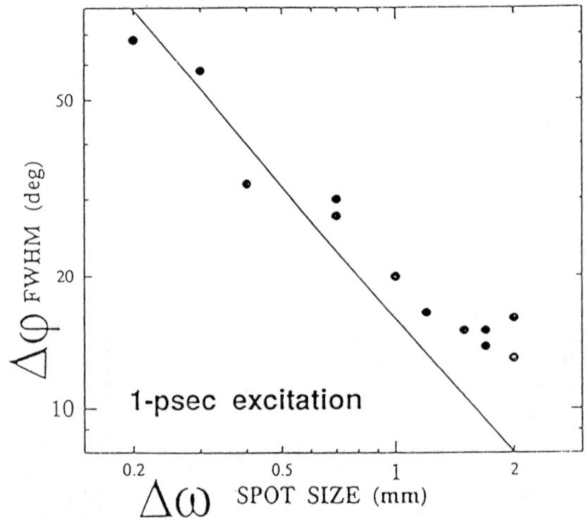

QThG35 **Fig. 2.** Excitation-spot-size dependence of the radiation beam divergence (1-ps excitation-pulse duration). The beam-divergence angle (FWHM) depends on the inverse of the excitation spot size. This dependence clearly shows the diffraction effect.

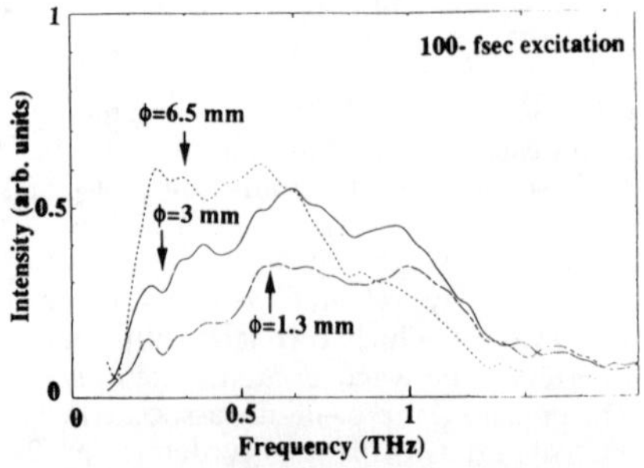

QThG35 **Fig. 3.** Diffraction effect in the spectral shape (100-fs excitation-pulse duration). As the excitation spot size became smaller, the longer-wavelength side was more easily diffracted and the shorter-wavelength component was emphasized.

In conclusion, properties of terahertz radiation from GaAs excited by ultrashort laser pulses were examined. Most properties can be understood by assuming a surface-surge-current model rather than optical rectification. We also found that the diffraction effects in the spatial and spectral properties of terahertz radiation should be considered carefully for future application as white probe light for time-resolved far-infrared spectroscopy.

*Department of Applied Physics, Tokyo Institute of Technology

**Department of Applied Physics, Tohoku University

†Research Institute for Scientific Measurements, Tohoku University

1. D. H. Auston, Appl. Phys. Lett. **43**, 713 (1983).
2. P. C. M. Planken, M. C. Nuss, W.

H. Knox, D. A. B. Miller, and K. W. Goossen, Appl. Phys. Lett. **61**, 2009 (1992).
3. S. L. Chuang, S. Schmitt-Rink, B. I. Greene, P. N. Saeta, and A. F. J. Levi, Phys. Rev. Lett. **68**, 102 (1992).
4. N. Sarukura, Z. Liu, Y. Segawa, S. Koshihara, Y. Kondo, Y. Shibata, T. Takahashi, and M. Ikezawa, in *22nd International Conference on the Physics of Semiconductors*, 1994, Canada, paper TuP-117.

QThG36

Femtosecond coherent spectroscopy of bulk ZnSe and ZnCdSe/ZnSe quantum wells

A. J. Fischer, D. S. Kim, J. Hays, W. Shan, J. J. Song, D. B. Eason,* J. Ren,* J. F. Schetzina,* H. Luo,** J. K. Furdyna,** Z. Q. Zhu,† T. Yao,† *Center for Laser Research and Department of Physics, Oklahoma State University, Stillwater, Oklahoma 74078*

We have found very strong nonlinear diffracted signals in femtosecond (100-fs) two-beam self-diffraction experiments on a thin (300-Å total thickness) ZnCdSe/ZnSe multiple-quantum-well (MQW) sample whose band gap is in the blue region (2.71 eV). The second harmonic of a self-mode-locked Ti:sapphire laser was tuned to the heavy-hole exciton resonance at low temperatures. Extremely-high-order nonlinear signals, up to 13th order, are visible to the naked eye at time delay $T \approx 200$ fs.[1] Even higher-order diffracted signals are very likely in thicker MQW samples.

As a function of T, the higher-order diffracted signals decrease faster. However, the spectrally resolved diffracted signals show essentially the same sharp excitonic features (Fig. 1). This implies that although the higher-order diffracted signals decrease faster as a function of time delay, the temporal shape of the higher-order diffracted signals in real time remains broad. It is also clear the nonlinearities of all orders are dominated by the excitonic resonances in both bulk ZnSe and ZnCdSe/ZnSe MQWs.

We also found four-wave mixing (FWM) signals in bulk ZnSe at negative time delays (see Fig. 2), which suggests that coherent exciton–exciton interaction[2] is important in the nonlinearity of these wide-band-gap materials.

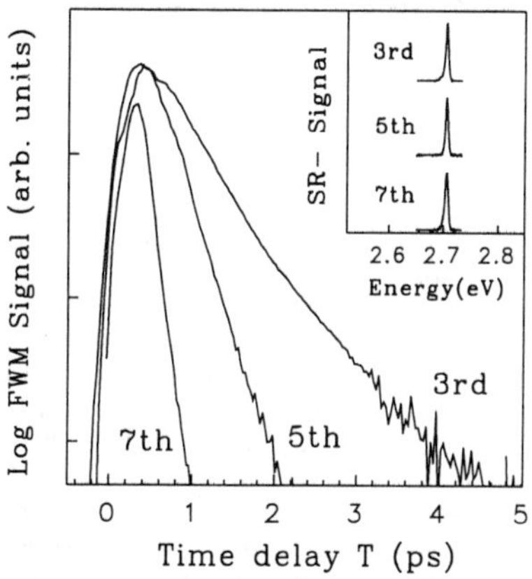

QThG36 Fig. 1. Semilogarithmic plot of third-, fifth-, and seventh-order diffracted signals as a function of time delay T. The spectrally resolved (SR) diffracted signals, shown in the inset, retain sharp excitonic features for all orders of nonlinearity.

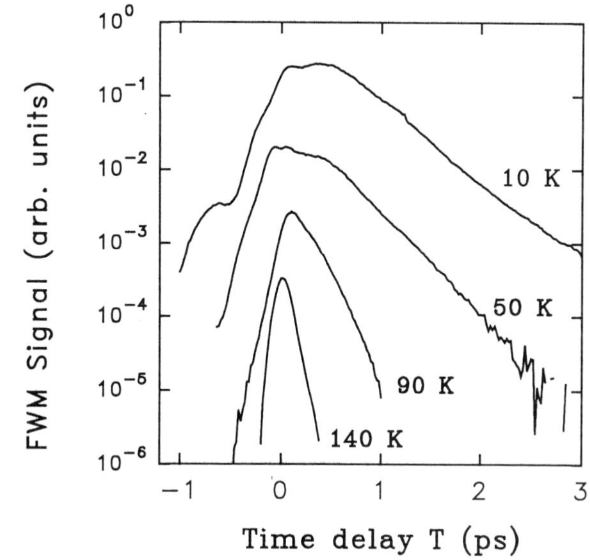

QThG36 Fig. 2. Four-wave-mixing (FWM) signal near heavy-hole excitonic resonances at 10, 50, 90, and 140 K for a ZnSe epilayer. Negative-time-delay signals are clearly visible.

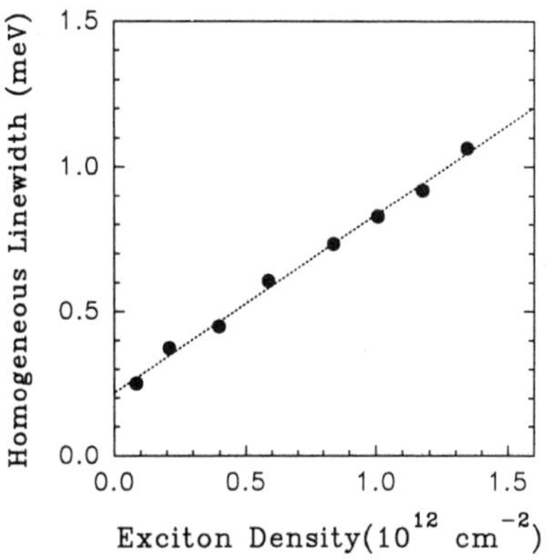

QThG36 Fig. 3. Homogeneous linewidth as a function of exciton density for a ZnCdSe/ZnSe MQW sample. The homogeneous linewidth of this sample at low exciton density is about 0.2 meV.

Exciton–phonon interactions are studied in bulk ZnSe and a ZnCdSe MQW by using FWM (Fig. 2); exciton–acoustic-phonon and exciton–longitudinal-optical-phonon coupling constants (γ and Γ_{LO}) are determined by assuming[3]

$$\Gamma_{ph} = \gamma T_L + \Gamma_{LO}/[\exp(E_{LO}/k_b T_L) - 1], \quad (1)$$

where Γ_{ph}, T_L, E_{LO}, and k_b are homogeneous linewidth of the exciton, the lattice temperature, the longitudinal-optical-phonon energy, and Boltzmann's constant, respectively.

The homogeneous lifetime for this MQW sample was studied as a function of exciton density (Fig. 3). The exciton–exciton interaction rate was determined by a fit to[4]

$$\Gamma(N_{ex}) = \Gamma_0 + \alpha N_{ex}, \quad (2)$$

where N_{ex} is the exciton density, α is the exciton–exciton rate, and Γ_0 represents any residual phase coherence that decays through scattering mechanisms other than exciton–exciton scattering.

* *Department of Physics, North Carolina State University, Raleigh, North Carolina 27695*
** *Department of Physics, University of Notre Dame, Notre Dame, Indiana 46556*
† *Institute of Material Research, Tohoku University, Miyagi 980, Japan*

1. A. J. Fischer, D. S. Kim, J. Hays, W. Shan, and J. J. Song, Phys. Rev. Lett. **73**, 2368 (1994).
2. C. Stafford, S. Schmitt-Rink, and W. Schäfer, Phys. Rev. B **41**, 10000 (1990).
3. J. Lee, E. Koteles, and M. Vassel, Phys. Rev. B **33**, 5512 (1986).
4. L. Shultheis, J. Kuhl, and A. Honold, Phys. Rev. Lett. **57**, 1635 (1986).

QThG37

Hot-phonon effects in femtosecond luminescence spectra of electron–hole plasmas in CdS

P. Haring Bolivar, V. Klimov, H. Kurz, *Institute for Semiconductor Technology II, Rhine–Westphalia Technical University, 52056 Aachen, Germany*

The carrier intraband relaxation in CdS is studied with subpicosecond time resolution by using luminescence upconversion. The cooling of a highly dense plasma (0.3–1.4 × 10^{18} cm^{-3}) is significantly slowed down with respect to the calculated energy-loss rates predicted by the dominant[1] long-range Fröhlich interaction involving longitudinal optical (LO) phonons. Even under the assumption of strong screening[2,3] of the carrier–LO-phonon interaction, the measured relaxation times cannot be reproduced. The retardation of the cooling is attributed to the nonequilibrium buildup of high populations (heating) of certain modes of both LO and acoustic phonons. The slowed cooling through the heating of zone-center optical phonons (optical-phonon bottleneck) has been reported previously.[4-6] Here, for the first time to our knowledge, observation of the development of acoustic-mode disturbances is reported. The disturbances evolve on a subpicosecond time scale and are observed through the retardation of the optical-phonon decay (acoustic-phonon bottleneck).

The samples under investigation are CdS monocrystal platelets with the c axis in the platelet plane. The samples are excited at 3.1 eV by frequency-doubled 100-fs pulses from a mode-locked Ti:sapphire laser. The luminescence is then spectrally and time resolved in a standard upconversion setup.[7] All measurements are performed at room temperature. Figure 1 shows typical photoluminescence spectra of CdS taken at various delay times (Δt = 0.2–10 ps) and an excitation density of $n_e = 1.4 \times 10^{18}$ cm^{-3}. There is one broad luminescence band at 2.46 eV with exponential slopes to high and low energies from the maximum. We concentrate on the high-energy part of the spectra. This spectral region is located above the fundamental band gap and reflects the thermal distribution of the electron–hole plasma. We fit an electron temperature T_e by a Fermi–Dirac distribution of the already-thermalized electrons and holes, assuming a common carrier temperature. The energy-loss rates depicted in Fig. 2 are obtained in a differential mode, dividing the temperature change by the time difference of two consecutive spectra. The measured data are compared with various theoretical predictions. In the time regime between 300 and 800 fs, the data can be well fitted by assuming the LO-phonon–acoustic-phonon interaction restricts the energy loss of the plasma (LO-phonon bottleneck). It is interesting that in this case the carrier loss rate is independent of the excitation density. After 800 fs the observed further reduction of the cooling rate can be explained by the heating of specific acoustic-phonon modes coupled to the zone-center LO phonons. The wave vector q_a of these acoustic modes is specified by $\hbar\omega(q_a) = 1/2 \cdot \hbar\omega_{LO}$. The generation of the nonequilibrium acoustic phonon populations leads to the retardation of the carrier cooling (by 3 orders of magnitude over simple expectations) by slowing down the optical-phonon decay. This assumption is confirmed by the modeling of the temperature dynamics of the plasma, LO phonons, and acoustic phonons (Fig. 3).

V. K. acknowledges support from the Alexander von Humboldt Foundation. P. H. B. is supported by the Heinrich Hertz Foundation.

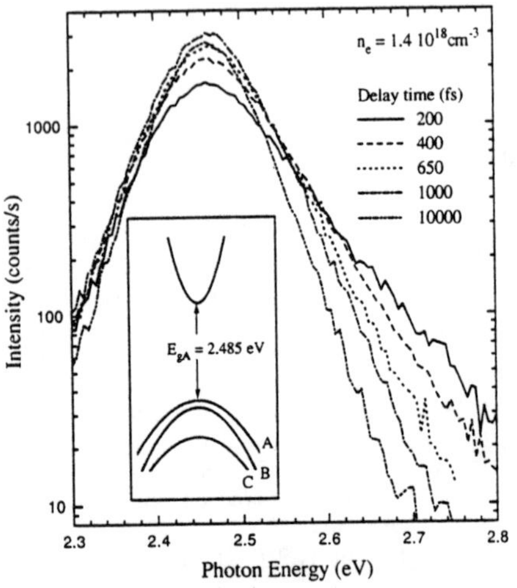

QThG37 Fig. 1. Time-resolved photoluminescence spectra of CdS (300 K) measured at an excitation density of 1.4×10^{18} cm^{-3}. Inset schematically depicts the band structure of CdS in the center of the Brillouin zone.

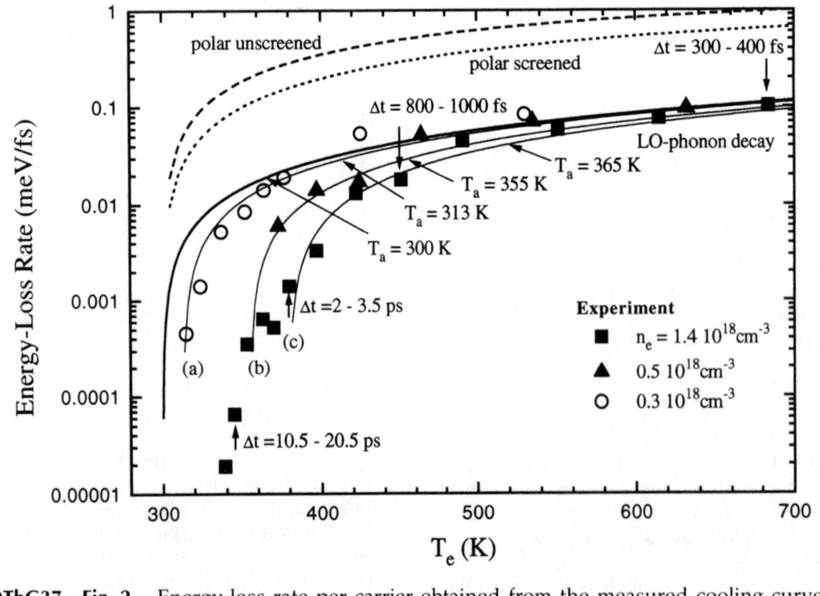

QThG37 Fig. 2. Energy-loss rate per carrier obtained from the measured cooling curves (symbols) and calculated for various energy-loss mechanisms: unscreened (dashed curve) and statically screened (dotted curve) polar-carrier–LO-phonon interaction and LO-phonon decay (thick and thin solid curves). The cooling rates for the latter mechanism are calculated by assuming various temperatures of acoustic phonons as indicated.

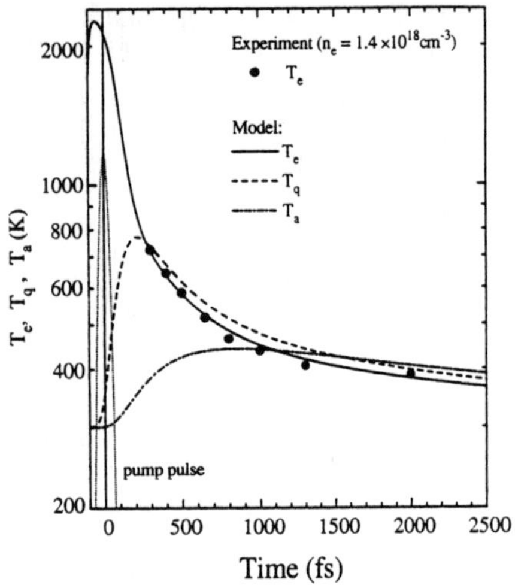

QThG37 Fig. 3. Measured dynamics of carrier cooling (circles, $n_e = 1.4 \times 10^{18}$ cm^{-3}) in comparison to calculation, with LO- and acoustic-phonon heating taken into account. Temperature dynamics in electron–hole-plasma, LO phonons ($q = 5.7 \times 10^6$ cm^{-1}), and acoustic phonons are shown by solid, dashed, and dashed-dotted curves, respectively.

1. See, for example, J. Shah, Solid State Electron. **21**, 43 (1978).
2. M. Pugnet, J. Collet, and A. Cornet, Solid State Commun. **38**, 531 (1981); J. Collet, J. L. Oudar, and T. Amand, Phys. Rev. B **34**, 5443 (1986).
3. E. J. Yoffa, Phys. Rev. B **23**, 1909 (1981).
4. J. Shah, IEEE J. Quantum Electron. **22**, 1728 (1986).
5. W. Pötz and P. Kocevar, Phys. Rev. B **28**, 7040 (1983).
6. S. E. Kumekov and V. I. Perel, Sov. Phys. JETP **67**, 193 (1988).
7. J. Shah, IEEE J. Quantum Electron. **24**, 276 (1988).

QThG38

Electron-density measurement in subpicosecond-laser-produced plasmas

M. A. Mazing, A. P. Shevelko, E. A. Yukov, A. McPherson,* C. K. Rhodes,* *Optical Division, P. N. Lebedev Physics Institute, Russian Academy of Sciences, Leninsky Prospect 53, 117924 Moscow, Russia*

Plasma was created by irradiating transparent solid targets with a high-power subpicosecond laser pulse. The KrF laser beam,[1] with λ = 248 nm, was focused onto flat targets, and intensities as high as ≈10^{19} W/cm^2 were produced at the target surface. The laser pulse duration was ≈300 fs. Transparent solid targets of LiF, MgF_2, CaF_2, and BaF_2 were used. Earlier, an electron density of the order of 10^{23} cm^{-3} in these laser-produced plasmas was evaluated by measuring the density enhancement of the dielectronic satellite spectra emitted by lithiumlike Mg and Si ions.[2] An electron density of the order of 10^{23} cm^{-3} was also estimated from a comparison of experimental x-ray spectra in the 9–15-Å band from a solid BaF_2 target and from a theoretical model[3] (x-ray emission from a 650-fs laser-produced barium plasma).

Here the results of Stark-broadening measurements are reported. Spectra of hydrogen- and heliumlike Mg and F ions were investigated. The x-ray spectra were recorded by using a von Hamos crystal spectrograph.[3] The spectral resolution observed was not worse than 1000. Spectra were recorded in 1–3 laser shots.

The Stark widths of Mg and F ion lines were measured for $1s-np$ and $1s^2-1snp$ transitions with $2 \leq n \leq 4$. Using the calculated Stark widths, we obtained the electron densities in the interval $2 \times 10^{22} \leq N_e \leq 5 \times 10^{22}$ cm^{-3}.

For the CaF_2 target, the electron density was deduced also from measuring and comparing with theory the intensity ratio for the $2^1P_1-1^1S_0$ resonance and the $2^3P_1-1^1S_0$ intercombination lines of heliumlike calcium. This method gave $3 \times 10^{22} \leq N_e \leq 4 \times 10^{22}$ cm^{-1} as a result.

Some specific peculiarities of spectra obtained are discussed. A comparison with similar spectra emitted by the same targets irradiated by nanosecond laser pulses is also presented.

* *Department of Physics, University of Illinois, Chicago, Illinois 60680*

1. T. S. Luk, A. McPherson, G. Gibson, K. Boyer, and C. K. Rhodes, Opt. Lett. **14**, 1113 (1989).
2. A. Zigler, V. L. Jacobs, D. A. Newman, P. G. Burkhalter, D. J. Nagel, T. S. Luk, A. McPherson, K. Boyer, and C. K. Rhodes, Phys. Rev. A **45**, 1569 (1992).
3. W. H. Goldstein, A. Zigler, P. G. Burkhalter, D. J. Nagel, A. Bar-Shalom, J. Oreg, T. S. Luk, A. McPherson, and C. K. Rhodes, Phys. Rev. E **47**, 4349 (1993).
4. M. A. Mazing and A. P. Shevelko, in *X-Ray Plasma Spectroscopy and the Properties of Multiply-Charged Ions*, Vol. 179 of the Proceedings of the Lebedev Physics Institute, I. I. Sobelman, ed. (Nova, Commack, N.Y., 1987), p. 1.

QThG39

Direct measurement of prepulse suppression by using a plasma shutter

D. M. Gold, *Lawrence Livermore National Laboratory, L-251, P.O. Box 808, Livermore, California 94550*

Generation of intense, ultrashort optical pulses with a high contrast between the main pulse energy and prepulse energy is essential to creating hot solid-density plasmas that emit ultrafast x-ray pulses.[1] Prepulse energy focused to intensities above 10^{12} W/cm^2 may cause ionization before the femtosecond pulse arrives at the target surface. Significant hydrodynamic expansion of preionized plasma changes the nature of the femtosecond pulse interaction to a laser–plasma coupling with cooler, low-density plasma. Attempts to improve laser pulse contrast by compensating these effects include direct pulse phase control, spectral or temporal windowing, and intensity-dependent harmonic generation, saturable absorption, or optical switching. The plasma-shutter technique has important advantages, including prepulse suppression following all amplification stages,[2] spatial pulse cleaning,[3] passive self-action, wide bandwidth acceptance, and extension to high-repetition-rate laser systems.[4]

Successful prepulse suppression by using a plasma shutter has been inferred in earlier experiments,[2] but none gave a direct measurement of the amount of suppression achievable with this technique. We quantify plasma-shutter suppression by using a high-dynamic-range second-order autocorrelation with an SiO_2 target placed at Brewster's angle and irradiated by 130-fs 616-nm pulses focused to ~10^{16} W/cm^2. We also report on spatial, spectral, and temporal changes in the pulse after reflection from the self-induced plasma.

The plasma shutter (Fig. 1) is a self-induced, plasma-based reflective optical switch created by focusing an intense, ultrashort laser pulse onto an initially transmissive target material. The target transmits the low-intensity prepulse before ionizing and reflecting the high-intensity femtosecond pulse. A low-noise laser autocorrelation was first taken without any plasma shutter. The contrast ratio was measured to be ~10^4. The *P*-polarized beam was then focused in vacuum onto a raster-scanned SiO_2 target placed at Brewster's angle, 55.6° ± 1°. The autocorrelation was repeated with the beam reflected from the plasma shutter. Figure 2 shows that the contrast ratio was improved by a factor >625.

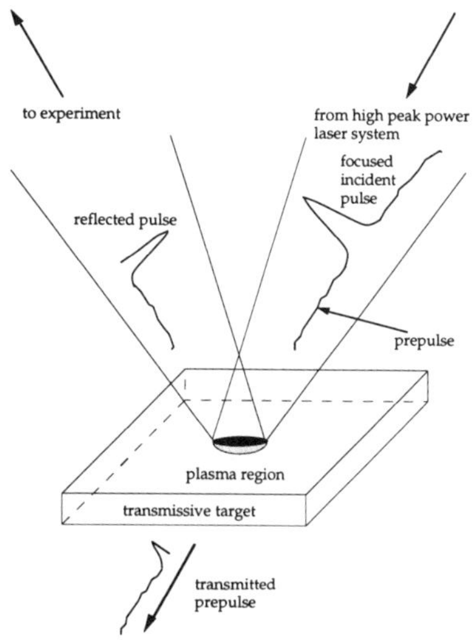

QThG39 Fig. 1. Schematic diagram illustrating how plasma shutter improves the contrast ratio by passing the low-intensity prepulse and reflecting the high-intensity femtosecond pulse.

Single-shot autocorrelations of the incident and reflected pulses showed a 7–10% reduction in pulsewidth with a similar spectral broadening. A spectral Doppler blue shift of several angstroms characterized a plasma expansion velocity of the order of the sound speed, 10^7 cm/s. The reflected-pulse spatial profile indicated a highly specular reflection and reflective spatial filtering at high incident intensities.

We have demonstrated an enhancement in the contrast ratio of high-intensity, ultrashort pulses by a factor >625 with a passive, generally applicable method that reduces prepulse, regardless of its source, without significant deleterious effects on achievable pulse intensity. The resulting pulse contrast ratio is competitive with that of the other prepulse reduction techniques mentioned. More generally, the plasma shutter is a plasma-based optical element that demonstrates a unique wide-bandwidth, ultrafast switching capability for intense subpicosecond laser pulses.

1. M. M. Murnane, H. C. Kapteyn, and R. W. Falcone, Phys. Rev. Lett. **62**, 155 (1989).
2. H. C. Kapteyn, M. M. Murnane, A. Szoke, and R. W. Falcone, Opt. Lett. **16**, 490 (1991); S. Backus *et al.*, Opt. Lett. **18**, 134 (1993); A. Zigler, Opt. Lett. **16**, 1261 (1991).
3. N. K. Moncur, Appl. Opt. **16**, 1449 (1977).
4. S. Backus *et al.*, Opt. Lett. **18**, 134 (1993).

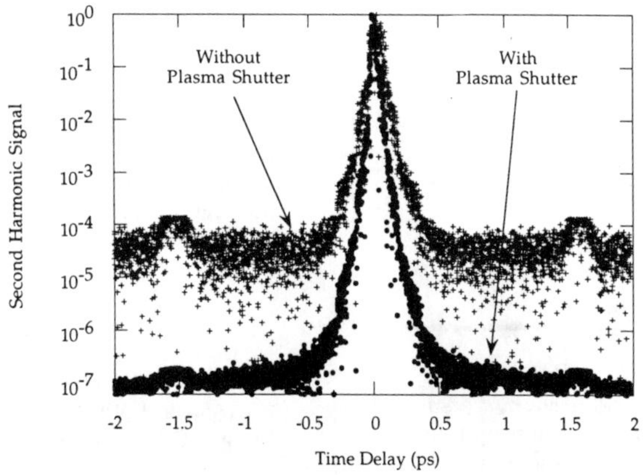

QThG39 Fig. 2. High-dynamic-range autocorrelation showing that the plasma-shutter prepulse suppression at Brewster's angle improves the contrast ratio by a factor >625.

QThG40

Time-resolved reflectivity measurements of electron heat conduction and hydrodynamics in an ultrashort-pulse plasma

D. M. Gold, R. Walling, *Lawrence Livermore National Laboratory, L-251, P.O. Box 808, Livermore, California 94550*

In plasmas produced by ultrashort-pulse lasers, electron heat transport down steep thermal gradients carries absorbed laser energy away from the plasma and into cold bulk target regions. Supersonic electron conduction is predicted to be the limiting factor for both the peak and the duration of the initial plasma heating.[1] Therefore, a quantitative understanding of electron-energy transport rates and heat-wave details is important to studying these plasmas as high-energy-density, strongly coupled matter and to their application as high-brightness x-ray sources. Hydrodynamic expansion, at the speed of sound, follows most electron conduction, lowering the plasma temperature and density and obscuring the experimental reflection signal. The transition between these two primary plasma energy channels exists on a several-picosecond time scale, which can be probed only by femtosecond pulses.

We present pump–probe reflectivity measurements, spatially and temporally resolving plasma reflectivity with a nonintrusive (low-intensity) probe pulse. The reflectivity signal and simulations are used to infer evidence of a transition between rapid heat transport away from the plasma through electron conduction and significant hydrodynamic expansion.

Plasma was created from an SiO_2 (fused quartz) target by the intense (~10^{16}-W/cm^2), ultrashort (130-fs) laser pump pulse.[2] A nonintrusive fraction of the pump beam was used to probe the plasma. A temporal resolution of 130 fs (probe-pulse-width limited) was attained by measuring reflectivity as a function of the time delay betwen the arrival of the pump pulse and the probe pulse. A spatial resolution of ~3 μm was achieved by aperturing magnified images of the reflected and transmitted probe beams at the target-interaction region.

The time-resolved reflectivity signal (Fig. 1) indicates a sharp reflectivity increase on plasma formation, followed by two distinct relaxation times. The initial (~100-fs) reflectivity drop is associated with electron conduction to the bulk target, lowering the surface plasma temperature and density. The second reflectivity relaxation appears to be coincident with expected significant hydrodynamic expansion and has a time scale comparable to this expansion. Simulations integrating beam propagation and the LASNEX hydrodynamic plasma code point to these correlations. The experimental electron conduction signal is not seen at intensities $<10^{16}$ W/cm^2 or in impure forms of SiO_2-based materials. In both cases, a smaller temperature gradient reduces the reflectivity signature for electron conduction below reflectivity losses due to hydrodynamic expansion.

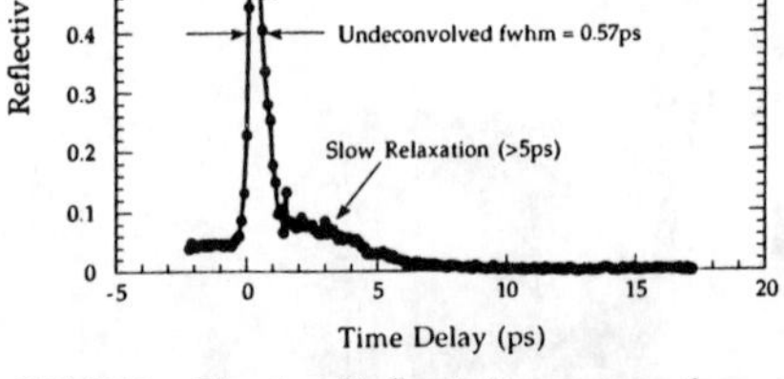

QThG40 Fig. 1. Reflectivity versus time in the central plasma region with an SiO_2 target. Incident pump intensity is 1×10^{16} W/cm^2. The probe is S-polarized at a 30° angle of incidence, and the pump is normally incident.

1. R. M. More, *Laser Absorption in the Ultra-Short Pulse Domain* (Lawrence Livermore National Laboratory, Livermore, Calif., 1993).
2. W. E. White, H. Nathel, W. Tulloch, and D. M. Gold, "A compact laser system for producing sub-100fs laser pulses with millijoule energies," in *American Physical Society: Division of Plasma Physics Annual Meeting* (American Physical Society, New York, 1990).

QThG41

Guiding a one-dimensional nanometer-diameter optical beam

Tetsuro Kobayashi, Hiroki Taki, Akihiro Morimoto, Masahiro Kitagawa, *Engineering Science, Osaka University, Toyonaka, Osaka 560, Japan*

The thinnest limit of the diameter of a light beam propagating in a dielectric waveguide is of the order of the wavelength in free space. This fact makes high-density integration of optical devices difficult. If there is any way to decrease the beam diameter without any limitation by wavelength, it will make a great contribution to realization of the high-density integration of optical devices as well as of high-spatial-resolution optical heads.

Here we propose a new kind of one-dimensional (1-D) optical waveguide that makes it possible to guide a very thin (nanometer-diameter) optical beam.

As far as an optical beam is 3-D (having 3-D propagation phase constants), the above limitation in the beam diameter is inevitable. We consider, for example, an optical beam that propagates in the z direction with a phase constant β in a dielectric guide composed of a single core and a single clad. The transverse components of the phase constants k_x and k_y in the core and of the attenuation constants κ_x and κ_y in the clad (evanescent wave) must satisfy

$$\beta^2 + k_x^2 + k_y^2 = \omega^2\varepsilon_{\text{core}}\mu_0,\ \beta^2 - \kappa_x^2 - \kappa_y^2 = \omega^2\varepsilon_{\text{clad}}\mu_0. \quad (1)$$

In ordinary dielectric guides in which $\varepsilon_{\text{core}} > \varepsilon_{\text{clad}} > 0$, both k_x and k_y are real for the 3-D wave. Accordingly,

$$\beta,\ k_x,\ k_y,\ \kappa_x,\ \kappa_y \le \omega\sqrt{\varepsilon_{\text{core}}\mu_0} = \frac{2\pi n_{\text{core}}}{\lambda_0}. \quad (2)$$

As a result, the thinnest beam thickness is limited to the effective wavelength as follows:

$$d_x,\ d_y \ge \frac{\lambda_0}{4n_{\text{core}}}. \quad (3)$$

Equation (3) corresponds to a kind of uncertainty relation between k and r. On the other hand, if the core is a negative dielectric, i.e., $\varepsilon_{\text{core}} < 0$, at least one of k_x and k_y is imaginary and now Eqs. (2) and (3) are invalid. In other words, for a 1-D or 2-D guided wave with a negative dielectric, the beam diameter might be free from the limitation imposed by the wavelength. In fact, 2-D waveguides have been realized by using a metal as the negative dielectric.[1,2] In these guides 2-D waves can propagate as a surface plasmon and a surface polariton. Particularly for the surface polariton, the beam thickness can be decreased without any limitation by the wavelength,[2] although the beam width is still limited by the wavelength.

A 1-D waveguide, proposed here for the first time to our knowledge, is formed by rolling a 2-D waveguide up into a thin cylinder, as shown in Fig. 1. Here we consider the simplest case: the lowest-order TM mode of a metal-pin guide [Fig. 1(a)]. It is easily found that the fields in a metal core ($r < a$) with a dielectric constant ε_1 (<0) and in the clad ($r > a$) with ε_2 (>0) are given by

$$E_z = AI_0(\gamma_1 r),\gamma_1 = (\beta^2 - \omega^2\varepsilon_1\mu)^{1/2},$$

$$H_\phi = \frac{j\omega\varepsilon_1}{\gamma_1}AI_1(\gamma_1 r),\ E_r = \frac{\beta}{\omega\varepsilon_1}H_\phi\ (r < a), \quad (4)$$

$$E_z = BK_0(\gamma_2 r),\ \gamma_2 = (\beta^2 - \omega^2\varepsilon_2\mu)^{1/2},$$

$$H_\phi = -\frac{j\omega\varepsilon_2}{\gamma_2}BK_1(\gamma_2 r),\ E_r = \frac{\beta}{\omega\varepsilon_2}H_\phi\ (r > a), \quad (5)$$

respectively, where I_j and K_j are the jth order modified Bessel functions. From the boundary conditions for E_z and H_ϕ at $r = a$, we obtain

$$-\frac{\varepsilon_2}{\varepsilon_1} = \frac{\gamma_2 I_1(\gamma_1 a)K_0(\gamma_2 a)}{\gamma_1 I_0(\gamma_1 a)K_1(\gamma_2 a)}. \quad (6)$$

The phase constant β and the beam radius can be calculated as functions of core radius a by using Eqs. (4)–(6). Typical results are

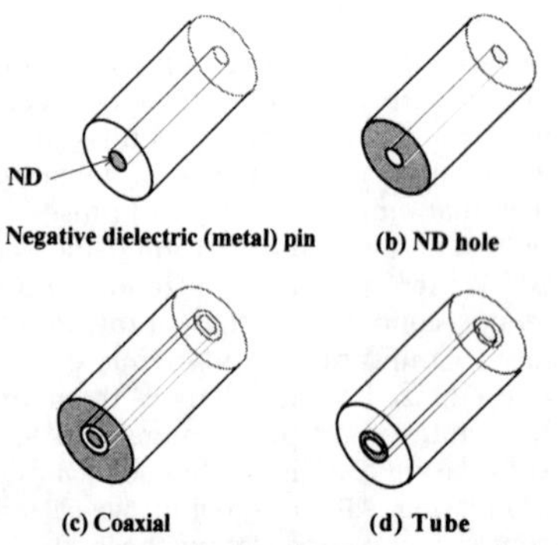

QThG41 Fig. 1. Typical examples of 1-D waveguides. For (a) and (b), outer media are not always necessary. Whether a 1-D wave exists or not depends on $\varepsilon_{\text{core}}/\varepsilon_{\text{clad}}$.

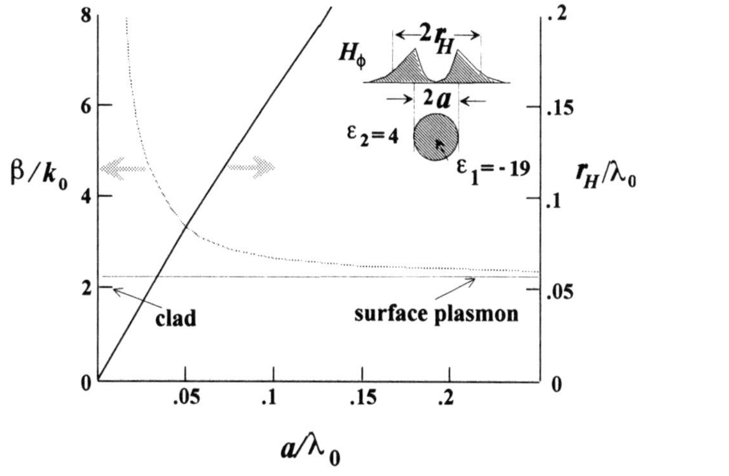

QThG41 Fig. 2. Beam diameter and propagation constant of a 1-D wave as functions of core radius {for the lowest TM mode of a negative dielectric-pin waveguide [Fig. 1(a)]}.

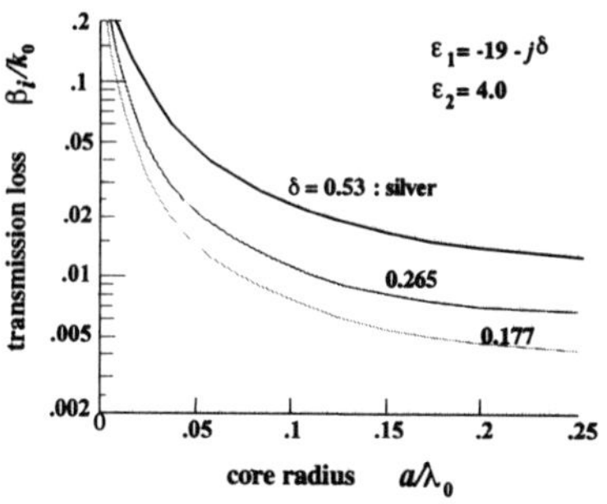

QThG41 Fig. 3. Transmission loss of the negative-dielectric (silver) pin guide.

shown in Fig. 2. As the core radius decreases, β increases and the beam radius decreases without limitation by the wavelength, and transmission cutoff never appears. In actual metals, however, dielectric constants are not real but are complex with a large negative real part and a small negative imaginary part that gives transmission loss. The calculated propagation losses are shown in Fig. 3. The losses are much higher than those of an ordinary optical waveguide and are slightly larger than those of a surface plasmon. By using 1-D waveguides we can obtain a very thin optical beam at the cost of transmissible length. Looking on the bright side of this fact, we can say that these waveguides are rather suitable for micro-optical-device applications because the loss effect is small. One of the important applications is an optical head with high spatial resolution for optical recording, manipulation, and processing and photon STM. The following are typical expected performances of the optical head: core, 20-nm-diameter silver cylinder; clad, normal dielectric with $n_{clad} = 2$; wavelength, $\lambda_0 = 633$ nm; beam diameter, 33 nm; transmission loss, 3 dB/410 nm; spatial data density, ~100 Gbit/cm^2.

There are many other attractive applications as EO/OE conversion devices with Pockels dielectrics and as OEIC devices with semiconductors. At the same time, there are many complex problems to be solved to realize the above applications, e.g., fabrication of such a thin metal pin, reduction of transmission loss, coupling with a normal waveguide as an optical fiber, and evaluation of the evanescent field from the end of the 1-D guide.

1. T. Takano and J. Hamasaki, IEEE J. Quantum Electron. **QE-8**, 206 (1972).
2. J. J. Burke, G. I. Stegeman, and T. Tamir, Phys. Rev. B **33**, 5186 (1986).

QThG42

Resonance Raman spectroscopy and electron–phonon coupling in PbS quantum dots

Todd D. Krauss, Frank W. Wise, *Department of Applied Physics, Cornell University, Ithaca, New York 14853*

Zero-dimensional semiconductor nanostructures, or quantum dots, showing confinement effects in the electronic states have generated considerable interest recently.[1] Although some information exists on the vibrational modes and electron–phonon coupling in II–VI quantum dots (CdS_xSe_{1-x}),[2,3] virtually no corresponding information exists for IV–VI quantum dots. The lead chalcogenides provide unique access to the strong-confinement regime because nanocrystals with $R/a_B \sim 0.1$ can be fabricated.[4] One predicted effect of strong exciton confinement is decreased exciton–phonon coupling.[3,5] This is important because the exciton–phonon coupling fundamentally determines the linewidth, and thus the strength, of resonant optical properties.

We have measured Raman spectra of 4-nm PbS quantum dots with the excitation varying across the 1s–1s exciton resonance (570 to 633 nm). Figure 1 shows a typical Raman spectrum with peak assignments based on the bulk phonon energies. Interestingly, we observe TO and LO + TO peaks that have not been seen in previous Raman studies of semiconductor quantum dots.[2,3,6] This spectrum provides information about phonon confinement in the nanocrystals and about the exciton–phonon coupling.

Phonon confinement manifests itself in Raman spectra as a shift of the eigenfrequency from the bulk value and as a broadening of the phonon linewidth.[7,8] The shift of the LO frequency from 207 to 216 cm^{-1} and the broadening of the peak indicate confinement of the LO-phonon wave function. Figure 2 shows the fundamental LO Raman peak, with a calculation accounting for phonon confinement,[7,8] as well as a peak calculated with the same fit parameters and the bulk phonon frequency. The best fit is obtained with a bulk phonon linewidth of 17.5 cm^{-1} and a phonon-confinement wave function that goes to zero at the dot diameter of 4 nm.

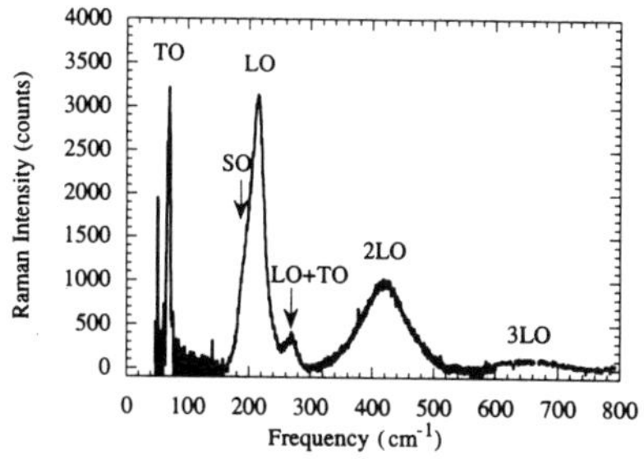

QThG42 Fig. 1. Resonant Raman spectrum of 4-nm-diameter PbS quantum dots obtained at 5 K. The excitation wavelength was 581 nm.

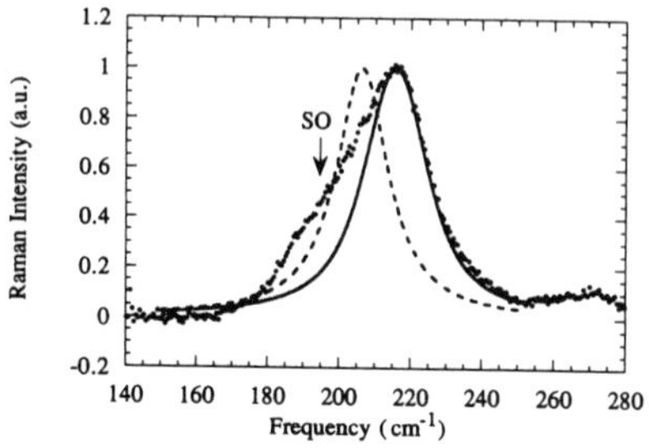

QThG42 Fig. 2. LO Raman peak at $\omega = 216$ cm^{-1}. The dots are the experimental data, and the solid curve is the theoretical fit obtained by using a phonon-confinement model. The dashed curve is a calculated bulk PbS Raman peak obtained by using the same parameters used in theoretical fit to the data. We believe the deviation from the theoretical fit to be caused by the SO modes.

Using an offset harmonic-oscillator model of electron–phonon coupling,[9] we find the Huang–Rhys parameter $S \sim 0.3$ with an electronic dephasing rate $\Gamma \sim 30$ cm^{-1}. The electron–phonon coupling in the nanocrystals is reduced from the calculated bulk value $S = 3.1$. The value of S that we obtain for PbS nanocrystals is comparable to that measured with CdS_xSe_{1-x} nanocrystals of similar radii.[2,3,5] However, this is somewhat unexpected because $R/a_B \sim 1$ in the CdS_xSe_{1-x} nanostructures. Possible explanations for the surprisingly large electron–phonon coupling include the following: (1) because Raman spectroscopy is a steady-state measurement, we could be measuring a state with charge trapped at the surface; (2) the assumption of an infinite confining potential is incorrect.

Interestingly, we find a systematic variation in S as the exciting wavelength is tuned across the exciton resonance. The exciton–phonon coupling decreases mono-

Thursday Posters

tonically from $S = 0.6$ at 581 nm to $S = 0.1$ at 610 nm. This trend in S has not, to our knowledge, been reported previously and is not predicted by the offset-harmonic-oscillator model of electron–phonon coupling. We are currently investigating whether this could be related to variations in particle size across the inhomogeneously broadened 1s–1s resonance.

1. L. E. Brus, J. Phys. Chem. **90**, 2555 (1986).
2. M. C. Klein, F. Hache, D. Ricard, and C. Flytzanis, Phys. Rev. B **42**, 11123 (1990).
3. J. J. Shiang, S. H. Risbud, and A. P. Alivisatos, J. Chem. Phys. **98**, 8432 (1993).
4. J. L. Machol, F. W. Wise, R. C. Patel, and D. B. Tanner, Phys. Rev. B **48**, 2819 (1993).
5. S. Schmitt-Rink, D. A. B. Miller, and D. S. Chemla, Phys. Rev. B **35**, 8113 (1987).
6. D. Bersani and P. P. Lottici, Phys. Status Solidi B **174**, 575 (1992).
7. H. Richter, Z. P. Wang, and L. Ley, Solid State Commun. **39**, 625 (1981).
8. I. H. Campbell and P. M. Fauchet, Solid State Commun. **58**, 739 (1986).
9. R. Merlin, G. Güntherodt, R. Humphreys, M. Cardona, R. Suryanarayanan, and F. Holtzberg, Phys. Rev. B **17**, 4951 (1978).

QThG43

Josephson plasma resonance in $YBa_2Cu_3O_{7-\delta}$ with intrinsic weak links

Ch. Jaekel, G. Kyas, H. G. Roskos, H. Kurz, W. Prusseit,* H. Kinder,*
Institute for Semiconductor Technology II, Rhine–Westphalia Technical University, Sommerfeldstrasse 24, 52056 Aachen, Germany

The growth of high-T_c superconductors on silicon wafers is generally hampered by the existence of a critical thickness d_c at which microcracks appear within the film. The critical thickness d_c for $YBa_2Cu_3O_7$ on YSZ (yttria-stabilized zirconia) on silicon is known to be about 50–70 nm. Recent experimental results suggest that the density of microcracks or weak links increases rapidly at thicknesses well below the established value of the critical thickness.[1] Our aim was to determine whether these weak links can work as intrinsic Josephson junctions. Highly resistive (>1000-Ω·cm) silicon wafers are evaporated with 40-nm-thick buffer layers of YSZ and *c*-axis-oriented epitaxial $YBa_2Cu_3O_{7-\delta}$ films whose thickness varies between 25 and 50 nm. The critical current density j_c is several 10^5 A/cm², and T_c values between 74 and 83 K are determined.

With a time-domain terahertz-spectroscopy arrangement we measure the frequency-dependent transmissivity of extremely thin superconductor films. From the ratio of the transmission coefficients of the sample and a reference sample (substrate without $YBa_2Cu_3O_{7-\delta}$ film), we can extract both the real and the imaginary parts of the terahertz conductivity without using the Kramers–Kronig relation.

One of the characteristic parameters of a Josephson junction is the plasma frequency f_{pl}, which is given by

$$f_{pl} = \sqrt{\frac{I_c}{2\pi\Phi_0 C}} = \sqrt{\frac{j_c t}{2\pi\Phi_0 \varepsilon\varepsilon_0}}, \quad (1)$$

where C is the junction capacitance, ε is the dielectric constant, and t is the barrier thickness. To date, the plasma resonance has been directly observed in Josephson-tunnel junctions made of low-T_c superconductors.[2] Here we report on similar resonances probably originating from weak links.

In Fig. 1 the Fourier-transformed transmission data are plotted for a 50-nm-thick film at various temperatures. Below T_c, an absorption at about 300 GHz appears and shifts to higher frequencies with decreasing temperature. The frequency at which the plasma-induced attenuation occurs, shown in Fig. 2 as a function of temperature, is closely related to the critical current den-

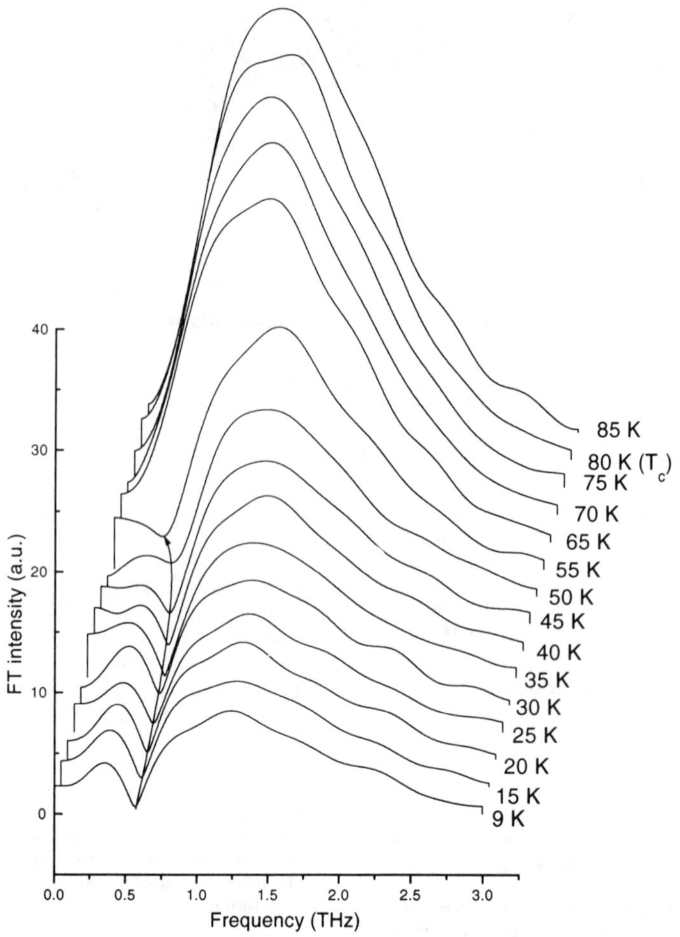

QThG43 Fig. 1. Fourier transform of the transmitted terahertz pulse. At 15 K below the transition temperature, a distinct absorption is observed.

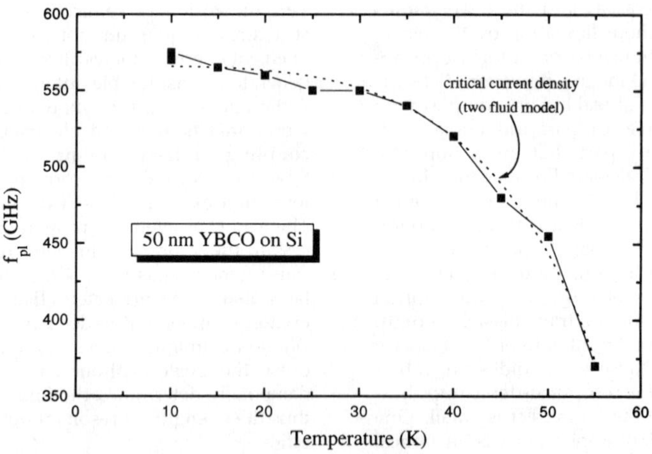

QThG43 Fig. 2. Resonance frequency as a function of temperature. The dashed curve shows the temperature behavior of the critical current density.

Thursday Posters

sity, as expected from Eq. (1). We discuss these results as evidence for Josephson plasma resonances at weak links present in these films at a high density.

This work was supported by German Federal Ministry for Research and Technology grant 13N6288.

*Department of Physics, E10, Munich Technical University, James-Franck Strasse 1, D-85747 Garching, Germany

1. C. Jaekel, C. Waschke, H. G. Roskos, H. Kurz, W. Prusseit, and H. Kinder, Appl. Phys. Lett. **64**, 3326 (1994).
2. A. J. Dahm, A. Denenstein, T. F. Finnegan, D. N. Langenberg, and D. J. Scalapino, Phys. Rev. Lett. **20**, 859 (1968).

QThH 2:30 pm Hyatt A

Coherent Control of Atomic and Molecular Dynamics

Douglas Shumacher, *University of Michigan, Presider*

QThH1 (Invited) 2:30 pm

Controlling quantum dynamics with tailored femtosecond light pulses

Bern Kohler, Vladislav V. Yakovlev, Jianwei Che, Jeffrey L. Krause, Michael Messina, Kent R. Wilson, Nikolaus Schwentner,* Robert M. Whitnell,** YiJing Yan,† *Department of Chemistry, University of California, San Diego, La Jolla, CA 92093-0339*

Quantum dynamics can be controlled, at least in principle, by tailoring the frequencies and phases (i.e. coherence) of a light filed which interacts with atoms or molecules. In this experiment we use a tailored, ultrashort pump pulse to control the vibrational dynamics of iodine molecules. Our objective is to excite a wave packet in the bound region of the B electronic state of I_2, which maximally overlaps a minimum-uncertainty target state localized near the outer turning point with a negative momentum. Theory is used to compute the light field which best achieves the target. Except for limiting the duration of the field to a maximum of 0.5 ps, no restrictions are made on the form the electric field can assume. As described in detail elsewhere,[1] the optimal light field for achieving the target is then a single pulse with a negative frequency chirp.

We synthesize an experimental pulse with the required pulse shape with a novel optical parametric amplifier (OPA).[2] The amplified OPA signal pulse is tuned to a center wavelength of 571 nm and then shaped by propagation through a prism compressor. By translating one of the prisms into and out of the optical path, the pump pulse can be both positively and negatively chirped. The precise pulse shape is measured by frequency-resolved optical gating (FROG).[3]

Two pump pulses are compared. One is negatively chirped to approximately match the theoretical pulse shape for producing our target. The second excitation pulse is identical with the first, except that the sign of the frequency-dependent phase is reversed. Theory predicts[1] that such a positively chirped pulse gives rise to distinctly different wave packet dynamics. To detect the resulting wave packet evolution, a 45-fs, 399-nm probe pulse is used to excite the wave packet to a higher-lying electronic state, and the total laser-induced fluorescence (LIF) is collected from this state as a function of delay time between the pump and probe pulses.

The FROG images and the retrieved time-dependent fields of the excitation pulses are shown in Fig. 1. The two pulses have nearly identical intensity profiles versus time and versus frequency; they differ only in their coherence.The LIF signals recorded with these two pulses are shown in Fig. 2. Clearly, changing the excitation pulse coherence gives rise to different vibrational dynamics, and we are able to control the focusing of matter, varying the height of the second peak. The different LIF peak heights and widths are reasonably reproduced by full quantum simulations based on the measured electric fields of Fig. 1, the results of which are also shown in Fig. 2. This initial experiment demonstrates that control of nonstationary states of matter is possible with tailored light fields. Related experiments should allow the control of the products of chemical reactions.

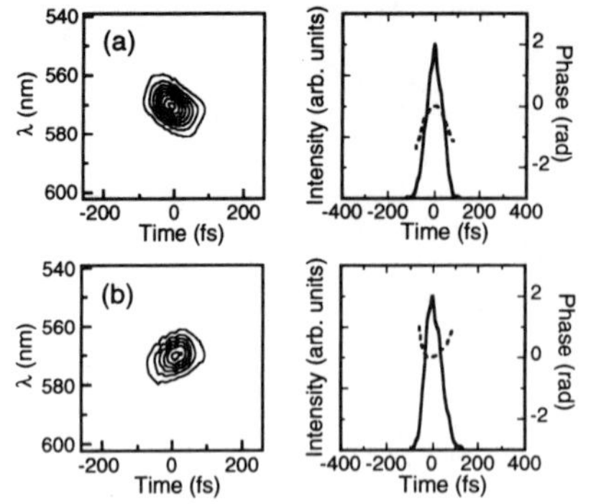

QThH1 Fig. 1. The FROG signals (polarization-gate geometry) recorded for the negatively (a) and positively (b) chirped pump pulses used to generate the LIF signals of Fig. 2. The center wavelength of both pulses is 571 nm. The electric fields (intensity and phase) as a function of time computed from the FROG signals are shown on the right.

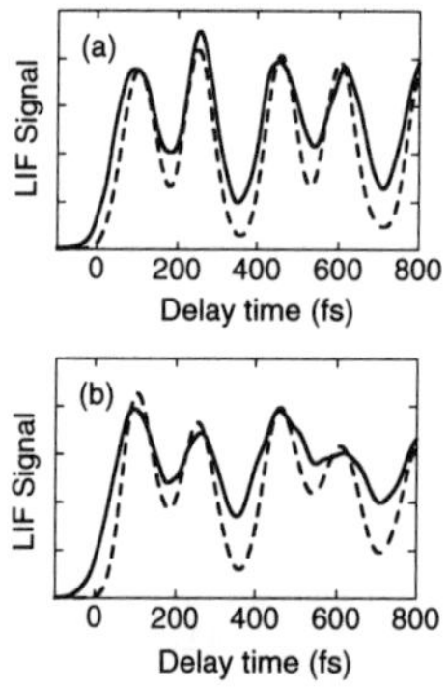

QThH1 Fig. 2. (a) The LIF signal (solid curve) recorded with the negatively chirped pump pulse of Fig. 1(a). (b) The LIF signal (solid curve) recorded with the positively chirped pump pulse of Fig. 1(b). The dashed curves are quantum simulations using the actual electric fields measured by FROG.

*Institute for Experimental Physics, Free University of Berlin, Germany

**Guilford College, Greensboro, North Carolina 27410

†Department of Chemistry, Hong Kong University of Science and Technology, Hong Kong

1. J. L. Krause, R. M. Whitnell, K. R. Wilson, Y-J. Yan, and S. Mukamel, J. Chem. Phys. **99**, 6562 (1993).
2. V. V. Yakovlev, B. Kohler, and K. R. Wilson, Opt. Lett., accepted for publication.
3. R. Trebino and D. J. Kane, J. Opt. Soc. Am. A **10**, 1101 (1993).

QThH2 3:00 pm

Phase control in the two-color photoionization of HD and the photodissociation of HD^+

B. Sheehy, B. Walker, L. F. DiMauro, *Department of Chemistry, Brookhaven National Laboratory, Upton, New York 11973*

An intense (~2 × 10^{13}-W/cm^2) two-color laser field with a variable relative phase between the two colors is used in a study of the photoionization of HD and the subsequent photodissociation of the molecular ion HD^+. Both processes show a sensitive phase dependence. The photoionization shows a larger-than-expected modulation in yield, with a period of π in the relative phase between the two colors. in the photodissociation of HD^+, either the H or D fragment will be charged, and an asymmetry in the angular distribution of either charged fragment, modulated with a period of 2π in the phase, is observed.

The 527-nm component of the two-color field is produced by doubling the 1053-nm fundamental in a BBO crystal; the doubled

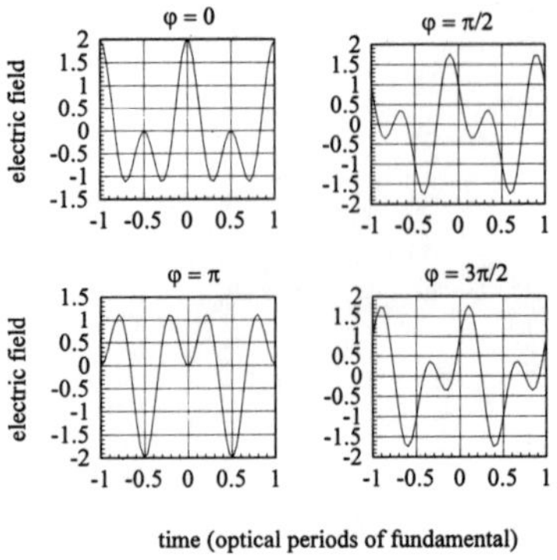

QThH2 Fig. 1. Time variation of the two-color laser field.

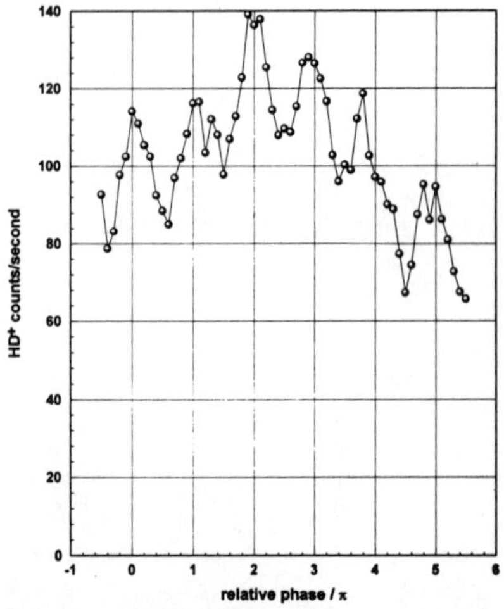

QThH2 Fig. 2. Production of HD^+ as a function of relative phase.

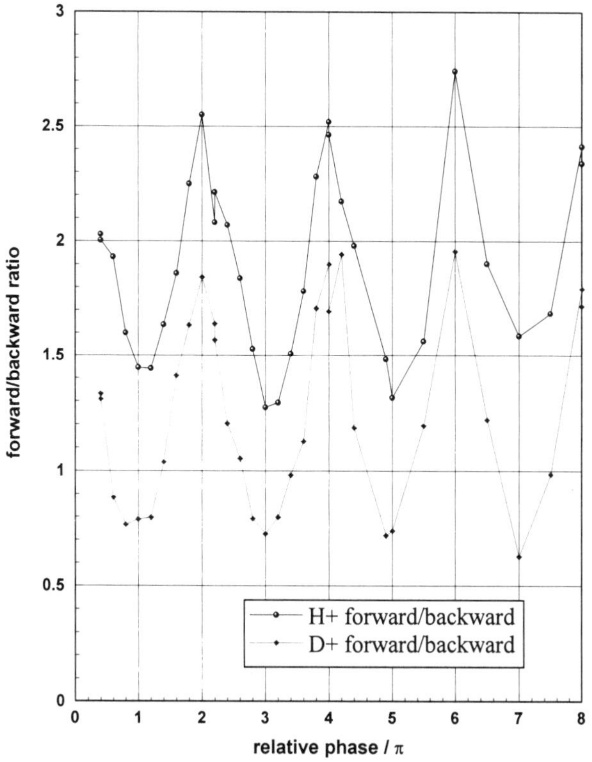

QThH2 Fig. 3. Forward/backward ratio of H^+ and D^+ as a function of relative phase.

light has a well-defined phase relative to the fundamental, which may then be varied by passing the collinear beams through a dispersive medium of variable length. The shape of the field in time for several values of the phase is shown in Fig. 1. At $\phi = 0$ and $\phi = \pi$ the field amplitude is asymmetric and the maximum points are in the positive and negative directions, respectively, whereas the waveforms at $\phi = \pi/2$ and $\phi = 3\pi/2$ have equal amplitudes in both directions but are time-reversed images of each other. The light is focused into the sample gas at the center of a time-of-flight mass spectrometer. An extraction field is present and allows ions emitted toward or away from the detector to be distinguished.

The HD^+ yield is plotted as a function of phase in Fig. 2. The modulation period of π suggests that the ionization depends on the maximum magnitude of the instantaneous electric field occuring in an optical cycle, which oscillates with the same period. This would be expected for multiphoton ionization in the tunneling regime, i.e., when the Keldysh parameter $\gamma < 1$. For the data shown, $\gamma = 1.9$, but the modulation depth is 30%.

In Fig. 3 the ratio of the number of atomic ions emitted toward/away from the detector (forward/backward) is plotted for both H^+ and D^+. This asymmetry is modulated with a period of 2π in the relative phase, again consistent with a rate dependent on the maximum instantaneous field strength, in this case both the direction and the magnitude of the maximum being important. The depth of modulation of these curves is a maximum for a 50/50 intensity mix of 527/1053-nm light. A phase-dependent isotopic preponderance of the charged dissociation fragment cannot be excluded, but the identical phase of the modulations in the H^+ and D^+ forward/backward asymmetries, and their similar depths of modulation, suggest that it is small in the experimentally accessible parameter range.

This work was supported by the U.S. Department of Energy.

QThH3 **3:15 pm**

Direct measurement of wave-packet dynamics

E. Constant, H. Stapelfeldt, P. B. Corkum, *National Research Council of Canada, Ottawa, Ontario K1A 0R6, Canada*

We present a new technique for measuring wave-packet dimensions and dynamics. It uses a weak pump pulse to create a vibrational wave packet in I_2 and a delayed intense femtosecond pulse to rapidly ionize the molecule to a highly charged state. We follow the dynamic of the wave packet by measuring the kinetic energy release from the resulting Coulomb explosion as a function of the pump–probe delay.

The experiment is illustrated in Fig. 1. A 200-fs blue pulse centered at 485 nm, with an energy of 200 nJ, excites a dissociative wave packet of iodine molecules. The internuclear distance, initially well defined, can then freely evolve in the new molecular dissociative state (linear combination of $B\ ^3\Pi_u^+$ and $^1\Pi_u^+$). After a chosen time we irradiate these molecules (both pulses are focused to the same 10-μm diameter and at the same place) with a powerful femtosecond pulse (50 μJ, 100 fs, 625 nm) that causes rapid ejection of several electrons and gives molecular ions an additional Coulomb potential energy inversely proportional to their internuclear distance. This energy is then released as kinetic energy, and its measured value gives the internuclear distance.

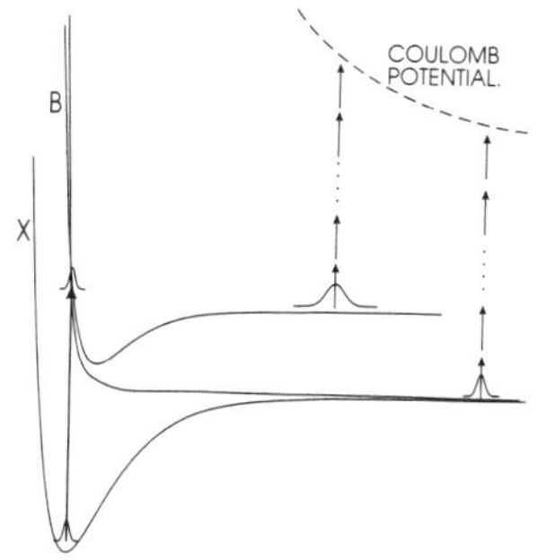

QThH3 Fig. 1. Schematic energy diagram.

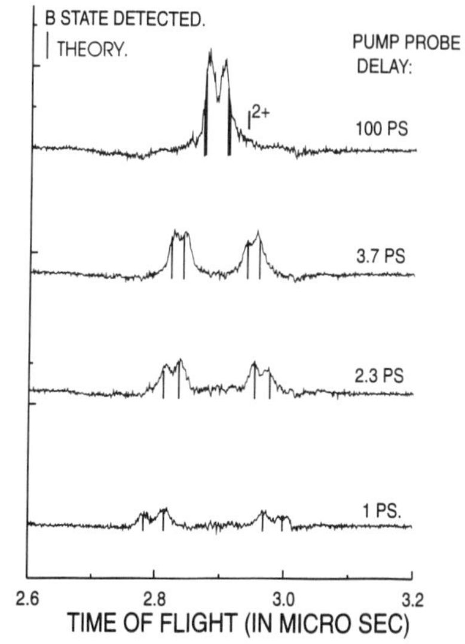

QThH3 Fig. 2. Time-of-flight spectra for I^{2+}.

We measure this energy in a time-of-flight spectrometer. The recorded time-of-flight spectrum is mostly determined by the ions' species, but its fine structure is a direct signature of their initial velocities. A velocity directed toward the detector will lead to a time of flight shorter than that produced by one in the opposite direction.

Apertures in the spectrometer ensure that only molecules that are aligned with the axis are detected. Because such a molecule will eject two ionic fragments in opposite directions, the recorded time-of-flight spectra will basically consist of two peaks. The time difference between these two peaks is determined only by the repelling electric field, the mass, and the charge of the species and their initial velocity: $\Delta T = 2mV/(qE)$. Consequently, ΔT measures the explosion velocity and therefore the internuclear distance. Furthermore, the peak width reflects the wave packet size.

Two methods are used to reject signal from unexcited molecules: (1) An exploding pulse polarized perpendicularly to the time-of-flight axis is used. It induces a torque[1] only on these highly polarizable molecules and tends to align them along the polarization. Thus it reduces the number of unexcited detected molecules. (2) The spectra due to these remaining molecules are simply recorded without any blue pulse and then are subtracted from the total time-of-flight spectra.

Figure 2 shows our experimental results for the I^{2+} ions when the 485-nm polarization is parallel to the time-of-flight axis. The expected positions of the peaks for the explosion channels $I^{2+}-I^{2+}$ and $I^{2+}-I^{+}$ are noted. Agreement between experimental results and the model are discussed, as well as the potential of this method for characterizing dissociation products and measuring the shape of molecules produced by tailored light pulses.[2]

1. P. Dietrich, D. T. Strickland, M. Laberge, and P. B. Corkum, Phys. Rev. A **47**, 2305 (1993).
2. J. L. Krause *et al.*, J. Chem. Phys. **99**, 6562 (1993).

QThH4 **3:30 pm**

Quantum control in a finite anharmonic ladder of states

Z. D. Carter, K. Mochizuki,
M. R. Matthews, J. R. Gardner,
D. J. Heinzen, *Department of Physics, University of Texas, Austin, Texas 78712*

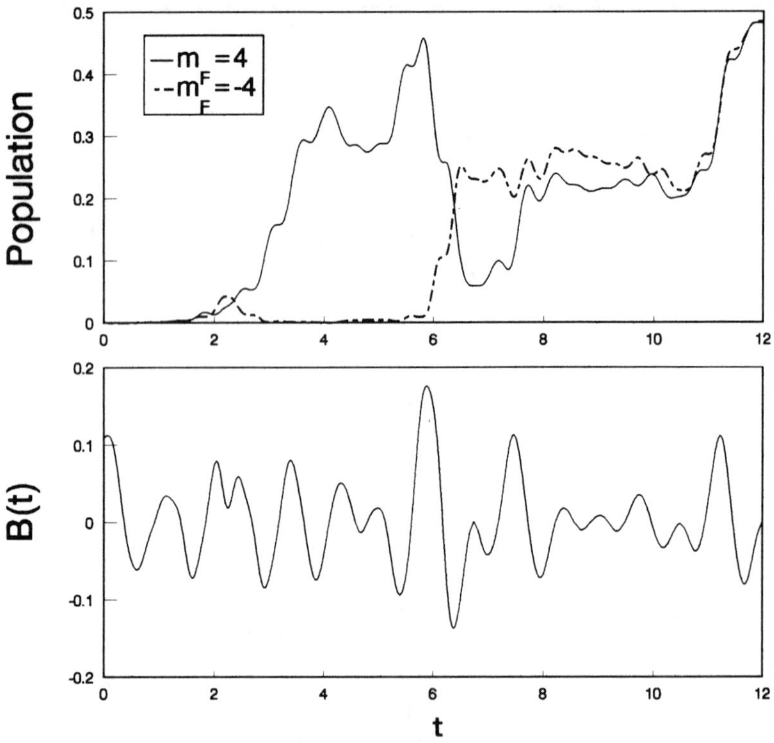

QThH4 Fig. 1. Example of the quantum control of a spin wave function. (a) Probability for a Cs atom to be in the $|F = 4, M_F = +4\rangle$ state (solid curve) and the $|F = 4, M_F = -4\rangle$ state (dashed curve) as a function of time, starting from the $|F = 4, M_F = 0\rangle$ state and evolving under the Hamiltonian H. The final state is a very close approximation to the desired $|\psi(t_f)\rangle$. (b) Form of $B(t)$ that results in the evolution shown in (a). Time is measured in units of the Larmor period, and $B(t)$ is measured in units of B_0.

Control of the quantum state of a system is of intense current interest. Although much attention has been directed toward the control of electronic or nuclear translational wave functions, there is significant motivation to explore the control of a spin system. First, a spin system has only a finite number of states. Second, a spin system may be manipulated by radio-frequency magnetic fields, and this makes experiments with complex perturbing pulse shapes more feasible. Finally, unique quantum states that are of interest for precision-measurement of quantum-optics applications may be prepared.

We have explored this concept for the $|F = 4, M_F\rangle$ manifold of Zeeman levels of the Cs atomic ground state. Simultaneous application of a uniform static magnetic field, a time-dependent transverse magnetic field, and a uniform optical field results in a Hamiltonian $H = -\boldsymbol{\mu}\cdot\mathbf{B}(t) + m_F^2\omega_E|m_F\rangle\langle m_F|$, where $\mathbf{B}(t) = B_0\hat{z} + B(t)\hat{x}$ and $\boldsymbol{\mu}$ is the atomic magnetic-dipole operator. The second term in this Hamiltonian represents the light shift induced by the optical field. With $B(t)$ set to zero, it yields a finite anharmonic ladder of quantum states labeled by M_F. Because of the anharmonicity, $B(t)$ induces motions of the spin that are not simply rotations of the initial state. Rather, it is possible by judicious choice of $B(t)$ to produce an arbitrary spin state.

A specific example of this is shown in Fig. 1. We assume that we want to produce the transformation $|\psi(t = 0)\rangle = |F = 4, M_F = 0\rangle \rightarrow |\psi(t_f)\rangle = |F = 4, M_F = -4\rangle + |F = 4, M_F = +4\rangle$. We assume that $\omega_E = 0.02\omega_L$, where ω_L is the Larmor frequency. Using the method of Shi and Rabitz,[1] we find that this may be accomplished with $B(t)$, as shown in Fig. 1(b). As may be seen in Fig. 1(a), $B(t)$ produces in 12 Larmor periods a state that is almost exactly equal to the desired $|\psi(t_f)\rangle$. By application of other forms of $B(t)$, other quantum states may be prepared. This unique ability to control the spin state is interesting for several reasons. First, squeezed spin states that may lead to improved signal-to-noise ratios in certain experiments may be prepared.[2] Other interesting spin states, such as Schrödinger-cat states, could be prepared. Also, this spin wave function may be transferred to an optical field, so that arbitrary field states could be prepared.[3]

We report on our calculations and on an apparatus to demonstrate this behavior in an experiment with spin-polarized laser-cooled Cs atoms released from a magneto-optical trap. In preliminary experiments we have observed Cs Zeeman resonances narrower than 200 Hz. Finally, a method to monitor the quantum state is required. We discuss a method, based on Ramsey's method of separated oscillatory fields, to accomplish this.

We gratefully acknowledge the support of the National Science Foundation, the National Aeronautics and Space Administration, and the Texas Advanced Research Program.

1. S. Shi and H. Rabitz, J. Chem. Phys. **92**, 364 (1990).
2. D. J. Wineland, J. J. Bollinger, W. M. Itano, and D. J. Heinzen, Phys. Rev. A **50**, 67 (1994).
3. A. S. Parkins, P. Marte, P. Zoller, and H. J. Kimble, Phys. Rev. Lett. **71**, 3095 (1993).

QThH5 **3:45 pm**

State-specific atomic excitation by using strong-field, short-pulse lasers

Mark Edwards,* Charles W. Clark, *National Institute of Standards and Technology, Room A253, Building 221, Gaithersburg, Maryland 20899*

In recent experiments involving high-intensity, short-duration laser pulses tightly focused into an atomic-vapor cell, atoms surviving the pulse in excited states have been observed.[1–6] This raises the possibility of using such laser pulses to efficiently produce large numbers of atoms in selected excited states.

We have used a one-dimensional model atom interacting with a realistic high-intensity ($I \sim 10^{13}–10^{15}$ W/cm^2), short-pulse ($\tau \sim 100–600$ fs) focused laser beams to study the effects of the peak pulse intensity, the pulse duration, and the focal peak-intensity distribution for the optimal production of atoms surviving the pulse in specific bound states. We have investigated these effects by using two methods: (1) direct numerical integration of the time-dependent Schrödinger equation (TDSE) and (2) Floquet analysis of the atom interacting with a constant-amplitude laser field for many field amplitudes in conjunction with Landau–Zener theory.

We have found that the results of the direct integration of the TDSE regarding the model atom's probability of surviving a pulse in a particular bound state can be accurately predicted by using few-Floquet-

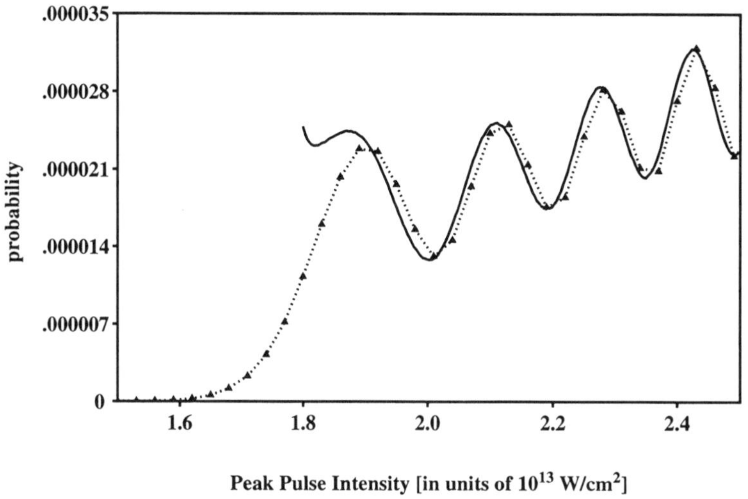

QThH5 Fig. 1. Population of level E_5 after the passage of the pulse as a function of peak pulse intensity. The triangles show the results of direct numerical integration of the time-dependent Schrödinger equation, and the solid curve shows the results of the double-crossing Landau–Zener theory. The model system consists of an electron moving in a one-dimensional, long-range potential of the form $V(x) = -V_0 a/(x^2 + a^2)^{1/2}$ whose parameters are fixed to replicate the ionization energy of Xe, as in Ref.[1] Level E_5, to which the graph refers, lies at the same energy as the Xe $4f$ state.

state models, such as Landau–Zener theory whose parameter values are obtained from complex Floquet eigenvalues. Figure 1 shows the single-atom probability for the model atom to survive a Gaussian pulse in a particular excited state in comparison with the Landau–Zener/Floquet predictions. This circumstance makes computation of such probabilities sufficiently efficient for the performance of a wide-ranging, systematic study of the effects of the pulse duration and the spatial distribution of peak intensities over the laser focus.

Individual atoms can survive the laser pulse in a bound state if the peak intensity at the location of the atom within the laser focus exceeds the resonant intensity of the state. When the instantaneous laser intensity equals the state's resonant intensity, resonant excitation occurs and the excited state and the state is populated. This process also occurs on the trailing pulse edge, and the state will be repopulated at this time. Several processes limit the final amount of excited-state population. The two resonant excitations can exhibit quantum interference, causing the final state population to oscillate as a function of peak pulse intensity (see Fig. 1). Furthermore, between the two resonance times the excited atoms will undergo decay into the continuum, and this will also tend to decrease the survival probability of the atom. The amount of population that is transferred on the leading and the trailing pulse edges depends on three factors: (1) the population of the ground state at excitation, (2) the ground-state-to-excited state coupling at excitation, and (3) the time the two states spend near resonance.

We find that the total yield of surviving atoms is optimal for neither very long nor very short pulses. For short pulses, the level crossings are essentially diabatic and very little excitation occurs. For long pulses, the crossings are nearly adiabatic and large population transfer occurs. Most of this population decays into the continuum, however.

Also with Department of Physics, Georgia Southern University, Statesboro, Georgia 30460-8031

1. M. P. de Boer and H. G. Muller, Phys. Rev. Lett. **68**, 2747 (1992).
2. G. N. Gibson, R. R. Freeman, and T. J. McIllrath, Phys. Rev. Lett. **69**, 1904 (1992).
3. J. G. Story, D. I. Duncan, and T. F. Gallagher, Phys. Rev. Lett. **70**, 3012 (1993).
4. R. B. Vrijen, J. H. Hoogenraad, H. G. Muller, and L. D. Noordam, Phys. Rev. Lett. **70**, 3018 (1993).
5. M. P. de Boer, L. D. Noordam, and H. G. Muller, Phys. Rev. A **47**, R45 (1993).
6. R. R. Jones, D. W. Schumacher, and P. H. Bucksbaum, Phys. Rev. A **47**, R49 (1993).

QThI — 2:30 pm, Hyatt B

Microcavity Effects II

Richart E. Slusher, *AT&T Bell Laboratories, Presider*

QThI1 (Invited) — **2:30 pm**

Three-dimensional optical lattices

Gilbert Grynberg, Philippe Verkerk, *Laboratoire Kastler-Brossel, Department de Physique de l'Ecole Normalc Supérieure, 24 rue Lhomond, F-75231 Paris cedex 05, France*

The interference pattern of several beams of the same frequency generally leads to a space dependent variation of the total intensity and of the local polarization. The light-shift of an atom in such a standing wave is thus space dependent and appears as an optical potential for the atomic motion. The occurrence of quantized vibrational states for the external degrees of freedom was first demonstrated in the 1D case on the cesium $6S_{1/2}(F = 4) - 6P_{3/2}(F = 5)$ transition.[1] The occurrence of narrow lines and long damping times associated with the evolution of the external degrees of freedom was interpreted.[2] More recent experiments showed that similar results are obtained in the 2D and 3D cases.[3]

Several beam geometries were used to realize these optical lattices leading to different patterns for the location of cold atoms. We describe experiments performed with beams aligned along a regular or a stretched tetrahedron, which lead to optical lattices having different symmetries depending on the beams geometry. Two types of lattices were considered: the "ferromagnetic" lattices where all the occupied sites correspond to the same polarization of light and the "antiferromagnetic" lattices where the atoms have the same probability of occupying sites of opposites circular polarization. In this last case, we have studied the variation of the relative occupation probability with a static magnetic field and have showed how to measure a spin temperature in these lattices.

Finally, we describe a few problems (density, atom-atom interaction etc . . .) that are encountered in these lattices and we show that it is possible to overcome some of these difficulties by developing a new kind of lattices working on a transition connecting two levels for which there is a *dark state* in absence of motional and magnetic couplings.[4] Preliminary observations obtained on the cesium $6S_{1/2}(F = 3) - 6P_{3/2}(F = 2)$ transition shows that the atoms can be significantly colder in these "gray" lattices.

1. P. Verkerk *et al.*, Phys. Rev. Lett. **68**, 3861, 1992; P. Jessen *et al.*, Phys. Rev. Lett. **69**, 49 (1992).
2. J. Y. Courtois and G. Grynberg, Phys. Rev. A **46**, 7060 (1992).
3. G. Grynberg *et al.*, Phys. Rev. Lett. **70**, 2249, 1993; A. Hemmerich and T. W. Hansch, Phys. Rev. Lett. **70**, 410 (1993); A. Hemmerich *et al.*, Europhys. Lett. **22**, 89 (1993); P. Verkerk *et al.*, Europhys. Lett. **26**, 171 (1994).
4. G. Grynberg and J. Y. Courtois, Europhys. Lett. **27**, 41 (1994).

QThI2 — **3:00 pm**

Cascaded optical cavities with two-level atoms

J. Gripp, J.-Y. Jeong, S. L. Mielke, L. A. Orozco, *Department of Physics, State University of New York, Stony Brook, New York 11794-3800*

The present investigation is directed toward understanding the behavior of an open quantum optical system driven by light produced by another quantum optical system.[1-3] We want to study the response of a quantum system to nonclassical light. The field driving the second system has nonclassical fourth-order correlation functions.

The theoretical treatment of this class of problems has to go beyond the Markovian approximation.

We have experimentally investigated the steady-state characteristics of a system that consists of two single-mode optical cavities traversed by a collimated beam of optically pumped rubidium atoms. The rubidium behaves as a collection of two-level atoms. The two cavities are cascaded: a coherent source drives the first cavity, the output of which is coupled through a unidirectional device into the second. The characteristic parameters of the apparatus are in the intermediate-coupling regime of cavity quantum electrodynamics. The coupling constant of a single atom to the electromagnetic field is comparable to the decay rates of the polarization and the cavity.

We study the steady-state output of the cavities as a function of the system parameters to find regions of optical bistability. For certain parameters, double bistability can exist in the input–output characteristics of the system. To permit us to perform quantitative comparisons with experiments, our model includes the Gaussian transverse profile of the cavity mode, the homogeneous broadening of the two-level atoms, the mistunings of the excitation source with respect to the resonances of the cavities and the atoms, and the efficiency of the coupling into the second cavity.

For the first cavity we have measured the input and output intensities at the switching points of the bistable curve as a function of the atomic density. With the first cavity bistable, we have measured the input intensity necessary to switch the transmission of the second cavity. We have seen good quantitative agreement between our measurements and the theoretical predictions of the model.[4]

With the steady-state behavior understood, we are now proceeding to the characterization of the transmission of the coupled system at very low intensity. We want to look for modifications in the transmission due to the higher-order correlations present in the output of the first cavity.

This work is supported by the National Science Foundation.

1. H. J. Carmichael, Phys. Rev. Lett. **70**, 2273 (1993).
2. M. I. Kolobov and I. V. Sokolov, Opt. Spectrosc. **62**, 69 (1987).
3. C. W. Gardiner, Phys. Rev. Lett. **70**, 2269 (1993).
4. A. T. Rosenberger, L. A. Orozco, H. J. Kimble, and P. D. Drummond, Phys. Rev. A **43**, 6284 (1991).

QThI3 **3:15 pm**

Dynamics of nonlinear vacuum Rabi splitting in semiconductor microcavities

J.-K. Rhee, R. Lai, T. B. Norris, Y. Arakawa,* M. Nishioka,* *Center for Ultrafast Optical Science, University of Michigan, 2200 Bonisteel Boulevard, 1006 IST, Ann Arbor, Michigan 48109-2099*

Recent observations of vacuum-field Rabi splittings (VRS) in strongly coupled semiconductor microcavities[1–4] have generated considerable interest in the dynamics of microcavity exciton–polaritons. We have previously reported on the dynamics of time-domain vacuum Rabi oscillations in the low-excitation regime,[2] in which the VRS is independent of incident light intensity. As the light intensity is increased, however, we may expect two different nonlinear responses: First, higher-lying states of the Jaynes–Cummings ladder might be observable. Second, at high intensities phase-space filling and Coulomb screening of the exciton states result in a reduced oscillator strength that leads to a reduced VRS.

The $Al_xGa_{1-x}As$-based five-quantum-well microcavity sample used in this work is the same as that used for our previous publications, showing 6-meV VRS at 10 K in the low-excitation regime.[1] We first investigated nonlinear response in time-integrated reflectance spectra of the sample with impulsive on-resonant excitation. Figure 1 shows the excitation-dependent VRS and the corresponding normalized reflectance spectra at various excitation intensities. The reduced VRS at higher optical fluences is the manifestation of a reduced excitonic-transition oscillator strength in the microcavity. The VRS is given by[2]

$$\hbar\Omega = e\hbar\sqrt{\frac{N}{2m\varepsilon L_C}f_S}, \qquad (1)$$

where f_S is the areal excitonic-transition-oscillator strength per quantum well, N is the number of quantum wells, and L_C is the effective cavity length. The saturation of oscillator strength can be modeled as given in Ref. 5. Using these two models, we find a saturation optical fluence of 9×10^{11} photons/cm^2. The corresponding exciton saturation density is about 3×10^{10} excitons/cm^2, in good agreement with the theory (2.2×10^{10} excitons/cm^2).[5,6] The saturation optical fluence is reduced by factor of 4 with respect to the bare QW due to the cavity.

To study the dynamics of nonlinear VRS, we performed pump–probe experiments with on- and off-resonant excitation. The on-resonance-pump experimental results (Fig. 2) show the initial fast recovery of the reduced VRS. This fast recovery is due to coherent polariton radiation, so that most energy stored in the exciton–cavity system is coherently reradiated within the polari-

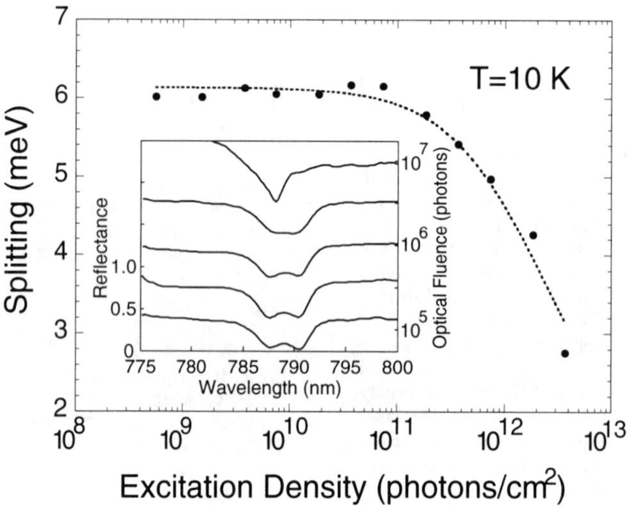

QThI3 Fig. 1. Normalized reflectance spectra at high optical fluences (inset) and excitation-intensity-dependent VRS, 70-fs laser pulses from a Ti:sapphire oscillator were tuned to the exciton–cavity resonance.

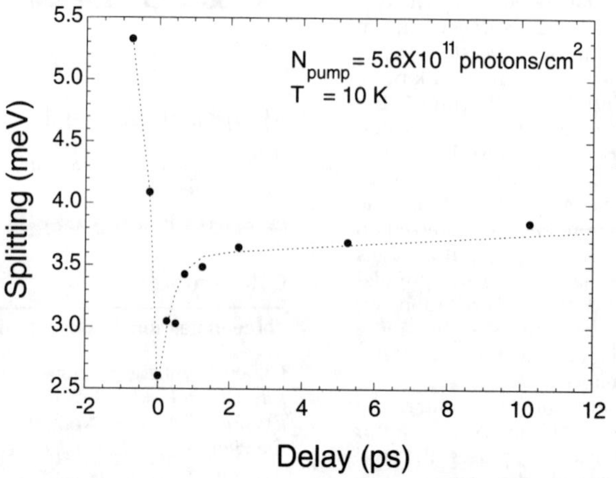

QThI3 Fig. 2. Time-resolved VRS with the on-resonance pump near zero time delay. VRSs are extracted from normalized reflectance spectra of the probe. The pump optical fluence was 5.6×10^{10} photons/cm^2. The dashed curve is a guide to the eye.

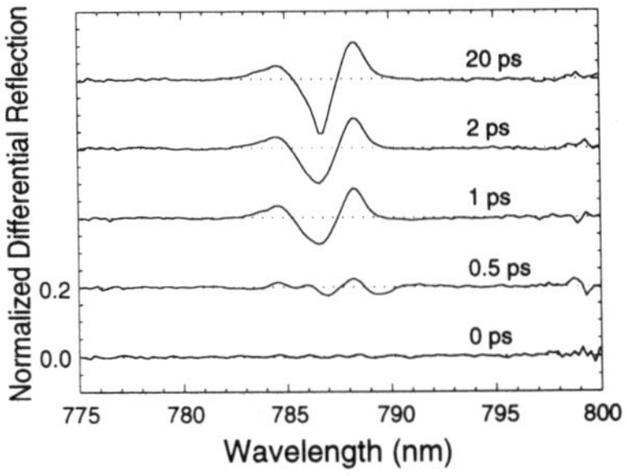

QThI3 Fig. 3. Time-resolved differential reflectance spectra with the off-resonant pump. Each curve is normalized with the incident probe spectrum. The probe was tuned to exciton–cavity resonance. The pump spectrum was selected at 750 nm with a 10-nm bandwidth from a white light continuum so as to excite the $Al_{0.2}Ga_{0.8}As$ barrier region. The excitation fluence was 2.7×10^{13} photons/cm^2. The sample temperature was 14 K.

ton decay time (260 fs).[2] At longer time delays, the VRS then recovers with a time constant of about 35 ps as the remaining incoherent exciton population decreases because of radiative decay and possibly because of ionization of excitons.

In the off-resonant-pump experiment (Fig. 3), the pump energy was selected far above the band gap, so that we excite EHP in both barriers and quantum wells. Free carriers excited in the quantum wells relax to the band edge within about 2 ps and bleach the exciton oscillator strength. Subsequently, carriers in the barrier region are captured in the quantum wells and slightly increase the degree of saturation in about 50 ps. The relaxation of the saturation in this case is now governed by energy relaxation and the carrier lifetime, unlike in the resonant-pump case. No evidence was observed in either resonant or off-resonant excitation of higher-lying states of the Jaynes-Cummings ladder.

The work at the University of Michigan was supported by the National Science Foundation through the Center for Ultrafast Optical Science under grant STCPHY 8920108.

*Institute of Industrial Science, University of Tokyo, 7-22-1 Roppongi, Minato-ku, Tokyo 153, Japan

1. C. Weisbuch, M. Nishioka, A. Ishikawa, and Y. Arakawa, Phys. Rev. Lett. **69**, 3314 (1992).
2. T. B. Norris *et al.*, Phys. Rev. B **50**, in press.
3. R. Houdré *et al.*, Phys. Rev. B **49**, 16761 (1994).
4. J. M. Jacobson *et al.*, in *International Quantum Electronics Conference*, 1994 Technical Digest Series, Vol. 9 (Optical Society of America, Washington, D.C., 1994), p. 50.
5. S. Schmitt-Rink, D. S. Chemla, and D. A. B. Miller, Phys. Rev. B **32**, 6601 (1985).
6. K.-H. Schlaad *et al.*, Phys. Rev. B **43**, 4268 (1991).

QThI4 **3:30 pm**

Strong-coupling-induced enhancement of luminescence-decay rate in a room-temperature semiconductor microlaser

R. Jin, M. S. Tobin, H. M. Gibbs,* G. Khitrova,* R. P. Leavitt, G. J. Simonis, D. Boggavarapu,** *Microphotonics Devices Branch, Army Research Laboratory, 2800 Powder Mill Road, Adelphi, Maryland 20783*

Modifications of spontaneous-emission properties of atoms in cavity quantum electrodynamics (QED) are generally classified into the weak-coupling regime, characterized by enhanced or inhibited spontaneous emission, and the strong-coupling regime, characterized by energy exchange between atoms and photons (e.g., vacuum Rabi oscillations). Progress in semiconductor microstructures has prompted great interest in studying similar effects in solid-state systems, with potential applications in microphotonics and optoelectronics.[1] Recently, evidence of both weak and strong cavity QED effects, such as spontaneous-emission modification and vacuum Rabi splitting, have been found in semiconductor microcavities[2] and microlasers.[3] Here we report a quantitative study of the luminescence decay rate of a GaAs/AlGaAs microlaser by time-correlated photon counting. Our result shows that the strong cavity–medium coupling can cause a significant enhancement of the luminescence-decay rate in a room-temperature semiconductor microlaser.

The high-reflectivity mirror of the microlaser structure has 22/28 quarter-wave periods in the front/back of a one-wavelength spacer layer. Each low-index layer consists of 73.56 nm of AlAs, and each high-index layer consists of 12 periods of 2.5-nm/2.5-nm $GaAs/Al_{0.3}Ga_{0.7}As$ superlattice. The spacer layer contains a 100-nm layer of GaAs between two 75.52-nm layers of $Al_{0.3}Ga_{0.7}As$. The structure is grown by molecular-beam epitaxy on a semi-insulating GaAs substrate. The wafer was not rotated during growth, giving sufficient variations in the cavity resonance positions relative to the luminescence peak of GaAs for studying the effect of coupling strength on the decay rate. For resonant-cavity microlasers, the active GaAs layer is located at the antinode of the laser field.

The luminescence lifetimes of the microlasers are measured by using a time-correlated photon-counting technique with carrier densities ranging from below transparency to above threshold. The pump source is a cavity-dumped picosecond styryl-9M dye laser synchronously pumped by a mode-locked, frequency-doubled Nd:YAG laser. The pump-pulse length is less than 10 ps at 810 nm, and the laser wavelength varies from 850 to 875 nm, with thresholds ranging from 30 to 60 pJ. The impulse response of the entire system is about 100 ps. The measurement is taken under the same experimental conditions for samples from the same wafer with substrates removed. Figure 1 shows a drastic difference in the measured luminescence lifetime depending on whether the microlaser cavity is on (870 nm) or off (850 nm) resonance with the GaAs luminescence peak (about 875 nm). The luminescence-decay rate is consistently faster in the resonant cavity microlaser, even at pump powers of only 2% of the threshold. Theoretical analysis indicates that the large enhancement of the decay rate is due to strong-coupling cavity-QED effects in the resonant cavity microlaser.

*Optical Sciences Center, University of Arizona, Tucson, Arizona 85721
**Department of Physics, University of Oregon, Eugene, Oregon 79403

1. E. Yablonovitch, Phys. Rev. Lett. **58**, 2059 (1987).
2. C. Weisbush *et al.*, Phys. Rev. Lett. **69**, 3314 (1992).
3. R. Jin *et al.*, Phys. Rev. A **49**, 4038 (1994).

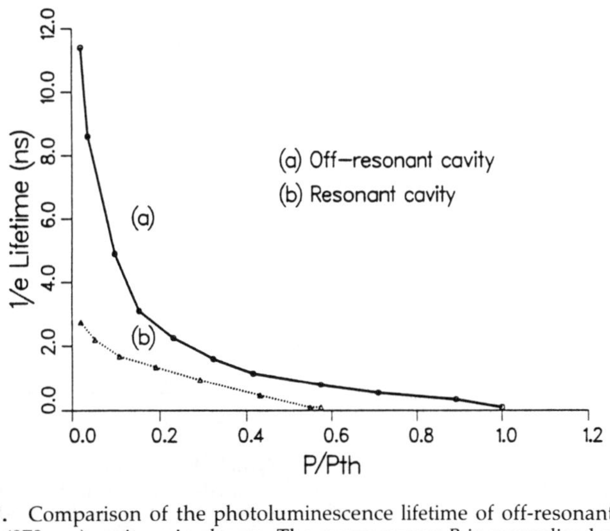

QThI4 Fig. 1. Comparison of the photoluminescence lifetime of off-resonant (850-nm) and resonant (870-nm) cavity microlasers. The pump power P is normalized with respect to the threshold power P_{th} of the off-resonant cavity microlaser. The normalized threshold of the resonant cavity microlaser is 0.574.

QThI5 **3:45 pm**

Coherent oscillations in semiconductor microcavities

Jagdeep Shah, Hailin Wang, T. C. Damen, W. Y. Jan, J. E. Cunningham, *AT&T Bell Laboratories, Holmdel, New Jersey 07733*

In this paper we present coherent nonlinear optical studies of a composite exciton–cavity system. By observing coherent oscillations in spectrally resolved transient four-wave mixing (FWM), we show, for the first time to our knowledge, the excitation of a coherent superposition of the two normal modes of the composite system, and we demonstrate directly in the time domain the coherent energy exchange between the exciton and the cavity.[1] Deep coherent oscillations with a period that varies with the normal-mode splitting persist even when detuning between the cavity and the exciton is much greater than the exciton–cavity coupling rate.

In general, there are two distinct dynamical regimes for the exciton–cavity interactions. In the perturbative (weak-coupling) regime, effects of the coherent coupling between the exciton and the cavity are negligible. Enhancement or suppression of spontaneous-emission rates is expected. In the nonperturbative (strong-coupling) regime, the coherent exciton–cavity coupling becomes dominant over dissipative decay processes. The persistence of the coherent oscillation when the cavity is detuned far from the exciton shows, surprisingly, that the dynamics of the nonlinear response are in the nonperturbative regime, whereas the dynamics of the linear response remain in the perturbative regime, leading to a unique situation in which the linear and the nonlinear responses remain in different dynamical regimes.

The microcavity consists of 15 and 20.5 stacks of $\lambda/4$ $Al_{0.11}Ga_{0.89}As/AlAs$ as the top and bottom mirrors, respectively. Two 150-Å GaAs quantum wells are placed at the center (antinode) of a λ cavity. Small variations in the cavity thickness allow us to tune the cavity resonance by moving the laser spot across the sample while the exciton energy remains unchanged. For FWM measurements (see the inset of Fig. 1 for geometry), 100-fs pulses from a Ti:sapphire laser are centered at the heavy-hole (hh)

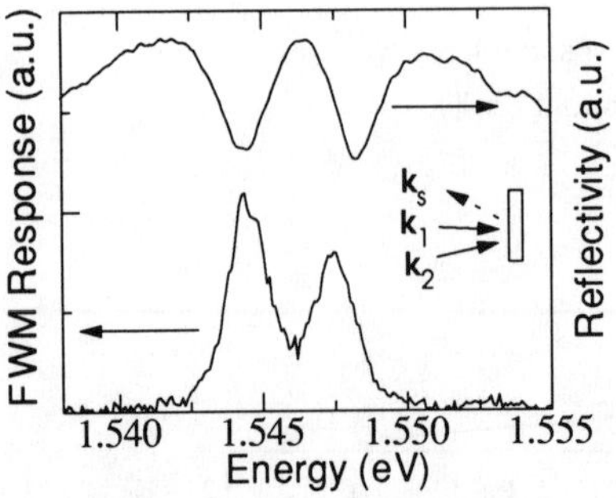

QThI5 Fig. 1. Spectra of both FWM emission (at zero delay) and the reflected laser pulse. The photon flux used for the FWM measurement is $10^{11}/cm^2$/pulse. The reflectivity is obtained at a much lower photon flux.

exciton energy. All measurements are performed at 10 K.

Figure 1 shows spectra of the reflected laser pulse and the FWM emission at zero delay when the cavity is nearly resonant with the hh exciton. Both spectra show a normal-mode splitting that disappears rapidly with increasing intensities because of saturation of the exciton transition. Figure 2 shows the dependence of FWM spectra on detunings between the cavity and exciton. Two resonances are observed, corresponding to the excitonlike and the cavitylike modes in the reflectivity spectrum (see inset).

Figure 3 shows the FWM response as a

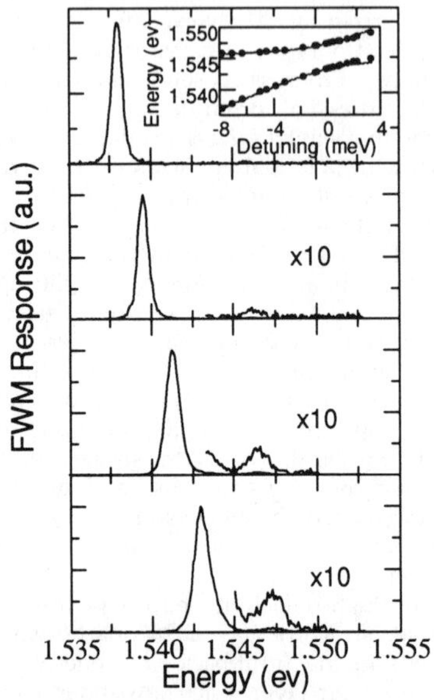

QThI5 Fig. 2. FWM spectra at zero delay with various exciton–cavity detunings. The inset shows the position of the normal modes in the reflectivity spectra as a function of the detuning. The fit (curves) gives an exciton–cavity coupling rate of 2.1 meV.

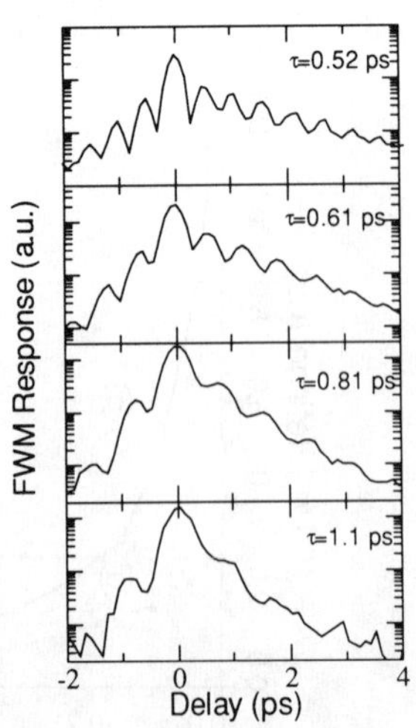

QThI5 Fig. 3. FWM signal as a function of the delay between the two incident pulses. The data are obtained at the peak of the corresponding cavitylike resonance in Fig. 2. τ is the period of the oscillation.

function of the delay between the two pulses. The data are obtained by first sending the FWM signal through a spectrometer centered at the energy of the corresponding cavitylike mode in Fig. 2. Periodic oscillations are observed in the transient response, and the period of the oscillation follows exactly the energy separation between the two normal modes.

Because we detected only one frequency component of the FWM signal, the oscillations in Fig. 3 are quantum beats, in contrast to the polarization interference observed in linear measurements. The oscillations are due to the excitation of a coherent superposition of the two normal modes and directly reflect the coherent energy exchange between the cavity and exciton.

Very deep oscillations in FWM persist when the cavity is detuned far from the exciton resonance. The persistence of the deep oscillation is very surprising because in this regime the coupling between the cavity and the exciton is very weak and the effects of the coherent coupling are negligible for linear measurements. The dynamics of the FWM response are still dominated by the coherent exciton–cavity coupling because the nonlinear response results from only the excitonic part of the normal-mode wave function. The extreme sensitivity of FWM to the coherent coupling leads to a unique situation in which the linear and the nonlinear responses remain in different dynamical regimes.

The decay dynamics of the normal modes and a theoretical description of the nonlinear response of the coupled system are also presented.[1]

1. H. Wang, J. Shah, T. C. Damen, W. Y. Jan, and J. E. Cunningham, submitted for publication.

QThJ — 2:30 pm — Hyatt C/D

Four-wave Mixing in Semiconductors

Rita Mahon, *U.S. Naval Research Laboratory, Presider*

QThJ1 (Invited) **2:30 pm**

Broadband four-wave mixing in semiconductor waveguides

H. Q. Le, J. P. Donnelly, S. H. Groves, E. A. Swanson, A. Darwish,* G. Lenz,* E. P. Ippen,* *Lincoln Laboratory, Massachusetts Institute of Technology, P.O. Box 73, Lexington, Massachusetts 02173-9108*

Four-wave mixing (FWM) near the semiconductor band edge is efficient owing to an enhancement of the third-order nonlinear susceptibility $\chi^{(3)}$. The enhancement is caused by the one-photon-resonance mechanism which can be stronger than the two-photon absorption mechanism that is dominant in the region far below the band edge; the mechanisms are illustrated in Fig. 1. For two frequency inputs ω_1 and ω_2, the conversion efficiency at $2\omega_1 - \omega_2$ or $2\omega_2 - \omega_1$ outputs is proportional to $|\chi^{(3)}|^2/(\alpha^2 + \beta_2^2\Delta\omega^4)f(L)$, where $\Delta\omega = \omega_1 - \omega_2$, α is linear loss, β_2 is the group-velocity dispersion, and $f(L)$ is a function of the device length L: $f(L) \approx 4e^{-\alpha L}\sin^2(\beta_2\Delta\omega^2 L/2) + (1 - e^{-\alpha L})^2$. For an optimal device length L, the maximum conversion efficiency for a given pump power is proportional to $|\chi^{(3)}|^2/(\alpha^2 + \beta_2^2\Delta\omega^4)$, defined as the figure of merit (f.o.m.). As α and β_2 are proportional to $\chi^{(1)}$, in quantum well structures with sharp and high density-of-state excitonic resonances, the enhancement of $\chi^{(3)}$ in certain range can be larger than that of $\chi^{(1)}$ by virtue of its higher

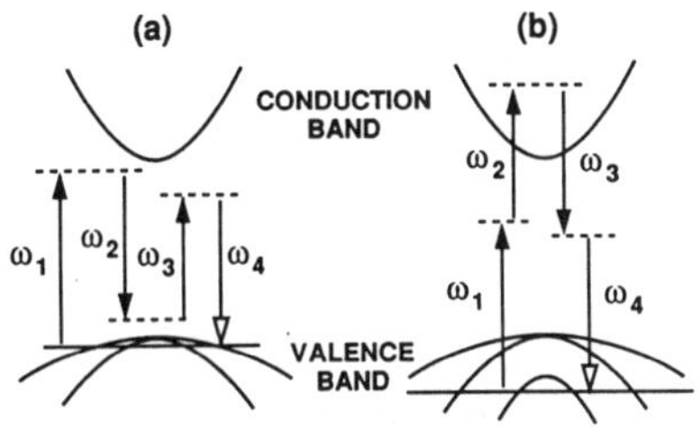

QThJ1 Fig. 1. Four-wave mixing processes via (a) one-photon resonance mechanism, and (b) two-photon resonance (absorption) mechanism.

order, resulting in a net enhancement of the f.o.m.

The $\chi^{(3)}$ enhancement has been demonstrated in GaAs/AlGaAs QW waveguide[1] at 0.8 μm and InGaAsP/InP QW waveguides at 1.5 μm.[2,3] Both picosecond pulse and cw studies produced similar results because the nonlinearity originates from ultrafast electronic polarization. Figure 2 illustrates the enhancement observed in a 1.2-mm long GaAs/AlGaAs QW waveguide. In Fig. 2(a), the diagonal components $\chi^{(3)}_{1111}$ and $\chi^{(3)}_{2222}$ in the region below the excitonic resonance are degenerate and assume the two-photon absorption $\chi^{(3)}$ value. Toward the band edge, both components increase in correlation with the two excitonic resonances (heavy hole and light hole), and at a rate faster than the corresponding linear absorption increases. The result on the f.o.m. of $\chi^{(3)}_{1111}$ is intrapolated and shown in Fig. 2(b) (dashed line), indicating a ~9-dB increase. The observed conversion efficiency was also shown (squares) exhibiting a ~13 dB enhancement.

Similar studies were conducted for InGaAsP/InP QW waveguides.[2,3] Here, QW design with high uniformity and high potential barrier was employed to achieve the needed sharp excitonic resonances. In the regime where the conversion efficiency is limited by dispersion rather than loss, a $\chi^{(3)}$ enhancement was also observed with a rate

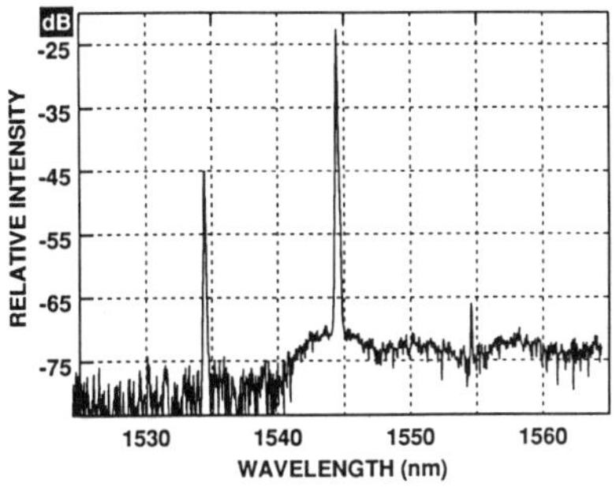

QThJ1 Fig. 3. A cw four-wave mixing spectrum in an InGaAsP/InP QW waveguide.

higher than that of β_2, resulting in a higher f.o.m. For a wavelength shift up to 20 nm, the conversion efficiency was ~ −22 dB with 60 mW cw pump[2] (Fig. 3) and −11 dB with picosecond pulses.[3] These conversion efficiencies and wavelength shifts are within the range for wavelength conversion devices.

This FWM approach to wavelength conversion has desirable properties for WDM networks that simultaneously require the capabilities of ultrafast, low latency, broadband, multi-channel, format independence, and high-dynamic range.

**Department of Electrical Engineering and Computer Science, Massachusetts Institute of Technology, Cambridge, Massachusetts 02139*

1. H. Q. Le and S. Di Cecca, Opt. Lett. **16**, 901 (1991); H. Q. Le and S. Di Cecca, IEEE Photon. Tech. Lett. **4**, 878 (1992).
2. J. P. Donnelly, H. Q. Le, E. A. Swanson, S. H. Groves, A. Darwish, and E. P. Ippen, Proceedings of LEOS Conference (IEEE, Piscataway, NJ), p. 330 (1994).
3. A. Darwish, G. Lenz, E. P. Ippen, H. Q. Le, J. P. Donnelly, S. H. Groves, and E. A. Swanson, CLEO '95.

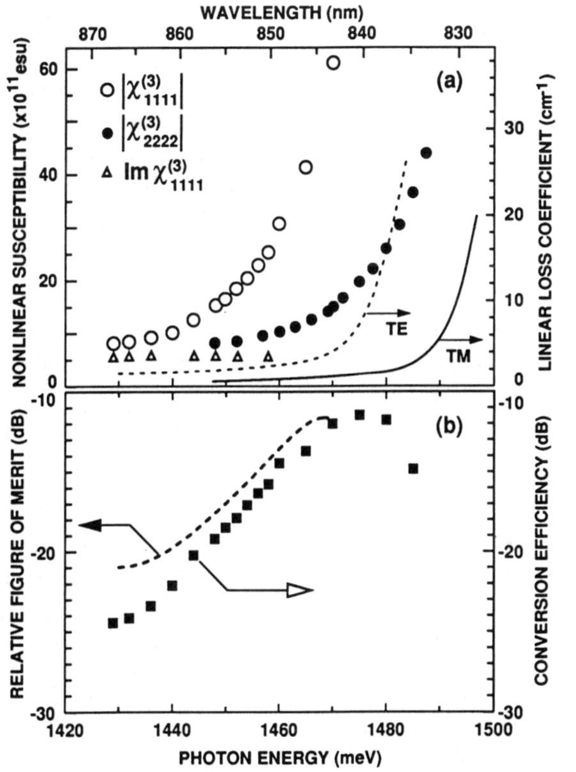

QThJ1 Fig. 2. (a) Magnitude and dispersion of the nonlinear susceptibility tensor components $|\chi^{(3)}_{1111}|$ and $|\chi^{(3)}_{2222}|$, measured for AlGaAs/GaAs QW waveguides,[1] exhibiting correlation with the heavy-hole (TE) and light-hole (TM) exciton resonances, respectively. Also shown is $\mathrm{Im}\chi^{(3)}_{1111}$ which is nearly constant, indicating that the increase in $\chi^{(3)}_{1111}$ is mostly real. (b): dash line shows the calculated figure of merit based on interpolated results in (a), and squares show the observed conversion efficiency with a constant pump power and wavelength detuning. The figure of merit is higher than the observed conversion efficiency at low photon energy indicates that the 1.2 mm device length employed was not optimal for this photon energy range.

QThJ2 **3:00 pm**

Frequency conversion in passive InGaAsP/InP waveguides

A. M. Darwish, G. Lenz, E. P. Ippen, H. Q. Le,* J. P. Donnelly,* S. H. Groves,* E. A. Swanson,* *Department of Electrical Engineering and Computer Science and Research Laboratory of Electronics, Massachusetts Institute of Technology, Cambridge, Massachusetts 02139*

Four-wave mixing (FWM) is an important technique for optical frequency conversion in WDM systems. It allows format-independent frequency conversion over large frequency separations[1,2] and can be used for dispersion compensation by conjugating the signal at midway.[3] Several recent reports [4,5] have described experiments performed in active semiconductor amplifiers. We have studied FWM at $\lambda = 1.5$ μm in *passive* waveguides. In passive waveguides[6] the nonlinearity relies on virtual transitions and does not suffer from limitations imposed by the carrier-modulation time constant or the carrier-heating time constant. Moreover, signal-to-noise ratio (SNR) can

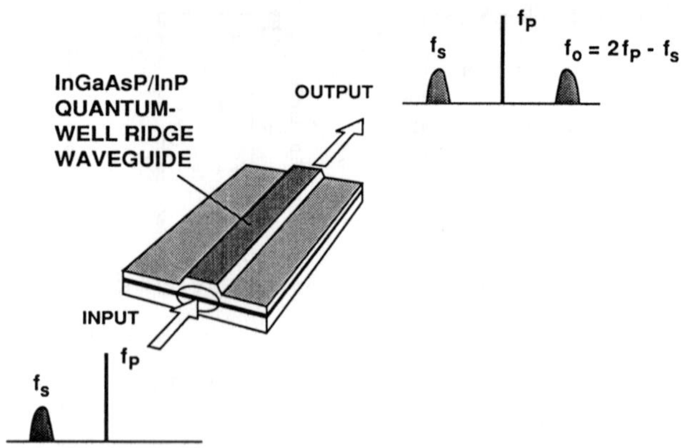

QThJ2 Fig. 1. Schematic illustration of wavelength conversion by means of FWM in a passive InGaAsP/InP single-quantum-well waveguide.

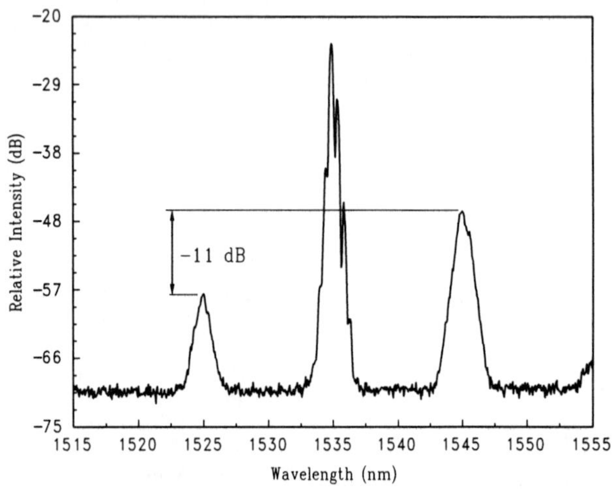

QThJ2 Fig. 2. Four-wave mixing with picosecond pulses.

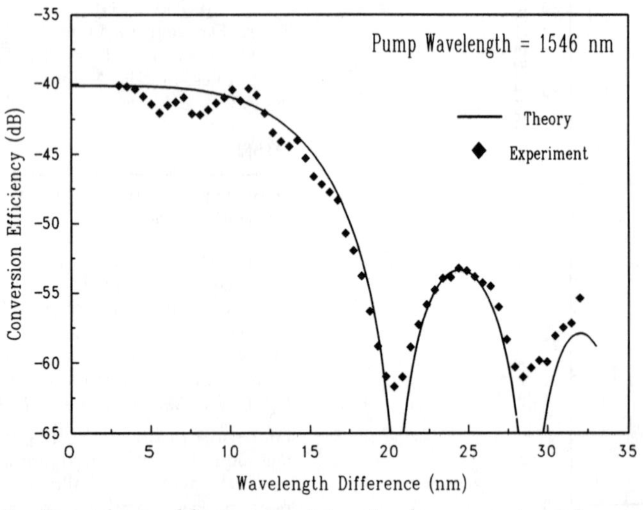

QThJ2 Fig. 3. Dependence of four-wave-mixing signal on pump-input-frequency separation. Phase-matching nulls determine the GVD coefficient. A GVD of 3200 ps^2/km is inferred from the curve.

be improved over that in active waveguides because of the absence of amplifier spontaneous emission.

The experimental setup is shown in Fig. 1. Two color-center lasers, KCl:Tl and NaCl:OH, are used for pump and probe. Both lasers are synchronously pumped by an actively-mode-locked Nd:YAG laser with a 100-MHz repetition rate. The KCl:Tl laser, tunable from 1460 to 1560 nm, produces 10-ps pulses. The NaCl:OH laser, tunable from 1530 to 1660 nm, produces 3–10-ps pulses. In a 7.5-mm-long passive InGaAsP/InP single-quantum-well waveguide with a band gap of QW ~ 1490 nm, a conversion efficiency of −11 dB was accomplished for a frequency conversion of 20 nm (~2.7 THz) with the pump, 10-ps KCl:Tl pulses, centered at 1535 nm and the input signal, 3-ps NaCl:OH pulses, at 1545 nm, as shown in Fig. 2. The imaginary part of $\chi^{(3)}$, the two-photon-absorption (TPA) coefficient β, imposes limitations on the conversion efficiency if the pump intensity approaches the waveguide saturation intensity, $I_{sat} = 1/\beta L$, where L is the waveguide length. Short pulses provide us with the high intensities required to investigate TPA effects. The tuning characteristic of the FWM signal was studied by tuning the input away from the pump wavelength in ~0.5-nm steps. With a cw pump centered at 1546 nm, the input was tuned from 1515 to 1543 nm. The resulting conversion efficiency, along with the theoretical fit, is shown in Fig. 3. Phase-matching nulls, due to group-velocity dispersion (GVD), are clearly evident. This provides a general measurement technique for GVD. For our case a GVD of 3200 ps^2/km is deduced. Because the nonlinearity does not have a characteristic time constant, in contrast to active waveguides, the main bandwidth-limiting mechanism is GVD, as Fig. 3 indicates. Hence, a flat response over large bandwidth should be observed in short devices. To demonstrate the broadband nature of the nonlinearity, we downconverted an input signal at 1507 nm to 1637 nm with a −47-dB output conversion efficiency (low efficiency is the result of phase mismatch).

In summary, we have studied nondegenerate FWM in passive waveguides at 1.5 μm. Our results suggest that a large conversion efficiency is achievable over a broad bandwidth. Bandwidth is mainly limited by GVD. We present measurements of $\chi^{(3)}$, real and imaginary parts as a function of wavelength, close to the band gap, and we show the resonant nonlinearity enhancement of n_2. In addition, we discuss TPA limitations on FWM-conversion efficiency.

**MIT Lincoln Laboratory, Lexington, Massachusetts 02173-9108*

1. M. C. Tatham, G. Sherlock, and L. D. Westbrook, IEEE Photon. Technol. Lett. **5**, 1303 (1993).
2. R. Schnabel, W. Pieper, M. Ehrhardt, M. Eiselt, and H. G. Weber, Electron. Lett. **29**, 2047 (1993).
3. R. M. Jopson, A. H. Gnauck, and R. M. Derosier, Electron. Lett. **29**, 576, (1993).
4. K. Kikuchi, M. Amano, C. E. Zah, and T. P. Lee, Appl. Phys. Lett. **64**, 548 (1994).
5. A. Uskov, J. Mork, J. Mark, M. C. Tatham, and G. Sherlock, Appl. Phys. Lett. **65**, 944 (1994).
6. J. P. Donnelly, H. Q. Le, E. A. Swanson, A. Darwish, and E. P. Ippen, in *Proc. 7th Annual Meeting of the Lasers and Electro-Optics Society* (Institute of Electrical and Electronics Engineers, New York, 1994), p. 330; also H. Q. Le and S. D. Cecca, IEEE Photon. Technol. Lett. **8**, 879 (1992).

QThJ3 3:15 pm

Physical and nonphysical divergences of semiconductor optical susceptibilities

C. Aversa, J. E. Sipe, M. Sheik-Bahae,* E. W. Van Stryland,** *Department of Physics and Ontario Laser and Lightwave Research Centre, University of Toronto, 60 St. George Street, Toronto, Ontario M5S 1A7, Canada*

Calculations of nonlinear-optical-response coefficients of semiconductors are often plagued by nonphysical divergences as the frequencies of the incident light approach zero. We present a formalism to derive, in the independent-particle approximation, expressions for the electronic contributions to semiconductor nonlinear-optical susceptibilities that do not suffer from these unrealistic divergences. Our approach is much simpler and more transparent than another recently described scheme.[1] In addition to practical advantages for obtaining explicit susceptibility expressions, this simplification allows us to more easily understand the origin of nonphysical divergences in model calculations and how they may be avoided.

We have applied our expressions to a two-band-model calculation of the nondegenerate nonlinear absorption and refraction described by $\chi^{(3)}(-\omega; \omega, \Omega, -\Omega)$. Our resulting dispersions correct previous results that suffered from certain nonphysical divergences. This allows us to unambiguously obtain the two-band-model predictions of the nonlinear refractive index and two-photon absorption ($\omega = \Omega$), the dc Kerr effect and third-order contributions to the Franz–Keldysh effect ($\Omega = 0$), and the below-band-gap virtual photoconductivity ($\omega = 0$). All of these processes are described by a finite $\chi^{(3)}$, and each predicts a finite $\chi^{(3)}(0; 0, 0, 0)$.

Finally, we show that under certain resonant conditions quite physical divergences can occur in the semiconductor susceptibilities. In particular, this allows us to understand how the coherent control of current, which has previously been understood by an interference between one- and two-photon absorption paths, can be described within the more conventional framework of nonlinear optics by realistic divergences that arise in $\chi^{(3)}(0; \omega, \omega, -2\omega)$. These realizations, and our well-behaved susceptibilities, allow us to consider other situations in which physical divergences occur, and we discuss a coherently controlled current effect that apparently has not been previously described.

**Department of Physics and Astronomy, University of New Mexico, Albuquerque, New Mexico 87131*

***Center for Research and Education in Optics and Lasers and Departments of Physics and Electrical Engineering, University of Central Florida, Orlando, Florida 32826*

1. J. E. Sipe and E. Ghahramani, Phys. Rev. B **48**, 10755 (1993).

QThK 4:30 pm
Hyatt A

Molecules in Strong Laser Fields

Frederick H. Mies, *National Institute of Standards and Technology, Presider*

QThK1 4:30 pm

Above-threshold dissociation in CO^{2+} at high frequencies

D. L. Hatten, J. Zhu,* W. T. Hill III,* Y. Cui,** J. Goldhar,** *Department of Physics, Box 167, Rose-Hulman Institute of Technology, 5500 Wabash Avenue, Terre Haute, Indiana 47803*

One of the more conspicuous nonlinear effects associated with atoms and molecules

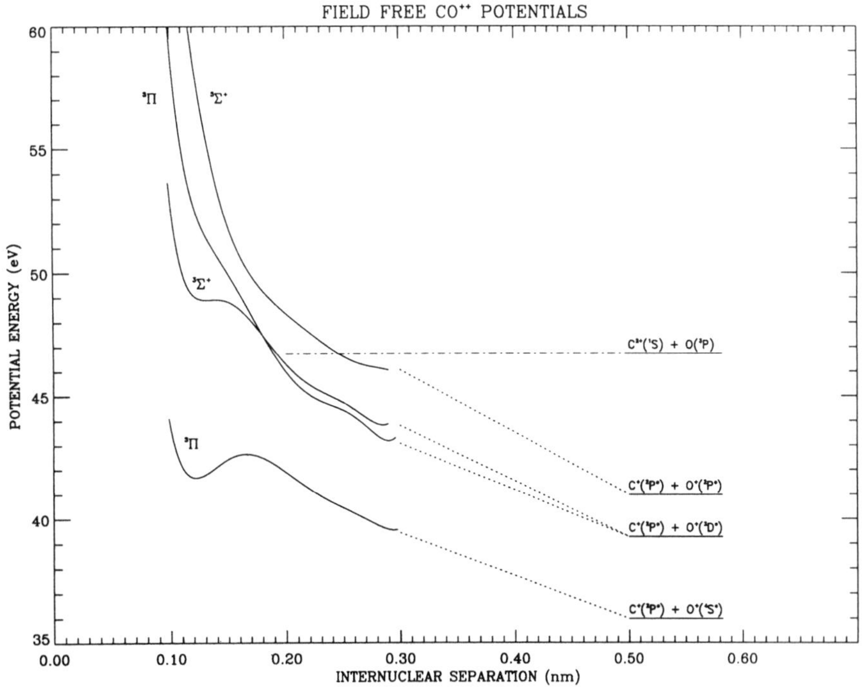

QThK1 Fig. 1. Selected diabatic CO^{2+} potential curves.

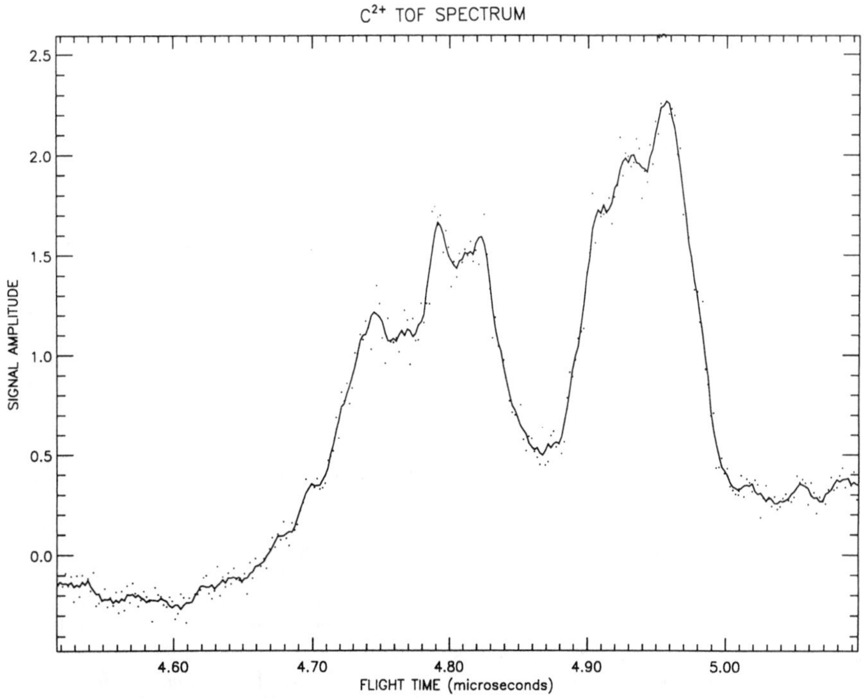

QThK1 Fig. 2. A portion of the CO dissociative ionization spectrum centered on the C^{2+} peaks. The peak that begins near 4.9 μs corresponds to ions initially directed away from the TOF detector.

in the presence of intense fields is excess photon absorption in the continuum. Excess photon absorption manifests itself in two very prominent ways, above-threshold ionization (ATI) and above-threshold dissociation (ATD). Similar to its electronic counterpart, ATD corresponds to the situation in which the diatomic system absorbs more photons than the minimum necessary to dissociate. This gives rise to an ion spectrum punctuated with peaks spaced by a fixed fraction of the photon energy, which depends on the relative mass of two departing fragments. Although ATI has been observed in a variety of atomic systems and even in molecular systems, ATD has previously been reported only in H_2^+. In this paper we describe what we believe to be the first observation of ATD in a system other than H_2^+, i.e., CO^{2+}. ATD is structure sensitive and, hence, wavelength sensitive. Although apparently not seen at lower frequencies, ATD is observable at high frequency, $\lambda = 193$ nm (6.4 eV). As is discussed, ATD is supported by a strong asymmetric dissociation channel, and by asymmetric we mean

$$AB^{(p+q)+} \rightarrow A^{p+} + B^{q+},$$

with $|p - q| \geq 2$.

Our experiments were performed at peak intensities ranging from 10^{14} to 10^{15} W/cm², and a Wiley–McLaren-type time-of-flight detector (TOF) was used to determine both the energy and the species of the ejected ions. A small sample (pressure of about 5×10^{-8} Torr) of CO was driven to the $^3\Pi$ metastable state of CO^{2+} by seven 193-nm photons. By this time, the molecular structure of CO^{2+} is fully dressed by the intense field, opening multiple dissociation channels. Specifically, there is strong radiative coupling between the ground $^3\Pi$ state and the $^3\Sigma^+$ state as well as the lowest excited $^3\Pi$ state. The excited $^3\Pi$ and $^3\Sigma^+$ states of CO^{2+} are perturbed by potentials that asymptotically approach the $C^{2+} + O$ threshold. Consequently, as shown in Fig. 1, by crossing of the diabatic curves near 46.5 eV, the equivalent adiabatic curves will lead to one- and two-photon asymmetric dissociation, ATD. The large peak centered at 4.80 µs and the step centered at 4.76 µs in Fig. 2 are the ATD peaks in the C^{2+} spectrum.

**Institute for Physical Science and Technology, University of Maryland, College Park, Maryland 20742*

** *Department of Electrical Engineering, University of Maryland, College Park, Maryland 20742*

QThK2 (Invited) 4:45 pm

Stabilization of molecules in strong laser fields

D. Normand, M. Schmidt, C. Cornaggia, *C.E.A.—DSM/DRECAM/SPAM, Bât. 422, Centre d'Etudes de Saclay, 91191 Gif-sur-Yvette Cedex, France*

Diatomic molecules in strong laser fields get aligned along the field polarization and stabilized in laser-induced states. The stabilization appears robust against variation of field-parameters. The transient species explode at the end of the interaction.

The multielectron dissociative ionization (MEDI) of molecules with strong laser fields in excess of 10^{15} W/cm² have been studied. The molecules explode in energetic ions which are emitted along the laser polarization axis. To understand the behavior of molecules in such intense fields, we have analyzed the fragment kinetic energy releases E_{exp}. The conclusions of our experimental study are the following:

(1) Fragmentation proceeds only via few channels with characteristic E_{exp} values depending on the molecule. These energies do not significantly depend on the laser parameters (wavelength, peak intensity and pulse duration).

(2) The E_{exp} values correspond to the explosion of the molecule from an effective internuclear distance R_{eff} larger than the equilibrium internuclear separation of the neutral molecule R_e. We find the empirical law, $R_{eff} = 2.3\sqrt{R_e}$, where R_e and R_{eff} are expressed in angstroms [2]. This law applies for all molecules investigated so far (H_2, N_2, CO, O_2, CO_2, C_2H_2, Cl_2, I_2, etc.) and for all observed fragmentation channels (e.g., up to Cl_2^{10+}).[1]

(3) In MEDI experiments performed with 2 ps laser pulses, the subsequent ionization of the fragments by the laser, called post-dissociation ionization (PDI), is not observed. The absence of PDI strongly suggests that the fragments are formed at the end of the pulse.

(4) We performed a control experiment where the 2-ps laser pulse is replaced by two 130-fs laser pulses delayed by 1.5 ps. Having a relative intensity of 1:4, the two 130-fs pulses are inscribed in the 2-ps pulse envelope, simulating a 2-ps pulse with a ⟨⟨hole⟩⟩ in he leading edge. Although the peak intensity of the double-pulse is the same as that of the 2-ps pulse, we now clearly observe that the second pulse post-ionizes the fragments created by the first one.[2] This experiment proves that if the fragments were formed in rising edge of the 2-ps laser pulse, the PDI probability would be high enough to be detected. The absence of PDI signals in the experiments performed with picosecond Gaussian pulses implies that the fragmentation cannot occur during the pulse. This confirms the hypothesis of molecular stabilization.

Our results contradict the intuitive model in which the MEDI interaction is supposed to occur in the leading edge of the pulse, giving rise to competition betweeen ionization and dissociation of the molecules. We propose an alternative model in which valence electrons are strongly coupled to the laser field giving rise to light-induced bound states of the transient molecule species. In this approach, the field forces the electrons to oscillate on an internuclear scale, which compensates for the Coulomb repulsion between the ions. The molecules are stabilized at an effective internuclear separation, characteristic of the molecule, and explode into fragments only at the end of the laser pulse below some threshold intensity.

1. M. Schmidt, D. Normand, and C. Cornaggia, Phys. Rev. **A50**, 5037, 1994.
2. D. Normand, M. Schmidt, and C. Cornaggia, to be published.

QThK3 5:15 pm

Critical distance for explosive ionization of diatomic molecules

M. Y. Ivanov, T. Seideman, P. B. Corkum, *Steacie Institute for Molecular Sciences, National Research Council of Canada, M-23A, Montreal Road, Ottawa, Ontario K1A 0R6, Canada*

The present work focuses on the role of nuclear coordinates in intense-field ionization of diatomic molecules and their ions. The major conclusion of our study is that the intense-field electron dynamics in a double-well potential of a diatomic molecule (molecular ion) are qualitatively different from the electron dynamics in a single-well potential of an atom. This qualitative difference leads to a dramatic quantitative consequence. Namely, there is a critical window of internuclear distances in which the ionization rates for a diatomic molecular ion are many (e.g., 4) orders of magnitude higher than those at large internuclear distances, $R \rightarrow \infty$, and those near the equilibrium distance R_{eq}).

At $R \approx R_{eq}$ the two wells of the molecular potential almost merge into one deep well and ionization is very much like atomic ionization. When R increases, the internal barrier between the two wells rises [Fig. 1(a)] and electron tunneling between the wells becomes difficult. In intense fields tunneling is suppressed even more strongly. As a result, even at $R \sim 3$–4 Å, the electron gets localized in one of the potential wells for a time longer than the laser cycle. Let us assume that the electron is localized in the left well. During the laser cycle the laser field biases the total potential $V(x, t) = V_{mol}(x) - xE \cos \omega t$ by moving the left well up and the right well down [Fig. 1(b)], and the electron rises with the ground state of the left well. (In contrast, if the electron is not localized, it adjusts adiabatically to the changing potential and slips into the ground state of the right well.)

The most efficient region of ionization corresponds to internuclear separations for which the top of the external barrier becomes suppressed as compared to the internal barrier between the wells [typically, $R \sim 3$–5 Å; Fig. 1(b)]. The electron localized in the left well can now very easily tunnel through the thin and low internal barrier directly to the continuum [Fig. 1(b)]. Tunneling from the ground state of the right well—a process very much like conventional intense-field ionization of atoms—is much more difficult [Fig. 1(b)].

As the internuclear separation increases further, the internal barrier between the wells becomes wider and higher, and it becomes more difficult to penetrate. As a result, the ionization rate starts to decrease rapidly and becomes atomiclike at $R \rightarrow \infty$.

Figure 2 shows some results of our calculations for various charged states of a model molecule M_2. The total ionization probability is shown for an optical pulse with duration of 30 fs, a characteristic time scale for a nuclear motion in heavy diatomics (such as I_2). First, there is a pronounced peak in the ionization probability around $R \approx 4$–5 Å. Second, the ionization rate of a triply charged molecular ion M_2^{3+} in this region is higher than the ionization rate of

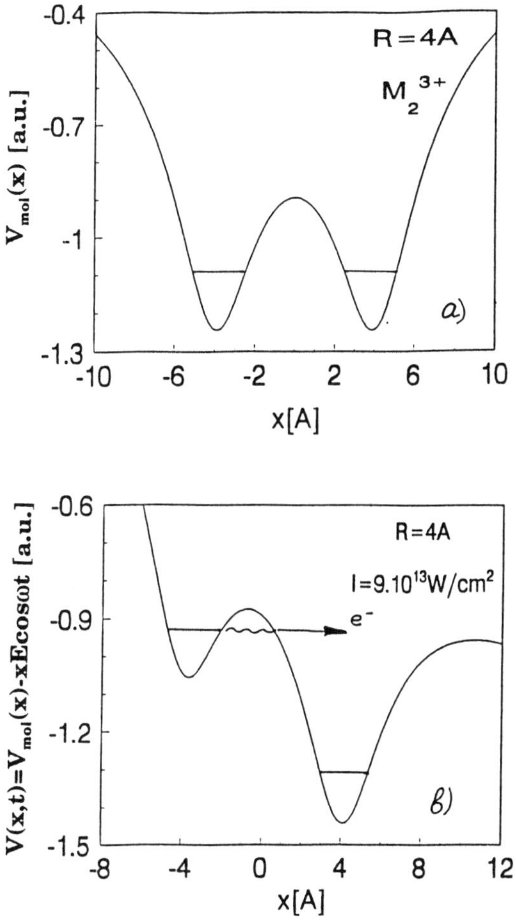

QThK3 Fig. 1. Pictorial scheme for the ionization of a diatomic molecule. (a) Field-free potential for electron motion in a model diatomic molecule; internuclear distance, $R = 4$ Å. (b) Full potential molecule plus laser field; $V(x, t) = V_{mol}(x) - xE \cos \omega t$ at $\cos \omega t = 1$.

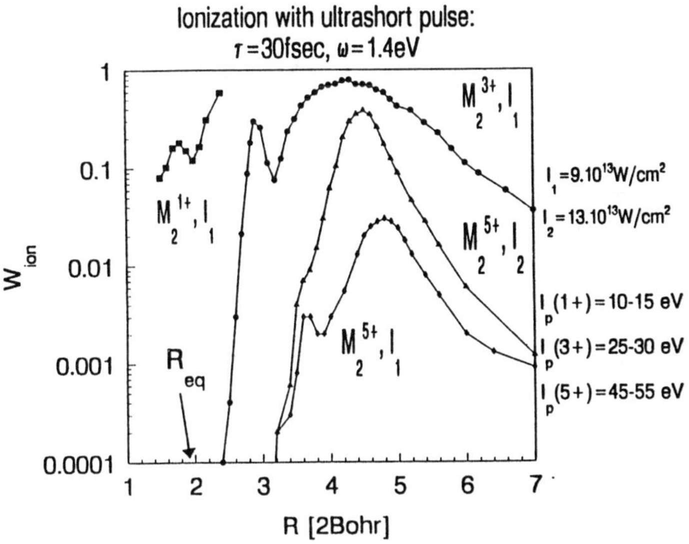

QThK3 Fig. 2. Ionization of variously charged model molecular ions by an ultrashort optical pulse.

a stable molecule M_2^{1+} at $R \approx R_{eq}$, and the ionization of M_2^{5+} in the critical region is almost as efficient as the ionization of M_2^{1+} at R_{eq}. This is in spite the ionization potentials of M_2^{3+} and M_2^{5+} being much higher than that of M_2^{1+} at $R \approx R_{eq}$.

These results suggest the following scenario of multielectron ionization of a diatomic molecule with picosecond optical pulses. First, several electrons are stripped until the molecule reaches the first dissociating charged state (M_2^{2+} or M_2^{3+}). The molecule starts to dissociate and enters the critical region of internuclear distances, where ionization is highly efficient. The intensity needed to reach the dissociating state at R_{eq} is sufficient to rapidly ionize several more electrons in the critical region, until a highly charged state is reached. At this point the time τ required for the nuclei to pass the critical region is too short to permit significant ionization to the next charged state, $\Gamma_{ioniz}\tau \ll 1$. The nuclei fly apart, with the kinetic energy given by the same fraction of the Coulomb energy at R_{eq}, independent of the charged state and determined only by R_{eq}/R_{crit}. Not only does this scenario solve the experimental puzzle raised in Ref. 1, but also our results open new opportunities in molecular science, which are discussed.

1. M. Schmidt, D. Normand, and C. Cornaggia, "Laser-induced trapping of chlorine molecules with pico- and femtosecond pulses," submitted to Phys. Rev. A.

QThK4 **5:30 pm**

Dissociation, ionization, and Coulomb explosion of H_2^+ with short, intense laser pulses

André D. Bandrauk, S. Chelkowski, Osman Atabek,* *Faculty of Sciences, University of Sherbrooke, Sherbrooke, Quebec J1K 2R1, Canada*

Numerical solutions of the time-dependent Schrödinger equation have been obtained by using a split-exponential-operator method[1] for H_2^+ in intense, short laser pulses and taking into account dissociation and ionization simultaneously. The exact time-dependent wave function thus obtained allows one to examine both electronic and proton evolution during and after the pulse. Competition between ionization and dissociation are shown to occur in view of the short time scale for the proton motion, $\tau_p \sim 10$ fs, which is comparable to the ionization time scale at intensities $I \geq 10^{14}$ W/cm^2. At lower intensities, suppression of dissociation has been predicted to occur for specific initial vibrational levels and intensities[2] due to laser-induced avoided crossings between electronic states. The effect of ionization on such laser-induced stabilization of the nuclear states is examined in detail.

At intensities $I \geq 10^{14}$ W/cm^2, ionization is accompanied by Coulomb explosion of the protons. The kinetic-energy distribution of the protons has the signature of the laser-induced avoided-crossing mechanism. Because present models of high-order harmonic generation invoke a recollision mechanism between ionizing electrons and nuclei, a search for the influence of Coulomb explosion on harmonic generation has been made. Illustrations of time-dependent proton and electronic wave packets are used to assess the coupling between particles and the influence of such couplings on the highly nonlinear response of a H_2^+ system to external intense laser pulses.

**Laboratory of Molecular Photophysics, University of Paris-South, Orsay, Cedex 91405, France*

1. Phys. Rev. A **46**, R5342 (1992).
2. A. D. Bandrauk, *Molecules in Laser Fields* (Dekker, New York, 1993).

QThK5 **5:45 pm**

Dissociation dynamics of H_2^+ and D_2^+ in a strong 10-μm laser field

F. A. Ilkov, S. Turgeon, T. D. G. Walsh, S. L. Chin, *Center for Optics, Photonics, and Lasers, Department of Physics, University of Laval, Québec, Québec G1K 7P4, Canada*

In the spectral region from the UV to the near infrared, the processes of multiphoton ionization and dissociation of H_2 (D_2) have been thoroughly studied and are well understood.[1] The situation is quite different in the mid-infrared spectral range. Experiments were performed to study the behavior of H_2 and D_2 in the presence of an intense 10.6-μm (CO_2) laser field. Both molecular and atomic ions were observed, with the latter possessing some initial value of kinetic energy. Ion yields and kinetic-energy dependences as functions of the laser intensity were measured.

By considering the ion-yield results (Figs. 1 and 2), it is clear that H_2 dissociates faster than does D_2. For example, Fig. 1 shows that at the lowest intensity for which a signal can be observed, the atomic-ion yield is approximately an order of magnitude higher than the molecular-ion yield, indicating that even at a relatively low intensity (i.e., the appearance intensity of H_2^+) most of the molecular ions dissociate. Referring to Fig. 2, we see that this is not the case for deuterium. The D_2 molecule, which is twice as heavy, dissociates much more slowly than its lighter counterpart (i.e., H_2) with the atomic-ion signal overtaking the molecular-ion signal only after some increase (above the appearance intensity for D_2^+) in the laser intensity.

Figure 3 shows that the dissociative tunnel ionization of H_2 (D_2) results in the creation of protons (deuterons) with a kinetic energy that rises from 0.37 eV (0.32 eV) to 0.76 eV (0.6 eV) as the laser intensity increases from 5×10^{13} to 1×10^{14} W/cm^2.

The consideration and analysis of these experimental results have led us to regard the dissociative tunnel ionization as a three-step process: (1) tunnel ionization of the neutral molecule ($A_2 \rightarrow A_2^+$), (2) dissociation of the molecular ion ($A_2^+ \rightarrow A + A^+$), and (3) associative tunnel ionization of the neutral fragment ($A \rightarrow A^+$). The first of these steps has previously been discussed.[2] The dissociation of the molecular ion (step 2) has been considered within the context of the barrier-suppression and chaotic dissociation theories. The former model, proposed by Dietrich and Corkum,[3] provides a good prediction for the dissociation threshold intensity. However, it cannot adequately explain the experimentally observed kinetic-energy dependence. Chaotic dissociation theory,[4] on the other hand, correctly predicts the intensity threshold for dissociation as well as negligible fragment kinetic energy. This then allows for the ever-present laser field to further tunnel-ionize the dissociating molecular ion ($A \rightarrow A^+$), resulting in a Coulomb repulsion of the two atomic ions and producing the observed kinetic energy. Numerical simulations have been performed and are in good agreement with the experimental results.

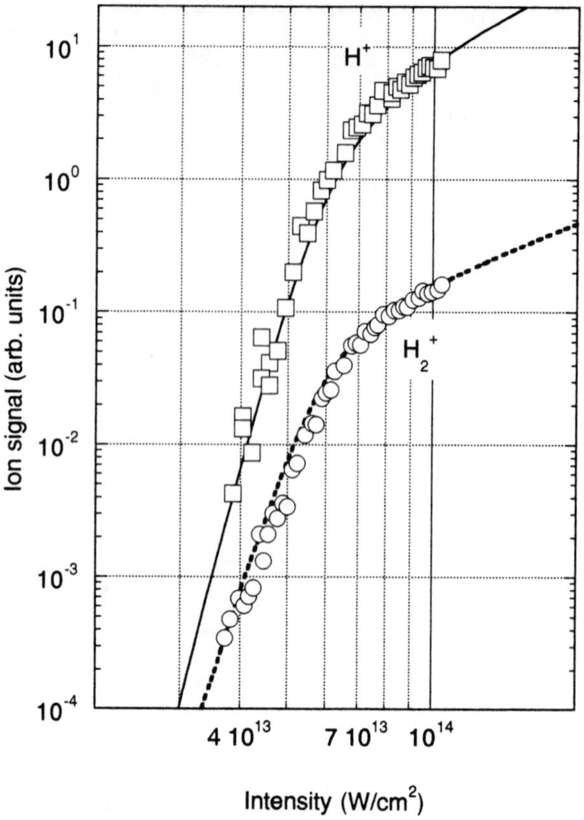

QThK5 **Fig. 1.** Ion yield versus CO_2 laser intensity for both molecular (H_2^+) and atomic (H^+) ions of hydrogen.

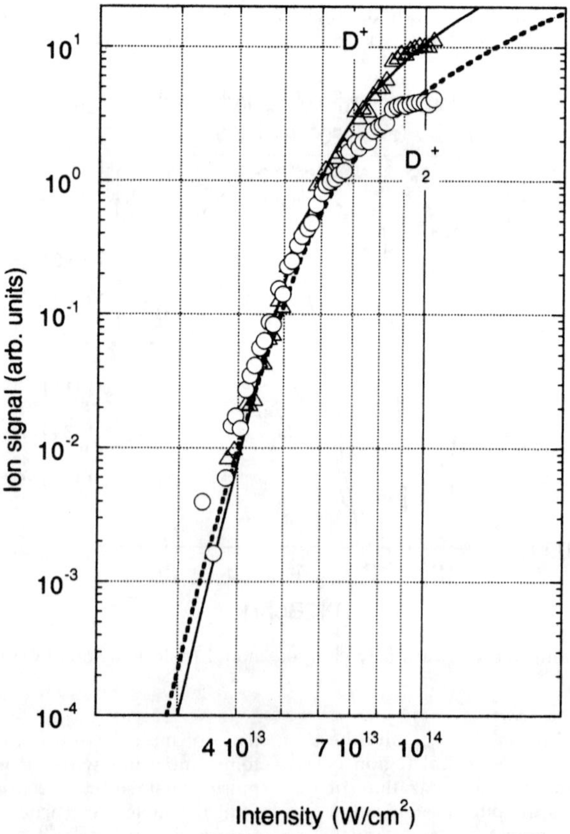

QThK5 **Fig. 2.** Ion yield versus CO_2 laser intensity for both molecular (D_2^+) and atomic (D^+) ions of deuterium.

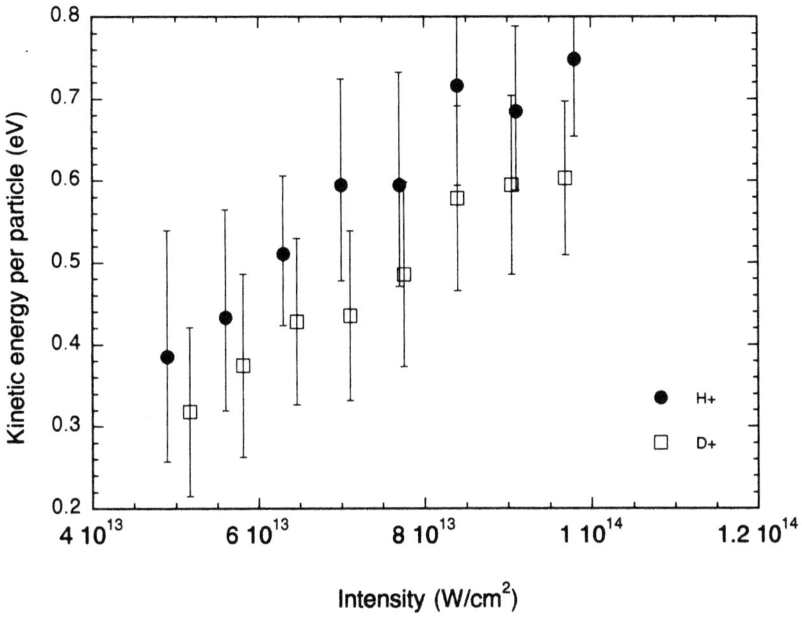

QThK5 Fig. 3. Kinetic energy of dissociation (per proton, deuteron) as a function of the incident CO_2-laser intensity.

1. A. D. Bandruk and S. C. Wallace, eds. *Coherence Phenomena in Atoms and Molecules in Laser Fields*, Vol. 287 of NATO ASI Series B: Physics (Plenum, New York, 1992).
2. S. L. Chin *et al.*, Laser Physics **3**, 298 (1993).
3. P. Dietrich and P. B. Corkum, J. Chem. Phys. **97**, 3187 (1992).
4. M. E. Goggin and P. W. Milonni, Phys. Rev. A **37**, 796 (1988).

QThL — 4:30 pm, Hyatt B

Semiconductor Nanostructures and Microcavities

Klaus Streubel, *Photonics Microwave Engineering, Presider*

QThL1 (Invited) — **4:30 pm**

Quantum nanostructures formed by self-ordering on patterned nonplanar substrates

Eli Kapon, *Institute of Micro and Optoelectronics, Department of Physics, Swiss Federal Institute of Technology, 1015 Lausanne, Switzerland*

Self-ordering of compound semiconductor nanostructures is an attractive approach towards the realization of quantum structures. The self-ordering process obviates the need for lithography in defining the nanometer-sized structures and can in principle improve their uniformity. These attributes are particularly important in the context of low-dimensional quantum structures, e.g., quantum wires (QWRs) and quantum dots (QDs), for which dimensions in the 10-nm range and uniformity of better than 10% are desirable in order to achieve clear quantum confinement effects. Several approaches utilizing spontaneous self-ordering of nanostructures have been proposed and studied. In that case, however, the random formation sites of the wires or dots prevents the precise registration of arrays of such quantum structures. Such arrays are particularly useful in applications of such low-dimensional structures in optoelectronics.

This talk will review recent progress in the realization of quantum well (QW) and QWR heterostructures formed via self-ordering on patterned, nonplanar substrates.[1-3] The approach relies on the self-limiting surface profiles which evolves during organometallic chemical vapor deposition of GaAs/AlGaAs heterostructures on corrugated substrates. For corrugations oriented along the direction on a (100) substrate, extremely sharp corners are obtained between AlGaAs quasi-(111)A planes. Deposition of thin (a few nm) GaAs layers in these grooves then yield crescent-shaped QWRs of effective widths in the 10-nm range. Similarly, AlGaAs vertical QW structures are formed at the center of these grooves due to segregation of group III atoms. The size and shape of these self-ordered QW and QWR structures depend only on the growth conditions and the layer composition, and thus can be closely controlled. Furthermore, the grooves act as seeds determining the sites of the self-ordered wells and wires, hence it is possible to achieve ordered arrays of these nanostructures. Detailed growth experiments and modeling of the self-ordered phase, which helps to reduce size and composition fluctuations.

The QWRs exhibit absorption spectra revealing a quasi-one-dimensional (1-D) subband structure in agreement with model calculations; subband separations up to 50 meV have been observed. Strong polarization anisotropy is found in absorption spectra obtained with light polarized parallel or perpendicular to the wire axis. Time resolved luminescence spectra evidence fast carrier capture into the QWRs via the self-ordered QWs which grow connected to the wires. Diode lasers incorporating the self-ordered QWs and QWRs operate at room temperature. The vertical-QW the well. The QWR lasers show gain spectra which are peaked at the quasi-1-D subbands, and operate with sub-mA threshold currents.

1. E. Kapon, M. C. Tamargo, and D. M. Hwang, Appl. Phys. Lett. **50**, 347 (1987).
2. E. Kapon, D. M. Hwang, and R. Bhat, Phys. Rev. Lett. **63**, 430 (1989).
3. E. Kapon, "Lateral Patterning of Quantum Well Heterostructures by Growth on Nonplanar Substrates," *Epitaxial Microstructures*, A. C. Gossard, Ed., Semiconductors and Semimetals, Vol. 40 (Academic, Boston, 1994), p. 259.

QThL2 — **5:00 pm**

Lasing in self-ordered vertical-quantum-well heterostructures

E. Kapon, B. Dwir, H. Pier, A. Gustafsson, J. M. Bonnard, D. M. Hwang,* E. Colas,* *Institute of Micro and Optoelectronics, Department of Physics, Swiss Federal Institute of Technology, 1015 Lausanne, Switzerland*

Quantum-well (QW) semiconductor lasers exhibit interesting and useful properties owing to the quantum confinement in their active regions. Traditionally, these structures are obtained by epitaxial growth of thin (<20-nm) semiconductor films lying in the substrate plane. The QW potential distribution is thus achieved by controlling the thickness and composition of the grown layers. Here, we describe the lasing properties of a new type of QW, formed by segregation of an AlGaAs alloy grown on a nonplanar substrate. The vertical orientation of this self-ordered QW structure leads to lasing with linear polarization oriented normal to the substrate plane.

The vertical AlGaAs vertical-QW (VQW) structures are grown by organometallic chemical-vapor deposition in $[01\bar{1}]$-oriented V-grooves etched on (100) n^+-GaAs substrates.[1-3] The active layer consists of a 0.3-μm-thick undoped $Al_{0.3}Ga_{0.7}As$ layer sandwiched between $Al_{0.7}Ga_{0.3}As$ p and n layers. Growth in the groove results in formation of an extremely narrow (less than 20-nm-wide) region of AlGaAs lying in the (011) plane, in which the Al mole fraction is lower than that on both lateral sides. This lateral variation in Al mole fraction is clearly visible in transmission electron micrographs, showing a contrast between different $Al_xGa_{1-x}As$ compositions (Fig. 1), as well as in energy dispersive x-ray-emission spectra of the VQW region.

Room-temperature cross-sectional cathodoluminescence (CL) spectra of the VQW laser structure allow the identification of the different effective band gaps around the VQW (Fig. 2). On both sides of the V-groove, the planar active layer emits at 660 nm, which corresponds to an Al mole fraction of $x = 0.36$. Excitation of the VQW

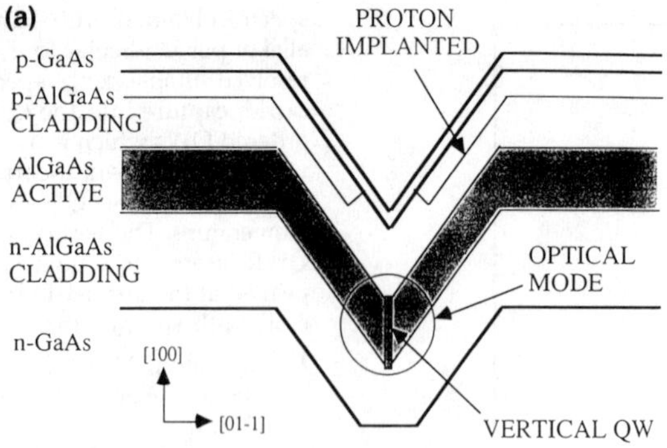

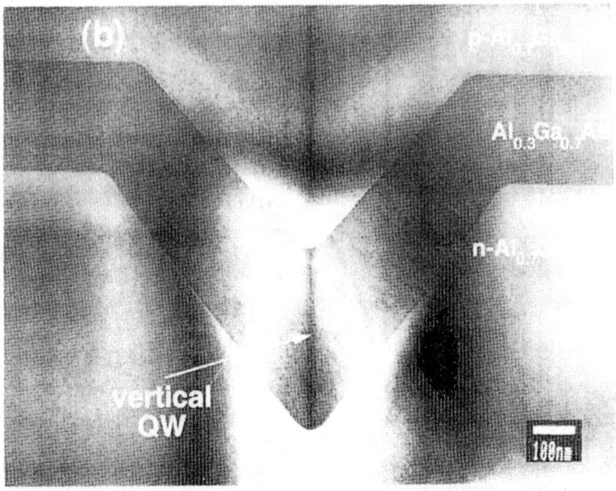

QThL2 Fig. 1. Vertical QW laser structure: (a) schematic cross section; (b) TEM cross section.

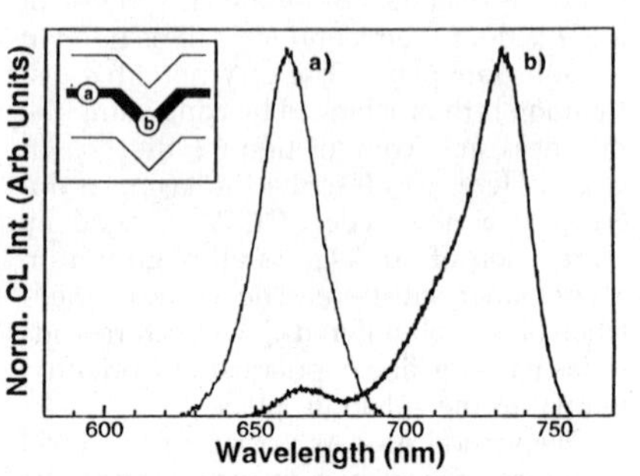

QThL2 Fig. 2. Spot-mode room-temperature cathodoluminescence spectra of the vertical QW laser of Fig. 1 at two positions: (a) AlGaAs layer; (b) groove center.

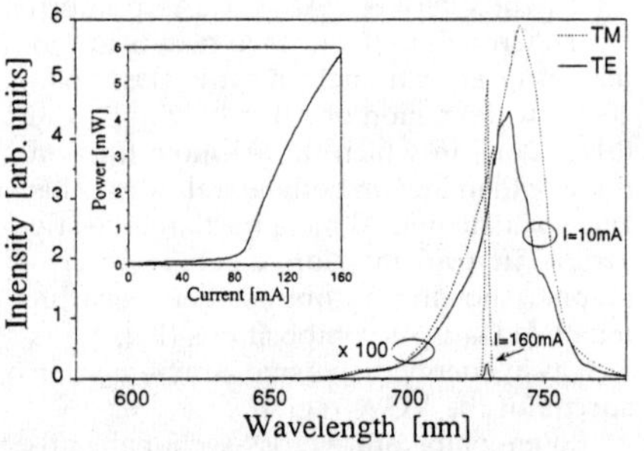

QThL2 Fig. 3. Polarized amplified-spontaneous-emission and lasing spectra of the vertical QW laser of Fig. 1 for analyzer oriented in the [011] (TE) and [100] (TM) directions. Inset shows the light-versus-current characteristic; I_{th} = 87 mA.

region gives rise to a stronger line at 730 nm and a weaker one at 665 nm, corresponding to recombination in the AlGaAs near the VQW (equivalent $x = 0.35$) and to recombination at the VQW (equivalent $x = 0.22$), respectively. The relatively weak emission line from the background AlGaAs ($x = 0.35$) active layer may be partly due to efficient carrier capture into the VQW region.

The p-side contact of the VQW laser is fabricated by using proton implantation to selectively excite the VQW region (see Fig. 1). Lasers with a cavity length of $L = 1$ mm operate at room temperature (pulsed) with a threshold current of 87 mA. The near-field pattern consists of a ≈1-μm-diameter circular spot, centered on the VQW region. Polarized emission spectra of the VQW laser, measured below and above threshold, are shown in Fig. 3. Below threshold, spectra similar to the CL ones are obtained, indicating efficient electroluminescence (EL) from the VQW. Recombination takes place mainly at the VQW, although some emission from the background AlGaAs composition is also visible. Above threshold, lasing occurs in the VQW only, at 727 nm; the blue shift relative to the EL peak is probably due to band filling. The beam is polarized (particularly above threshold) in the [100] direction (TM-like), consistent with the polarization-selection rules for electron–heavy-hole transitions in a QW.

The VQW configuration is particularly useful for realizing QW lasers with extremely narrow (<<1 μm) QW active regions, because the VQW width is determined with high accuracy by the thickness of the AlGaAs layer. The width of the VQW in our case is only 400 nm. This design flexibility is important for minimizing the threshold current[4] and for the realization of AlGaAs quantum-wire lasers.[3]

* *Bellcore, Red Bank, New Jersey 07701*

1. M. Walther *et al.*, Appl. Phys. Lett. **60**, 521 (1992).
2. E. Kapon *et al.*, Superlattices and Microstructures **12**, 491 (1992).
3. G. Vermeire *et al.*, J. Cryst. Growth **124**, 513 (1992).
4. E. Kapon, Opt. Lett. **15**, 801 (1990).

QThL3 **5:15 pm**

Microcavity surface-emitting laser with a 133-μA threshold current

K. L. Lear, K. D. Choquette, R. P. Schneider, Jr., S. P. Kilcoyne, K. M. Geib, *Department of Photonics Research 1312, Sandia National Laboratories, MS 0603, P.O. Box 5800, Albuquerque, New Mexico 87185-0603*

The prospect of thresholdless lasing is tantalizing from the aspects of both complete modal confinement and low-threshold laser applications. Vertical-cavity surface-emitting lasers afford straightforward longitudinal mode control through single-wave cavities and can be defined with small areas for large lateral mode spacings as well. Here we report on studies of a microcavity surface-emitting laser with a 2 × 2-μm² area and, to our knowledge, the lowest room-temperature current threshold yet reported for any diode laser.

The laser has been realized by producing a lateral index step and current aperture due to conversion of AlGaAs to an oxide by selective lateral oxidation.[1] The small lateral dimensions of the unoxidized cavity result in a calculated in-plane mode spacing of 70 nm in the region of the 980-nm lasing wavelength for the triple-InGaAs-quantum-well structure. Although the in-plane cavity has a low *Q*-factor due to the moderate reflectivity of the oxide/AlGaAs interface, proper lateral mode positioning could contribute to a partial suppression of in-plane amplified spontaneous emission. The oxide structure also serves as an intracavity step-index lens for focusing the vertical mode.

Subthreshold (100-μA) emission spectra are shown in Fig. 1 for both horizontal di-

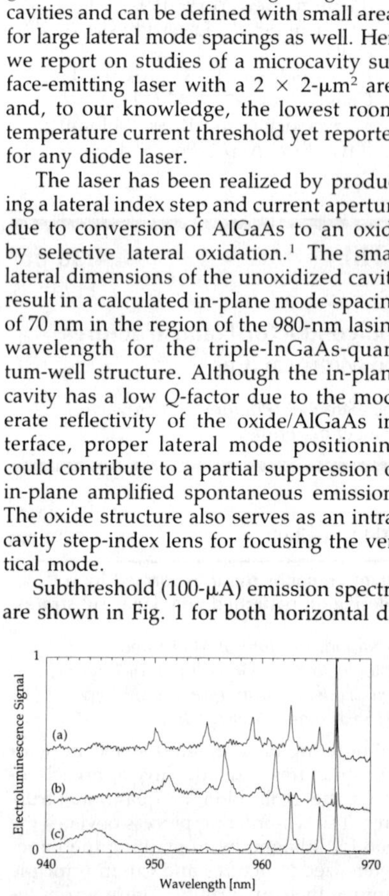

QThL3 Fig. 1. Subthreshold (100-μA) emission spectra for a microcavity VCSEL in the two orthogonal horizontal directions (a, b) and in the vertical direction (c).

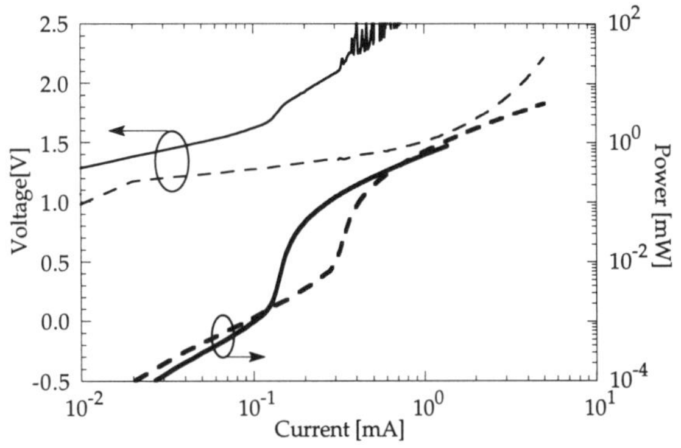

QThL3 Fig. 2. Light–current (heavy curves) and voltage–current (light curves) dependencies for a 2 × 2-μm^2 (solid curves) and 7 × 7-μm^2 (dashed curves) laser.

rections and the vertical direction. A strong, widely spaced modal pattern is evident in all directions at wavelengths shorter than 967 nm, at which lasing in the fundamental mode occurs. The multiple peaks in the vertical emission correspond to various higher-order modes. The average transverse mode spacing is approximately 2 nm, with larger-area lasers showing much denser spectra. The horizontal-emission spectra also contains several peaks with much higher finesse and smaller spacing than expected for the small low-Q lateral cavity. Some of these lines correspond to scattered vertical emission, but other strong lines appear where there is no vertical-emission peak. Details of three-dimensional spectral measurements and associated modal analysis are presented.

The light–current dependence is presented in a log-log plot in Fig. 2 for the 2 × 2-μm^2 laser as well as for a larger 7 × 7-μm^2 one. The two lasers have approximately equivalent offsets in the intercepts between the spontaneous and stimulated emission lines, indicating comparable beta-factors despite the difference in size. The voltage curve for the small laser does show an unexpected increase in voltage at threshold, which is under further investigation. The room-temperature cw threshold current for the small laser is 133 μA, smaller than the previously reported lowest threshold current of 165 μA for a high-reflectivity-coated 125-μm-long edge-emitting laser.[2] The VCSEL reported here has a calculated cavity lifetime of 0.7 ps, compared to the 20 ps for the edge-emitting laser. The long cavity lifetime contributes to large internal-absorption losses, making this edge emitter's 0.4% external quantum efficiency smaller than the 53% external efficiency of the VCSEL by a factor of more than 100. The combination of short cavity lifetime and low threshold currents also make this new microcavity laser very promising for future dynamical studies.

1. K. D. Choquette *et al.*, in *International Semiconductor Laser Conference*, 1994, postdeadline paper 4; submitted to Electron. Lett.
2. T. R. Chen *et al.*, in *International Semiconductor Laser Conference*, 1994, paper 26.

QThL4 **5:30 pm**

Lasing modes in semiconductor Fabry–Perot microcavities with intracavity dielectric spheres

K. E. Meissner, P. L. Gourley, T. M. Brennan, B. E. Hammons, *Sandia National Laboratories, M/S 0350, Albuquerque, New Mexico 87185*

Microcavity lasers are important both for the study of basic optical physics and as new sources of coherent light. Here, we present a new microcavity with a fundamental geometry consisting of a dielectric sphere within a Fabry–Perot resonator. In this case the Fabry–Perot resonator provides longitudinal confinement and the dielectric sphere provides lateral confinement of the optical field. We study the lasing modes (spectral and spatial) produced in this idealized hybrid microcavity resonator with the intent of extending the study to the case of arbitrarily shaped objects for device application.

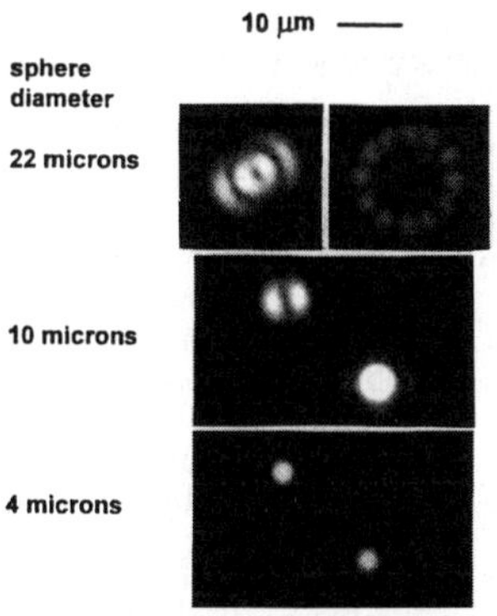

QThL4 Fig. 1. Images of lasing modes for various sizes of polystyrene spheres in the microcavity.

The resonator consists of an MBE-grown mirror and gain region on a GaAs wafer. This is basically an open-cavity vertical-cavity surface-emitting laser (VCSEL). The polystyrene sphere is placed on the wafer and is covered with a glass dielectric mirror that acts as an output coupler. The semiconductor gain region is excited with 750-nm laser light. The polystyrene sphere provides lateral optical confinement, increases the cavity Q, and thus encourages lasing action.

When polystyrene spheres of various sizes are placed in the microcavity and the gain region pumped with cw laser light, the structure begins to lase above a threshold. The onset of lasing is easily recognized because of a narrowing of the emitted light. Lasing modes from a collection of spheres are shown in Fig. 1. As expected from simple diffraction theory, the larger spheres are capable of supporting more higher-order modes than are the smaller spheres. These higher-order modes include whispering-gallery modes. However, under concentric excitation, the lower-order modes are found to be prevalent in all sizes of spheres in which lasing action occurred. Figure 2 shows the lasing threshold versus sphere size. The curve indicates that the structure has a threshold minimum around 6 μm. Below this, the threshold increases dramatically until the spheres under 4 μm in diameter no longer show lasing action. This is most likely due to larger diffraction losses that are incurred as the spheres become smaller. To increase the available gain region and prove the mode distribution, we excited the semiconductor with femtosecond pulses. This permitted more than one mode to lase at a time. Figure 3 shows the resulting spectrum from such an experiment on a 6-μm sphere. The two lowest-

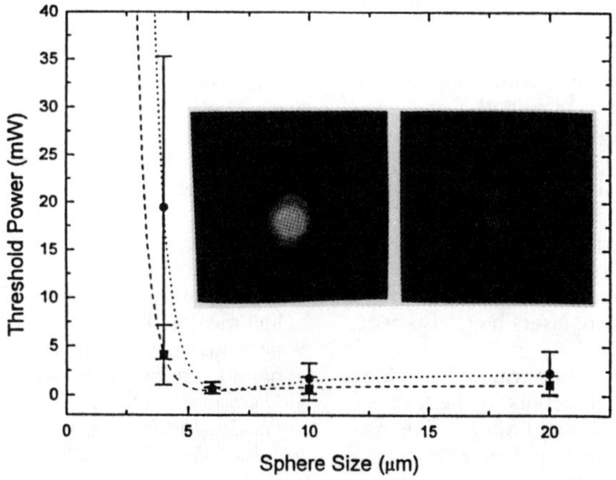

QThL4 Fig. 2. Threshold power for the TEM_{00} (squares with dashed curve) and the TEM_{01} mode (circles with dotted curve). The cavity was pumped with a 25-μm-FWHM spot. Inset shows images of the TEM_{00} mode (left) and the TEM_{01} mode (right).

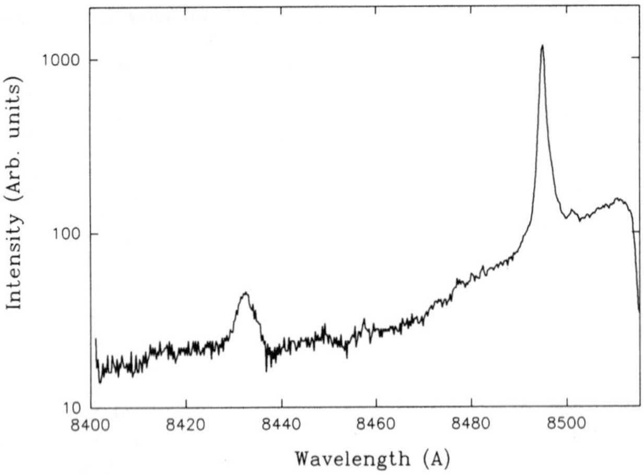

QThL4 Fig. 3. Spectrum from 6-μm polystyrene sphere pumped with 100-fs, 750-nm laser light in a 25-μm-FWHM spot. The lowest-order mode is on the right.

order modes are oscillating in the cavity and are clearly seen in the spectrum. From the separation of the modes, properties of the sphere can be ascertained.

We have observed lasing in polystyrene microspheres contained within a semiconductor microcavity laser under both cw and femtosecond excitation. Lower-order modes were found to dominate, and the structure demonstrated a preferred sphere diameter. Additionally, we were able to observe higher-order modes, including whispering-gallery modes in larger spheres. Development of this method may provide a useful way to probe the structure of small (~10-μm), arbitrarily shaped particles.

QThL5 **5:45 pm**

Polymer microdisk and microcylinder lasers

M. Kuwata-Gonokami,* A. Dodabalapur, R. H. Jordan, H. E. Katz, M. L. Schilling, R. E. Slusher, S. Ozawa,** *AT&T Bell Laboratories, Murray Hill, New Jersey 07974*

Optical microresonators are of interest for low-threshold microlasers and LEDs and for studies of quantum-electrodynamic modification of spontaneous emission. As the optical mode volume of a microresonator approaches a cubic wavelength, the number of modes interacting with the optically active spectral region in the microresonator approaches unity. This can result in a single lasing mode within the spectral gain width and in very strong coupling between the optical field and the gain medium. Very-high-Q optical microresonators with strong optical confinement to the gain region can be obtained by using the whispering-gallery modes located around the edge of a disk or cylinder. Microlasers based on these resonators have previously been demonstrated in semiconductor systems.[1,2] Polymer-microsphere lasers have also been demonstrated.[3]

We have fabricated optical microcavities by using polymer materials in the form of microdisks and microcylinders. The microdisks are patterned in polymer films spun on silicon or glass substrates. The gain material is typically a rhodamine or pyrromethene dye dissolved in a polymer host, typically polystyrene. The optical fields are confined to the gain layer by dielectric layers of poly(vinly alcohol) that are spun on above and below the polystyrene gain layer. These encapsulating layers have a dielectric constant lower than that of the polystyrene and effectively confine the optic fields to the gain layer. The layer thicknesses are typically a few hundred nanometers. Pulsed optical pumping in the green resulted in lasing near 570 nm in microdisks with diameters near 20 μm and with a pyrromethene-dye-doped gain layer. The lasing mode spacings are consistent with whispering-gallery modes around the edge of the disks.

Polymer-microcylinder optical resonators were formed by dipping small-diameter optical fibers in a dye-doped polymer solution and allowing the resulting polymer droplets to cure. As the droplets cure, cylindrical resonators self-assemble around the glass core by surface-tension and adhesion effects. Similar to the microdisk, the cylindrical-resonator optical fields are confined to the polymer layer by dielectric contrast of the higher refractive index of the polymer relative to that of the glass core. These cylindrical resonators potentially have very high Q values because the surfaces that define the resonant modes at the cylinder edge can be extremely smooth. These resonators, for various dye-doped polymers, also lased in the orange and red portions of the spectrum when they were pumped in the green by a pulsed dye laser. The lasing-mode spacing is complex for fibers with diameters greater than 20 μm, but it consists of only a few evenly spaced modes for fibers with diameters in the 10-μm range. These modes also can be identified as whispering-gallery modes at the outer surface of the cylindrical polymer beads. Many interesting phenomena, including output coupling of the lasing modes, lifetimes of the lasing-dye molecules, and new fabrication and self-assembly techniques, are being studied.

Permanent address: Department of Applied Physics, Faculty of Engineering, University of Tokyo 7-3-1, Hongo, Bunkyo-ku, Tokyo 113, Japan

***Department of Applied Physics, University of Tokyo, 7-3-1 Hongo, Bunkyo-ku, Tokyo 113, Japan*

1. S. L. McCall, A. F. J. Levi, R. E. Slusher, S. J. Pearton, and R. A. Logan, Appl. Phys. Lett. **60**, 289 (1992).
2. A. F. J. Levi, R. E. Slusher, S. L. McCall, T. Tanbun-Ek, D. L. Coblentz, and S. J. Pearton, Electron. Lett. **28**, 1010 (1992).
3. M. Kuwata-Gonokami, K. Yakeda, H. Yasuda, and K. Ema, Jpn. J. Appl. Phys. Lett. **31**, 99 (1992).

QFA 8:00 am

Hyatt B

Molecular Dynamics

Robert R. Alfano, *City University of New York, Presider*

QFA1 8:00 am

Coherent properties of large- and small-size J-aggregates of pseudoisocyanine

Kazuhiko Misawa, Takayoshi Kobayashi, *Department of Physics, University of Tokyo, 7-3-1 Hongo, Bunkyo-ku, Tokyo 113, Japan*

J-aggregates of cyanine dyes are attracting the interest of many scientists because of large second-order[1] and third-order[2,3] optical nonlinearities. The excitonic transition of the J-aggregates is characterized by a sharp absorption band, called the J-band, below the monomer transition. The optical nonlinearity is expected to be enhanced because of the delocalized exciton.

Recently, we were successful in making oriented J-aggregates in a polymer film by a novel method called vertical spin coating. In the present paper we describe the determination of the absorption spectra of coherent mesoaggregates and fluid dynamically oriented macroaggregates by this method.

An aqueous solution of pseudoisocyanine bromide (PIC-Br, Japanese Research Institute for Photosensitizing Dyes) and polyvinyl alcohol (PVA, Kanto Chemicals) was spun on a substrate attached to the shaft of an electric motor along the radial direction of rotation by centrifugal force. This vertical-spin-coating method is described in detail elsewhere.[4]

Figure 1 shows the dichroic absorption spectra with several dye amounts in 80 mg of PVA. At a weight ratio of 9 mg/80 mg (PIC-Br/PVA), a large dichroic ratio of 5–10 is observed, suggesting the formation of highly oriented J-aggregates. As the weight ratio is decreased, the dichroic ratio gradually decreases. At weight ratios of 7/80, 6/80, and 4/80, small aggregates randomly oriented are simultaneously obtained in a sample film, in addition to large J-aggregates well oriented along one axis.

The remarkable result in the present paper is that J-aggregates become isotropically dispersed around a ratio of 3/80. At a ratio lower than 2.80, the J-band disappears and only the monomer spectrum can be observed. From the concentration dependence of the dichroism, an oriented film of J-aggregates is concluded to be composed of the small-size coherent aggregates dispersed isotropically in the sample and large-scale rod-shaped crystallites formed by nucleation. The former, called mesoaggregate, is characterized by coherently delocalized excitation over the mesoscopic size, and the latter, called macroaggregate, can be oriented by macroscopic fluid dynamical centrifugal force during the vertical-spin-coating process. This macroaggregate is considered to be an ensemble of inhomogeneously sized distributed mesoaggregates.

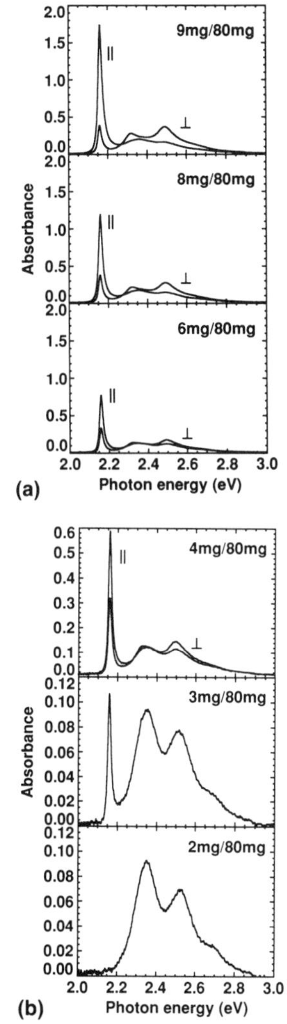

QFA1 Fig. 1. Dichroic absorption spectra with 9, 8, 6, 4, 3, and 2 mg of PIC in 80 mg of PVA.

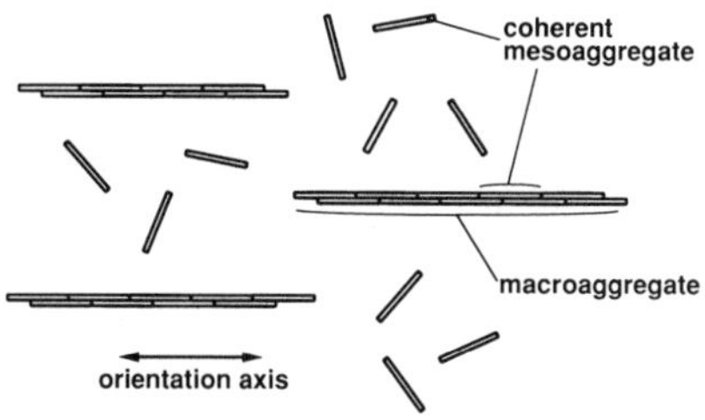

QFA1 Fig. 2. Coherent mesoaggregates and oriented macroaggregates. The former is characterized by coherent excitation over the aggregate, and the latter is an inhomogeneous ensemble of mesoaggregates.

Comparing the absorption spectrum of 4/80 with that of 3/80, we conclude that mesoaggregates start to form a macroaggregate around these weight ratios. Figure 2 shows a sketch of the two types of aggregates.

Finally, we separated the dichroic absorption spectra of the oriented J-aggregates into the spectra of the isolated mesoaggregates and macroaggregates (Fig. 3). The former band is sharp and nearly isotropic, whereas the latter has broad width

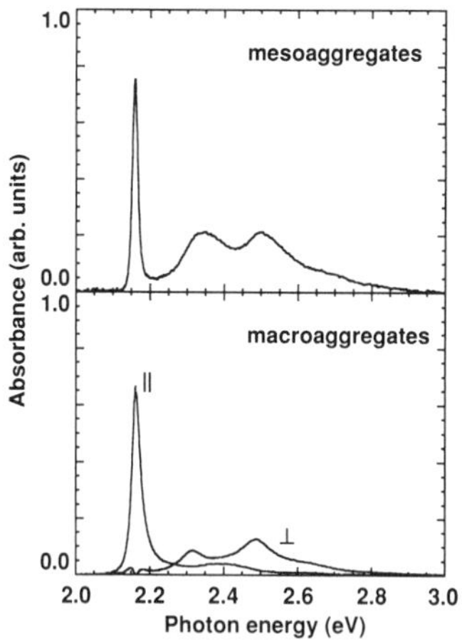

QFA1 Fig. 3. Absorption spectra of isolated coherent mesoaggregates and oriented macroaggregates.

with a higher energy shoulder and is strongly dichroic. The J-band for the perpendicular polarization is due to the mesoaggregates, whereas the J-band for the parallel polarization is a superposition of those of the mesoaggregates and oriented rod-shaped macroaggregates.

1. Y. Wang, Chem. Phys. Lett. **126**, 209 (1986).
2. Y. Wang, J. Opt. Soc. Am. B **8**, 981 (1991).
3. K. Misawa, K. Minoshima, H. Ono, and T. Kobayashi, Chem. Phys. Lett. **220**, 251 (1993).
4. K. Misawa, H. Ono, K. Minoshima, and T. Kobayashi, Appl. Phys. Lett. **63**, 577 (1993).

QFA2 8:15 am

Time-resolved fluorescence of aggregated chlorophylls

H. Z. Wang, X. G. Zheng, W. D. Mao, Z. X. Yu, Z. L. Gao, S. Y. Yang,* *Institute for Laser and Spectroscopy, Zhongshan University, Guangzhou 510275, China*

Investigating delayed light emission and continued oxygen release, two phenomena that occur long after illumination in a photosynthetic system, is important for understanding the primary photosynthetic process. It is well known that a large amount of the chlorophylls in chloroplasts are in aggregate. We have investigated delayed light emission from aggregated chlorophylls. By means of time-resolved spectroscopy, a nonemissive energy-storage state in aggregated chlorophylls has been discovered for what we believe to be the first time. The results, which reveal the physical mechanism of delayed light emission, can be used to explain the continued oxygen release in photosynthetic systems.

When aggregated chlorophylls are injected into acetone, deaggregation occurs and the energy stored in the nonemissive state is released as monochlorophyll fluorescence. The experimental results show that absorbed light energy can be stored in the nonemissive state for 20 min.

The nonemissive state is an aggregated delocalized state. Shortly after the pump-

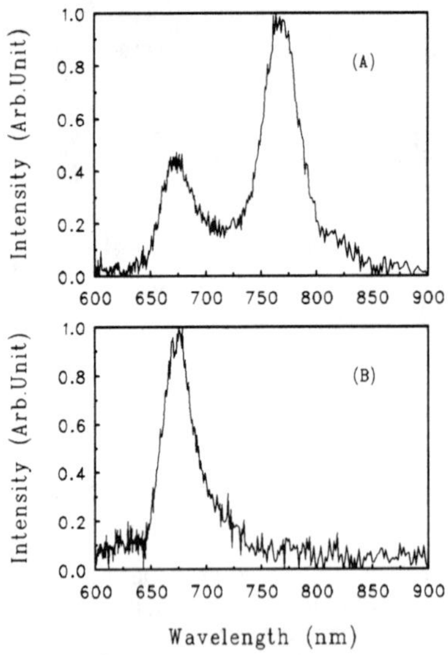

QFA2 Fig. 1. Fluorescent spectra of aggregated chlorophylls at room temperature: (A) during the first 100 ps of decay and (B) 10 ns to 2 min after the pump pulse.

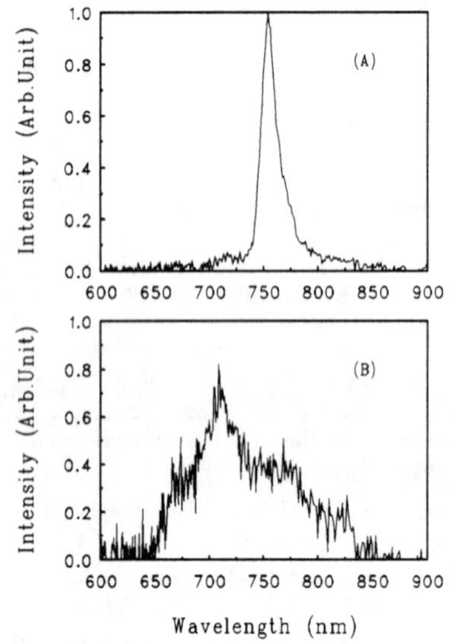

QFA2 Fig. 2. Fluorescent spectra of aggregated chlorophylls at 77 K: (A) during the first 100 ps of decay and (B) 10 ns to 2 min after the pump pulse.

ing of a high-density pulse, the nonemissive state is fully occupied and leads to short-duration emission (at 750 nm). Only when an exciton in the nonemissive state interacts with thermal phonons and then transits into the monomolecular state is a photon emitted as monochlorophyll fluorescence. Therefore, at room temperature there are two fluorescent components at the monochlorophyll fluorescent wavelength, one with a lifetime of 4.4 ns and the other with a much longer lifetime. Figures 1 and 2 show the fluorescent spectra of the first 100 ps of the decay and at elapsed times from 10 ns to 2 min.

As shown in Figs. 1(B) and 2(B), within 10 ns to 2 min after the pump pulse, the emission at room temperature is shorter in wavelength and higher in intensity than that at 77 K because there are more phonons and higher phonon energies at room temperature than at 77 K.

This work reveals that chlorophylls are important not only for light harvesting and energy transfer but also for energy storage in photosynthetic systems.

**Shanghai Institute of Plant Physiology, Chinese Academy of Sciences, Shanghai 20032, China*

QFA3 **8:30 am**

Femtosecond transfer dynamics from π-conjugated polymers into dopant-dye molecules

P. Haring Bolivar, G. Wegmann, R. Kersting, M. Deussen,* U. Lemmer,* R. F. Mahrt,* H. Kurz, H. Bässler,* E. O. Göbel,* *Institute for Semiconductor Technology II, Rhine-Westphalia Technical University–Aachen, 52056 Aachen, Germany*

Conjugated polymers have attracted much attention because of their potential application as very effective, cheap, and easy-to-produce luminescent devices.[1,2] From the technological point of view, there is a strong interest in being able to easily adjust the emission wavelength of a polymer LED in order to make polychromic displays feasible. One possibility for controlling the emission wavelength is doping the polymer with other molecules.[3,4] In this paper we report on our investigations of the energy transfer from conjugated soluble polymers poly-(phenyl phenylene vinylene) (PPPV) to the admixtured dye molecule DCM (4-dicyanomethylene-2-methyl-6-*p*-dimethylaminostyryl-4H-pyran). Varying the dye concentration drastically changes the dynamics of the transfer process, allowing us not only to observe the transport into the DCM molecules, but also to estimate the total trap density and to give further insight into the transport between polymer chromophores.

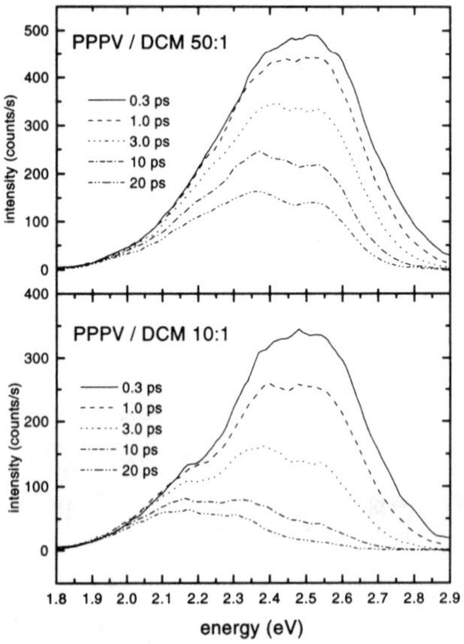

QFA3 Fig. 1. Time-resolved luminescence spectra for two samples with different DCM contents at various time delays after excitation.

All experiments were performed with a Kerr-lens mode-locked Ti:Al_2O_3 laser (76-MHz repetition rate) in a luminescence upconversion setup.[5] The samples, which were excited with 100-fs light pulses at 3.1-eV, consisted of a polycarbonate matrix (80-wt. % PC) blended with a mixture (20 wt. %) of PPPV and DCM with various weight ratios of PPPV-to-DCM content (1:0, 100:1, 50:1, 20:1, 10:1, and 4:1). The samples were kept at 10 K in a closed cycle cryostat to prevent degradation at the high excitation intensities (I_{CW} = 13 W/cm^2) required by this experimental technique.

In Fig. 1 we show the time evolution of the time-resolved luminescence spectra for two samples with different DCM concentrations. The first sample had a PPPV-to-DCM ratio of 50:1 (corresponding to a DCM concentration of 1×10^{19} cm^{-3}), and the second had a ratio of 10:1. Both samples clearly show a relaxation from the PPPV emission (centered at 2.55 eV) into the DCM states (emission at 2.05 eV). The change in the relaxation dynamics is evident. Figure 2 shows the time evolution of the PPPV luminescence at 2.55 eV for varying DCM concentrations. In Fig. 3 we summarize all measurements, depicting the mean emission energy of time-resolved luminescence

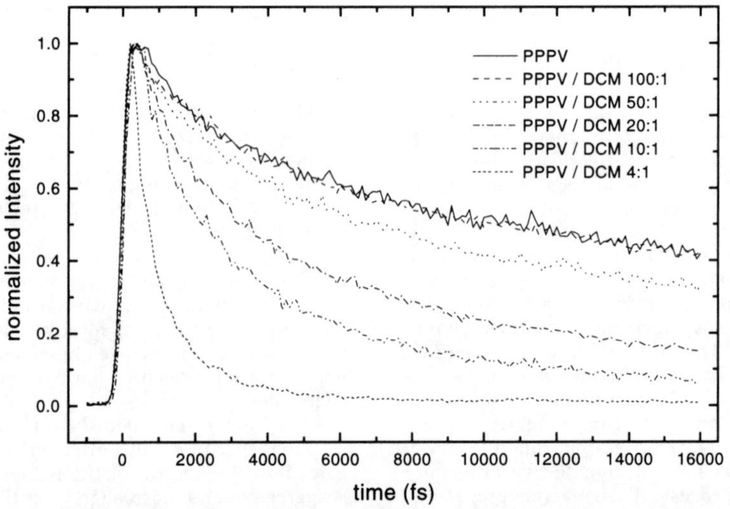

QFA3 Fig. 2. Time evolution of the luminescence at 2.55 eV (PPPV) for various DCM concentrations.

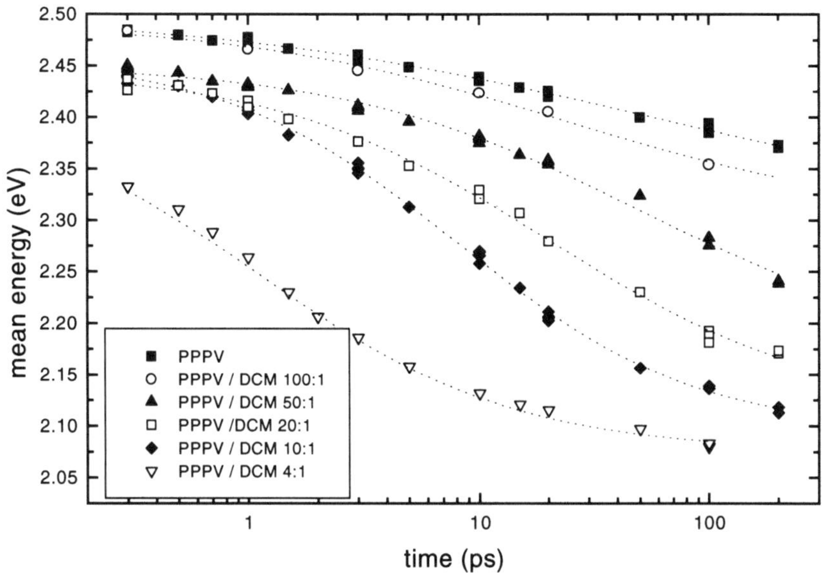

QFA3 Fig. 3. Mean energetic position of the luminescence as a function of time delay after excitation for various DCM concentrations. Curves connecting the data of the samples are included as a guide to the eye.

spectra as a function of time for all DCM concentrations studied. By extrapolating the DCM-concentration dependence of the decay dynamics we conclude a rough estimate for the trap density of pure PPPV of 5×10^{18} cm^{-3}.

We gratefully acknowledge financial support by the German Research Association and the Heinrich Hertz Foundation.

Department of Physics and Department of Physical Chemistry and Center for Materials Science, Philipps University–Marburg, Renthof 5, 35032 Marburg, Germany

1. J. H. Burroughes, D. D. C. Bradley, A. R. Brown, R. N. Marks, K. Mackay, R. H. Friend, P. L. Burns, and A. B. Holmes, Nature **347**, 539 (1990).
2. M. Schwörer, Phys. Blätter **50**, 52 (1994).
3. M. Uchida, Y. Ohmori, T. Noguchi, T. Ohnishi, and K. Yoshino, Jpn. J. Appl. Phys. **32**, L921 (1993).
4. A. Ochse, U. Lemmer, M. Deussen. J. Feldmann, A. Greiner, R. F. Mahrt, H. Bässler, and E. O. Göbel, Mol. Cryst. Liq. Cryst., in press.
5. J. Shah, IEEE J. Quantum Electron. **24**, 276 (1988).

QFA4 **8:45 am**

Femtosecond transient configuration change of a polydiacetylene backbone

T. A. Pham, A. Daunois, J.-Y. Bigot, J. Le Moigne,* *Nonlinear Optics and Optoelectronics Group, Strasbourg Institute of Physics and Chemistry of Materials, National Center for Scientific Research Unité Mixte 380046, ULP-EHICS, 23 rue du Loess, F-67037 Strasbourg Cedex, France*

The polydiacetylenes (PDAs) are particularly attractive materials as models of quasi-one-dimensional electronic systems with a high degree of crystallinity. Important nonlinear optical and electro-optical properties have been obtained in these low-dimensional structures.[1]

The stable PDA backbone is well known to be an acetylenic one (=CR–C≡C–CR′)$_n$, and its excited configuration is expected to exhibit a butatrienic structure (–CR=C=C=CR′)$_n$. In this paper we report the study of this configuration's dynamics in a PDA by using photon echo and pump–probe techniques with a 10-fs temporal resolution.

The samples are highly oriented poly[1.6-di(n-carbazolyl)-2.4 hexadiyne] (pDCH) films grown by epitaxy on a potassium acid phtalate crystalline substrate.[2] They are polymerized up to 98%. The room-temperature absorption spectra of a 120-nm-thick film are shown in Fig. 1(a) for light polarized parallel and perpendicular to the polymer-chain axis. The high dichroic ratio (>20) on the main excitonic resonance at 1.919 eV indicates an important degree of orientation. Femtosecond experiments have been performed on this sample with 60-fs CPM pulses at 620 nm, which are amplified with a 5-kHz-repetition-rate copper-vapor laser and are compressed to 10 fs with the fiber-grating-prism compression technique.

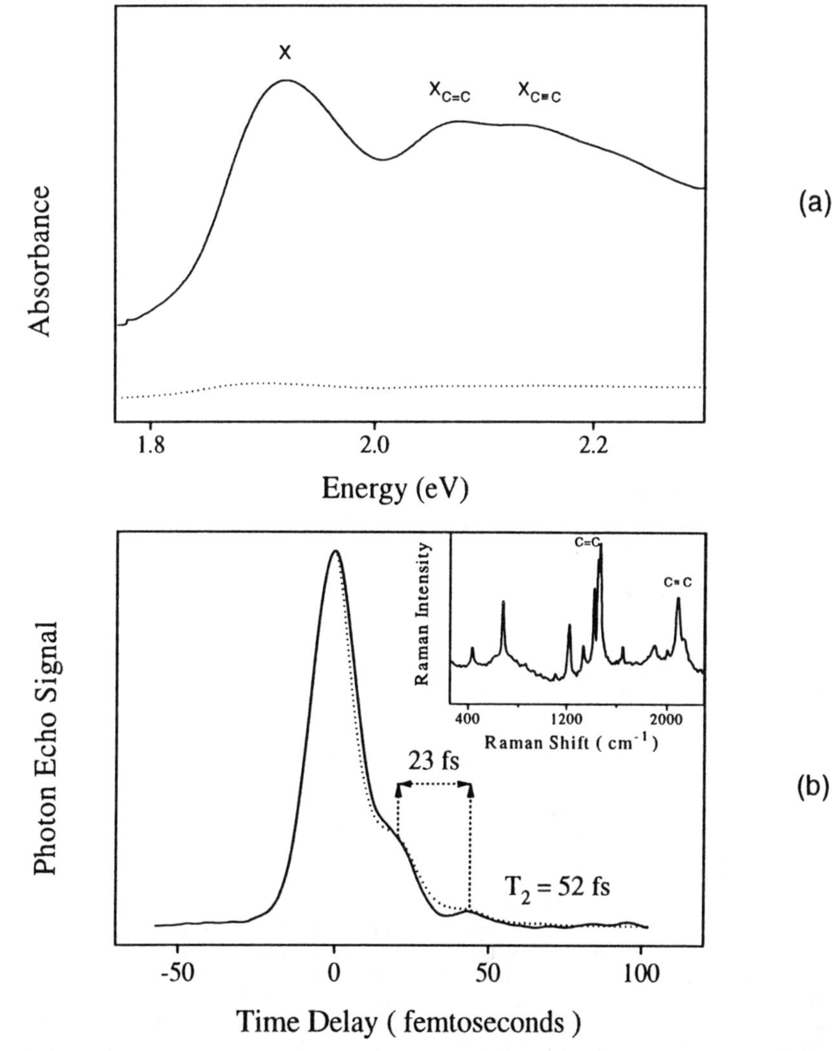

QFA4 Fig. 1. (a) Absorbance spectra of 120-nm epitaxial pDCH thin films for light polarization parallel (solid curve) and perpendicular (dotted curve) to the chain axis. (b) photon-echo decay obtained with 10-fs pulses (solid curve) and corresponding fit (dotted curve). The inset displays the Raman spectrum of pDCH excited at 632.8 nm.

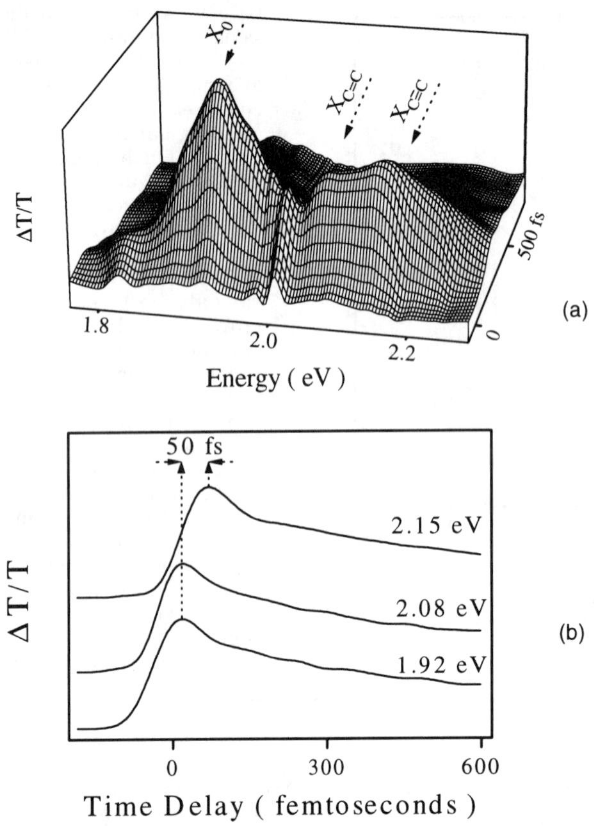

QFA4 Fig. 2. Pump–probe results from 60-fs pump and 10-fs probe pulses. (a) Three-dimensional plot of differential transmission spectra for delays between −100 and +800 fs. (b) Time-resolved differential transmission spectra for three photon energies of the probe pulse, 1.92, 2.08, and 2.15 eV, corresponding to the *X*, *X*↔C=C, and *X*↔C≡C vibronic states, respectively.

Photon-echo experiments performed with two 10-fs pulses with wave vectors k_1 and k_2 are measured in the direction $2k_1 - k_2$ as a function of delay τ between the two pulses. In Fig. 1(b) we have plotted the amplitude of a photon-echo signal obtained with an energy of 400 pJ/pulse (~5 GW·cm^{-2}). The signal reveals a transverse excitonic relaxation time T_2 of (52 ± 5) fs as well as 23-fs-period quantum beats, which we attributed to the C=C stretching mode (180 meV) of the PDA chain [inset of Fig. 1(b)]. The dashed curve is a fit using a simplified three-level system (ground state and two excited vibronic states: $|2\rangle$, $|3\rangle$) interacting with an ultrashort pulse. The photon-echo decay behavior is proportional to $[1 + \alpha\cos(\omega_{23}t)]^2 \exp(-4t/T_2)$. The fitting parameters are $\omega_{23} = 0.273$ rad fs^{-1}, corresponding to a 180-meV energy splitting between levels $|2\rangle$ and $|3\rangle$, $T_2 = 52$ fs, and $\alpha = 0.11$, corresponding to the ratio of the dipole moments for the transitions from the ground state to the two excited states. The observed quantum beats suggest that most of the dephasing is related to the C=C stretching of the PDA backbone through exciton *X*↔C=C phonon interactions. The *X*↔C≡C coupling, with a 16-fs period, is delayed and does not show up during the dephasing process. The corresponding quantum beats due to the C≡C stretching mode [inset of Fig. 1(b)] should be observed with our 10-fs pulses.

To investigate this observation further, we performed pump–probe experiments (with 60-fs pump and 10-fs probe pulses). Figure 2(a) displays a three-dimensional plot of differential transmission spectra $\Delta T/T$ for various delays between the pump and probe pulses. These spectra reveal a spectral hole burning at the pump energy and population bleaching of the exciton (*X*) and vibronic (*X*↔C=C, *X*↔C≡C, *X*↔carbon-skeleton planar shear modes) states, as well as excited-state formation. The most important feature is a delay between the *X*↔C=C and *X*↔C≡C. In Fig. 2(b) three time-resolved differential transmissions, measured with a 4-fs steps, are displayed. They correspond to the exciton E_x, $E_x + F_{C=C}$ (double-bond vibronic), and $E_x + E_{C≡C}$ (triple-bond vibronic) resonances. The observed signals reveal a 50-fs delay between the two vibronic sidebands.

The observed dephasing and population dynamics suggest a femtosecond transient configuration change of the pDCH backbone.[3] Under excitation with the ultrashort pulse, the electronic distribution of the initial stable acetylenic chain of the pDCH undergoes a quasi-equiprobable redistribution along the chain, leading to the formation of a butatrieniclike backbone configuration. During the dephasing process, within 52 fs an energy transfer along the chain by means of the system vibration mode occurs: quantum beats corresponding to the C=C modes are observed. Later, an electronic localization due to the formation of self-trapped excitons (within 120 fs) along the chain takes place, leading to a return to the acetylenic configuration.

* *Organic Materials Group, Strasbourg Institute of Physics and Chemistry of Materials, National Center for Scientific Research Unité Mixte 380046, ULP-EHICS, 23 rue du Loess, F-67037 Strasbourg Cedex, France*

1. M. Schott and G. Wegner, in *Nonlinear Optical Properties of Organic Molecules and Crystals*, D. S. Chemla and J. Zyss, eds. Academic, New York, 1987, Vol. **2**, p. 3.
2. J. Le Moigne, F. Kajzar, and A. Thierry, Macromolecules **24**, 2622 (1991).
3. T. A. Pham, A. Daunois, J.-C. Merle, J. Le Moigne, and J.-Y. Bigot, Phys. Rev. Lett., in press.

QFA5 **9:00 am**

Ultrafast vibronic relaxation of large molecules in solution studied on a 20-fs time scale

T. Hasche, S. Ashworth, E. Riedle, M. Woerner, T. Elsaesser, *Max Born Institute for Nonlinear Optics and Ultrafast Spectroscopy, Rudower Chaussee 6, D-12489 Berlin, Germany*

Vibronic relaxation processes of large molecules in the condensed phase, which occur on an ultrafast time scale, have been the subject of numerous time-resolved experiments in the regime between 100 fs and several picoseconds.[1] Recently, high-repetition-rate solid-state lasers have become available for the generation of pulses of less than 30-fs duration,[2,3] allowing the investigation of relaxation steps in time intervals of several tens of femtoseconds, i.e., comparable to vibrational periods of the molecules. In this paper we describe a study of the relaxation of high-lying singlet S_n states in organic-dye molecules in two-color pump–probe measurements with an unprecedented time resolution of about 20 fs, providing new insight into femtosecond material excitations and ultrafast internal conversion processes.

In Fig. 1 the absorption and emission spectra of the dye IR125 and the spectral profiles of the excitation pulses at 425 nm and the probe pulses at 850 nm are shown. The probe spectrum overlaps with both the S_0–S_1 absorption and the S_1–S_0 emission bands. Near-infrared pulses of 17-fs duration are generated in a mode-locked Ti:sapphire laser providing an output power of 450 mW. Excitation pulses are derived from the laser output by frequency doubling in a thin BBO crystal and subsequent recompression (duration, 30 fs; average power, 40 mW). The very high time resolution is evident from the cross-correlation trace plotted in Fig. 2(a).

In Fig. 2 the increase of transmission observed with the full bandwidth of the probe pulses of 50 nm is plotted as a function of delay time for (a) the first 150 fs and (b) up to 3.5 ps. We find a very fast rise of the signal within 60 fs followed by a much slower increase extending into the picosecond regime. The latter is mainly due to

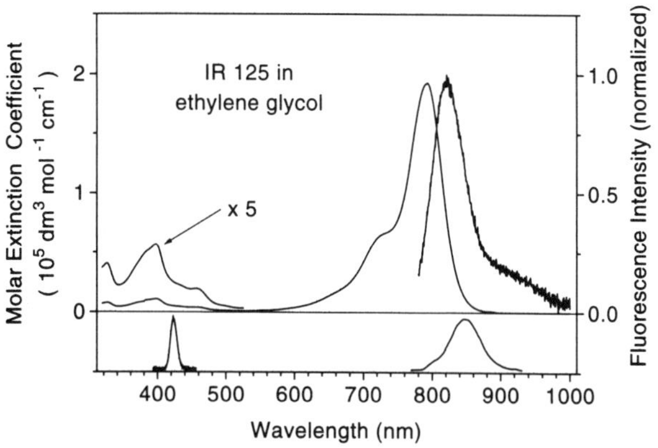

QFA5 Fig. 1. Absorption (left-hand side) and emission (right-hand side) spectra of the dye IR125 dissolved in ethylene glycol. The spectral profiles of the excitation pulses at 425 nm (bandwidth, 11 nm) and of the probe pulses at 850 nm (width, 50 nm) are shown on the abscissa.

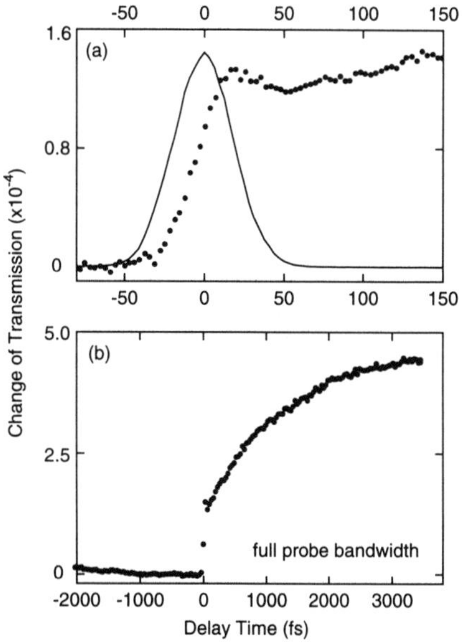

QFA5 Fig. 2. Time-resolved increase of transmission of IR125 measured with the full bandwidth of the probe pulses at 850 nm. The transmission change $\Delta T = T - T_0$ is plotted as a function of delay time (a) for the first 150 fs after excitation at 425 nm and (b) for times up to 3.5 ps (points T_0, T respectively denote transmission of the sample before and after excitation). The solid curve represents the cross correlation of the pump and the probe recorded with a 100-μm BBO crystal at the location of the sample.

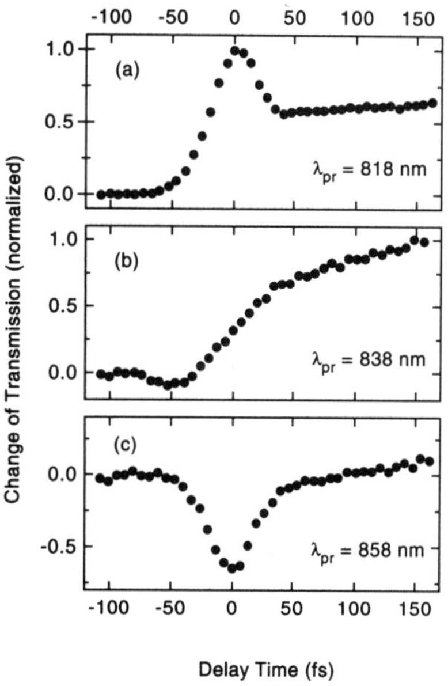

QFA5 Fig. 3. Time-resolved data for different spectral positions (spectral width, 11 nm) within the overall bandwidth of the probe pulses.

polar solvation in ethylene glycol, resulting in a red shift of the emission band.

To provide insight into the different components of the femtosecond response, the probe pulse was spectrally resolved after interaction with the sample. In Fig. 3 we present results for three spectral positions (probe bandwidth, 11 nm). The fast rise of transmission shows peak structures and/or oscillatory features related to the coherent vibronic polarization and its evolution on a time scale comparable to the vibrational periods. It is important to note that this coherent response is observed even for the large energy separation of pump and probe pulses. The rapid increase of transmission at short wavelengths is mainly due to bleaching of the electronic ground state. At longer wavelengths, the signal is dominated by the onset of stimulated S_1–S_0 emission, revealing a fast population of S_1 states through S_n relaxation in the 50–100-fs regime. The microscopic mechanisms underlying the ultrafast transients are discussed in detail.

1. For a review, see T. Elsaesser and W. Kaiser, Annu. Rev. Phys. Chem. **42**, 83 (1991).
2. P. F. Curley, Ch. Spielmann, T. Brabec, F. Krausz, E. Wintner, and A. J. Schmidt, Opt. Lett. **18**, 54 (1993).
3. S. Backus, M. T. Asaki, C. Shi, H. C. Kapteyn, and M. M. Murnane, Opt. Lett. **19**, 399 (1994).

QFA6 **9:15 am**

Terahertz time-domain reflection spectroscopy of liquid water

L. Thrane, P. Uhd Jepsen, R. H. Jacobsen, S. R. Keiding, *Institute of Chemistry, University of Aarhus, Langelandsgade 140, DK-8000 Aarhus C, Denmark*

In this paper we report an investigation of the temperature dependence of the far-infrared spectrum of liquid water in the range from 2 to 40 cm^{-1}. To do this we have used a new experimental technique based on ultrashort electromagnetic pulses (terahertz pulses) radiated from small semiconductor antennas driven by femtosecond laser pulses.[1] The bandwidth of the terahertz pulses is very broad, extending from 50 GHz to several terahertz (1 THz = 33.356 cm^{-1}). By reflecting these pulses off a silicon-water interface and measuring the frequency-dependent change in phase and amplitude of the pulses, the optical constants, the index of refraction, and the absorption coefficient of liquid water can be obtained if knowledge of the optical constants of silicon is assumed. The pulse sequence obtained from the terahertz time-domain reflection spectrometer is shown in Fig. 1. By Fourier-transforming the two pulses and taking their complex ratio, one directly obtains the optical constants for water, as shown in Fig. 2. We observe good agreement with the literature for both index and absorption at room temperature.

In order to measure the properties of liquid water as functions of temperature, the sample cell was placed on a 70-W Peltier element. This allowed us to change the temperature from the supercooled region (270 K) up to 314 K. Index and absorption were measured for 10 different temperatures. All absorption data resemble the data at 292 K, a smooth absorption curve increasing approximately as $\nu^{1/2}$ with the absorption decreasing with temperature.

To represent the temperature-dependent absorption, we have chosen to plot the logarithm of the integrated absorption from 0.06 to 1 THz as a function of T^{-1}, an Arrhenius plot, as shown in Fig. 3. The integrated absorption at 314 K was used to normalize the plot. The rationale for this plot comes from the simplest description of a polar liquid, the Debye model. In this model it can be shown the integrated absorption is approximately proportional to τ^{-1}, where τ is the orientational relaxation time of the molecular dipoles. Furthermore, τ is expected to follow an exponential temperature dependence, scaled by the activation enthalpy of the relaxation process:

$$\tau = \tau_0 \exp\left(\frac{\Delta H}{kT}\right).$$

With these assumptions, we obtain an activation enthalpy of 2.4 kcal/mol, or 870 cm^{-1}, as given by the solid line in Fig. 3. The dashed curve represents the temperature dependence of the relaxation time as observed in a recent Raman experiment by Tominanga *et al.*[2]

Below 283 K the absorption remains more or less constant. It is interesting that the observed ΔH is significantly smaller than

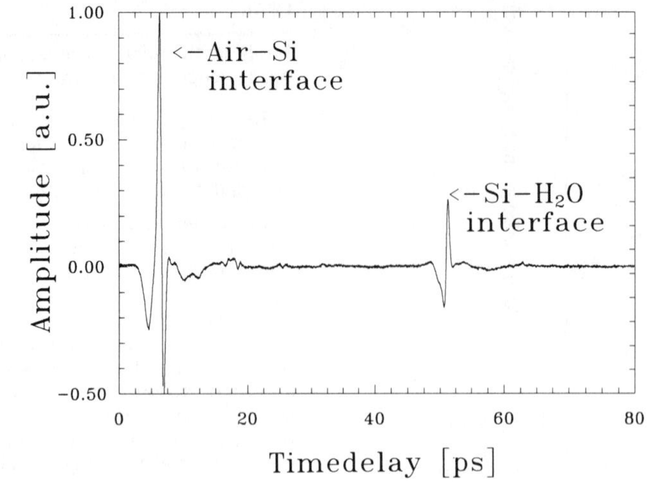

QFA6 Fig. 1. Terahertz pulses reflected from the air/silicon and silicon/water interfaces. The amplitude corresponds to an induced current of 0.4 nA, and the time constant of the detection system is 30 ms.

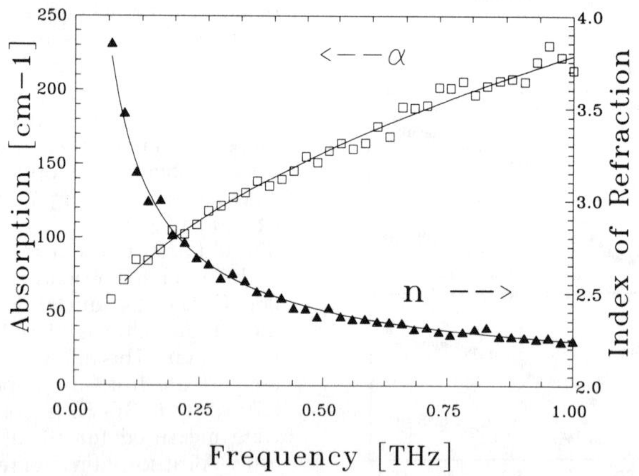

QFA6 Fig. 2. Index of refraction (filled triangles) and absorption coefficient (open squares) for liquid water at 292 K.

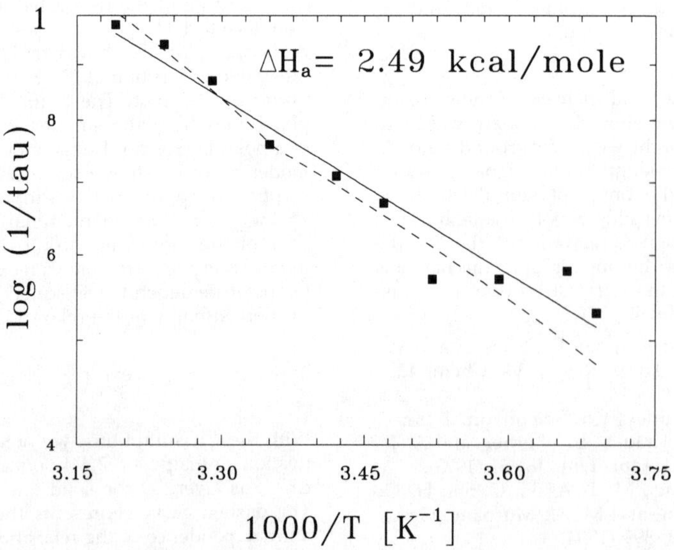

QFA6 Fig. 3. Integrated absorption from 0 to 1 THz (Proportional to $1/\tau$) on a log scale versus the inverse temperature. The solid line is a fit corresponding to an activation energy of 2.49 kcal/mol, and the dashed curve is the temperature dependence of the inverse relaxation time obtained from Raman experiments in Ref. 2.

the ΔH inferred from dielectric relaxation in the microwave region. This could indicate that an additional orientational relaxation takes place in water without breaking the hydrogen-bond network.

1. M. van Exter, Ch. Fattinger, and D. R. Grischkowsky, Opt. Lett. **14**, 1128 (1989).
2. K. Mizoguchi, Y. Hori, and Y. Tominanga, J. Chem. Phys. **97**, 1961 (1993).

QFA7 **9:30 am**

Femtosecond vibrational dynamics of adsorbates near the desorption threshold

J. P. Culver,* M. Li,** Z.-J. Sun,*,** R. M. Hochstrasser,** A. G. Yodh,*
Laboratory for Research on the Structure of Matter, University of Pennsylvania, Philadelphia, Pennsylvania 19104

Femtosecond optical excitation of surfaces creates a unique environment for molecular adsorbates that can initiate surface chemistry. Of particular importance to femtosecond photoprocesses on metal surfaces are the energy-transfer mechanisms in highly perturbed hot adsorbate/substrate systems. Several experiments have used intense femtosecond excitation of the substrate to desorb diatomic molecules from metal surfaces.[1,2] Complementary experiments of energy-transfer rates with low-power excitation pulses have also been performed.[3,4] However, when molecules are excited near the desorption threshold, anharmonic couplings between different adsorbate modes and altered electronic potentials will effect energy-transfer processes. Coupling rates measured at low excitation therefore may not be valid in the desorption regime.

We present time-resolved IR spectroscopic studies of CO on Cu(111) following visible excitation pulses with fluences capable of desorbing approximately 10^{-7} of a monolayer per pulse (i.e., just at the desorption threshold). The measurements were made with absorbed fluences up to 0.75 mJ/cm^2 (see Ref. 2), and we find that a single-rate model adequately describes the data. In particular, the coupling rates are constant within 20%. This is, to our knowledge, the first real-time, state-specific evidence that low-excitation coupling rates can account for the vibrational temperatures of adsorbates on metal surfaces near their desorption threshold.

In this experiment the transient vibrational dynamics of a low-frequency mode are probed by monitoring its degree of excitation *indirectly* through the frequency shift it induces in a higher-frequency mode. This technique has been used to study the dynamics of CO on Cu(111),[3] Cu(100),[4] and Pt(111)[4] metal surfaces. By using a gated upconversion technique, the time evolution of the CO internal-stretch vibration (ω_0 = 2073.5 cm^{-1}, Γ_0 = 4.5 cm^{-1}) is followed after femtosecond heating of the substrate. For a top-bound CO on Cu(111), the dephasing mode is frustrated translation. From the time response of the CO stretch-mode absorption (Fig. 1), the coupling rates of the frustrated translation to the electrons γ_e and the phonons γ_l can be extracted. Figure 2 shows the temperatures

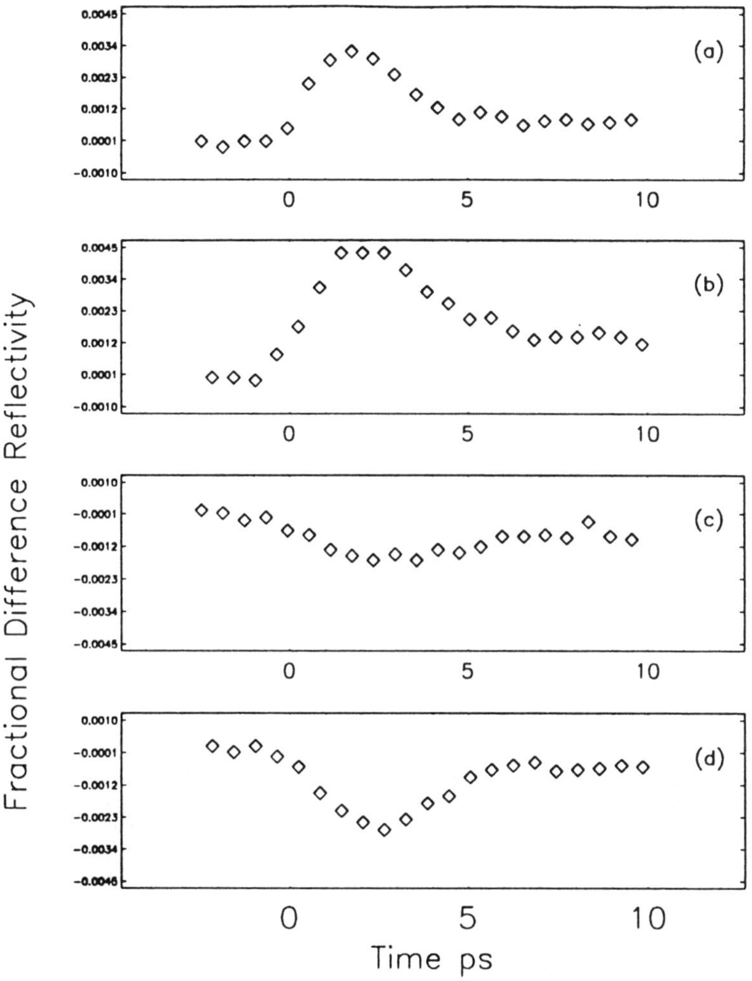

QFA7 **Fig. 1.** Transient fractional difference reflectivity of CO on Cu(111) for a fluence absorbed of F_{abs} = 0.5 mJ/cm^2. The probe laser was tuned near the CO-stretch Lorentzian absorption at (a) 2077.5, (b) 2075.0, (c) 2068.0, and (d) 2066.0 cm^{-1}.

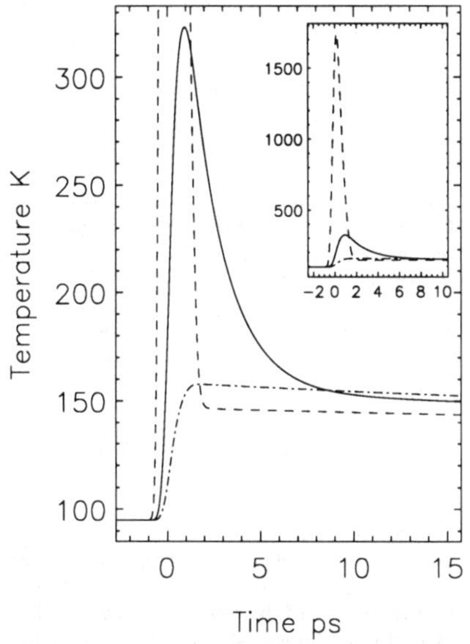

QFA7 **Fig. 2.** Calculations of electron (dashed curve) and phonon (dotted-dashed curve) reservoir temperatures as a function of time for F_{abs} = 0.75 mJ/cm^{-2}. The frustrated-translation temperature (solid curve) was calculated by using the measured coupling rates γ_e and γ_l.

of the frustrated translation, T_{ft}, the electrons, T_e, and the lattice, T_l, resulting from a 500-fs visible pulse with an absorbed fluence of 0.75 mJ/cm^2.

By examining the vibrational response of CO at adsorbed pump fluences in the range 0.3–0.75 mJ/cm^2, the excitation dependence of the frustrated-translation coupling to the substrate can be determined (Fig. 3). Fitting the measured coupling rates to a linear dependence on the fluence absorbed, we find γ_e = (184 + $\alpha_e F_{abs}$) GHz, with α_e = 34 ± 41 GHz·cm^2/mJ, and γ_l = (316 + $\alpha_l F_{abs}$) GHz, with α_l = 58 ± 78 GHz·cm^2/mJ. It is evident that up to excitation levels sufficient to induce desorption, fluence-independent (or weakly dependent $\delta\gamma/\gamma < 20\%$) rates account for the measured coupling of the frustrated translation to the substrate electrons and phonons. Using molecular-dynamics calculations that incorporate electronic friction, Tully and coworkers predicted temperature dependent coupling rates for the CO/Cu(100) system.[5] We compare our measurements to these simulations and discuss the relevance of adsorbate/substrate coupling rates to femtosecond-laser induced desorption.

*Also with Department of Physics, University of Pennsylvania, Philadelphia, Pennsylvania 19104

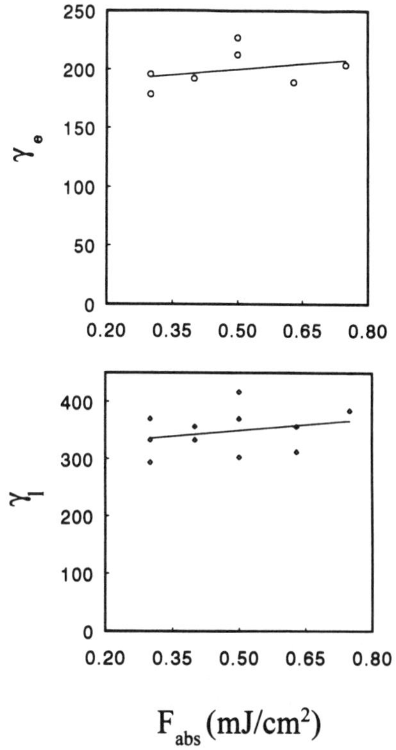

QFA7 **Fig. 3.** Effective coupling rates γ_e and γ_l as a function of substrate excitation. Lines are linear fits described in the text.

**Also with Department of Chemistry, University of Pennsylvania, Philadelphia, Pennsylvania 19104

1. R. R. Cavanaugh, D. S. King, J. C. Stephenson, and T. F. Heinz, J. Phys. Chem. **97**, 786 (1993) and references therein.
2. A desorption rate of 1% monolayer per pulse with a F_{abs} = 2.2 mJ/cm^2 was previously found for CO on Cu(111); J. A. Prybyla, H. W. K. Tom, and G. D. Aumiller, Phys. Rev. Lett. **68**, 503 (1992).
3. J. P. Culver, M. Li, L. G. Jahn, R. M. Hochstrasser, and A. G. Yodh, Chem. Phys. Lett. **214**, 431 (1993); J. P. Culver, M. Li, L. G. Jahn, R. M. Hochstrasser, and A. G. Yodh, in *Laser Spectroscopy and Photochemistry on Metal Surfaces*, W. Ho and H.-L. Dai, eds. (World Scientific, Singapore), in press.
4. T. A. Germer, J. C. Stephenson, E. J. Heilweil, and R. R. Cavanagh, J. Chem. Phys. **101**, 1702 (1994); T. A. Germer, J. C. Stephenson, E. J. Heilweil, and R. R. Cavanagh, Phys. Rev. Lett. **71**, 3327 (1993); T. A. Germer, J. C. Stephenson, E. J. Heilweil, and R. R. Cavanagh, J. Chem. Phys. **98**, 9986 (1993).
5. J. C. Tully, M. Gomez, and M. Head-Gordon, J. Vac. Sci. Technol. A**11**, 1914 (1993).

QFB 8:00 am

Hyatt C/D

Coherent Phenomena in Semiconductors I

Makoto Kuwata-Gonokami, *University of Tokyo, Presider*

QFB1 8:00 am

1.3-fs oscillation of the quantum coherence in a GaAs multiple quantum well: observation of the two-photon coherently excited biexciton

K. B. Ferrio, Nicolas Bonadeo, Duncan G. Steel, Makoto Kuwata-Gonokami,* *Harrison M. Randall Laboratory of Physics, University of Michigan, Ann Arbor, Michigan 48109*

Theoretical and experimental studies have revealed the complex, inherently many-body nature of the transient resonant nonlinear optical response of excitons in GaAs multiple quantum wells (MQWs). The established model of these systems, the semiconductor Bloch equations (SBE), shows the importance of Coulomb and exchange interactions. But studies of the polarization dependence show behavior not predicted by the established SBE when they are extended to include the vector nature of the electromagnetic interaction. There have, however, been numerous experimental indications and theoretical predictions that biexcitons are important in the transient nonlinear response of MQWs.[1-3] These measurements rely on the polarization-selection rules for biexcitons and time ordering of the applied fields, which are different for single-photon and two-photon excitations.

The coherently excited biexciton, however, has a unique feature that distinguishes it from other electronic excitations: the two-photon coherent-excitation pathway, in contrast to stepwise two-photon excitation, does not involve a population of the intermediate (single exciton) state but, rather, creates a true quantum coherence between the ground and biexciton states. This coherence oscillates with frequency $\omega_B = 2\omega_X - \Delta$, where $\hbar\omega_X$ and $\hbar\Delta$ are the single-exciton energy and biexciton binding energy, respectively. In a four-wave-mixing experiment, it is possible to directly interrogate the strength and time dependence of this coherence as shown[4] and recently demonstrated[5] for biexcitons in CuCl.

In this paper we report direct observations of this quantum coherence oscillation in a 10-period GaAs/AlGaAs MQW with 98-Å wells and 98-Å barriers at liquid-helium temperatures. The interferometric four-wave-mixing experiment is shown schematically in Fig. 1. The temporal quantum oscillation is observed in the phase $\omega_B\tau$ of the emitted field, where τ is the delay between the pump field $E_{pump}(t)$ and the probe field $E_{probe}(t - \tau)$, both tuned to the single-exciton resonance.

The observation of the two-photon coherence occurs for so-called negative delay ($\tau < 0$), for which E_{pump} arrives before E_{probe}. (Positive delay corresponds to the usual two-level nonlinear optical response). Using linearly cross-polarized fields for E_{pump} and E_{probe} to suppress contributions from spin-independent nonlinear optical responses, we observe the oscillation of Fig. 2(a), corresponding to 1.3 fs, approximately half the inverse excitation frequency, as expected for a quantum coherence of the biexciton and ground states. The phase of the laser is shown in Fig. 2(b) for direct comparison.

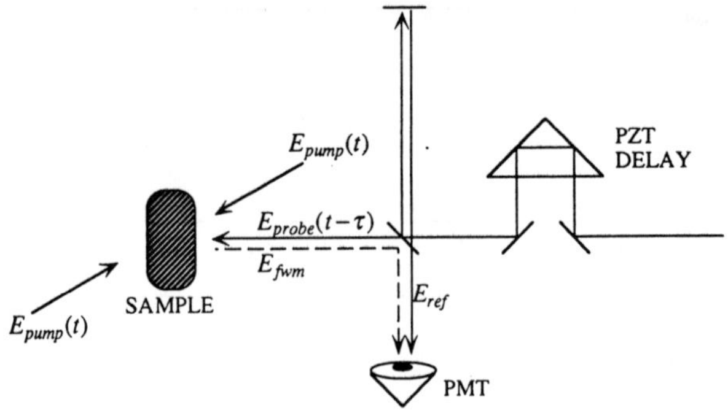

QFB1 Fig. 1. Schematic representation of the phase-conjugate interferometer for direct observation of the ground-state biexciton coherence.

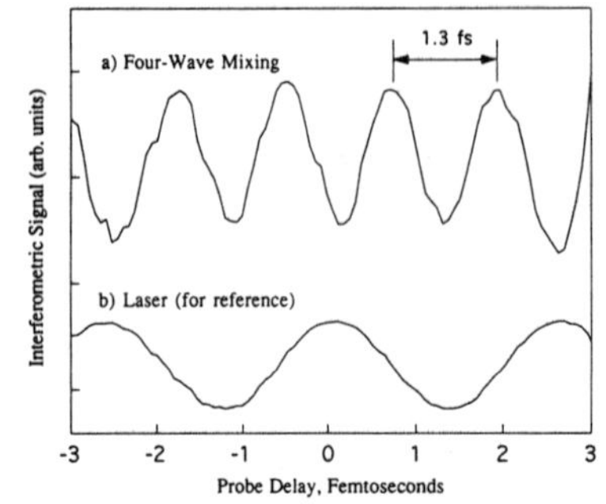

QFB1 Fig. 2. Interferometrically detected four-wave-mixing signal, showing a period half that of the laser displayed for reference in the lower panel.

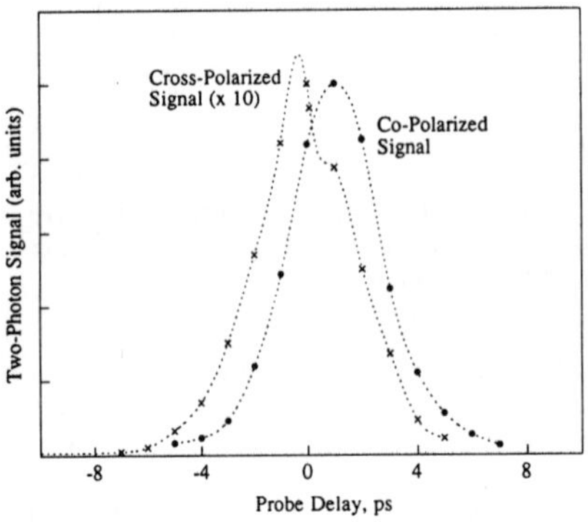

QFB1 Fig. 3. Strength of the two-photon-coherence for co- and cross-linearly polarized fields. Dashed curves are guides for the eye only.

Figure 3 shows that the signal is a highly asymmetric function of τ for cross-polarized excitation. For $\tau < 0$ the oscillation arises from a true quantum coherence, manifesting the nonlinear response of a coherent two-photon excitation. The reduced signal for $\tau > 0$ shows that the nonlinear optical response due to spin-dependent interactions is weaker, in agreement with earlier predictions.[6] Copolarized excitation, for comparison, gives a signal arising for $\tau > 0$, as expected for single-exciton physics.

This work was supported by the U.S. Air Force Office of Scientific Research and the U.S. Army Research Office.

* *Department of Applied Physics, Faculty of Engineering, University of Tokyo, Hongo, Bunkyo-ku, Tokyo 113, Japan*

1. B. F. Feuerbacher, J. Kuhl, and K. Ploog, Phys. Rev. B **43**, 2439 (1991).
2. H. Wang, J. Shah, T. C. Damen, and L. N. Pfeiffer, Solid State Commun. **91**, 869 (1994).
3. K. Bott, O. Heller, D. Bennhardt, S. T. Cundiff, P. Thomas, E. J. Mayer, G. O. Smith, R. Eccleston, J. Kuhl, and K. Ploog, Phys. Rev. B **48**, 17418 (1993).
4. I. M. Beterov and V. P. Chebotaev, in *Progress in Quantum Electronics*, J. H. Sanders and S. Stenholm, eds. (Pergamon, Oxford, 1974).
5. T. Ogasawara, K. Ema, and M. Kuwata-Gonokami, in *Quantum Electronics and Laser Science*, 1993 Technical Digest Series, Vol. 12 (Optical Society of America, Washington, D.C., 1993), p. 10.
6. E. Hanamura, Solid State Commun. **12**, 951 (1973).

QFB2 8:15 am

Quantum beats from biexcitons in narrow quantum wells: polarization and density dependence

T. F. Albrecht, A. Schulze, M. Koch, T. Meier, K. Bott, J. Feldmann, W. Stolz, J. Kuhl,* E. J. Mayer,* S. W. Koch, E. O. Göbel, S. T. Cundiff, *Department of Physics and Scientific Center for Materials Sciences, Philipps- University, D-35032 Marburg, Germany*

Transient four-wave mixing (TFWM) has been extensively used during the last few years to study exciton dynamics in semiconductors and in semiconductor heterostructures. These studies have provided significant insight into exciton dynamics in both ordered and disordered systems. There are, nevertheless, still certain aspects of the excitonic TFWM response that are not completely understood. In particular, a complete understanding of the polarization and intensity dependences of the signal strength,

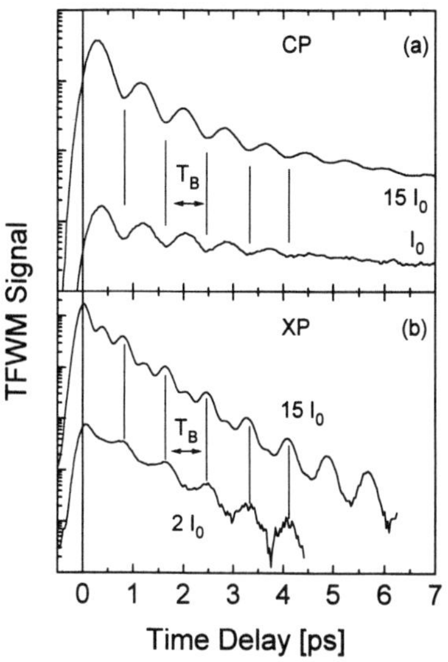

QFB2 Fig. 1. TFWM signal as a function of delay for (a) CP and (b) XP cases. In both cases the results at two intensities are shown (vertical displacement for clarity). The equally spaced vertical lines mark the beat minimums in (a) and the beat maximums in (b).

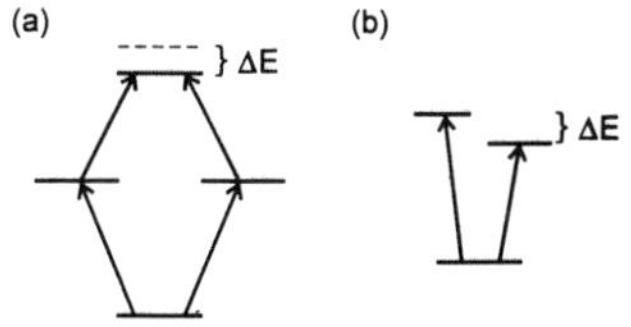

QFB2 Fig. 2. Level diagrams under consideration for (a) a diamond configuration and (b) a V configuration.

the decay of the time-integrated signal, and the temporal characteristics of the time-resolved signal has proven elusive. Recent theoretical and experimental work has indicated that exciton–exciton interactions are important.[1,2] Such interactions include incoherent processes, such as excitation-induced dephasing, and the formation of biexcitons. Additionally, some of the phenomena result from disorder, making them sample dependent.

We present here the results of TFWM experiments in multiple quantum wells. The data were obtained from a GaAs/AlAs superlattice with 4.1-nm wells and 3.0-nm barriers. We have obtained similar results from five other samples, including four samples consisting of unstrained InGaAs/InP. The key features that connect all of these samples are the very narrow well widths and the resulting strong inhomogeneous broadening arising from well-width fluctuations.

For colinearly polarized (CP) incident pulses in a two-pulse experiment we observe a clear quantum beating [Fig. 1(a)] in the time-integrated signal. This beating is independent of intensity and corresponds to an energetic splitting of 5.2 meV. It does have the striking characteristic of starting with a minimum at zero delay. For cross-linearly polarized (XP) pulses we observe quantum beats with the same period, but starting with a maximum [Fig. 1(b)]. Additionally, the XP beats have a very peculiar intensity dependence; at high intensity they exhibit beats with half the period of those present in the low-intensity or CP cases. For co-circularly polarized pulses there are no beats (data not shown).

All of these features are qualitatively reproduced in a model that includes, in addition to biexcitonic contributions [as in Ref. 2 and Fig. 2(a)], both inhomogeneous broadening and fifth-order contributions. The inhomogeneous broadening is essential for the manifestation of biexcitonic beating in the time-integrated signal; in a homogeneously broadened sample the beating occurs in the time-resolved signal but not the time-integrated one. The strong disorder also localizes the excitons, serving to strongly enhance the biexcitonic contributions to the optical properties[3] and the biexcitonic binding energy. The minimum at zero time delay for CP pulse means that the states that are interfering to produce the beats are in a diamond configuration [Fig. 2(a)], because a V configuration [Fig. 2(b)] always produces a maximum at zero delay. The intensity dependence suggests that terms beyond third order are contributing, and, indeed, simply extending the calculation to include fifth-order terms qualitatively reproduces the observed effect, a halving of the beat period at high intensity in the XP configuration and no change with intensity in the CP case.

*Max Planck Institute for Solid State Research, Heisenbergstrasse 1, D-70506 Stuttgart, Germany

1. H. Wang, K. Ferrio, D. G. Steel, Y. Z. Hu, R. Binder, and S. W. Koch, Phys. Rev. Lett. **71**, 1261 (1993).
2. K. Bott, O. Heller, D. Bennhardt, S. T. Cundiff, P. Thomas, E. J. Mayer, G. O. Smith, R. Eccleston, J. Kuhl, and K. Ploog, Phys. Rev. B **48** (1993).
3. K. Brunner, G. Abstreiter, G. Böhm, G. Tränkle, and G. Weimann, Phys. Rev. Lett. **73**, 1138 (1994).

QFB3 8:30 am

Photon echoes from the absorption continuum of a semiconductor

S. Glutsch, S. Bar-Ad, M.-A. Mycek, D. S. Chemla, U. Siegner,* *Department of Physics, University of California at Berkeley, and Materials Sciences Division, Lawrence Berkeley Laboratory, 1 Cyclotron Road, MS 2-300, Berkeley, California*

The theory of photon echoes has been intensively investigated for interaction-free particles since 1979.[1] Numerical calculations for semiconductors, excited at or below the band edge, have been reported.[2] However, a very important configuration for ultrashort-pulse experiments corresponds to excitation in the continuum or simultaneous excitation of bound and continuum states, as shown in Fig. 1. To the best of our knowledge, this configuration has never been studied theoretically, despite the fact that one can expect marked differences between the responses of bound and continuum states.

We have solved the semiconductor Bloch equations,

$$i\hbar \frac{\partial}{\partial t} n(\mathbf{r}, t) = [H(\cdot, t) *, n(\cdot, t)](\mathbf{r}) + i\hbar \left(\frac{\partial n(\mathbf{r}, t)}{\partial t}\right)_{\text{deph}},$$

$$H(\mathbf{r}, t) = \begin{pmatrix} -\frac{\hbar^2}{4m}\Delta & -\mu E(t) \\ -\mu^* E^*(t) & -\frac{\hbar^2}{4m}\Delta \end{pmatrix} \delta(\mathbf{r}) - V(\mathbf{r})[n(\mathbf{r}, t) - n(\mathbf{r}, -\infty)],$$

$$m = \frac{m_e m_h}{m_e + m_h},$$

$$V(\mathbf{r}) = \frac{e^2}{4\pi\varepsilon_0 \varepsilon r},$$

$$n(\mathbf{r}, -\infty) = \text{diag}(0, 1)\delta(\mathbf{r}),$$

$$\left(\frac{\partial n_{ij}(\mathbf{r}, t)}{\partial t}\right)_{\text{deph}} = \begin{cases} 0 & \text{for } i = j, \\ -\varepsilon n_{ij}(\mathbf{r}, t) & \text{for } i \neq j, \end{cases}$$

up to the third order in the field for a four-wave-mixing (FWM) geometry for various conditions of the excitation. The pulses are assumed to have a Gaussian form, $E(t) = \bar{E}(2\pi\sigma^2)^{-1/2} \exp(t^2/2\sigma^2)\exp(-i\bar{\omega}t)$, and the second pulse is delayed by Δt.

For all explicit results, excitonic units are used (GaAs: 1 Ry = 7.4 meV; $\hbar$/Ry = 120 fs), and we assume a phenomenological homogeneous linewidth $\hbar\varepsilon = \hbar/T_2 = 0.2$ Ry. Figure 1 shows the linear spectrum and the power spectra $|\bar{E}(\omega)|^2$ of the laser pulse for various excitation conditions.

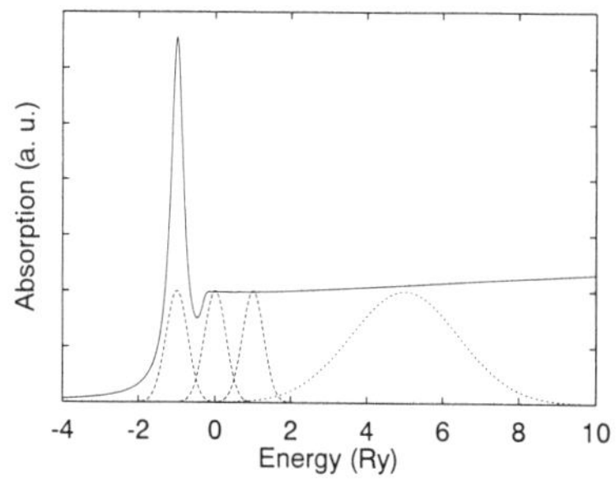

QFB3 Fig. 1. Absorption spectrum (solid curve) and power spectra for excitation conditions $\sigma = 2.5\ \hbar$/Ry, $\hbar\omega = -1$, 0, and +1 Ry (dashed curve) and $\sigma = 0.5\ \hbar$/Ry, $\hbar\omega = +7$ Ry (dotted curve).

In Fig. 2 time-integrated FWM signals $\int_{-\infty}^{+\infty} dt |P^{(3)}(t)|^2$ are plotted on a logarithmic scale versus time delay Δt for $\hbar\bar{\omega} = -1$, 0, and +1 Ry and $\sigma = 2.5\ \hbar$/Ry (dashed curves in Fig. 1). For excitation at the exciton, the signal is proportional to $\exp(-2\varepsilon\Delta t)$ for $t \to +\infty$ and $\exp(4\varepsilon\Delta t)$ for $t \to -\infty$. The behavior for negative time delay is due to polarization interference.[3] In the continuum case, the signal shows a decay with $\exp(-4\varepsilon\Delta t)$ for $t \to +\infty$ that indicates that the signal is delayed by $2\Delta t$.

In Fig. 3 the creation of a photon echo is demonstrated for excitation high in the continuum with $\hbar\omega = 7$ Ry, $\sigma = 0.5\ \hbar$/Ry (dotted curve in Fig. 1), and $\Delta t = 4\ \hbar$/Ry. The time-resolved signal $|P^{(3)}(t)|^2$ is plotted versus time t. A photon echo that has ap-

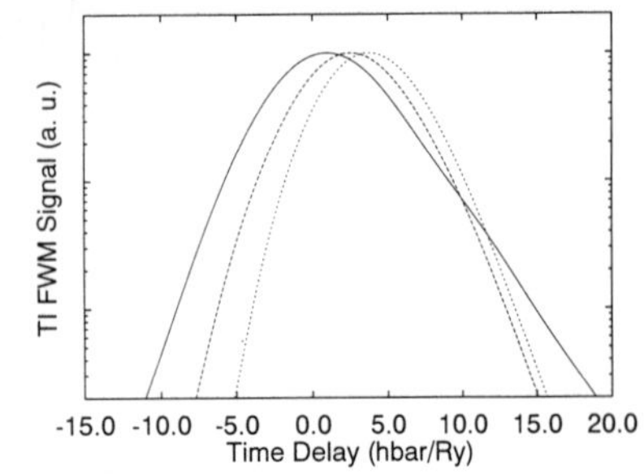

QFB3 Fig. 2. Time-integrated FWM signal on a logarithmic scale versus time delay for $\sigma = 2.5\ \hbar/\text{Ry}$ and $\hbar\omega = -1$ (solid curve), 0 (dashed curve), and +1 Ry (dotted curve).

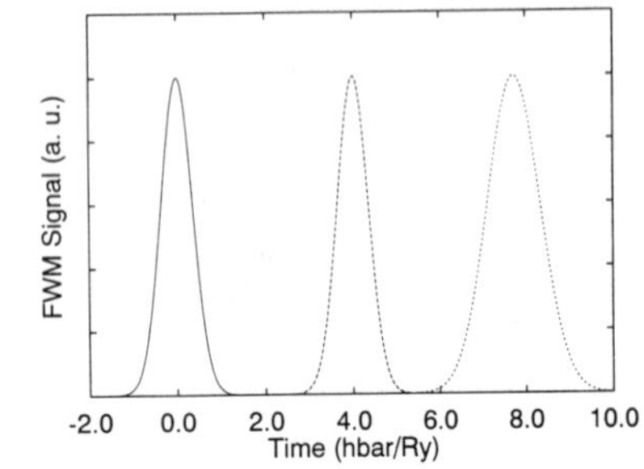

QFB3 Fig. 3. Intensities $|E_1(t)|^2$ (solid curve) and $|E_2(t)|^2$ (dashed curve) and time-resolved FWM signal (dotted curve) on a linear scale for $\sigma = 0.5\ \hbar/\text{Ry}$, $\hbar\omega = +7$ Ry, and $\Delta t = 4\ \hbar/\text{Ry}$.

proximately a Gaussian shape and a maximum at $t = 2\Delta t - 4\varepsilon\sigma^2$ appears.

In conclusion, we have provided convincing theoretical evidence for a photon echo in a homogeneously broadened semiconductor excited in the absorption continuum.

This work was supported by the Director, Office of Energy Research, Office of Basic Energy Sciences, Division of Material Sciences of the U.S. Department of Energy under contract DE-AC03-76SF00098. Two of the authors (S. G. and U. S.) wish to thank the German Research Association for financial support.

* *Department of Physics, University of Marburg, Renthof 5, D-35032 Marburg, Germany*

1. T. Yajima and Y. Taira, J. Phys. Soc. Jpn. **47**, 1620 (1979).
2. M. Lindberg, R. Binder, and S. W. Koch, Phys. Rev. B **45**, 1865 (1992).
3. K. Leo *et al.*, Phys. Rev. Lett. **65**, 1340 (1990).

QFB4 **8:45 am**

Coherent dynamics of resonantly excited quantum-well excitons

N. T. Pelekanos, E. J. Mayer, M. Hübner, J. Kuhl, W. W. Rühle, R Hay,* K. Ploog,*
Max Planck Institute for Solid State Research, Heisenbergstrasse 1, D-70506 Stuttgart, Germany

The dynamics of resonantly excited excitons in quantum wells (QWs) is an active field of research.[1–3] Resonant excitation offers the advantage of generating and probing the time evolution of an exciton population with a fixed ($k_\parallel \approx 0$) wave vector. Here, we present resonant time-resolved photoluminescence (PL) experiments on a high-quality 200-Å GaAs/$Al_{0.3}Ga_{0.7}As$ single QW, with 0.25-meV PL linewidth at lower temperature and less than 0.05 meV of Stokes shift.

In Fig. 1 we show the resonant PL decay at $T = 15$ K, for two different circular-polarization configurations: $\sigma^+\sigma^+$ and $\sigma^-\sigma^+$ (i.e., excite with σ^+ or σ^- and detect σ^+, respectively). The PL decay was measured by using a streak camera with a temporal resolution of 2.5 ps (2/e decay of the laser pulse). For times longer than a few tens of picoseconds the two configurations are equivalent, indicating that stationary equilibrium between the exciton spin states has been established. Focusing in the first tens of picoseconds, during which nonequilibrium excitons are on display, we observe a fast initial decay τ for the $\sigma^+\sigma^+$ and a distinct rise after the laser pulse (visible in the curve) for the $\sigma^-\sigma^+$. In the latter case the observed delay of 25–30 ps for the PL to reach its maximum gives a first-order estimate for the spin relaxation in our QW.

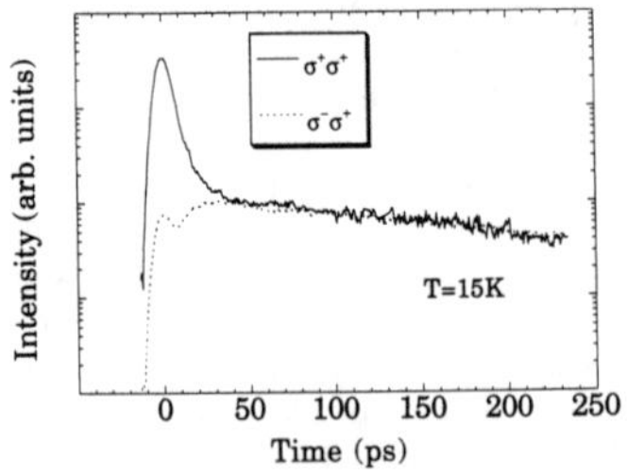

QFB4 Fig. 1. PL decay for resonant excitation on a 200-Å GaAs/AlGaAs single QW for two different polarization configurations: $\sigma^+\sigma^+$ and $\sigma^-\sigma^+$.

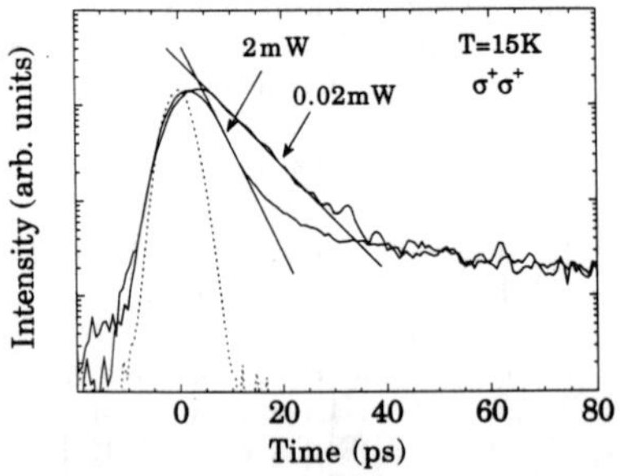

QFB4 Fig. 2. Resonant PL decay in the $\sigma^+\sigma^+$ configuration for various power levels. 2 mW correspond to 6×10^9 excitons/cm². The temporal response to a 1-ps laser pulse is also indicated (dotted curve).

The fast decay time τ in the $\sigma^+\sigma^+$ configuration strongly depends on the exciton density. In Fig. 2 we show the PL decay for two different power levels. The system response to a 1-ps laser pulse is also indicated. Clearly, τ decreases with increasing exciton density. This behavior can be understood by considering that τ reflects the rapid depletion of the $k_\parallel = 0$ optically active state by (1) scattering with acoustic phonons and other excitons into the $k_\parallel > k_0$ optically inactive state (k_0 being the photon wave vector), (2) radiative recombination, and (3) spin flip into other exciton spin states. Increasing the exciton density makes the exciton–exciton scattering more efficient and hence reduces τ. In Fig. 3 we plot τ as a function of exciton density n_x, and we compare it with $T_2/2$ independently measured in a degenerate-four-wave-mixing (DFWM) experiment. Clearly, τ is larger than $T_2/2$ and $d(T_2/2)/dn_x$ is larger than $d\tau/dn_x$. We explain this qualitatively by the following consideration: in the case of an exciton with $k_\parallel < k_0$, exciton–acoustic-phonon and exciton–exciton scattering can leave the exciton either inside or outside the $k_\parallel < k_0$ optically active range. Both processes contribute to $T_2/2$, whereas only the second contributes to τ, explaining the larger values of τ and its weaker dependence on exciton density.

Finally, by measuring the long thermalized PL decay time to be $\tau_{therm} = 580$ ps and the homogeneous linewidth to be $\Gamma_h = 0.45$ meV from the DFWM measure-

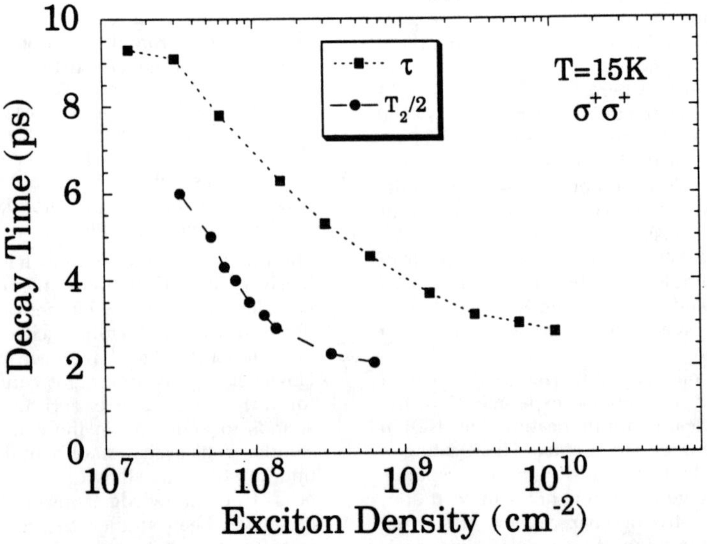

QFB4 Fig. 3. Comparison of the fast PL decay τ and $T_2/2$ as a function of exciton density. For estimating the dephasing time we assumed homogeneously broadened transitions.

ments for the same exciton density and temperature $T = 15$ K, we can apply the following equation[3]:

$$\tau_{\text{therm}} = \frac{3\Gamma_h}{4E_0}\tau_0\frac{1 + \exp(\Delta/kT)}{1 - \exp(\Gamma_h/kT)},$$

where $E_0 = 0.1$ meV is the exciton kinetic energy for $k = k_0$ and $\Delta = 0.1$ meV is the spin exchange splitting. The above equation yields the radiative constant at $k_\parallel = 0$, $\tau_0 = 24$ ps, which agrees very well with the theory[4] and implies that τ is not limited by radiative recombination.

**Paul Drude Institute, Hausvogteiplatz 5-7, D-10117 Berlin, Germany*

1. B. Deveaud, *et al.*, Phys. Rev. Lett. **67**, 2355 (1991).
2. R. Eccleston *et al.*, Phys. Rev. B **44**, 1395 (1991).
3. A. Vinattieri *et al.*, Solid State Commun. **88**, 189 (1993).
4. L. C. Andreani *et al.*, Solid State Commun. **77**, 641 (1991).

QFB5 9:00 am

Bloch oscillations in semiconductor superlattices with nonresonant optical excitation

Christian Waschke, Hartmut G. Roskos, Heinrich Kurz, Klaus Köhler,* *Institute for Semiconductor Technology II, Rhine-Westphalia Technical University, Sommerfeldstrasse 24, D-52056 Aachen, Germany*

With time-resolved tetrahertz emission spectroscopy, Bloch oscillations of optically excited wave packets in semiconductor superlattices have been observed. All time-resolved experiments conducted up to now with other measurement techniques have concentrated on wave packets generated by resonant excitation of Wannier–Stark-ladder (WSL) states evolving from the first electron miniband. In this paper we present evidence for Bloch oscillations in the WSL at nonresonant excitation as high as 80 meV above the fundamental band gap. This is, to our knowledge, the first observation of coherent charge oscillations after excitation of high-lying states, in which ultrafast dephasing has been assumed to suppress such oscillations.

The measurements are performed on semiconductor superlattices with miniband widths of 7 to 50 meV for the first electron miniband. A reverse-bias field is applied between the doped substrate and a semitransparent Cr-Schottky contact on the sample surface. Low-temperature photocurrent and electro-reflectance measurements reveal clearly the expected excitonic features characteristic for formation of WSLs at the lowest electron/hole minibands. Additionally, we identify much weaker structures in the spectra, indicative of WSLs evolving from higher minibands.

In Fig. 1, detected terahertz transients are plotted for various photon energies of the excitation pulses. The wavelength is tuned from low energies beneath the band gap (1.531 eV) to high energies (1.642 eV) above the second electron miniband. At the band edge the signal consists of a single-cycle transient emitted during excitation of

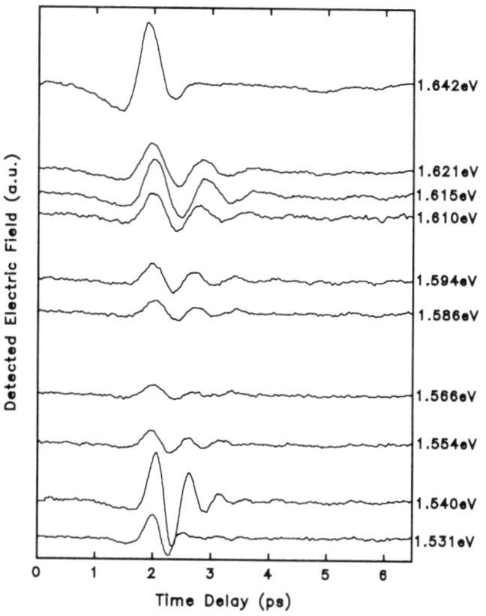

QFB5 Fig. 1. Terahertz transients of the Bloch oscillation as a function of the excitation energy.

the superlattice (time-delay zero: around 1.5 ps). It is attributed to instantaneous polarization. Additionally, for excitation in the WSL of the first miniband around 1.540 eV, a strong oscillation is observed. When the laser is tuned to shorter wavelengths, the amplitude first decreases but increases again above 1.58 eV. Around 1.61 eV, at which the second miniband of the electrons is expected, the amplitude of the oscillations has a second maximum. The oscillations disappear with shorter excitation wavelength, leaving a single-cycle transient at 1.642 eV.

The oscillating terahertz radiation emitted after excitation of the WSL evolving from the first miniband is known to result from Bloch oscillations of electron wave packets. A feature characteristic of Bloch oscillations is the electric-field tunability of the oscillation frequency with a characteristic linear field dependence.[1,2] At nonresonant excitation, the emission of electromagnetic radiation exhibits this bias dependence, proving that the emission after excitation of high-lying states originates from Bloch oscillations.

It is remarkable that the dephasing of the Bloch oscillations for nonresonant excitation is only slightly faster than that for resonant excitation at the band gap. This is completely unexpected from the dc-photocurrent, electro-reflectance, and time-resolved four-wave-mixing measurements. These techniques probe optical conduction-to-valence-band transitions.[3] The linewidth and time resolution, respectively, are determined by the optical dephasing time. In contrast to these experiments, terahertz emission spectroscopy is sensitive to the intraband polarization in either the valence band or the conduction band, in which dephasing does not depend on the optical frequencies.

**Fraunhofer Institute for Applied Solid State Physics, Tullarstrasse 72, D-79108 Freiburg, Germany*

1. C. Waschke, H. G. Roskos, R. Schwedler, K. Leo, H. Kurz, and K. Köhler, Phys. Rev. Lett. **70**, 3319 (1993).
2. H. G. Roskos, *Festkörperprobleme*, Vol. 34 of Advances in Solid State Physics (Vieweg Verlag, Braunschweig), in press.
3. Recently, a direct comparison of the intraband and interband dynamics of Bloch oscillations for resonant excitation has been performed (Dekorsy *et al.*, submitted for this conference).

QFB6 9:15 am

Bloch oscillations of excitonic and continuum states in superlattices

T. Dekorsy, P. Leisching, K. Köhler,* H. Kurz, *Institute for Semiconductor Technology II, Rhine-Westphalia Technical University, D-52056 Aachen, Germany*

We report on the simultaneous performance of four-wave-mixing[1] (FWM) and transmittive electro-optic-sampling[2] (TEOS) experiments to investigate Bloch oscillations in biased GaAs/$Al_xGa_{1-x}As$ superlattices. These techniques allow a comparative study of the coherent dynamics of the inter- and the intraband polarization, because they can be performed in exactly the same experimental arrangement. FWM signals are well known to have their strongest contributions from excitonic resonances at low excitation densities.[4] The contribution of continuum states can be neglected because of the strong dephasing of their *interband* coherence, which is typically below 100 fs in GaAs quantum wells and bulk GaAs, even at 10 K. In contrast to FWM, TEOS enables us to monitor directly the dynamics of the *intraband* dipole moment induced by the spatial motion of coherent electronic wave packets in amplitude and phase.[2] Thus, in TEOS experiments we directly measure the source of coherent terahertz radiation emitted by Bloch oscillations.[3] A careful comparison of TEOS and FWM data under exactly the same excitation conditions reveals significant differences in the frequency and dephasing of Bloch oscillations. These specific differences show that Bloch oscillations observed in FWM are based on excitonic quantum beats, whereas Bloch oscillations observed in TEOS are mainly performed by electrons in continuum states. The observation of Bloch oscillations performed by continuum states is of prime importance for the realization of electronically driven Bloch oscillations and of Bloch oscillations at room temperature. The results reported here are supported by recent results in terahertz emission experiments, in which Bloch oscillations were observed for excitation well above the lowest optical transitions in the superlattice.[5]

The reason for the frequency differences observed is illustrated in Fig. 1. At higher fields, the initial-wave-packet results from the coherent superposition of (heavy-hole 0)-to-(electron 0) (h_0–e_0) and (heavy-hole 0)-to-(electron −1) (h_0–e_{-1}) WS states. If Coulomb interaction between electrons and holes is neglected, the frequency/field relation of the Bloch oscillations performed by electrons in continuum states is $\nu_{BO} = eFd/h$, where F is the electric field and d is the superlattice period. For excitonic wave packets the exact oscillation frequency *also*

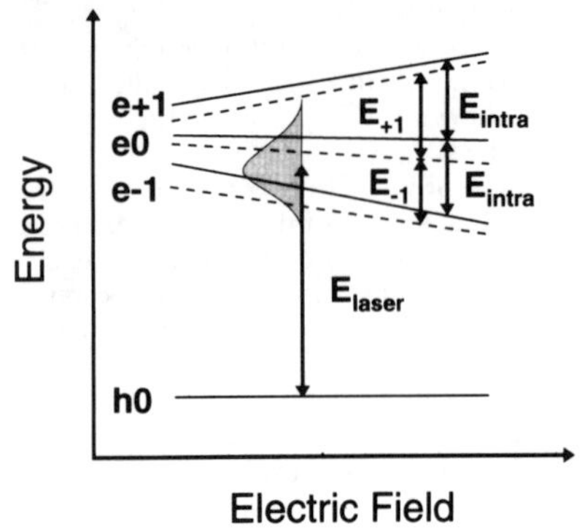

QFB6 Fig. 1. Illustration (not to scale) of the field dependence of the h_0–e_{-1}, h_0–e_0, and h_0–e_{+1} Wannier-Stark continuum (solid lines) and exciton transitions (dashed lines) at higher fields. The energy splitting between continuum states is denoted as E_{intra}, between the h_0–e_0 and h_0–e_{-1} excitons as E_{-1}, and between the h_0–e_0 and h_0–e_{+1} excitons as E_{+1}. E_{laser} is the laser energy. $E_{-1} < E_{intra} < E_{+1}$ at higher electric fields.

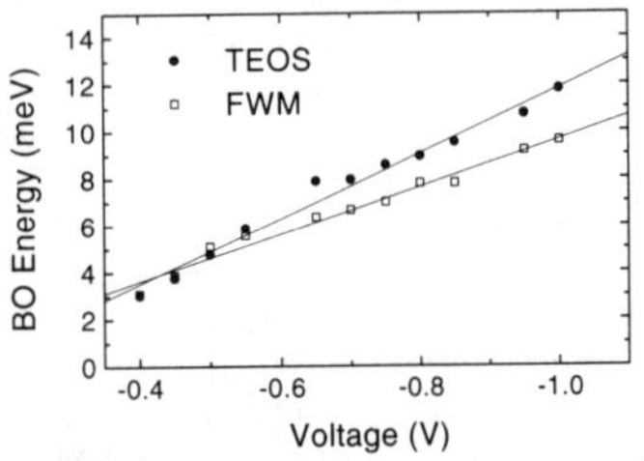

QFB6 Fig. 2. Energy of the Bloch oscillations (BO) as a function of the applied voltage obtained from the FWM and TEOS data. Straight lines are linear fits to the data.

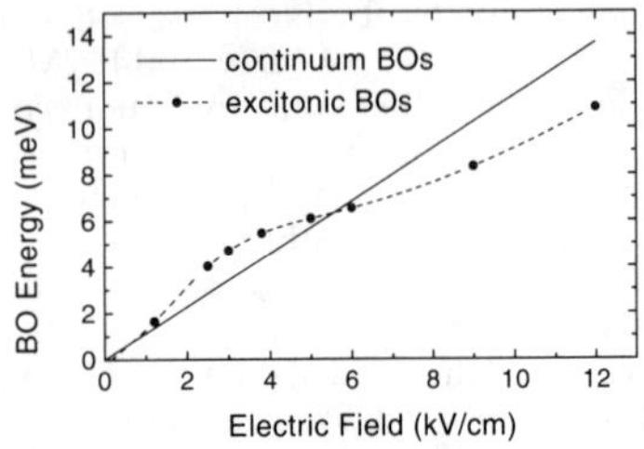

QFB6 Fig. 3. Energies of Bloch oscillations (BOs) calculated from the field dependence of the excitonic binding energies (filled circles) and the semiclassical frequency dependence $h\nu_{BO} = eFd$ (straight line).

depends sensitively on the field dependence of the exciton binding energies of the transitions involved. At fields $F \approx \Delta/ed$, where Δ is the width of the miniband, the electronic WS wave functions become strongly localized, leading to an increase of the absorption strength and the exciton binding energy $E_B(h_0, e_0)$ of the h_0–e_0 transition, whereas the h_0–e_{-1} transition becomes indirect, with a reduced exciton binding energy $E_{B(h0,e-1)}$. Therefore, the excitonic Bloch oscillation frequency is expected to rise less with the field than the semiclassical frequency/field relation expected for Bloch oscillations performed by electrons in continuum states. The maximum deviation expected from $\nu_{BO} = eFd/h$ in FWM is the difference in exciton binding energies $E_{B(h0,e0)} - E_{B(h0,e-1)}$ at high fields.

The sample is a GaAs/$Al_{0.3}Ga_{0.7}As$ superlattice that has 35 periods of 97-Å-wide GaAs wells and 17-Å-wide barriers and is externally biased. The excitation pulse derived from a Ti:sapphire laser is centered at 798 nm between the h_0–e_0 WS transition and the h_0–e_{-1} WS transition. Figure 2 shows the oscillation energies obtained from simultaneously performed FWM and TEOS experiments as a function of the reverse bias. Pronounced energy differences are clearly observed at higher electric fields. The maximum energy difference measured at −1.0 V is 2.2 meV. In addition, distinct differences in the dephasing time of the interband (FWM) and intraband (TEOS) polarization are observed in the time-resolved data.

We performed numerical calculations of the exciton binding energies of the direct and indirect transitions. In Fig. 3 the Bloch-oscillation frequency obtained from these calculations is compared to the semiclassical frequency/field relation. The numerical results are in quantitative agreement with the experiments. Thus we conclude that we observe Bloch oscillations performed by electronic wave packets in continuum states in TEOS, whereas excitonic quantum beats are observed in FWM. The results are compared to recent theoretical predictions for differences in the inter- and intraband dynamics obtained from the solution of the semiconductor Bloch equations.[6]

This work is supported by the German Research Association and the Volkswagen foundation.

*Fraunhofer Institute for Applied Solid State Physics, D-79108 Freiburg, Germany

1. J. Feldmann *et al.*, Phys. Rev. B **46**, 7252 (1992); K. Leo *et al.*, Solid State Commun. **84**, 943 (1992). P. Leisching *et al.*, Phys. Rev. **50** (November 15, 1994).
2. T. Dekorsy *et al.*, Phys. Rev. B **50**, 8106 (1994).
3. C. Waschke *et al.*, Phys. Rev. Lett. **70**, 3319 (1993).
4. D.-S. Kim *et al.*, Phys. Rev. Lett. **68**, 1006 (1992).
5. C. Waschke, H. G. Roskos, *et al.*, submitted for this conference.
6. T. Meier *et al.*, Phys. Rev. Lett. **73**, 902 (1994).

QFB7 (Invited) **9:30 am**

Coherent dynamics at the semiconductor band edge

M. Wegener, *Institut für Physik, University Dortmund, D-44221 Dortmund, Germany*

The pioneering experiments on the coherent dynamics of excitons[1] and continuum states[2] in semiconductors have opened a new field of research. Employing transient four-wave mixing (FWM) they were able to determine the phase relaxation time, i.e. T_2 of excitons and continuum states, respectively. In this paper we review the progress in refined experimental techniques as well as in the understanding of the microscopic physics of the underlying scattering processes.

When a semiconductor is illuminated with a very short pulse, hence with a very large bandwidth, it is not at all clear which states will actually contribute to the measured FWM-signal. This problem, which gives rise to a common criticism of experiments with "too short"-pulses, can, however, be avoided if the diffracted wave is spectrally resolved. The pulses still excite a coherent superposition of many quantum mechanical states; the dynamics of all individual constituents, however, can clearly be separated in the FWM-spectra. This has enabled us to distinguish between interference effects and dephasing in the coupled coherent dynamics of excitons and continuum states at low temperatures.[3] Here we have used 140-fs pulses and bulk Germanium as a model substance.

A much more detailed picture of the underlying microscopic physics can be drawn from experiments that employ the dependence of the FWM-spectra on the polarization of the two incident beams. It was first realized in Ref. 4 (for the case of excitons) that certain contributions of carrier-carrier scattering can be switched on and off by choice of the polarization. This enabled a sensitive test of current state-of-the-art many body theories.[5] Suprisingly, we find that at low carrier densities ($\approx 10^{15}$ cm^{-3}) so-called polarization scattering dominates over scattering with occupied states, which is usually considered the dominating scattering mechanism.

In a second set of experiments we employ the same techniques and concepts to GaAs at low temperatures, taking advantage of the recent progress in generation of $\approx$10 fs-pulses with self-modelocked Ti:Sapphire lasers. Experimental data in a wide range of carrier densities (9 10^{14} cm^{-3} – 2 10^{18} cm^{-3}) are compared with model calculations. From analyzing the FWM-spectra (as a function of time-delay) and their polarization dependence we obtain detailed insights into the **k**-dependence of dephasing and the influence of memory-effects[6] (non-Marcovian dephasing).

1. L. Schultheis, J. Kuhl, A. Honold, and C. W. Tu, Phys. Rev. Lett. **57**, 1635 (1986).
2. P. C. Becker, H. L. Fragnito, C. H. Brito-Cruz, R. L. Fork, C. E. Cunningham, J. E. Henry, and C. V. Shank, Phys. Rev. Lett. **61**, 1647 (1988).
3. T. Rappen, U. Peter, M. Wegener, and W. Schäfer, Phys. Rev. B **48**, 4879 (1993).
4. H. Wang, K. B. Ferrio, D. G. Steel, Y. Z. Hu, R. Binder, and S. W. Koch, Phys. Rev. Lett. **47**, 1261 1993).
5. T. Rappen, U. Peter, M. Wegener, and W. Schäfer, Phys. Rev. B **49**, 10774 (1994).
6. M. Wehner, D. Steinbach, T. Marschner, W. Stolz, and M. Wegener, unpublished.

QFC 10:30 am Hyatt B

Atoms in Strong Laser Fields

Howard R. Reiss, *American University, Presider*

QFC1 (Invited) 10:30 am

Above-threshold ionization beyond the high harmonic cutoff

L. F. DiMauro, *Brookhaven National Laboratory, Department of Chemistry, Upton, New York 11973*

Experimental studies of atoms in intense laser fields are often complicated by the large dynamic range of signal covered over an intensity range that meaningful displays the full evolution of the dynamics: from the perturbative to the tunnel ionization regime. We have developed kilohertz repetition rate laser systems which allow us to reach peak intensities of 10^{16} W/cm^2, and detect photoionization yields over 12 orders of magnitude, and so effectively span the relevant intensity and signal ranges. Theoretical models offering a unified description of various strong field phenomena: optical harmonic generation (OHG), above-threshold ionization (ATI), and multiple ionization, may now be critically tested. Semiclassical trajectory calculations suggest that the rescattering of the field-driven electron with the core can result in OHG, high energy ATI, and multiple ionization. Evidence for rescattering is observed in the angular distributions for high-order ATI peaks as off-axis scattering rings. These structures depend strongly on the intensity and scale approximately as $8U_p$. A quantum-mechanical model based on the single active electron approximation (SAE), which predicts these structures and their intensity dependence, will be compared with the semiclassical model. Strong-field multiple photoionization of helium has been studied using a 100 fs, 780 nm kilohertz laser system. The He^+ yields are well predicted by the SAE model; however an enhanced double ionization called nonsequential (NS) ionization, is observed over an intensity range in which the single ionization dynamics evolves from multiphoton to the tunneling regime. However, the NS double ionization yield scales only with the ac tunneling rate for single ionization. The origin of the nonsequential mechanism will be discussed in terms of the rescattering model.

QFC2 11:00 am

Coherence effects in multiphoton ionization studied by using femtosecond pulses

R. R. Jones, *Department of Physics, University of Virginia, McCormick Road, Charlottesville, Virginia 22901*

Ultrashort laser pulses with large pulse energies have been used to study high-intensity multiphoton ionization of atoms with unfocused laser beams. Intensity-dependent coherence effects, which were masked in previous experiments by spatial-intensity variations at the laser focus, are observed.

Specifically, four-photon ionization of Na has been studied by using 150-fs 777-nm laser pulses with energies up to 20 mJ. During the laser pulse, transient resonances between the ground state and several excited states occur because of intensity-dependent ac Stark shifts.[1] These resonances occur twice in each pulse, once on the rising-intensity edge and again on the falling-intensity edge. The transition probability during the second resonance interval depends on the relative phase difference $\delta\phi$ between the two states and on the population in each state. In general, $\delta\phi$ depends on the peak pulse intensity and the time interval between the two resonances. At the intensities studied, ionization out of the excited states dominates. Therefore, interferences in the excited-state population are reflected in the total ion yield.

When a single pulse is used to ionize the Na atoms, modulations in the total ionization signal are observed as a function of the peak intensity in the pulse. At certain intensities, the ionization probability actually *decreases* with increasing laser intensity.

The coherence established between the ground state and the excited states during one pulse can be exploited to control the ionization yield from a second, temporally delayed phase-coherent pulse. In this configuration, the phase difference $\delta\phi$ between two resonant states is acquired during the time interval between the two laser pulses and is directly proportional to the relative pulse delay. Two identical, temporally delayed pulses are created from a single pulse in a Michelson interferometer. The total ionization yield from the two pulses oscillates as a function of the delay between them. The peaks of these oscillations show an ionization enhancement of as much as a factor of 20 over the single-pulse yield. In addition, the Fourier transform of the ionization signal versus time delay yields the transition frequencies that contribute to the excited-state population.[2] Therefore, not only does the two-pulse technique allow significant control over multiphoton ionization, but it can also identify the bound states that contribute to the process.

This work was supported by the Office of Naval Research.

1. R. R. Freeman, P. H. Bucksbaum, H. Milchberg, S. Darack, D. Schumacher, and M. E. Geusic, Phys. Rev. Lett. **59**, 1092 (1987).
2. R. R. Jones, C. S. Raman, D. W. Schumacher, and P. H. Bucksbaum, Phys. Rev. Lett. **71**, 2575 (1993).

QFC3 11:15 am

Intense-field photoionization of calcium: influence of coupled states and electron correlation

Barry Walker, B. Sheehy, L. F. DiMauro, *Brookhaven National Laboratory, Upton, New York 11973*

Electron correlation and its influence on photoionization spectroscopy are long-standing interests in atomic physics. A scenario for investigating final-state electron correlation involving intense-field photoionization of a neutral atom at a frequency that couples and Rabi-splits ion-core levels (Fig. 1) has been described.[1] Electron correlation is predicted to cause the energy spectrum of the free electron to mirror the splitting of the ion levels. We have investigated this process experimentally and present the results.

The light for ionization and ion coupling is produced by frequency-doubling regeneratively amplified Ti:sapphire mode-locked pulses. The light pulses were better than twice the $\Delta\nu\Delta\tau$ transform limit, had an FWHM of 180 fs, and tuned from 380 to 406 nm. The species studied was calcium produced in an atomic beam.

The interaction of the light with calcium gives perturbative, nonresonant ionization when it is detuned from ion-core resonances, as verified by a measured I^2 intensity dependence. The photoelectron spectrum (PES) for the ionization of the neutral as a function of wavelength is shown in Fig. 2. In the low-intensity limit [Fig. 2(a)] the experimental peak values agree well with expected values based on the two-photon excitation to an unperturbed-ion ground state. Although the low-field limit shows spectra with clear peaks at the expected excess photon energy, the more-complicated high-field spectra [Fig. 2(b)] cannot be understood without consideration of the light-induced coupling in the Ca^+-ion core.

When the laser is tuned to an ion-core resonance ($4s$–$4p$), the PES as a function of intensity shows that the high-intensity

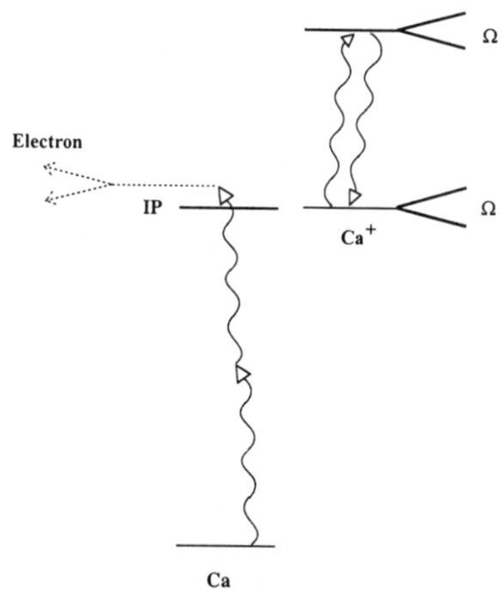

QFC3 Fig. 1. Simplified diagram of calcium showing the neutral ionization, ion-core resonance, Rabi splitting Ω, and the proposed impression of the splitting on the free electron.

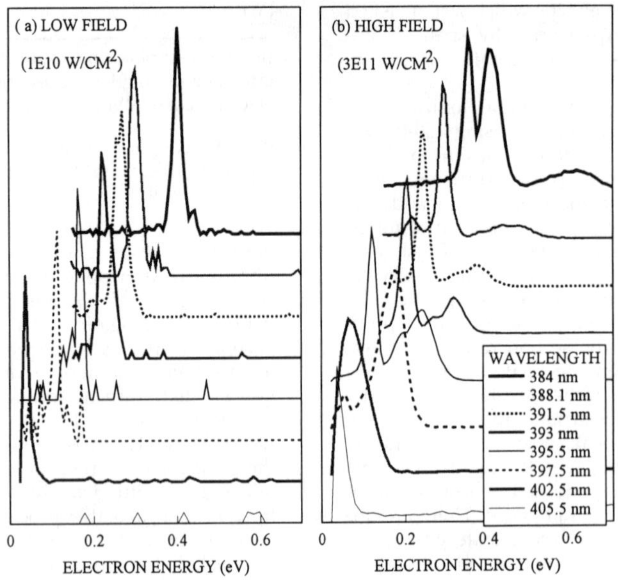

QFC3 Fig. 2. PES for calcium at several wavelengths in the low-intensity and high-intensity limits.

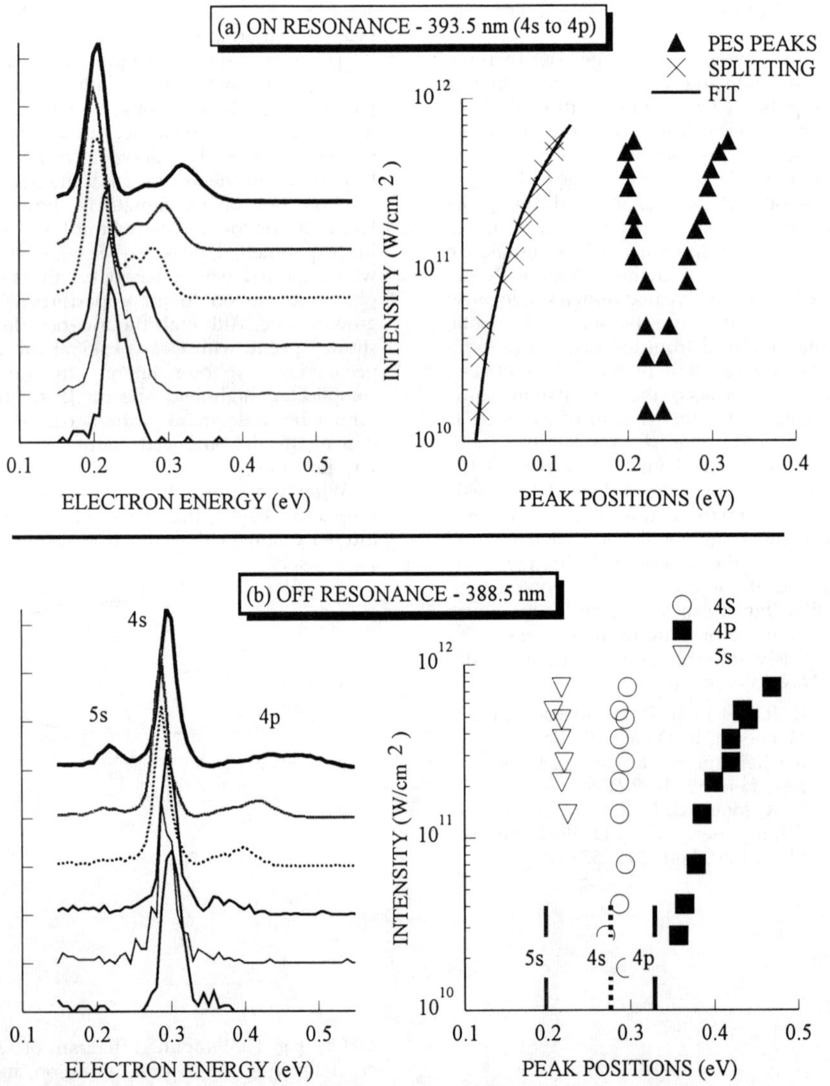

QFC3 Fig. 3. PES for calcium (a) on and (b) off of an ion-core resonance. A plot of the peak positions as a function of intensity accompanies the spectrum.

structure is the result of a peak splitting that fits well to a linear electric-field dependence [Fig. 3(a)], consistent with a Rabi splitting. When the laser is tuned off the resonance [Fig. 3(b)], the ion final-state peaks are identified in the neutral-ionization spectrum by extrapolation to field-free values and by laser polarization dependence. Though off resonance the 4*s* ground state of the ion is the dominant feature in the spectrum, excited-ion-state-population and peak shifting are still significant. The 4*p* state, for example, is observed to shift similarly to the splitting in the on-resonance case, despite a large detuning (330 cm^{-1}).

The splitting of the neutral-ionization peaks when an ion-core resonance is excited expresses some features of a simple Rabi picture, but the effect is more complex. The peak splittings are asymmetric in form on resonance, and off resonance the shifting is also asymmetric. Generally, the exchange between the ion core and the free electron through electron correlation is observed to strongly influence the outgoing electron final-state distribution.

This work was supported by the U.S. Department of Energy.

1. R. Grobe and J. H. Eberly, Phys. Rev. A **48**, 623.

QFC4 **11:30 am**

Observation of the direct triple ionization of argon atoms

S. Augst, A. Talebpour, S. L. Chin, Y. Beaudoin,* M. Chaker,* *Department of Physics, Pav. Vachon, University of Laval, Quebec, Quebec G1K 7P4, Canada*

The occurrence of direct double ionization of atoms that are exposed to intense laser fields has been known for over a decade.[1] By "direct double ionization" we mean that doubly charged ions are created directly from neutral atoms rather than through a sequential process of singly charged ions being ionized to created doubly charged ions. Recently, interest in this topic has been renewed, with the introduction of high-repetition-rate lasers, making it possible to perform statistical studies on the ionization signal and to derive various theoretical models to explain this behavior.[2–5]

We have studied this phenomenon in argon by using a 600-fs 1053-nm laser. The laser was focused with an *f*/30 lens. A large *f*-number was used to produce a very large focal volume, which permits the detection of a signal even for very low rates of ionization. Figure 1 shows the ion signal as a function of laser intensity. Each data point corresponds to a single laser shot. Direct double ionization is clearly demonstrated by the appearance of the second charge state at intensities lower than the saturation intensity of the first charge state. At low intensities the second-charge-state production is dominated by direct double ionization. When the first charge state reaches saturation, there are no more neutral atoms at the focus and direct double ionization is no longer possible. As the laser intensity is increased, the second charge state production is dominated by sequential ionization.

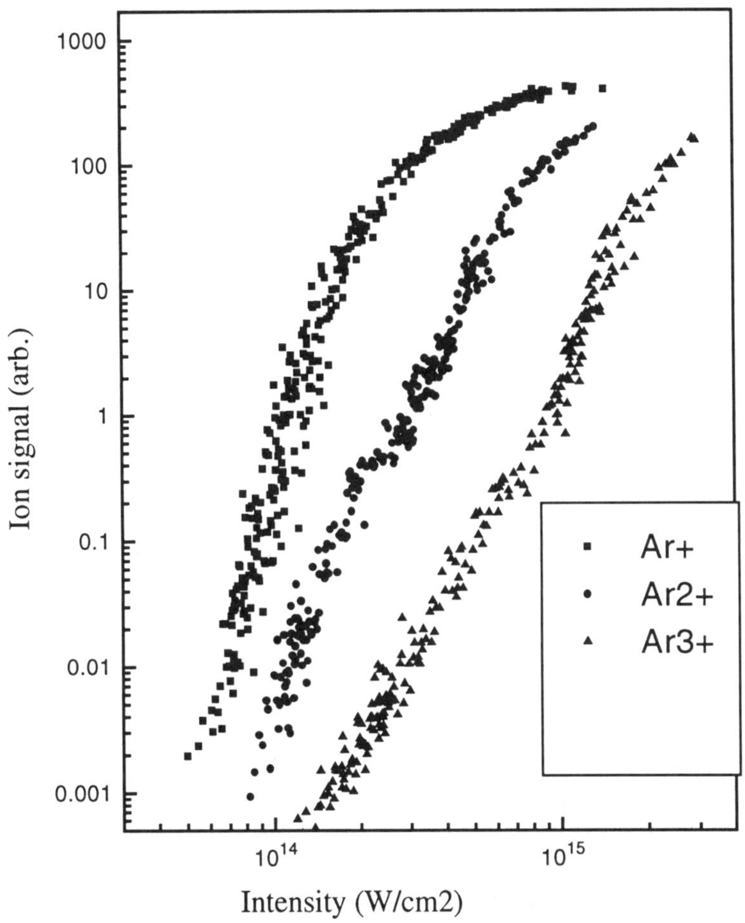

QFC4 Fig. 1. Argon-ion signal as a function of incident laser intensity for a 600-fs 1053-nm laser pulse.

An unexpected feature is evident in the third charge state. A significant triply charged ion signal is present at laser intensities lower than the saturation intensity of both the second and first charge states. This indicates the appearance of triply charged ions at intensities for which there are still neutral atoms remaining in the laser focus. We discuss this phenomenon and review how it relates to recently proposed theoretical models.

**INRS–Energie et Materiaux, Montee St. Julie, Varennes, Quebec J3X 1S2, Canada*

1. A. L'Huillier, L. A. Lompre, G. Mainfray, and C. Manus, J. Phys. B **16**, 1363 (1983).
2. D. N. Fittinghoff, P. R. Bolton, B. Chang, and K. C. Kulander, Phys. Rev. Lett. **69**, 2642 (1992).
3. P. B. Corkum, Phys. Rev. Lett. **71**, 1994 (1993).
4. D. N. Fittinghoff, P. R. Bolton, B. Chang, and K. C. Kulander, Phys. Rev. A Brief Reports **49**, 2174 (1994).
5. B. Walker, B. Sheehy, L. F. DiMauro, P. Agostini, K. J. Schafer, and K. C. Kulander, Phys. Rev. Lett. **73**, 1227 (1994).

QFC5 **11:45 am**

Appearance intensities for multiply charged ions in a strong laser field

Miroslaw Brewczyk,*
Kazimierz Rzazewski,** Charles W. Clark, *Electron and Optical Physics Division, National Institute of Standards and Technology, Technology Administration, U.S. Department of Commerce, Gaithersburg, Maryland 20899*

Many of the phenomena observed[1] in a strong-field laser–atom interaction, such as high-harmonic generation, above-threshold ionization, and stabilization, can be obtained in simple models that invoke a picture of one electron under the influence of an effective potential; the observed phenomena are also found in numerical experiments that integrate the time-dependent Schrödinger equation for one-electron systems under fairly realistic conditions.

The production of multiply charged ions by laser irradiation of neutrals is, on the other hand, of an essentially many-electron character. It has not yet been possible to perform numerical experiments with a realism comparable to that which has been attained in the one-electron case, and it seems unlikely that in the near future any such treatment will be applicable to an atom with more than two electrons: the two-electron system in the nonperturbative regime is just beginning to be explored,[2,3] although there has been extensive work on a model system of two electrons moving in one dimension.[4] However, with the increase of attainable laser intensity, experiments have moved into a regime of unprecedently high ionization: recently, experiments[5] at close to 10^{19} W/cm^2 have exhibited ion-charge states of more than 40. Thus there is a need to develop a theoretical framework within which large numbers of electrons can be treated.

We treat multiple ionization within the time-dependent Thomas–Fermi approximation. The atomic electrons are represented by a smoothly varying density function $n_e(\mathbf{r}, t)$, which satisfies a hydrodynamical equation of motion. We solve this equation numerically. In the case of radiation polarized linearly along the z axis, this amounts to integrating a time-dependent partial differential equation for the density $n_e(p, z, t)$ as a function of the two cylindrical polar coordinates p and z. This was accomplished by discretizing space on a grid of 0.1-a.u. spacing and solving the equation of motion for the difference of $n_e(p, z, t)$ from the static Thomas–Fermi value (since charge near the nucleus is not disturbed even in very strong fields). We used a combination of Lax and overrelaxation methods in the numerical solution.

Results are reported here for the case of Xe ($Z = 54$) and Nd:YAG laser radiation. For intensities up to 10^{16} W/cm^2, we find that the final charge states are relatively

QFC5 Table 1. Comparison with experiment of the appearance intensities, for Xe ionized by a Nd:YAG laser, computed with the help of various statistical models: TF, present work; static TF, static Thomas–Fermi model of Ref. 6; TFDJ, static Thomas–Fermi–Dirac model of Ref. 6; Experiment, Ref. 7. Intensity is in units 10^{13} Wcm^{-2}.

Charge state	TF	static TF	TFDJ	Experiment
1	-	0.	0.15	7
2	-	0.14	1.83	18
3	-	2.12	9.19	40
4	84	11.3	30.7	92
5	169	37.7	79.5	160
6	406	98.2	181.5	214
7	710	220	367	672
8	1024	442	683	802

insensitive to the temporal profile of the pulse. The appearance intensity, i.e., the intensity at which a given ionic species appears, exhibits reasonable agreement with experiments, as shown in Table 1. Early calculations relevant to the experimental conditions[5] show much more dependence on the pulse structure. However, we generally confirm the estimates of Ref. 5 that the maximum charge state that can be produced by direct laser–atom interaction is about 20.

*Filia Uniwersytetu Warszawskiego, ul. Lipowa 41, 15-424 Bialystok, Poland
**PAN Center for Theoretical Physics and College of Science, Al. Lotnikow 32/46, 02-668 Warsaw, Poland

1. B. Piraux, A. L'Huillier, and K. Rzazewski, eds., *Super-Intense Laser-Atom Physics*, NATO ASI Series (Plenum, New York, 1994).
2. X. Tang, H. Rudolph, and P. Lambropoulos, Phys. Rev. A **44**, R6994 (1991); P. Lambropoulos and X. Tang, in Ref. 1, p. 403.
3. S. Blodgett-Ford, J. Parker, and C. W. Clark, in Ref. 1, p. 391.
4. J. H. Eberly and R. Grobe, in Ref. 1, p. 445.
5. A. McPherson, B. D. Thompson, A. B. Borisov, K. Boyer, and C. H. Rhodes, Nature **370**, 631 (1994).
6. M. Brewczyk and M. Gajda, J. Phys. B **21**, L383 (1988); M. Brewczyk and K. Rzazewski, J. Mod. Opt. **38**, 1883 (1991); for a review see M. Brewczyk, M. Gajda, and K. Rzazewski, Laser Phys. **3**, 343 (1993).
7. S. Augst, D. Strickland, D. D. Meyerhofer, S. L. Chin, and J. H. Eberly, Phys. Rev. Lett. **63**, 2212 (1989).

QFC6 **12:00 m**

High-harmonic generation in diatomic molecular gases by using intense subpicosecond laser pulses

Y. Liang, S. Augst, A. Talebpour, P. A. Golovinski, S. L. Chin, Y. Beaudoin,* M. Chaker,* *Center for Optics, Photonics, and Lasers, Department of Physics, University of Laval, Quebec, Québec G1K 7P4, Canada*

High-harmonic generation in molecular gases has been investigated recently. Krause *et al.*[1] have calculated the harmonic spectrum of a stretched hydrogen molecule in a linearly polarized laser field. Theories of high-harmonic generation from diatomic molecular ions have also been developed.[2,3] In this paper we present an experimental study of high-harmonic generation in the diatomic molecular gases H_2, D_2, N_2, and O_2 by using a 600-fs 1053-nm Ti:sapphire/Nd:glass laser and a 160-fs 800-nm Ti:sapphire laser. The rare gases Ar and Xe are also used in the experiments for comparison.

Wahlström *et al.*[4] have shown that a harmonic signal from atoms varying with laser intensity can be divided into three regions. It increases at first rather steeply in the cutoff region, it then increases much more slowly in the plateau region, and finally it saturates when the gas becomes ionized. First, we concentrate our study of harmonic generation in diatomic molecules in the cutoff region with relatively low laser intensities by using a 600-fs 1053-nm Ti:sapphire/Nd:glass laser. The harmonic-conversion efficiencies are compared by choosing various gas pairs. The results show that for a fixed laser frequency and in the absence of static multiphoton resonances, the ionization potential and mass are not sensitive parameters for harmonic efficiency but the static polarizability is a key factor. If the atoms or diatomic molecules have the same static polarizabilities, they should produce harmonics with approximately the same efficiency at a fixed pump laser wavelength[5] even though the ionization potentials are different. We discuss a recently proposed theoretical model of the polarizability effect on the high harmonic efficiency.[6]

At relatively high laser intensity, the harmonics reach the plateau and the saturation region and the influence of ionization and dissociation of molecules becomes important. Experiments have been performed in the gases Ar, N_2, O_2 by using a 160-fs 800-nm Ti:sapphire laser. These three gases have nearly the same polarizabilities. By measuring the 9th and 13th harmonics, it is shown that at relatively low laser intensity the harmonic efficiencies are nearly the same among the three gases, as described before. However, with increased laser intensity, the harmonics in O_2 saturate before those in Ar and N_2 and show a lower efficiency than others. This could happen because O_2 has a relatively lower ionization potential than do Ar and N_2. The earlier ionization of O_2 reduces the probability of the harmonics from neutral molecules, so that the efficiency decreases. This means that if the atoms or diatomic molecules have nearly the same static polarizabilities, the one with higher ionization potential can produce harmonics with higher efficiency with a relatively higher laser intensity. Ar and N_2 behave in a similar way because they have nearly the same ionization potential. Other details of dissociation have been observed and are discussed.

*INRS–Énergie et Matériaux, Varennes, Québec J3X 1S2, Canada

1. J. L. Krause, K. J. Schafer, and K. C. Kulander, Chem. Phys. Lett. **178**, 573 (1991).
2. M. Y. Ivanov and P. B. Corkum, Phys. Rev. A **48**, 580 (1993).
3. T. Zuo, S. Chelkowski, and A. D. Bandrauk, Phys. Rev. A **48**, 3837 (1993).
4. C. G. Wahlström, J. Larsson, A. Persson, T. Starczewski, S. Svanberg, P. Salières, Ph. Balcou, and A. L.'Huillier, Phys. Rev.A **48**, 4709 (1993).
5. Y. Liang, S. Augst, S. L. Chin, Y. Beaudoin, and M. Chaker, J. Phys. B, in press.
6. S. L. Chin and P. A. Golovinski, submitted to J. Phys. B.

QFC7 **12:15 pm**

Harmonic generation with elliptically polarized light

W.-C. Liu,* C. W. Clark, *National Institute of Standards and Technology, Room A253, Physics Building, Room A253, Gaithersburg, Maryland 20899*

Recent experiments[1] on atomic gases interacting with elliptically polarized light have revealed that the high-harmonic spectrum decreases with increasing driving field ellipticity and that the ellipticities of harmonics exhibit complex dependence on harmonic order and the intensity and the ellipticity of the driving fields. We look at this problem in two model systems.

First, we discuss a V-type three-level atom with degenerate upper levels ($m = \pm 1$) to simulate the ground and first excited states of an atom interacting with a laser. This is the simplest model system that can show effects of elliptical polarization. The harmonic spectra of this system have been calculated numerically by using the Floquet scheme. We find that the harmonic spectra have an interesting two-plateau structure, and we obtain simple analytic formulas for the plateau cutoffs. The axes of harmonics have significant rotations only around the points at which the Floquet eigenenergies are near an avoided crossing (Fig. 1). Away from crossings, the axes of the polarization ellipse of the harmonic radiation are parallel to those of the driving field but the ellipticity of the harmonic radiation still exhibits complex dependence.

Second, to investigate the effects of the ionization continuum, we follow the theory proposed by Lewenstein *et al.*[2] and describe a model atom in terms of low-frequency, elliptically polarized fields. If we assume that (1) the atomic ground state directly couples with the continuum, (2) the effects

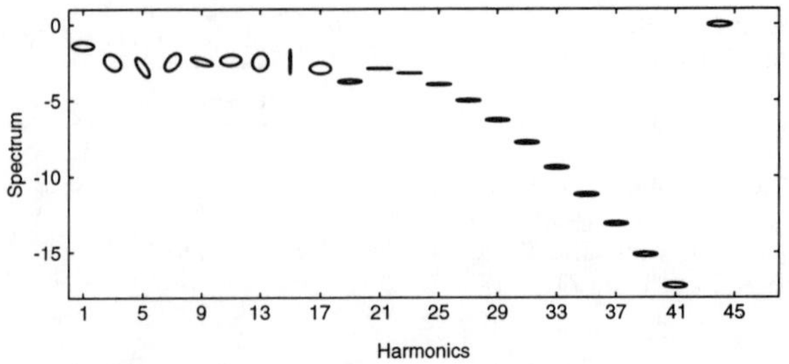

QFC7 Fig. 1. Harmonic spectrum of a three-level system with a driving-field strength of 0.475 a.u. and an ellipticity of 0.2. This is the typical spectrum when Floquet eigenenergies are near an avoided crossing. The polarization ellipse of the driving field is shown in the upper-right-hand corner; the polarization ellipses of the harmonics are as indicated. Intensity of the spectrum is displayed on a logarithmic scale.

of all other bound states are negligible, and (3) after excitation to the continuum, the electron can move freely in the radiation field, unaffected by the atomic potential, we can obtain analytic formulas for dipole moments and the harmonic spectra. In the low-frequency limit, we use a saddle-point method to evaluate the integral and simplify the results. With this model we can further explore phase differences between harmonics in orthogonal directions.

**Also with Institute of Physical Science and Technology, University of Maryland, College Park, Maryland 20742*

1. K. S. Budil, P. Salieres, A. L'Huillier, T. Ditmire, and M. D. Perry, Phys. Rev. A **48**, R3437 (1993); F. A. Weihe, S. K. Dutta, and P. H. Bucksbaum, in *OSA Annual Meeting*, 1993 Technical Digest Series, Vol. 16 (Optical Society of America, Washington, D.C., 1993); H. H. Burnett and P. B. Corkum, in *OSA Annual Meeting*, 1993 Technical Digest Series, Vol. 16 (Optical Society of America, Washington, D.C., 1993); F. A. Weihe, S. K. Dutta, and P. H. Bucksbaum, joint April Meeting of American Physical Society and AAPT (American Physical Society, New York, 1994).
2. M. Lewenstein, Ph. Balcou, M. Yu. Ivanov, A. L'Huillier, and P. B. Corkum, Phys. Rev. A **49**, 2117 (1994).

QFD — 10:30 am — Hyatt C/D

Coherent Phenomena in Semiconductors II

Jason B. Stark, *AT&T Bell Laboratories, Presider*

QFD1 (Invited) — 10:30 am

Rabi flopping in semiconductors

S. T. Cundiff,* A. Knorr, J. Feldmann, S. W. Koch, E. O. Göbel, H. Nickel,** *Fachbereich Physik und Wissenschaftliches Zentrum für Materialwissenschaften, Philipps-Universität, Renthof 5, D-35032 Marburg, Germany*

Many coherent optical experiments that were originally performed on atomic or molecular systems[1] are now being performed in semiconductors.[2] This has been made possible by the current ready availability of lasers with pulse widths on the order of the dephasing time in semiconductors (100 fs). Because coherent phenomena can be strongly modified by many-body interactions, these experiments provide insight into the many-body physics present in semiconductors. Optically induced transient nutation, also known as Rabi flopping,[1,3] where strong resonant incident optical fields drive the excitation from the ground state to the excited state and back again, is a coherent experiment of particular interest. This interest arises due to the theoretical prediction that the flopping frequency can be strongly modified by many-body contributions.[4]

We present the results of experiments that allow us to observe Rabi flopping in semiconductors[2]. The observation of the Rabi flopping in semiconductors is hindered not only by the short dephasing rates, but also by the significant shortening of the dephasing times at the elevated excitation densities produced by the strong resonant incident fields. The experimental technique is based on observing the pulse shape modifications induced on a femtosecond pulse by propagation through a thin semiconductor film, similarly to the first studies in molecules.[6] Due to the short time scales involved, the pulse-shape modifications are observed by cross-correlation with a reference pulse. Additionally the pulse-shape modification due to propagation through an unexcited sample is compared to that through a weakly excited sample. This is achieved by a weak copropagating prepulse, which is chopped, and performing differential detection with lockin amplifier.

Analysis of this experiment shows that the signal is proportional to the polarization, which itself is proportional to the temporal derivative of the induced density.[5] Hence, a zero crossing in the differential cross-correlation signal corresponds to a density maximum. The analysis shows that a zero crossing that appears above a critical density, shifts to earlier times with increasing intensity, and is independent of prepulse intensity, is a signature of Rabi flopping.

We performed this experiment on a InGaAs/GaAs 2-period multiple quantum well. The observed signal displays precisely these characteristics. Numerical solutions of the coupled Semiconductor-Maxwell-Bloch-Equations, corresponding to the experimental conditions, provide good agreement with the experiment. The calculations confirm that the observations are a manifestation of Rabi flopping in semiconductors.

**Current address: AT&T Bell Laboratories, Crawfords Corner Road, Holmdel, New Jersey 07733*

***Deutsche Bundespost Telekom Fernmeldetechnisches Zentralamt, D-64295, Germany*

1. L. Allen and J. H. Eberly, *Optical Resonance and Two-Level Atoms* (Wiley, New York, 1975).
2. N. Peyghambarian, S. W. Koch, and A. Mysyrowicz, *Introduction to Semiconductor Optics* (Prentice Hall, Englewood Cliffs, 1993); F. Henneberger, S. Schmitt-Rink, and E. O. Göbel, eds., *Optics of Semiconductor Nanostructures* (Akadamie Verlag, Berlin, 1993).
3. I. I. Rabi, Phys. Rev. **51**, 652 (1937); I. I. Rabi, J. R. Zacharias, S. Millman, and P. Kusch, Phys. Rev. **53**, 318 (1938).
4. R. Binder, S. W. Koch, M. Lindberg, N. Peyghambarian, and W. Schäfer, Phys. Rev. Lett. **65**, 899 (1990); T. Östreich, A. Knorr, Phys. Rev. B **48**, 17811 (1993).
5. S. T. Cundiff, A. Knorr, J. Feldmann, S. W. Koch, E. O. Göbel, and H. Nickel, Phys. Rev. Lett. **73**, 1178 (1994); S. T. Cundiff, A. Knorr, J. Feldmann, S. W. Koch, E. O. Göbel, and H. Nickel, Phys. Stat. Sol. B **188** (1995).
6. G. B. Hocker and C. L. Tang, Phys. Rev. Lett. **21**, 591 (1968).

QFD2 — 11:00 am

Dark states in semiconductors

M. Lindberg, R. Binder,* *Institute for Physics, Åbo Akademy, Porthansgatan 3, 20500 Åbo, Finland*

In atomic systems the concept of dark states is already well known.[1] Essentially, dark states are special dressed superposition states of the excited states that are not dipole coupled to the ground state of the atom even if the excited states themselves are. Therefore, they remain inactive during the interaction. Atoms with V- or Λ-type energy levels allow for the creation of dark states. Usually defined in terms of stationary light fields coupled to separate transitions with a common ground state, a dark state is created if the detunings of the two light fields from their respective transition frequencies in a V or Λ configuration are the same.

In a semiconductor the level structure of the optical transition is more complex and the transitions are strongly influenced by the Coulomb interaction (excitonic effects). However, transitions to the bound exciton states have characteristics similar to those of a single atom. In a GaAs quantum well, at least four bound exciton transitions are involved. By using the selection rules, the number of resonant transitions can be reduced to two, which form a V configuration. Because of the fast scattering, relaxation, and dephasing processes, one has to explore the dark states under pulsed-excitation conditions. In this project we show that the signature of the dark states can be obtained if a semiconductor quantum well is dipole coupled to two oppositely circularly polarized subpicosecond light pulses.

In a GaAs quantum well there are two heavy-hole (hh) and two light-hole (lh) excitons. For both kinds of excitons, one exciton couples to right-handed circularly polarized light σ_+ and the other couples to left-handed circularly polarized light σ_-. The multiband semiconductor Bloch equation[2] (SBE) includes these selection rules along with the linear and nonlinear Coulomb interaction contributions, which give rise to excitonic features, band shifts, and induced optical fields.

We have solved the SBE for a hh – lh splitting of 40 meV, with a linear absorption spectrum shown in Fig. 1(a), with two strong pulses.

To show the existence of dark states, we compute the differential absorption of a 200-fs σ_+ pulse in the presence of a σ_- pulse. The center frequency of the σ_+ pulse is chosen to be 10 meV (5 meV) below the lh-exciton resonance. In Fig. 1(b) we show the differential absorption of the σ_+ pulse as a function of the center frequency of the σ_- pulse. Whenever the detuning of the σ_- pulse from the hh-exciton resonance equals that of the σ_+ pulse from the lh-exciton resonance, we observe a minimum in the differential absorption. The mini-

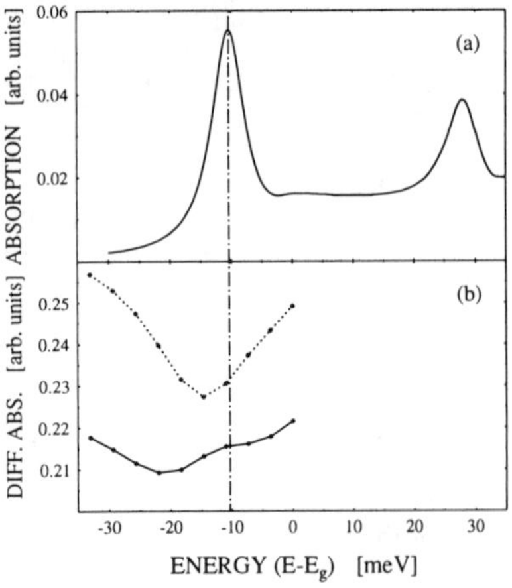

QFD2 Fig. 1. (a) Linear absorption spectrum with hh- and lh-exciton resonances. (b) Differential absorption of a σ_+ pulse, centered at 18 meV (solid curve) and 23 meV (dashed curve) versus the center frequency of a σ_- pulse.

mum shown in Fig. 1(b) is a manifestation of the optically induced intervalence-band coherence, i.e., the coherence between the hh and lh valence bands. Without the intervalence-band coherence, the minimum does not appear. Incoherent bleaching or saturations effects would not require such a coherence.

**Optical Sciences Center, University of Arizona, Tucson, Arizona 85721*

1. B. W. Shore, *The Theory of Coherent Atomic Excitation* (Wiley, New York, 1990), p. 824.
2. M. Lindberg, R. Binder, Y. Z. Hu, and S. W. Koch, Phys. Rev. B **49**, 16942 (1994).

QFD3 **11:15 am**

Transition to a multiphoton excitonic response in GaAs

A. C. Schaefer, N. H. Bonadeo, D. G. Steel, *Harrison M. Randall Laboratory of Physics, University of Michigan, Ann Arbor, Michigan 48109*

Recently, the source of the optical nonlinearities near the band edge in bulk GaAs has been the subject of much study. At high density, it is well known that phase-space filling and screening dominate. At moderate densities, excitation-induced dephasing (EID) becomes a major source of nonlinearity.[1] The role of the biexciton has also been suggested by recent data.[2] We find, however, that below exciton densities of about 3×10^{14} cm^{-3} a well-defined transition from behavior dominated by EID to a completely different nonlinear response characterized by induced absorption, spin-dependent interactions, and an excitation-induced line narrowing (EIN) is observed. The data suggest a multiphoton nonlinear response and may indicate collective effects.

The data are obtained from measurements in high-quality bulk GaAs (lh–hh degeneracy lifted by strain) by using frequency- and time-domain differential transmission (DT) and four-wave mixing (FWM) in an excitation-density regime of 10^{13}–10^{15} cm^{-3}. As shown in Fig. 1, the cw DT reveals induced absorption across the whole exciton line at the lh(1*s*) resonance[3] that grows superlinearly with excitation density. This behavior is unexpected in the context of the semiconductor Bloch formalism for these systems and is indicative of a nonlinear response involving multiple photons. Time-resolved DT reveals that this response is immediate (within 200-fs resolution). The fact that the response is immediate, has a decay time comparable to that of the free exciton (≈2 ns), is resonant with the exciton, and is qualitatively the same for several samples (with different strain and luminescence characteristics) suggests that this behavior is intrinsic. Polarization-sensitive FWM measurements show that the behavior derives nominally from spin-dependent interactions in accordance with the dipole selection rules.

In the same regime, the 2*s* states show a strong nonlinear response, which indicates an excitation-induced line narrowing,

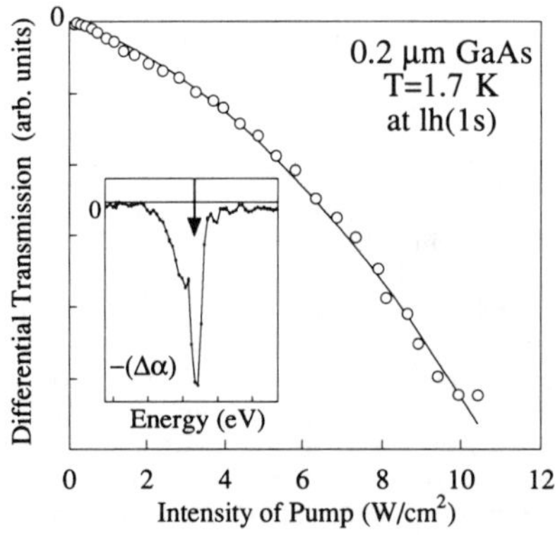

QFD3 Fig. 1. Dependence of DT signal at lh(1s) exciton on pump intensity, showing superlinear growth. Fit is a sum of linear and quadratic terms. Inset: spectrum of lh(1s) DT, showing complete induced absorption. Arrow indicates position of lh absorption resonance.

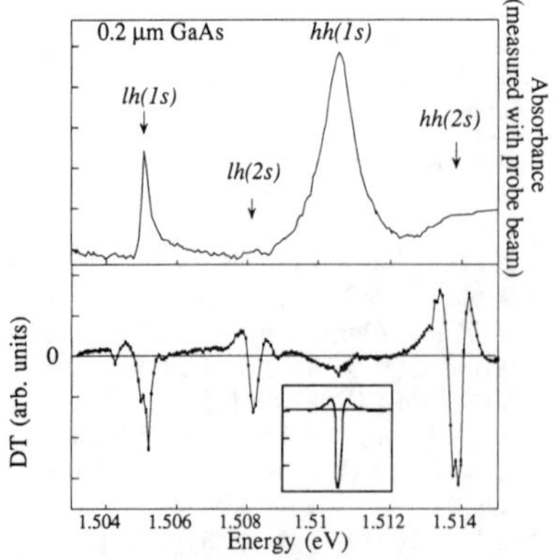

QFD3 Fig. 2. Top: typical absorption spectra. Note non-Lorentzian line shape at lh(1s), possibly indicating polariton effects. Bottom: DT spectrum at an excitation density of approximately 5×10^{13} cm^{-3}, showing strong EIN behavior at 2*s* states. Note complete induced absorption at hh(1s). A corresponding signal is detected at these resonances in FWM. Inset: line shape predicted for EIN.

as shown in Fig. 2. The inset in Fig. 2 shows the standard line shape expected in transmission for a harmonic oscillator, where the dephasing is given by $\gamma = \gamma_0 - \beta I$, with $\beta > 0$. This observation, coupled with the transition at higher densities to behavior dominated by an increase in dephasing (EID), raises questions about the fundamental mechanisms of the dephasing at these densities.

The observation of a multiphoton response at low densities indicates behavior that has yet to be understood. In the standard formalism, it is expected that at these densities the nonlinear response should be third order and linear with any one of the excitation fields. Currently, a model based on collective effects (excitation-induced enhancement of excitonic oscillator strength)[4] is being considered. Also, the role of polaritons is suggested from measurements of transport properties and also from the deviation of the lh absorption line from a Lorentzian profile[5] (see Fig. 2).

This work is supported by the U.S. Army Research Office and the U.S. Air Force Office of Scientific Research.

1. H. Wang, *et al.*, Phys. Rev. Lett. **71**, 1261 (1993).
2. M. Jiang, *et al.*, in *International Quantum Electronics Conference*, 1994 Technical Digest Series, Vol. **9** (Optical Society of America, Washington, D.C., 1994).
3. A. C. Schaefer, *et al.*, in *International Quantum Electronics Conference*, 1994 Technical Digest Series, Vol. **9** (Optical Society of America, Washington, D.C., 1994).
4. Hanamura, unpublished; I. E. Perakis and D. S. Chemla, Phys. Rev. Lett. **72**, 3202 (1994).
5. G. Czajkowski and A. Tredicucci, Nuovo Cimento **14D**, 1203 (1992); A. Tredicucci, *et al.*, Phys. Rev. B. **47**, 10348 (1993).

QFD4 **11:30 am**

Transient time-irreversible optical interaction in zinc-blende semiconductors

A. R. Bungay, H. Loh, A. Malinowski, S. V. Popov, N. I. Zheludev, V. F. Kamalov, E. V. Slobodchikov,* *Department of Physics, University of Southampton, Highfield, Southampton S017 1BJ, U.K.*

Recently we have found a class of media, namely, zing-blende semiconductors (e.g., GaAS and InSb), in which polarization-sensitive interaction with light exhibits time-irreversible behavior.[1,3] This phenomenon is due to relativistic **q**-linear spin splitting (**q** is the wave vector of the electron) in the band structure. Here we report what we believe to be the first results of a femtosecond pump–probe study of light-induced time-irreversible birefringence/dichroism.

Zinc-blende-type materials, such as $A^{III}B^{V}$ semiconductors, provide a nearly ideal ground for observation of optical *T*-odd effects. Let us consider a simple two-band structure in GaAs. In zinc-blende materials, accounting for the spin–orbit interaction leads, in particular, to small **q**-linear terms in carrier dispersion that lift the spin de-

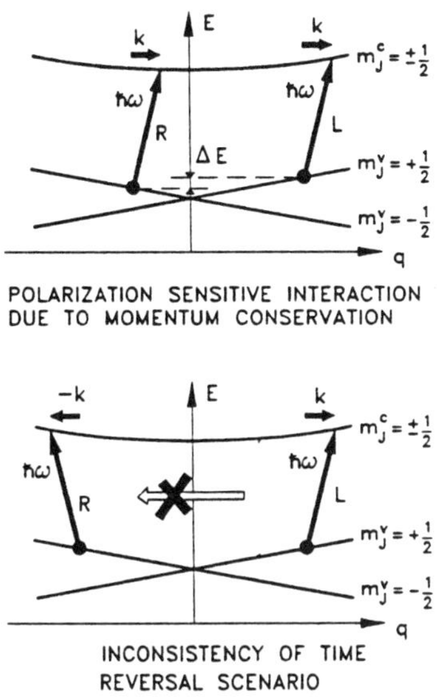

QFD4 Fig. 1. Optical interaction with GaAs near the center of the Brillouin zone (**k** is the projection of the photon wave vector on the **q**-vector direction, where the splitting occurs).

generacy. The **q**-linear splitting is on a scale of only a few milli-electron-volts and may often be ignored. However, it is crucial when polarization-sensitive optical effects are considered. For clarity, we take account of this splitting [$E(\mathbf{q}, m_J = 1/2) \neq E(\mathbf{q}, m_J = -1/2)$] only in the light-hole valence band, and we presume that the conductivity band is parabolic and is degenerate with respect to the projection of the total angular momentum $\mathbf{J}$:$E(\mathbf{q}, m_J = 1/2) = E(\mathbf{q}, m_J = -1/2)$ (see Ref. 2 for more details). When the finite value of the photon wave vector is considered, the transitions promoted by left- and right-hand-polarized light (L and R in Fig. 1) of the same quantum energy $\hbar\omega$ begin and end at levels that differ by ΔE on the dispersion curve, leading to differential right/left absorption/refraction. This absorption/refraction is time-irreversible, as can be seen by reversing the scenario ($T \Rightarrow -T$:$\mathbf{k} \Rightarrow -\mathbf{k}$; $\mathbf{q} \Rightarrow -\mathbf{q}$, $m_J \Rightarrow -m_J$): the transition that is time conjugated to $m_J^v = 1/2 \Rightarrow m_J^c = -1/2$ cannot be consistently performed because it requires R-polarized light for promotion rather than L-polarized as in the time-direct process. However, time reversal does not affect the handedness of light polarization. Because of the peculiar anisotropy of the splitting, this results in T-odd birefringence/dichroism for linearly polarized light, which manifests itself in nonreciprocal transmission and reflection experiments.[1] If the crystal is stimulated with light, the excitation creates free carriers, the band gap re-normalizes (reduces), the probe wave promotes transitions with higher **q**, and T-odd birefringence/dichroism increases. This is the nature of the nonlinear pump–probe time-irreversible interaction reported here. Specifically, we observed pump-induced enhancement of T-odd birefringence/dichroism in reflective configurations.

With a 2-ps Ti:sapphire laser (λ = 750 nm) and with a 90-fs Cr:forsterite laser (λ = 1260 nm) we clearly observed pump-induced correction to the natural time-irreversible effect in GaAs and InSb (see Figs. 2 and 3, respectively). In the experiment with GaAs, the pump and the probe waves were identically linearly polarized. Light-induced T-odd birefringence/dichroism in this case competes with the specular nonlinear anisotropic effect due to band-structure undulation,[4] but it may be easily separated by measuring the dependence of the effect on pump–probe delay and on crystal orientation (Fig. 2). The anisotropic effect gives rise to the sharp peak at zero pump–probe delay, and the tail is due to the T-odd induced birefringence/dichroism slowly decaying with carrier recombination rate. In the experiments with InSb, the probe wave was linearly polarized and the pump wave was circularly polarized. Here, light-induced T-odd birefringence/dichroism competes with the specular inverse Faraday effect due to magnetization of the sample with the pump pulse.[5] The reflected probe-pulse-polarization rotation is presented as a function of pump–probe delay and crystal orientation (Fig. 3). The specular inverse Faraday effect produces the sharp peak at zero pump–probe delay, and the tail is due to T-odd birefringence/dichroism. Also, the specular-inverse-Faraday-effect rotation does not depend on crystal orientation.

We analyze potential use of the effect in solid-state spectroscopy, project its prospective device applications in nonreciprocal valves, and discuss its thermodynamic meaning in the context of the inapplicability of the Onzager principle to nonequilibrium systems.

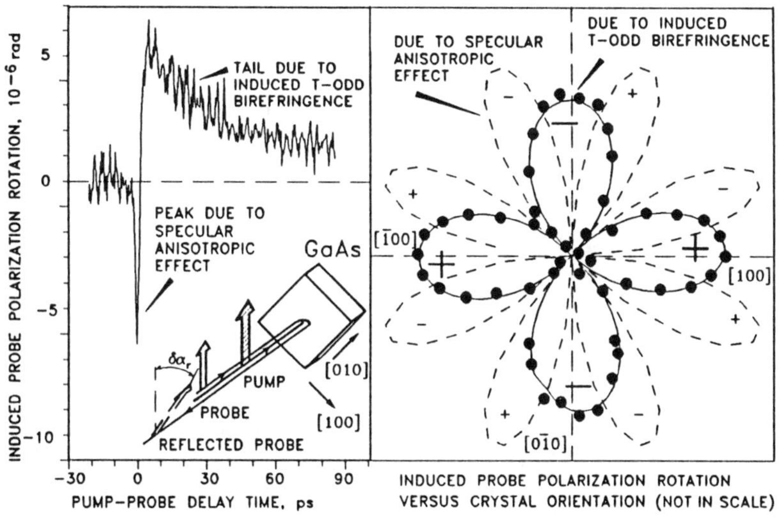

QFD4 Fig. 2. Manifestation of light-induced T-odd birefringence/dichroism in GaAs.

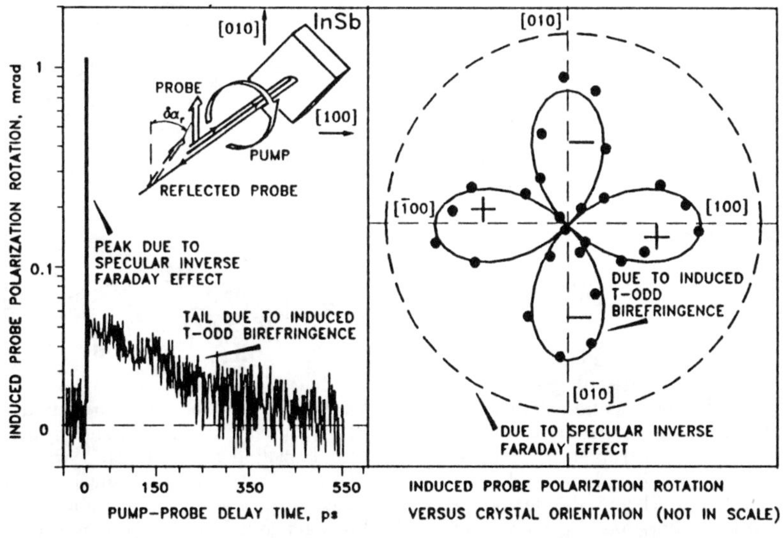

QFD4 Fig. 3. Manifestation of light-induced T-odd birefringence/dichroism in InSb.

Institute of Chemical Physics, 4 Kosigina Street, Moscow 117977, Russia

1. A. R. Bungay, S .V. Popov, Yu. P. Svirko, and N. I. Zheludev, Chem. Phys. Lett. **217**, 249 (1994).
2. N. I. Zheludev, S. V. Popov, Yu. P. Svirko, A. Malinowski, and D. Yu. Paraschuk, Phys. Rev. B **50** (October 1994).
3. A. R. Bungay, Yu. P. Svirko, and N. I. Zheludev, Phys. Rev. B **47**, 16141 (1993).
4. N. I. Zheludev, A. R. Bungay, S. V. Popov, and Yu. P. Svirko, in 1994 Technical Digest Series, Vol. 7 (Optical Society of America, Washington, D.C., 1994), paper WC6-1, also Opt. Lett., in press.
5. S. V. Popov, N. I. Zheludev, and Yu. P. Svirko, Opt. Lett. **19**, 13 (1994).

QFD5 **11:45 am**

Degree and state of polarization of coherent four-wave-mixing signals from GaAs/AlGaAs multiple quantum wells

J. A. Bolger, Shekhar Patkar, A. E. Paul, W. Sha, Arthur L. Smirl, *Laboratory for Photonics and Quantum Electronics, 100 IATL, University of Iowa, Iowa City, Iowa 52242*

Degenerate four-wave mixing (FWM) has been shown to be a powerful tool for studying coherent transient processes in semiconductors, and recent studies[1,2] indicate that the polarization-selection rules can be very useful in discriminating among various proposed models for these processes, particularly in delineating the roles of excitation-induced dephasing (EID), the local-field correction (LFC), and biexcitons. Here, we describe the measurement of all four Stokes parameters that completely characterize the polarization state and degree of polarization of the FWM signal in GaAs/AlGaAs multiple-quantum-well samples under a variety of experimental conditions. The ellipticity, the azimuthal angle of the ellipse, and the degree of polarization vary dramatically with experimental conditions and the sample type. Nevertheless, we show that all of the observed tendencies agree qualitatively with models that include EID and LFC phenomenologically. Homogeneous or inhomogeneously broadened two or three-level models are invoked as appropriate for the experimental conditions and sample type.

The measurements were performed by using 100-fs pulses obtained from a tunable mode-locked Ti:sapphire laser and the conventional two-pulse geometry shown in Fig. 1(a). A quarter-wave plate and analyzer were used to select out the six intensity components used to obtain the Stokes parameters[3] for various angles θ_{12} between $\mathbf{E}_1$ and $\mathbf{E}_2$ [see Fig. 1(b)]. From the Stokes parameters we determined the fraction P of the light that was polarized ($P = 1$ denotes completely polarized light), and for the polarized portion we extracted the ellipticity angle ε and the azimuthal angle θ_{sig} for the polarization ellipse, as illustrated in Fig. 1(c). Each of these parameters was determined as a function of (1) the angle θ_{12} between the two linearly polarized input pulses, (2) the pump fluence, (3) the time delay τ_{12} between $\mathbf{E}_1$ and $\mathbf{E}_2$, and (4) the detuning between the light- and heavy-hole excitons for two temperatures (8 and 80 K) and for a sample that was homogeneously broadened (at 80 K) and for one that was inhomogeneously broadened.

Typical experimental results for a homogeneously broadened sample at 80 K with the wavelength tuned just below the heavy-hole resonance (so that negligible light holes were excited) are summarized in Fig. 2 for three selected fluences. Several features are worth noting. The contrasting behaviors of the azimuthal angle at high and low fluences, as shown in Fig. 2(a), confirm previous observations.[1,3,5] However, also notice that the FWM signal is initially fully and linearly polarized when the input pulses are polarized parallel to each other, but the degree of ellipticity increases [Fig. 2(b)] and the degree of polarization decreases [Fig. 2(c)] as the angle between the input polarizations is increased. Also, the ellipicity decreases with increasing fluence, but, surprisingly, the degree of polarization is only weakly dependent on fluence. The polarization ellipses for each fluence and for selected angles are sketched in Fig. 2(d) to illustrate these trends.

For the conditions depicted in Fig. 2, a homogeneous system that includes two two-level heavy-hole transitions and that phenomenologically includes EID and LFC qualitatively reproduces the principal trends. An example of our extensive parametric simulations for such a 2 × 2 model (which illustrate the dependence of the azimuthal angle, the ellipticity, and the degree of polarization on the input-pulse polarizations and on the EID parameter for a nonzero level of LFC) are presented in Fig. 3. Key results of this 2 × 2 homogeneous model are that (1) without EID (contrary to observation), no ellipticity, no depolarization, and a linear dependence on θ_{12} are obtained regardless of the fluence and regardless of the level of LFC included and (2) without LFC (also contrary to observation), no ellipticity is obtained regardless of the fluence or the level of EID included.

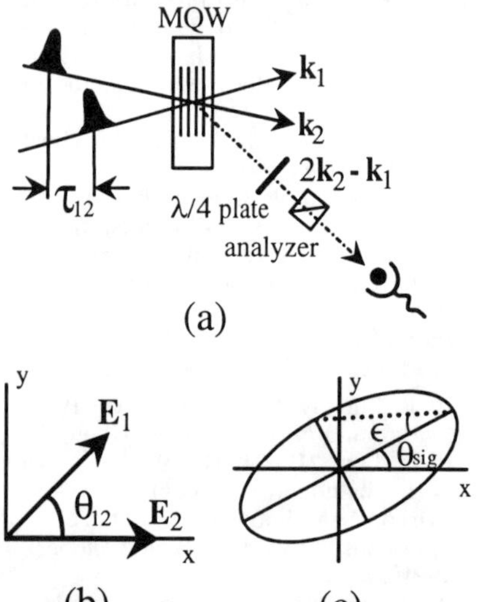

QFD5 Fig. 1. (a) Experimental geometry for measuring the Stokes parameters of the degenerate four-wave-mixing signal as a function of (b) the angle between the polarizations of the two pump pulses. (c) Notation used to describe the polarization state, where θ_{sig} denotes the azimuthal angle for the elliptical polarization and ε denotes the ellipticity angle.

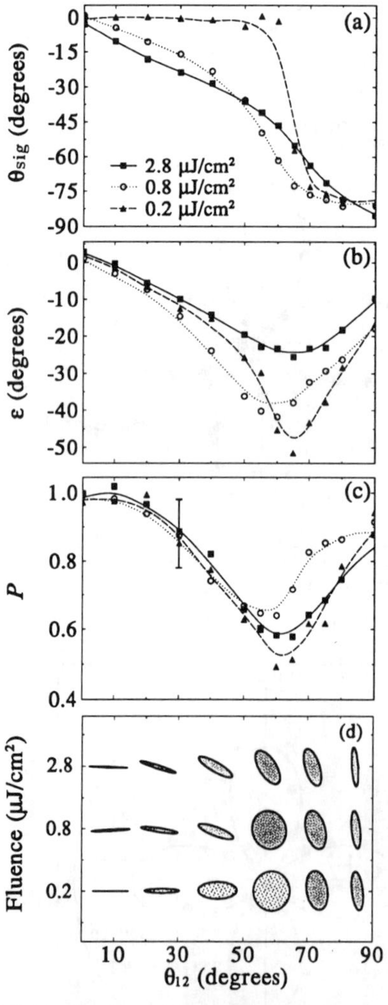

QFD5 Fig. 2. (a) Azimuthal angle, (b) ellipticity angle, c) degree of polarization, and (d) resulting polarization state as a function of incident polarizations for three different incident fluences. For the measurements described here the laser was tuned 2.5 meV below the heavy-hole exciton and the time delay τ_{12} was set to zero.

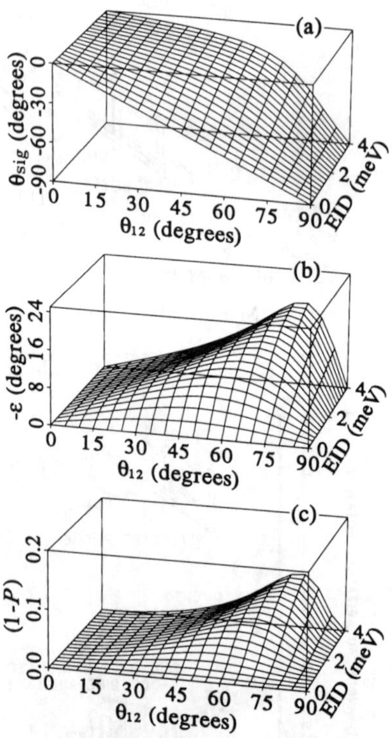

QFD5 Fig. 3. Calculated (a) azimuthal angle, (b) ellipticity angle, and (c) degree of polarization as a function of the relative angle between input linear polarizations and increasing EID parameter for a homogeneous 2 × 2 system and for a fixed local-field strength of 1 meV.

Hence, both EID and LFC are necessary. Corresponding results are discussed for inhomogeneously broadened two-level (2 × 2) and homogeneously and inhomogeneously broadened three-level (2 × 4) systems.

1. R. Eccleston, J. Kuhl, D. Bennhardt, and P. Thomas, Solid State Commun. **86**, 93 (1993).
2. H. Wang, K. Ferrio, D. G. Steel, Y. Z. Hu, R. Binder, and S. W. Koch, Phys. Rev. Lett. **71**, 1261 (1993).
3. K. Bott, O. Heller, D. Bennhardt, S. T. Cundiff, P. Thomas, E. J. Mayer, G. O. Smith, R. Eccleston, J. Kuhl, and K. Ploog, Phys. Rev. B **48**, 17418 (1993).
4. For a detailed description, see R. M. A. Azzam and N. M. Bashara, Ellipsometry and Polarized Light (North-Holland, Amsterdam, 1988).
5. Y. Z. Hu, R. Binder, S. W. Koch, S. T. Cundiff, H. Wang, and D. G. Steel, Phys. Rev. B **49**, 14382 (1994).

QFD6 12:00 m

Polarization dependence of quantum beats between light-hole and heavy-hole excitons in GaAs quantum wells

J. Kuhl, E. J. Mayer, G. O. Smith, K. Bott,* V. Heuckeroth,* T. Meier,* P. Thomas,* S. W. Koch,* R. Hey,** K. Ploog,**
Max Planck Institut for Solid State Research, Heisenbergstrasse 1, D-70569 Stuttgart, Germany

The appearance of quantum beats (QBs) between heavy-hole (hh) and light-hole (lh) excitons in GaAs quantum wells (QWs) has been observed in many degenerate-four-wave-mixing (DFWM) experiments with subpicosecond optical pulses.[1,2] Although the variation of the beat amplitude and phase with polarization of the incident pulses was recognized early on, a consistent theoretical interpretation of exciton coherence for the case of simultaneous excitation of hh and lh excitons is still missing.

Here we present a theoretical and experimental investigation of the polarization-selection rules for DFWM and quantum beating between hh and lh excitons in a GaAs QWs by using a novel three-pulse configuration. Three 400-fs pulses (bandwidth, 50 cm^{-1}) from a mode-locked Ti:sapphire laser tuned to a frequency between the hh and lh exciton transition were applied to excite the backward-reflected DFWM signal on a high-quality GaAS single QW [hh photoluminescence (PL) linewidth, 0.3 meV; no Stokes shift between the PL and PL excitation spectra]. The temporal separation τ_{12} between the first pulse (wave vector $\mathbf{k}_1$) and the second pulse (wave vector $\mathbf{k}_2$) is continuously varied, whereas the third pulse (wave vector $\mathbf{k}_3$) always coincides with the second ($\tau_{23} = 0$). Figure 1 shows the time-integrated DFWM signal diffracted in the direction $\mathbf{k}_1 + \mathbf{k}_2 - \mathbf{k}_3$ for three different combinations of circularly polarized pulses. These signals reveal the following remarkable properties: (1) under the above excitation conditions, we observe a decay time of 3.0 ps for $\tau_{12} \geq 0$ in all polarization configurations. This decay time can be attributed to the hh exciton dephasing time. (2) For $\tau_{12} \leq 0$ the $\sigma^+\sigma^-\sigma^+$ signal rises with a time constant of 3.8 ps, much shorter than the time constant of the rise time of the other two curves. The rise time of the other two curves is determined by the incoherent lifetime of the hh excitons. (3) The $\sigma^+\sigma^+\sigma^+$ and the $\sigma^+\sigma^-\sigma^+$ signals display hh/lh exciton beats, whereas the $\sigma^-\sigma^+\sigma^+$ signal exhibits no modulation. (4) The signal for the $\sigma^+\sigma^+\sigma^-$ configuration (not shown in Fig. 1) is about 100 times smaller and is probably due to fifth- and higher-order effects.

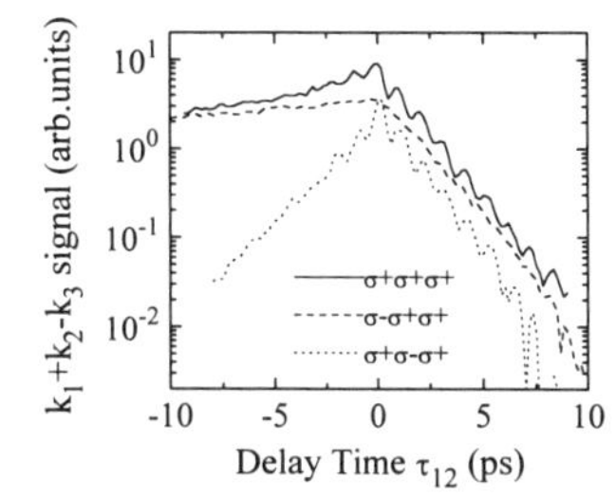

QFD6 Fig. 1. Measured time-integrated DFWM signal detected in the $\mathbf{k}_1 + \mathbf{k}_2 - \mathbf{k}_3$ direction versus the delay time τ_{12} for various circular polarizations of the incident excitation fields.

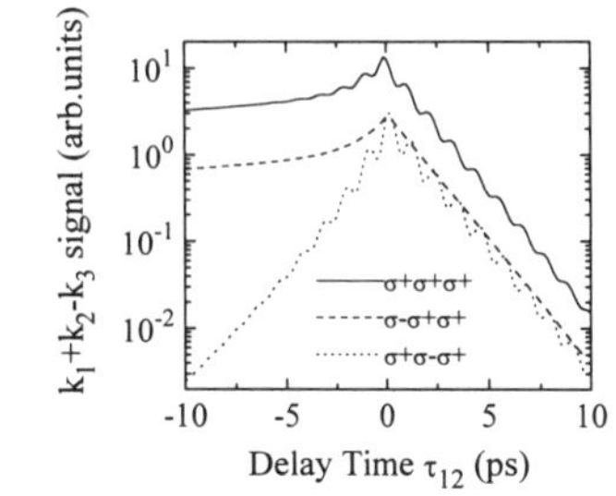

QFD6 Fig. 2. Calculated time-integrated DFWM signal versus the time delay τ_{12} for the same circular polarization configurations as in Fig. 1. The pulse width and energy of the excitation field are the same as for the experimental data (Fig. 1).

Theoretical calculations of these signal curves have been accomplished by solving the optical Bloch equations for appropriate model systems. For two independent three-level systems, which represent noninteracting hh and lh excitons, it is impossible to create any modulation of the signal diffracted into the $\mathbf{k}_1 + \mathbf{k}_2 - \mathbf{k}_3$ direction for the $\sigma^+\sigma^+\sigma^+$ geometry. For a correct description of the beating features, it is important to take into account exciton/exciton Coulomb interaction, which can manifest itself through local-field effects (LFE), excitation-induced dephasing (EID), or biexciton formation (BIF). In our approach the many-body-interaction effects have been included phenomenologically by describing the exciton excitation in the framework of a 10-level system consisting of the exciton ground state, four single-exciton states, and five two-exciton states. This phenomenological inclusion of two-exciton states has proven its great potential for the modeling of excition–exciton Coulomb coupling for many experimental situations.[3,4]

Numerical solutions of the optical Bloch equations, calculated to third order in the electric field for this 10-level system (see Fig. 2), reproduce all of the mentioned characteristics of the experimental curves in Fig. 1. Similar agreement between experimental data and theoretical results is achieved if linearly, instead of circularly, polarized pulses are used.

Our study demonstrates that the occurrence of hh–lh exciton beats in DFWM signal traces depends on the polarization geometry and that two-exciton states are important in the generation of quantum beats.

* *Department of Physics and Materials Science Center, Philipps University, Renthof 5, D-35032 Marburg, Germany.*
** *Paul Drude Institute for Solid State Electronics, Hausvogteiplatz 5-7, D-10177 Berlin, Germany*

1. B. F. Feuerbacher, J. Kuhl, R. Eccleston, and K. Ploog, Solid State Commun. **74**, 1279 (1990).
2. K. Leo, J. Shah, T. C. Damen, A. Schulze, T. Meier, S. Schmitt-Rink, P. Thomas, E. O. Göbel, S. L. Chuang, M. S. C. Luo, W. Schäfer, K. Köhler, and P. Ganzer, IEEE J. Quantum Electron. **28**, 2498 (1992).
3. K. Bott, O. Heller, D. Bennhardt, S. T. Cundiff, P. Thomas, E. J. Mayer, G. O. Smith, R. Eccleston, J. Kuhl, and K. Ploog, Phys. Rev. B **48**, 17418 (1993).
4. E. J. Mayer, G. O. Smith, V. Heuckeroth, J. Kuhl, K. Bott, A. Schulze, T. Meier, D. Bennhardt, S. W. Koch, P. Thomas, R. Hey, and K. Ploog, Phys. Rev. B **50** (1994); E. J. Mayer, G. O. Smith, V. Heuckeroth, J. Kuhl, K. Bott, A. Schulze, T. Meier, S. W. Koch, P. Thomas, R. Hey, and K. Ploog, Phys. Rev. B, in press.

QFD7 12:15 pm

Polarization studies of the free polarization decay on GaAs: evidence for excitonic coherence transfer

N. H. Bonadeo, M. Jiang, V. Subramaniam, D. G. Steel, *Harrison M. Randall Laboratory of Physics, University of Michigan, Ann Arbor, Michigan 48109*

When a system is excited by an ultrashort pulse of light, a macroscopic polarization is created. After the excitation, the coherently prepared dipoles radiate in the forward direction while losing coherence because of dephasing by scattering, diffusion, etc., resulting in an exponential decay in time of the macroscopic polarization. Thus, the signal in the forward direction, the free polarization decay (FPD), has a decay time related to this polarization decay time (T_2), and in the limit of delta-function pulses it has no rise time.

In a system that conserves angular momentum the FPD is expected to conserve the polarization of the excitation beam. However, if the system is in the presence of some coupling leading to scattering, there is a possibility of a transfer of the dipole coherence from one transition to another, permitting the coherent emission of radiation with a polarizatiton other than the initial one (see Fig. 1). Such an effect is predicted in the original theory of Bloch–

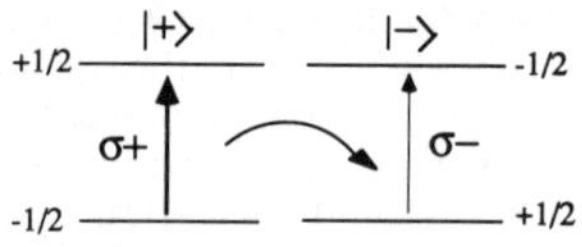

QFD7 Fig. 1. Energy diagram for the light hole of GaAs corresponding to the magnetic subtracts that are connected through the interaction.

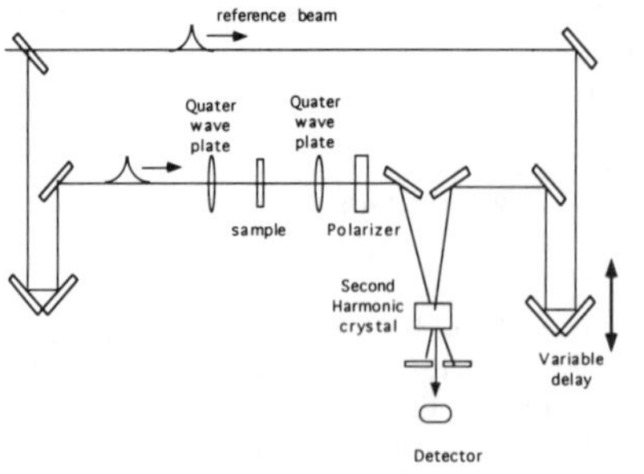

QFD7 Fig. 2. Experimental configuration. The signal is mixed with a reference beam in the second-harmonic crystal in order to achieve time resolution. The last quarter-wave plate is set such that both together add 0 or π phase shift. In this manner it is possible to select the polarization of the excitation beam or the opposite polarization.

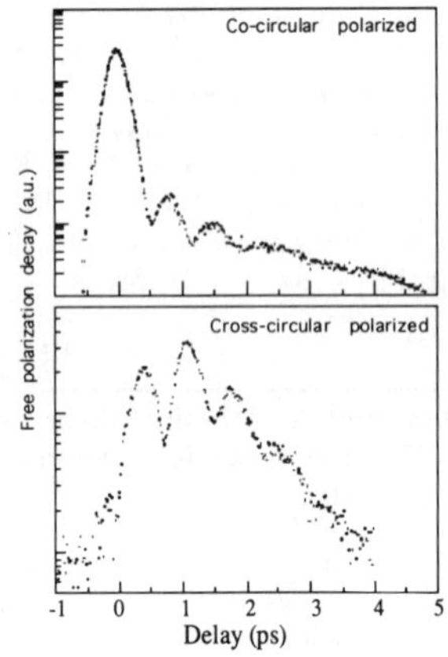

QFD7 Fig. 3. Top panel shows the copolarized free polarization decay. The signal around zero is the transmission of the beam and is followed by the FPD with a decay determined by the dephasing rate. Bottom panel shows the orthogonally polarized FPD. A clear rise time with a maximum at 1.2 ps can be clearly seen. Quantum beats are observed in both free polarization decays because of coupling to the heavy hole.

Redfield–Wangles for NMR (Ref. 1), with evidence in atomic systems just recently reported.[2] Indeed, it would be expected that the phase shift accompanying the scattering process would, on average, destroy the induced coherence. However, in the presence of symmetry breaking, it can be shown that such transfer of quantum coherence may exist.[3]

To look for this system, we examined the polarization decay after a transient of excitation of the exciton in GaAs at 4 K. The appropriate energy-level diagram and corresponding dipole selection rules are shown in Fig. 1. After excitation with a 130–180-fs σ_+-polarized pulse, the emitted signal is upconverted in a second-harmonic crystal with a reference beam (see Fig. 2). With the use of a quarter-wave plate and polarizer in front of the crystal, it is possible to select between the corotating (σ_+) and counterrotating (σ_-) polarization. In this way it is possible to independently obtain data on the free polarization decay as a function of time for the two circular polarizations.

The preliminary results are shown in Fig. 3. The top curve shows the upconverted coherent radiation polarized in the direction of the incident pulse. The data show a peak centered at $t = 0$, corresponding to the transmitted pulse. However, after this there is additional emitted radiation that decays exponentially with the dephasing time. This is the usual free polarization decay. The spectrum of the laser is centered at the light hole, but it also covers the heavy hole, leading to the observed quantum beats. As indicated above, in the absence of coherence transfer, no emission would be expected with polarization σ_-. The lower curve in Fig. 3 shows, however, that there is indeed radiation emitted with the opposite polarization (relative to the incident field) and that the radiation is not instantaneous but, rather, peaks after the incident pulse with a rise time of 1.2 ps. The strength of the field intensity is of the order of 1/40 the copolarized FPD. These data are in agreement with the concept of coherence transfer. Simple band mixing is clearly not involved because such coupling would lead to an exact replication of the copolarized FPD with no rise time. In addition, simple incoherent processes such as spin flips that do lead to luminescence are also not involved because not only do they not lead to a coherent electromagnetic polarization, but they also occur on a much longer time scale. We note that in our picture of coherent coupling between the orthogonally polarized states, one might expect a faster rise time for the orthogonal polarization after excitation of the heavy hole. Indeed, this is in agreement with measurements. Related effects in the four-wave-mixing response have been observed elsewhere.[4]

This work is supported by the U.S. Army Research Office and the U.S. Air Force Office of Scientific Research.

1. C. P. Slichter, Principles of Magnetic Resonance (Springer-Verlag, Berlin, 1990), 3rd ed., p. 655.
2. A. M. Bacon, H. Z. Zhao, P. J. Laverty, L. J. Wang, and J. E. Thomas, Phys. Rev. A **49**, 379 (1994).
3. Marcus Lindberg and Stephan Koch, personal communication, 1994.
4. J. Kuhl, personal communication, 1994.

QFE 2:00 pm
Hyatt B

Laser Driven X-Ray Sources

Luiz Da Silva, *Lawrence Livermore National Laboratories, Presider*

QFE1 (Invited) 2:00 pm

Demonstration of a 10-Hz, femtosecond-pulse-driven XUV laser at 41.8 nm in Xe IX

G. Y. Yin, B. E. Lemoff, C. L. Gordon, III, C. P. J. Barty, S. E. Harris, *Edward L. Ginzton Laboratory, Stanford University, Stanford, California 94305*

We recently demonstrated a new class of XUV lasers in which an intense circularly polarized femtosecond laser pulse is used to tunnel ionize a gaseous target species while simultaneously producing the hot electrons necessary to collisionally excite the species.[1] The first specific systems collisionally excited in this way were proposed by Lemoff *et al.*,[2] who gave calculations for lasing in Ar IX(Ne-like), Kr IX(Ni-like), and Xe IX(Pd-like). We have observed lasing at 41.81 nm on the $4d^95d\ ^1S_0$-$4d^95p\ ^1P_1$ transition in Xe IX. In contrast to previous XUV lasers operating by collisional excitation that required solid density targets and pump laser pulse energies of several hundred joules, a 10-Hz, 800-nm laser pulse with an energy of ~70 mJ and time duration of ~40 fs is longitudinally focused in a differentially pumped cell containing 5 to 12 torr of Xe gas. Laser amplification was observed with an estimated gain coefficient of 13 cm^{-1} and a total gain of exp(11).

A partial energy level diagram of Xe IX is shown in Fig. 1. Experimentally observed transitions agree to published Xe IX spectral data[3] within the instrumental limit, and all levels are in good agreement with the RCN/RCG atomic physics code calculation.[4] The laser transition is marked with a bold arrow bar. The Xe^{+8} ions are created by the optical pulse through tunneling ionization. With an estimated intensity of 3×10^{16} W/cm^2 at the focal region, the calculated tunneling rate[5] from neutral Xe to Xe IX is 1.5×10^{17} sec^{-1}. The ionization rate from Xe IX to Xe X is 5×10^5 sec^{-1}, and is thus inconsequential. The 40-fs pulse produces a nearly uniform plasma of Xe IX in the focal region. For circularly polarized laser field ionization, as a result of the conservation of angular momentum[6] the electrons will retain a kinetic energy which is equal to the quiver energy, $\varepsilon = e^2E^2/4m\omega^2$, at the time of ionization. The last four electrons ionized for each Xe^{+8} created are expected to have energies greater than 104 eV, thus contributing to collisional excitation of the upper laser level. Because the electron ion relaxation time is on the order of 10 ns, the ions remain very near to room temperature and the Doppler broadening is negligible during the period we are interested in.

Lasing at 41.81 nm is evident in the dependence of the signal on both pressure

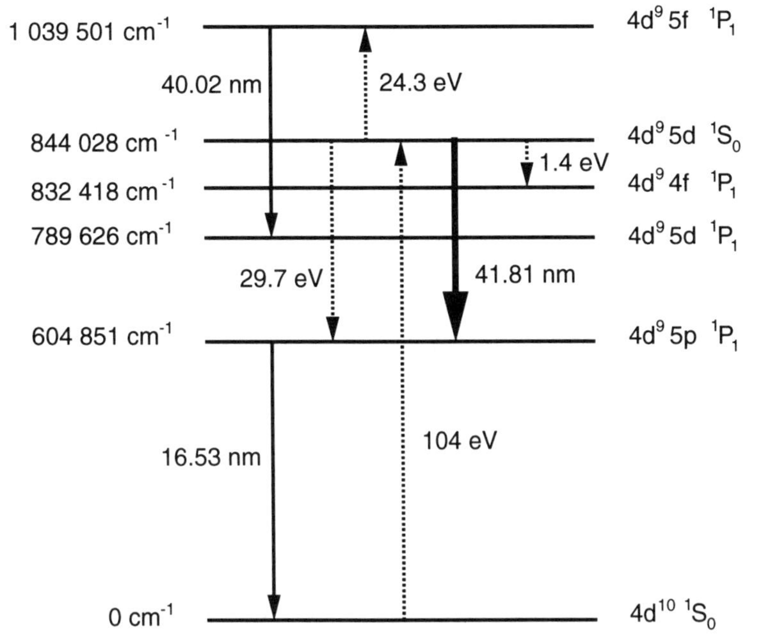

QFE1 Fig. 1. Partial energy level diagram of Xe IX, showing observed lines at 16.53 nm, 40.02 nm, and the laser line at 41.81 nm. The dashed lines indicate the dominant collisional excitation and de-excitation channels of the upper laser level.

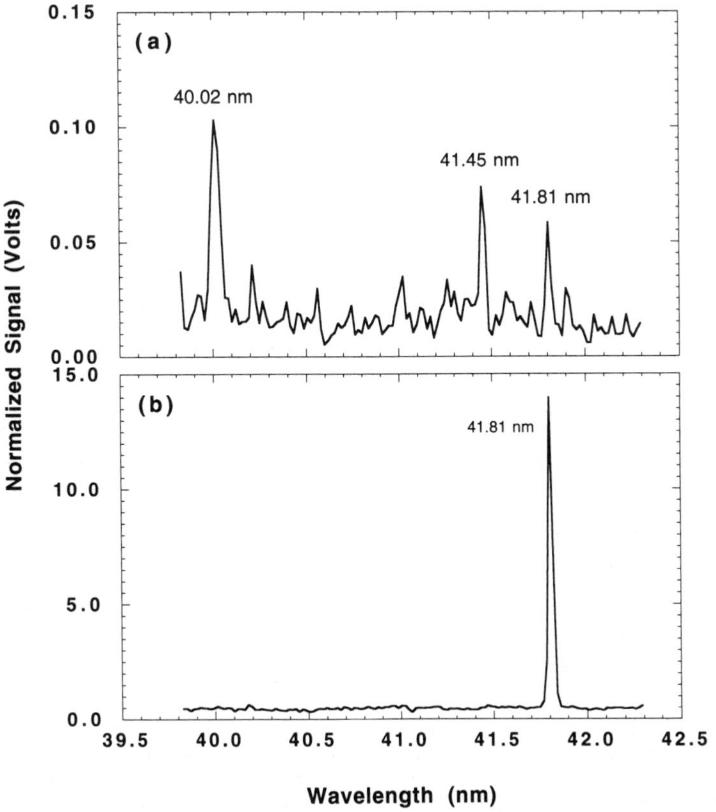

QFE1 Fig. 2. Observed emission spectrum at an estimated length of 8.4 mm and a pressure of (a) 3 torr, and (b) 12 torr. Note the factor of 100 scale change between the two plots.

and length. Figures 2(a) and 2(b) show spectral scans in the vicinity of 41.8 nm and a length of 8.4 mm at Xe pressures of 3 and 12 torr, respectively. Figure 3 shows the 41.81-nm signal at a pressure of 12 torr and several lengths ranging from 3.9 to 8.4 mm. A conservative exponential gain fit yields the differential gain of 13.3 cm^{-1}. At an effective length of 8.4 mm, the total length gain is exp(11.2).

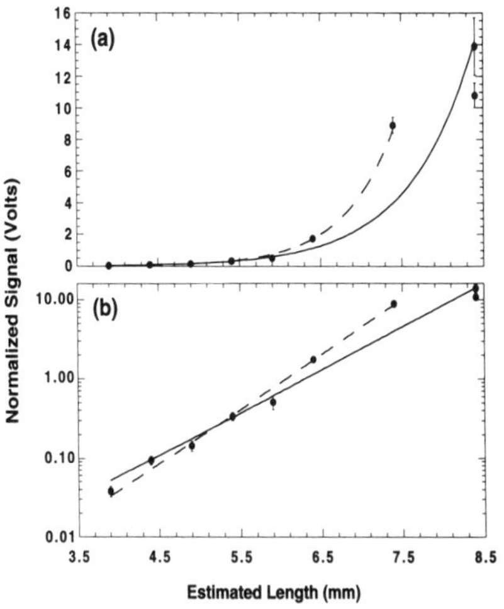

QFE1 Fig. 3. Length dependence at a pressure of 12 torr of the observed 41.81 nm line emission. Linford curves are shown fit to all the data (solid line, 13.3 cm^{-1}) and 3.9 through 7.4 mm only (dashed line, 16.8 cm^{-1}): (a) linear scale, (b) logarithmic scale.

1. B. E. Lemoff, G. Y. Yin, C. L. Gordon, III, C. P. J. Barty, and S. E. Harris, Phys. Rev. Lett., to be published.
2. B. E. Lemoff, C. P. J. Barty, and S. E. Harris, Opt. Lett. **19**, 569 (1994).
3. J. Sugar and V. Kaufman, Physica Scripta **26**, 419 (1982).
4. R. D. Cowan, The Theory of Atomic Structure and Spectra (UC–Berkeley, 1981), Secs. 8-1, 16-1, and 18-13.
5. N. H. Burnett and P. B. Corkum, J. Opt. Soc. Am. B **6**, 1195 (1989).
6. P. B. Corkum, N. H. Burnett, and F. Brunel, Phys. Rev. Lett. **62**, 1259 (1989).

QFE2 **2:30 pm**

High-order frequency conversion in a plasma waveguide

Charles G. Durfee III, James Lynch, Howard M. Milchberg, *Institute for Physical Science and Technology, University of Maryland, College Park, Maryland 20742*

High-order frequency conversion is one of several applications of a recently demonstrated plasma waveguide.[1] The plasma waveguide has a mode structure similar to that found in optical fibers[2] but may guide pulses of much higher intensity. Furthermore, in the limit that the refractive index is dominated by the free electrons and the ionic contribution may be neglected, the plasma-waveguide mode structure is independent of wavelength.[3] Because the waveguide may support multiple beams (without any threshold for the channeled beams, as in self-focusing), wave-mixing phenomena and harmonic generation at very high intensities may be studied and exploited. The plasma waveguide avoids the phase shift associated with the passage of the beam through a focus, which is an important limiting factor in harmonic-generation experiments using gaseous targets. Instead, there is a contribution to the phase velocity resulting from the mode structure of the waveguide. This phase variation with distance is linear because each eigenmode phase front is flat. There are, therefore, three sources of dispersion that must be considered in phase matching: the free electrons, the bound electrons, and the waveguide contribution. For harmonic generation to moderately high order (such that the harmonic light may fall near an atomic or ionic resonance), phase matching may be obtained through an appropriate mix of species. This has been done previously for focused beams, but guiding the beam over many Rayleigh lengths should allow for much higher conversion efficiency.

For very-high-order frequency generation, the ionic resonances are less important and phase matching may be obtained through high-order difference-frequency generation.[4] For example, phase matching for the process $\omega_3 = m\omega_1 - n\omega_3$ must satisfy the condition $mk_1 - nk_2 - k_3 = 0$, where the wave vectors k include the refractive-index and waveguide phase terms. It can be shown that the dispersion of the plasma waveguide phase term is the same as that of the plasma refractive index and that the phase-matching condition becomes $m\lambda_1 - n\lambda_2 - \lambda_3 = 0$. Phase-matched XUV frequency generation should be possible by choosing $\lambda_1 << \lambda_2$ and $m >> n$. Essentially, the phase mismatch created in generating the frequency $\omega_3 \approx m\omega_1$ is compensated by subtracting photons of a lower frequency, whereas the output frequency is shifted only slightly.

1. C. G. Durfee III and H. M. Milchberg, Phys. Rev. Lett. **71**, 2409 (1993).
2. C. G. Durfee III, J. Lynch, and H. M. Milchberg, Opt. Lett. **19**, 1937 (1994).
3. H. M. Milchberg, C. G. Durfee III, and J. Lynch, J. Opt. Soc. Am. B, in press.
4. P. L. Shkolnikov, A. E. Kaplan, and A. Lago, in *Short Wavelength V*, M. D. Perry and P. B. Corkum, eds., OSA Proceedings Series, Vol. 17 (Optical Society of America, Washington, D.C., 1993), p. 137.

QFE3 **2:45 pm**

Possible soft-x-ray lasing to the ground states in oxygen and carbon ions

B. Wellegehausen, A. Egbert, H. Eichmann, C. Momma, S. Nolte, Boris N. Chichkov,*
Institute for Quantum Optics, University of Hannover, Welfengarten 1, D-30167 Hannover, Germany

The recent availability and rapid progress of intense short-pulse laser systems has resulted in growth of interest in the development of high-repetition-rate soft-x-ray lasers driven by these intense laser pulses. One of the most popular soft-x-ray-laser approaches is based on tunnel ionization of atoms (molecules) in an intense laser field.[1] This ionization mechanism is very selective (because of the exponential dependence of the ionization rate on laser intensity and ionization potential) and permits stripping of all atoms to a necessary, definite ionization stage. In the case of a linearly polarized laser field, the free electrons can be left relatively cold, which allows the creation of a transient population inversion between excited and ground ion states during the three-body recombination cascade. To date, theoretical studies of a possible x-ray lasing to the ground ion state have concentrated on H- and Li-like ions.[1] Recent experimental results[2] have demonstrated evidence of lasing in H-like Li ions at 135 Å on the L_α transition.

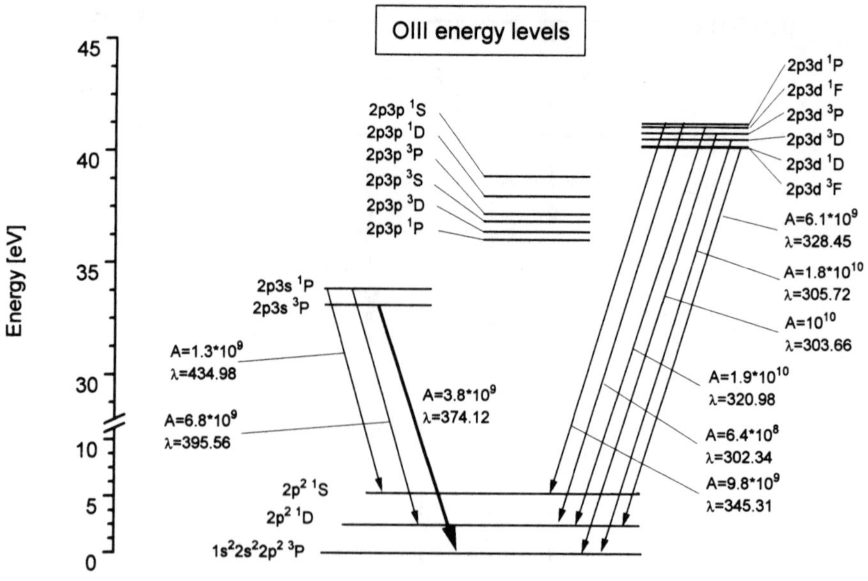

QFE3 Fig. 1. Energy levels, wavelengths, and radiative-decay probabilities for the O III ion.

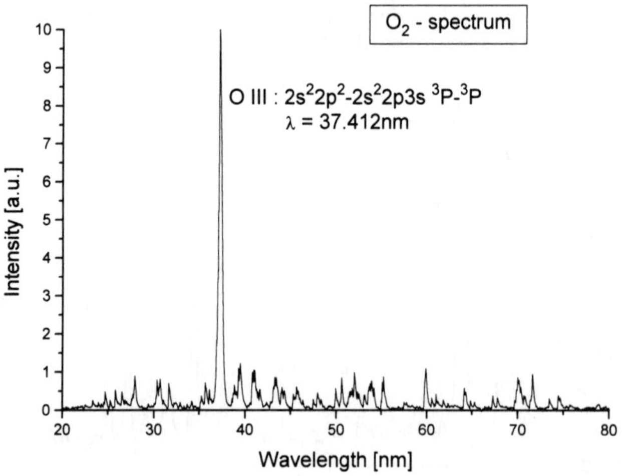

QFE3 Fig. 2. Example of a longitudinal spectrum of a plasma produced in an oxygen-gas jet.

Our theoretical analysis allows us to conclude that conditions necessary for soft-x-ray lasing to the ground electron configuration can be realized more easily with more complex ions. A trivial argument for this conclusion follows immediately from the expression for the population inversion density, $\Delta N = N_1 - N_0 g_1/g_0$, where g_0 and g_1 are statistical weights for the lower and upper laser levels, respectively. It is clear that more-favorable conditions for the population inversion will be obtained between levels with g_1/g_0 as low as possible ($g_1/g_0 \to 0$). For the L_α transition in H-like ions, $g_1/g_0 = 3$. We consider, as an example, Be-, C-, and N-like ions and present experimental results obtained with O II, O III, and C III ions for transitions with $g_1/g_0 \leq 1$. Figure 1 shows the energy levels, wavelengths, and radiative decay probabilities for transitions in O III. Figure 2 shows an example of an O III spectrum obtained with a commercial Ti:sapphire laser (100 mJ, 150 fs) focused with *f*/12 optics into an oxygen-gas jet (diameter, ~1 mm). Spectra were recorded in the direction of beam propagation. As can be seen, the $2p3s\ (^3P)$–$2p^2\ (^3P)$ line at $\lambda = 374.12$ Å is absolutely dominating in the spectrum. Detailed studies of this anomalous radiation are reported. Preliminary investigations of dependence on pressure and length variations support a suggestion of lasing on this line.

An additional advantage of lasing to the ground electron configuration in a complex ion is related to its splitting (due to electrostatic interaction) on various LS terms. Therefore, the lower laser level (LS term) belonging to the ground electron configuration can be located above the actual ground state of this ion. This creates an exit channel out of the lower laser level due to electron-impact deexcitation, and it simplifies the threshold conditions for lasing to the ground electron configuration. This is the case for the O II and C III ions. Investigations of possible soft-x-ray lasing in these ions are also discussed.

P. N. Lebedev Institute of Physics, Leninsky Prospect 53, Moscow 117924, Russia

1. N. H. Burnett and P. B. Corkum, J. Opt. Soc. Am. B **6**, 1195 (1989); N. H. Burnett and G. D. Enright, IEEE J. Quantum Electron. **26**, 1797 (1990); B. M. Penetrate and J. N. Bardsley, Phys. Rev. A **43**, 3100 (1991); D. C. Eder *et al.*, Phys. Rev. A **45**, 6761 (1992).
2. Y. Nagata *et al.*, Phys. Rev. Lett. **71**, 3774 (1993); D. C. Eder *et al.*, Phys. Plasmas **1**, 1744 (1994); T. D. Donnelly *et al.*, in *High Field Interactions and Short Wavelength Generation*, 1994 Technical Digest Series, Vol. 16 (Optical Society of America, Washington, D.C., 1994), p. 157.

QFE4 (Invited) **3:00 pm**

Multiphoton-induced x-ray emission from clusters

C. K. Rhodes, *Laboratory for Atomic, Molecular, and Radiation Physics, Department of Physics, University of Illinois at Chicago, Chicago, Illinois 60607-7059*

The ability to apply power densities controllably at or above vigorous thermonuclear levels ($>10^{19}$ W/cm^3) in materials is the basic issue for achieving efficient amplification of x-rays. Recent experimental and theoretical findings concerning the multiphoton production of x-rays from clusters and high-intensity modes of channeled propagation in plasmas indicate an entirely new method for producing the conditions necessary for strong amplification in the multi-kilovolt range. These two new nonlinear phenomena are being united to produce and control the imperative power compression; the multiphoton mechanism serves to establish the condition locally while the confined propagation provides the required spatial organization. The present work, which experimentally demonstrates the first to our knowledge combined expression of these two complex nonlinear processes through direct x-ray imaging of Xe(M) emission (~1 keV) in stable self-trapped channels, (a) reveals the exceptional compatibility of their mutual scaling for realizing the necessary power density, (b) provides confirming evidence for the action of a superstrong coherent multi-electron intense-field interaction in the X-ray generation from the Xe clusters, and (c) furnishes new detailed information on the dynamics of the radial intensity distributions associated with the channeled propagation. The resulting knowledge of the scaling relations underlying these phenomena enables the optimum conditions for amplification to be specified up to a quantum energy of ~5 keV. The harmonious use of these new nonlinear processes is expected to lead to an advanced generation of extraordinarily bright x-ray sources in the multi-kilovolt region having a peak brightness of $\sim 10^{31}$ – 10^{33} γs^{-1} $(mrad)^{-2}$ $(mm)^{-2}$ $(0.1\%\ BW)^{-1}$, a level sufficient for biological holographic imaging capable of providing a high resolution visualization of the molecular anatomy of cells, tissues, and organisms in the natural state.

QFE5 **3:30 pm**

Reduction of pumping energy and wavelength scaling of Ni-like soft-x-ray lasers

H. Daido, Y. Kato, K. Murai, S. Ninomiya, R. Kodama, H. Takabe, F. Koike* *Institute of Laser Engineering, Osaka University, 2-6 Yamada-oka, Suita, Osaka 565, Japan*

We report efficient production of Ni-like soft-x-ray-laser media[1] with a curved-slab target[2,3] and multiple laser pulses,[4] as shown in Fig. 1. With this technique we have obtained lanthanide soft-x-ray lasers with a spectral range between 6 and 8 nm.[4] The bend angle of 10 mrad for a 2.5-cm-long target, corresponding to a 2.5-m radius of curvature, provided the most intense lasing for the Ni-like Nd soft-x-ray laser. A laser beam of 1.053-μm wavelength was focused to a line of 2.8-cm length and approximately 80-μm average width. The laser pulse was composed of two pulses with the individual pulse widths of 100 ps, separated by 400 ps. The laser energy was approximately 150 J for each 100-ps pulse, corresponding to an average irradiance at focus of 7.9×10^{13} W/cm^2. A gain coefficient of 3.1 cm^{-1} and a gain–length product of 7.8 for Nd have been achieved at 7.97 nm with a pumping-laser energy of 250 J on a 2.5-cm-length target, which is an order of magnitude lower than that for the previous Eu-laser experiment.[1]

Figure 2 shows the spectra of the Ni-like Nd ($Z = 60$), Sm ($Z = 62$), Gd ($Z = 64$), and Tb ($Z = 65$) lasers recorded with an x-ray streak camera, where Z is the atomic number. Each spectrum is composed of two $J = 0–1$ lines; a longer-wavelength line due to the (3/2, 3/2)-to-(5/2, 3/2) transition and a shorter-wavelength line due to the (3/2, 3/2)-to-(3/2, 1/2) transition are indicated by the arrows. The peak of the x-ray-laser pulse is 80 ps before the peak of the second pumping pulse.

The pumping energy could be further reduced by reducing the line-focus width. One-dimensional hydrodynamic simulation code in a cylindrical geometry was used to simulate the two-dimensional expansion. The result shows that a line-focus width of 50 μm will result in almost the same gain region as that with the focus width of more than 100 μm. The first pumping pulse may need a slightly higher fluence to compensate for the expansion loss. Our experimental results suggest that a similar gain may be obtained by using a second pumping pulse of shorter duration and much smaller energy. Consequently, a significant reduction of the pumping energy required to obtain the same gain–length product can be achieved with better spatial coherence because of the smaller source size. We report the required pumping conditions for extending the wavelength range, as shown in Fig. 3, with this technique by using atomic-physics (GRASP2) and plasma-hydrodynamics (ILESTA) codes.

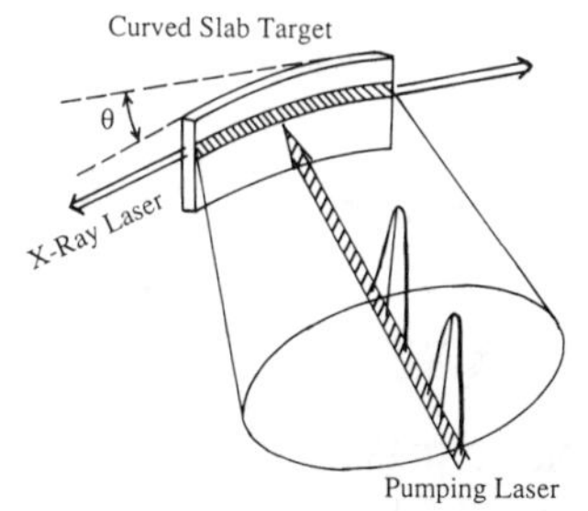

QFE5 Fig. 1. Schematic diagram of a curved-slab target pumped with a double-pulse laser.

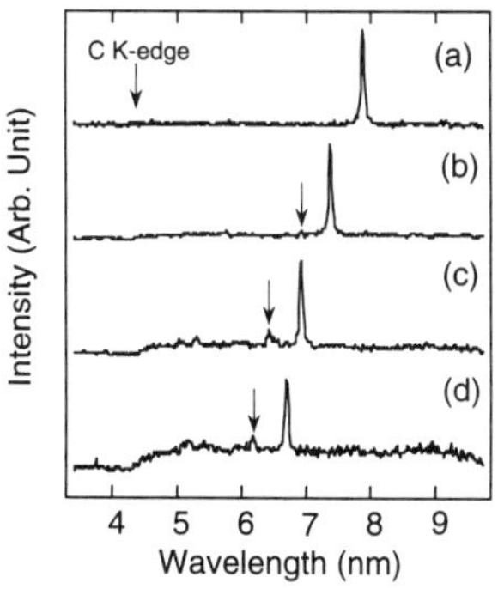

QFE5 Fig. 2. Typical spectra of (a) Nd, (b) Sm, (c) Gd, and (d) Tb recorded with a flat-field spectrometer and an x-ray streak camera. The shorter-wavelength $J = 0–1$ lines are indicated by the arrows. The relative peak intensities of the longer-wavelength $J = 0–1$ lines are 8.5, 2.8, 1.5, and 1.0 for Nd, Sm, Gd, and Tb, respectively. The gain lengths are 2.5 cm.

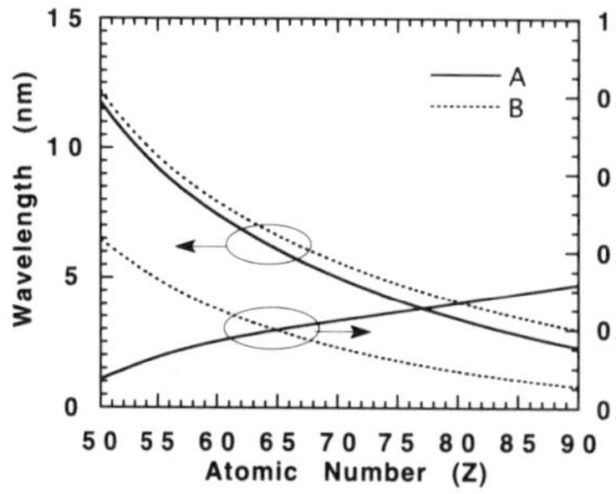

QFE5 Fig. 3. Calculated shorter- and longer-wavelength Ni-like-ion $J = 0–1$ transitions, which agree well with experiment, and corresponding oscillator strengths, as a function of atomic number.

Physics Laboratory, School of Medicine, Kitasato University, 1-15-1 Kitasato, Sagamihara, Kanagawa 228, Japan

1. B. J. MacGawan *et al.*, Phys. Fluids B **4**, 2326 (1992).
2. J. G. Lunney, Appl. Phys. Lett. **48**, 891 (1986).
3. R. Kodama *et al.*, Phys. Rev. Lett., in press.
4. H. Daido *et al.*, Opt. Lett., in press.

QFE6 **3:45 pm**

Vacuum laser acceleration by second- or third-order ponderomotive force

N. B. Baranova, B. Ya. Zel'dovich, *Center for Research and Education in Optics and Lasers, University of Central Florida, Suite 400, 12424 Research Parkway, Orlando, Florida 32826-3271*

Many schemes for laser acceleration of elementary particles have been suggested and discussed; see, e.g., Ref. 1 and references therein. For the inverse Čherenkov effect schemes (which allow for acceleration of the first order in the laser field strength **E**) one must focus a powerful laser beam inside a medium with a nonzero deviation of the refractive index from the vacuum one; optical breakdown is an evident obstacle. Other first-order schemes require the delivery of laser radiation quite near (about a wavelength distance) to the gratings, holes in mirrors, glass tubes, etc., making these schemes also vulnerable to the optical breakdown.

We discuss in the present paper schemes[2] that allow laser acceleration of charged particles in true vacuum (also called VLA) at a large distance from any refracting or reflecting substances. The driving force for those acceleration processes was chosen to be either second-order ponderomotive force $\mathbf{F}^{(2)}$ or third-order ponderomotive force $\mathbf{F}^{(3)}$. The expression for the second-order force for a charged particle in its own rest frame[3,4] is well known:

$$\mathbf{F}^{(2)} = -\nabla U^{(2)},$$

$$U^{(2)} = \frac{e^2}{2mc^2}\langle \mathbf{A}\cdot\mathbf{A}\rangle = \frac{e^2}{4m\omega^2}\mathbf{E}\cdot\mathbf{E}^*. \quad (1)$$

Here $\langle\cdot\rangle$ denotes time averaging and $\mathbf{A}(\mathbf{r}, t)$ denotes the real vector potential, so that for the gauge choice with $\varphi = 0$ one has

$$\mathbf{E}_{\text{real}}(\mathbf{r}, t) = \frac{1}{2}\mathbf{E}(\mathbf{r})\exp(-i\omega t) + c.c.$$

$$= -\frac{1}{c}\frac{\partial \mathbf{A}}{\partial t}. \quad (2)$$

The third-order force requires that the time-averaged cube of the field strength be nonzero at the particle's frame. That force was calculated in Ref. 5:

$$F_i = -\frac{\partial U^{(3)}}{\partial x_i}, \quad (3)$$

$$U^{(3)} = -\frac{e^3}{2m^2c^3}\left\langle \frac{\partial(A_j^2)}{\partial x_k}\int_{-\infty}^{t} A_k dt'\right\rangle. \quad (4)$$

Reference 5 contains a suggestion that a quite steep gradient of $U^{(2)}$ synchronously moving with the particle may be produced through the interference of two optical beams, $E_1 \propto \exp(-i\omega t + i\mathbf{k}_1\mathbf{r})$ and $E_2 \propto \exp(-2i\omega t + i\mathbf{k}_2\mathbf{r})$, illuminating the path $z = ut$ at some specific angles θ_1 and θ_2 to the z axis so that

$$1 - \frac{u}{c}\cos\theta_1 = 2 - 2\frac{u}{c}\cos\theta_2. \quad (5)$$

As for the third-order acceleration, the important point is that one can get a nonzero time-averaged cube in the particle frame by using monochromatic illumination (with one frequency ω only) by the coherent waves satisfying the condition

$$\frac{u}{c}(\cos\theta_1 + \cos\theta_2 - \cos\theta_3) = 1. \quad (6)$$

The case in which the first and the second waves are represented by the same field is also a possibility. The particular case $\theta_3 = 90°$ is shown in Fig. 1; for simplicity, the wave $E_1 \equiv E_2$ with $\theta_1 = 60°$ is not depicted.

The acceleration rate, i.e., the growth of the electron's kinetic energy K, was calculated to be

$$\frac{dK}{dz} \approx G\frac{P}{P_0}\frac{mc^2}{K} \quad (7)$$

for the second-order force and

$$\frac{dK}{dz} \approx G\frac{P}{P_0}\left(\frac{mc^2}{K}\right)^{3/2} \quad (8)$$

for the third-order force. Here P (W/cm²) is the light-beam power density. At the wavelength $\lambda = 1$ μm the value of P_0 is approximately 1.4×10^{18} W/cm². At $P = P_0$ the quiver velocity of a nonrelativistic electron would be about c. The parameter G is approximately 5 GeV/cm for a 1-μm wavelength.

The second part of this paper deals with the combination[2] of two ideas: conical focusing[1] and a diffraction-grating device to produce a focal spot illuminating the target line at an angle $\theta \neq 0$ but with the spot moving along the target with controllable speed (including c).[6,7] The illumination of a conical diffraction grating by a broad, ultrashort pulse with the radially growing chirp allows us to do that: Fig. 1 shows an example of such focusing at an angle 90° toward the axis; the focal spot runs in the z-direction with the speed of light. This scheme allows us to obtain a high power density P accompanying the ultrarelativistic particle by means of a bandwidth-limited light pulse ($\delta\omega = 1/\tau_0$) of relatively moderate energy Q. If the interaction length is L, then the necessary value of pulse energy is

$$Q \approx 0.3PL\lambda\tau_0. \quad (9)$$

The same energy is required for the paraxial Gaussian copropagating pulse at given P, L, and τ_0. However, in the latter case extremely strong Doppler downshift occurs: $\omega' \approx \omega mc^2/K \ll \omega$. On the other hand, for Bessel-conical focusing with $\theta \sim 1$ rad (see Fig. 1) one obtains $\omega' \approx \omega K/mc^2 \gg \omega$ and the forward-radiated x-ray wave acquires the frequency

$$\omega_F \approx \omega\left(\frac{K}{mc^2}\right)^2. \quad (10)$$

These bremsstralung x rays will be very monochromatic, $\Delta\omega_F/\omega_F \approx \lambda_0/2\pi L$. The last two statements about the frequency and the bandwidth of forward-emitted x rays for the laser illumination of ultrarelativistic electrons at nonzero angle θ are well known. The emphasis of this paper is that they may be produced at a relatively low energy cost Q for the specially prepared ultrashort pulse accompanying the particle.

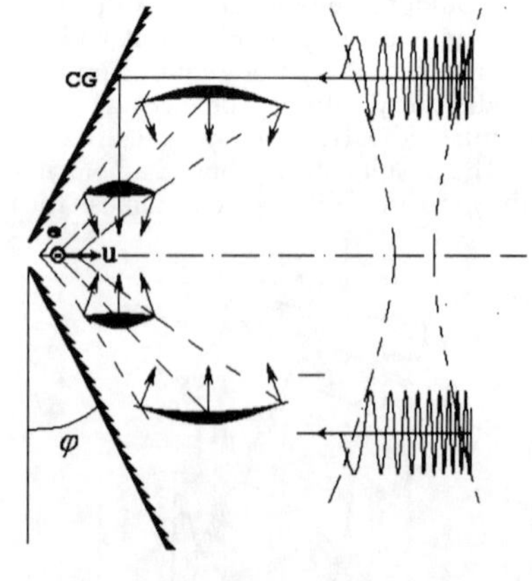

QFE6 Fig. 1. Conical grating (CG) scheme to direct photons at an angle ($\theta = 90°$ here) relative to an electron velocity $\mathbf{u} = u\mathbf{e}_z$, $u \approx c$, and to follow an electron along length $L \gg c\tau_0$.

1. L. C. Steinhauer and W. D. Kimura, J. Appl. Phys. **72**, 3237 (1992).
2. N. B. Baranova, B. Ya. Zel'dovich, and M. O. Scully, Sov. Phys. JETP **78**, 249 (1994).
3. A. V. Gaponov, M. A. Miller, Sov. Phys. JETP **7**, 168 (1958).
4. L. S. Brown, T. W. B. Kibble, Phys. Rev. **133**, A705 (1964).
5. N. B. Baranova, B. Ya. Zel'dovich, J. Am. Opt. Soc. B **8**, 27 (1991).
6. Zs. Bor, S. Szatmari, A. Muller, Appl. Phys. B **32**, 101 (1983).
7. M. H. Sher, J. J. Macklin, J. F. Young, S. E. Harris, Opt. Lett. **12**, 891 (1987).

QFF **2:00 pm**

Hyatt C/D

Dynamics of Electronic Excitations

Presider To Be Announced

QFF1 **2:00 pm**

Dependence of time-resolved pump–probe signals on probe spectral shape

J.-P. Likforman, M. Joffre, D. Hulin, *Laboratory of Applied Optics, ENSTA, Ecole Polytechnique, National Center for Scientific Research, URA 1406, F-91120 Palaiseau, France*

We demonstrate that in femtosecond pump–probe experiments the signal is not always independent of the probe-pulse spectral shape. This is in contradiction to the common interpretation of these experiments, in which the pump–probe spectra are simply assumed to represent time-dependent absorption spectra. Indeed, the absorption of a sample as measured by a single low-intensity optical field is independent of its spectral shape. For an optically thin sam-

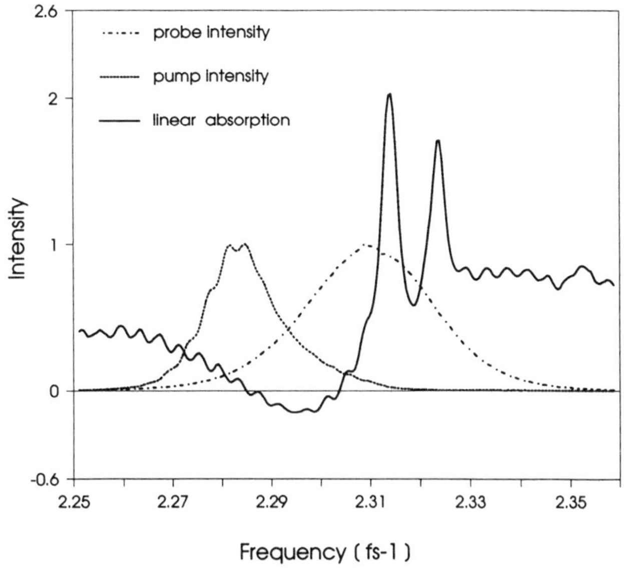

QFF1 Fig. 1. Experimental configuration: pump spectrum (dotted curve) is plotted together with the absorption profile of the sample (solid curve) and a probe spectrum centered on the excitonic resonance (dash-dotted curve).

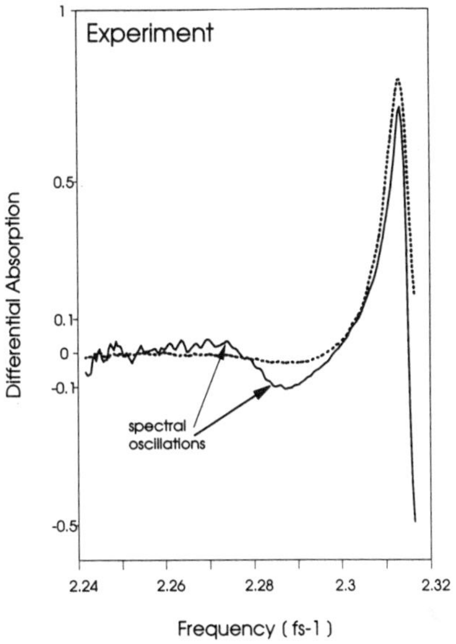

QFF1 Fig. 2. Comparison of experimental differential-absorption signals obtained with two different probe pulses in an optical-Stark-effect experiment. The experiment using a probe centered on the absorption line (solid curve) gives strong spectral oscillations at frequencies for which the probe is decreasing. The oscillations almost disappear in the experiment with the probe detuned to a lower frequency because of the higher amplitude of the probe at the corresponding frequencies.

ple, the absorption can be written

$$\alpha(\omega) \propto \ln\left(\frac{I_i}{I_t}\right) \propto \operatorname{Im}\left[\frac{P^{(1)}(\omega)}{\varepsilon_{pr}(\omega)}\right], \quad (1)$$

where I_i and I_t are the incident and transmitted probe intensities, respectively, $\varepsilon_{pr}(\omega)$ is the electric field of the optical pulse, and $P^{(1)}(\omega)$ is the induced linear polarization. The linear polarization at ω is proportional to the field at the same frequency, and the division by $\varepsilon_{pr}(\omega)$ makes the signal independent of the pulse spectrum. Similarly to relation (1), the pump–probe signal is measured as

$$-\Delta\alpha(\omega) \propto \ln\left(\frac{I_t}{I_t^{(p)}}\right) \propto \operatorname{Im}\left[\frac{P^{(3)}(\omega)}{\varepsilon_{pr}(\omega)}\right], \quad (2)$$

where $I_t^{(p)}$ is the transmitted probe intensity in the presence of the pump and $P^{(3)}(\omega)$ is the nonlinear polarization induced by both pump and probe pulses. However, in the nonlinear regime, the signal at frequency ω results from frequency mixing between the two pulses and is no longer proportional to the probe at ω. Therefore, dividing by $\varepsilon_{pr}(\omega)$ in relation (2) does not preclude the differential signal from depending on the probe spectral shape. All three terms of the third-order polarization[1] are affected by this phenomenon, not only population terms[2] but also induced-grating and perturbed-free-induction-decay terms.

To illustrate this discussion, we present an experiment performed with a femtosecond titanium:sapphire oscillator in an optical-Stark-effect configuration in which the pump is tuned below the excitonic resonance of a GaAs multiple-quantum-well structure (Fig. 1). We compare differential signals recorded with two different probe pulses, one centered on the excitonic absorption and the other one at a lower energy (Fig. 2). We find that the signal does indeed depend on the shape of the probe and that the spectral oscillations,[1] due to the abrupt perturbation of the linear polarization by the pump pulse, are enhanced at frequencies for which the probe field is less intense. In the case of a sharp absorption line, the probe $\varepsilon_{pr}(\omega)$ contributes to the nonlinear polarization $P^{(3)}(\omega)$ proportionally to its value at the resonance and gives rise to a signal at frequencies as far from the resonance as the pump spectral width. Thus, the division by a probe field of decreasing amplitude [relation (2)] strongly amplifies the spectral oscillations at the corresponding frequencies (solid curve in Fig. 2). On the contrary, if the probe is detuned from the exciton resonance, the spectral oscillations vanish entirely (dotted curve in Fig. 2). Finally, we would like to stress that the pump–probe signal depends not only on the power spectrum of the probe pulse but also on its spectral phase.

1. C. H. Brito-Cruz, J. P. Gordon, P. C. Becker, R. L. Fork, and C. V. Shank, IEEE J. Quantum Electron. **QE24**, 261 (1988).
2. E. Tokunaga, A. Terasaki, T. Wada, K. Tsunetomo, Y. Osaka, and T. Kobayashi, J. Opt. Soc. Am. B **10**, 2364 (1993).

QFF2 (Invited) 2:15 pm

Ultrafast spectroscopy of GaAs

Mary Ann Mycek, Uwe Siegner, Stephan Glutsch, Daniel S. Chemla, *Department of Physics, University of California at Berkeley, and Materials Sciences Division, Lawrence Berkeley Laboratory, 1 Cyclotron Road, MS 2-300, Berkeley, California.*

Surprising and novel results are obtained for both the linear and the nonlinear optical response of GaAs under magnetic field. Using a variety of spectroscopic techniques, we measure field dependent effects due to Coulomb correlation.

QFF3 2:45 pm

Time-resolved carrier capture in quantum-well separate-confinement heterostructures

C. Y. Sung, T. B. Norris, X. K. Zhang,* M. Sneed,* P. K. Bhattacharya,* *Center for Ultrafast Optical Science, University of Michigan, 2200 Bonisteel, Ann Arbor, MI 48109-2099*

Carrier-capture dynamics are crucial in the performance of quantum-well-based devices, such as quantum-well lasers. Some quantum-mechanical calculations predict strong resonance of the carrier capture as a function of well width due to wave-function overlay between the barrier and the well,[1] although other calculations, including carrier–carrier scattering and screening effects, indicate an important density dependence of the capture and thermalization in operating lasers.[2]

A number of experiments on time-resolve carrier capture into quantum wells have been performed; however, the transient distribution functions occurring in separate-confinement-heterostructure (SCH) laser structures have not been measured. The best way to do so is to perform time-resolved pump–probe differential transmission spectroscopy, which will allow the capture of electrons and holes to be determined separately, and to determine the distributions as well as the populations. Toward this end, we have performed initial investigations of capture in SCH structures.

The samples consist of 15 separate single GaAs quantum wells, each surrounded by 500-Å $Al_{0.1}Ga_{0.9}As$ barrier layers and 100-Å AlAs cladding layers. The substrates were removed by selective etching to allow transmission experiments. The measurements were performed in a standard pump–probe setup at room temperature. Micro-

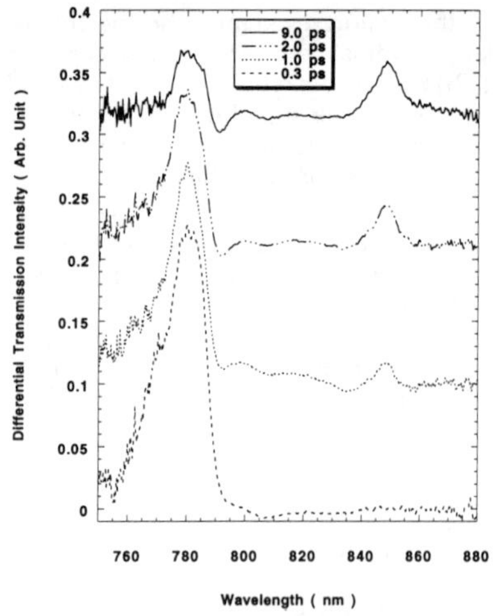

QFF3 Fig. 1. Differential transmission spectra at four different time delays from a sample with a 70-Å well width.

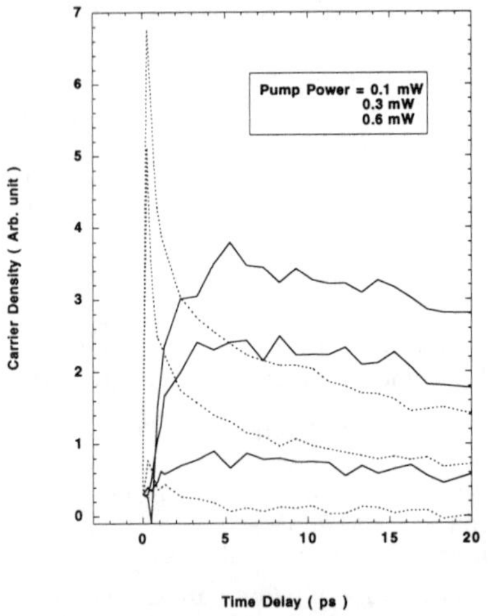

QFF3 Fig. 2. Temporal evolutions of carrier density: dashed curves are carriers on the interbarriers; solid curves are carriers in the wells. Three different pump pulse intensities are shown.

joule 100-fs pulses from a 250-kHz Ti:sapphire regenerative amplifier were used to generate a white-light continuum in a sapphire crystal. The pump energy was tuned to the $Al_{0.1}Ga_{0.9}As$ barrier region. The dispersion of the broadband probe (1.4 to 1.65 eV) was compensated in a double-pass prism pair, so that transmission spectra of the entire near-band-edge region could be obtained with 120-fs resolution.

Some typical differential transmission spectra are shown in Figure 1 for a sample with a 70-Å well width, clearly showing the capture of the carriers from the barrier into the well. Figure 2 shows spectrally integrated amplitudes of the well (solid curves) and barrier (dashed curves) transitions (which are to first order proportional to the populations) for different pump powers. The capture shows a fast initial capture on a single-picosecond time scale, followed by a slower capture that is essentially complete at approximately 5 ps. Over the limited range of quantum-well parameters and carrier densities investigated to date, only minor changes in the capture rate have been observed. Further work will investigate the distributions and capture under very high pumping and under lasing conditions.

* *Solid State Electronics Laboratory, University of Michigan, 1301 Beal Avenue, Ann Arbor, Michigan 48109-2122*

1. P. W. M. Bloom, C. Smit, J. E. M. Haverkort, and J. H. Wolter, Phys. Rev. B **47**, 2072 (1993).
2. Y. Lam and J. Singh, IEEE J. Quantum Electron. **30**, 1196 (1994).

QFF4 **3:00 pm**

Femtosecond hole thermalization in bulk GaAs

F. Vallée, P. Langot, R. Tommasi,*
Laboratory of Quantum Optics, National Center for Scientific Research, Ecole Polytechnique, F-91128 Palaiseau Cédex, France

Femtosecond spectroscopy has been extensively exploited to analyze elementary carrier interactions in semiconductors. Most measurements to date have been performed under conditions such that electron dynamics dominate the transient response; little is known about hole-relaxation dynamics in direct-gap semiconductors. By using time-resolved luminescence this problem has recently been addressed in n-doped GaAs, in which the high-density electron plasma introduced by doping constitutes an important thermal bath for the nonequilibrium holes.[1-3] We show, for the first time to our knowledge, that by using proper excitation and probing conditions, the heavy-hole (hh) thermalization dynamics can be selectively analyzed in intrinsic bulk GaAs by using two-wavelength femtosecond absorption-saturation spectroscopy.

Because of the large hh/electron mass ratio, the average momentum of a thermal hh distribution is much larger than that of electrons and, consequently, the two distributions are partly separated in *k*-space. In the nonequilibrium regime, the electrons can also be confined to small momentum by photoexciting them with an average momentum that is smaller than the thermal one (i.e., $\lambda_{pump} \geq 850$ nm at 300 K in GaAs). Under these conditions, for large wave vectors (i.e., large probe-photon excess energy) the transient absorption saturation ($\Delta T/T$) predominantly contains information on the filling of the hh states (for instance, the electron contribution to $\Delta T/T$ is about 1/7 the hole contribution for $\lambda_{probe} = 750$ nm), making possible a selective analysis of the hh thermalization dynamics.

The pump and probe pulses are created by frequency conversion of a femtosecond $Ti{:}Al_2O_3$ laser. The pulses are spectrally broadened in an optical fiber, and the two synchronized, independently tunable (in the range 890–720 nm) 70-fs pulses are selected by using grating-pair compressors. The high repetition rate and stability of our system permit a high-sensitivity ($\Delta T/T$ of the order of 10^{-6}) when a standard pump–probe setup is used.

Figure 1 shows the measured $\Delta T/T$ for a probe wavelength of 750 nm and a pump wavelength of 850 nm, corresponding to the injection of cold holes. Under these conditions, no instantaneous band filling is observed and the short time-delay signal is dominated by band-gap renormalization, which decreases the transmission. The subsequent rise is due to filling of the probed hh states as the initial distribution gains energy, and it indicates that after 500 fs, the holes are in thermal equilibrium with the lattice.

By assuming monoexponential hh dynamics, a characteristic thermalization time of 150 fs has been estimated. This time has been found to be independent of both the carrier density (in the investigated range of 3×10^{16} to 7×10^{17} cm^{-3}) and of the initial excess energy of the electron gas (λ_{pump}

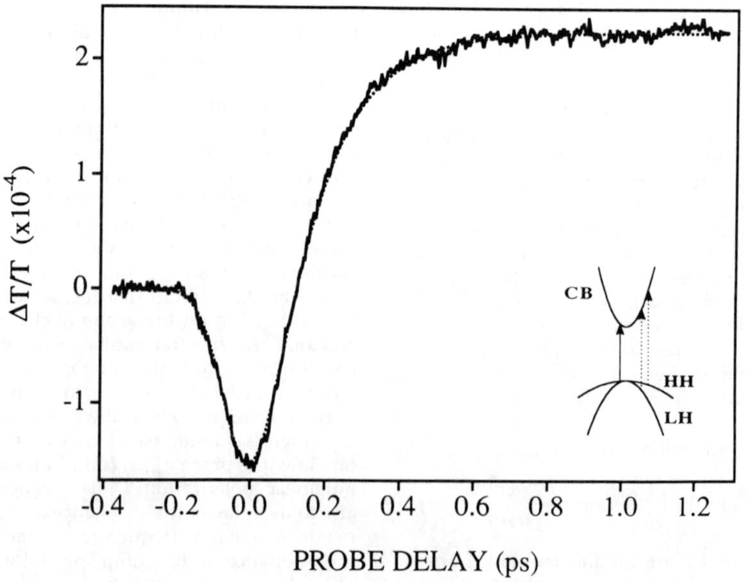

QFF4 Fig. 1. Transient absorption saturation in GaAs measured for a pump wavelength of 850 nm and a probe wavelength of 750 nm. The photoexcited carrier density is 1×10^{17} cm^{-3}. Inset indicates the optical transition for the pump (solid arrow) and the probe (dotted arrow).

varying in the range 870–850 nm), indicating that hole heating is dominated by hole–phonon interactions. Additional measurements were performed as a function of the probe and pump photon energies, yielding a precise description of the transient hole distribution and confirming the dominant role of the hole–phonon interactions. These measurements open up many possibilities for the investigation of hole interactions with their environment under various conditions in direct-gap semiconductors.

*Also with *Department of Physics, University of Bari, Via Orabona 4, I-70126 Bari, Italy*

1. X. Q. Zhou, K. Leo, and H. Kurz, Phys. Rev. B **45**, 3886 (1992).
2. A. Chébira, J. Chesnoy, and G. M. Gale, Phys. Rev. B **46**, 4559 (1992).
3. A. Tomita, J. Shah, J. E. Cunningham, S. M. Goodnick, P. Lugli, and S. L. Chuang, Phys. Rev. B **48**, 5708 (1993).

QFF5 **3:15 pm**

Direct mid-infrared spectroscopic measurements of hole relaxation in InGaAs/AlGaAs quantum wells

Z. Xu, P. M. Fauchet, C. W. Rella,* B. A. Richman,* H. A. Schwettman,* G. W. Wicks,** *Department of Electrical Engineering, University of Rochester, Rochester, New York 14627*

Measurements of infrared bleaching[1] and anti-Stokes Raman scattering[2] in n-type QWs have revealed an electron relaxation time ranging from less than 1 ps to 10 ps when the intersubband transition energy is larger than the LO phonon energy. The relaxation time depends on the excitation intensity and is influenced by intervalley scattering, LO-phonon screening, and hot phonons. In p-type quantum wells, the relaxation time is expected to be shorter because the density of states is larger and the band structure and the scattering mechanisms are different. Here, we report the first direct measurement of the relaxation time of holes in p-type quantum wells by the pump–probe technique using ~1-ps pulses generated by a mid-infrared free-electron laser.

The MBE-grown samples are 4-nm-wide strained $In_{0.5}Ga_{0.5}As$ quantum wells sandwiched between 8-nm-wide $Al_{0.5}Ga_{0.5}As$ barriers. The Be doping concentration is 10^{19} cm^{-3}, and the sheet density is thus 1.2×10^{13} cm^{-2} if all the carriers are transferred to the wells. The FTIR spectrum shows an absorption peak at 5.25 μm (220 meV) with a full width at half maximum (FWHM) of 50 meV (Fig. 1). The absorption peak is 1 order of magnitude wider than in n-type quantum wells because of the efficient hole-hole scattering, which is expected to produce a width of 20 to 30 meV.[3] The 5.25-μm absorption peak is due to the heavy-hole-to-light-hole transition, with the strain-induced energy splitting between heavy- and light-hole states $\Delta = 150$ meV and a quantum-confinement-induced energy splitting of 90 meV at $k = 0$.

Equal-wavelength pump–probe measurements are performed at room temperature by using the Stanford free-electron laser (FEL). The FEL pulses are < 1 ps long and are tunable from 4 to 6 μm. The peak pump intensity is varied from 0.3 to 10 GW/cm². The change of transmission, measured at 5.05 μm with a pump intensity of 2.5 GW/cm², is shown in Fig. 2 on a logarithmic scale. A relaxation time τ from ≤ 1 to 2 ps is obtained, depending on the wavelength and pump intensity. The data can be explained by assuming a homogeneously broadened two-level system with a saturation intensity of 3.2 GW/cm², which is significantly higher than in n-type quantum wells, in part because the absorption cross section for our p-type sample is an order of magnitude smaller than in n-type quantum wells. As is expected from a homogeneously broadened system, the relaxation time appears to be wavelength independent to within the accuracy of our measurements.

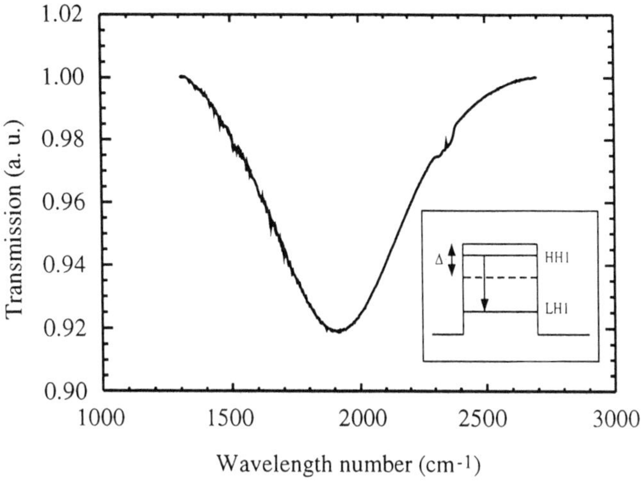

QFF5 Fig. 1. Infrared absorption of the QW sample. The strong absorption is due to the (heavy hole 1)-to-(light hole 1) transition, as indicated in the inset. Δ is the strain-induced energy splitting.

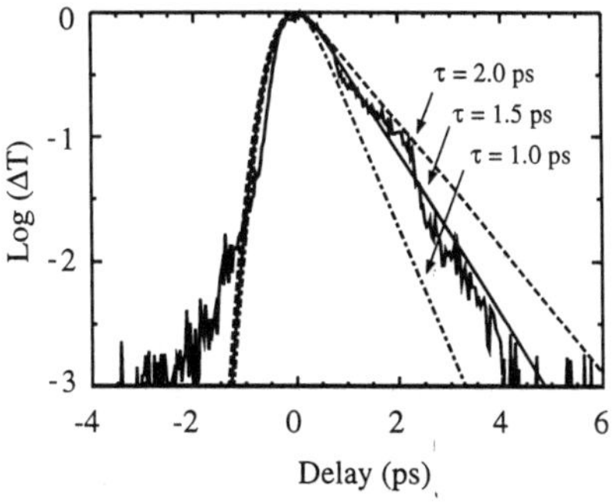

QFF5 Fig. 2. Pump-induced transmission change of the probe beam for a pump intensity of 2.5 GW/cm² measured at 5.05 μm as a function of the delay time, on a logarithmic scale. The fits assume an exponential relaxation time of 1, 1.5, and 2 ps.

In the valence band, the phonons and the holes can interact through the optical-deformation potential, which is more efficient and less easily screened than the Frohlich interaction, resulting in a faster hole scattering time. Our relaxation time does not reflect the initial intersubband scattering event[4] (dwell time) but rather the intrasubband cooling of the hot holes. The details are involved because, for example, hole–hole scattering is efficient and thus the hot holes can cool down quickly by giving energy to the cold-hole reservoir. The increase of the relaxation time at higher excitation intensity is consistent with hot-phonon reabsorption. Detailed fitting and analysis are presented.

We acknowledge support form the U.S. Office of Naval Research through contracts N00014-92-J-4063 and N00014-91-C-0170.

**Stanford Picosecond FEL Center, W. W. Hansen Experimental Physics Laboratory, Stanford University, Stanford, California 94305*

***Institute of Optics, University of Rochester, Rochester, New York 14627*

1. A. Seilmeier *et al.*, Phys. Rev. Lett. **59**, 1345 (1987); F. H. Julien *et al.*, Appl. Phys. Lett. **53**, 116 (1988); T. Elsaesser *et al.*, Appl. Phys. Lett. **45**, 256 (1989).
2. M. C. Tatham, J. R. Ryan, and C. T. Foxon, Phys. Rev. Lett. **63**, 1637 (1989).
3. M. Asada, IEEE J. Quantum Electron. **QE-25**, 2019 (1989).
4. J. F. Young, T. Gong, P. M. Fauchet, and P. J. Kelly, Phys. Rev., B **50**, 2208 (1994).

QFF6 **3:30 pm**

Ballistic charge oscillations in bulk GaAs

W. Sha, S. Patkar, Arthur L. Smirl, W. F. Tseng,* *Laboratory for Photonics and Quantum Electronics, 100 IATL, University of Iowa, Iowa City, Iowa 52242*

Advances in the growth and fabrication of semiconductor structures with features having dimensions comparable to (or less than) typical mean-free-path lengths and advances in the development of femtosecond laser sources with temporal resolutions less than the mean scattering times have made possible the investigation of coherent and ballistic transport in semiconductors. To date, for example, three-terminal (launcher-base-detector) devices have been used to study transport in the near-ballistic regime,[1,2] and velocity overshoot has been reported in transient-photocurrent experiments.[3] More recently, the nearly ballistic acceleration of electrons has also been reported on a femtosecond time scale.[4,5] In addition, coherent oscillatory electronic wave packets have been observed in coupled quantum wells,[6] and Bloch oscillations have been observed in superlattices.[7,8]

Here, to best of our knowledge, we report the first direct observation in a bulk semiconductor of coherent charge oscillations in the presence of a constant built-in field. These oscillations are the result of ballistic charge transport that occurs on a time scale short compared to the separately measured dephasing time. In these experiments, electrons and holes are suddenly generated bulk material in the presence of a constant field. For times short compared to the mean scattering time, the electrons and holes are initially accelerated in opposite directions by the presence of the constant field. As they move apart, a space-charge field develops as a result of the Coulomb attraction between the oppositely charged species. This restoring force causes the phasma to coherently oscillate until collisions dampen the oscillations.

Our measurements were performed at 80 K in a GaAs vertical p-i-n structure with a 3.5-kV/cm built-in field and a 400-nm-thick intrinsic region. The electrons and holes were photogenerated in the intrinsic region by a 100-fs pump pulse tuned to the GaAs band edge. The ballistic charge oscillations were then monitored with a time-delayed probe by measuring the change in the near-band-edge electroabsorption as the oscillating field screened the built-in p-i-n field. A representative trace of the time derivative of the change in transmission induced by these oscillations is shown in Fig. 1. Several features are worth noting. First, notice that clear oscillations are observed over a range of fluences. These oscillations decay in a few hundred femtoseconds, consistent with the expected mean time between collisions at this temperature in GaAs. Finally, the frequency of oscillation clearly increases with increasing fluence.

In Fig. 2 we plot the oscillation frequency for range of fluences. In obtaining these data, we verified that the optical absorption was linear over this range of fluences; therefore, this figure illustrates that the frequency of oscillation has a square-foot dependence on carrier density, consistent with phasma oscillations: $\omega_p^2 = ne^2/\varepsilon m^*$, where ω_p is the plasma frequency, n is the carrier density, e is the electronic charge, ε is the dielectric constant, and m^* is the effective mass. In fact, by estimating the carrier densities, we find that the measured frequencies deviate from the calculated plasma frequencies by less than 25% for the entire fluence range. Also, from a simple coherent-plasma model, we estimate the oscillation amplitude and the maximum kinetic energy gained in these experiments. The results (shown in Fig. 3) indicate that for low fluences the carriers move ~100 nm and gain more than 10 meV in energy before collisions disrupt their motion.

To verify that our oscillations were indeed the result of coherent charge motion, we also measured the carrier dephasing in an electric-field-free sample under other-

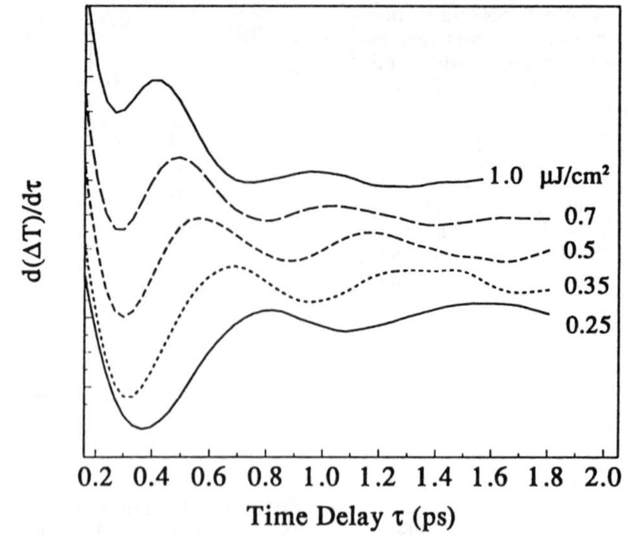

QFF6 Fig. 1. Time derivative of the change in electroabsorption versus pump–probe time delay for several fluences.

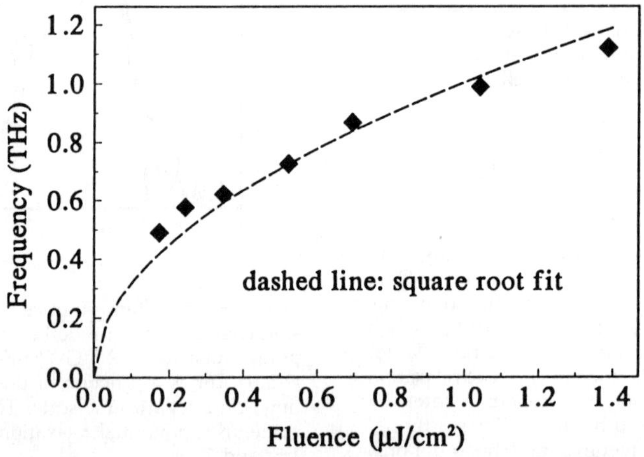

QFF6 Fig. 2. Frequency of the oscillating electric field (as extracted from the electroabsorption measurements) as a function of incident optical fluence. Note that because the electroabsorption senses only the magnitude of the electric field, a factor of 2 must be taken into account when the frequencies are extracted from the observed oscillations. The dashed curve is a least-squares fit to a square-root function.

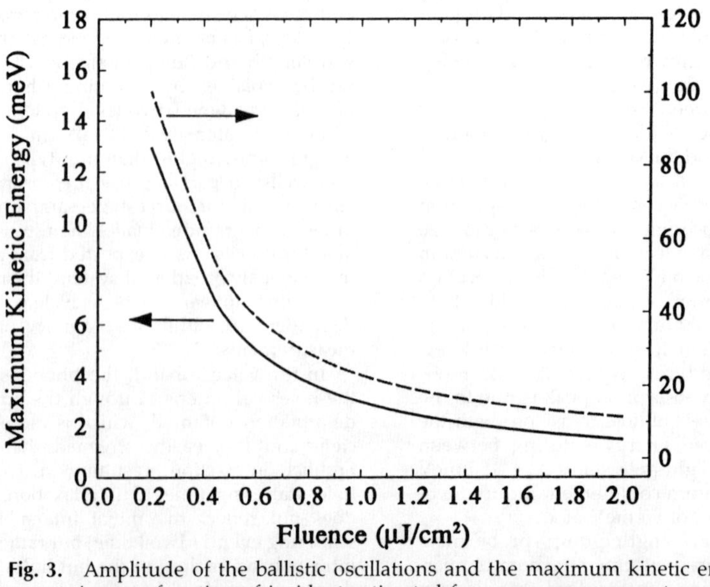

QFF6 Fig. 3. Amplitude of the ballistic oscillations and the maximum kinetic energy acquired per carrier as a function of incident estimated from our measurements.

wise identical conditions by using transient degenerate four-wave-mixing techniques. The dephasing times extracted from the four-wave signals are consistent with the decays of the oscillations shown in Fig. 1. Finally, to exclude the possibility that the oscillations are the result of intervalley scattering between the Γ and L valleys, we note that the maximum energy gained per carrier, as shown in Fig. 3, is of the order of 15 meV, which is much lower than the ~300 meV needed for intervalley transfer (and for Gunn oscillations).

Consequently, we have verified that we have directly observed, for the first time to our knowledge, coherent plasma oscillations of electrons and holes in a bulk semiconductor and that the charge oscillations occur in the ballistic-transport regime on a subpicosecond time scale.

*National Institute of Standards and Technology, Gaithersburg, Maryland 20899

1. A. F. J. Levi, J. R. Hayes, and R. Bhat, Appl. Phys. Lett. **48**, 1609 (1985).
2. M. Heiblum, M. I. Nathan, D. C. Thomas, and C. M. Knoedler, Phys. Rev. Lett. **55**, 2200 (1985).
3. K. E. Meyer, M. Passot, and G. A. Mourou, Appl. Phys. Lett. **53**, 2254 (1988).
4. W. Sha, T. B. Norris, W. J. Schaff, and K. E. Meyer, Phys. Rev. Lett. **67**, 2553 (1991).
5. B. B. Hu, E. A. DeSouza, W. H. Knox, M. C. Nuss, and J. E. Cunning, in *Ultrafast Phenomena*, 1994 Technical Digest Series, Vol. 7 (Optical Society of America, Washington, D.C., 1994), postdeadline paper 12.
6. K. Leo, J. Shah, T. C. Damen, A. Schulze, T. Meier, S. Schmitt-Rink, P. Thomas, E. O. Göbel, S. L. Chuang, M. S. C. Luo, W. Schäfer, K. Köhler, and P. Ganser, IEEE J. Quantum Electron. **28**, 2498 (1992).
7. T. Dekorsy, P. Leisching, K. Köhler, and H. Kurz, Phys. Rev. B **50**, 8106 (1994), and references therein.
8. J. Feldmann, K. Leo, J. Shah, D. A. B. Miller, J. E. Cunningham, T. Meir, G. von Plessen, A. Shulze, P. Thomas, and S. Schmitt-Rink, Phys. Rev. B **46**, 7252 (1992).

QFF7 **3:45 pm**

Electron dynamics near the plasmon resonance of copper metallic nanoparticles

J.-Y. Bigot, O. Crégut, A. Daunois, J.-C. Merle, J. Guille,* *Nonlinear Optics and Optoelectronics Group, Strasbourg Institute of Physics and Chemistry of Materials, National Center for Scientific Research, Unité Mixte 380046, ULP-EHICS, 23 rue du Loess, F-67037 Strasbourg Cedex, France*

The investigation of electron dynamics near the Fermi level of metals by using ultrashort optical pulses permits study of the thermalization of a hot-electron distribution to the lattice temperature. Previous pump–probe experiments, performed on Cu (Refs. 1 and 2) or on Au (Ref. 3) thin films close to the interband transitions form the *d* band to the Fermi level ($d \rightarrow E_F$) have

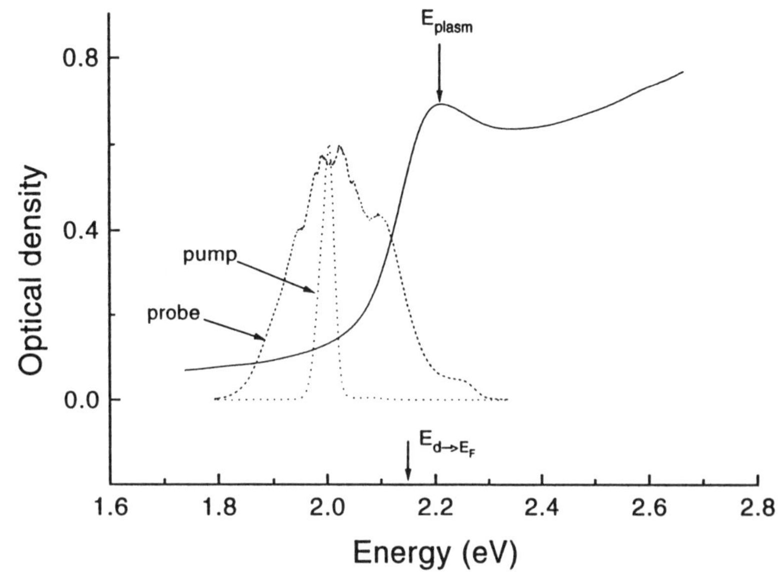

QFF7 **Fig. 1.** Optical density of a 200-μm-thick sample with 5-nm copper nanoparticles and pump and probe spectra.

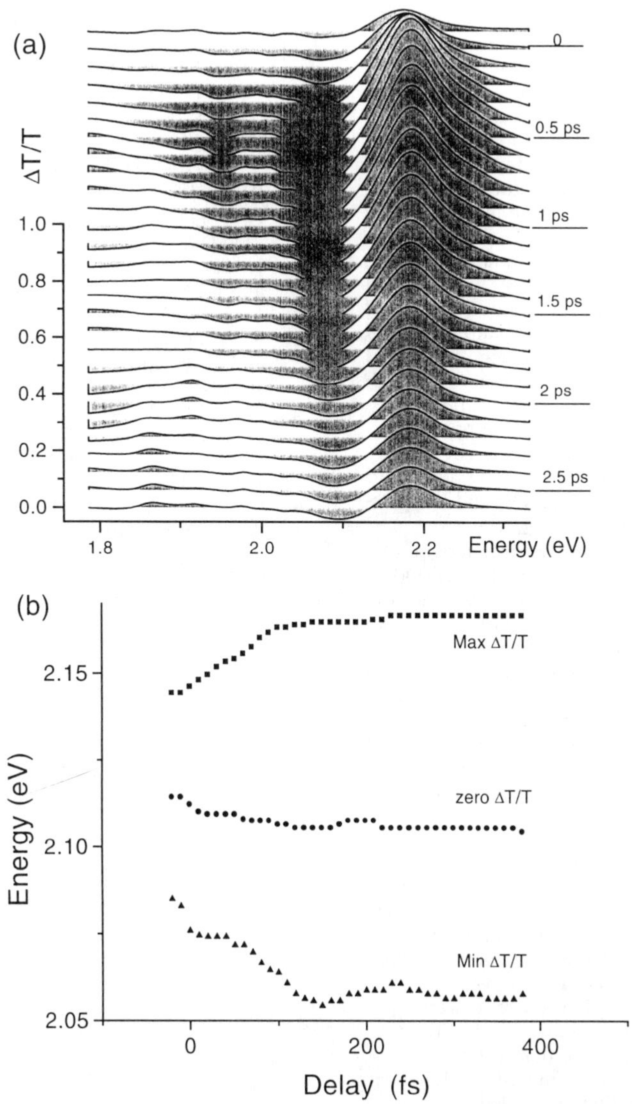

QFF7 **Fig. 2** (a). Temporal evolution of $\Delta T/T$ spectra measured for 100-fs delays and with a pump energy of ~30 nJ/pulse. (b) Time evolution of the spectral positions of the maximum, zero, and minimum signals $\Delta T/T$ measured with 10-fs delays.

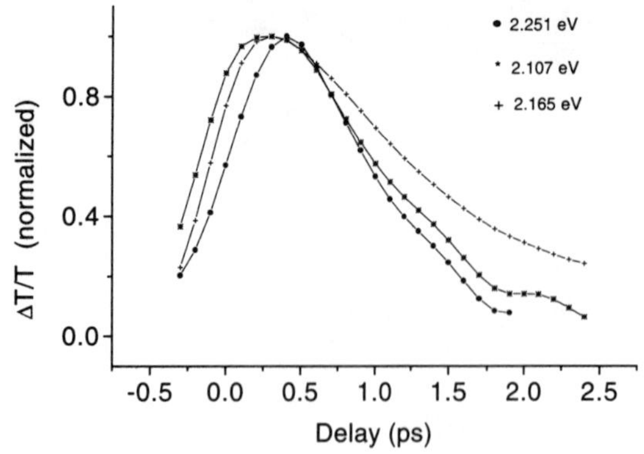

QFF7 Fig. 3. Thermalization process in the vicinity of the phasmon resonance for energies 2.251, 2.107, and 2.165 eV.

revealed intensity-dependent relaxation times varying between a few hundreds of femtoseconds to a few picoseconds. These studies could well be interpreted as a time-dependent electronic temperature relaxing to the lattice temperature. For small metallic nanoparticles embedded in glass, the electron dynamics can be expected to behave differently because of the existence of surface plasmon modes in the spectral region of dielectric confinement. This plasmon resonance, which corresponds to the index matching of the transparent matrix and the metal, is at the origin of the enhancement of the third-order susceptibility, as observed by several authors.[4,5]

In the present work we investigate the electron → lattice relation in Cu nanoparticles embedded in a glass matrix by using transmission pump–probe femtosecond spectroscopy. Figure 1 shows the optical density of a 200-μm-thick sample containing Cu particles of an average diameter of 5 nm together with the pump–pulse (80-fs) and probe–pulse (10-fs) spectra obtained from a compressed copper-vapor-laser-amplified CPM cavity. The arrows indicate the positions of the interband transition ($E_{d \to E_F}$ = 2.15 eV) and the plasmon resonance (E_{plas} = 2.21 eV).

Fig. 2(a) shows the spectrotemporal variation of the induced transmission $(\Delta T/T)(\hbar\omega, t)$ obtained with 100-fs delays between the pump pulses and the probe pulses. It shows the overall heating of electrons on both sides of the $d \to E_F$) transition and the subsequent thermalization to the lattice temperature. A closeup of the early dynamics reveals the detailed heating process. Figure 2(b) shows the spectral positions of $\Delta T_{max}/T$, $\Delta T/T = 0$, and $\Delta T_{min}/T$ measured with 10-fs delays during the first 300 fs. The interband transition ($\Delta T/T = 0$) remains unshifted because the Fermi level is temperature independent. The maximum (minimum) signal shifts to higher (lower) energies during the heating of the electron population. The heating is accomplished after about 200 fs. The measurement of temporal cross sections for various energies indicates a longer thermalization time τ_{th} above the Fermi edge than below it: τ_{th} = 3.5 ps at 2.2 eV as compared to 1.5 ps at 1.95 eV. In addition, a detailed study of τ_{th} in the vicinity of the plasmon resonance shows a slowing of the electron cooling at resonance (see Fig. 3).

These effects correspond to a thermalization process for the surface plasmons that is different from that for the quasi-particle states near the Fermi level. We discuss their origin, which we attribute to an inefficient energy exchange between the surface phasmons and the phonon bath. The resonant character of these modes is considered in a way similar to that for localized modes in ionic crystalline films.[6]

*Inorganic Materials Group, Strasbourg Institute of Physics and Chemistry of Materials, National Center for Scientific Research Unité Mixte 380046, ULP-EH1CS, 23 rue du Loess, F-67037 Strasbourg Cedex, France

1. G. L. Eesley, Phys. Rev. Lett. **51**, 2140 (1983).
2. H. Elsayed-Ali, T. Norris, M. Pessot, and G. Mourou, Phys. Rev. Lett. **58**, 1212 (1987).
3. R. Schoenlein, W. Lin, J. Fujimoto, and G. Eesley, Phys. Rev. Lett. **58**, 1680 (1987).
4. F. Hache, D. Richard, and C. Girard, Phys. Rev. B **38**, 7990 (1988).
5. L. Yang *et al.*, J. Opt. Soc. Am. B **11**, 457 (1994).
6. K. L. Kliewer and R. Fuchs, Phys. Rev. **150**, 573 (1966).